ADVANCES IN THE STUDY OF MAMMALIAN BEHAVIOR

SPECIAL PUBLICATIONS

This series, published by the American Society of Mammalogists, has been established for papers of monographic scope concerned with some aspect of the biology of mammals.

Correspondence concerning manuscripts to be submitted for publication in the series should be addressed to the Editor for Special Publications, Hugh H. Genoways (address below).

Copies of Special Publications of the Society may be ordered from the Secretary-Treasurer, Gordon L. Kirkland, Jr., Vertebrate Museum, Shippensburg State College, Shippensburg, Pennsylvania 17257.

Price of this issue $45.00

COMMITTEE ON SPECIAL PUBLICATIONS

ADVANCES IN THE STUDY OF MAMMALIAN BEHAVIOR

EDITED BY

JOHN F. EISENBERG

AND

DEVRA G. KLEIMAN

SPECIAL PUBLICATION NO. 7

THE AMERICAN SOCIETY OF MAMMALOGISTS

PUBLISHED 11 MARCH 1983

Library of Congress Catalog Card No. 82-73762

ISBN No. 0-934612-06-3

PARTICIPANTS IN THE FRONT ROYAL CONFERENCE ON MAMMALIAN BEHAVIOR, 16–19 AUGUST 1980*

Front row (left to right): Gail Michener, Edwin Gould, Don Wilson.

Second row: John Fentress, Ilan Golani, Katherine Ralls, Robert Tamarin, Devra Kleiman, Harry Jerison.

Third row: Tim Clutton-Brock, Patricia Moehlman, Ian Stirling, Paul Harvey, Peter Marler.

Fourth row: Robert Brownell, Brian Bertram, Theodore Grand, Christen Wemmer, John Vandenbergh, Jon Rood, Dietland Müller-Schwarze.

Fifth row: Karl Kranz, John F. Eisenberg, Jon Ballou, Bryan McNab, Kent Redford, William Franklin, Tom Grant, Hubert Hendrichs, John Seidensticker, Mdme and Pierre Charles-Dominique.

* Frank and Mrs. Bronson attended but missed the group photograph.

We dedicate this volume to Professor Heini Hediger and the late Professor R. F. "Griff" Ewer, pioneers in the study of mammalian behavior.

J.F.E. and D.G.K.

CONTRIBUTORS

BRIAN BERTRAM
The Zoological Society of London
Regent's Park
London, England NW1 4RY

F. H. BRONSON
Department of Zoology
The University of Texas
Austin, Texas 78712

PIERRE CHARLES-DOMINIQUE
Museum National D'Histoire Naturelle
Ecologie Generale
Brunoy, France

T. H. CLUTTON-BROCK
King's College Research Center
King's College
Cambridge, England CB2 1ST

JOHN F. EISENBERG
Florida State Museum
University of Florida
Gainesville, Florida 32611

JOHN FENTRESS
Department of Psychology
Dalhousie University
Halifax, Nova Scotia
Canada

WILLIAM FRANKLIN
Department of Animal Ecology
Iowa State University
Ames, Iowa 50011

ILAN GOLANI
Department of Zoology
Tel-Aviv University
Tel-Aviv, Israel

EDWIN GOULD
Department of Mammalogy
National Zoological Park
Smithsonian Institution
Washington, D.C. 20008

THEODORE I. GRAND
Oregon Regional Primate Research Center
Beaverton, Oregon 97005

T. R. GRANT
Department of Zoology
The University of Wollongong
Wollongong, New South Wales 2500
Australia

PAUL HARVEY
Department of Biological Sciences
University of Sussex
Sussex, England BN1 9QG

HUBERT HENDRICHS
Universitat Bielefeld
4800 Bielefeld 1, West Germany

HARRY JERISON
Departments of Psychiatry and Biobehavioral Sciences
University of California Medical School
760 Westwood Place
Los Angeles, California 90024

DEVRA G. KLEIMAN
Department of Zoological Research
National Zoological Park
Smithsonian Institution
Washington, D.C. 20008

PETER MARLER
Field Research Center
The Rockefeller University
Millbrook, New York 12545

BRIAN C. MCNAB
Department of Zoology
University of Florida
Gainesville, Florida 32611

Gail Michener
Department of Biological Sciences
The University of Lethbridge
Lethbridge, Alberta
Canada T1K 3M4

Patricia Moehlman
School of Forestry
Yale University
New Haven, Connecticut 06520

Greg Moran
Department of Psychology
University of Western Ontario
London, Ontario
Canada

D. Müller-Schwarze
College of Environmental Science and Forestry
State University of New York
Syracuse, New York 13210

Jon Rood
Conservation Center
National Zoological Park
Front Royal, Virginia 22630

Ian Stirling
Canadian Wildlife Service
Environment Canada
Edmonton, Alberta
Canada T6H 3S5

Robert H. Tamarin
Biological Science Center
Boston University
Boston, Massachusetts 02215

John Vandenbergh
Department of Zoology
North Carolina State University
Raleigh, North Carolina 27650

Christen Wemmer
Department of Conservation
National Zoological Park
Smithsonian Institution
Front Royal, Virginia 22630

Don E. Wilson
Bird and Mammal Laboratories
U.S. National Museum of Natural History
Smithsonian Institution
Washington, D.C. 20560

PREFACE AND ACKNOWLEDGMENTS

In 1978, Hugh Genoways suggested that a volume of readings in mammalian behavior would serve a useful purpose in summarizing recent advances in behavioral analyses for students in mammalogy and non-behaviorists, as well as for professionals in that field of study. At his request, we sought to organize a symposium bringing together workers from various sub-disciplines of behavioral studies to effect a synthesis of modern thought. To this end, a conference was organized at the National Zoological Park's Conservation and Research Center in Front Royal, Virginia. The participants met between 16 and 19 August 1980. The presentations represent a cross section of interests within the discipline of animal behavior and should present to the reader a balanced account of our efforts and highlight problem areas of pressing concern.

Studies of mammalian behavior have gained great popularity in the last 25 years. The emerging science of sociobiology has provided a theoretical focus for interpretations of the manner in which social behavior can evolve. Studies of migratory behavior, habitat utilization, and feeding ecology contribute directly to our understanding of how ecosystems are organized and how wildlife populations may be managed. Studies of social organization and breeding behavior of particular species uncover yet another subtle layer of organization within a community which permits planners and managers to understand how populations are organized and how they regulate themselves through time. There is hardly a facet of conservation biology that is not touched upon by studies of behavior. It is our hope that this set of readings will introduce students to the various sub-disciplines of behavioral studies and highlight their relevance to applied problems. Aside from applications, this treatment of problems in behavior analysis has intrinsic theoretical interest of its own. We hope that latent interest in students can be sparked as they sample the selections here offered.

The conference was supported by a generous grant from the Friends of the National Zoo. Organizational details of the conference were ably handled by Ms. Tabetha Carpenter and Mr. Jack

Williams. Mrs. Wyotta Holden graciously served as editorial assistant and Sigrid (James) Bruch redrew many of the final figures. Karl Kranz, Kent Redford, and Jon Ballou served as conference aides. John Seidensticker and Katherine Ralls assisted as moderators and discussants. Professional discussions were enlivened by Melvin Sunquist, Larry Collins, Georgina Mace, and Robert Brownell. Special thanks are due the staff of the Conservation Center for numerous courtesies. We wish to thank once again Dr. Theodore H. Reed, Director of the National Zoo, for his support in helping to make this conference and volume a reality.

J.F.E. and D.G.K.

CONTENTS

PART I. STRUCTURE, DEVELOPMENT, AND FUNCTION

INTRODUCTION

ANIMALS are characterized by their ability to independently initiate or respond to environmental changes by movement. Movement in higher vertebrates is accomplished by means of contractile tissue (muscles) which are attached to moveable structures and/or associated with a skeletal system permitting forces to be exerted against a supportive substrate. Movement is coordinated in the short term by a nervous system while the propensity to exhibit certain forms of movement is mediated by a neuroendocrine system more responsive to long-term environmental changes. Thus, behavior in higher vertebrates is the overt manifestation of neuromuscular and endocrine system activity. What a mammal does is in no small measure what its muscular and skeletal systems permit it to do, when coordinated by the nervous system which modulates sensory input and alters responsiveness to external stimuli. Physiological mechanisms of animal behavior patterns have been investigated for centuries and will only be touched on briefly in this series of essays.

Grand discusses how body structure constrains the form of behavior patterns exhibited within the framework of development by comparing and contrasting the trunk, limb, and body proportions of an infant rhesus macaque (*Macaca mulatta*) with those of a juvenile, a subadult, and finally an adult. He highlights the fact that behavior is constrained within the limits of body structure but that, when considered from a developmental standpoint, the behavior patterns expressed change dramatically through processes of maturation so that the whole form of the animal is altered in adulthood, creating new possibilities and new limits for behavioral expression. His essay not only brings home the reality of form and function analysis for the interpretation of behavior but also points out that living organisms change through time as the processes of development and growth unfold. The repertoire of an infant is far different from that of an adult, both with respect to its reactions to the environment and the form of social behaviors which it can show toward conspecifics in the same social unit. Grand's essay then appropriately leads to a consideration of behavioral ontogeny, namely how behavior patterns unfold through time as an organism matures from birth to adulthood and finally passes from adulthood to senescence.

In the essay by Fentress, ontogeny is the central theme but he immediately highlights some of the most persistent problems plaguing all behavioral analyses. Because behavior generally consists of movement or transitions between static postures, the description of behavior patterns becomes a central issue. How does one describe how an animal locomotes? How does one extend a description of locomotion to include the movements of two interacting animals relative to one another. Fentress does not propose a single descriptive framework but notes that the problems of description and methods for adequately describing and categorizing mammalian behaviors need to be broadened. Fentress' essay suggests that behavioral analysis must emerge from the "descriptive ethogram" and categorization of behaviors into "functional" groups to a new plane of sophistication. The complex scheme of development, whereby individual motor units are combined and recombined to form whole functional behavior units and there is interaction of behavioral expression with context, needs a deeper level of analysis.

Fentress' essay addresses other important problems still plaguing students of behavior, such as the "nature-nurture" dichotomy. Behavior patterns have a developmental history; thus important problems concerning the degree to which environment influences developmental processes must always be fully investigated. Some behavior patterns, such as face-washing in rodents, may be very resistent to environmental modification and indeed suggest that the neuromuscular patterns for the exhibition of such behaviors are programmed rather deeply within the genetic system of the animal. On the other hand, the development of interindividual relationships in the process of ontogeny may be subject to a great deal of environmental monitoring. Some interaction patterns may be idiosyncratic to the dyad in question whereas other forms of social behavior may also be very resistent to environmental modification. For example, regurgitation of food by adult wolves to their offspring may appear without prior experience with the sudden appearance of offspring and, in conformity with the face-washing behavior of rodents, may be very resistent to modulation by the external environment.

In the essay by Golani and Moran, the application of the Eshkol-Wachmann Movement Notation technique designed for choreographic description has been applied to the interactions of wolves. Their essay highlights the semantic difficulties encountered by earlier investigators in determining which pair of an interacting set was submissive or dominant. The approach by Golani and Moran makes description far more rigorous and suggests that classical categorizations of behaviors ignored the all important transitions from stable configurations varying qualitatively in their degree of predictability.

Golani's and Moran's behavioral analysis emphasizes the importance of the partner's role and distance in the understanding of the progress of a dyadic encounter. Each animal's role can be described quantitatively and cannot be understood without reference to the partner. Two points of importance emerge: 1) changes in the social relationships within a group may result in a change of roles, with significant alterations in the behavioral expressions of individuals; and 2) patterns of social interaction can be traced back to non-social behavior, showing the conservatism of behavioral expression.

Ultimately, an individual's capacity to respond appropriately to its environment as it develops is dependent on physiological events. The endocrine system develops at discrete stages during ontogeny and the activities of the endocrine system are strongly regulated by external events. Photoperiod controls gonadal activity but the social milieu, including the number of adult males and females present in the vicinity, may have strong effects on the rate of development.

Vandenbergh discusses some recent findings which highlight 1) how sensitive the mammalian organism is to the social environment, even *in utero,* and 2) how laboratory findings may relate to function in natural populations.

Jerison addresses the mammalian brain as an information processing system. The mammalian brain increases in absolute size as body size increases, but the relationship is negatively allometric. When the class Mammalia is considered as a whole, it can be noted that larger mammals have relatively smaller brains than do most small mammals. There is great scatter in such a plot of contemporary species and Jerison explores the nature of such variation.

By employing endocasts of fossil brains, he can track the rate of increase in relative brain size within different vertebrate lineages.

The folding of the vertebrate brain, which is pronounced in species with relatively large brains, can be viewed as an anatomical process designed to increase the surface area of a solid body increasing in mass. The surface of the brain is the area where the sensory-motor projection areas lie and folding of the surface can be seen as an adaptation to increase projection areas as sensory and motor neurones increase in number.

Jerison concludes his essay with a discussion of the human brain and its evolutionary history. Clearly the information storage and processing capacities of humans are enormous relative to other mammalian species, but the evolution of relatively large brains has occurred many times in distantly related mammalian families. The exact nature of the selective forces which direct the evolution of large-brained species remains an enduring subject for study.

J.F.E. and D.G.K.

THE ANATOMY OF GROWTH AND ITS RELATION TO LOCOMOTOR CAPACITY IN MACACA

THEODORE I. GRAND

Abstract

FETAL, infant, juvenile, and adult male and female macaques differ from one another in two anatomical ways: 1) the proportional composition of tissues (for example, brain, muscle, bone, viscera), and 2) the distribution of body mass among the segments (for example, head, trunk, limbs). For example, the brain declines as a percentage of weight from 10% at birth to 1% at maturity, whereas muscle rises from 24% at birth to at least 40% during the first six postnatal months. The head is 25% of weight at birth but declines in adulthood to 6 or 7%. At the same time the thigh increases from 3.5 to more than 7% of total weight. Regional increases in musculature are sequential and functional. The masticatory muscles (and mandible) continue to grow from nutritional independence to sexual maturity, whereas the brain and eyes decline in relative size after birth. Early growth of the forearms supports the clinging newborn; later growth of shoulder and upper arm correlates with locomotor independence. Increased body fat is a phenomenon of maturity. Comparable shifts in tissue proportions, growth rates, and the segmental distribution of body mass in other mammals and birds reflect their particular evolutionary and adaptive strategies.

Introduction

In this paper I bring the attitudes and methods of an anatomist to the problems of growth. In the past I have compared adult forms of many genera from various size classes that differ in their locomotor patterns (Grand, 1977*a*, 1978). Here I concentrate on rhesus macaques (*Macaca mulatta*) including several size classes and varying levels of locomotor skill.

The dynamic and functional expressions of tissues (muscles and bones) and of segmental masses (head, thigh, foot) grade naturally into another facet of my research, that is, locomotor skill (Grand, 1972, 1976). Although it is a rather pure, non-social expression, locomotor behavior ties together all other behavioral functions. It brings the organism to a food source or to a mate; it permits seasonal migration from one place to another. The capacities for speed and acceleration or for endurance are built into muscle, bone, and adipose tissue, and thereby, open up or shut off the adaptive range of the individual (Harkness, 1976). Thus, from the shifting center of body mass in the growing individual to the jump of the adult male macaque at a terminal branch crossing, the growth of the musculoskeletal tissues is the bottom line for the analysis of detail and the top line for a behavioral perspective in all that I discuss. Each of the locomotor stages that is considered is directly related to changes in tissue composition and segmental distribution of mass.

I make no pretense that these data or my interpretations of them can fully satisfy the growth specialist. The usual investigations are longitudinal, in that they follow individuals year after year (Kowalski and Guire, 1974; Tanner, 1969). The growth spurt, for example, is a sharply accelerated change in weight and length of the individual. However, when growth rates for a number of individuals of the same age (or weight) are averaged, the individual peaks are smoothed out and the growth spurt blurs. I used cross-sectional sampling, because the data were obtained from dead animals. The animals arranged by increasing body weight in natural logs represent increasing age. Cross-sectional techniques have been used to study the growth of muscle groups in domestic animals (Berg, 1968; Butterfield and Johnson, 1968; Fowler, 1968; Widdowson, 1968) and in wild birds (Austin and Ricklefs, 1977; O'Connor, 1975, 1977; Ricklefs, 1968, 1973; Ricklefs and Weremuick, 1977; Ricklefs et al., 1978).

In such research, what should interest the student of behavior? Motor skill is one key to survival. As a consequence of the primacy of locomotion, almost 70% of a typical mammal is devoted to the locomotor apparatus (Grand, 1977*a*, 1978). Motor skills reflect adaptations within a given habitat, and a certain fit must exist between the level of skill an animal attains and where and how it moves in searching for food and mates and avoiding predators. In a troop of macaques, for example, one sees a distribution of skill levels: 1) a young animal cannot jump across a gap between terminal branches because the animal is too small and too weak; 2) a mature male cannot cross the same gap because the branches cannot bear his weight. I want ethologists to recognize and appreciate the anatomical bases of such motor problems.

Mason (1979) and Vaillant (1977) stress that both sociality and psychological stability proceed through a number of distinct phases; the individual threads come together at different rates to create the fabric of a reproductively and socially mature animal. Similarly, tissue proportions and gross bodily changes proceed at different rates; the maturation of each tissue and region is synchronized, as it should be, to a particular functional demand at a particular stage of the individual's life-history.

Methods

Precise methods have evolved for the study of adult forms (Grand, 1977*a*) and have been focused more recently on the developmental process (Grand, 1977*b*). Tissue decomposition or "fractionation" and segmentation data are interconvertible with studies by human anatomists, physical educators, and the Ross "School" of Kinanthropometrists. In one technique, skin, muscle, bone, and viscera are successively dissected and weighed to determine the proportion of each tissue to live weight. In the second, the body is dismembered at each of the major segments to create a picture of the segmental distribution of body mass.

One aspect of growth is increasing body weight, and weight is the baseline in this cross-sectional sample. Here I demonstrate that specific tissues and segments grow slower or faster than one another with respect to increasing body weight as the animal develops through time.

The basis for all biomechanical analyses is the location of the center of gravity (Palmer, 1944). To locate the center of gravity in young and adult animals, I used the same linear proportions but included measured differences in segmental mass. A calculation of the product of moments (Miller and Nelson, 1973) determined the anteroposterior position of the center of gravity.

Results

The transformation of the macaque body from a 250 g fetus to a 6 to 12 kg adult is presented step-by-step.

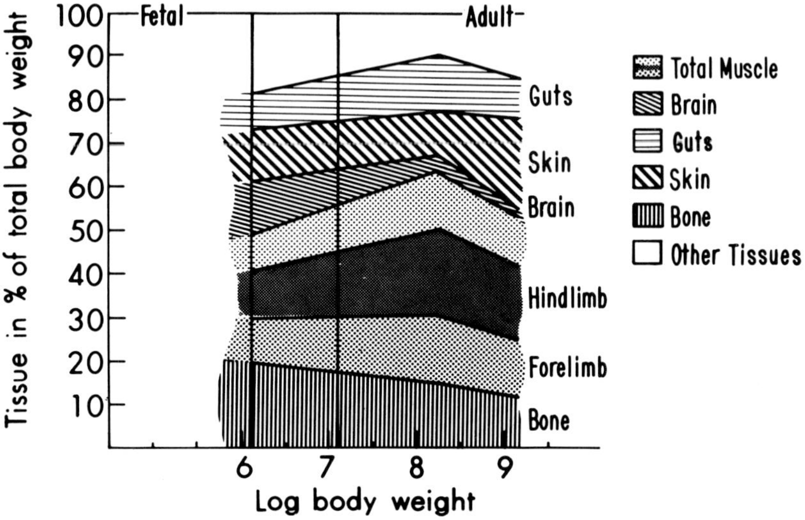

FIG. 1. Changes in tissue proportions from fetus to adult in *Macaca mulatta* (log body weight in g).

Tissue composition: proportional changes in bone, muscle, brain, skin, and viscera (Fig. 1).—The brain achieves mature size early in life. Muscle, heart, and bone grow with locomotor demand and thus account most directly for later body weight increases. Regional prominance of the nuchal, forearm, and calf muscles is evident in the newborn; other groups, for example, shoulder, hip, and masticatory muscles, grow later on (Fig. 2).

Segmental distribution: the limb segments.—The hand, foot, tail, and head are relatively heavier at birth than they are later in life; the upper arm and thigh grow heavier with age (Grand, 1977*b*) (see Fig. 3).

Infants have forearms that are disproportionately heavier than their upper arms. However, after 800 to 1,000 g the upper arms grow at an accelerated rate (see Fig. 4A).

The same rate change holds for the hind limb. The foot slows up first, the calf next, while the thigh continues to grow with size and locomotor independence (see Fig. 4B).

Segmental distribution: the trunk and head.—The mechanically important regions of truncal mass do not change uniformly, nor do their component tissues change in similar ways from one segment

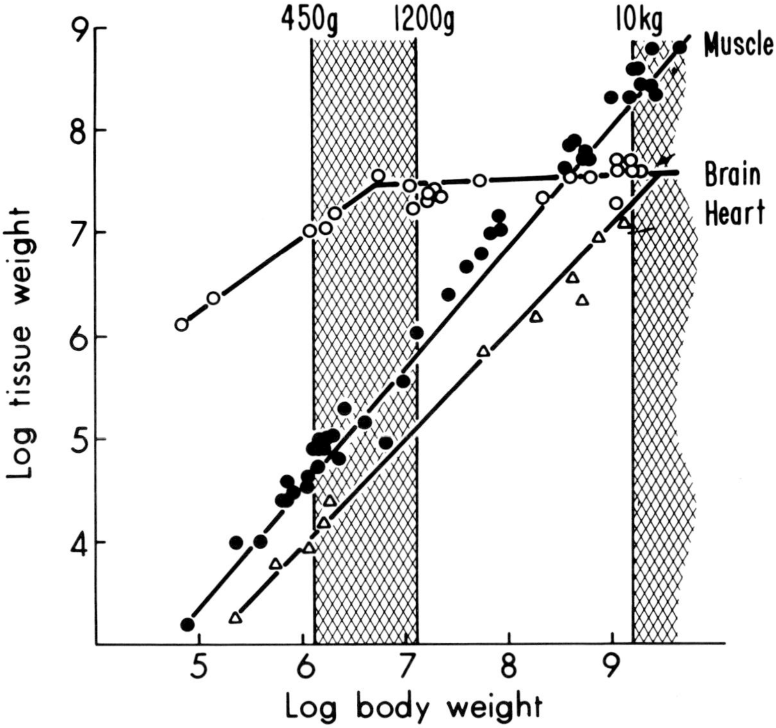

FIG. 2. Growth rates of tissues (muscle, brain, heart) in relation to body weight in g.

to the next. The head constitutes about 25% of birth weight, but drops to 6 or 7% of weight in the adult (Figs. 5 and 6).

Three changes in the composition of the head are of great interest. First, the brain achieves full growth by 1,000 g body weight. Second, the eyes reach mature size at about 3 kg body weight. Third, at birth the mandible and masticatory muscles are absolutely smaller than the other tissue complexes. However, as the animal becomes nutritionally independent, as the teeth grow, and as the mandible is subjected to the compressive strains of chewing, the masticatory muscles grow at an accelerated rate (see Fig. 7). Ultimately, of course, muscle grows more in mass than bone. A dimorphism in the musculature is also evident—the percentage of the head that is muscle is twice as great in the adult male as in the adult female.

Sexual dimorphism in the mass of the head results from these

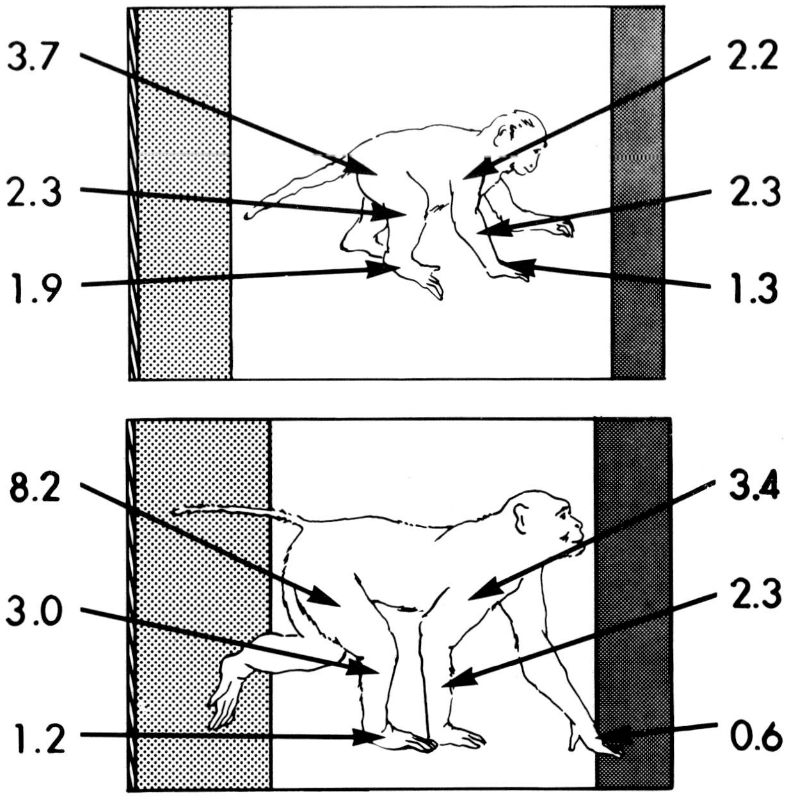

FIG. 3. Proportions of limb segments in relation to body weight in a newborn (top) and adult male macaque (bottom). Numbers along sides of figures are percentage weights of each limb segment (thigh, calf, foot, upper arm, forearm, hand). The stippled rectangles show the relative proportions of both hindlimbs (light stipple) and both forelimbs (heavy stipple).

tissue differences. In 3 to 8 kg animals, the head of males is heavier than that of females. In males, the brain, masticatory musculature, skull, and mandible are absolutely heavier so that the head mass is greater than that of females. However, as males grow beyond 8 kg the remainder of the body outgrows the head and its component tissues and the head continues to drop in relative weight. Thus, the male head becomes smaller.

Each shoulder represents 3 to 3.5% of body mass of the infant (Fig. 6) but rises to more than 5% in the adult male. The greatest increase is in propulsive musculature. The lumbar region represents about 17% of weight at birth, but increases to more than 20% in adulthood. The proportional composition of tissues within the segment changes. The viscera decline from 50 to 35% of the segment, whereas the back extensors rise from almost 6.5 to 9.5%. The pelvic region increases from 7 to 11% of total body weight. The pelvic viscera decline in relative weight, but the muscle increases significantly. The back musculature rises from 2 to 4.5% of the segment, whereas hip muscles rise from 23 to 30%. The relative growth of the major propulsive groups in percent of total weight is shown in Fig. 8.

Biomechanics.—To demonstrate the shifting center of gravity in the growing individual is logically impossible from a cross-sectional sampling technique. Nevertheless, infant and adult macaques, drawn to the same scale with their mass differences assigned, show a significant posterior shift in the center of gravity, about 10% of body length (Fig. 9). This point in combination with body weight and the relative strength of muscle groups is the key to locomotor capacity and changing efficiency.

Discussion

The weight increases in macaques from 400 g at birth to over 8 kg in maturity are accompanied by non-uniform changes in the tissues and segments. The brain and eyes mature early in life, muscle and bone later, and adipose tissue accumulates still later. Fowler (1968) pointed out that the components mature in relation to functions and those elements tied by one functional demand grow at the same rate. The macaque counterpart to Fowler's domestic animal sequence is as follows: 1) immediate sensory processing and information storage; 2) emerging nutritional independence; 3) locomotor independence; 4) the dimorphism of reproductively mature females and males.

At 400 to 800 g and up to 4 months of age, infants must process both visual and social information. Consequently, the brain and eyes attain mature size shortly after birth. The head is 25% of body weight, the brain 50% of head weight. By contrast, the masticatory muscles weigh less than 1% of the head, the mandible slightly more. Because of the physical immaturity, with the body less than 25% muscle, the newborn clings to his mother in order to be carried from

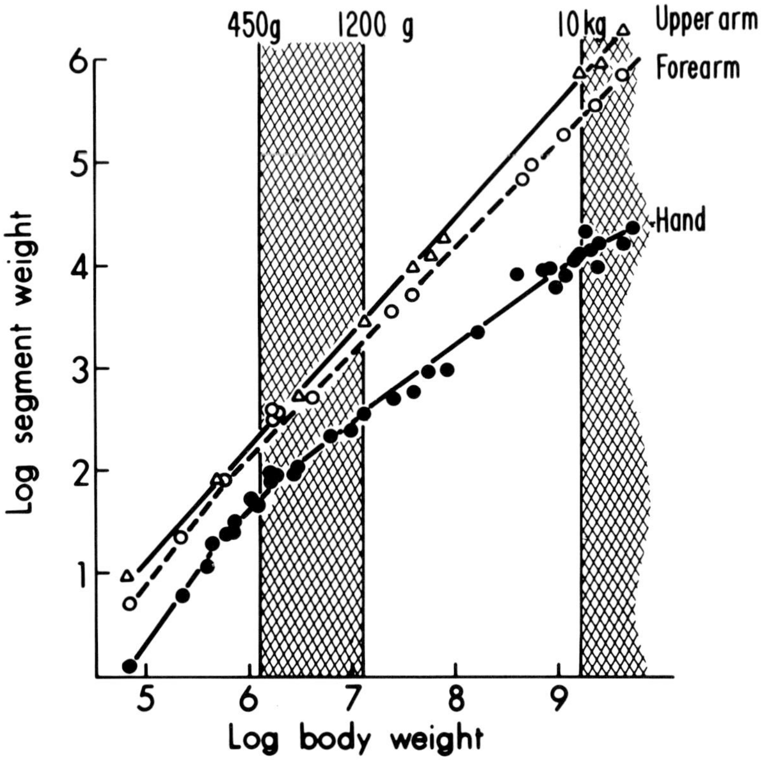

FIG. 4. Growth rates of the limb segments and head in relation to body weight. A (above), forelimbs; B (opposite page), hindlimbs and head.

place to place. The nuchal extensor muscles, which support the head, and the forearm and calf muscles, which flex the fingers and toes, are disproportionately large. Segmental "disproportions" reflect this muscular immaturity. As a result of the massive head and the light hindlimbs, the animal is top- or front-heavy, and the forelimbs do more pulling in quadrupedal walking than the hindlimbs do in pushing (Kimura et al., 1979). This is a dependent, rather inefficient locomotor stage with no significant sexual dimorphism, and only the earliest exploration of physical capacity.

At 1,000 g and 6 months of age, the juvenile macaque processes its own food and moves independently. Mandible and masticatory muscles have begun to grow. Total muscle rises to 40 or 45% of

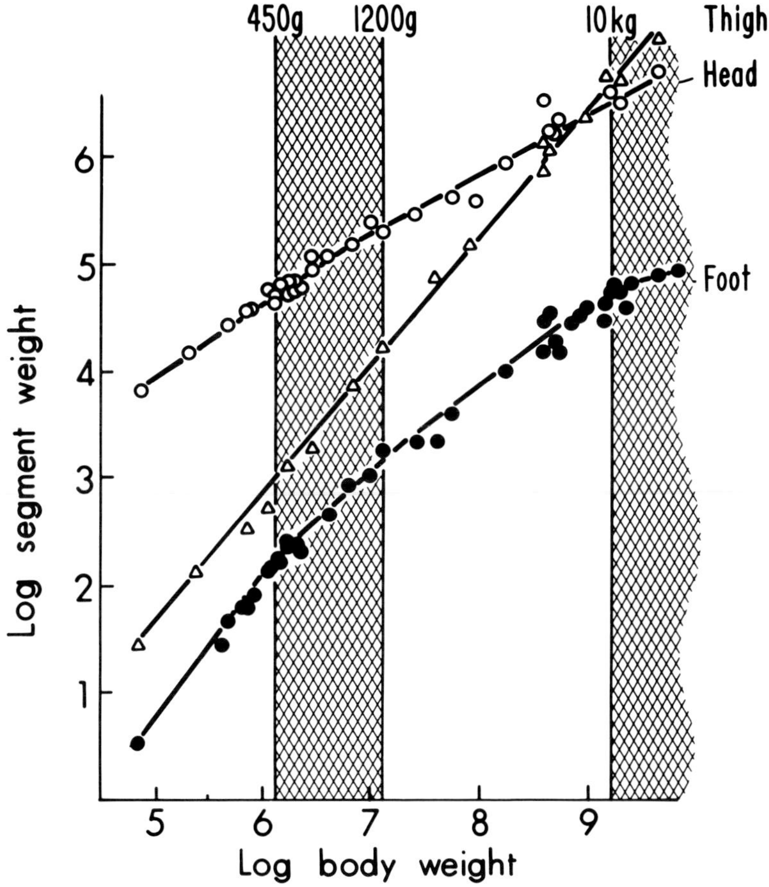

total weight and the juvenile is exceptionally strong. Propulsive muscles of the thigh, hip, back, upper arm, and shoulder grow disproportionately. The heart grows with the musculature and there is increased cardiovascular demand. The center of gravity moves posteriorly; the hindlimbs thrust or push the body more effectively (Fig. 9). There is marked exploitation of physical skill and experimentation. As juveniles grow towards sexual maturity, bodily dimorphism emerges.

At 3 to 8 kg, reproductively mature females may begin to accumulate a higher proportion of fat than males. In any case, this leads

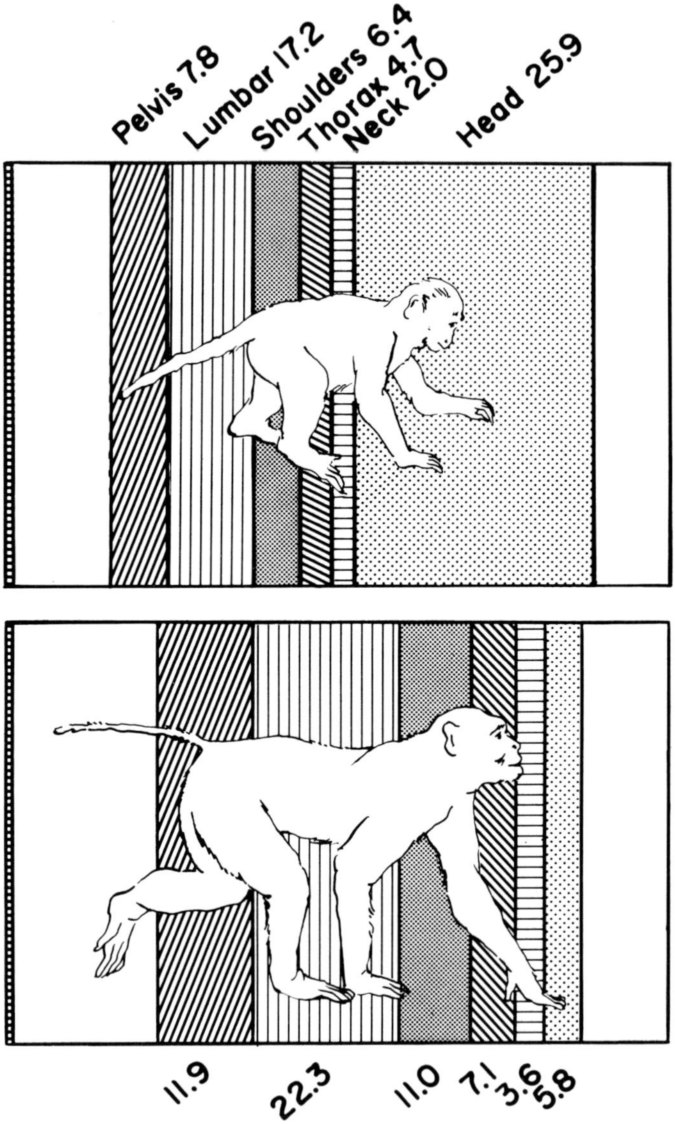

FIG. 5. Segmental distribution of masses in the trunk of a newborn (top) and an adult male (bottom). The numbers along the top and bottom of the figure represent the percentage of that segment in relation to body weight.

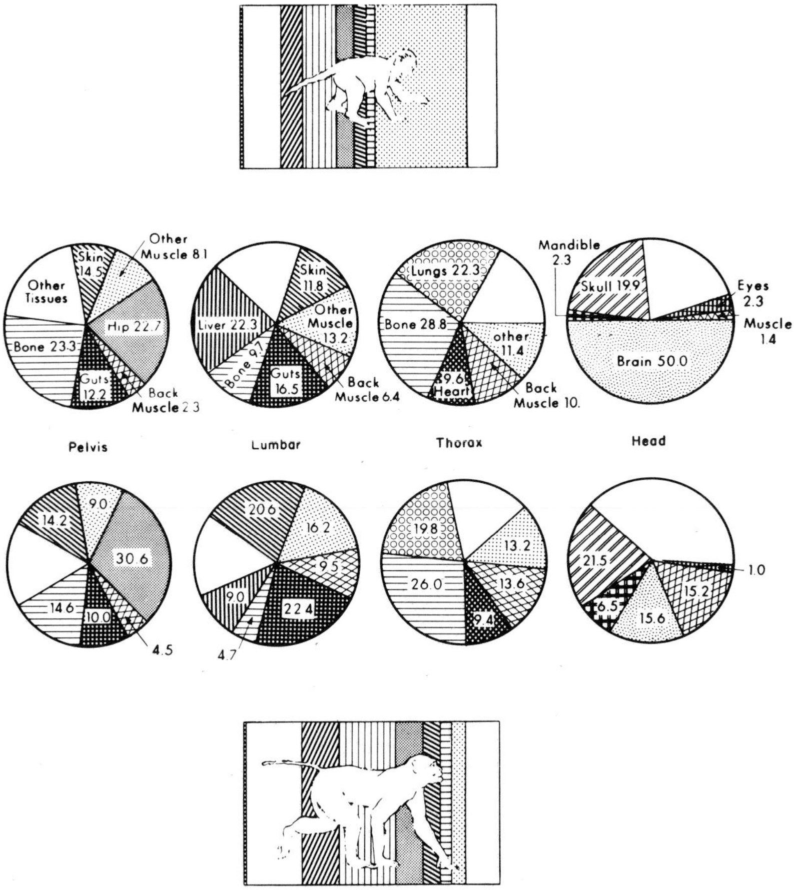

FIG. 6. Tissue decomposition of the truncal masses in infant and adult.

to a slight drop in the proportion of muscle to body weight. The head becomes relatively smaller and the hindlimbs two times larger than the forelimbs. The center of gravity continues to shift posteriorly. In this size class, locomotor activities become somewhat restricted and play declines. Sexual dimorphism increases.

Above 8 kg the males have become reproductively and socially mature. The head is relatively smaller, the thighs still more massive than before. The male brain is absolutely larger than the female

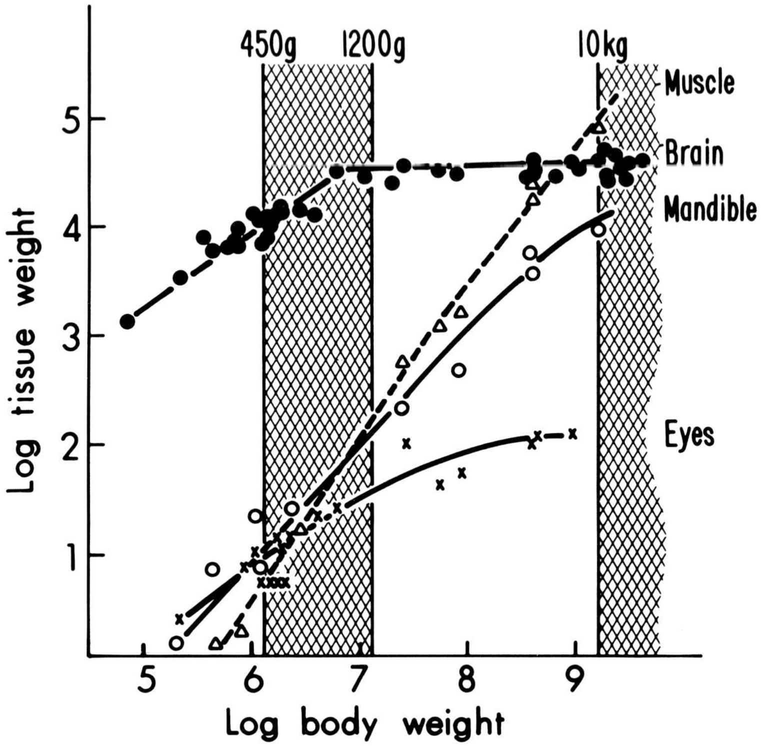

FIG. 7. Growth rates of tissues (brain, eyes, masticatory muscles, mandible) in relation to body weight.

brain and the male masticatory muscles may be twice as large. In free-ranging semi-arboreal animals great body size is a locomotor inconvenience. For example, the largest males of a troop make detours that other troop members need not make; they must avoid slender terminal branches and large gaps because of the manner in which their greater weight deforms the branches.

It is evident that each stage differs from that which precedes it, yet each stage and its capacities are integrated into the life of the macaque troop—the mother accomodates to her helpless newborn; the juvenile explores and exploits its relatively great strength; heavy-bodied adults become quite conservative both in their activity and energy expenditure. This range of skills found within a single social unit reminds one of the dramatic, if more attenuated, locomotor

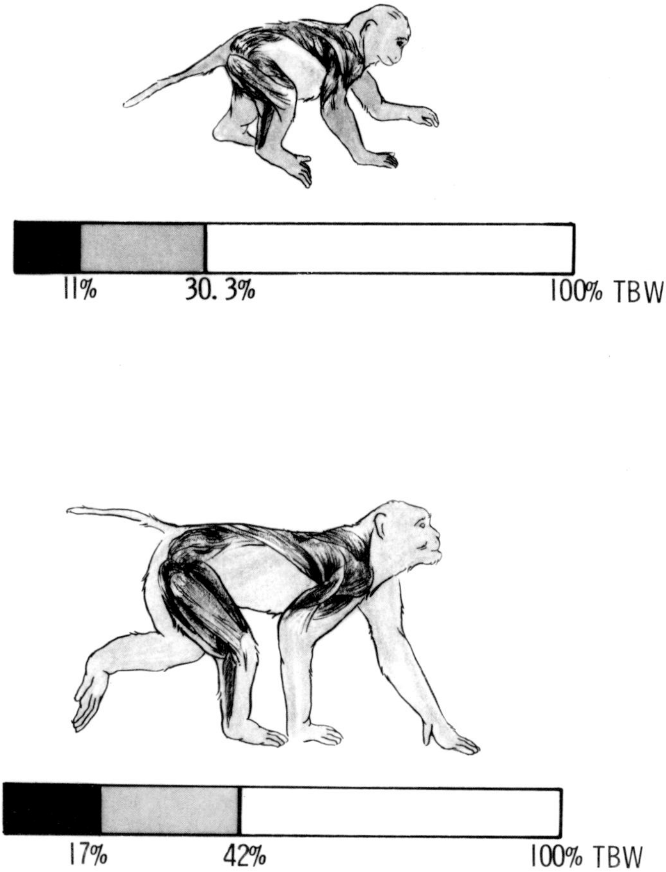

FIG. 8. Propulsive muscles in relation to total muscle in infant (top) and adult male (bottom). Black bar identifies propulsive muscle and grey bar the remainder of muscle; TBW is total body weight).

development of the human infants—the inefficient stagger of the 1 to 2 year old, mature walking at 7 to 9 years. Along with changes in musculature, distribution of mass, and the center of gravity, the human infant must learn the reciprocal discharge of some muscle groups and the suppression of others, the coordinated swing of the contralateral limbs, and the efficient weight transfer from head to toe. Five to 6 years are required for this pattern to mature (Okamoto, 1973).

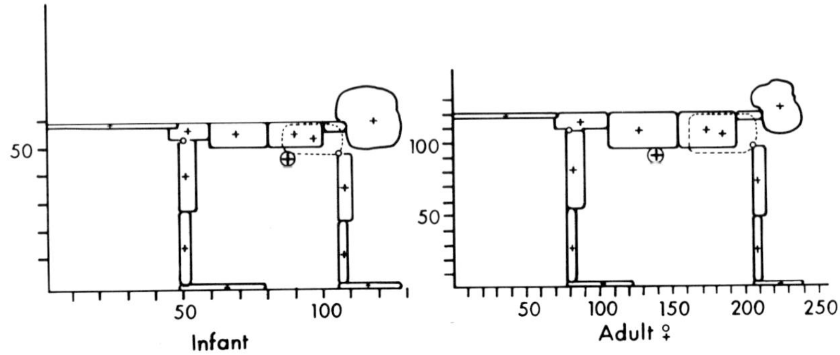

FIG. 9. Positional shift in the center of gravity in infant (left) and adult male (right) reduced to the same vertical and horizontal size. Symbols are: +, the location of the center of mass of each segment; ++, the locations of the centers of mass of the thorax (posterior +) and shoulder (anterior +, in dotted outline); ⊕, center of gravity of the whole body.

Mason (1979) and Vaillant (1977) described the complexities of psychosocial development from birth to reproductive maturity in terms and images strikingly similar to those of anatomical and locomotor development. Thus, Mason (1979) said the following: [Evolutionary purpose] "gives no clue as to how an adaptive outcome is actually achieved by a given individual. . . . These are empirical questions and they constitute a major task for ontogenetic studies (p. 4). . . . At every stage in its development the individual is an organized entity. . . an ongoing enterprise with established needs and functions, constrained by its existing organization . . . in active commerce with its environment (p. 4). . . . Ontogeny . . . is a continuous process, extending from conception to senescence, and is subject to natural selection from the beginning to the end of the life-cycle. It is the organism's entire life history that evolves (p. 8)." And Vaillant (1977) stated that: "ego mechanisms of defense describe unconscious and sometimes pathological mental processes that the ego uses to resolve conflict among the four lodestars of our inner life: instincts, the real world, important people, and the internalized prohibitions provided by our conscience and our culture (p. 9). . . . A central thesis of this book is that if we are to master conflict gracefully and to harness instinctual strivings creatively, our adaptive styles must mature . . . . The Berkeley investigators observed that, over time, behaviors reflecting reaction formation and fantasy declined, but that those reflecting altruism and suppression increased. . . . The middle years of the adult life cycles revealed not only increasing career commitment and responsibility for others but also a progressive maturation of adaptive modes . . . . The Grant Study men (as adolescents) were twice as likely to use immature defenses as mature ones; but as young adults they were twice as likely to use mature mechanisms as immature ones; and finally, in middle life they were four times as likely to use mature as immature defenses (pp. 329–330)."

Both writers pointed out the connection between internal (endogenous) and external (exogenous) worlds. Mason (1979), however, dealt directly with the phylogenetic meanings of these inner and outer worlds. When the environment is stable or predictable, animals may mature quickly and show automated or wired-in responses. When the environment is variable or semi-erratic, special demands are placed upon the organism and behavioral "openness" becomes more adaptive. This openness or flexibility has been one major trend of mammalian evolution and prolonged immaturity one of its driving mechanisms.

The macaque external world shows an interesting dichotomy between structural objects like the forest and social objects like other animals. The components of the former are relatively stable; that is, they are fixed in space and relatively permanent in time, but they are emotionally neutral. The latter objects, as Mason (1979) noted, are moody, capricious, unpredictable, but sources of social value. An anatomical and developmental dichotomy parallels this separation of the world. The locomotor tissues have built-in limitations (joint surfaces, muscle architecture, bony leverages) for the world of trees and branch crossings. These tissues mature slowly and the social group takes care of the individual. By contrast, the nervous system has many more degrees of freedom and its neuronal populations are particularly sensitive to early experience and other social factors. As a consequence, the neural tissues attain mature size early to gain maximum stimulation. Lengthened immaturity enhances social plasticity. This dichotomy in the external world mirrors growth rate differences of the internal world.

Nevertheless, despite the psychosocial advantages of lengthened immaturity, the growing animal is exposed to other exogenous hazards. The longer the time of development, the greater the chance that tissue maturation will be deflected by pertubations of the food supply. Food reduction may delay the development of the brain early in ontogeny. Later in life food reduction may delay bone and muscle growth; still later in time, fat deposits are altered (Berg, 1968; Widdowson, 1968). But whether these nutritional shifts impair or simply delay tissue maturation is debated. Malina and Zavaleta (1980) have recently shown that, whereas people in recent generations of all industrialized nations have grown taller and heavier, Mexican-Americans have not grown as much. Presumably, this discrepancy is due to a lower plane of nutrition.

Once an animal reaches adulthood, the stresses of famine, migration (or hyperactivity), or childbearing may be buffered by stores of adipose tissue. Several independent lines of evidence exist. First, stored fat allows Canadian geese to undertake their annual migration but it also permits females to initiate nesting before food is freely available. Depletion of energy reserves for the eggs may cause the female to stop laying. Whatever the precise mechanism, clutch size in geese is closely attuned to food supply (Raveling, 1979). Second, Laudrie (1975), the French historian, found a peculiar demographic phenomenon in birth and death records of humans over the past three hundred years. Correlated with the predictably high death rates during famines and sieges, he found a drastic reduction in births. As grain prices peaked and as deaths from starvation and epidemic rose, the number of conceptions fell and women complained increasingly of cessation of menstruation. This relation between famine and sterility he calls "famine amenorrhea."

Frisch puts the explanation into physiological perspective (Howell, 1979). The human female stores on the average almost twice as much adipose tissue as the male. If a woman falls below a critical level of fatness (17 to 20%), reproductive function is impaired; she may stop menstruating and become temporarily sterile. This has recently been noted among female endurance athletes, particularly marathon runners, during heavy training. Do these fat and energy dimorphisms found in macaques and other mammals serve as survival mechanisms for the reproductively mature during periods of reduced nutrition?

Conclusions

Within the limits of my method, these functional interpretations are probably correct. However, whether this method would be adequate for other analyses depends far more upon the number and breadth of hypotheses generated. My introduction to the anatomy of macaque locomotor development should be adequate, if the questions are limited to ones such as the following. 1) What are the adaptive values of anatomical and locomotor maturity? 2) What are the effects of reduced levels of nutrition for each developmental stage? 3) How is energy intake at each stage of the life history balanced with metabolic demands for growth, for maintenance, and for activity? 4) As a troop of monkeys moves through the canopy what are the locomotor problems that each skill class has? By extension, as the motor capacity of an animal changes, how does it vary its route to keep up? Do different skill classes exploit different parts of the habitat (Temerin et al., 1980)? How does an animal judge how far it can jump? 5) Do macaque females possess more body fat than males and does the additional fat buffer the animal against the energetic demands of gestation and lactation?

Acknowledgments

I want to express my thanks to several friends and colleagues with whom I discussed this analysis: Neal Cross, Chuck Kowalski, Alan Martin, Charlie Menzel, and Bill Ross. In addition, I had the assistance of many staff members of the Oregon Primate Center: Jim Palotay, who cooperated in numberless ways and made animals available to me; Janina Ely, who tightened, then polished versions of the manuscript; Bob Coffin, Melissa Morrow, Wayne Borum, who taught me the mechanics of editing on a computer terminal; Kathy Kerr and Jeanette Cissman, who did the charts and figures; Harry Wohlsein, whose films have put me close to the biomechanics of macaque development. The work described in this chapter, Publication No. 1223 of the Oregon Regional Primate Research Center, was supported in part by USPHS grant No. 00163-23.

Literature Cited

Austin, G. T., and R. E. Ricklefs. 1977. Growth and development of the rufous-winged sparrow (*Aimophila carpalis*). Condor, 79:37–50.

Berg, R. T. 1968. Genetic and environmental influences on growth in beef cattle. Pp. 429–450, *in* Growth and development of mammals (G. A. Lodge and G. E. Lamming, eds.). Plenum Press, New York, 527 pp.

Butterfield, R. M., and E. R. Johnson. 1968. The effect of growth rate of muscle in cattle on conformation as influenced by muscle-weight distribution. Pp. 212–223, *in* Growth and development of mammals (G. A. Lodge and G. E. Lamming, eds.). Plenum Press, New York, 527 pp.

FOWLER, V. R. 1968. Body development and some problems of its evaluation. Pp. 195–211, *in* Growth and development of mammals (G. A. Lodge and G. E. Lamming, eds.). Plenum Press, New York, 527 pp.

GRAND, T. I. 1972. A mechanical interpretation of terminal branch feeding. J. Mamm., 53:198–201.

———. 1976. Differences in terrestrial velocity in *Macaca* and *Presbytis*. Amer. J. Phys. Anthrop., 45:101–108.

———. 1977*a*. Body weight: its relation to tissue composition, segment distribution, and motor function. I. Interspecific comparisons. Amer. J. Phys. Anthrop., 47:211–240.

———. 1977*b*. Body weight: its relation to tissue composition, segment distribution, and motor function. Amer. J. Phys. Anthrop., 47:241–248.

———. 1978. Adaptations of tissues and limb segments to facilitate moving and feeding in arboreal folivores. Pp. 231–241, *in* The ecology of arboreal folivores (G. G. Montgomery, ed.). Smithsonian Institution Press, Washington, D.C., 547 pp.

HARKNESS, R. D. 1976. Locomotion. Pp. 134–162, *in* Environmental physiology of animals (J. Bligh, J. L. Cloudsley-Thompson, and A. J. McDonald, eds.). Halsted Press, John Wiley and Sons, New York, 456 pp.

HOWELL, N. 1979. Demography of the Dobe Kung. Academic Press, New York, 389 pp.

KIMURA, T., M. OKADA, AND M. ISHIDA. 1979. Kinesiological characteristics of primate walking: its significance in human walking. Pp. 297–312, *in* Environment, behavior, and morphology: dynamic interactions in primates (M. E. Morbeck, H. Preuschoft, and N. Gomberg, eds.). Gustav Fischer, New York, 410 pp.

KOWALSKI, C. J., AND K. E. GUIRE. 1974. Longitudinal data analysis. Growth, 38:131–169.

LAUDRIE, E. L. 1975. Famine amenorrhea (seventeenth-twentieth centuries). Pp. 163–178, *in* Biology of man in history (R. Forster and O. Ranum, eds.), Johns Hopkins Univ. Press, Baltimore, Maryland, 224 pp.

MALINA, R., AND R. N. ZAVALETA. 1980. Secular trend in the stature and weight of Mexican-American children in Texas between 1930 and 1970. Amer. J. Phys. Anthrop., 52:453–462.

MASON, W. 1979. Ontogeny of social behavior. Pp. 1–28, *in* Handbook of behavioral neurobiology, Vol. 3: Social behavior and communication (P. Marler and J. G. Vandenbergh, eds.). Plenum Press, New York, 410 pp.

MILLER, D. I., AND R. C. NELSON. 1973. Biomechanics of sport. A research approach. Lea and Febiger, Philadelphia, 265 pp.

O'CONNOR, R. J. 1975. The influence of brood size upon metabolic rate and body temperature in nestling blue tits, *Parus caeruleus*, and house sparrows, *Passer domesticus*. J. Zool., 175:391–403.

———. 1977. Growth strategies in nestling passerines. Living Bird, 16:209–238.

OKAMOTO, T. 1973. Electromyographic study of the learning process of walking in 1- and 2-year old infants. Pp. 328–333, *in* Medicine and sport. Vol. 8: Biomechanics (S. Cerquilini, A. Venerando, and J. Wartenweiler, eds.). Karger, Basel, 511 pp.

PALMER, C. E. 1944. Studies of the center of gravity in the human body. Child Devel., 15:99–180.

RAVELING, D. G. 1979. The annual cycle of body composition of Canada geese with special reference to control of reproduction. Auk, 96:234–252.

RICKLEFS, R. E. 1968. Patterns of growth in birds. Ibis, 110:419–451.

———. 1973. Patterns of growth in birds. II. Growth rate and mode of development. Ibis, 115:177–201.

RICKLEFS, R. E., AND S. WEREMUIK. 1977. Dynamics of muscle growth in the starling and Japanese quail: a preliminary study. Comp. Biochem. Physiol., 56A:419–423.

RICKLEFS, R. E., D. C. HAHN AND W. A. MONTEVECCHI. 1978. The relationship between egg size and chick size in the laughing gull and Japanese quail. Auk, 95:135–144.

TANNER, J. M. 1969. Growth at adolescence. Blackwell Sci. Publ., Oxford, 326 pp.

TEMERIN, L. A., B. P. WHEATLEY, AND P. S. RODMAN. 1980. The influence of body size on foraging in primates. Am. J. Phys. Anthrop., 52:280–281 (abstract).

VAILLANT, G. E. 1977. Adaptation to life. Little, Brown and Co., Boston, 396 pp.

WIDDOWSON, E. M. 1968. The effect of growth retardation on postnatal development. Pp. 224–233, *in* Growth and development of mammals (G. A. Lodge and G. E. Lamming, eds.). Plenum Press, New York, 527 pp.

A VIEW OF ONTOGENY

JOHN C. FENTRESS

Abstract

THIS essay explores thematic issues relevant to the understanding of behavioral ontogeny in mammals. First, methods of behavioral taxonomy that are rigorous, yet broad in perspective, must be sought. Both the dynamic and relational features of behavior deserve special consideration. One can then seek rules of relation among defined control systems and their surrounds, along with tendencies for these control systems to become self-organized. There is some evidence for a hierarchical progression in ontogeny where systems move from a state of loose structure to simple well-defined structure to more elaborate combinations of action. Extrinsic influences may play many roles in ontogeny, with strict forms of instruction (as in learning theory) being just one. These extrinsic influences combine with the organism's genetic heritage to bridge past, present, and future. Self-organizing tendencies in behavior may not only contribute importantly to initial differentiation (in cooperation with the surround) but act to buffer the developing organism from various perturbations in the surround. Both individual action patterns and socially coordinated forms of expression may yield to similar methods of analysis, thus giving a more satisfactory conceptual grasp on the critical features of ontogenetic change. These issues are illustrated through research on rodent movement patterns and canid social organization. Recent models in developmental biology and biophysics may prove useful in guiding certain lines of future research in behavioral ontogeny.

"An organism is merely a transition, a state between what was and what will be."

F. Jacob, 1973

Introduction

Ontogeny is nature's way of joining an organism's evolutionary history to the present moment, thereby gaining some assurance for access to the future. The central view I wish to express in this chapter is that our models of behavioral development depend upon the slices we make in time, between actions, and across levels of organization. Yet it is clear that our initial distinctions dictate the subsequent rules of relation we find, and that it is these relations which generate our explanations about how behavior develops. In short, the initial problem is one of gaining an appropriate taxonomy of the behavioral phenomena we wish to investigate.

Problems and Approaches

Taxonomy of change.—The problem of obtaining a satisfactory taxonomy of behavioral development is magnified over that which might seem adequate for adult animals, because development involves change. Our taxonomies must thus be sufficiently fluid to deal with change in various dimensions while still being sufficiently stable to structure our thought. The problem of behavior categories is never easy to deal with (for example, see Fentress, 1973*a*, 1976*a*), but at least it should be recognized in explicit terms. Wilson (1975) summarized the issue well with his statement: "If the construction of categories is hopeless, it is also profitable" (!). It is a matter of treating categories in a reasonable manner, and also recognizing that a variety of complementary schemes for classification can be used with profit (Golani and Moran, this volume).

Compartments and relations.—A second major problem is how to evaluate the rules of relation among events both intrinsic and extrinsic to the system we are investigating. This, of course, is basic to all ethology (Hinde, 1970; Marler and Hamilton, 1966). It may indeed be the major problem in all of natural science, for somehow we must deal with a world of definable pieces that are also interconnected. If interactions were the only feature we would end up with so much homogenous soup, whereas if extreme compartmentalization were the rule there would be no way to obtain organized action. The ancient battles of nature *versus* nurture in discussions of behavioral ontogeny reveal how very difficult it is to think about. Somehow we must gain better tools to think about nature *plus* nurture (and *vice versa*!).

Diversity.—The third problem is that the diversity of animal species is reflected through the diversity of behavioral patterns these species perform, and a probable diversity of mechanisms. How are we to generalize? As Hinde (1972) stated, it is critical for the biologist to maintain a sense of "humility in the face of diversity." Yet we also want to do something more than just assemble an animal stamp collection. This issue is relevant not only to cross-species comparisons, but also to different classes and levels of activity within a single organism. In science we seek organizational themes that transcend diversity rather than fly in the face of it (Holton, 1978). Darwin's theory of evolution, based as it was explicitly upon the fact of diversity, is biology's primary claim to fame in this respect (Mayr, 1970). In the study of development, we have not even begun to approach this stage of sophistication (for example, see Bateson, 1976; Fentress, 1981*a*, 1981*b*).

Approach.—Given these three basic problems (taxonomy of change and relation, evaluation of interactive and self-ordering processes, and diversity), where should one begin—and how? I start with description by concentrating upon two quite different systems of behavior that have interested me, my students, and colleagues for several years. These are: 1) the organization of relatively simple movement sequences in rodents (*Mus musculus*), and 2) the organization of social actions in canids (*Canis lupus, C. latrans*). This will certainly provide diversity! And it should also lead to thoughts about how one might relate quite different levels of behavioral organization (that is, action patterns of the individual and socially integrated behaviors). My next task will be to look both "upward" and "downward" in search for possible constraints in development by 3) exploring social support systems in canids, and 4) evaluating experience in the context of genetic potentialities. I hope to show that some common themes can be found in each of these instances. Two are basic. The first is that rules of relation among processes may be a more important focus than a concentration restricted to processes in isolation. The second is that experience may promote self-organizational tendencies in mammalian behavior rather than just impose the details of ontogeny, from "without" so to speak (as, for example, implied by most learning models of development). My emphasis throughout is upon problems of description and analysis, for these are the prerequisites to solutions.

Dissecting the Action Stream

It is common for ethologists to speak, with some sense of awe, about the "stream of behavior" (Hinde, 1970). This implies a continuity in expression that defies simple segregation into component parts. Perhaps most obvious to those who work with mammals, actions appear not only multidimensional but also often blend together with no clear lines of demarcation. How many notes are there in the howl of a wolf; how many postures does a wolf show in communicating with its comrades; and where do we draw the line between "communication" and something else (Fentress, 1978*a*, 1978*b*; Golani, 1976; Moran and Fentress, 1979; Schenkel, 1947; Schleidt, 1974)? Often "units" appear to be useful fictions at best—convenient (indeed necessary) for analysis, but not be accepted in the more absolute sense of "really real."

Toward a Behavioral Taxonomy: Perspectives and Problems

I can illustrate the themes of behavioral taxonomy and its relevance to ontogeny through the "sometimes convenient fiction" of movement components in the facial grooming of rodents.

Rodent grooming: form and function.—Few observers of adult rodents would have difficulty deciding when the subject is grooming its body surface, for there is a characteristic series of head, body, and (most obviously) forepaw movements that are directed toward the body surface. However, the behavioral category of grooming has as its primary referent the presumed function of attending to peripheral irritants rather than to a detailed decomposition of the movement itself. This is equally true of subcategories of grooming, such as grooming of the head, belly, and back (Eisenberg, 1963; Fentress, 1972). The observer certainly makes some mental note of the distinctive features of movement used to gain access to these various body surface regions, but the categories most commonly refer to the *summed consequences* of individual movement components.

The distinction between form and function can be critical to developmental analysis. What if we observe an infant mouse making forepaw movements that are reminiscent of those used in adult grooming, but in which the forepaws fail to contact the face? Is this "grooming?" Most observers would say "yes," or at least, "it's a precursor to grooming"—a response which I have confirmed by showing films to several classes of students. What these students do, as we all do, is to mix criteria for classifications—picking and choosing amongst them as the situation or our biases dictate. This is often adequate for initial inquiry, as form and function necessarily have a certain cohesion. But this cohesion does not have to be perfect because animals can achieve the same end result through variable means (Golani, 1976; Lashley, 1951). Such facts can cause havoc with our categories, and most critically to the inferences we draw from them.

Other issues in description.—In addition to the multiple descriptive criteria that we use to form categories, there are at least four additional issues of importance. First is the degree of *discontinuity* we accept in separating one category from another. For example, if notes blend together in a wolf howl, do we speak of one behavioral unit or many imperfectly separable units? The second is the degree of *heterogeneity* we will accept within a category. The broader our categories, the more heterogenous they will appear at refined levels of inquiry. The third issue is the most *appropriate level* to make the categories for our problem. Thus broad features of behavior, such as fighting, grooming, howling, or mating, may contain many distinguishable features, but for some problems the broader categories are of primary interest. The critical issue is how to relate these potentially diverse levels of inquiry to one another. The fourth issue is how to deal with categories whose features may *change,* as they do in ontogeny. Obviously these issues are not completely separable; they are related features of the broader problem of behavioral taxonomy.

Perceptual context.—Human perceptual systems are very good at making categories, which is both a bonus and a limitation to observational research. For example, in speech perception we can segregate relatively continuous sounds into discrete phonemes with normally little ambiguity (Lieberman, 1977). Change the context within which particular sounds are embedded, however, and confusions can, and often do arise. This means that we do not interpret events in isolation. The issue can be particularly important for developmental analyses. If a young child articulares the sound "ba," I may, quite unambiguously, decide that it refers to a "ball" or to the sound of a sheep, depending on the context. But the lack of ambiguity is not attributable to the sound per se; it is a function of a broader, usually implicit, contextual analysis on my part. It is an interesting exercise to splice bits of animal behavior film together into distorted sequences and find out what happens to the observer's categories (another useful classroom exercise)!

Clusters versus single dimensions.—Fentress and Stilwell (1973) found that it was possible to subdivide facial grooming in mice into a variety of stroke types, without ambiguity, if the observer combined various dimensions, such as movement amplitude, point of contact on the face, and coupling between the forelimbs. These actions were quite discrete as *clusters.* Woolridge (1975) confirmed this, but also demonstrated that when single dimensions of movement—such as velocity or amplitude of limb movement, and area of face covered—were analyzed quantitatively, the distribution of data points appeared much more continuous. Similarly, Machlis (1977) found that quantitative analyses of perceptually distinct bouts of pecking in chicks yielded intervening durations that had not been appreciated previously.

Ontogenetic perspective.—Because the behavior of animals varies across numerous dimensions, these elementary considerations can be extremely important to developmental analyses. For example, Stilwell and I (see Fentress, 1978*c*) found that it was not possible to classify grooming strokes with the criteria derived from adult mice until the animals were approximately 10 days of age. Only later Golani and I (pers. observ.) separated out basic dimensions such as the details of kinematic articulation in various limb segments and the resulting contact pathways between the forepaws and face, and a fuller picture of early grooming ontogeny emerged. Thus infant mice might show characteristically rhythmical forearm movements with only fortuitous contact between the forepaws and the face; each of these contributions to the total pattern could be tracked independently.

There were still some basic decisions, if one wanted to continue to use the broad (and originally functional) category of grooming (versus something else). Thus young mice in the filming chamber would occasionally "struggle" with rhythmic leg movements as well as "groom" with rhythmic leg movements. The distinction was that in "grooming" as opposed to "struggling," the head always turned to the midline and the forepaws turned perpendicular to the ground rather than horizontal to it. By tracking these individual dimensions, and looking at their rules of combination in time, one could view the processes of ontogeny in a more systematic and precise manner than was possible through the original broad category of "grooming." We still use these broad terms when conversing, but they do not dictate unwanted constraints in our analyses.

Pattern formation: differentiation and integration.—The points I have made have important ramifications for the question of pattern formation in behavioral ontogeny because the behavior of young animals often appears much less distinct (that is, less clearly differentiated) than that of adults. Stilwell and I found that young mice exhibit certain strokes (called "X" strokes; Fentress, 1978*c*) that are intermediate in their characteristics between two adult stroke types, an expected result if behavioral systems differentiate progressively from common origins as do many structural systems in biology (Ede, 1978). Yet it is a perspective that has received relatively little attention in the behavioral literature. Many of these issues can be isolated quantitatively even at the level of interactions *between* animals, such as in Havkin's (1977) findings of a progressive differentiation of highly symmetrical and asymmetrical interactions from earlier stages of loose symmetry during social ontogeny in wolves. The issues raised in rodent grooming have their ramifications elsewhere.

Heterogeneity.—A related issue of potential importance is that of heterogeneity in classes of behavior defined for young animals. In classifying the individual facial movements in young mice, there can be considerable looseness of expression from one moment to the next within a "category." This of course also relates closely to the problem of distinctions between categories, as we have found with wolf vocalizations. Adult categories such as "howl," "squeak," "bark," and "growl" are certainly heterogeneous but also distinct (Fentress, 1967). As shown by Field (1978, 1979), individual criteria, such as frequency and syllable duration within a category, may be even more variable for young animals than for adults. This may be of little interest if the categories remain distinct, but they do not always do so. Thus early "howls" and "whines" are much more similar in infant wolves than in adults (Harrington and Mech, 1978), perhaps suggesting a developmental linkage between the two. Even if one can preserve the distinction between behavioral categories in developmental analyses, precise measurements of individual dimensions in terms of their distinctiveness are necessary preludes to more satisfactory evaluations of the processes of behavioral ontogeny.

Higher-order patterns.—So far I have treated the broad categories of behavior as more or less isolated events. To do so is an error. Grooming in rodents, for example, is but part of still higher-order patterns of expression. Thus adult rodents commonly groom during transition between active and inactive states; these states and the expression of grooming are not truly independent.

Rules of transition between behavioral categories are as basic as are rules of composition within these categories. Simply put, young animals—like young children—often surprise us with respect to *when* they do things (that is, what context), as well as *how* they do things.

Thematic Illustration: "Motor Traps" and "Hierarchies"

A young mouse may be crawling across a table and suddenly start to groom its face. If we are used to observing adult animals only, we are surprised. A young wolf may suddenly break away from a rough and tumble interaction with a peer and start chewing on a bone or stick. Such findings are common. Recent research in our laboratory has suggested reasons for these "unexpected" transitions. We summarize these under the general heading of *motor traps.*

Action combinations and function.—It is usually possible to arrange the performance of even a single class of behavior into a descriptive hierarchy (Dawkins, 1976; Gallistel, 1980; Tinbergen, 1951). For example, Fentress and Stilwell (1973) found that the facial grooming strokes could be arranged into higher-order "units" of expression on the basis of differential probabilities throughout a grooming sequence. These units in turn are combined sequentially into still higher-order "phrases," etc., much in the manner of the letters, words, and phrases of the English language (see also Fentress, 1972, 1977*a*, 1980). Grooming itself is embedded through sequencing rules between actions such as walking and sitting still.

At the level of broad functional behavior categories, similar motor patterns may be used to achieve quite distinct endpoints. Thus a wolf can bite food, a rival, or even its mate; or animals can locomote toward or away from a variety of objects in their environments.

"Traps": through similarities in form.—Young mammals often get "trapped" by movement, or body posture, and use the similarity of movements/postures in different functional categories of behavior as a basis for making transitions between these categories. Golani and I demonstrated this by setting infant mice up into a posture

commonly used by adults when grooming. Finding themselves thusly situated, the animals frequently commence to groom (that is, make the rhythmic forelimb movements we are calling, for convenience, grooming). And if they slow down or doze off, they can be made to "groom" again merely by pinching the tail lightly!

In a non-manipulative context these young mice also commence "grooming" movements if, by chance, the forepaws contact the face during locomotion (such as when attempting to navigate a turn). Wolf pups who are "playing" roughly with one another may find themselves biting a stick or food object rather than the partner, and commence feeding. This very common class of observation is reminiscent of Lind's (1959) earlier analyses of "transitional actions," but it has not yet entered the developmental literature and is only revealed precisely through detailed analyses of "whole patterns" at different levels of complexity. It suggests that there may be shifting priorities in levels of control with development—early stages being "trapped" at lower levels of motor similarity (Fentress, 1981*a*, 1981*b*; Teitelbaum, 1977).

Foci of flexibility.—These observations are complemented by data we have recently obtained on motor flexibility in reference to a *single goal.* Adult animals can cause havoc with single-criterion behavioral categories because they are able to adjust particular actions to achieve a constant endpoint. If a cuff is placed over one forelimb of an adult mouse, it can (under most circumstances) successfully facial groom by adjusting various other body segments. Similarly, Golani and I have found that mice at approximately 10 days (when the facial grooming categories of Fentress and Stilwell, 1973, become useful) can adjust facial grooming in the nest if they happen to have one forelimb weighted by a littermate. I have noted preliminarily that very young mice (for example, 5 to 6 days) are unable to make these adjustments—or at least they do not as often do so.

Hypothesis of ontogenetic hierarchies.—Our data suggest that younger animals may be operating at lower hierarchical levels than are older animals. More direct evidence for this contention stems from several sources: 1) mice at approximately 4 to 5 days of age can locomote across a table for brief distances. If one taps a single hindleg of the mouse, this hindleg will often "run away," leaving the progression of the other three limbs more or less intact (Fentress, 1972). 2) Newborn mice will swim if placed into a warm bath. Templeman and I (see Fentress, 1978*c*) found that the coupling between the individual forelegs and hindlegs was very loose in these young animals, so that only one limb would "swim" at any given time or a single limb would be left behind while the others swam. Bekoff and Trainer (1979) have confirmed this for rats, and conclude that in young animals each limb may have its own neural pattern generator (that is, run by lower hierarchical levels). 3) Havkin (in litt.) has documented several cases in wolf pups where a single limb gets "left behind" causing the animal to trip over an obstacle—again suggesting partially independent limb control. Although young mammals do not always show this independence of movement components, they are more likely to do so than are adults. Thus one must speak in "more or less" rather than "all or none" terms.

Form and orientation in behavior.—A different perspective with a similar point is the separation of form and orientation in mammalian movement patterns. As I mentioned, young mice may go through many of the individual kinematic articulations of facial grooming without making successful functional contact with the face; this is sometimes due to the failure of one limb segment to adjust for variations in the other. Also there are times when *each* limb segment (or even individual trajectory components within a limb segment) appears to be slightly out of tune. For many years ethologists have noticed that the successful orientation of movement patterns may follow the identifiable expression of these movements. Thus, Eibl-Eibesfeldt (1956) reported that young squirrels may articulate recognizable movements of nut opening without joining them into a well-oriented nut-opening sequence, and Fentress (1967) found that a young wolf could be made to pounce in response to a moving object without orienting the pounce toward the object. What these observations indicate clearly is 1) that different dimensions of integrated performance may have different developmental courses, and 2) that the synthesis of these dimensions into higher-order units of expression may often be a later developmental event than is the perfection of these dimensions in isolation.

Pieces and Their Relations: The Differentiation of Integration

The picture can occasionally be more subtle than I have implied. For example, Stilwell and Fentress (Fentress, 1978*c*) found that the synthesis of mouse grooming strokes into higher-order units *and* the proper sequencing of these units all emerged at approximately 14 to 16 days of age. While perfect generalizations about ontogenetic sequencing are thus not possible at this stage, a careful dissection of behavioral dimensions within an ontogenetic perspective may lead to the establishment of formal rules for which separable mechanisms may be discerned.

Diffuse and modular properties.—There is an additional issue that has the appearance of paradox. Infantile (young) behavioral patterns are often less clearly differentiated than those of adults, and young animals also frequently exhibit behavioral fragments that are not fully integrated into broader functional wholes. Both diffuse and modular properties can be seen in young mammals.

The diffuse nature goes beyond undifferentiated form, applying also to processes that initiate actions and affect their intensity (that is, "motivational processes"—see Fentress, 1973*a,* 1976*a,* and 1978*a,* for fuller discussion). Thus mild pinches of the tail, which can generate grooming in young mice in the appropriate "sitting" posture, are much less effective in older animals. Analogously, Hall (1979) showed that direct injections of food into the stomachs of infant rats may lead to "diffuse activation," with grooming behavior as a common consequence. As the animals get older they are more likely to restrict responses to food injections to feeding activities. Infant wolves as well as infant mice can be made to feed by the application of various peripheral stimuli; in adult animals one can still do this but only in a more statistical and fragile sense.

Thus, in the behavior of young mammals, higher-order (for example, "motivational") processes are frequently quite unstructured (for example, watch a human infant smile in response to gas pains as well as social stimuli), with the structure being provided by the motor set of the creature at the moment (very easily shown in young children). Once the "set" action is initiated, then it may be expressed in splendid isolation, that is, without in turn generating other, descriptively distinct, actions that in older animals join in the service of a common function.

So the apparent paradox potentially resolves itself *if* one can assume (or better yet, gain evidence!) that action systems at higher levels of organization are often diffusely organized at the same time that lower order elements are relatively discrete and independent from one another (for example, see Gallistel, 1980). Of course, "higher" and "lower" are relative; at any given level there might be a progression from 1) imperfectly structured (that is, not clearly definable) to 2) clearly structured but unintegrated (in a broader functional sense) to 3) both structured and integrated. Recent data we have collected support this set of hypotheses.

Stages of Structure: Clarification and Elaboration

In this section I give two examples that differ both in terms of species and level of complexity. One involves behavioral coordination of limb segments for the individual organism. The second involves social interactions between animals.

Limb coordination.—In the ontogeny of grooming in inbred house mice (*Mus musculus*), three quite separable stages can be discerned. In the first (postnatal days 1 to 4) the animals perform a variety of movements that are relatively independent in terms of sequential connection, are difficult to distinguish on precise morphological grounds, and achieve competent function only on a fortuitive basis. In the second stage (postnatal days 5 to 9), the movement details become better coordinated with one another, take on a clearly distinguishable taxonomic basis, and achieve reliable functional contacts with the face. At the same time these movements become simplified in terms of their total expression. Next (postnatal days 10+) individual grooming movements become differentiated *and* coordinated, thus capturing *both* the complexity and order of earlier stages. Even at this third stage, however, the integration between grooming as a total functional complex and other actions is still not fully adult-like. Grooming can still be elicited by a variety of higher-order stimuli, if the animal is in an appropriate behavioral set, and also the predictability of grooming in behavioral transitions extrapolated from adults is quite imperfect. The ontogeny of rodent grooming thus shows 1) an initial differentiation of component acts, regardless of their total functional utility, 2) a clarification, and also simplification, of individual behavioral components, and 3) a resynthesis of these, now differentiated, behavioral components into higher-order combinations. Put simply, this involves a process of clarification (simplification), then elaboration (Hines, 1942).

Social coordination.—In a study of early social interactions, Havkin (in litt.) has obtained evidence that around four weeks of age wolf pups perform a variety of loosely coupled actions that are often relatively independent of the actions of the other animal. By approximately eight weeks the interactions become more clearly defined, with the two animals tightly coupled (for example, knowledge of the action of one animal becomes a good predictor of the second). At approximately 12 weeks the animals show a much lower proportion of action that most observers would unhesitatingly classify as "play." The individual animals are obviously attending to one another in that they are normally found in close proximity, but the details of their individual actions are more separable and elaborate than at eight weeks of age. Thus, rather like "social behavior" among the forelimbs and head of an individual mouse, two wolf pups show ontogenetic transitions between a variety of loosely defined movements leading to fortuitous contact between them, to a simplification and clarity of interactive structure, to a subsequent re-elaboration into what one would label as more "mature" interaction.

Analogous transitions are being revealed for human children. Reigel (1976) has argued that interactions of infants at 4 to 6 weeks of age appear as "collective monologues," whereas by 10 to 12 weeks they are more like "simple dialogues," implying a progressive restriction in freedom of the behavior of the child as a function of its partner. MacCombie (summarized by Cairns, et al., 1980) has shown that children go through an intermediate stage in play in which they tend "to entrap each other in common activities." This implies a close symmetry in the performance of the children. Later the children escape from a simple synchrony in their behavior; that is, their performance is more complex and elaborate.

A Cautionary Note on Hierarchical Interpretation

My wife and I have had the pleasure of watching the development of language in our two daughters. While it is certainly true that words become identifiably structured before sentences, these are not totally independent events. For example, when "higher-order" grammatical structures are clearly being practiced, the articulation of individual words may deteriorate, and then become re-perfected. This suggests that hierarchical levels in expression are not fully independent, and the perspective about "which level" becomes perfected first can depend upon the particular time frame emphasized. An analogous phenomenon of momentary deterioration of "lower levels" of expression may also occur during the mastery of socially integrated movements, as suggested by recent observations of Havkin (in litt.) on wolf social ontogeny. Thus, the simple view that lower levels achieve perfection before, and independently of, higher levels must be judged cautiously. Development obviously proceeds in two directions at once; that is, the perfection of higher levels and the refinement of lower levels of expression represent a cooperative enterprise involving the entire intact organism at any time.

The stages of simplification and then elaboration that follow loose structure may permit the perfection of progressively larger and larger blocks of action, and indeed in later ontogeny may be triggered by momentary deteriorations as just mentioned. Certainly child psychologists (for example, see Bruner, 1973) as well as investigators concerned with the perfection of motor skills in adults (for example, see Bernstein, 1967; Posner, 1978) commonly speak of the progressive elaboration of higher and higher units of expression, where the participating elements gain a progressive automaticity in their expression. Whether or not there is a momentary deterioration, then re-perfection, appears to be a relatively unexplored issue.

The extent to which the rules of social entrapment bear formal similarities to the "motor traps" defined earlier for individual animals should also be pursued. In each case the behavioral consistencies are based to a large measure on similarities of form. Indeed, it may be this production of form that contributes to higher-order states in young animals as well as the reverse. Thus we have found that by forcefully engaging a young wolf in some movements used in "play" (for example, rhythmically tugging upon the forepaws), it is often possible to reduce the animal's tendencies to flee. This does not work so well with older animals! However, the form of movement, even imposed movement, in young animals can indeed switch them into quite different functional "sets."

Toward Categories of Social Ontogeny

I have already started considering social relations and their ontogeny by introducing such intuitively based constructs as social "play." Three important points should be highlighted. First, one may be able to abstract rules of ontogeny and performance and apply them to social relations of animals as a "superorganism" (Moran et al., 1981) as well as to coordinated actions within the individual, such as in the stages of loosely elaborate, simply structured, and elaborately structured. Such measures can in principle be quantified. Second, terms such as "play" are most often imperfectly and intuitively defined, and merely set the stage for subsequent dimensional analyses rather than providing either unitary descriptions or explanations (Symons, 1978). Third, the appearance and disappearance of certain styles of behavioral organization at particular stages of life adds a caution to thinking about unidirectional trends in ontogeny as well as to viewing all infant and juvenile behavior as leading to a perfected adult state. Selection operates at each phase of an animal's life, and there are special needs that the animal progresses through as a function of its life phase. Little animals are much more than imperfect big animals (Oppenheim, 1980).

"Play"

Definitions and their limits.—In the canid literature the construct of "play" has been used frequently (for example, see Bekoff, 1974; Fox, 1971), but it has not been analyzed for its dynamics of organization (Fagen, 1978, 1981; Havkin, 1977). Rather, categories such as "wrestling," "playbow," and "ear posture" tend to be treated both statistically and as logical equivalents. This form of description is reminiscent of the typological categories of early evolutionary biology (see Mayr, 1970, for a valuable critique of such descriptions). And there are certain classificatory criteria that are of questionable analytical validity, one of which is "exaggerated movement." What does "exaggerated" mean precisely—larger amplitude, faster excursion of limbs, poor coordination, or some (undefined) combination of these? The available literature does not yield satisfactory answers (Henry and Herrero, 1974; Symons, 1978).

That observers can agree that animals are playing is but a first step, not a solution. White's (1977) comment that "it looks like fun"(!) is a refreshing escape from some of the pseudoprecision of previous definitional attempts (see also Fagen, 1978; Welker, 1971). Symon's (1978:121–122) description of play as behavioral patterns that "differ from similar patterns in the species repertoire, and either lack immediate function or, at least, lack the function of the patterns they resemble" highlights some of the most important issues for ontogenetic analysis (though perhaps embarassingly appears to apply quite well to immature forms of rodent grooming discussed above!).

Ontogenetic perspectives.—Descriptive, causal, and functional analyses can easily become confounded in discussing constructs such as play (White, 1977), and it is not always clear what level of order is most relevant to revealing properties of "dynamic programming" (Fagen, 1978) within which play constructs might in the future be translated. Although little animals are more than just imperfect adults, some of the movements that appear to characterize play (description) might indeed be produced (causation) by immaturity of the sensory-motor and higher neural apparatus, and contribute to (function) the maturation of this apparatus (Bekoff et al., 1980). For example, the stumbling movements of young wolf pups that often give them the appearance of lack of seriousness of intent, may reflect immaturity of the animals' motor apparatus, and head-bobbing might be due in part to imperfect visual resolution (compare head-bobbing of human adults when they have difficulty making out a visual stimulus, especially one that is moving). But these are hypotheses that can only be resolved through more adequate *descriptions* of what animals actually do at different ages.

Symmetry, Order, and Relations

Havkin's documentation of a progression from loose symmetry in social contact of young wolves to forms of interaction that are either highly symmetrical *or* asymmetrical demonstrates the value of such descriptions (Havkin, 1977). Symmetry is an abstract concept that can be applied to a variety of natural (as well as artificial) phenomena (Shubnikov and Koptsik, 1974), and it permits an escape from the blinders imposed by traditional typological categories.

A complementary approach to the problem of *ordering principles* in wolf social behavior has been developed by Moran (1978; Moran et al., 1981) in his analysis of supplanting interaction in adult animals (see also Golani and Moran, this volume). Moran selected three physical dimensions involved in such interactions: 1) distance between the animals; 2) angle of orientation between the animals; and 3) point of closest opposition of one animal on the body of a second. He then examined the relations among these three dimensions within an abstracted "interaction space." He found that during supplanting behavior the animals maintained relatively fixed configurations in but two regions of the total interaction space, and also moved between actions within these regions through a small set of theoretically possible trajectories.

The relational perspective.—The analyses of Havkin and Moran are important to social behavior and its ontogeny because the actions of one animal are defined with explicit reference to the actions of another. Thus symmetry can only be defined in terms of relations between two (or more) animals; the same holds with inter-animal distance, orientation, point of nearest opposition, and the synthesis of these (Golani, 1976; Golani and Moran, this volume). It is thus possible to document that animals indeed do maintain relational boundaries to their actions, that they subject individual movements to the maintenance of these relational boundaries, and that both these relational and individual features of behavioral order can be tracked systematically throughout ontogeny.

Ontogeny of relations.—Havkin (1981) has been able to track shifts in strategy involved in turning of the head by one animal in the context of its relations to a second. This type of canid movement in adults led to such ideas as "submissive posturing" and the controversy surrounding these ideas (for example, see Schenkel, 1967). From about 6 to 8 weeks of age, wolf pups typically turn their heads up in response to contact (for example, pushing) by another pup, resulting in the former losing balance and falling on its back. At about 8 weeks of age, the animal begins to turn its head *down* and to the side, and often does so without (or prior to) direct contact by the partner. This becomes combined with biting of the forelegs of the partner, etc.—it is thus an active strategy of interaction. The animal that turns its head toward the ground also appears much less likely to fall over in simple response to its partner. Through preliminary biomechanical analyses, Havkin (1981) has supported the contention that the shift in style of head turning leads to a more stable support base and is accompanied by the adoption of configurations by the "superior" animal (that is, one that will end up on top in a fall), which both maximize the force it can exert on its partner and minimize the force it will receive from its partner. For example, characteristic "T" formations develop, where the animal at the stem of the "T" has a demonstrable biomechanical advantage over that at the top of the "T." The animals are becoming more sophisticated in techniques that may later serve in agonistic encounters. However, "play," "submission," and "dominance" do not have to be invoked in these analyses (Aldis, 1975; Bekoff, 1974; Symons, 1978).

Through related analytical approaches, Havkin (in litt.) has also documented a shift in the tendency of frequent role reversals (where one animal will "give up" its demonstrable "wrestling" advantages over another) to longer follow-throughs of this biomechanical advantage. The behavior becomes less "playful" in its appearance, but the boundary between play and non-play does not become a major concern. (Indeed, I am not sure that the wolves themselves could make such a distinction!). Typology is replaced by systematic documentation of the dynamic relations among social participants. One can envision the future isolation of mechanisms in such a research approach, something that is almost inconceivable when the original multidimensional constructs are treated in a unitary and typological (static) manner.

Prospective.—Although it is too early to state precisely how the ontogenetic observations of Havkin (1977, 1981) may lead to the more stable asymmetric interactions between wolves documented by Moran (1978, 1982), and how these in turn can be related to traditional concepts, such as dominance and submission, the outlook is promising. For example, recent data of Havkin (in litt.) suggest that by the time wolf pups are 6 to 8 weeks of age, individual characteristics can be discerned, such as probability of initiations of contact, persistence of contact, and strategies of positioning during contact, which may predict long lasting social asymmetries in adulthood. Two additional features of this approach are that 1) it permits the systematic tracking of individual differences at the same time that general (shared) strategies of development are abstracted, and 2) the chasm between social behavior and individually coordinated movements (above) is partially bridged by the application of analytical perspectives equally relevant to each (Golani and Moran, this volume).

Interestingly, the adult animal which consistently supplanted others in Moran's (1978) study (and thus appeared "dominant" by traditional criteria) often assumed positions in the relational configuration that Havkin has shown to be biomechanically *disadvantageous*. Supplanting animals thus may have a greater freedom of movement in comparison to those they displace, and the security of their status may be witnessed through their willingness to accept a disadvantaged position (Schenkel, 1967). Also, Moran (in litt.) has found that the displaced animal in a supplanting interaction is most often the one that initiates movement from one stable configuration to another, as if actively seeking a more advantageous position.

Although these last speculations move some distance away from the objective analyses performed by Moran and Havkin, they do indicate how such analyses may yield broad insights into social organization that would otherwise pass unnoticed.

Individuals and Their Groups

The differentiation and integration of motor actions defined at the level of the individual does have a relevance for social and other broader behavioral contexts (Moran et al., 1981). It is a matter of slicing behavior in different ways and at different levels, depending upon the particular set of principles being sought. Certainly one can question the extrapolation from neonatal wolf puppies made by Scott and Nagy (1980) that "there are no behavior patterns present that are similar to those in adults." One has to define explicitly the criteria of pattern formation being used, not to mention the concept of similarity itself! However, these authors do raise the interesting metaphor of "behavioral metamorphosis" as a warning that different developmental phases may be both descriptively and causally separable (see also Oppenheim, 1980).

In summary, I believe that future investigations of behavioral ontogeny will need to pay attention both to the successive differentiation and incorporation of individual actions (or more accurately, dimensions of expression) into higher-order units, and an awareness that such hierarchies themselves are not unidimensional; at each phase of life we investigate, the animal has special skills and limitations (and is subjected to special selective pressures). It is much less easy to suggest precisely how one can achieve this goal.

Many authors have suggested that individual actions (dimensions) in young animals are relatively unconnected—at least by the rules found in adults. Thus Leyhausen (1965) found that the movements of predation (watching, crouching, stalking, pouncing) are performed separately by young kittens, as well in varied combinations that do not form a cohesive "predatory" whole. They are also frequently connected to action components found in other higher-order systems as defined for adult cats. These general points are carried further by the analysis of nursing activities in rat pups by Hall et al. (1977). Newborn rat pups show a sequence of actions that consist of 1) rooting, 2) grasping the nipple, and 3) sucking—more or less as a chain reflex. However, during the first week of life, nursing is not directly affected either by nutritional needs (for example, deprivation time) or success (for example, dry nipple). By 8 to 10 days there is nutritional control over attachment, but the animals will still continue to suckle a dry nipple. Only by 14 to 16 days do the rat pups respond to a dry nipple by moving off. Thus it was shown that a reliable sequence (all parts of which must be performed for the pups to survive) *can* be fractionated differentially at different ages through systematic manipulations of particular variables. The feeding system which might appear unitary in infant rats is actually decomposable. And, of course, as the rats get still older the nursing sequence is eliminated.

Group Structure and Ontogeny

These last observations on feeding are reminiscent of Mason's (1979) distinction between *trophic* and *informational* systems in social ontogeny. In order to understand behavioral ontogeny in any altricial mammalian species, one certainly must appreciate the support provided to the infant by others in its social group, whether parents, aunts and uncles, siblings, or cousins. These other animals may provide a wide range of specific "instructions" to the infant mammal, but they are also necessary suppliers of nutrition, shelter, and the like. The beneficial effects for the offspring may be direct or indirect, and vary in their specificity (Kleiman and Malcolm, 1981).

I next discuss some observations on the nexus of social systems in captive wild canids collected in collaboration with J. Ryon, N. Bennett, F. Harrington, H. Parr, and L. White. Possible roles of experience in the elaboration of integrated behavioral patterns will be treated with an emphasis on support (care providing) systems per se, and their relevance to the constraints in ontogeny.

The main points are that 1) many members of the social group within which coyotes and wolves are born may participate directly and indirectly in their rearing, and 2) given even relatively normal social rearing conditions, a variety of sophisticated behavioral patterns can develop without having to be learned directly. The two points merge in that predispositions toward pup care can be seen in adult animals with minimal previous experience that might be assumed to provide necessary specific prerequisites.

The cohesion among group members is as sophisticated as that among the individually defined actions discussed previously, which reinforces my argument that it is important to seek rules of order in behavior at a variety of levels of expression. Further, in each case it is often the *relations* among distinguishable components of the behavioral system under investigation that is of primary interest. These relations provide the functional cohesiveness in behavior, and can be achieved in various ways (Fentress, 1980).

What we see also argues for the importance of intrinsic ordering in the development of these relations, that is, the ontogeny of even complex forms of mammalian behavior can be viewed as being self-organized (Nicolis and Prigogine, 1977), supported by the environment but not enslaved in its full details. The view that emerges is subtle and does not argue, for example, for simple notions of pre-specification. Clearly all behavior is molded in time. On the other hand, one can easily over-evaluate the instructive nature of the environment, a perspective that Edelman (1978:54) alluded to in a quite different context by stating that it "puts the brain too much at the mercy of the outside world."

Support Networks in Canids

We have documented social support systems in both captive coyotes (*Canis latrans*) and wolves (*Canis lupus*) since 1972 and 1973 respectively, including one season (1973) in which two wolf pups were given to an unmated adult pair, and four seasons following the last birth of coyote pups. The details of these studies are published elsewhere (Bennett, 1979; Fentress and Ryon, in press); thus only highlights will be abstracted here. Both species exhibit clear patterns of distributed pup care, as well as provisioning of the mothers. Further, the animals are able to perform the various associated activities even when they have been reared outside of a normal social group (that is, without the opportunity to "learn" directly from other pack members; see Figs. 1 and 2).

Initial observations.—When two 3.5-month female pups were given to an unbred ten-year-old male and four-year-old female, three dens were dug *by the male* and within two days he regurgitated food to the pups. The female also fed the pups by regurgitation and food carrying, but less frequently than the male (40 versus 111 observations). The adult female was also observed to solicit food from the male on five occasions, although he never fed her. The following year the two adults bred, and we have had successfully-

FIG. 1. Wolf pup soliciting feeding from an adult.

FIG. 2. Adult (left) provisioning food to half-grown wolf pups.

reared litters every subsequent year except for 1975 (Fentress, 1978*a*).

Roles of males and females in pup care.—The adult males actively participated in den construction (reviews by Kleiman and Malcolm, 1981; Vehrencamp, 1979; Wittenberger, 1979), pup feeding, the feeding of mothers, and feeding of subadult pack members (Kleiman and Eisenberg, 1973; Kleiman and Malcolm, 1981). They are solicited by each group member, although adult males typically do not solicit feeding nor are they fed.

Adult female non-mothers also feed pups by carrying and/or regurgitating solid food items. Some of these females also feed mothers and subadults. This behavior can include feeding mothers with whom they have had previous antagonistic interactions. In contrast, mothers of pups do not feed other mothers, nor are they as likely to feed subadults. Adult females do not feed adult males.

Selectivity.—The feeding of pups by adults is surprisingly indiscriminate, but pups from our different captive litters are always amalgamated into a single group (Murie, 1944). The feeding of mothers by other adults, however, is selective, even after the pups are amalgamated. (Certain females may even share dens prior to pup amalgamation, that is, during the births of their litters.) Subadults are also fed in a selective manner. Subadults are in many ways intermediate animals; they feed pups (but not adults) and are fed by adults. Feedings of mothers by adult males can begin several days prior to parturition. The feeding of other animals continues for several months, with a hiatus occurring during the subsequent breeding season (February and March) (see recent reports on pup care in canids by Allen, 1979; Bekoff, 1978; Camenzind, 1978; Frame and Frame, 1976; Gier, 1975; MacDonald, 1979; Moehlman, 1979, this volume; Peterson, 1977; Zimen, 1976).

Solicitation.—Although animals may actively solicit feeding, the correlation between feeding and solicitation is far from perfect. There can, for example, be a drop in feeding behavior during the autumn even though solicitations continue. Also, in subsequent seasons a given animal may become more willing to feed with little solicitation, so that the overall ratio of solicitation to feeding is quite variable. Further, once an animal has established a strong feeding tendency it may continue to display seasonal feeding of an adult female even in the absence of pups. For example, we had one adult coyote male that continued to feed his mate even though, because of his vasectomy, there had been no new pups in four seasons. It is also not uncommon to see an adult exhibit feeding intention movements to the pups of a female, only to be driven away by the mother. (Of course, expressions of willingness to feed may help an animal retain or regain full integration into the social group, but such considerations are beyond my scope.)

Social nexus.—These observations indicate that canid pups are exposed to, and supported by, most older members of the pack from a very early age. As soon as pups emerge from the dens (3 to 5 weeks), they receive considerable attention from older pack members, including, if I may say so, "play." What the pups "learn" from these interactions is not known, but the opportunities for diverse social experience are many. It is thus striking how normally animals can develop when separated from these diverse social supports (Ginsburg, 1978). Critical assessments of deviations, or at least limitations, in social performance under conditions of limited experience must await future research.

Experience and the Phenotypic Boundaries of Behavior

We have data implying that previously isolated animals lose in not having an awareness of the social intricacies of the social group to which they are later introduced. For example, handreared pups may behave toward other pack members in an indiscriminant manner—a poor strategy, given the realities of pack life. These pups also appear prone to the "spoiled brat" syndrome; for example, they can be too bold in asserting their rights over food items (see Rogers, 1973, on the role of socialization in eliminating idiosyncratic behaviors in chimpanzees, *Pan troglodytes*). This can be detrimental, but we have not pursued the matter systematically because the reintroduced animals are often endangered. The pack, too, must learn about pups at an appropriate age.

The full development of behavior within the context of the social group does not terminate after infancy, but continues throughout an animal's lifetime. For example, handreared pups that are "tamed" to close human contact will revert to a shyness of people even after five or six months of age if they can be integrated successfully into a social group of "untamed" animals (Fentress et al., 1978). Our preliminary data suggest that tendencies (Hinde, 1970) to behave in one way versus another may remain much more sensitive to differential experience than do the mechanisms that underlie the detailed patterning of actions.

Intrinsic biases and experience.—It is much easier for wolves to regain timidity than to retain tameness to humans. Also the animals frequently do not generalize from known handlers to other humans. In a word, there are intrinsic biases toward certain styles of interaction (for example, approach versus avoidance) that can only be overcome with the maintenance of special conditions. Scott and Nagy (1980) have recently argued that in *Canis* the consequences of experience at one stage of development may be minimally expressed at a later stage because of developmental discontinuities. This is also reminiscent of Alder's (1975) study showing that docility in Skomer voles (*Clethrionomys glareolus*) could be produced and maintained only in animals that were *both* genetically predisposed and had proper environmental support. Similarly, Wecker (1963) showed that *Peromyscus maniculatus* retained a genetic bias toward grassland habitats that could only be altered through continuing support of a woodland habitat. Although neither study systematically tracked the duration of altered tendencies of behavioral expression, they each do indicate clearly that there are intrinsically organized predispositions that can be modulated only so far by reasonably altered experience.

One factor, shown particularly nicely by Stevenson's (1967) research on song selection by chaffinches, is that animals frequently seek out sources of environmental stimulation that reinforce already existing behavioral biases. Many investigators of human development have begun to stress the role infants have in shaping their own experiences, including those provided by the parents (for example, see Bell, 1968; Cairns et al., 1980). These *reciprocal relations* between the developing infant and its environment can be appreciated by watching the ontogeny of behavior of wolf pups within a social group. Through self-initiated interactions with other pack members, they gain access to a wide range of experiences that guarantee normal social behavior.

Role flexibility.—One consequence of this genetic and developmental richness is that wolves within a social group can perform a variety of different roles appropriately if circumstances dictate. Thus, shifts in the "social hierarchy" can lead each participant immediately to perform a variety of actions that were previously exhibited by another animal which had occupied that position (see Brown, 1978, on "variance enhancement" and "role switching" in communal birds). Such transitions can be dramatic and not only include the range of actions normally classified as "dominant" or "submissive." Other transitions can be seen in care-giving actions, such as when one adult female feeds another as a function of her own breeding success for that season. Thus, two females may switch roles as a function of which animal happens to breed (Fentress and Ryon, in press). And these switches can be reversed (several times). The animals obviously have a lot of behavioral competencies that they do not necessarily exhibit. The precise interplay between genetic foundations and various forms of social (and other) experiences in setting up these competencies can only be sorted out through systematic future research.

These observations point to the fluidity, multidimensionality, and relational nature of wolf social behavior throughout ontogeny. Classical static concepts, such as unidimensional dominance hierarchies, miss the very essence of this "dynamic programming" (Fagen, 1978; Moran, 1982) within the total network of social interactions, and alternative conceptual frameworks will have to be developed in the future (Lockwood, 1979; Rowell, 1974). For example, young wolves not only go through frequent role reversals with their peers, but these relations are also context dependent. Indeed, a normally passive individual in play may defend food items with particular vigor, an action which shows little resemblance to any simple linear hierarchical framework. Indeed, unlike previous authors (for example, see Bekoff, 1978), we have found that constructs involving the hierarchical ordering of social behavior in young wolves and coyotes are often of limited value (Bennett et al., pers. comm.).

Mechanisms of Phenotype Construction

In this section I explore Tinbergen's (1963) fundamental challenge and seek ways to synthesize the temporal perspectives of evolution, development, and short term production of behavior in terms of the combined functional consequences. I emphasize two themes: 1) the different roots of connectivity among actions that one sees during ontogeny; and 2) the diverse ways in which experience during an organism's lifetime might be expected to interact with its genetic structure.

Actions and Connections

Young animals frequently make behavioral transitions on the basis of motor similarities (see the concept of "motor traps" outlined above) whereas later the transitions between actions are more commonly based upon a common function (for example, biting and kicking in a fight). Indeed, the early stages of ontogeny are often marked by a *failure* to achieve integrated function in the adult sense. Thus infant rodents may swipe near the faces with grooming movements without making contact between the forepaws and face; young male wolves may show pelvic thrusting movements in an apparent functional vaccuum, and many carnivorous species often pounce in response to moving objects without properly orienting toward these objects (for example, see Fentress, 1967).

Such observations are compatible with the idea that lower levels in the "behavioral hierarchy" develop prior to the more abstractly integrated higher levels (Dawkins, 1976; see Gallistel, 1980, for related discussion). Although one should not accept too literally an interpretation of hierarchically-ordered ontogeny, one might expect ramifications upon pattern formation of experience in early ontogeny to follow the lines of physical similarity in movement, whereas later the ramifications might operate within more abstractly defined and "higher" functional units that can contain action of quite diverse form. To my knowledge this hypothesis has not been subjected to an explicit test, although it would not be difficult to do so by taking actions that are found in different functional groupings (for example, biting food, biting an opponent, biting a mate) and asking whether developmental modifications of one of these actions, through enhancement or retardation, had measureable ramifications either when the same action was used in a different functional context or across the performance of qualitatively different actions in the same functional context.

A potential difficulty, however, is that we cannot yet anticipate *how* differential experience interacts with the organism's prior makeup. The most common model is that there is a nearly isomorphic relationship (that is, one-to-one physical matching) between what goes in and what comes out. We thus learn to speak a foreign language by copying a tutor.

Roles and Limits of Experience

The role of instructive experience in ontogeny may be overplayed. Thus, although we have often tried to alter the basic temperament of wolves through various handrearing procedures, our success, at best, has been minimal (which is quite a different issue from statistical significance; Henderson, 1980). Our wolves howl, dig dens, exhibit a variety of social postures under appropriate circumstances, and care for pups with considerable efficiency without ever being taught. More dramatically, perhaps, infant mice display remarkably normal face grooming movements even after limb amputation at birth which restricts the normal routes of sensory feedback (Fentress, 1973*b*).

This is not to say that changes in experience during ontogeny do not have their measurable consequences—only that *certain* behavioral features of the developing organism can remain remarkably consistent under superficially different circumstances.

Embryological considerations.—The early mammalian embryo consists of undifferentiated blastomeres (embryonic cells) that are clumped together into a solid ball or *morula.* Only later does a modified blastula (*blastocyst*) appear with the subsequent clear differentiation of body parts and functions. Now the differentiated somatic cells that make up, say, an eye or an ear are genetically equivalent to one another (Davidson, 1976; Ede, 1978). They are phenotypically different only because they have, in some sense, undergone differential experience. They, of course, do not "learn" to become an eye or an ear; rather, different combinations of genetic machinery are activated due to the environment in which the different cells reside. The same happens to brain tissue (Jacobson, 1978) and, we might suppose, the behavioral actions that result from brain tissue (Chomsky, 1980). The extent of compartmentalization in embryonic development, as in behavior, is imperfect (Fentress, 1976*a*, 1980); that is, *mosaic* and *regulative* development are balanced depending upon the extent as well as form of communication between participating cell groups. This is directly analogous to Bateson's (1976) thoughtful discussion of specific and non-specific factors in behavioral ontogeny.

Genes and experience.—Several fundamental ideas can be abstracted from such considerations: 1) experience, in some form, is *always* necessary for differentiation (Lehrman, 1970; Marler, 1976); 2) experience might activate (select from) pre-existing potentialities rather than merely add "new" information by instructional processes (Jerne, 1967); 3) there may be important buffers in a developing system that permit basically normal ontogeny under quite varied conditions (Bateson, 1976); and 4) the degree and form of specificity of consequences of experience at any one age may depend importantly upon the integrative pathways that are established at that age (see above discussion and Bateson, 1976).

Prospectus.—How, then, do we perform more satisfactory developmental analyses? First, we must provide more detailed ontogenetic descriptions of behavioral dimensions and their *various* rules of interconnection (Golani, 1976). Second, we must examine explicitly the range of ramifications (and, conversely, buffers) when particular routes of experience are added during particular ontogenetic phases. Third, we must consider not only that experience can have such diverse roles as maintenance, facilitation, and induction (Gottlieb, 1976), but also that specification may take place by giving the organism access to information stores already intrinsic to it (for example, "selection" in the sense of Jerne, 1967) in addition to its obvious ancillary role of providing instructions per se.

The literature on behavioral development lacks any systematic exploration of these fundamental considerations that lead to predictable hypotheses, which in turn might stimulate more productive approaches to developmental problems in the future. In this way we may succeed in obtaining a better understanding of just how, in Jacob's (1973) terms, the organism plays its critical role as "transition" between past, present, and future.

Synoptic Overview: Ontogeny through Past, Present, and Future

The major issues and themes that I have introduced in this essay may be summarized under two major headings: 1) the differentiation of integration in behavior (Table 1), and 2) the progressive specification of developing systems (Table 2).

One of my goals has been to stress the intimacy through which past, present, and future are woven together in all forms of mammalian behavior. A second goal has been to view phenomena and problems at quite different levels of complexity to see whether any thematic commonalities can be discerned. I now focus on what

TABLE 1
DIFFERENTIATION OF INTEGRATION

I. Functional Configurations (Dimensions and Relations)
A. Relations among action dimensions
B. Relations among stimulus dimensions
C. Hierarchical order and generalization gradients
II. Specifications of Experience and Generalization
A. Range, completeness, perfection of inputs
B. Range, completeness, perfection of outputs
C. Variations in quality, strength, timing
III. Roles of Experience
A. Modulation (accentuate/diminish)
B. Specification (instruct/select)
C. Preservation (maintain/remove)
IV. Self-Organizational Tendencies
A. Unspecified to differentiated
B. Compartmentalized to integrated
C. Progressive higher-order autonomies (based upon history of prior interactions)

is perhaps *the* major issue in all of development, namely ways to dissect rules of relation between interactive and self-organizing tendencies of participating systems. Although this has already been alluded to in several places, it will perhaps be helpful to recapitulate the main points herewith. The aim is to suggest some areas of future research.

Differentiation of Integration

Functional configurations (dimensions and relations).—The first argument is a behavioral taxonomy that is both dynamic and relational. The dynamic aspect is important in order to capture the very essence of development (that is, change), and the relational aspect is important because actions and their precursors in animals form functionally unified configurations. To understand these functional configurations, the investigator needs to examine explicitly relations among dimensions of action that contribute to the same endpoint (for example, mating, fighting, grooming), and relations among stimulus features that give rise to one or more of these action components. This is a combined process of division and synthesis and permits the ordering of behavior into descriptive hierarchies (Dawkins, 1976; Fentress, 1972). The investigator can then ask at what level and by what set of rules a given experience generalizes (that is, what are its ramifications?). We might expect, for example, generalization to follow gradients other than merely on the basis of simple physical similarity. Thus, physically quite distinct actions or stimulus components might be influenced by an experience restricted to a subset of these actions or stimulus components. A compelling biological reason to suggest this is simply that it is *together* that these components form a functional whole. Future experiments could be designed to test this prediction, but at present systematic data are not available.

Specifications of experience.—One set of investigations is to evaluate the consequences of manipulating a broad range of inputs that can vary in completeness and/or distortion in comparison to those in normal experience. This is analogous to the separable issues of stimulus "poverty" and "degeneracy" discussed in models of human speech development (for example, see Chomsky, 1980). The empirical question is: To what extent can an incomplete or imperfect (that is, distorted) set of experiences facilitate normal ontogeny (for example, in comparison to more severe deprivation conditions)? The measures would be the range, completeness, and perfection of one or more behavioral outputs. The most adequate data would involve studies in which variations in the qualitative features of a given experience are combined with variations in the amount, or strength, of the experience as well as the time frame over which it is made available (see Fentress, 1976*a*, 1976*b*, 1977*b*, 1980, for related discussions of integrative networks in behavior).

Roles of experience.—Most models of behavioral development are based at least implicitly upon learning models (that is, "instructions" from the environment). As Gottlieb (1976) has noted, an embryological perspective suggests that experience can serve to facilitate and maintain as well as induce developmental changes in behavior. I suggest an even finer breakdown where: 1) modulation can be measured either by accentuating developmental tendencies (Gottlieb's facilitation) or diminishing (for example, retarding) these tendencies; 2) specification (Gottlieb's induction) can either act through an instructive mechanism (that "adds" information from the outside, so to speak) or a selective mechanism (one that gives the organism access to latent information stores, such as through gene activation (Davidson, 1976) or immunological selection (Jerne, 1967; Edelman, 1978); 3) preservation has two sides, the maintenance of systems, as suggested by Gottlieb, or the removal of behavioral circuitry, as stressed in many neurobiological models (Cowan, 1978; Jacobson, 1978). The purpose is first to suggest we liberalize our ideas of what roles experience may play in development, and second, to make explicit that experience can work in two directions (for example, such as by removing circuitry as well as by preserving it). This latter point is important when we recall that different classes of behavior can be mutually antagonistic; therefore, developmental events that, for example, facilitate one set of developmental pathways, might do so at the expense of others. Even perfection within a pathway may depend upon removing unwanted connections (Cowan, 1978).

Additionally, a model of the effects of a particular experience can depend on the level of analysis employed. Thus, instruction at a phenomenological level may be mediated by mechanisms that would be judged as operating through selection (Jerne, 1967). Future experiments would include an evaluation of the relative importance of various possible consequences of experience as a function of the strength, timing, etc., of the experience. For example, would moderate amounts of experience *X* accentuate pre-existing developmental biases even if these do not "match" *X* precisely, whereas at a higher threshold *X* would *respecify* the behavior? Dual action of this type is common in complex integrative systems (Fentress, in press) but has not been explicitly explored in the context of ontogeny.

Embryological models of development emphasize that the consequences of experience may not be obvious. Thus cellular precursors to a limb or eye may depend upon stem cells being embedded in an appropriate environment which need not resemble the final differentiated structure. Chomsky (1980) has emphasized that even complex forms of human language capacity may differentiate in a similar way; that is, the formation of brain circuitry depends upon the environment but the environment does not map isomorphically onto the final differentiated structure. Further, Bellugi (1971) has shown how children may make characteristic errors in speech that deviate systematically from any detailed models they are given by adults. Such "failures" to map in a one-to-one way with the details

TABLE 2
PROGRESSIVE SPECIFICATION THEMES IN BEHAVIORAL ONTOGENY

1) Initially system undifferentiated and incompletely specified.
2) Differentiation and specification made possible through interactions with surround.
3) These interactions need not specify in strict or exhaustive sense, but may also generate self-organizational action (that is, capitalize upon intrinsic capabilities).
4) The inputs that are relevant need not be superficially obvious, but may be a product of conditions through which animal normally passes.
5) Item 4 is different from issue of specificity of "inducer;" non-obvious influences may be either specific or non-specific, and further tests are needed.
6) With time there is an increased compartmentalization in the dual sense of 1) clarification of descriptive boundaries and 2) reduced influence of extrinsic influences.
7) Compartments, in the senses of 6, become coordinated into higher-order functional units, thus generating a hierarchical order.
8) The generalization of the consequences of experience follow, in part at least, functional configurations that are operative at the time.
9) A possible dynamic is that increased exposure to an extrinsic influence increases the functional specificity of its consequences, thus moving from the modulation of existing biases to specification, or even re-specification. Temporal parameters must also be considered here (for example, changing sensitivities of target).
10) Future research should concentrate on functional configurations of individually defined acts/systems, where relations are given equal emphasis to the components by which they are composed. A bi-directional association between "pieces" and "relations" is suggested, each dictating in part the properties of the other.
11) Functionally distinct systems may antagonize the development of one another, just as functional synergies may facilitate mutual development. Many more studies of behavioral antagonisms and synergies during ontogeny are needed.
12) It is also important to bear in mind that animals have distinct needs, limitations, and potentialities at different periods of development; continuities of effect between developmental phases must be tested for rather than assumed.

of experience, and thus the construction of "errors" as judged on the basis of this experience, argue strongly for intrinsic organizational processes (see Rosch, 1973 on the development of "natural categories" in human perception). In animal behavior the search for systematic deviations from the dictates of experience might provide useful insights into the nature of intrinsic ordering principles in ontogeny.

Self-organizational tendencies.—A very common observation in developmental biology is that systems become differentiated through interactions with the surround, after which they become progressively less easily influenced by alterations of the surround (Cowan, 1978; Ede, 1978; Edwards and Palka, 1976; Kimmel and Eaton, 1976). Similar phenomena occur in behavioral development (Bateson, 1976; Marler, 1976). This common sequence suggests that systems become progressively more self-organized (see Fentress, 1976*a*, 1980, for discussion in terms of integrative processes), which promotes intrinsically-ordered developmental changes and protects the differentiated system from dissipative influences. Selective stabilization of synapses during ontogeny, such as through differential activity, is now a widely accepted phenomenon in developmental neuroscience (Changeaux and Danchin, 1976). Functionally adequate circuitry is preserved; non-functional circuits remain sensitive to disruption.

Mathematical models of self-organizing systems in development have recently been developed with analogy to lateral inhibition in sensory physiology (Meinhardt, 1979). These and related models attempt a quantitative account of the dynamics of pattern formation, where pattern can be produced only by some extrinsic influence yet maintained only by a certain insulation from extrinsic influence (reviews in Haken, 1977; Katchalsky et al., 1974; Nicolis and Prigogine, 1977). This very exciting line of thought completely avoids all simple dichotomies between intrinsic and extrinsic influences by emphasizing balanced degrees and types of cooperative action at different time periods.

Although a detailed extrapolation of these concepts is not possible, the fields of dissipative structures and related areas may suggest ways to improve at least our conceptualizations of the processes of development. For the present purposes, I limit possible applications of these thoughts to three issues we have examined: 1) the transition from unspecified (for example, "homogeneous") to differentiated systems; 2) the transition from compartmentalized components of behavior to their integrated expression; and 3) the formation of progressively higher-order blockings ("units") of action, each of which is maintained in a partially autonomous state (that is, having but circumscribed interactions with the surround). Each of these proposed phases is based upon the history of prior interactions within and between the systems in question.

One clear implication is that the developing organism develops in terms of complex rules that may transcend the operation of particular factors, whether intrinsic or extrinsic in origin (Bateson, 1976). A thematic inquiry is equally important to genetic considerations as it is to considerations of experience. For example, the human nervous system has approximately 10^7 neurons per gene(!), a fact that makes simple models of one-to-one specification totally unrealistic (Wyman, 1976).

Themes in Progressive Specification

Twelve of the most salient points on the theme of progressive specification of behavioral systems during ontogeny are listed in Table 2.

Clearly, there are many gaps in our knowledge about both the phenomena and mechanisms of behavioral ontogeny. For the student of mammalian behavior, these gaps loom large indeed. And I am certain that there are many gaps we do not yet even perceive. Yet this is precisely what gives the study of behavioral ontogeny in mammals its fascination. Present and future investigators, as their colleagues in the past, will certainly trip often along the way, as I suspect I have unwittingly done here. However, the combination of precise description and liberalization of our present concepts will certainly pay off in long-range benefits. The issues go beyond abstract intellectual challenge, for they eventually tell us something about ourselves as well as the other creatures with whom we share the class Mammalia.

Acknowledgments

I acknowledge the many useful comments from my colleagues and students on an earlier draft of this manuscript. Many of their constructive suggestions were not incorporated adequately in the present text due to my own limitations of expression. They did, however, lead to some notable improvements. Conversations and correspondence with Pat Bateson, Nancy Bennett, Noam Chomsky, John Eisenberg, Ilan Golani, Zvika Havkin, Devra Kleiman, Lee Machlis, Peter Marler, Greg Moran, Heather Parr, Jenny Ryon, and Lilyan White were particularly helpful. I also thank all of the participants of the Front Royal Symposium, and particularly its organizers, for the atmosphere of free yet critical intellectual inquiry that they promoted. Mary MacConnachie typed various scribbles and drafts, and deserves much credit for bringing the present manuscript to completion. I also thank Dalhousie University and the Natural Sciences and Engineering Research Council (A9787) for their support.

Literature Cited

ALDER, E. M. 1975. Genetic and maternal influences on docility in the skomer vole (*Clethrionomys glareolus skomerensis*). Behav. Biol., 13:251–255.

ALDIS, O. 1975. Play fighting. Academic Press, New York, 319 pp.

ALLEN, D. L. 1979. Wolves of Minong: their vital role in a wild community. Houghton Mifflin, Boston, 499 pp.

BATESON, P. P. G. 1976. Rules and reciprocity in behavioral development. Pp. 401–421, *in* Growing points in ethology (P. P. G. Bateson and R. A. Hinde, eds.). Cambridge University Press, Cambridge, 548 pp.

BEKOFF, A., AND W. TRAINER. 1979. The development of interlimb coordination during swimming in postnatal rats. J. Exp. Biol., 83:1–11.

BEKOFF, M. 1974. Social play and play-soliciting by infant canids. Amer. Zool., 14:323-340.

———. 1978. Behavioral development in coyotes and eastern coyotes. Pp. 97–126, *in* Coyotes: biology, behavior, and management (M. Bekoff, ed.). Academic Press, New York, 384 pp.

BEKOFF, M. J. A. BYERS, AND A. BEKOFF. 1980. Prenatal motility and postnatal play: functional continuity? Dev. Psychobiol., 13:225–228.

BELL, R. Q. 1968. A reinterpretation of the direction of effects in studies of socialization. Psychol. Rev., 75:81–95.

BELLUGI, U. 1971. Simplification in children's language. Pp. 95–119, *in* Language acquisition: models and methods (R. Huxley and E. Ingram, eds.). Academic Press, New York, 310 pp.

BENNETT, N. 1979. Some aspects of cooperative pup-rearing in a pack of captive wolves (*Canis lupus*). Unpublished M.A. thesis, Dalhousie Univ., Halifax, 62 pp.

BERNSTEIN, N. A. 1967. The coordination and regulation of movement. Pergamon Press, Oxford, 197 pp.

BROWN, J. L. 1978. Avian communal breeding systems. Ann. Rev. Ecol. Syst., 9: 123–155.

BRUNER, J. 1973. The organization of early skilled action. Child Devel., 44: 1–11.

CAIRNS, R. B., J. A. GREEN, AND D. J. MACCOMBIE. 1980. The dynamics of social development. Pp. 79–106, *in* Early experiences and early behavior: implications for social development (E. C. Simmel, ed.). Academic Press, New York, 217 pp.

CAMENZIND, F. J. 1978. Behavioral ecology of coyotes on the National Elk Refuge, Jackson, Wyoming. Pp. 267–294, *in* Coyotes: biology, behavior and management (M. Bekoff, ed.). Academic Press, New York, 384 pp.

CHANGEAUX, J.-P., AND A. DANCHIN. 1976. Selective stabilization of developing synapses as a mechanism for the specification of neuronal networks. Nature, 264:705–712.

CHOMSKY, N. 1980. Rules and representations. Columbia Univ. Press, New York, 299 pp.

COWAN, W. M. 1978. Selection and control in neurogenesis. Pp. 59–79, *in* The neurosciences: fourth study program (F. O. Schmitt and F. G. Worden, eds.). M.I.T. Press, Cambridge, Massachusetts, 1185 pp.

DAVIDSON, E. H. 1976. Gene activity in early development. Academic Press, New York, 2nd ed., 452 pp.

DAWKINS, R. 1976. Hierarchical organization: a candidate principle for ethology. Pp. 7–54, *in* Growing points in ethology (P. P. G. Bateson and R. A. Hinde, eds.). Cambridge Univ. Press, Cambridge, 548 pp.

EDE, D. A. 1978. An introduction to developmental biology. John Wiley and Sons, New York, 246 pp.

EDELMAN, G. M. 1978. Group selection and phasic reentrant signaling: a theory of higher brain function. Pp. 51–100, *in* The mindful brain (G. M. Edelman and V. B. Mountcastle, eds.). M.I.T. Press, Cambridge, Massachusetts, 100 pp.

EDWARDS, J. S., AND J. PALKA. 1976. Neural generation and regeneration in insects. Pp. 167–186, *in* Simpler networks and behavior (J. C. Fentress, ed.). Sinauer Associates, Sunderland, Massachusetts, 403 pp.

EIBL-EIBESFELDT, I. 1956. Uber die ontogenetische Entwicklung der Technik des Nüsseöffnen vom Eichhörnchen (*Sciurus vulgaris* L.). Z. Säugetierk., 21: 132–134.

EISENBERG, J. F. 1963. The behavior of heteromyid rodents. Univ. California Publ. Zool., 69:1–114.

FAGEN, R. M. 1978. Evolutionary biological models of animal play behavior. Pp. 385–404, *in* The development of behavior: comparative and evolutionary aspects (G. M. Burghardt and M. Bekoff, eds.). Garland Press, New York, 429 pp.

———. 1981. Animal play behavior. Oxford Univ. Press, New York, 684 pp.

FENTRESS, J. C. 1967. Observations on the behavioral development of a handreared male timber wolf. Amer. Zool., 7:339–351.

———. 1972. Development and patterning of movement sequences in inbred mice. Pp. 83–132, *in* The biology of behavior (J. Kiger, ed.). Oregon State Univ. Press, Corvallis, 180 pp.

———. 1973*a*. Specific and nonspecific factors in the causation of behavior. Pp. 155–224, *in* Perspectives in ethology (P. P. G. Bateson and P. H. Klopfer, eds.). Plenum Press, New York, 336 pp.

———. 1973*b*. Development of grooming in mice with amputated forelimbs. Science, 179:704–705.

———. 1976*a*. Dynamic boundaries of patterned behavior: interaction and self-organization. Pp. 135–169, *in* Growing points in ethology (P. P. G. Bateson and R. A. Hinde, eds.). Cambridge University Press, Cambridge, 548 pp.

———. 1976*b*. (ed.). Simpler networks and behavior. Sinauer Associates, Sunderland, Massachusetts, 403 pp.

———. 1977*a*. The tonic hypothesis and the patterning of behavior. Ann. New York Acad. Sci., 290:370–395.

———. 1977*b*. Opening remarks: constructing the potentialities of phenotype. Ann. New York Acad. Sci., 290:220–225.

———. 1978*a*. Conflict and context in sexual behavior. Pp. 579–614, *in* Biological determinants of sexual behavior (J. Hutchison, ed.). John Wiley and Sons, New York, 822 pp.

———. 1978*b*. Defining behavioral representations. Behav. Brain Sci., 1:20.

———. 1978*c*. Mus musicus: the developmental orchestration of selected movement patterns in mice. Pp. 321–342, *in* The development of behavior: comparative and evolutionary aspects (G. M. Burghardt and M. Bekoff, eds.). Garland Press, New York, 429 pp.

———. 1980. How can behavior be studied from a neuroethological perspective? Pp. 263–283, *in* Information processing in the nervous system (H. M. Pinsker and W. D. Willis, Jr., eds.). Raven Press, New York, 366 pp.

———. 1981*a*. Order in ontogeny: relational dynamics. Pp. 338–371, *in* Behavioral development: the Bielefeld Interdisciplinary Project (K. Immelman, G. W. Barlow, M. Main, and L. Petrinovich, eds.). Cambridge Univ. Press, New York, 754 pp.

———. 1981*b*. Sensory-motor development. Pp. 293–318, *in* The development of perception: Psychobiological perspectives, Vol. 1 (R. N. Aslin, J. Alberts and M. R. Peterson, eds.). Academic Press, New York, 463 pp.

———. In press. Ethological models of hierarchy and patterning of species-specific behavior. *In* Handbook of neurobiology: motivation (E. Satinoff and P. Teitelbaum, eds.). Plenum Press, New York.

FENTRESS, J. C., AND C. J. RYON. In press. A long term study of distributed pup feeding and associated behavior in wolves. *In* Portland wolf symposium (P. C. Paquet and J. O. Sullivan, organizers).

FENTRESS, J. C., AND F. P. STILWELL, 1973. Grammar of a movement sequence in inbred mice. Nature, 244:52–53.

FENTRESS, J. C., R. FIELD, AND H. PARR. 1978. Social dynamics and communication. Pp. 67–106, *in* Behavior of captive wild animals (H. Markowitz and V. J. Stevens, eds.). Nelson Hall, Inc., Chicago, 314 pp.

FIELD, R. 1978. Vocal behavior of wolves (*Canis lupus*): variability in structure, context, annual/diurnal patterns, and ontogeny. Unpublished Sc.D. Thesis, The Johns Hopkins Univ., Baltimore, 243 pp.

———. 1979. A perspective on syntactics on wolf vocalizations. Pp. 182–205, *in* The behavior and ecology of wolves (E. Klinghammer, ed.). Garland Press, New York, 588 pp.

FOX, M. W. 1971. Behavior of wolves, dogs and related canids. Harper and Row, New York, 220 pp.

FRAME, L. H., AND G. W. FRAME. 1976. Female African wild dogs emigrate. Nature, 263:227–229.

GALLISTEL, C. R. 1980. The organization of action: a new synthesis. Lawrence Erlbaum Associates, Hillsdale, New Jersey, 432 pp.

GIER, H. T. 1975. Ecology and social behavior of the coyote. Pp. 247–259, *in* The wild canids: their systematics, behavioral ecology and evolution (M. W. Fox, ed.). Van Nostrand Rheinhold, New York, 508 pp.

GINSBURG, B. E. 1978. The genetics of social behavior. Pp. 1–15, *in* Perspectives in ethology (P. P. G. Bateson and P. H. Klopfer, eds.). Plenum Press, New York, 3:1–263.

GOLANI, I. 1976. Homeostatic motor processes in mammalian interactions: a choreography of display. Pp. 69–134, *in* Perspectives in ethology (P. P. G. Bateson and P. H. Klopfer, eds.). Plenum Press, New York, 2:1–340.

GOTTLIEB, G. 1976. Conceptions of prenatal development: behavioral embryology. Psychol. Rev., 83:215–234.

HAKEN, H. 1977. Synergetics. An introduction. Nonequilibrium phase transitions in physics, chemistry and biology. Springer-Verlag, New York, 355 pp.

HALL, W. G. 1979. Feeding and behavioral activation in infant rats. Science, 205: 206–208.

HALL, W. G., C. P. CRAMER, AND E. M. BLASS. 1977. Ontogeny of suckling in rats: transition toward adult ingestion. J. Comp. Physiol. Psychol., 91: 1141–1155.

HARRINGTON, F. H., AND L. D. MECH. 1978. Wolf vocalization. Pp. 109–132, *in* Wolf and man (R. L. Hall and H. S. Sharp, eds.). Academic Press, New York, 210 pp.

HAVKIN, G. Z. 1977. Symmetry shifts in the development of interactive behaviour of two wolf pups (*Canis lupus*). Unpublished M.A. thesis, Dalhousie Univ., Halifax, 40 pp.

———. 1981. Form and strategy of combative interactions between wolf pups (*Canis lupus*). Unpubl. Ph.D. dissert., Dalhousie Univ., Halifax, 126 pp.

HENDERSON, N. D. 1980. Effects of early experience upon the behavior of animals: the second twenty-five years of research. Pp. 39–77, *in* Early experiences and early behavior (E. C. Simmel, eds.). Academic Press, New York, 217 pp.

HENRY, J. D., AND J. M. HERRERO. 1974. Social play in the American black bear: its similarity to canid play and an examination of its identifying characteristics. Amer. Zool., 14:371–389.

HINDE, R. A. 1970. Animal behaviour: a synthesis of ethology and comparative psychology. Second ed. McGraw-Hill, New York, 876 pp.

———. 1972. Social behavior and its development in sub-human primates. Univ. Oregon Press, Eugene, 76 pp.

HINES, M. 1942. The development and regression of reflexes, postures and progression in the young macaque. Contrib. Embryol., Carnegie Instit. Washington, 30:153–204.

HOLTON, G. 1978. The scientific imagination: case studies. Cambridge Univ. Press, Cambridge, 382 pp.

JACOB, F. 1973. The logic of life: a history of heredity. Random House, New York, 348 pp. (First published in France, 1970, *La Logique du Vivant: Une Histoire de L'Hérédité*. Gallimard, Paris.)

JACOBSON, M. 1978. Developmental neurobiology. Second ed. Plenum Press, New York, 562 pp.

JERNE, N. K. 1967. Antibodies and learning: selection versus instruction. Pp. 200–205, *in* The neurosciences: a study program (G. C. Quarton, T. Melnechuk, and F. O. Schmitt, eds.). The Rockefeller Univ. Press, New York, 1:1–962.

KATCHALSKY, A. K., V. ROWLAND, AND R. BLUMENTHAL. 1974. Dynamic patterns of brain cell assemblies. Neurosciences Res. Program Bull., 12:3–187.

KIMMEL, C. B., AND R. C. EATON. 1976. Development of the Mauthner cell. Pp. 186–202, *in* Simpler networks and behavior (J. C. Fentress, ed.). Sinauer Associates, Sunderland, Massachusetts, 403 pp.

KLEIMAN, D. G., AND J. F. EISENBERG. 1973. Comparisons of canid and felid social systems from an evolutionary perspective. Anim. Behav., 21:637–659.

KLEIMAN, D. G., AND J. R. MALCOLM. 1981. The evolution of male parental investment in mammals. Pp. 347–387, *in* Parental care in mammals (D. J. Gubernick and P. H. Klopfer, eds.). Plenum Press, New York, 459 pp.

LASHLEY, K. S. 1951. The problem of serial order in behavior. Pp. 112–136, *in* Cerebral mechanisms in behavior (L. A. Jeffress, ed.). John Wiley and Sons, New York, 311 pp.

LEHRMAN, D. S. 1970. Semantic and conceptual issues in the nature-nurture problem. Pp. 17–52, *in* Development and evolution of behavior (L. R. Aronson, E. Tobach, D. S. Lehrman, and J. S. Rosenblatt, eds.). W. H. Freeman, San Francisco, 656 pp.

LEYHAUSEN, P. 1965. Über die Funktion der relativen Stimmungs-hierarchie (Dargestellt am Beispiel der phylogenetischen und ontogenetischen Entwicklung des Beautefangs von Raubtieren). Z. Tierpsychol., 22:412–494.

LIEBERMAN, P. 1977. The phylogeny of language. Pp. 3–25, *in* How animals communicate (T. A. Sebeok, ed.). Indiana Univ. Press, Bloomington, 1128 pp.

LIND, H. 1959. The activation of an instinct caused by a "transitional action." Behaviour, 14:123–135.

LOCKWOOD, R. 1979. Dominance in wolves: useful construct or bad habit? Pp. 225–244, *in* The behavior and ecology of wolves (E. Klinghammer, ed.). Garland Press, New York, 588 pp.

MACDONALD, D. W. 1979. 'Helpers' in fox society. Nature, 282:69–71.

MACHLIS, L. 1977. An analysis of the temporal patterning of pecking in chicks. Behaviour, 63:1–70.

MARLER, P. 1976. Sensory templates in species-specific behavior. Pp. 314–329, *in* Simpler networks and behavior (J. C. Fentress, ed.). Sinauer Associates, Sunderland, Massachusetts, 403 pp.

MARLER, P., AND W. J. HAMILTON. 1966. Mechanisms of animal behavior. John Wiley and Sons, New York, 771 pp.

MASON, W. 1979. Ontogeny of social behavior. Pp. 1–28, *in* Handbook of behavioral neurobiology. Vol. 3: social behavior and communication (P. Marler and J. G. Vandenbergh, eds.). Plenum Press, New York, 411 pp.

MAYR, E. 1970. Populations, species and evolution. Belknap Press, Cambridge, Massachusetts, 453 pp.

MEINHARDT, H. 1979. The random character of bifurcations and the reproducible processes of embryonic development. Ann. New York Acad. Sci., 316:188–202.

MOEHLMAN, P. D. 1979. Jackal helpers and pup survival. Nature, 277:382–383.

MORAN, G. 1978. The structure of movement in supplanting interactions in the wolf. Unpublished Ph.D. dissert., Dalhousie Univ., Halifax, 211 pp.

———. 1982. Long-term patterns of agonistic interactions in a captive group of wolves (*Canis lupus*). Anim. Behav., 30:75–83.

MORAN, G., AND J. C. FENTRESS. 1979. A search for order in wolf social behavior. Pp. 245–283, *in* The behavior and ecology of wolves (E. Klinghammer, ed.). Garland Press, New York, 588 pp.

MORAN, G., J. C. FENTRESS, AND I. GOLANI. 1981. A description of relational patterns of movement during 'ritualized fighting' in wolves. Anim. Behav., 29:1146–1165.

MURIE, A. 1944. The wolves of Mt. McKinley. U.S. Natl. Park Service Fauna Service, 5:1–238.

NICOLIS, G., AND I. PRIGOGINE. 1977. Self-organization in nonequilibrium systems. John Wiley and Sons, New York, 491 pp.

OPPENHEIM, R. W. 1980. Metamorphosis and adaptation in the behavior of developing organisms. Devel. Psychobiol., 13:353–356.

PETERSON, R. O. 1979. Social rejection following mating of a subordinate wolf. J. Mamm., 60:219–221.

POSNER, M. I. 1978. Chronometric explorations of mind. Lawrence Erlbaum Associates, Hillsdale, New Jersey, 271 pp.

REIGEL, K. F. 1976. The dialectics of human development. Amer. Psychol., 31: 689–70.

ROGERS, C. M. 1973. Implications of a primate early rearing experiment for the concept of culture. Pp. 185–191, *in* Precultural primate behavior (E. W. Menzel, Jr., ed.). S. Karger, New York, 258 pp.

ROSCH, E. H. 1973. Natural categories. Cognitive Psychol., 4:328–350.

ROWELL, T. E. 1974. The concept of social dominance. Behav. Biol., 11:131–154.

SCHENKEL, R. 1947. Ausdrucks-Studien an Wölfen. Behaviour, 1:81–129.

———. 1967. Submission: its features and functions in the wolf and dog. Amer. Zool., 7:319–330.

SCHLEIDT, W. M. 1974. How "fixed" is the fixed action pattern? Z. Tierpsychol., 36:184–211.

SCOTT, J. P., AND Z. M. NAGY. 1980. Behavioral metamorphosis in mammalian development. Pp. 15–37, *in* Early experiences and early behavior. Implications for social development (E. C. Simmel, ed.). Academic Press, New York, 217 pp.

SHUBNIKOV, A. V., AND V. A. KOPTSIK. 1974. Symmetry in science and art. Plenum Press, New York, 420 pp. (Translated from original Russian by D. Harker.)

STEVENSON, J. G. 1967. Reinforcing effects of chaffinch song. Anim. Behav., 15: 427–432.

SYMONS, D. 1978. Play and aggression: a study of rhesus monkeys. Columbia Univ. Press, New York, 246 pp.

TEITELBAUM, P. 1977. Levels of integration of the operant. Pp. 7–27, *in* Handbook of operant behavior (W. K. Honig and J. E. R. Staddon, eds.). Prentice-Hall, Englewood Cliffs, New Jersey, 689 pp.

TINBERGEN, N. 1951. The study of instinct. Clarendon Press, Oxford, 228 pp.

———. 1963. On aims and methods of ethology. Z. Tierpsychol., 20:410–433.

VEHRENCAMP, S. L. 1979. The roles of individual, kin, and group selection in the evolution of sociality. Pp. 351–394, *in* Handbook of behavioral neurobiology. Vol. 3: Social behavior and communication (P. Marler and J. G. Vandenbergh, eds.). Plenum Press, New York, 411 pp.

WECKER, S. C. 1963. The role of early experience in habitat selection by the prairie deer mouse, *Peromyscus maniculatus bairdi*. Ecol. Monogr., 33:307–325.

WELKER, W. I. 1971. Ontogeny of play and exploratory behaviors: a definition of problems and a search for new conceptual solutions. Pp. 171–228, *in* The ontogeny of vertebrate behavior (H. Moltz, ed.). Academic Press, New York, 500 pp.

WHITE, L. 1977. Play in animals. Pp. 15–32, *in* Biology of play. J. W. Lippincott Co., Philadelphia, 217 pp.

WILSON, E. O. 1975. Sociobiology: the new synthesis. Harvard Univ. Press, Cambridge, 697 pp.

WITTENBERGER, J. F. 1979. The evolution of mating systems in birds and mammals. Pp. 271–349, *in* Handbook of behavioral neurobiology. Vol. 3: Social behavior and communication (P. Marler and J. G. Vandenbergh, eds.). Plenum Press, New York, 411 pp.

WOOLRIDGE, M. W. 1975. A quantitative analysis of short-term rhythmical behaviour in rodents. Unpublished Ph.D. dissert., Oxford Univ., 201 pp.

WYMAN, R. J. 1976. A simpler network for the study of neurogenetics. Pp. 153–166, *in* Simpler networks and behavior (J. C. Fentress, ed.). Sinauer Associates, Sunderland, Massachusetts, 403 pp.

ZIMEN, E. 1976. On the regulation of pack size in wolves. Z. Tierpsychol., 40: 300–341.

A MOTILITY-IMMOBILITY GRADIENT IN THE BEHAVIOR OF THE "INFERIOR" WOLF DURING "RITUALIZED FIGHTING"

ILAN GOLANI AND GREG MORAN

Abstract

THE structure and function of the behavior of the "inferior" wolf during "ritualized fighting" was the subject of a dispute between Lorenz (1949) and Schenkel (1967). This behavior is reexamined by studying the coordination and integration of nine kinematic variables which characterize it. We show that so-called inferiority involves a coordinated transition along a kinematic gradient, from motility to relative immobility and arrest and vice versa. As the superior wolf shifts closer to the forequarters of the inferior one, the inferior wolf shifts from forward progression to circling, pivoting, and squat-arrest. Lateral bending, arching of torso, and tucking in of hindquarters increase as well, and leg flexion and ground contact increase from rear to front. The same morphogenetic process unfolds in the opposite direction as the rival wolf shifts to oppose the hindquarters of the inferior at a distance.

Detailed comparisons show that a remarkably similar kinematic process is found in the transition in and out of relative arrest in a variety of contexts, mammalian species, and preparations. Versions of it are found during the circling and pivoting leading to defecation in canids; during the turning around in a circle several times before lying down to sleep in, for example, wolves (*Canis lupus*), martens (*Martes martes*), and raccoons (*Procyon lotor*); and in mammals under the influence of apomorphine, a psychoactive drug which induces forward progression, circling, pivoting, and ultimately relative arrest. The opposite version is found in the emergence from arrest in infant rodents and carnivores who proceed out of arrest by pivoting, then circling, then forward progression. It is also found in exploratory behavior of hedgehogs which emerge out of arrest following the same orderly transition, and in recovery from neurological lateral hypothalamic akinesia in rats.

In all the pathological preparations, the structure of movement reflects the organization of the mediating dopaminergic system. The so-called display of inferiority during ritualized fighting presumably reflects the same fundamental organization.

Introduction

This paper presents a systematic formal analysis of the behavior of the so-called "inferior" wolf (*Canis lupus*) during "ritualized fighting" (Schenkel, 1947, 1967). As described by Schenkel, during such interactions the "inferior" wolf alternates between approach and withdrawal, flight and aggression, inferiority and non-inferiority, and varying symptoms of social stress, such as inhibited locomotion, inhibited biting, and bending of the tail between the legs. In contrast to the behavior of the "inferior," the "superior" wolf assumes postures which appear as a challenge to the inferior; the superior's neck is often "exposed" in front of the inferior's jaws, the latter showing a strong inhibition to bite. At such times, the two wolves align in parallel or at a 45° angle, the jaws of the inferior facing the shoulder, neck, or cheek of the superior. In another commonly observed configuration the superior's jaws point at the inferior's hindquarters; the latter then circles in order to withdraw the unprotected part of the body and to parry the attack with his jaws.

The whole interaction appears as a preparatory phase for severe fighting. It is interpreted by Schenkel (1947, 1967) as ritualized fighting over a "privilege," in which the inferior momentarily "gives up the claim," thus blocking the aggression in the superior but shortly afterwards tries again. Such interactions follow severe fights (Schenkel, 1967), precede them, and most frequently occur in the absence of any fighting at all (Moran et al., 1981).

As vivid as Schenkel's description is, it suffers from a confounding of functional, motivational, and formal criteria (Purton, 1978). Also, it does not provide a comprehensive representation of the *changing* spatiotemporal relations between the wolves. As long as the animals assume relatively static postures or social configurations—these are illustrated and described, but the changes between these postures and configurations are mostly ignored—presumably because of the lack of an appropriate technology of description. In a way, this paper may be regarded as a response to a plea for a more dynamic and explicitly relational approach to wolf social behavior, made by Fentress et al. (1978).

This report and its predecessor (Moran et al., 1981), are not intended to replace Schenkel's description, but rather to complement it so that the rich and complex spatiotemporal maneuvering of the interactants around each other and in the environment become apparent and visible even to the casual observer. Also, by providing the actual formal structure of the interaction, students of behavior should have a more appropriate data base for deciding to what extent the behavior of the "inferior" wolf is "ritualized," as Schenkel suggests it is, or is merely another instance of adaptation to the complex environment of a moving rival.

Since ethology's early days, the procedure most commonly used in the description of behavior involves an initial categorization of behavioral units, and subsequent search for rules which explain their sequencing. Behavior is conceived as a chain of detail combined by seemingly external constraints—motivational, stochastic, or environmental. The "taxonomic" phase comes first, followed by the analysis of serial order. Quantitatively oriented approaches to the study of behavior have established a rigor in the assessment of variance *within* categories of behavior (Barlow, 1977; Schleidt, 1974), and in the analysis of serial order (Colgan, 1978; Hazlett, 1977; Todt, 1975). However, the crucial initial act of categorization of behavioral units—the qualitative distinction between the presumed building stones of behavior—remains to this day basically implicit.

One alternative to this approach—and again, only an alternative—is to regard behavior as structurally whole and study the organization of its structure. Not unlike a living tissue, it may be examined from a variety of perspectives, each yielding a specific plan of organization. The categorization of behavior into discrete events is not taken for granted. Instead, continuous variables are isolated and recorded simultaneously. The coordination and integration of these variables may or may not yield discrete categories. Using this approach, both serial order and behavioral taxonomy become aspects of the same fundamental behavioral process (Golani, 1981).

Our method for behavioral description involves basically a rigorous recording of kinematic variables in several spherical coordinate systems (Eshkol and Wachmann, 1958), with a subsequent search for geometrical invariance (Ganor and Golani, 1980; Golani, 1976; Golani et al., 1979). Detailed Eshkol-Wachmann movement notation motor scores, which represent the relevant behaviors and substantiate the description, were used as the basis for the present report.

The present paper presents a structural analysis of a central class of motor interaction sequences in wolves from two separate perspectives—social and individual. The analysis yields two separate plans of organization. Although the individual plan is finely coordinated with the social plan, it is also performed outside of the interactional context—in exploratory behavior during early development and in adulthood, in pathology, and under the effect of drugs. The interplay between the two plans accounts for the morphogenesis of the interaction.

The rationale for presenting natural and pathological behaviors within the same framework is that the structure of movement in the pathological preparations reflects the organization of the mediating dopaminergic system; the occurrence of similar behaviors in natural contexts presumably reflects the same fundamental organization.

The social plan and the components of the individual plan have already been described in some detail by Moran et al. (1981). The present paper ties together for the first time the individual components into a broader plan of organization. It also shows how the individual plan is coordinated with the social plan.

Extended Supplanting Interactions in Wolves

The terms "ritualized fighting," "inferiority," and "superiority" determine a priori the nature and significance of the interactions which were the very subject of the present study. In a study based on kinematic observables such terms can at most become working hypotheses which can be either established or refuted at the end of the study. Furthermore, many of the interactions observed by us consisted of relatively calm, seemingly indifferent involvement of the two wolves, not resembling a fight at all. Therefore, in order to concentrate on observables and avoid theory-loaded terms, the so-called superior wolf was designated as the *Supplanter S,* the so-called inferior was designated as the *Displaced* wolf *D,* and the class of interactions was designated as *extended supplanting.*

The Social Plan

The changing spatiotemporal relations between two interacting wolves can be described unequivocally by using three continuous "interactional" variables—the distance between the two closest body parts of the two interactants, the angle between the longitudinal axes of their bodies, and a so-called "relationship of opposition" (Golani, 1976; Moran et al., 1981). The relationship of opposition (Eshkol, 1980) is defined by the two body parts of the two respective interactants which are in close proximity to each other (for example, muzzle to hindquarters). Opposition can be recorded continuously, thus being represented by opposition "pathways" on the body surfaces of the two interactants.

During supplanting interactions the wolves establish and maintain for extended periods of time two qualitatively distinct relationships of opposition. These relationships are maintained steadily in spite of, or rather as a result of, the extensive movement of both wolves.

The first interaction involves an opposition between the hindquarters of *D* and the forequarters of *S.* It occurs in two interactional configurations. In the first, labelled "Follow," it may be maintained for up to half an hour as *S* walks behind *D* persistently (Fig. 1).

The second interactional configuration, labelled "Circle," also involves an opposition between the hindquarters of *D* and the forequarters of *S.* This configuration is also maintained for extended periods as the two wolves circle each other (Fig. 2).

The second relationship of opposition involves the forequarters of *D* and the shoulder of *S.* It also occurs in two configurations and is maintained as *S* either encircles *D* (Fig. 3), in a configuration labelled "Twist and Turn," or in a very brief configuration which involves, as in Twist and Turn, an opposition between *D*'s forequarters and *S*'s shoulder and, simultaneously, between *D*'s torso and *S*'s pelvis. It occurs as *S* hip thrusts *D* and is labelled accordingly (Fig. 4).

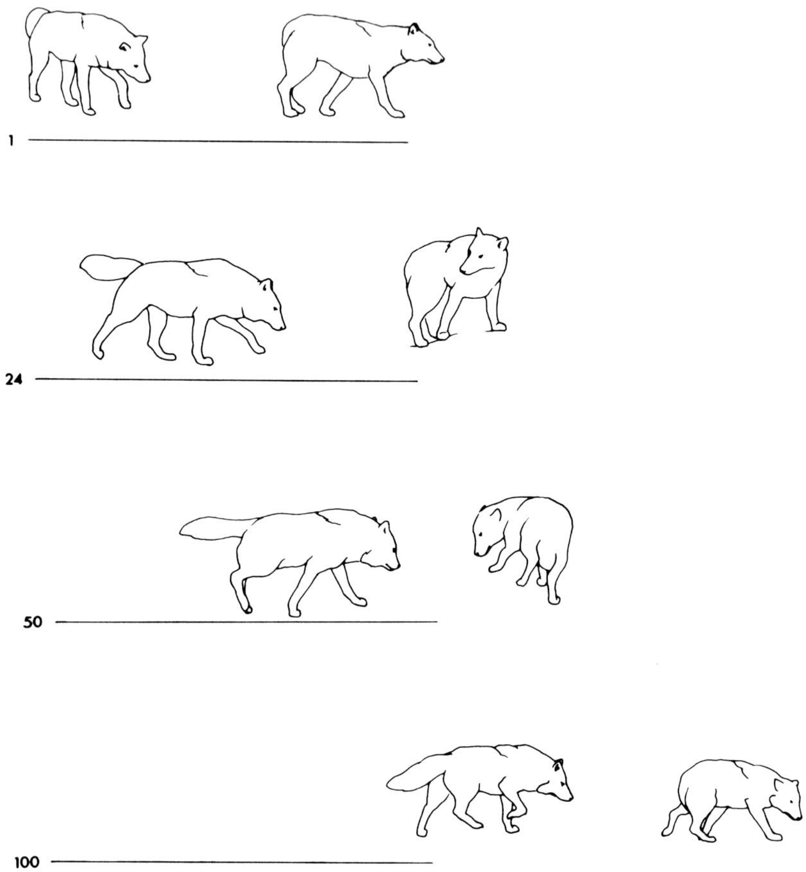

FIG. 1. Relatively stable configuration labelled "Follow" in wolves (*Canis lupus*). Numerals indicate frame numbers of film taken at 24 fPS.

As soon as the two animals move into the proximity of each other, they are seemingly "sucked into" one of these two relationships of opposition which are then maintained until the partners shift to the other stable relationship. The maintenance of a specific relationship of opposition acquires a priority over the maintenance of other behavioral invariants. While almost all behavioral variables change, opposition is invariant. The two animals move in unison as though connected by an imaginary "joint" which "forces" them to move in particular ways (Eisenberg and Golani, 1977; Golani, 1976, 1981).

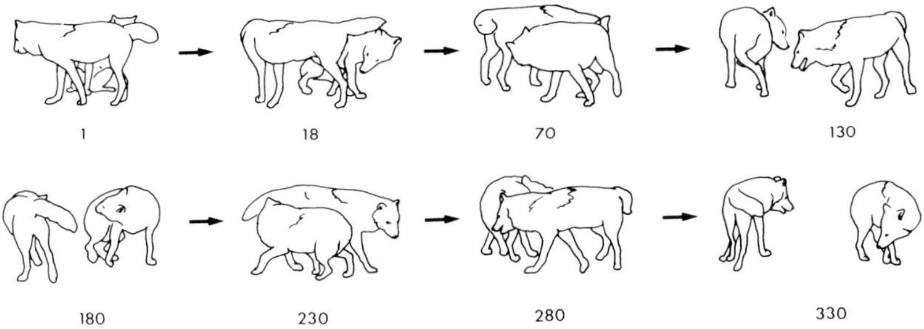

FIG. 2. Relatively stable configuration labelled "Circle."

The freedom for movement within the "joint" boundaries is disclosed by the other interactional variables of interanimal distance and relative orientation between partners. For each joint, specific values of distance and relative orientation are maintained temporarily and then change to new values which are again maintained, but always within a specified range typical to the joint. Because each joint involves a narrow *range* of oppositions, orientations and distances, these unitary processes have been described as "relatively stable configurations" (Moran et al., 1981).

When opposition is maintained at the far rear of *D*, the distance between the wolves is maintained at half a body length or more as in Follow (Fig. 1). When it is maintained more rostrally, on the side of *D*'s hindquarters, distance is typically maintained at approximately one half-wolf length, as in Circle (Fig. 2). When opposition is maintained still more rostrally at *D*'s forequarters, as in Twist-

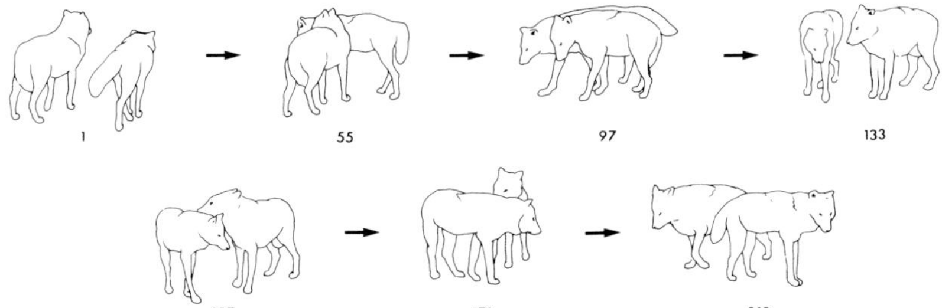

FIG. 3. Relatively stable configuration labelled "Twist-and-Turn."

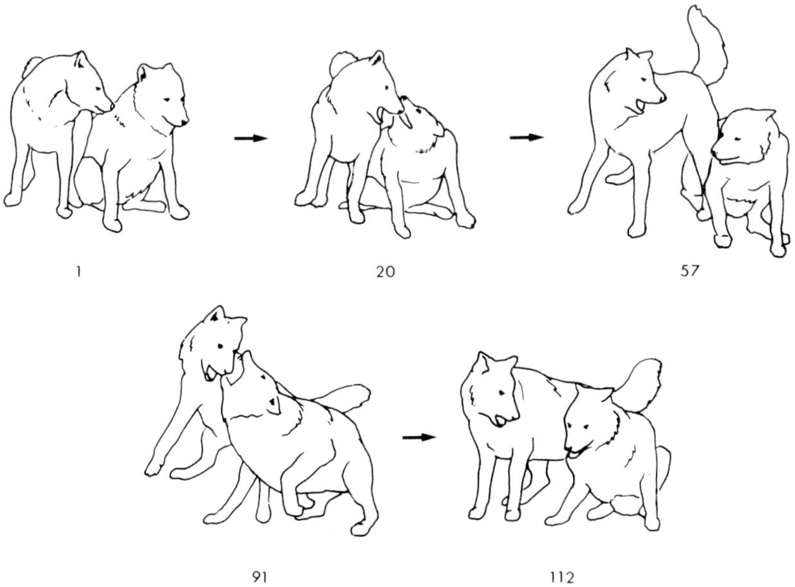

FIG. 4. Relatively stable configuration labelled "Hip Thrust."

and-Turn (Fig. 3) and Hip Thrust (Fig. 4), distance is maintained at contact or near contact. Thus, the maintenance of the largest distances is coupled to the maintenance of the most caudal oppositions, and the maintenance of smallest distances is coupled to the maintenance of rostral oppositions.

Relatively stable configurations have also been characterized by non-overlapping ranges of relative orientation between partners (Moran et al., 1981). Although orientation is a necessary part of the specification of configurations, it will not be discussed in the present paper, because its exposition does not add to the establishment of the coordination between interactional and individual variables suggested in this study.

Transitions between Relatively Stable Configurations

Transitions from one relatively stable configuration to the next involve a monotonic change along the relational variables. Monoton-

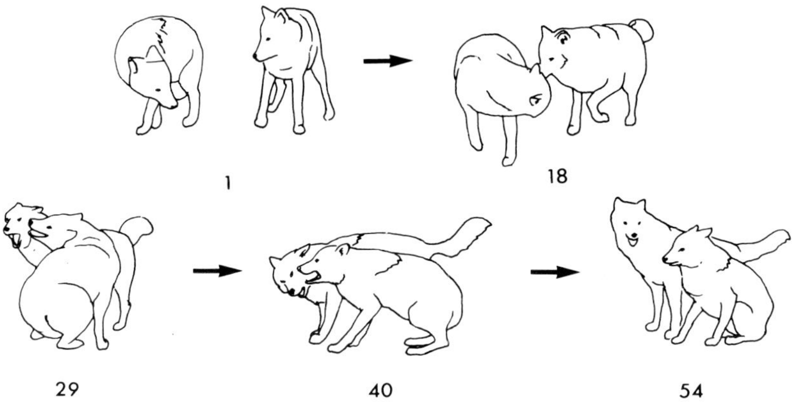

FIG. 5. The "Lunge/Swivel" transition.

icity is used here in the sense of a unidirectional, possibly intermittent change, but never in two opposite directions within the same transition. Opposition shifts monotonically on *D*'s body surface either clockwise or counterclockwise; distance is either increased or diminished monotonically; and relative orientation shifts monotonically for either $\alpha°$ or $360 - \alpha°$ degrees. The coordinated simultaneous transition along all three relational variables yields five complex patterns of transition which involve a change along all the variables and four simple transitions which involve a change only along the variable of relative orientation, as between Follow (Fig. 1) and Circle (Fig. 2). These transitions account for all logically possible combinations between monotonic changes of the three variables which can be performed between the four available relatively stable configurations. The configurations and transitions account for about 90% of the observed behavior and provide a partitioning of supplanting interactions in wolves (Moran et al., 1981).

The five complex transitions are illustrated in Figs. 5–9. *A shift of opposition on* D *from rear to front involves reduction of inter-animal distance.* This occurs in Lunge/Swivel (Fig. 5) and Walk-Up/Stop (Fig. 7) which involve a transition from Follow to Twist-and-Turn or Hip Thrust, and in Swivel/Stand Across (Fig. 6) which involves a transition from Circle to Twist-and-Turn. *A shift of opposition on* D *from front to rear involves an increase in inter-animal distance.* This occurs in Turn-to-Rear/Turn (Fig. 8) which involves a transition from Twist-and-Turn to Follow and in Walk

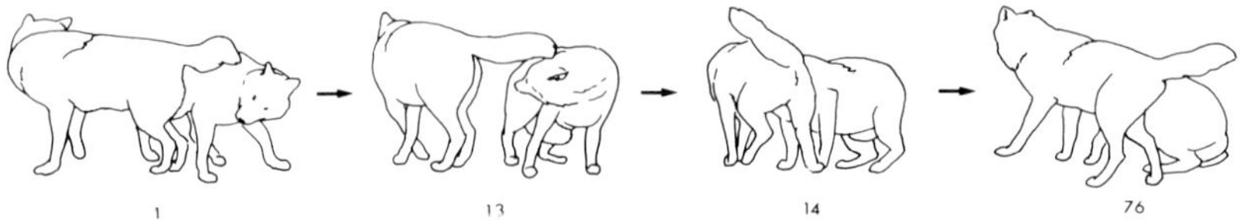

FIG. 6. The "Swivel/Stand Across" transition.

Away/Walk (Fig. 9) which also involves a transition from Twist-and-Turn to Follow.

To sum up, the social plan comprises four relatively stable configurations, five three-dimensional, and four unidimensional transitions between configurations. Transitions involve a monotonic change along the interactional variables. A shift of opposition on *D* from rear to front involves a reduction in distance and a shift of opposition on *D* from front to rear involves an increase in distance. *S* is shifted on *D*—because of *D*'s locomotion, *S*'s locomotion, or the locomotion of both wolves—from far and behind to close and in front, then back, from close and in front to far and behind.

The Individual Plan

The locomotion of an individual animal can be described in terms of the direction of its progression in relation to the longitudinal axis of its own body and in terms of the orientation of this axis in relation to the environment. The variable which describes progression may involve, for instance, forward, backward, and sideways progression. The variable that describes the change in orientation of the longitudinal axis of the body is called in the Eshkol-Wachmann (E-W) movement notation terminology "shift of front," and involves clockwise and counterclockwise changes of orientation of specified angular displacements (Eshkol, 1980; Eshkol and Wachmann, 1958). The term "shift of front" replaces the commonly used term

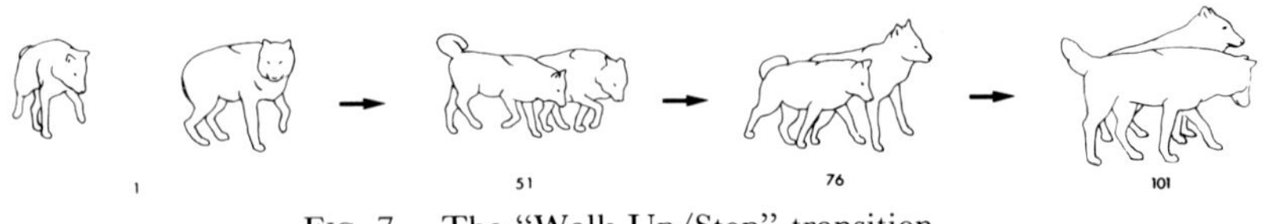

FIG. 7. The "Walk Up/Stop" transition.

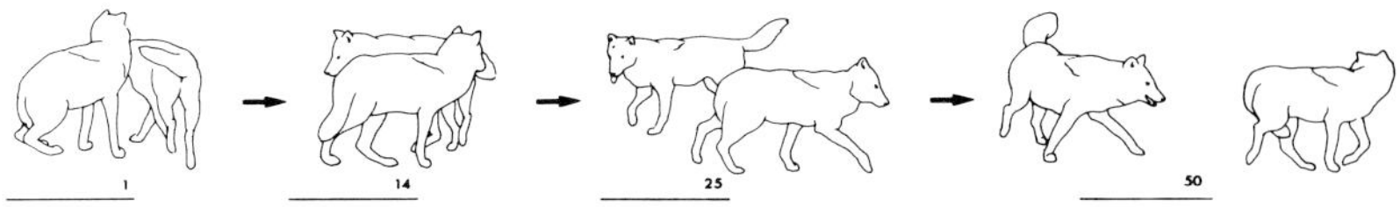

FIG. 8. The "Turn-to-Rear/Turn" transition.

"turning," in order to avoid ambiguity (in every day language turning refers to changes in orientation, which may also involve progression). A description in terms of these two simultaneous geometrical variables specifies both the direction of progression of an animal in relation to its own body and its orientation in the environment. For instance, forward progression alone generates a straight trajectory in the environment; forward progression with a shift of front generates circling. The tightness of the circling is determined by the ratio between the amount of forward progression and the amount of shift of front. With extensive forward progression and small change in orientation (small shift of front), the wolf describes a very wide circle; as the rate of forward progression diminishes and the rate of shift of front increase, the circling becomes tighter and tighter. Ultimately, a shift of front alone generates pivoting in place.

It was extremely difficult, if not impossible, to find a regularity in the coordination of these two variables when they were examined in isolation without reference to the context of the interaction. However, when examined in relation to interactional variables, a fundamental plan of coordination was also isolated at the level of the individual.

Individual Movements

D'*s forward walking and interanimal distance.*—During supplanting interactions, when *D* progresses in the environment it does

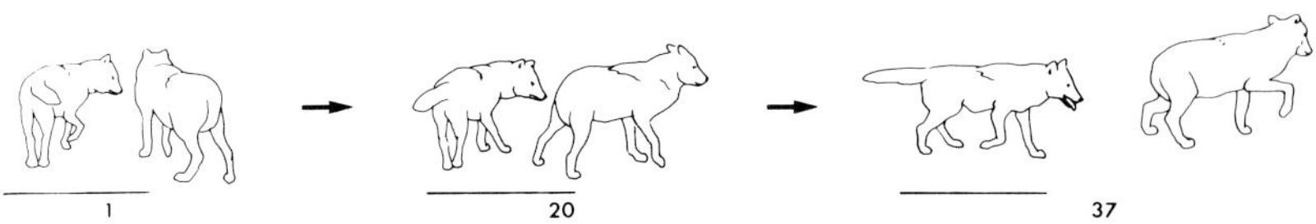

FIG. 9. The "Walk Away/Walk" transition.

so mostly by forward walking. It never walks backwards, and walks or jumps sideways only during Walk Away/Walk.

Forward walking and interanimal distance are tightly coordinated with each other. Increasing or large distances involve increasing or extensive forward walking by *D*. Decreasing or small distances involve diminishing, little, or no forward walking.

Thus, during the relatively stable configuration of Follow, as interanimal distance amounts to half-wolf length or more, *D* shows extensive forward progression (Fig. 1). During Circling, as distance is typically reduced, forward progression is minimal, locomotion comprising mostly a shift of front (Fig. 2); as distance is reduced to near contact or contact during Hip Thrust and Twist and Turn, no forward walking is observed (Figs. 3 and 4). Similarly, during transitions which involve a reduction in distance, forward walking is diminished and ultimately eliminated, although distance is still being reduced. During Lunge/Swivel, the extensive reduction in distance involves a cessation of forward walking at the initiation of the transition (Fig. 5). In Swivel/Stand Across, which also involves a reduction in distance, *D* performs no forward locomotion at all (Fig. 6). In Walk Up/Stop as distance is reduced, *D* either stops or slows down and then stops (Fig. 7).

During transitions which involve increasing distance, forward progression appears only toward their termination, when distance becomes sufficiently large. During Turn-to-Rear/Turn it appears only after an initial shift of front away from *S*, involving an increase in distance (Fig. 8). During Walk Away/Walk it appears after an initial progression sideways and away from *S* or after an initial shift of front away from *S* after distance increases (Fig. 9). Even during the "Follow" configuration, where forward locomotion is observed, *D* sometimes halts spontaneously forward progression and shifts its front. After such spontaneous momentary shifts of front of the fore-quarters backwards toward *S*, forward progression is resumed.

D*'s forward walking and opposition.*—Shift of opposition on *D* to hindquarters or the maintenance of opposition at hindquarters involves increasing or large amounts of forward progression respectively. Thus, during Follow when opposition on *D* is at the far rear, forward progression is the most extensive (Fig. 1). In Circle as opposition shifts somewhat forward on *D* to the side of its hindquarters, forward walking is reduced to a minimum. When opposition is at the torso or shoulder during Hip Thrust and Twist and Turn, there is no forward walking at all (Figs. 3 and 4).

Transitions of opposition from rear to front involve reduction and subsequent elimination or immediate elimination of forward walking, as in Walk Up/Stop (Fig. 7), Lunge/Swivel (Fig. 5), and Swivel/Stand Across (Fig. 6). Front to rear transitions of opposition involve appearance of forward walking, as in Turn-to-Rear/Turn (Fig. 8) and Walk Away/Walk (Fig. 9).

D'*s shift of front and distance.*—The coupling of shift of front to distance is as tight as that of forward walking, but in the opposite direction. Increasing and large interanimal distances involve diminishing with little or no shifts of front, whereas diminishing and small distances involve increasingly more extensive shifts along ever tighter circles, with ultimate arrest at contact or near contact.

During the Follow configurations momentary shifts of front which involve lateral bending of the torso and large lateral movements of head and neck backwards toward *S* occur (Fig. 1). With reduction of interanimal distance in Circle, shifts of front with only minimal forward progression are performed as *D* mostly performs a tight circle. As distance is reduced to near contact or contact, *D* performs pure shifts of front and merely pivots, as in Twist-and-Turn (Fig. 3).

During transitions from larger to smaller distances, the likelihood of shifts of front increases as does their amplitude. When such reduction in distance is slow, as in Walk Up/Stop, it may involve no shift of front at all or a shift of front which results in a Twist-and-Turn. Reduction in distance involves a slow, pure shift of front of almost 180° during Swivel/Stand Across (Fig. 6) and a fast pivot around hindquarters of almost 360° during Lunge/Swivel (Fig. 5).

Transitions from smaller to larger distances involve a shift of front at their initiation, before distance is increased sufficiently, as in Turn-to-Rear/Turn (Fig. 8). In Walk Away/Walk a sideways progression away from *S* often substitutes for an initial shift of front away from *S* (Fig. 9).

D'*s shift of front and opposition.*—Shifts of opposition to rear, and caudally maintained oppositions, involve a reduction in the likelihood of shift of front, an increase of the diameter of the circle, and a reduction in the amount of continuous shifts of front.

Shifts of opposition to front and frontally maintained oppositions involve an increasing likelihood of shifts of front along tighter circles.

Caudal opposition in Follow involves only occasional shifts of front of forequarters (Fig. 1); somewhat more rostral opposition in Circle involves a continuous shift of front along relatively large circles; frontal opposition in Twist-and-Turn involves pivoting around the hindquarters.

Transitions of opposition to rear involve an initial shift of front while opposition is still frontal. The shift of front disappears when opposition shifts to the rear (see Turn-to-Rear/Turn, Fig. 8, and Walk Away/Walk, Fig. 9).

Shift of opposition from rear to front is coupled to increasing shift of front as in Swivel/Stand Across (Fig. 6), Lunge/Swivel (Fig. 5), and the simple transition from Follow to Circle.

Several additional individual variables which describe ground contact and kinematic relations between body segments show also a coupling to distance and opposition.

D*'s ground contact, amount of leg flexion, arching and bending of torso, and tucking-in of hindquarters in relation to interanimal distance.*—Increasing distances typically involve a reduction in foot and body contact with ground, higher gait, unarching of torso, less lateral bending of torso and minimal or decreasing tucking-in of hindquarters and tail, proceeding from front to rear. Sometimes, however, increasing distances are not coupled to these changes, or these changes occur only intermittently as distance increases. Nevertheless, with increasing distance *D*'s ground contact is never increased, gait never becomes lower, torso never becomes more arched and more laterally bent, and tucking-in of hindquarters never increases. In other words, the changes along all these kinematic variables follow a principle of monotonic change in that they may change intermittently or not at all, but if they do change, they do so unidirectionally in the above specified directions.

Diminishing distances typically involve changes in all the above mentioned kinematic variables, but in the opposite direction, and proceeding from rear to front. A principle of monotonicity also applies here.

In *D* large distances in Follow typically involve a low digitigrade gait, minimal or no arching of torso, and minimal tucking-in of tail and pelvis. As distance is reduced the gait becomes progressively and monotonically lower and more plantigrade; arching and tucking-in are monotonically increased (Fig. 1). As distance is reduced to contact or near contact, *D* either stops without changing the other individual variables, as in Walk Up/Stop (Fig. 7), or shifts monotonically to plantigrade, flexes the hindlegs, arches and tucks in its hindquarters; there is a reluctance to release hindleg contact with the ground resulting in squatting and arrest. The latter is typically maintained during the relatively stable configuration of Hip Thrust and Stand Across (Fig. 10), which is a static version of Twist-and-Turn. In transitions, it is typically seen in Lunge/Swivel, and less drastically in Swivel/Stand Across. Transition to larger distances typically involves progressive monotonic release of ground contact from front to rear, extensions of the legs (also from front to rear) reduction of arching, and the monotonic reversal of tucking-in. It is shown in Turn-to-Rear/Turn and Walk Away/Walk (Figs. 8 and 9).

D*'s ground contact, amount of leg flexion, arching of torso, and tucking-in of hindquarters in relation to opposition.*—The same rules that apply for distance also apply for opposition. Shifts of opposition from rear to front typically involve monotonically increasing ground contact from rear to front, flexion of legs along the same gradient, arching, and tucking-in of hindquarters. Opposition shifts from front to rear involve monotonic changes in the opposite direction along all the above-mentioned individual variables.

Immobility properties are attributed to squatting and crouching (for example, see Fig. 10) because 1) during the assumption of these postures, *D* often performs large amplitude movements of its forequarters without recruiting its hindquarters for action (as though its hindquarters were immobilized); 2) when hip thrusted by *S* during the assumption of this posture, *D* may tumble while rigidly maintaining the tucking-in of the hindquarters and the flexed position of the hindlegs, again giving the impression of partial immobility; 3) *D* is never observed to walk forward out of this posture while *S* is in close proximity. If progressing at all, *D* may do so by jumping sideways or "galloping" forwards with seemingly paralyzed hindquarters and rigidly-flexed hindlegs.

Discussion of D*'s behavior*

During extended supplanting, distance reduction and shift of opposition from rear to front on *D* are coupled to each other, as

FIG. 10. Squatted arrest during a "Stand Across."

well as to: 1) reduction and ultimate elimination of forward progression; 2) increase in likelihood of shifts of front of increasing amplitudes; 3) increasing lateral bending of torso and wavering from side to side; 4) increasing arching of torso; 5) increasing tucking-in of hindquarters and tail; 6) rear to front progression of leg flexion; and 7) rear to front increase of ground contact, involving first digitigrade, then plantigrade stance, and ultimately continuous pelvis and hindleg contact with ground.

The opposite course of change occurs with increasing interanimal distances and shifts of opposition from front to rear. Thus, as *S* is shifted on *D* back and forth between "far behind" and "close in front," *D* keeps shifting back and forth along the above specified multidimensional gradient. Such shifts do not necessarily involve the *whole* gradient from arrest to extensive forward progression. *D* can "plug in" at any values along the gradient and change the direction of the shift at any values. But the direction of change is always common to all the variables, one way or another. In other words, *D* does not necessarily shift from complete arrest to highest motility, and back to complete arrest—it may start anywhere along the multidimensional gradient and move unidirectionally along it back and forth without reaching the extreme ends of the gradient. It is as though *D* is "sucked" in and out of squatted arrest through an imaginary whirlpool, seemingly having to work its way in and out by circling and pivoting. When "sucked in" it becomes progressively more arched, with tucked-in hindquarters, lower stance and increasing ground contact. Upon coming out, it becomes progressively less arched, with straight body and higher digitigrade stance.

The inter- and intra-individual variability which the wolves show with respect to the *absolute* values taken along the above described kinematic variables is large. Not only do wolves vary with respect to the distances and oppositions which precipitate a specific absolute amount of arching, bending, squatting, arrest, ground contact, and others, but also the same wolf may vary with respect to these behaviors at different times. For example, it may walk plantigrade at one time, when still at a distance from *S*, and at another time it may walk in a low digitigrade stance at the same distance. When such differences are observed between wolves they are usually explained in terms of differences in individual "character" and "experience." When observed in the same wolf, they are explained as differences in "mood." To predict quantitatively the correlation between distance, opposition, and the seven individual kinematic variables, it would be necessary to account quantitatively for the effects of individual characteristics and experience. The construction of such a quantitative model is obviously far in the future. The present study is therefore qualitative, in the sense that it merely specifies the direction of changes and their coordination with one another, without regard to their absolute values.

The morphogenesis of *D*'s behavior during supplanting is accounted for by a coupling between two relational and seven individual variables, all of which change monotonically one way or the other, but always in the above specified directions. At least three arguments could be raised against the existence of coupling between relational and individual variables: 1) by describing different aspects of the same behavior, a redundant representation is produced (redundancy is fallaciously interpreted as "a coupling"); 2) in many instances, the mere physical environment of the partner imposes certain forms of behavior (at such times, it is unnecessary to invoke other morphogenetic forces); 3) in most cases, the performance of progression and/or shift of front constitutes the easiest and most straightforward way to achieve the specific management of opposition and distance. Therefore, a principle of least action could account for the observed individual behavior.

The answer to the argument of presumed descriptive redundancy is that the representation of the same behavior in terms of multiple variables is only partly redundant. For instance, the particular shift in opposition and reduction of distance accomplished during a Lunge/Swivel could theoretically be performed by either *S* or *D* or by both in a variety of combinations of progressions and shifts of front; in reality, it is performed only by *D*. Similarly, a shift of front by *D* could be performed, for example, after a Follow in Lunge/Swivel in a variety of directions in relation to *S*. The fact that such a shift is always performed by *D* alone, and that the shift is almost always finely regulated to fit the particular interactional constraints, is not obvious. *D* performs large or small shifts of front in relation to the environment, but the interactional consequences are invariant. The invariance and the fitness in the quantitative aspects of the individual and interactional variables suggest a coupling between them. Tacitly, it implies a coupling between visual input which regulates distance and opposition, and visual, vestibular, and kinesthetic inputs which regulate the shift of front. However, because we have no access to input variables within our methodological framework, and because results are formulated exclusively in terms of observables, control is only implied.

Could it be that the physical environment imposed by *S* during the interactions provides a sufficient explanation for *D*'s behavior? For instance, it could be that *D* performs some shifts of front only because *S* is standing in its way, thus making forward walking physically impossible, as in Turn-to-Rear/Turn (Fig. 8) and Twist-and-Turn (Fig. 3). Similarly, some shifts of front may constitute the most straightforward way to reduce distance and shift opposition monotonically as in Lunge/Swivel (Fig. 5) or Swivel/Stand Across (Fig. 6). In other words, the very fine coordination of interactional and individual variables could explain the sufficiency of the interactional constraints, the principle of monotonic change, and some hypothetical principle of least action, which together account for the observed behavior.

To disprove this argument it is necessary to show that the same coupling between individual variables is generated in the absence of physical contraints and without direct functional consequences, that is, "in Vacuo." Such instances can indeed be observed. During Follow, *D* performs spontaneous lateral wavering of the head, neck, and torso, which may occasionally increase in amplitude to involve circling, pivoting, and squatting at a large distance from *S*. In the field, such coupling was observed when a young golden jackal (*Canis aureus*) sighted a mature jackal galloping toward it across a field. The young jackal pivoted into a squat-arrest while still at a large distance from the attacking jackal, and thereby did not escape (Golani, pers. observ.). Domestic dog owners are certainly familiar with this phenomenon, commonly observed in so-called "submissive" dogs upon approaching a reprimanding master.

Finally, a principle of least action does not account for the consistent differences in individual movements employed by *S* and *D* during supplanting.

The above described multidimensional kinematic coupling is not restricted to the behavior of *D* during supplanting interactions. Versions of it, involving some or most of the individual variables, can be observed in a variety of species and contexts—during development and in adulthood, and under both normal and pathological conditions. The various versions are similar in some respects and different in others. In this first report we shall, however, focus on similarities, not contrasts.

Ontogeny

Infant Exploratory Behavior

When infants of some quadrupedal mammals (for example, black rats, *Rattus rattus*; Norway rats, *R. norvegicus*; wildcats, *Felis libica*; and European badgers, *Meles meles*) are placed outside their nest, they first arrest and then display a regular buildup and spread of activity. This behavioral process was called "warm-up" (Golani et al., 1979). The process involves 1) a repetition of increasing movement trajectories of the head (either due to its own movements on the neck or due to movements of the neck, forelegs, torso, pelvis, and hindlegs), within specific dimensions of movement, and 2) an orderly incorporation of additional dimensions of movement. The "warm-up" sequence undergoes in ontogeny a series of behavioral transformations (Golani et al., 1981). However, in all of the species examined and at all days during the neonatal period, infants never walked forward before performing lateral movements, and never performed lateral movements before being arrested. The process is best illustrated in a "full blown" sequence of warm-up in a laboratory rat (*Rattus norvegicus*) at day 11 (Fig. 11). After prolonged arrest the rat performs increasingly larger lateral head trajectories, progressively recruiting more caudal limb and body segments. These culminate in pivoting and only then does the rat shift to forward walking, which is then followed by head-raising in the air away from the substrate. Support is regained cephalocaudally and ground contact is reduced from forequarters to hindquarters. It is as though to move out of an arrested position the rat has to repeatedly move its front limb and body segments, thus recruiting more caudal segments for movement in increasing amplitudes. Also, it is as though it cannot walk forward before performing increasingly larger shifts of front. Later in development, at day 15 or so, repetition is eliminated and the rat shifts directly from arrest to a large shift of front of some 90°–180° and then proceeds to walk forward. At this time its behavior resembles most closely the behavior of the *D* wolf as it moves out of arrest during supplanting interactions.

Adult Exploratory Behavior

The performance of a large shift of front in the transition from arrest to forward walking can be observed in adult laboratory rats when placed outside their cage in an unfamiliar environment (Nau and Golani, pers. observ.). But the orderly transition from initial arrest to extensive forward walking is performed in a much more striking form in the hedgehog, *Hemiechinus auritus* (Moualem and Golani, pers. observ.). Figure 12 illustrates a typical "warm-up" sequence in a normal adult. After pronounced arrest which involves arching of torso, tucking-in of hindquarters, and flexion of all four legs, the hedgehog performs increasingly larger shifts of front which are then transformed into forward walking. Concurrently, it reduces the torso arch and tucks out its pelvis gradually, extends forelegs and then hindlegs, steps with forelegs first, then incorporates hindlegs, and in general recruits limb and body segments for movement cephalocaudally. As exploration proceeds, it incorporates increasingly longer forward trajectories and eliminates pivoting and then circling. Some hedgehogs revolve like a top for several rounds of 360° each before proceeding to walk forward.

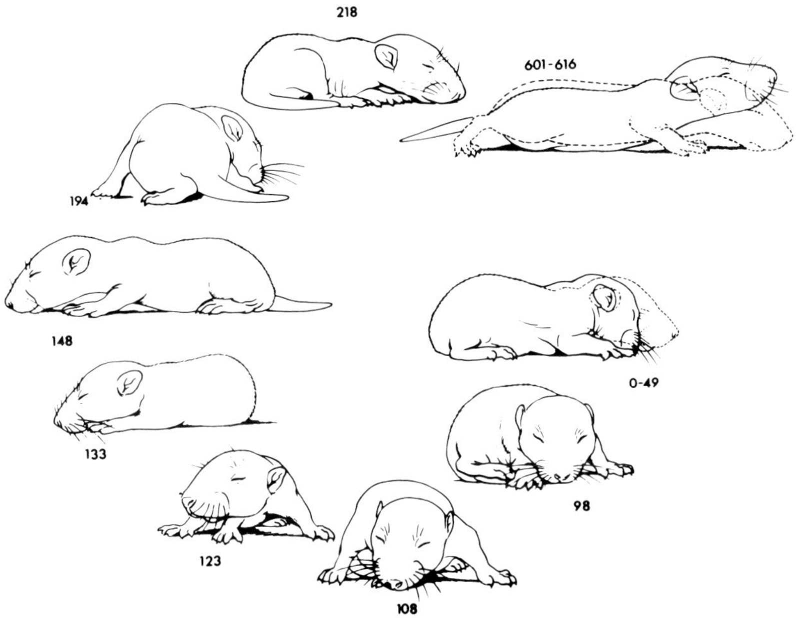

FIG. 11. Warm-up sequence in an 11-day-old infant laboratory rat (*Rattus norvegicus*). Tracings were made from film taken at 24 fPS. Numerals stand for frame numbers. Tracings should be followed in a clockwise direction starting from frame "0." Where possible, tracings show the position reached at the maximal extent of a movement along a dimension. After being placed on a horizontal surface outside the nest (frame 0, interrupted line tracing), the infant performs several small lateral head and neck movements, not represented in drawing. Increasingly larger lateral head trajectories follow as the infant recruits first head, neck, and one foreleg in a sideways step (frame 49, continuous line tracing); then head, neck and both forelegs in sideways stepping, and the ipsilateral hindleg in backward stepping (frame 98); then all four legs—forelegs to its right, hindlegs to its left (frames 108, 123). Pure minimal head-raising in air then occurs (frame 123), and after lowering of head, infant continues to pivot (frames 133–218). Movement forward, along the longitudinal dimension then sets in as infant walks for several steps, stops (frame 601, interrupted line tracing), and then raises head away from surface (frame 616 continuous line tracing) (from Golani et al., 1981).

Toilet Behavior and Lying Down to Sleep

In two situations which involve transition from motility to relative arrest an opposite sequence of change is observed. Before defecation, female dogs and puppies establish continuous snout contact with the ground, then proceed from forward locomotion to circling, to pivoting, to arrest, concomitantly arching their back and flexing

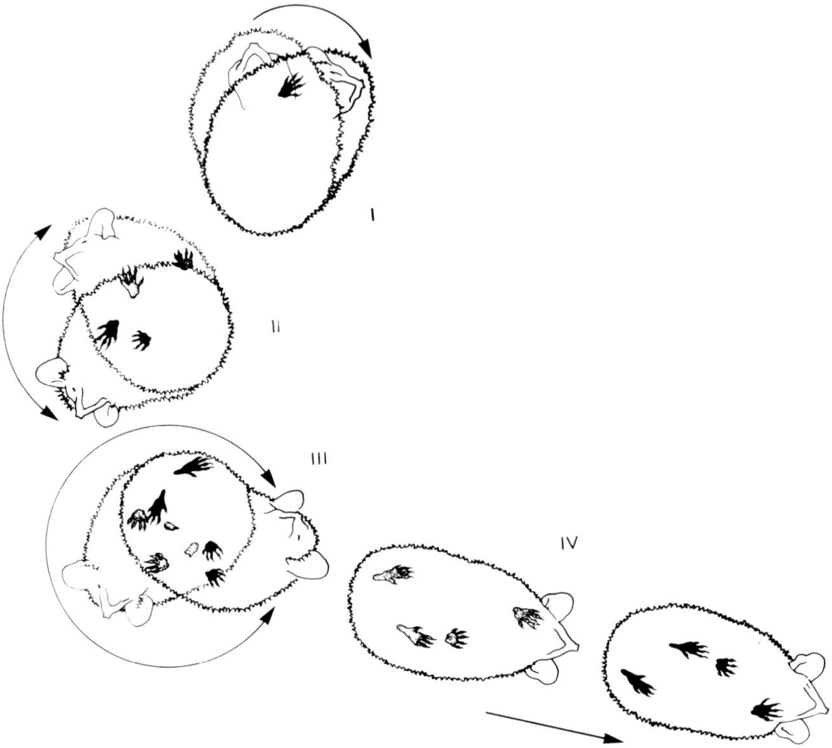

FIG. 12. Bottom view of a "warm-up" sequence in hedgehog exploratory behavior (*Hemiechinus auritus*). Tracings were made from film taken at 24 fPS; the hedgehog was placed on glass and filmed from underneath to show progressive cephalocaudal leg extension and increasing ground contact, reverse of tucking-in of hindquarters, and reduction of arching of torso (seen as stretching longitudinally). Roman letters stand for four successive stages. Dotted feet stand for earlier, dark feet for later, positions of feet.

their hindlegs. During defecation, the dog is relatively immobile. If snatched up in the air, it maintains the basic crouched posture. Most owner manuals advise dog owners to use nose contact, circling and pivoting as clues for a "forthcoming accident" (Wilmhurst, 1955). The other situation involves "turning around in a circle" several times before lying down to sleep. Darwin referred to this behavior, citing it as an instinctive action performed "in Vacuo," without any obvious function. Konig (cited in Trumler, 1973) reported that the amount of circling is positively correlated with the amount of forward locomotion practiced by the dog prior to lying down. Konig caused wolves, jackals, foxes (*Vulpes vulpes*), martens (*Martes martes*), minks (*Mustela vison*), and raccoons (*Procyon lotor*) to run on a treadmill and then watched how they lay down. After short runs, one turn-around preceded lying. After a long one, the animal turned around several times.

Not all dogs show this behavior, and in general, animals clearly do stop and locomote, without going through this behavior. Nevertheless, transitions are often observed. It is possible that the above described bizzare and heretofore unexplained behaviors reflect aspects of the same basic transition toward arrest observed during supplanting.

Pathology and Influence of Drugs

The "warm-up" structure is also observed in neurological recovery from severe bilateral lateral hypothalamic (LH) damage in rats as they shift from total immobility to high motility (Golani et al., 1979). The opposite process, or "warm-up" *in reverse,* is seen under the influence of a psychoactive drug called apomorphine (Szechtman et al., 1980).

Before going into the description of these two opposite processes, a few words are necessary about the nature of the two preparations. LH akinesia is presumably part of what has been described as the "dopamine deficiency syndrome" (Hornykiewicz, 1972). It is produced by destroying the nigrostriatal bundle (Marshall et al., 1974; Stricker and Zigmond, 1974; Ungerstedt, 1971*a*), which supplies the striatum with dopamine (Ungerstedt, 1971*b*). LH lesions are, therefore, assumed to involve *deficient* stimulation of the striatum with subsequent recovery due to the presumed development of a so-called denervation supersensitivity. Apomorphine, on the other hand, is a drug considered to be a dopamine receptor agonist (Anden et al., 1967; Ernst, 1967). Its administration involves, therefore, *excessive* stimulation of the striatum with presumed subsequent gradual shut down of synaptic transmission in the striatum. After LH lesions, rats become immobile and then gradually recover motility. Under apomorphine, they first become "hyperactive" and then gradually subside to relative arrest.

After severe LH damage, exploratory behavior recovers along several relatively independent dimensions of movement which appear successively. The pattern of recovery, which takes days or even months to complete, is reiterated in every sequence of activity that the rat performs after pronounced arrest. Every such sequence culminates at the stage of recovery that the animal has reached. It is this sequence of reiterations which was originally termed "warm-up" (Golani et al., 1979).

Although the process of warm-up takes seconds or minutes to complete in an LH-lesioned rat, the process of reverse warm-up under apomorphine takes 1 to 2 hours. Nevertheless, within this longer time span the rat's behavior undergoes similar transformations along the same variables, but in reverse and in the opposite direction. When two complete sequences of warm-up with LH damage and apomorphine-induced behavior are compared, the following symmetries are observed (Table 1). Apomorphine-treated rats exhibit exaggerated licking and gnawing (Ernst, 1967). Frequent licking movements in the air are also performed by *D*, the displaced wolf, during supplanting. Figure 13 illustrates four stages in the process of orderly regression from motility to relative arrest under apomorphine. It can be seen how forward walking gives way to circling, which in turn gives way to revolving and then to pivoting. The two extremes of the process—the initial extensive rearing and the final arrest of the hindquarters—are not illustrated. Note the concomitant change along the other relevant variables. Thus, both after LH damage and under apomorphine, rats show very extended versions of the same basic transition from motility to arrest and vice versa (including repetition and extension in time), which also is performed by *D* during supplanting interactions.

Whether in recovery or under apomorphine, the rat's exploratory behavior is finely coupled to the physical structure of the environment. While it abides by the structural rules of either "warm-up" or the reverse, the rat's behavior is also molded by environmental constraints, as is the behavior of the *D* wolf during extended supplanting. When an LH-lesioned rat is placed in a corner, it still performs the warm-up sequence, but upon encountering a wall early in the process, it does not perform a vertical scan on it. Then, during successive encounters with the same wall it performs increasingly larger vertical scans. Similarly, it does not walk forward when first encountering the open side of the corner, but during successive encounters, it performs increasingly larger forward progressions which eventually lead it out of the corner. The rat appears to explore the corner, but at any one moment it only employs the

TABLE 1
SYMMETRICAL CONTRAST IN BEHAVIOR OF LABORATORY RATS AFTER LH LESIONS AND APOMORPHINE

Warm-up after LH damage (after Golani et al., 1979)	Exploratory behavior under apomorphine (after Szechtman et al., 1980)
From arrest to exaggerated motility.	From exaggerated motility to arrest of hindquarters.
Starts from arrest with movement along the lateral dimension, then incorporates forward walking, then vertical scans, then rearing away from substrate.	Starts with rearing away from substrate, then incorporates forward walking, then shifts of front (along the lateral dimension), then partial arrest.
Along each of these dimensions starts with small amplitude movements, then shifts to exaggerated amplitudes, then subsides.	Along each of these dimension starts with exaggerated amplitudes, then either subsides or eliminates the respective dimensions.
From maximal to minimal ground contact (plantigrade to digitigrade).	From minimal to maximal ground contact (digitigrade to plantigrade).
From flexion to hyperextension of legs (from front to rear).	From hyperextension to flexion of hindlegs.
From continuous snout contact to release of contact.	From release of snout contact to continuous snout contact.
Lateral bending of torso increases initially then decreases.	Lateral bending of torso is missing initially, then increases to a maximum, then decreases.

dimensions and amplitudes of movement available to it. The same sequence of movements is also seen in the open field in the absence of walls, and the corner only molds the individually constrained behavior (Golani et al., 1979). Similarly, under apomorphine, rats actually explore the environment, following its contours with their snouts, rearing along surfaces, and so forth (Ellinwood and Kilbey, 1975). In so doing, however, they only employ the dimensions available to them at the time (Szechtman et al., 1980).

A parallel to *D*'s behavior during extended supplanting is evident; in all the above mentioned instances and preparations an individual plan is carried out in coordination with environmental constraints—be they the contours of the physical environment or those of a social rival.

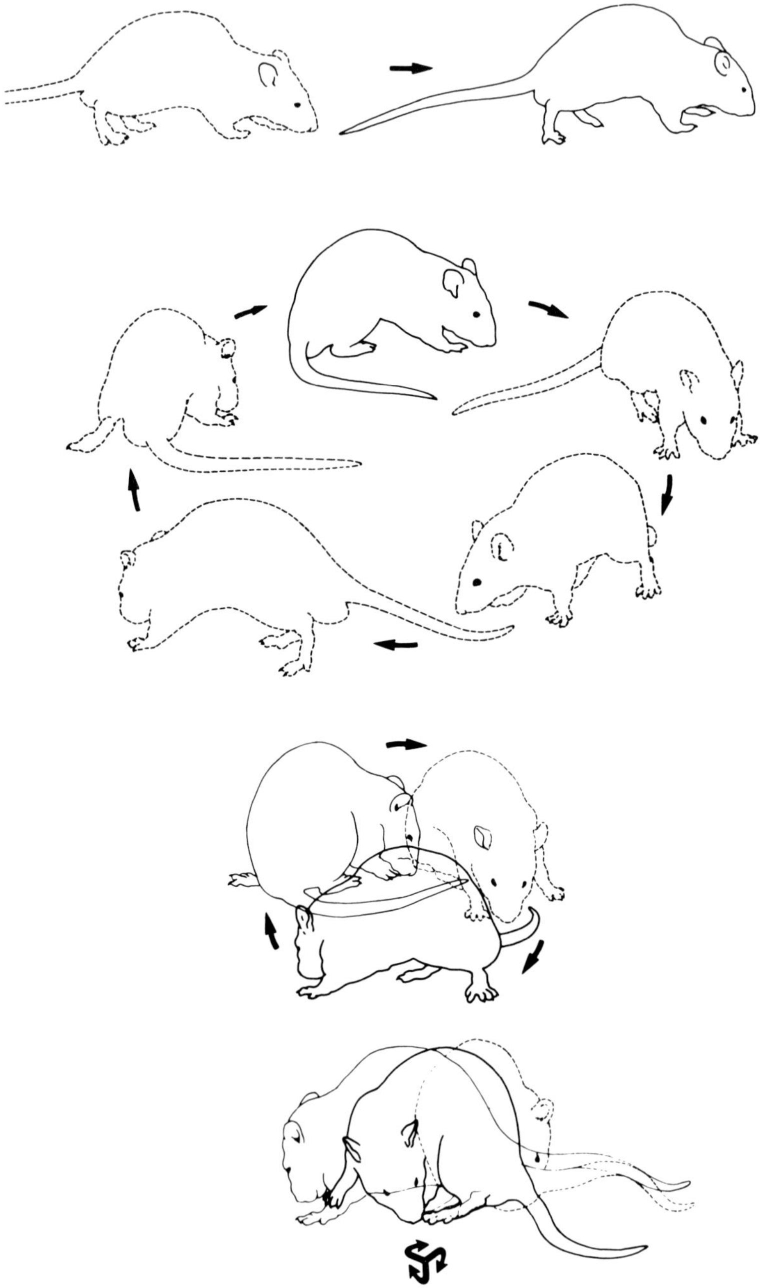

General Discussion

A highly ordered "Zeitgestalt" is involved in the transition from motility to arrest and vice versa (from arrest to motility) in a variety of mammals. Its synchronic properties are expressed in the coordination between many kinematic variables, including forward progression, shift of front, leg flexion, ground contact, dorsal arching of torso, lateral bending of torso, tucking-in of pelvis, rearing, and vertical scans. Different versions of this "Zeitgestalt" involve a coordinated change along some or most of these variables. Its temporal (diachronic) properties are expressed in the orderly transition between dimensions of movement. It is employed in development and adulthood, in exploration, and during social interactions. It is observed in neurological recovery and under the influence of a stimulant drug. It unfolds in seconds, minutes, or hours, in various versions, in two opposite temporal sequences. As it unfolds, it is finely coupled to environmental constraints. Like Lorenz's (1937) "instinct-tools," this structure can be employed in a variety of contexts, being finely adjusted to the situation. The classical ethological distinction between internally constrained "core" behaviors, which may or may not be superimposed with external orientational components, is partly applicable also here. However, the fallacious distinction between internal and external constraints gives way to a distinction between environmental or more particularly interactional variables, on the one hand, and individual variables which involve the kinematics of the individual animal regardless of its partner on the other.

Whereas a global principle was formulated for the behavior of *D*, we are still far from formulating an equivalent global principle for *S*. This is so, in spite of the fact that the components of *S*'s individual behavior during each of the interactional patterns are specified in Moran et al. (1981). Whatever the global structure of *S*'s behavior, *S* is always higher than *D* on the multidimensional

---

←

FIG. 13. Four stages in apomorphine-induced exploratory behavior in the laboratory rat. Illustrations were made from film, and represent (from top to bottom) forward walking, circling, revolving, and pivoting, of a rat during successive stages of drug action (from Szechtman et al., in prep.).

motility gradient during the relatively stable configurations: in Follow, it walks higher than *D* without arching, without tucking-in of hindquarters, and without performing spontaneous shifts of front; in Circle the forward progression component is larger for *S* than for *D* while *S* remains higher on all the other variables; in Twist-and-Turn, *S* walks forward as it encircles *D*, which then performs only pure shifts of front; and in Hip Thrust, *S* moves on all four legs near *D* whose hindquarters are collapsed. Because most transitions are initiated by *D*, *S* either keeps performing during their initiation regardless of its previous behavior, or stops momentarily and then responds with the appropriate interactional response, using forward progression and/or shift of front, as the immediate situation dictates. It may be concluded that the behavior of *D*, the so-called "inferior" animal, is both interactionally and individually constrained and that the individual constraints are part of a global structure. *S*, the so-called "superior" animal, is also interactionally constrained and its individual behavior is always higher than that of *D* with regard to the motility gradient. We thus end up with a global, formal definition of the behavior of the inferior wolf and with a formal definition of the difference between the simultaneous behaviors of the two rivals in terms of their level of motility.

The values of the various kinematic variables that *D* assumes with regard to motility are of crucial significance when *S* and *D* find themselves facing each other at contact or near contact. As was suggested by Schenkel (1967) and recently by Moran et al. (1981), all extended supplanting is so designed as to both avoid and provide the opportunity for this interactional configuration, which then leads to severe fighting and possibly into the crucial "dominance fight" described by Klinghammer (pers. comm.). When this configuration occurs, both *S* and *D* rear up on their hindlegs, placing their forelegs on each other's shoulders, leaning and pushing forward with all their weight. Once one wolf loses balance and falls on its side or back, the other commences to bite it extensively. Efficient pushing is obtained by rearing high up and darting forward with a straight torso so that both the wolf's weight and force act along a straight trajectory. It is at such time that *D*'s low motility plays against it. The plantigrade stance, the tucked in hindquarters and the laterally bent torso, result in a low, curved posture which cannot oppose the impact of *S*'s forward thrust. It is only by raising its own values on the variables of the motility gradient that *D* can push S over, thereby providing an option for a long-term reversal of social roles.

Acknowledgments

This study was supported by a grant to I. Golani from the United States-Israel Binational Science Foundation (B.S.F.) Jerusalem, Israel, and by grants to G. Moran and J. C. Fentress from the Canadian National Science and Engineering Research Council (NSERC).

Literature Cited

ANDEN, N., T. HÖKFELT, AND A. RUBENSON. 1967. Evidence for DA receptor stimulation by apomorphine. J. Pharm. Pharmacol., 19:627–629.

BARLOW, G. W. 1977. Modal action patterns. Pp. 98–134, *in* How animals communicate (T. A. Sebeok, ed.). Indiana Univ. Press, Bloomington, 1128 pp.

COLGAN, F. W. (ed.). 1978. Quantitative ethology. J. Wiley, New York, 364 pp.

EISENBERG, J. F., AND I. GOLANI. 1977. Communication in Metatheria. Pp. 575–599, *in* How animals communicate (T. A. Sebeok, ed.). Indiana Univ. Press, Bloomington, 1128 pp.

ELLINWOOD, E., AND M. KILBEY. 1975. Amphetamine stereotypy: the influence of environmental factors and prepotent behavioral patterns on its topography and development. Biol. Psychiat., 10:3–16.

ERNST, A. M. 1967. Mode of action of apomorphine and dexamphetamine on gnawing compulsion in rats. Psychopharmacology, 10:316–323.

ESHKOL, N. 1980. 50 lessons by Dr. M. Feldenkrais. Movement Notation Society for the Research Center for Movement Notation, Faculty of Fine Arts, Tel Aviv Univ., 160 pp.

ESHKOL, N., AND A. WACHMANN. 1958. Movement notation. Movement Notation Society, Weidenfeld and Nicholson, 203 pp.

FENTRESS, J. C., R. FIELD, AND H. PARR. 1978. Social dynamics and communication. Pp. 67–106, *in* Behavior of captive wild animals (H. Markowitz and V. Stevens, eds.). Nelson-Hall, Chicago, 200 pp.

GANOR, I., AND I. GOLANI. 1980. Coordination and integration in the hind-leg step cycle of the rat: kinematic synergies. Brain Res., 195:57–67.

GOLANI, I. 1976. Homeostatic motor processes in mammalian interactions: A choreography of display. Pp. 69–134, *in* Perspectives in ethology (P. P. G. Bateson and P. H. Klopfer, eds.). Plenum Publishing Corporation, New York, 2:1–340.

———. 1981. The search for invariants in motor behaviour. Pp. 372–390, *in* Behavioural development. The Bielefield Interdisciplinary Project (K. Immelmann, G. W. Barlow, L. Petrinovich, and M. Main, eds.). Cambridge Univ. Press, 754 pp.

GOLANI, I., D. L. WOLGIN, AND P. TEITELBAUM. 1979. A proposed natural geometry of recovery from akinesia in the lateral hypothalamic rat. Brain Res., 164:237–267.

GOLANI, I., et al. 1981. "Warm-up" along dimensions of movement in the ontogeny of exploration in rats and other infant mammals. Proc. Natl. Acad. Sci., 78:7226–7229.

HAZLETT, B. A., ed. 1977. Quantitative methods in the study of animal behavior. Academic Press, New York, 222 pp.

HORNYKIEWICZ, O. 1972. Biochemical and pharmacological aspects of akinesia. Pp. 127–149, *in* Parkinson's disease: rigidity, akinesia, behaviour (J. Siegfried, ed.). Hans Huber Publishers, Bern, Stuttgart, Vienna, 1:1–219.

LORENZ, K. 1937. The establishment of the instinct concept. Pp. 259–315, *in* Studies in animal and human behaviour, Vol. 1. Methuen & Co. Ltd. London, 403 pp.

———. 1949. Er redete mit dem Vieh, den Vögeln und der Fischen. Borotha, Wien, 254 pp.

MARSHALL, J. F., J. S. RICHARDSON, AND P. TEITELBAUM. 1974. Nigrostriatal bundle damage and the lateral hypothalamic syndrome. J. Comp. Physiol. Psychol., 87:808–830.

MORAN, G., J. C. FENTRESS, AND I. GOLANI. 1981. A description of relational patterns of movement during "ritualized fighting" in wolves. Anim. Behav., 29:1146–1165.

PURTON, A. C. 1978. Ethological categories of behaviour and some consequences of their conflation. Anim. Behav., 26:653–670.

SCHENKEL, R. 1947. Ausdrucks-studien an wölfen. Behaviour, 1:81–129.

———. 1967. Submission: Its features and function in the wolf and dog. Amer. Zool., 7:319–329.

SCHLEIDT, W. M. 1974. How "fixed" is the fixed action pattern? Z. Tierpsychol., 36:184–211.

STRICKER, E., AND M. J. ZIGMOND. 1974. Effects on homeostasis of intraventricular injections of 6-hydroxydopamine in rats. J. Comp. Physiol. Psychol., 86:973–994.

SZECHTMAN, H., et al. 1980. Pp. 511–517, *in* Enzymes and neurotransmitters in mental disease (E. Usdin, T. L. Sourkes, and B. H. Youdim, eds.). John Wiley & Sons Ltd., 650 pp.

TODT, D. 1975. Short term inhibition of outputs occurring in the vocal behaviour of blackbirds (*Turdus merula* m.L.) J. Comp. Physiol., 98:289–306.

TRUMLER, E. 1973. Understanding your dog. Faber and Faber, London, 264 pp.

UNGERSTEDT, U. 1971*a*. Adipsia and aphagia after 6-hydroxydopamine induced degeneration of the nigrostriatal dopamine system. Acta Physiol. Scand., suppl., 367:95–122.

———. 1971*b*. Stereotaxic mapping of the monoamine pathways in the rat brain. Acta Physiol. Scand., suppl., 367:1–48.

WILMHURST, C. 1955. Obedience training for your dog. Dover Publ., Inc., New York, 250 pp.

THE ROLE OF HORMONES IN SYNCHRONIZING MAMMALIAN REPRODUCTIVE BEHAVIOR

JOHN G. VANDENBERGH

Abstract

SUCCESSFUL reproduction in mammals depends on a chain of coordinated physiological and behavioral events. Coordinating events during three life periods are examined—early development, puberty, and adulthood.

Neonatal endocrine factors are modified by maternal experience and can change sexual orientation and reproductive performance in rats. Similarly, recent work in mice has shown that endocrine interactions between developing fetuses can influence later reproductive performance.

The age of puberty also is subject to environmental stimuli. Stimuli from adult males accelerates the onset of puberty in females and stimuli from females inhibits puberty in juvenile females. Pheromones have been identified as important components of this stimulus complex. Recent work with wild populations of mice is described, suggesting a possible role for such puberty-modifying pheromones in regulating populations.

The two-way interaction between hormones and behavior is described in adults. Work on primates has shown that intersexual stimuli can influence the seasonal onset of breeding activities and can modify the hormones regulating ovulation. Similarly, interactions between sexes at the time of copulation can influence ovulation and the course of pregnancy.

Introduction

Successful reproduction in mammals depends upon a chain of coordinated physiological and behavioral events. For example, the uterus must undergo elaborate changes in anticipation of ovulation but, if sperm are not available to meet a viable ovum in the oviduct, all of these preparations are useless. In almost every link of this chain hormones play a role in coordinating the physiological changes and synchronizing these with the required reproductive behaviors. The details of how hormones mediate specific physiological events is beyond the scope of this paper. In this paper I review the role played by hormones in synchronizing mammalian reproductive behavior and physiology. The individual links in the chain of coordinated behavioral and physiological events can be subdivided into units—events in early development, events during puberty, and events during adulthood. Each of these units are discussed separately.

Events in Early Development

A fertilized mammalian ovum will either develop as a male or as a female depending on whether it has been fertilized by a sperm bearing a Y chromosome or an X chromosome (Mittwoch, 1973). Once set in a direction toward maleness or femaleness, however, many aspects of sexual differentiation depend upon the presence or absence of steroid hormones during prenatal and early postnatal development.

A fundamental difference between females and males is that the activity of the female gonad peaks and wanes with a cycle specific to the species, whereas the activity of the male gonad is relatively constant. In both sexes, of course, seasonal and other factors can alter gonadal activity. The basis for this cyclicity has been best worked out in rodents and it is clear that cyclic gonadal activity, the feminine pattern, will occur in either sex unless testosterone is present during a specific period during neonatal development.

In addition to inhibiting cyclicity, testosterone is also necessary for the later appearance of masculine sexual behaviors. That differentiation of sexual behavior is dependent upon the presence of testosterone early in life was first clearly demonstrated by Phoenix et al. (1959) in the guinea pig (*Cavia porcellus*). Injection of mothers with testosterone during pregnancy resulted in the masculinization of the genitalia of female offspring. When these offspring were ovariectomized in adulthood and injected with estrogen and progesterone they showed less feminine behavior than did control females that had not been exposed to testosterone neonatally. In contrast, the genitalia and behavioral orientation of males exposed to testosterone *in utero* were not different from controls.

Work quickly shifted to the laboratory rat (*Rattus norvegicus*), a species with a shorter gestation period and one in which sexual differentiation continues for a short while postnatally. In the laboratory rat, castration could be performed immediately after birth and thus prior to the end of the process of sexual differentiation. Neonatally-castrated male rats resembled females when they reached adulthood in that they displayed lordosis and other feminine behaviors in pairing tests with stimulus males after injection with estradiol benzoate and progesterone (Grady et al., 1962).

These pioneering experiments spawned studies which have amplified, extended, and clarified the basic finding that in mammals, female characteristics develop in the absence of gonadal hormones and male characteristics depend upon the presence of androgens during specific periods of early development. The research supporting this conclusion has been thoroughly reviewed (Beach, 1971; Goy and Goldfoot, 1973; Whitsett and Vandenbergh, 1978). My purpose is not to reexamine this literature but rather to provide the background for some recent experiments which reveal the influence of factors affecting the prenatal environment and later reproductive performance.

Armed with the knowledge that adult male behavior is influenced by neonatal hormone levels in laboratory rats, Ward (1972) hypothesized that applying stress to pregnant females would induce the production of androstenedione, a weak adrenal androgen. The fetus would thus be exposed to a weak androgen which, because of its suppressive effect on gonadotropin release, would result in reduced levels of testosterone production. The result would be a reduction of male behavior and an enhancement of female behavior in the male offspring. She stressed rat mothers during days 14 to 21 of gestation by restraining them in brightly lit chambers for 45 min three times per day. Other mothers were left unhandled as controls. When the pups were born some of the newborn males of both the stressed mothers and the controls received 30 min of physical stress from day 1 to day 10 of life. This procedure yielded four groups of males—unstressed controls, prenatally stressed, postnatally stressed, and a group that were stressed both as fetuses and newborn pups. At 3 months of age all males were given mating tests with adult females. At the conclusion of these tests the males were castrated, injected with estradiol benzoate and progesterone, and paired with adult males. The results of these tests indicate that prenatal stress significantly reduced male sexual behavior and increased female sexual behavior in the male offspring. Stress delivered directly to the pup after it was born had a negligible effect on its later masculinity and there was only a slight increase in female behavior, defined as showing at least two lordotic responses to the male in a ten minute mating test. In contrast, the greatest increase in female receptive behavior occurred among the prenatally stressed males and the males stressed both pre- and postnatally. The results and more recent work by Ward and Weisz (1980) strongly suggest that stress applied to the mother can influence the hormonal environment of the fetus so that the male is demasculinized and feminized in its later behavior and responsivity to hormones.

Attention has also been given to the effects of prenatal stress on developing female rats. Herrenkohl (1979) stressed pregnant female rats from days 14 through 22 of gestation by exposing each animal to heat and bright light for three 45-min periods per day. Female pups born to stressed rats and those born to control rats were paired with stud males when they matured. The incidence of pregnancy was significantly lower in the prenatally stressed rats. Less than half of them delivered a litter compared to 78% of the control females. One explanation for this effect is that the stress applied to mothers while pregnant could have altered their maternal behavior or lactational ability and that these effects caused the decreased fertility of the offspring. In a subsequent experiment, Herrenkohl (1979) controlled for this by cross-fostering the pups between stressed and unstressed mothers. She found that a high proportion of non-stressed females reared by either stressed or non-stressed mothers became successful mothers themselves. However, females stressed prenatally showed severely impaired fertility whether they were cross-fostered or not. The prenatal stress therefore seems to affect later reproduction not by disrupting rearing conditions but by altering the fetus, probably through a hormonal mechanism.

From the examples given above and from other studies (Goy, 1970; Wilson, 1978) it is clear that hormonal manipulations during the neonatal period profoundly influence later reproductive behavior and performance. It will be important to determine normal hormone levels and fluctuations during the prenatal and early postnatal periods in males and females to understand the basis for normal development. Several such studies have appeared for a variety of mammals such as laboratory rats (Weisz and Ward, 1980), domestic pigs (*Sus scrofa*) (Raeside and Sigman, 1975), domestic cattle (*Bos taurus*) (Challis et al., 1974), rhesus monkeys (*Macaca mulatta*) (Resko, 1975), and humans (Winter et al., 1977).

In one fascinating series of studies the investigators took advantage of a natural variation in the prenatal endocrine environment. In both laboratory rats (Clemens et al., 1978) and house mice (*Mus musculus*) (vom Saal, 1979; vom Saal and Bronson, 1978, 1980*a*), female fetuses developing *in utero* between two male fetuses altered morphological, physiological, and behavioral reproductive characteristics. Females flanked by males had longer anogenital distances, more irregular estrous cycles as adults, and were more aggressive than females flanked by other females *in utero*. Recently, vom Saal and Bronson (1980*b*) found that female fetuses developing between two male fetuses have higher levels of testosterone in the blood and amniotic fluid than female fetuses flanked by sisters. Male fetuses have triple the level of circulating testosterone and presumably this is the source of the elevated testosterone levels in their female companions in the uterus.

Individual variations in reproductive performance seen among adult females could well be due to natural variations in the prenatal hormonal environment resulting from stress experienced by the mother during pregnancy or by the presence of varying numbers of male littermates. As information continues to be gathered we should have a rather complete picture of the profound dependence of successful reproduction on levels of hormones present very early in development and factors that influence the levels of these hormones.

After fetal and early postnatal development has been completed, endocrine functions related to reproduction remain relatively quiescent until puberty when a new burst of activity occurs.

Events during Puberty

Puberty in female mammals is thought to be the result of changes in estrogen receptivity of specific hypothalamic areas in the brain (Davidson, 1969; Moltz, 1975; Ramirez, 1973). This change in sensitivity of brain tissue to estrogen is then followed by the establishment of feedback mechanisms involving the production of releasing factors from neurosecretory areas of the hypothalamus and hormones from the anterior pituitary and the gonads. In females these feedback systems result in the onset of ovarian cyclicity and, in males, in the development of the testes. Studies have been more extensive on females than on males because of the greater ease and precision with which puberty can be detected in females. Puberty in females is signalled on a particular day by the appearance of changes in the vaginal epithelium or in some species by the onset of menstruation. In males, gonadal development is gradual and there is no specific event that signals puberty.

The onset of puberty is a developmental event that is subject to modification by environmental stimuli. In humans, for example, there has been a secular trend toward earlier puberty over the past century among women living in developed nations (Tanner, 1973; Zacharias and Wurtman, 1969). This trend has been attributed to improvements in nutrition, health care, and perhaps increasing levels of social stimuli (Vandenbergh, 1974; Zacharias and Wurtman, 1969). The stimuli influencing the chain of events leading to puberty are poorly understood in humans (Bongaarts, 1980), but we are beginning to get a better understanding of these events in rodents (Bronson, 1979; Vandenbergh, 1980). Intersexual stimuli play a prominent role in both accelerating and inhibiting puberty in several species of rodents.

Extensive studies on the house mouse (*Mus musculus*) reveal the important role that the male plays in hastening the onset of puberty in the female. Female mice reared in the presence of adult males attain puberty earlier than those reared in all female groups, as measured by vaginal opening (Castro, 1967) or the occurrence of first estrous (Vandenbergh, 1967). Much of the acceleratory effect of the male is transmitted through male urine which has been applied directly to prepubertal females (Colby and Vandenbergh, 1974; Cowley and Wise, 1972). A bioassay, based on uterine weight changes following exposure of prepubertal females to male urine (Lombardi et al., 1976) has permitted us to make preliminary steps in the chemical isolation of the active ingredient in male urine (Vandenbergh et al., 1976). Our results using column chromatography suggest that the pheromone is a small peptide or material bound to such a peptide. Using a similar separation procedure, Marchlewska-Koj (1980) has found that a similar fraction of male urine induces the pregnancy block or "Bruce" effect (Bruce, 1960). As yet it is not known whether the same compound or mixture of compounds induces acceleration of female puberty, the pregnancy block effect, and perhaps the estrous synchronization effect as well. This question can be answered only when one of the pheromones is identified and tested in all three effects.

An additional important finding is that the production of puberty accelerating pheromone by the male is dependent upon androgen levels (Lombardi et al., 1976). Castration of males results in the loss of pheromonal activity in their urine and replacement by testosterone proprionate injections results in restoration of pheromonal activity in a dose-dependent fashion. This suggests that social factors which may influence testosterone levels in males may have an effect on the male's ability to accelerate sexual development in females. Lombardi and Vandenbergh (1977) tested this notion by allowing a pair of adult male mice to fight for 7 days and then testing the urine from both the dominant and the subordinate mouse for its ability to accelerate puberty. Urine from the dominant male mouse was found to be effective in stimulating the onset of early puberty but the urine of subordinate males was no more effective than water. The loss of ability to stimulate puberty in females by subordinate males was presumed to be due to a decline in testosterone levels because organs such as seminal vesicles which are known to be testosterone dependent were reduced in size in the subordinate males compared to the dominant males. The results suggest that the social environment of the male can have an effect on the age at which juvenile females attain sexual maturity.

We have recently tested whether wild male mice living in natural populations would show a fluctuation in their ability to accelerate puberty in females as a function of season or population density (Massey and Vandenbergh, 1981). In this study, eight pairs of wild house mice were released on a highway island which was one quarter of a cloverleaf along an interstate highway near Raleigh, North Carolina. Highway islands contain natural populations of small rodents (Massey and Vandenbergh, 1980). We trapped and removed rodents from two of these islands and released the eight pairs of mice. Live trapping with Sherman traps was conducted at approximately monthly intervals and population estimates were obtained according to the Schnabel (1938) method. Each trap was fitted with a piece of filter paper to collect any urine excreted during the period of confinement. Filter paper impregnated with urine from adult males was then frozen and later exposed to juvenile females in the laboratory to test for its ability to accelerate the onset of puberty. Based on trapping data, the population grew to about 55 mice in December 1979. Urine was collected from males when the population was low in June 1979 and when it was at its peak in January 1980. Juvenile females were then exposed to filter paper impregnated with male urine from day 25 through day 31 of age and vaginal smears were taken daily to determine the age at which first estrous occurred. The results revealed that male urine collected at either time was equally capable of inducing accelerated sexual development. Under these levels of population density, the puberty accelerating pheromone in males did not vary in potency with season of the year or population density. It is possible that the densities obtained in the natural populations were insufficient to induce reductions in testosterone and subsequent changes in pheromonal potency. Nevertheless, this study did demonstrate that adult male mice living under natural conditions are capable of producing a urinary pheromone that accelerates puberty in females.

Another and perhaps more important pheromonal effect is that of puberty inhibition by female stimulation. Females reared in groups of four or more attained first estrous at a later age than females reared singly (Vandenbergh et al., 1972). Drickamer (1977) later showed that puberty in juvenile females can be delayed by exposure to the urine of grouped females. Grouping of females is essential because the urine of isolated females is without effect on puberty in juvenile females. Realizing that puberty inhibition may be an extremely effective method of population regulation, Massey and I (1980) conducted much the same study as described above on highway island populations of house mice except this time we tested the urine of adult females collected from high- and low-density populations at different seasons of the year.

Two highway islands were cleared of indigenous rodents and eight pairs of second generation wild-caught house mice were released in Spring 1978. By December, both populations had peaked—one at an estimated level of 16 mice and the other at 74 mice. Urine collected from females in these populations when they were at low density and when they were at high density was tested for its ability to inhibit puberty. Puberty inhibition was tested by exposing 25- to 31-day-old albino female mice in the laboratory to urine-impregnated filter paper. Daily vaginal smears revealed when first estrous occurred. Control data were provided by females exposed to clean pieces of filter paper. The onset of puberty was unaffected by urine from females in the sparse population at either its high or low point or from the more dense population at its lowest point. Only urine collected from the more dense population when it reached its peak delayed the onset of puberty (Fig. 1). The delay in pubertal onset was approximately one week, sufficient to have a major impact on population dynamics.

This finding agrees well with laboratory studies showing that only female mice reared in groups display the ability to produce a urinary substance which can inhibit puberty (Drickamer, 1974). Tactile contact among the females in the group is necessary. Overt fighting among the females in the strains of mice used in these studies is rare so it is unlikely that the pheromone production is related to a severely stressful situation. The factors influencing the ability of the female to produce the inhibitory pheromone remain obscure but probably revolve around endocrine changes.

The findings described above provide a good example of the theme of this chapter—the interactions between hormones and behavior that result in reproductive synchrony. The chain of interactive events can be followed schematically (Fig. 2). Interactions among females induce the production of a urinary pheromone which delays the onset of puberty in juvenile females. This delay, in turn, reduces the rate of population growth (see Cole, 1954, for a discussion of the importance of generation time to population dynamics). Slower population growth results in a reduction of females in the population and presumably a reduction in interactions among the females. Reducing the rate of interactions closes the chain of events by inducing a reduction in the production of the inhibitory pheromone.

The conclusions I have drawn from our work on populations of mice living on highway islands are speculative. To date we have only studied a few populations and we have not excluded other possible influences on the production of the puberty inhibiting pheromone. More detailed studies of vegetational changes are required as well as replicate studies of populations living in other habitats. A variety of factors have been identified that control the growth of rodent populations. Changes in the inhibitory influences of females on puberty may be but one of an array of mechanisms used to control populations of mice.

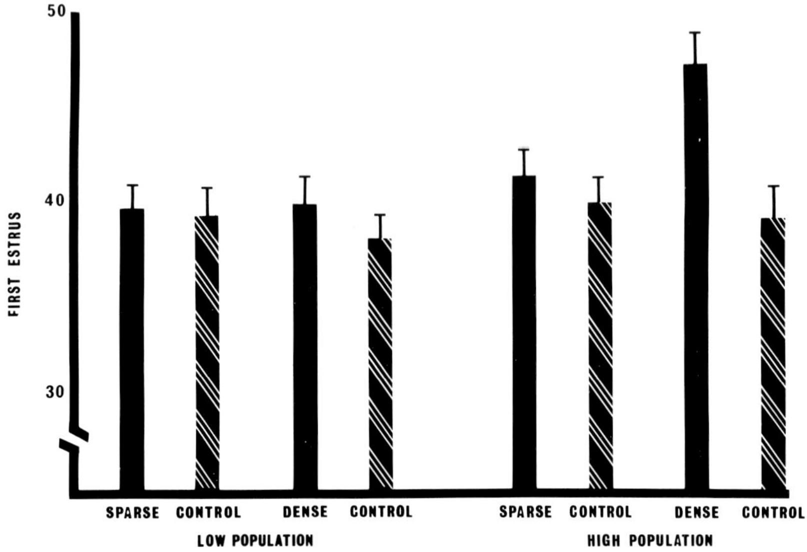

FIG. 1. The onset of puberty in house mice living on highway islands (data from Massey and Vandenbergh, 1980).

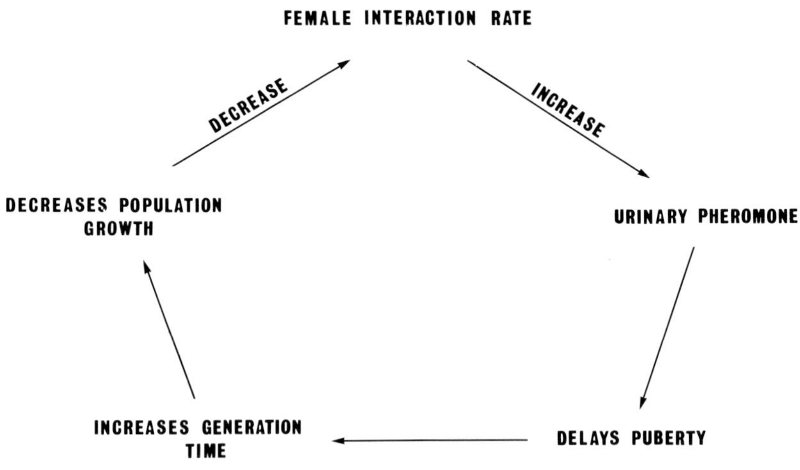

FIG. 2. A diagram of the possible role played by a urinary pheromone in regulating house mouse populations.

Events during Adulthood

A variety of synchronizing events must occur in adulthood, most of them mediated by hormones, to result in the production of the next generation. These synchronizing events can be thought of at three levels: 1) interactions between the organism and the physical environment to induce a state of fertility; 2) social interactions among individuals in a group to modulate fluctuating hormone levels; and 3) the fine tuning of interactions between a male and female at the time of copulation. Using only a few examples of many, I describe events at each of these levels with primary attention to the last two.

One or more fluctuations in the physical environment such as photoperiod, temperature, or vegetational changes induce seasonal breeding in a variety of mammals. A discussion of such factors is beyond the scope of this paper but I begin with an examination of how social factors can interact with the physical environment.

Rhesus monkeys which live in a free-ranging colony on an island off the coast of Puerto Rico near La Parguera breed with a seasonal rhythm (Vandenbergh and Vessey, 1968). Mating activities begin in September or October and end in February. At the time of the study, the colony consisted of several social groups; there were slight differences in the timing of mating among groups. We hypothesized that interactions between sexes within a group could "fine tune" the timing of the mating season. I first tested this (Vandenbergh, 1969) by removing adult males and females from a group prior to the onset of the mating period and pairing them in outdoor cages on the island. Males were paired with females before hormonal manipulations and then after the females had been artificially brought into estrus with estradiol benzoate injections. I was able to show a sharp increase in mounting and copulatory behavior by males exposed to sexually receptive females at a time of the year when such females are normally not present.

In addition to the behavioral change, these males showed a more rapid testicular recrudescence and increase in reddening of their sex skin. Sex skin is a testosterone dependent, secondary sexual characteristic. These results suggest that a male in the non-breeding condition is capable of responding to a sexually active female and that he can show both behavioral and hormonal recrudescence at an abnormal time of the year.

We then monitored the level of sexual behavior in three social groups of monkeys over a 2-year period in the field (Vandenbergh and Drickamer, 1974). After baseline data were collected in the first year, we implanted two females in one group with pellets of estradiol benzoate well before the normal onset of mating activity. This hormone treatment brought these females into a sexually receptive state and, within 2 weeks, we began to observe males in this group showing sexual interactions with the females. The timing of the mating season occurred at the usual time of the year in the other two groups, that is, October through February. In the experimental group containing the two hormone-implanted females showing sexual receptivity at an atypical time of the year, mating behavior appeared within the group about 3 months ahead of schedule. Interestingly, the two highest ranking males responded to the females first. The males not only showed behavioral changes due to the presence of sexually receptive females but they displayed reddening of the sex skin and apparently became fertile abnormally early because the births in this group occurred significantly earlier than in the control groups. Thus, the untreated females ovulated earlier than expected either due to stimuli from the sexually receptive females or from the males that became sexually active.

We have tested whether testosterone-treated males could stimulate female ovarian function during the non-mating season but were unsuccessful (Vandenbergh and Post, 1976). We implanted males with testosterone and induced both sex skin coloration changes as well as behavioral changes in these males. Although they directed high levels of sexual behavior to the females, we were unable to detect any changes in ovarian function. So, we assume that interactions among the females rather than stimuli from the males were coordinating ovarian function.

In addition to coordinating seasonal events among adults in social groups, the interaction between behavior and hormones operates at a more immediate level as well. Recently, Keverne (1979) housed talapoin monkeys (*Miopithecus talapoin*) in small, mixed-sexed groups of adults. The females were ovariectomized and implanted with estradiol in silastic tubes to maintain a low but normal level of estradiol in the circulation. Observations of the group disclosed which of the females were dominant and which were subordinate. Dominant females were the object of far less aggression from either males or females and received far more sexual behavior than subordinate females. Dominant and subordinate females were then given extra implants of estradiol to induce an ovulatory surge of luteinizing hormone (LH) from the pituitary. In primates an LH surge typically follows a sudden increase in estrogen levels and is thought to be responsible for ovulation. Only the dominant females exhibited the typical LH surge; subordinate monkeys given the same dose of estradiol showed no LH response. The lack of response to estradiol was due to prolactin levels twice as high in subordinate females than in dominant females. When prolactin levels were suppressed by administering a dopamine antagonist, bromocryptine, these subordinate females were again capable of displaying the LH surge after estradiol administration. These results demonstrate that social interactions between the members of a group can alter the female primate's neuroendocrine balance resulting in infertility among socially subordinate females.

The interplay between physical and social factors prepares each sex for mating. The internal fertilization process universal to mammals is an event which must be finely orchestrated so that active sperm meet a viable egg. To accomplish this, mammals have developed a variety of mechanisms. One of the most effective is the induction of ovulation by copulatory stimulation as is seen in a variety of mammals, such as the domestic cat (*Felis catus*), several mustelids, the domestic rabbit (*Oryctolagus cuniculus*), and a few species of mice. Adler (1974) described this process in a variety of species and Dewsbury (1975) provided an excellent review of the relationship between copulatory behavior and both ovulation and pregnancy maintenance in muroid rodents. The reader should refer to these papers and a wealth of other literature for further information on reflex and spontaneous ovulation.

In the remainder of this paper, I focus on how the male's behavior can facilitate the induction and maintenance of pregnancy in laboratory rats, as recently described by Adler (1979).

A consistent feature of the copulatory pattern in laboratory rats is a series of intromissions preceding each ejaculation (Dewsbury, 1975). The function of these pre-ejaculation intromissions was examined by allowing one group of pairs of rats to intromit at a normal level and restricting the number of intromissions in another set of pairs (Wilson et al., 1965). Twenty days after copulation 90% of the females receiving unrestricted intromissions were pregnant compared to only 20% of the females in the restricted group. Thus, intromissions prior to ejaculation were necessary for inducing or maintaining pregnancy.

In laboratory rats, unlike many other mammals such as primates, there is not a prolonged luteal phase after ovulation. The pattern of multiple intromissions shown by rats apparently triggers the progestational phase through a neuroendocrine mechanism. Adler (1969) showed that the proportion of females becoming progestational is directly related to the number of intromissions received. Only 10% of females receiving fewer than four intromissions became progestational compared to 85% of females receiving 13 to 16 intromissions. Measures of circulating progesterone levels showed that progesterone increased as a function of increased intromission frequency (Adler et al., 1970).

Further work relating the physiological consequences of copulatory behavior showed that denervation of the vaginocervical area by cutting the pelvic nerves of female rats resulted in a loss of responsivity to intromissions by the male (Spies and Niswender, 1971). Among these pelvic nerves, the pudendal enervates a field extending from the base of the clitoris to the base of the tail and laterally along the inner surface of the thighs. The size of the sensory field can increase in size by 30% following injections of estrogen into ovariectomized females (Komisaruk et al., 1972; Kow and Pfaff, 1973/1974). The enlarged sensory field caused by estrogen enhances the female's progestational response and may facilitate the female's motor activity to allow for intromission. Thus, the essential act of depositing the seminal fluid in the female's vagina is the result of an elegant set of complex interactions between hormones and neural mechanisms.

Conclusion

From the studies reviewed in this chapter dealing with the perinatal, pubertal, and adult phases of a mammal's life cycle it is clear that, in terms of behavior, hormones act as effective messengers. The messages they carry to a tissue, be it a part of the central nervous system or a peripheral part of the anatomy, can be accepted in all or in part depending upon the prior and current experiential history of the tissue. The intensity of the message hormones carry is only partly dependent upon their level in circulation.

In my opinion, two important questions remain. First, which of the many hormonal dependent behaviors we have noted are dependent on social or developmental factors? More research is needed to identify the factors which modulate the behavioral responses that have already been studied. Second, how can environmental factors such as the presence of socially dominant conspecifics or courting members of the opposite sex influence the manner in which hormones act upon target tissues? Our traditional concept that hormone "A" stimulates behavior "B" no longer holds. We must now know the prior history and current state of the animal exposed to hormone "A" if we are to be able to predict that behavior "B" is to occur. The findings I have discussed reveal some of the complex interactions between hormones, behavior, and the environment resulting in successful reproduction. We have made considerable progress in understanding how the chain of events operates within the brain and other organs of the body. But, only a beginning has been made in this area. As we look more deeply into the interactions between physiology and behavior, other complex and fascinating chains of events will be discovered.

Acknowledgments

I thank Drs. J. M. Whitsett, R. M. Grossfeld, and L. C. Ulberg for suggesting several important improvements in the manuscript. This work was supported in part by PHS grant MH 30577 and appears as paper 8493 of the Journal Series, North Carolina Agricultural Research Service, Raleigh, North Carolina 27650.

Literature Cited

ADLER, N. T. 1969. Effects of the male's copulatory behavior in successful pregnancy in the female rat. J. Comp. Physiol, Psych., 69:613–622.

———. 1974. The behavioral control of reproductive physiology. Pp. 259–286, *in* Reproductive behavior (W. Montagna and W. A. Sadler, eds.). Plenum Press, New York, 376 pp.

———. 1979. On the physiological organization of social behavior: sex and aggression. Pp. 29–71, *in* Handbook of behavioral neurobiology. Vol. 3: Social behavior and communication (P. Marler and J. G. Vandenbergh, eds.). Plenum Press, New York, 411 pp.

ADLER, N. T., J. A. RESKO, AND R. W. GOY. 1970. The effect of copulatory behavior on hormonal change in the female rat prior to implantation. Physiol. Behav., 5:1003–1007.

BEACH, F. A. 1971. Hormonal factors controlling the differentiation, development, and display of copulatory behavior in the ramstergig and related species. Pp. 249–296, *in* The biopsychology of development (E. Tobach, L. R. Aronson, and E. Shaw, eds.). Academic Press, New York, 593 pp.

BONGAARTS, J. 1980. Does malnutrition affect fecundity? A summary of evidence. Science, 208:565–569.

BRONSON, F. H. 1979. The reproductive ecology of the house mouse. Quart. Rev. Biol., 54:265–299.

BRUCE, H. M. 1960. A block to pregnancy in the mouse caused by proximity to strange males. J. Reprod. Fertil., 1:96–102.

CASTRO, B. M. 1967. Age of puberty in female mice: relationship to population density and the presence of adult males. An. Acad. Brasil. C., 39:289–291.

CHALLIS, J. R. G., ET AL. 1974. The concentration of androgens, oestrogens, progesterone and luteinizing hormone in the serum of foetal calves throughout the course of gestation. J. Endocrinol., 60:107–115.

CLEMENS, L., B. GLADUE, AND L. CONIGLIO. 1978. Prenatal endogenous androgenic influences on masculine sexual behavior and genital morphology in male and female rats. Horm. Behav., 10:40–53.

COLBY, D. R., AND J. G. VANDENBERGH. 1974. Regulatory effects of urinary pheromones on puberty in the mouse. Biol. Reprod., 11:268–279.

COLE, L. C. 1954. The population consequences of life history phenomena. Quart. Rev. Biol., 29:103–137.

COWLEY, J. J., AND D. R. WISE. 1972. Some effects of mouse urine on neonatal growth and reproduction. Anim. Behav., 20:499–506.

DAVIDSON, J. M. 1969. Feedback control of gonadotropin secretion. Pp. 343–388, *in* Frontiers in neuroendocrinology (W. F. Ganong and L. Martini, eds.). Oxford Univ. Press, New York, 430 pp.

DEWSBURY, D. A. 1975. Diversity and adaptation in rodent copulatory behavior. Science, 190:949–954.

DRICKAMER, L. C. 1974. Sexual maturation of female house mice: social inhibition. Devel. Psychobiol., 7:257–265.

———. 1977. Delay of sexual maturation in female house mice by exposure to grouped females or urine from grouped females. J. Reprod. Fertil., 51: 77–81.

GOY, R. W. 1970. Early hormonal influences on the development of sexual and sex-related behavior. Pp. 196–207, *in* The neurosciences: second study program (F. O. Schmitt, ed.). The Rockefeller Press, New York, 1068 pp.

GOY, R. W., AND D. A. GOLDFOOT. 1973. Hormonal influences on sexually dimorphic behavior. Pp. 169–186, *in* Handbook of physiology. Section 7: Endocrinology (R. O. Grepp, E. B. Astwood, and S. R. Geiger, eds.). American Physiological Society, Washington, D.C.

GRADY, L. K., C. H. PHOENIX, AND W. C. YOUNG. 1962. Role of the developing rat testis in the differentiation of the neural tissues mediating mating behavior. J. Comp. Physiol. Psychol., 59:176.

HERRENKOHL, L. R. 1979. Prenatal stress reduces fertility and fecundity in female offspring. Science, 206:1097–1099.

KEVERNE, E. B. 1979. Sexual and aggressive behavior in social groups of talapoin monkeys. Pp. 271–297, *in* Sex, hormones and behaviour. Ciba Foundation Symp., 62:1–382.

KOMISARUK, B. R., N. T. ADLER, AND J. B. HUTCHINSON. 1972. Genital sensory field: enlargement by estrogen treatment in female rats. Science, 178: 1295–1298.

KOW, L. M., AND D. W. PFAFF. 1973/1974. Effects of estrogen treatment on the size of receptive field and response threshold of pudendal nerve in the female rat. Neuroendocrinol., 13:299–313.

LOMBARDI, J. R., AND J. G. VANDENBERGH. 1977. Pheromonally-induced sexual maturation in females: regulation by the social environment of the male. Science, 196:545–546.

LOMBARDI, J. R., J. G. VANDENBERGH, AND J. M. WHITSETT. 1976. Androgen control of the sexual maturation pheromone in house mouse urine. Biol. Reprod., 15:179–186.

MARCHLEWSKA-KOJ, A. 1980. Partial isolation of pregnancy block pheromone in mice. Pp. 413–414, *in* Chemical signals in vertebrates and aquatic animals (D. Müller-Schwarze and R. M. Silverstein, eds.). Plenum Press, New York, 445 pp.

MASSEY, A., AND J. G. VANDENBERGH. 1980. Puberty delay by a urinary cue from female house mice in feral populations. Science, 209:821–822.

———. 1981. Puberty acceleration by a urinary cue from male mice in feral populations. Biol. Reprod., 24:523–527.

MITTWOCH, U. 1973. Genetics of sex differentiation. Academic Press, New York, 253 pp.

MOLTZ, H. 1975. The search for the determinants of puberty in the rat. Pp. 35–154, *in* Hormonal correlates of behavior (B. E. Eleftheriou and R. L. Sprott, eds.). Plenum Press, New York, 1:1–440.

PHOENIX, C. H., ET AL. 1959. Organizing action for prenatally administered testosterone propionate on the tissues mediating mating behavior in the female guinea pig. Endocrinol., 65:369–382.

RAESIDE, J. I., AND D. M. SIGMAN. 1975. Testosterone levels in early fetal testes of domestic pigs. Biol. Reprod., 13:318–321.

RAMIREZ, V. D. 1973. Endocrinology of puberty. Pp. 1–28, *in* Handbook of physiology. Section 7: Endocrinology (R. O. Greep, E. B. Astwood and S. R. Geiger, eds.). American Physiological Society, Washington, D.C.

RESKO, J. A. 1975. Fetal hormones and their effect on the differentiation of the central nervous system in primates. Fed. Proc., 34:1650–1655.

SCHNABEL, Z. E. 1938. The estimation of the total fish population in a lake. Amer. Math. Monthly, 45:348–352.

SPIES, H. G., AND G. D. NISWENDER. 1971. Levels of prolactin, LH and FSH in the serum of intact and pelvic-neurectomized rats. Endocrinol., 88:937–943.

TANNER, J. M. 1973. Trend towards earlier menarche in London, Oslo, Copenhagen, the Netherlands and Hungary. Nature, 243:95–96.

VANDENBERGH, J. G. 1967. Effect of the presence of a male on the sexual maturation of female mice. Endocrinology, 81:345–349.

———. 1969. Endocrine coordination in monkeys: male sexual response to the female. Physiol. Behav., 4:261–264.

———. 1974. Social determinants of the onset of puberty in rodents. J. Sex Res., 10:181–193.

———. 1980. The influence of pheromones on puberty in female mice. Pp. 229–241, *in* Chemical signals in vertebrates and aquatic animals (D. Müller-Schwarze and R. M. Silverstein, eds.). Plenum Press, New York, 445 pp.

VANDENBERGH, J. G., AND L. C. DRICKAMER. 1974. Reproductive coordination among free-ranging rhesus monkeys. Physiol. Behav., 13:373–376.

VANDENBERGH, J. G., AND W. POST. 1976. Endocrine coordination in rhesus monkeys: female responses to the male. Physiol. Behav., 17:979–984.

VANDENBERGH, J. G., AND S. VESSEY. 1968. Seasonal breeding of free-ranging rhesus monkeys and related ecological factors. J. Reprod. Fertil., 15:71–79.

VANDENBERGH, J. G., L. C. DRICKAMER, AND D. R. COLBY. 1972. Social and dietary factors in the sexual maturation of female mice. J. Reprod. Fertil., 28:397–405.

VANDENBERGH, J. G., ET AL. 1976. Chromatographic separation of puberty accelerating pheromone from male mouse urine. Biol. Reprod., 15:260–265.

VOM SAAL, F. S. 1979. Prenatal exposure to androgen influences morphology and aggressive behavior of male and female mice. Horm. Behav., 12:1–11.

VOM SAAL, F. S., AND F. H. BRONSON. 1978. In utero proximity of female mouse fetuses to males: effect of reproductive performance during later life. Biol. Reprod., 19:842–853.

———. 1980*a*. Length of the estrous cycle in mice due to former intrauterine proximity to male fetuses. Biol. Reprod., 22:777–780.

———. 1980*b*. Sexual characteristics of adult female mice are correlated with their blood testosterone levels during prenatal development. Science, 208: 597–599.

WARD, I. L. 1972. Prenatal stress feminizes and demasculinizes the behavior of males. Science, 175:82–84.

WARD, I. L., AND J. WEISZ. 1980. Maternal stress alters plasma testosterone in fetal males. Science, 207:328–329.

WEISZ, J., AND I. L. WARD. 1980. Plasma testosterone and progesterone titers of pregnant rats, their male and female fetuses and neonatal offspring. Endocrinology, 106:306–316.

WHITSETT, J. M., AND J. G. VANDENBERGH. 1978. Hormonal influences on brain and behavioral development. Pp. 73–106, *in* Studies on the development of behavior and the nervous system: early influences (G. Gottlieb, ed.). Academic Press, New York.

WILSON, J. D. 1978. Sexual differentiation. Ann. Rev. Physiol., 40:279–306.

WILSON, J. R., N. ADLER, AND B. LEBOEUF. 1965. The effects of intromission frequency on successful pregnancy in the female rat. Proc. Natl. Acad. Sci., 53:1392–1395.

WINTER, J. S. D., C. FAIMAN, AND F. I. REYES. 1977. Sex steroid production by the human fetus: Its role in morphogenesis and control of gonadotrophins. Pp. 41–58, *in* Birth defects: original article series, Vol. 13 (J. W. Littlefield and F. de Grauchy, eds.). The National Foundation.

ZACHARIAS, L., AND R. J. WURTMAN. 1969. Age at menarche. Genetic and environmental influences. New England J. Med., 280:868–875.

THE EVOLUTION OF THE MAMMALIAN BRAIN AS AN INFORMATION-PROCESSING SYSTEM

HARRY J. JERISON

Abstract

THE brain's work is to make sense of the thousands of millions of neurochemical events produced by the interaction of an organism and its environment. It is more than mere metaphor to describe that work as *creating* a real world that enables a life to be lived. Emphasizing data and issues of evolutionary biology, this chapter reviews some details of that "creative" work and the relations among species in the organization and structure of their brains to do the work. Details of organization as well as many of the variations in brain size may be understood cladistically as reflecting the evolution of commonly derived traits and the coupling (and occasional decoupling) of factors that determine brain and body size. Spectacular anagenetic ("progressive") trends, above the species level, have also been discovered in which distantly related groups have achieved comparable grades of "encephalization" (processing capacity independent of body size).

Introduction

The highest functions of the mammalian brain may be thought of as providing a representation of the world in which an animal lives and a representation of the animal in its world. There is a mosaic map in the brain that can be correlated with a real world. The map is peculiar, a misshapen "homunculus" or "animalculus" (Welker, 1976). In primates a large part of it is visual cortex and other neural tissue involved in vision, and that system is normally pictured as a map of what the animal sees—its visual field. There is another map of the "auditory field," a representation of a kind of piano keyboard—a tonotopic map in which the elements may be imagined as tuning forks responsive to the environment of sounds. Yet other maps reflect the odoriferous environment, the environment of things touched, and the environment of things tasted. And there is another map of an internal environment of muscles and skin that places the animal itself inside a space. The work of the brain is to make sense of the information (which is in the form of neural activity) that it has about these environments, and in ordinary experience that work is reflected in the construction of a "real world" in which a life is lived (Eccles, 1979; Jerison, 1973).

Thus, the brain is an information-processing system. That system can be described in many ways. A classic description is according to functions localized in the brain. Another description emphasizes the work of the brain as a whole or of large masses of tissue in the brain. The first view examines anatomically localized functions in brains—vision, hearing, and touch, for example, or the speech-functions in the human brain. Species can be compared with respect to the place of these functions in their normal behavior patterns and the corresponding representations of the functions in the structure of their brains. The second view arose historically in the biometric analysis of brain-body relations—gross brain weight and gross body weight. The approach leads to the allometric analysis of species differences in brain size; deviations from allometry are described as degrees of encephalization.

In evolutionary perspective the data on localization are most relevant for cladistic analyses, showing relationships among species, especially living species, with respect to specialized adaptations. The allometric approach and the doctrine of encephalization are especially useful for anagenetic analyses—the measurement of the progressive evolution of the brain as revealed in fossil endocranial casts, or "endocasts." It is only recently that the anatomical and physiological justification for the biometric approach has been developed, and I review that in a separate section.

The remainder of this chapter is divided into four substantive sections: 1) the evolution of localized functions and the correlated fissural patterns (shape); 2) the anatomical and physiological significance of brain size; 3) the interaction between brain size and shape; 4) the evolution of brain size and its implications for the evolution of processing capacity.

Evolution of Localized Functions

Shortly after Broca's discovery of a localized speech area in the human brain, studies of soldiers with head wounds in the Franco-Prussian War of 1870 led to the identification of functionally specialized and localized sensory and motor areas in their exposed brains (see Young, 1970, for a detailed history). If a surgeon stimulated a small area behind the central sulcus (Fissure of Rolando), the soldier reported, for example, a tingling sensation in the thumb or in some other part of the body, depending on the site of stimulation. This helped define a somatosensory area. Stimulation anterior to the central sulcus produced automatic, "forced" movements of parts of the body, and this helped define a motor area. Studies of partial blindness following other wounds, helped define the visual cortex. In the century that followed, a fairly complete map of the human brain was developed. Experimental studies by psychologists and physiologists, in which animals were trained to respond differentially to different sensory events while the electrical activity of the brain was recorded, provided maps of sensory fields in the brains of various mammalian species. The motor cortex was mapped by laborious studies involving stimulation of exposed brain and observation of movements of various parts of the body (Welker, 1976).

This research program has led to an extensive catalogue of brain maps, two of which are shown in Fig. 1. These were selected to show how the behavioral specializations of different species are reflected in the contribution to the brain as a whole of the localized cortical areas controlling those specializations. The brains show maps of the forepaw and of the snout areas in the somatosensory cortex of two procyonids, the raccoon (*Procyon lotor*) and the coatimundi (*Nasua narica*). The raccoon "fingers" its food with its forepaws as part of its normal behavior, and the coatimundi does this sort of thing with its snout. This behavioral difference between closely related carnivores is well reflected in the differential representation of the two sensory surfaces in the brain (Welker et al., 1964).

This kind of differentiation is universal in the organization of mammalian brain-behavior relations. It deserves a name, and I have called it the "principle of proper mass." *The mass of neural tissue controlling a particular function is appropriate to the amount of information processing involved in performing the function. In com-*

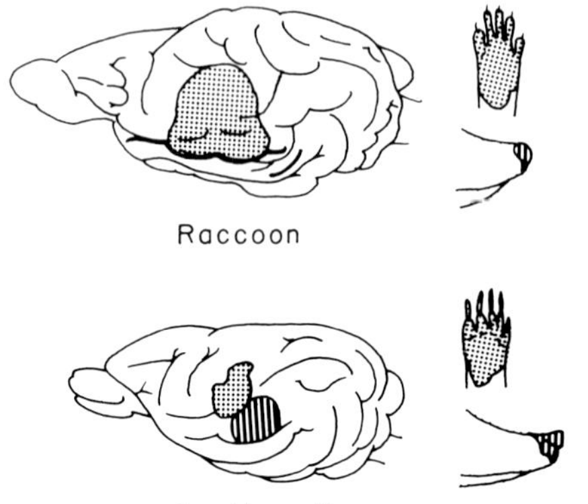

FIG. 1. Maps of hand and snout somatosensory areas in two procyonids. Snout area in raccoon is barely visible in lateral view as dark "line" at the anterior suprasylvian sulcus (adapted from Welker et al., 1964).

parisons among species the importance of a function in the life of each species is reflected by the absolute amount of neural tissue for that function in each species. Within a species the relative masses of neural tissue associated with different functions are proportional to the importance of the functions in the species (Jerison, 1973:8–9). This is a lot of verbiage for a principle. It boils down to the idea that the amount of brain tissue for a behavior is appropriate for the importance of the behavior. In the procyonid data of Fig. 1, the principle applies as is indicated because the raccoon is a "hand-toucher-and-feeler" and the coatimundi a "nose-toucher-and-feeler."

The preceding principle has held up wherever it has been examined. In the human brain the enlarged tongue area can be correlated with the motor control of the tongue in producing articulate speech. In bats (Suga and Jen, 1976), the principle is reflected in the enormous fraction of the brain devoted to an auditory cortex specialized for a finely tuned analysis of the 60 kHz frequency region that is the carrier of the sonar signal used in echolocation and sound ranging. In monkeys, apes, and humans, it is reflected in the enormous extent of the visual cortex relative to other functional areas of the brain (Kaas, 1980). Such analysis has been carried out in great detail, with the nature of the localized functions worked out behaviorally, anatomically, and physiologically (for example, see Diamond, 1979; Merzenich and Kaas, 1980; Welker, 1976; Zeki, 1978). The principle of proper mass is the basic theorem for interpreting the adaptive radiation, functional differentiation, and reorganization of the brain in living species of mammals.

It is possible to apply this approach to the fossil record by analyzing endocasts as if they were brains and assuming the uniformitarian hypothesis (that is, that the correlation between the fissural pattern and the functional map is approximately the same in the fossil relatives of living species). Such studies with primates (Martin, 1979; Radinsky, 1970, 1974), and with various families of carnivores (Radinsky, 1978, 1980) indicate that the kind of differentiation exemplified by Fig. 1 occurred in a progressive way. In the evolution of the raccoon from earlier less-differentiated procyonid species, for example, it is safe to assume that an increase in the forepaw area derived from an ancestral stage where a smaller absolute and relative size of the brain was involved in the control of forepaw movements.

Results as described above resolve into the following generalization: the progressive differentiation and elaboration of functional systems in phyletic sequences are typical. The localized functions that may be identified in endocasts are the same as those identified in studies of localization of function in living species. Uniformitarianism is an evolutionary principle that inevitably biases and constrains "discoveries" based on fossil evidence, however, and truly novel functions cannot be discovered merely by examining endocasts. Were the human brain available only as an endocast for inspection by some alien intelligence, for example, there could be no hint of the unusual human adaptations of speech and language, or even of the unusual functional (and slight structural) asymmetry of the human brain. The major achievement of the endocast approach in paleoneurology has been to have identified in many fossil species the same basic fissural patterns that can be observed in their living descendants. It has confirmed expectations about the evolution of specialized systems from more generalized ones.

In order to use information about fissural patterns for inferences concerning the localization of function in the brain, homologous fissures in ancestral and descendant species have to be identified. This is not a serious problem if brain size is the same. If it is not the same, however, there may be problems because there is a correlation between fissurization and brain size (see Jerison, 1982).

In the next section, where I deal with brain size, I provide a basis for understanding the correlation by explaining the role of the cortical surface area as a fundamental dimension of information-processing capacity. The section that follows, on "braininess and convolutedness," or the interaction of size and shape, shows how the surface-volume relationship in the brain is the basis for the quantitative form of the correlation. I conclude with a continuation of the present discussion by emphasizing the place of brain size in the analysis of the evolution of localized, functional systems.

Brain Size and Processing Capacity

The evolution of total neural information-processing capacity in mammals is equivalent to the evolution of brain size. The purpose of this section is to explain the equivalence. To develop the argument it is necessary, first, to agree on the unit for processing. In the past, the consensus has been that this should be the nerve cell, or neuron, with the connectivity of the neuron taken into account in an appropriate way. Although this approach leads to the conclusion that brain size is, indeed, an estimator (in the statistical sense) of processing capacity, the argument has to be based on limited data on the number of neurons and their average connectivity in different species (Jerison, 1973). Within the past decade there have been new insights into the structure and function of the mammalian neocortex that lead to a stronger argument with better data. In presenting these insights I am following discussions of the microscopic anatomy by Eccles (1979) and Szentágothai (1978; see Mountcastle, 1978), and somewhat less directly the discussions in Chow and Leiman (1970) and by Scheibel and Scheibel (1970). These studies imply a new and enormous significance for research on the gross anatomy of the brain as reported early in this century by Brodmann (1913) and more recently by Elias and Schwartz (1971), which I reanalyze here.

The central idea is that the mammalian neocortex, that great layer of cells that constitutes most of the visible brain in a typical dissection, is remarkably uniform in its organization with respect to the actual unit of processing. This unit consists of a column of cells in which the column works as a module containing about 2,000 neurons and is repeated again across the neocortex. Such functional columns have been identified throughout the brain, and their organization has been worked out in some detail (Szentágothai, 1978). The theory is that a column extends through the entire thickness of the cortex—0.5 to 3 mm or so, depending on the species, mouse through human or elephant or whale—and is 200 μm to 350 μm in diameter.

According to the new insights, the new theory of cortical function, such a column of cells is the *module* for information-processing in the mammalian cerebral neocortex. The size of a module is uniform across species as is the morphology from region to region within the brain beneath the cortical surface. The depth of the unit, counted in number of cells, is also remarkably constant in different species (Rockel et al., 1980). From the measure of their diameters, and according to Eccles (1979), there are about 2,000 modules per centimeter-squared of cerebral cortex. If we take the module as the unit of processing, it is clear that a fundamental measure of processing capacity in mammalian brains should be correlated with the area of the cerebral surface.

The equivalence of brain weight and processing capacity arises from the relationship between the cortical surface area and brain weight. Although a relationship between these measures has been recognized for more than a century (Baillarger, 1853), its mathematical form was not stated until a decade ago (Elias and Schwartz, 1971). This was done for a sample of 15 species derived from marsupial, carnivore, and cetacean brains. There was an impressive correlation between log brain volume and log cortical surface.

In correlation and regression analysis it is important to have a fairly uniform density of points throughout the range of values under analysis. It is, therefore, fortunate that Brodmann (1913) had presented data on cortical surface and brain weight in 33 additional species, representing 10 orders of mammals (including the orders, but not the species, sampled by Elias and Schwartz, 1971). The combined data on 48 species are presented in Fig. 2. The diagonal line is the equation:

$$\log S = .912 \log E + \log 3.7393 \tag{1}$$

Cortical surface S is in cm^2, and brain size E is weight in grams or, equivalently (specific gravity = 1.0), volume in cm^3.

The human brain is slightly below the mammalian average, or regression, in its surface area relative to its volume. The open circles in the small box in Fig. 2 shows the lightest, heaviest, and median

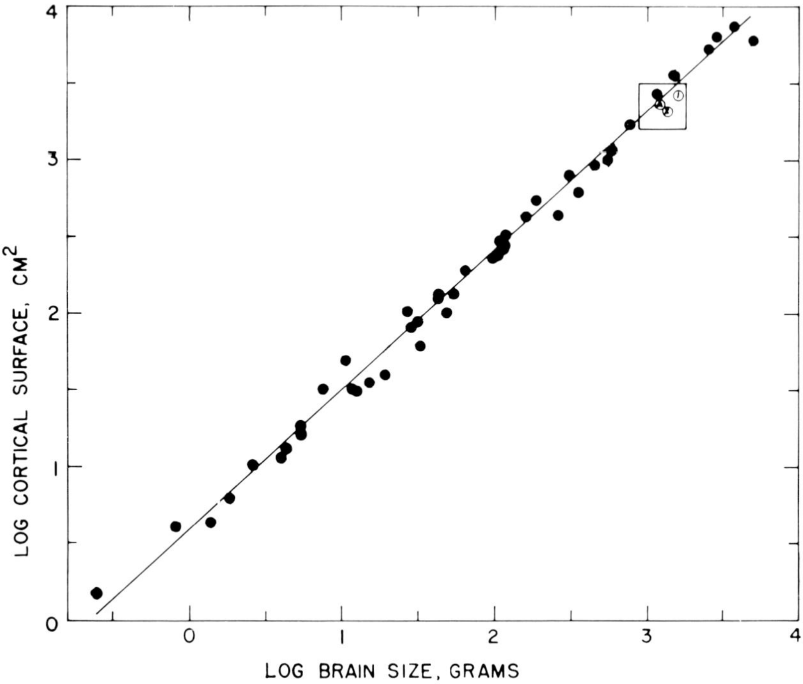

FIG. 2. Cortical surface and brain weight or volume in a sample of 48 mammalian species (r = 0.995). Box includes three data points for *Homo sapiens* and indicates minimum, median, and maximum brain volumes in a sample of 20 men and women, with the corresponding cortical surfaces. (Data from Brodmann, 1913, and Elias and Schwartz, 1971; graph from Jerison, 1982, copyright Plenum Press, New York.)

human brain weights. Of the 20 in the sample, four were above the regression and 16 below—$\chi^2 = 7.2$; $P = 0.007$.

The strong interspecific surface-volume relationship illustrated in Fig. 2 is almost deterministic in its force. It suggests a fundamental structural constraint on the brain's geometry. A full analysis is beyond the scope of this chapter, which emphasizes information-processing capacity and its evolution; I have discussed the analysis briefly elsewhere (Jerison, 1979*a*, 1982). The relationship is crucial because it enables one to estimate cortical surface from brain weight and, consequently, to relate information-processing capacity to brain weight.

Braininess and Convolutedness: The Interaction of Size and Shape

Larger brains are generally more convoluted, more fissured, than smaller brains. A quantification of the relationship is especially important now, because of the revival of interest in the analysis of the morphology of gyri and sulci (convolutions and fissures) in living and fossil mammalian brains (Falk, 1978, 1982; Holloway, 1974, 1976; Radinsky, 1979, 1980). If the pattern of fissurization (shape) is used to analyze the "proper mass" of brain tissue in the evolution of localized functional systems, how should the expected changes in convolutedness with the evolution of enlarged brains be treated?

The analysis can begin with a geometric argument underlying the interpretation of Fig. 2 in demonstrating that the brain changes its shape as its size changes. The argument is based on the geometry of similar solids. In all similar solids, surface area S is related to volume V by the equation:

$$S = kV^{2/3} \tag{2}$$

In a sphere, $k = (6^{2/3})(\pi^{1/3})$, or about 4.84. In a cube, $k = 6.0$. The shape of any solid can be *defined* by the value of k in Eq. 2.

Recall now Eq. 1, which is the allometric relation between cortical surface and brain weight or volume. This can be rewritten as:

$$S = 3.71E^{.91} \tag{3}$$

The exponent $.91 > 2/3$ in Eq. 3 implies a change in shape with a change in size, as noted earlier. The change is the sort of thing that would occur if a sphere expanded and became transformed into a cube. An equation like Eq. 3 could describe such a process, and the good fit of Eq. 3 to the data on mammalian brains indicates that the transformation of the shape of the brain was very orderly.

A simple geometric model for such a transformation is a sphere that develops fissures on its surface as its expands. Imagine the fissures as producing sectors through the center of the sphere, beginning with two hemispheres (a two-sectored sphere) and progressing to four, eight, and 16 sectors. At each grade of sectoring the solid would be described by Eq. 2. The two-sectored sphere would have $k = (1.5)(4.84)$; the four-sectored sphere, $k = 9.68$; the eight-sectored sphere, $k = 3 \times 4.84$, etc. Equation 3 can then be interpreted as an orderly progression between grades of sectoring. This is shown

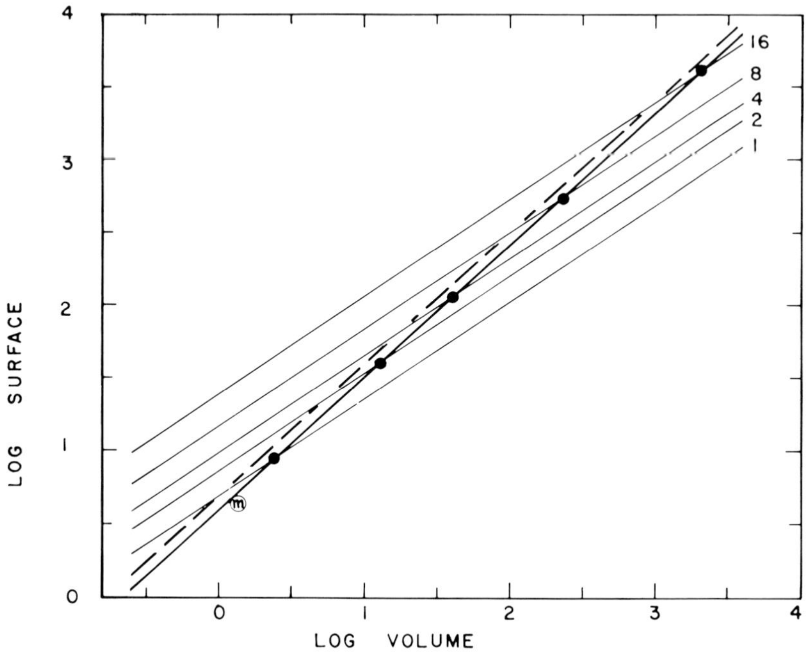

FIG. 3. "Regression" line from Fig. 6 (heavy solid line) and equivalent line from cortical surface-cerebral hemisphere regression superimposed on diagonal lines representing surface-volume relations in sectored spheres (see text).

in Fig. 3. The solid heavy diagonal in Fig. 3 is the regression line of Fig. 2. (The dashed diagonal is a regression line from an analysis of cortical surface as a function of the volume of the cerebral hemispheres rather than the whole brain and the results of which were almost identical to those for the whole brain although displaced as indicated in Fig. 3.) Small-brained species have smooth brains and are near the one-sectored sphere-lines. Humans, elephants, and whales have highly fissured brains and are between the eight-sectored and 16-sectored sphere-lines.

According to these results we should expect the amount of fissurization in the brains of different species to be determined almost entirely by the size of the brain. Elias and Schwartz (1971) measured the relationship, and their results are presented in Fig. 4. It is as orderly as anticipated; a quantitative analysis leads to additional insights into the evolution of fissurization and suggests some

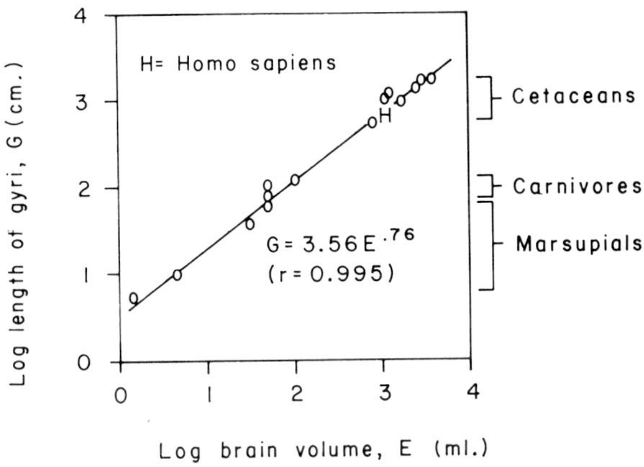

FIG. 4. Length of "external" gyri and brain size. "External" gyri are those visible at the surface of the undissected brain and correspond to gyri visible in fossil endocasts. (Data from Elias and Schwartz, 1971; graph modified from Jerison, 1982, copyright Plenum Press, New York.)

answers for the questions in the first paragraph of this section. The external envelope of the cerebral cortex and brain in mammals is fairly similar in shape in different species, and the surface of that envelope is related to brain size more or less according to Eq. 2 (with a suitable value for K). The regression of the length of the external gyri G versus brain size, presented in Fig. 4, is:

$$G = 3.56E^{.76} \tag{4}$$

The area of the external surface of the cerebral cortex is the product of the length of the gyri and their average width. A dimensional analysis of the relationship between Eq. 4 and Eq. 2, therefore, suggests that the average width of the external gyri is approximately constant. If we treat Eq. 4 as exact, then the average width would be proportional to the -0.09 power of brain size, which would imply that larger brains have slightly narrower gyri than do smaller brains. The negative exponent is probably an artifact of the manner in which the lengths were measured, however, and as a first approximation it is certainly correct to treat the average width of gyri W as a constant across species of mammals. Without further analysis, we also note that:

$$\text{Endocast surface} = WG \cong f(E^{2/3}) \tag{5}$$

These are important generalizations for comparative and evolutionary studies that relate localized functions in the brain to the pattern of fissures and gyri. They correct the misleading impressions that result from the scaling of homologous parts of the brains of different species to appear equal in size. An example of misleading scaling is in the right half of Fig. 5. This kind of illustration would lead one to identify "lower" species and "higher" species in terms of fissurization: the ringtailed lemur (*Lemur catta*) brain magnified to the size of the chimpanzee (*Pan troglodytes*) brain appears considerably less fissured, and the rhesus monkey's (*Macaca mulatta*) brain seems to be an intermediate grade.

A truer picture is indicated at the left in Fig. 5. Size is conserved as a dimension; the three brains are drawn to the same scale, indicated in centimeters below each brain. The analysis in Fig. 2 and Eq. 5 demonstrates that the amount of fissurization is related to the size of the brain. The width of the fissures is approximately the same in the three species. It is true that the chimpanzee has a more fissured brain than a rhesus monkey, and both have more fissured brains than the lemur, but the true relationship is suggested by Eq. 4 and demonstrates that the degree of fissurization is associated with the differences in the absolute sizes of the brains.

The total processing capacity is proportional to brain size, as indicated in the previous section, but the partitioning of processing capacity is according to functional areas and systems. There have been efforts to identify and localize these functional areas with respect to the fissural patterns in fossil "brains" (endocasts).

How should changes in shape such as the appearance of new convolutions be analyzed when these new fissures appear in larger brains? Exactly this question has arisen in the analysis of the endocast of the 2-million year old *Homo habilis,* KNM-ER 1470 (Holloway, 1974). The endocast is significantly larger than those from living great apes or from fossil australopithecines and, unlike the smaller endocasts, that of *H. habilis* has fissurization in the frontal lobe in the region of the left "third frontal convolution." This is Broca's speech area in living humans. Is this evidence for the appearance of a speech area in our 2 million year old ancestor?

The answer, unfortunately, is equivocal, because as we have seen new fissures had to appear in the larger hominid brains as a consequence of Eq. 4 and Fig. 4. Is the third left frontal convolution a speech area in the brain of *Homo habilis*? A positive response

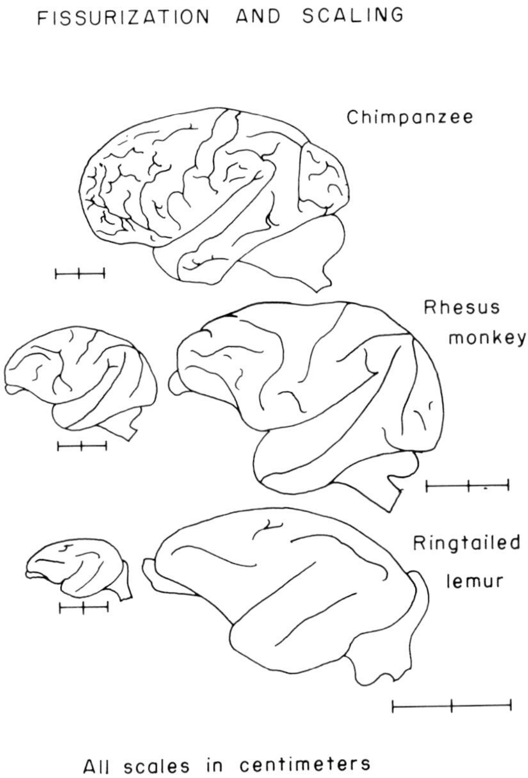

FIG. 5. Tracings of fissural patterns in three primate brains to illustrate the effect of scale on the judgment of fissurization.

would be equivalent to judging that living humans have speech areas in both hemispheres, rather than only in the dominant hemisphere, because this convolution system is present (and anatomically indistinguishable) in both hemispheres in living humans. A negative response, on the other hand, would be equivalent to denying a Broca's area where it is known to exist. The best we can do is to recognize that the basic evidence is about size, and that the *H. habilis* endocast is in the size range (supports an information processing capacity) that requires the presence of fissurization in a "Broca's area" of the brain. From the evidence of the endocasts alone, we will probably never know whether that is a speech area.

The only possibility for positive evidence would arise if asymmetries correlated with language are discovered in the external morphology of Broca's area in the living human brain. If such asymmetries are also found in a *H. habilis* endocast, this would be positive evidence. Research on the living brain (Connolly, 1950) has shown no morphological asymmetries visible on the surface.

We can put the analysis in another way. To the extent that the processing capacity of the living human brain is based on an enlarged frontal lobe, to the extent that the processing is associated with the evolution of language, and to the extent that the enlargement requires fissurization of a third frontal convolution in each hemisphere, we can judge that *H. habilis* had reached a language-potential grade of processing capacity. We cannot judge whether the capacity was used for language by *H. habilis,* though by reference to the brain of its living human descendant, it is reasonable to associate the capacity with some stage in the evolution of language (Holloway, 1976).

The association of convolutedness with braininess, that is, the confounding of information from fissural patterns by the allometric association of fissurization with brain size as given in Eq. 4, makes it difficult to partition information-processing by the evolving brain into functional units comparable to those discovered in the study of living brains. There is little doubt that the partitioning was according to those categories, but this can neither be verified nor falsified by evidence from fossil endocasts. The evidence is primarily derived by examining the adaptive radiation of the mammalian brain as represented in one stratigraphic population—the brains of living mammals. That evidence, with the help of the uniformitarian hypothesis, is so strong that conjectures from the evidence of fossil endocasts can have an important role in filling out our picture of the details of the history of the mammalian brain as a functional system.

Brain Size and Encephalization

The usual conclusion from the analysis of fissural patterns is that the present organization of the brain in a phyletic series was preceded by less elaborate specializations of the various functional systems. The biometric tradition, which involved the analysis of total brain size in relation to body size, leads to a similar conclusion for

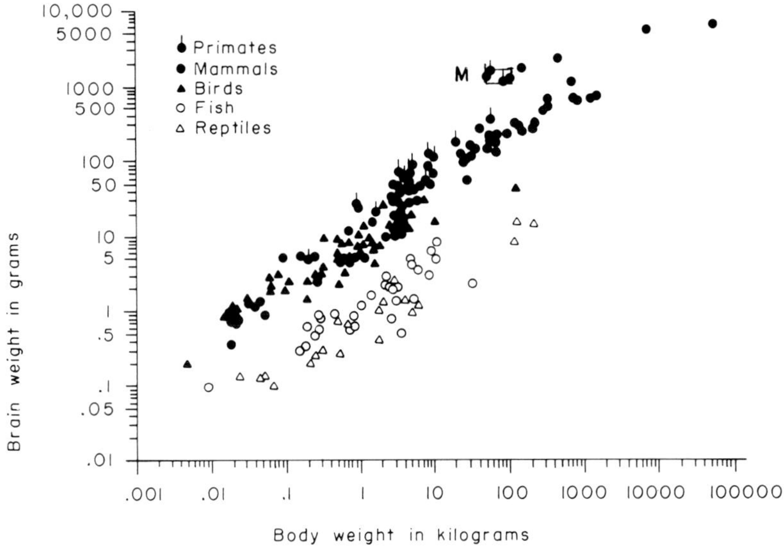

FIG. 6. Brain-body relations in 198 vertebrate species (data from Crile and Quiring, 1940; redrawn from Jerison, 1973).

the brain as a whole. It is as if the various directions of adaptation in different species were correlated quantitatively while maintaining their qualitative differences. An advance in visual capacities in some species would be mirrored by advances in tactile capacities in others. The story of the raccoon and the coatimundi as developed by Welker and his colleagues (Fig. 1) can apparently be extended to the brain as a whole. The interesting feature from the data of the brain as a whole is that it is possible to differentiate species according to grade, independently of direction of adaptation. This is the main contribution of the analysis of brain size and encephalization.

Five hundred million years of vertebrate evolution have produced an enormous variety of species with a corresponding variety of brains. The living species represented in Fig. 6 range in brain size from 0.069 g (a goldfish) to 6,800 kg (a blue whale). This is the range of information processing capacity that has evolved in vertebrates. In this section I review the fossil record of that evolution.

The data in Fig. 6 on about 200 species are from Crile and

Quiring (1940) and are discussed in more detail in Jerison (1973). The method of graphing, as a bivariate system in which each species is represented by both a brain weight and a body weight and with logarithmic scales for the axes, makes the major factors that control brain size clearly identifiable. Within each class of vertebrates, the most important of these is body size, evident in the oblique orientation of each cloud of points on the graph. Lines with a slope of about two-thirds can be drawn through the sets of points representing the "lower" and "higher" vertebrate classes (see Fig. 7, below). The line for the mammals is approximately

$$\log E = \tfrac{2}{3} \log P + \log .12 \qquad (6A)$$

which, in linear transformation, is the allometric function:

$$E = .12P^{2/3} \qquad (6)$$

E is brain weight and P is body weight in grams.

The factor differentiating lower from higher vertebrates is a vertical displacement of the points in "brain-body" space, that is, the encephalization factor. The two factors evident in Figs. 6 or 7 that control brain size in vertebrates are, thus, an allometric factor and an encephalization factor. A regression model identifies components of variance—the allometric factor is represented by the regression component and, within the mammals, it accounts for about 80% of the variance in brain size. Encephalization is the residual from the regression and accounts for the remaining 20% of the mammalian variance. The within-species variance, which is suggested only for the data of *Homo sapiens* ("M") in Fig. 6, represents error of measurement.

Before looking at the fossil record let us review the information on living species presented in Fig. 6. First, it is evident that error of measurement is a minor problem in this interspecific analysis. The rectangle marked "M" represents the entire range of values for 41 humans in the sample (brains 1,100–1,540 g; bodies 36–95 kg). The size and position of the circle drawn for each point is, thus, pretty much the right size to indicate a standard error of measurement for each species. Second, the orientation of the clouds of points with slopes of approximately ⅔ suggests a fundamental allometry of brain and body. The ⅔-exponent of Eq. 6 represents a mapping of the body's surface on the brain (Jerison, 1977). Third, although the differentiation between lower and higher vertebrates

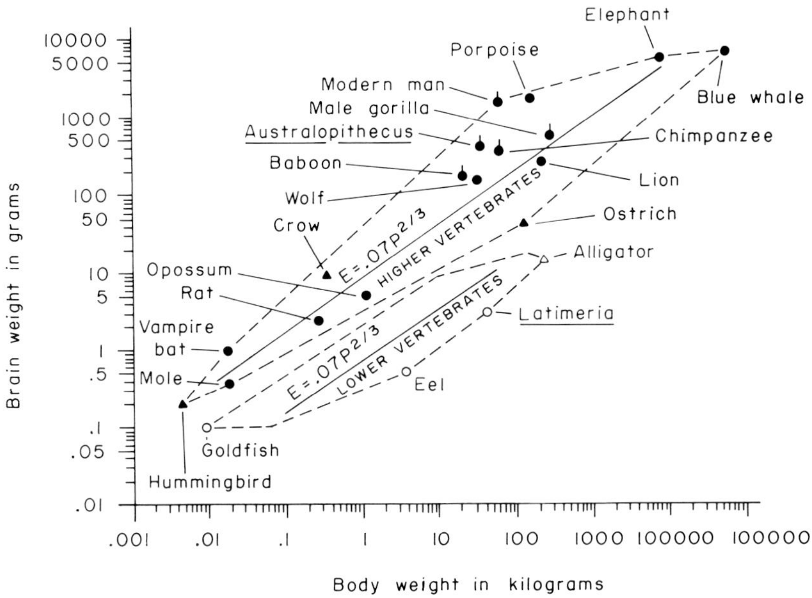

FIG. 7. Convex polygons and allometric lines at "slope" ⅔ for the discriminable groups of Fig. 6 (redrawn from Jerison, 1973).

in Fig. 6 is dramatic, it is equally interesting that within these groupings there is no clear differentiation of the classes. Fish and reptiles overlap one another, and so do birds and mammals. Other data suggest that this may be partly a matter of sampling and that there are differences within the groupings (Ebbesson and Northcutt, 1976; Hopson, 1980). Finally, Fig. 6 is a good illustration of the partitioning of brain size into an allometric component and an encephalization component.

A selected sample of the points in Fig. 6, analyzed somewhat differently, is shown in Fig. 7. Convex polygons are drawn about the data of the lower and higher vertebrates; allometric lines are drawn through each of these, and many of the points are identified to suggest the status of some familiar species. It is unfortunate that analysis with convex polygons has not been developed as a quantitative method, because a truer picture of the evolution of brain size and encephalization can be developed when it is pictured geometrically rather than numerically. A species is represented by a small region rather than a point in brain-body space, and its evolution could be pictured as movement in brain-body space.

Quantitative studies of the evolution of encephalization have depended on Eq. 6 or a similar equation to define an expected mammalian brain size. The general regression model can be used to illustrate such an analysis. Taking Eq. 6A as the general form of the equation, different degrees of encephalization in mammalian species appear as departures from the line. If Eq. 6A is taken to define an expected brain size E_e for a given body size, and E_m is the measured brain size, the regression model of Eq. 6A would have $\log E_m - \log E_e$ as the residual, which we may call $\log m$: Note

$$\log m = \log E_m - \log E_e \tag{7A}$$

$$m = E_m/E_e \tag{7}$$

The residual m is, thus, the encephalization quotient EQ, which is defined as the ratio of measured brain size to expected brain size (Jerison, 1973). This is evident if we use Eq. 6 to define E_e:

$$EQ = E_m/.12P^{2/3} \tag{8}$$

Within the mammals, $EQ = 1.0$ implies an exactly average amount of encephalization. $EQ = 2.0$ means that the brain is twice as large (twice as encephalized) as the average, the typical grade of higher primates, and $EQ = 0.5$ means that the brain is half as large as average, the typical condition for rodents.

The Fossil Record

The evolution of total processing capacity in mammals is the story of how the present diversity and pattern of their data in Fig. 6 came about. The answers can be read from data on critical fossils. The story begins about 200 million years ago with the evolution of mammals from their immediate ancestors, the mammal-like reptiles. What was the status of the reptiles at that time, and how did the earliest mammals compare with them? The evidence on the reptiles (Jerison, 1973; Hopson, 1977, 1980) indicates that living reptiles are good models for almost all known fossil reptiles with respect to encephalization. Adding dinosaurs, pterosaurs, and mammal-like reptiles to a picture like Fig. 6, and partitioning the data into appropriate convex polygons leads to an extended reptilian polygon parallel to and below that of living mammals and one that includes the

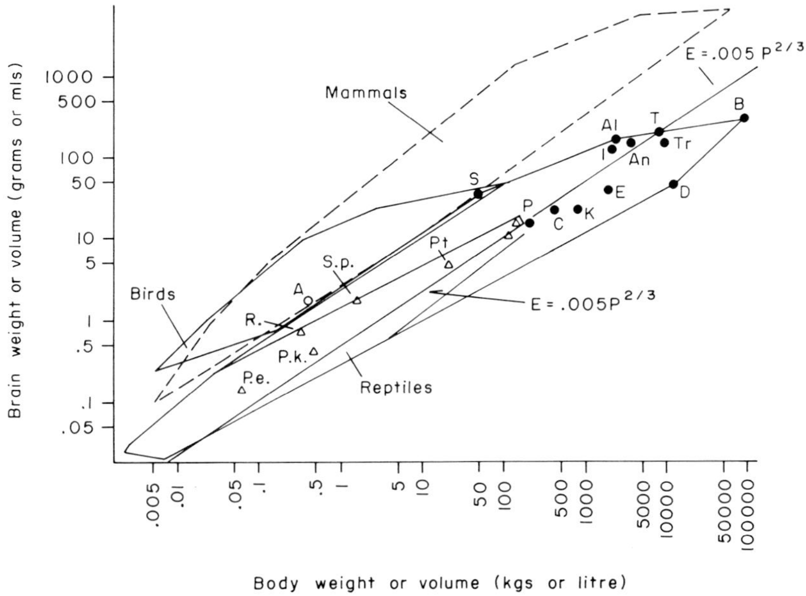

FIG. 8. Brain-body relations in various groups to illustrate the reptilian base from which mammalian encephalization was derived. Symbols are: open circle, *Archaeopteryx*; closed circles, dinosaurs; triangles, pterosaurs. (From Hopson, 1977; copyright Annual Reviews, Inc.).

living reptiles. Mammals evolved from that grade of encephalization. The data are portrayed in Fig. 8.

The earliest mammal with a known endocast is *Triconodon mordax* ("T" in Fig. 10) an Upper Jurassic species of about 150 million years ago. Its *EQ* was 0.28 (Jerison, 1973)—an approximate value, of course, but accurate enough to indicate that it was significantly more encephalized than its reptilian ancestors (typical *EQ* for reptiles relative to the mammalian average was and is about 0.06).

The story of the early mammals is important enough to be graphed with all of the archaic mammalian species. *Triconodon* (T) was evidently at about the same grade of encephalization as the living Virginia opossum (*Didelphis virginiana*) (D; *EQ* = .25 or so) and significantly above fossil and living reptiles as shown in Fig. 9. The other archaic mammals labeled in Fig. 9 are early Tertiary species that lived between 40 and 60 million years ago. The Mesozoic grade of encephalization in mammals persisted for at least 100 million years.

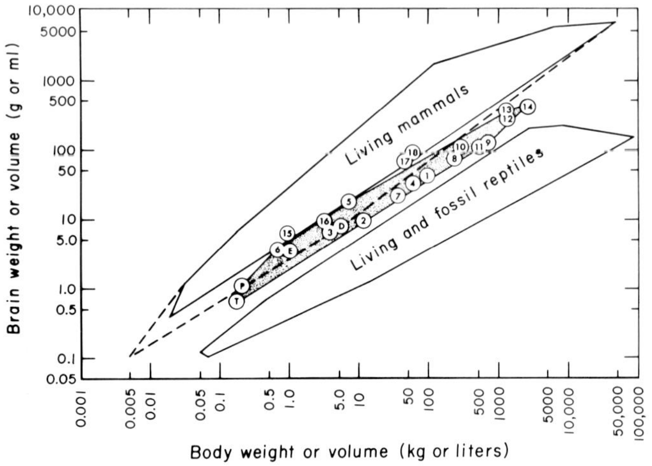

FIG. 9. Divergence of archaic mammals (shaded area) from the reptilian base. See Jerison, 1973, 1976, for additional information. (Modified from Jerison, 1973.)

The first increase in encephalization in mammals above the basal grade probably occurred in early Cenozoic primates—Paleocene *Plesiadapis,* Eocene and Oligocene omomyids, and Eocene adapids. I reviewed these data recently (Jerison, 1979*b*) and summarize them in Fig. 10A; the diagonal line is Eq. 6, and the polygon is that for living and subfossil prosimians. The seven fossils represent the beginnings of the adaptive radiation of primates with respect to encephalization. This includes *Aegyptopithecus,* the earliest anthropoid primate about which we have information.

From Fig. 10A, it appears that the Eocene adapids, *Smilodectes* and *Adapis,* were below the grade of encephalization of living prosimians. The Eocene omomyids, *Tetonius* and *Necrolemur,* were at that grade, and the Oligocene omomyid, *Rooneyia,* was above that grade and approached the grade of the living tarsier. If we treat the adapids as having been in the ancestry of the majority of living prosimians and the omomyids as having been in the ancestry of living tarsiers, the picture is that of increasing encephalization during the Eocene and Oligocene, a 20 million year interval. This is a different picture from the stasis suggested by the evidence of the earliest mammals. The earliest of the mammals that we know was already at a grade that occurs in living mammals. The earliest of

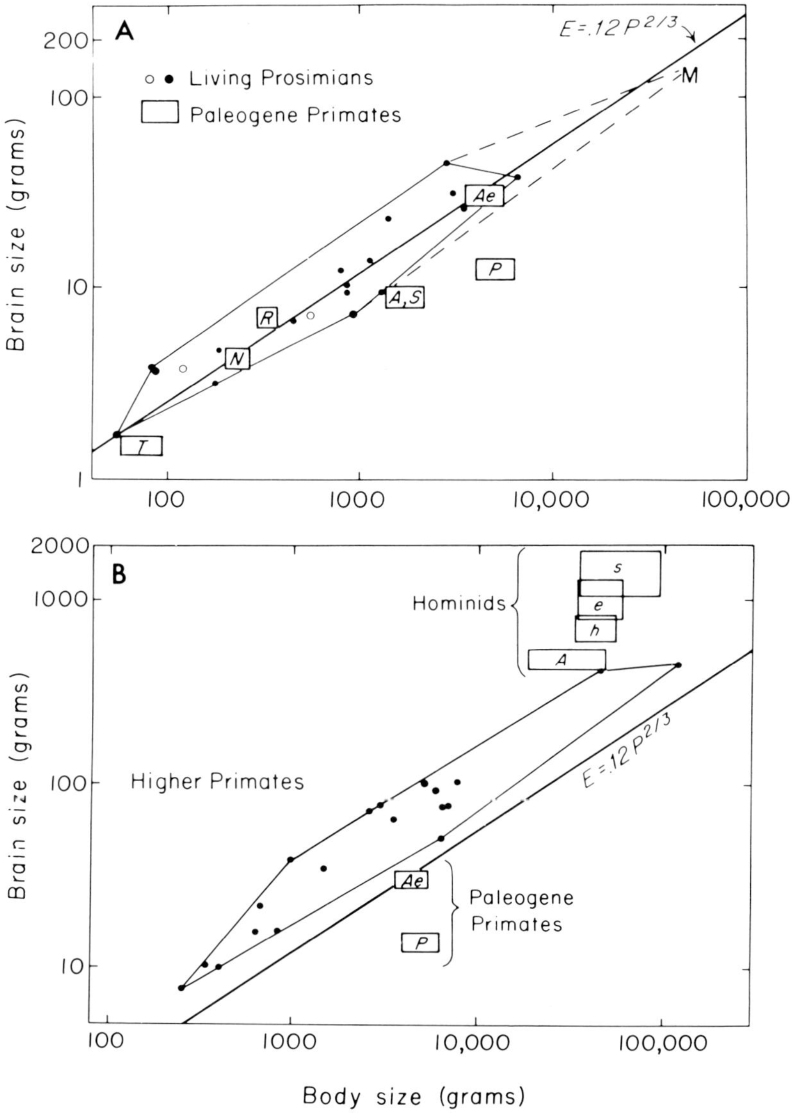

FIG. 10. Convex polygons enclosing primate data, shown relative to Eq. 6. Abbreviations are: A, *Tetonius* (*T*), *Necrolemur* (*N*), *Rooneyia* (*R*), *Adapis* (*A*), *Smilodectes* (*S*), *Plesiadapis* (*P*), and *Megalapis* (M); B, Paleogene primates *Aegyptopithecus* (*Ae*) and *Plesiadapis* (*P*), hominids *Australopithecus* (*A*), *Homo habilis* (*h*), *H. erectus* (*e*), and *H. sapiens* (*s*). (Adapted from Jerison 1973, 1979*b*.)

the primates, on the other hand, are intermediate between the basal mammalian grade and the grade of their closest and least encephalized relatives among the living primates.

The evidence from *Aegyptopithecus* is consistent with this picture of evolving (anagenetic) encephalization that was to continue in later species. This earliest of the anthropoids with known endocast is at a prosimian grade, well below the grade of any living "higher" primate. The relationship between *Aegyptopithecus* and the living higher primates in brain-body space is illustrated in Fig. 10B, which adds the data on this Oligocene species to those of living anthropoids. Although there is no question about assigning *Aegyptopithecus* to the anthropoids on the basis of other morphological characters, the evidence of the brain and encephalization indicates that this species was at a prosimian grade. Brain evolution lagged behind other characters.

The most complete history of the evolution of encephalization is available for carnivores and ungulates. The picture of ungulate evolution (Jerison, 1973, 1976) is probably correct as given in Fig. 11. The pattern is one of increasing encephalization with the passage of time and increasing diversity in the amount of encephalization. The picture of carnivore encephalization may be wrong and should be regarded as tentative because of new evidence.

Radinsky (1978) has recently analyzed a larger and better dated sample of species than those used for Fig. 11. His data are consistent with those presented above for the evolution of the ungulates, but he did not find the same relationship between progressive and archaic carnivores, nor did he find any significant differences between progressive ungulates and carnivores at any time in their histories which might have reflected an evolutionary competition. In general his analysis leads to a view of encephalization as being more weakly related to fitness than suggested by Fig. 11. There are some unresolved problems in the analysis of body weight in fossil species, however (see Jerison, 1979*a*); a technical problem centers around the choice between incompatible regression methods that Radinsky and I have used. The problem can easily be resolved by undertaking multiple regression analysis of body weight from skeletal measures in living species and applying these equations to the fossils. The required multiple correlation coefficient should be very high ($R > 0.99$) to permit accurate estimation of body weight from skeletal information available on fossils.

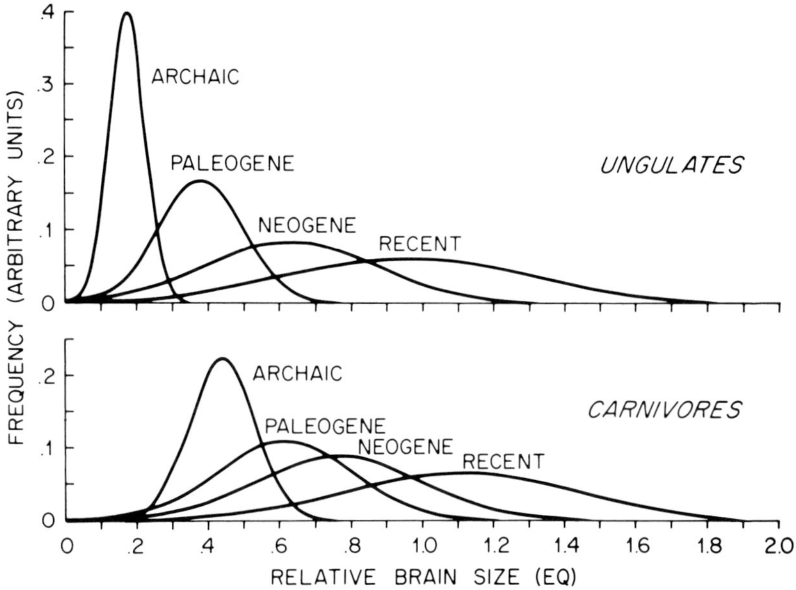

FIG. 11. Distributions of encephalization quotients in various assemblages of ungulates and carnivores. "Archaic" refers to orders that are now extinct: amblypods, condylarths, and creodonts. (From Jerison, 1973; copyright Academic Press, New York.)

Very Large Brains: Cetaceans and Hominids

In the diversification of the mammals, very large-brained species appear in three orders—proboscideans, cetaceans, and primates. The proboscidean brain size is known primarily on the basis of Pleistocene and Recent species, a rather incomplete record. A more complete story can be developed for the other two orders, with evolutionary evidence from an excellent fossil record.

The evolution of encephalization in the cetacean brain is presented in Fig. 12, utilizing the method of convex polygons (Jerison, 1978). Paleogene, Neogene, and Recent grades are compared with the archaic mammalian grade shown earlier in Fig. 8 and with the present hominid and pongid grades. There are two points to emphasize. First, as an order, the cetaceans had achieved their present grade of encephalization no later than the Lower Miocene epoch, about 18 million years ago. Second, within the order Cetacea, the

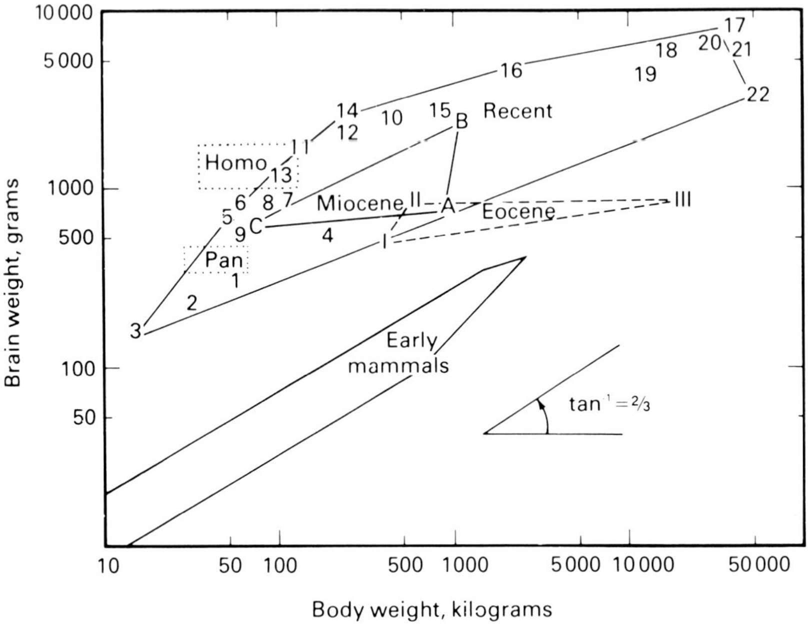

FIG. 12. Encephalization in Recent and fossil cetaceans, compared with that in man, chimpanzee, and early mammals (cf. Figure 9). Points 1–22 are Recent cetaceans, A–C are Miocene cetaceans, and I–III are Eocene cetaceans. (From Jerison 1978.)

convex polygon is very obviously oriented along a shallower slope than ⅔ in brain-body space. These conclusions are based on quantitative but non-numerical methods, which illustrate the utility of an analysis with convex polygons rather than *EQ*s.

Slopes (exponents in equations like Eq. 6) shallower than ⅔ occur often enough in the analysis of encephalization in land mammals to warrant a brief methodological digression on their significance. The shallow slopes can be understood with the help of a simple theory of encephalization (Jerison, 1977) that derives an equation only slightly more complex than Eq. 6 and does this from elementary neurological principles. The basic theoretical equation is:

$$E = .1mP^{2/3} + A \tag{9}$$

The new term A refers to an amount of tissue *added* to a brain independent of body size. Human speech and language areas are such systems of tissue. When Eq. 9 is graphed it can appear as a family of curves with m (that is, EQ) and A as parameters. The theory derives two parameters, or partitions, of encephalization—a multiplicative amplification factor m and the additive factor A, a necessary partition in view of the structure of the brain. According to the theory, shallow slopes result from trying to fit straight lines when the true shape of the "allometric function" is given by Eq. 9, which is graphed as a curve that is concave upward in log-log brain-body-space and has a ⅔ slope as its limit.

This is probably not a correct approach to cetacean data, however. The shallow slopes in Fig. 12 are predictable from selection pressures of a marine environment in which there are no gravitational restrictions on size. Large bodies could have evolved independent of the evolving brain, and the model for cetacean encephalization would then be quite different from that in land mammals. To explain, we need a bit more theory of encephalization.

The role of body weight is heuristic rather than fundamental for the theory. Body weight works because it provides an estimate of the surface of the brain. Brain surface area is conceptually independent of the body surface. The estimate of brain surface area provides a measure of the basic map at the periphery of the brain, a map of the sensory surfaces in the skin, retina, basilar membrane of the ear, olfactory epithelium, and kinaesthetic and proprioceptive receptors and of the motor systems controlling the body's musculature. The amplification term m is proportional to the number of times that the map is repeated in the brain, and the additive term represents processing capacity that is independent of that map. In order to apply the theory to cetacean encephalization it would be necessary to determine the P-term in Eq. 9 from something other than body size. In principle, the summed cross-sectional areas of the medulla and of the cranial nerves would replace $P^{2/3}$ in the equation. The m-term would then be equal, rather than merely proportional, to the amplification.

The model of cetacean encephalization should not involve gross body weight at all. A reasonable conjecture from Fig. 12 is that the cetaceans are highly encephalized—some are at a grade comparable to *H. sapiens* as measured by EQ—but that the encephalization of large cetacean species will almost always be underestimated by formulas like Eq. 8 or 9 because their body surface overestimates the size of the "basic map" in the brain and thus underestimates the amount of amplification undergone by the map. Cetacean species evolved anagenetically with respect to body size, but processed about the same amount of information with their brains. Lande (1979) analyzes the meaning of the allometric exponent in population-genetic terms, which quantifies the evolutionary decoupling of the genetics of brain and body.

Relative stability in body size in the hominids during 3 or 4 millions years of this history for which there is evidence on the brain simplifies the analysis of hominid encephalization. It is essentially a history of brain size, which increased from slightly above a pongid grade at about 500 cm^3 to almost three times that size at about 1,400 cm^3. Body size may have increased from about 30 kg to about 50 or 60 kg during that interval according to the available evidence (McHenry, 1975), but there are important human populations alive today in which adult body weights of 30 or 40 kg are typical. It is, therefore, not critical to control for body size in analyzing hominid encephalization.

The pattern of hominid encephalization can be visualized from the data in Fig. 12 by imagining appropriate overlapping rectangular regions between the living chimpanzee and human rectangles. Because of the similarities in body weight, encephalization in the hominid groups can be measured directly from absolute brain size; the adjustment for body size and allometry is approximately the same for the several groups. Figure 13 shows another way to present these data. Hominid data are in the right half of this illustration; the other data are included for an insight into the variability of brain size. The coefficient of variation is approximately 10% in all of the species that are shown. Figure 13 is thus evidence of a constancy in within-species diversity in brain size.

Our special interest in our own evolutionary history, and the rather good fossil record of hominid endocranial evolution justify yet another perspective for the hominid data. Fig. 14 puts the story into a framework of time, based on a summary of the data on the historic course of evolution of information-processing capacity in the hominids (Blumenberg, 1978). Blumenberg's taxonomic decisions were not quite the same as those used for the hominid data of Fig. 13, but the two patterns are in reasonable agreement. The links among the species in Fig. 14 are drawn to contrast a contin-

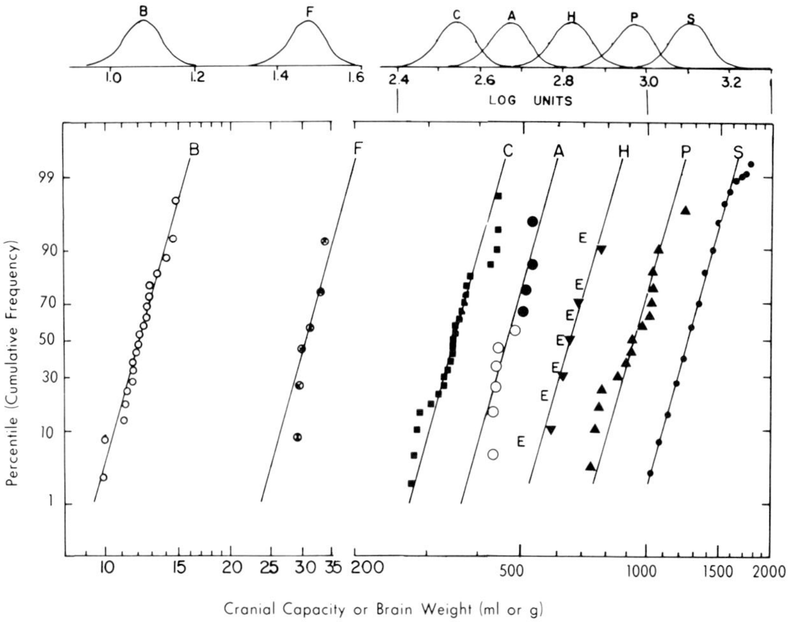

FIG. 13. Cumulative frequency distributions of brain size, with corresponding normal curves (top). Abbreviations are: B, *Bathygenys reevesi* (an Oligocene oreodont); F, *felis catus*; C, chimpanzee; A, *Australopithecus*; E, *Equus caballus*; H, *Homo habilis*; P, *H. erectus*; S, *Homo sapiens*. Parallel lines on "probability" paper with logarithmic abscissa imply equal coefficients of variation, in this instance about 10%. (From Jerison, 1979*a*.)

uous model, which would permit one to make inferences about the rate and pattern of evolution of brain size, and a discrete model that assumes stasis at each of several grades and rapid evolution between grades.

The continuous model is heuristic, permitting sophisticated mathematical development. Witness Lande's (1979) work that led to an alternate approach to the appearance of allometric exponents less than ⅔ (see the discussion here of cetacean encephalization), and his success in distinguishing short-term from long-term differentiation of brain-body evolution. He could describe short-term differentiation as reflecting primarily selection for body size, and long-term differentiation as allowing for independent evolution of encephalization.

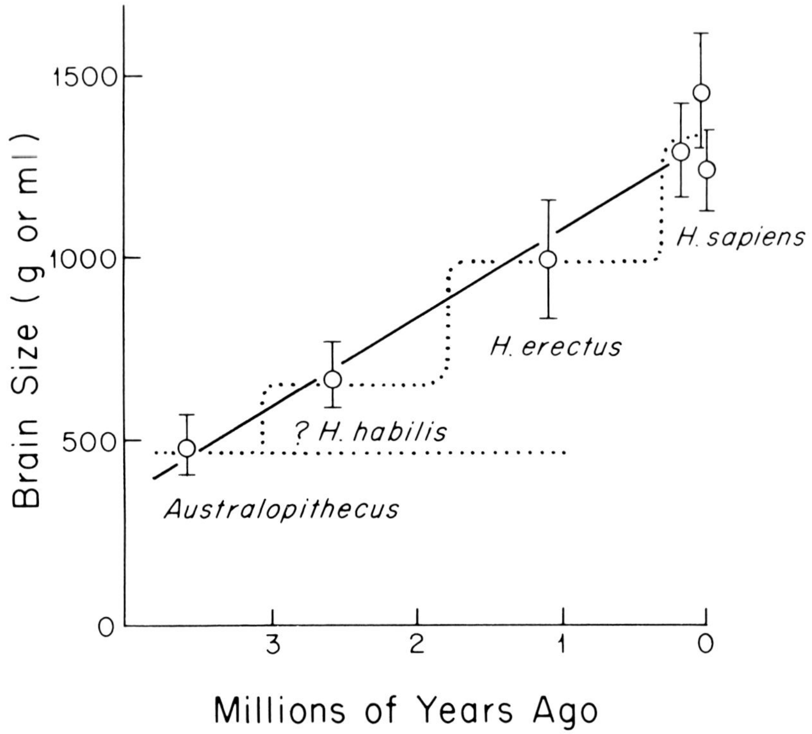

FIG. 14. Rate of hominid brain evolution. Diagonal is a continuous model. Steps represent discrete model of "punctuated equilibria." (Data from Blumenberg, 1978.)

The discrete model is probably correct, historically. The dotted steps in Fig. 14 are actually about as long as the lives of each taxon. *Australopithecus* and *Homo* apparently coexisted, and the persistence of the australopithecine grade is suggested in the dotted line of Fig. 14. The duration of *H. habilis* is unknown, though the indicated range of between 1.8 and 3 million years ago is consistent with current opinion among paleoanthropologists. *Homo erectus* is known for the entire indicated interval, from about 1.5 to 0.4 million years ago, and the earliest *H. sapiens* in Blumenberg's group (Steinheim man) is dated at about 0.3 million years ago (Kennedy, 1980). Kennedy would include Steinheim man with *H. erectus* and would put the earliest of *H. sapiens* that has been recovered at about 0.2 million years ago. This is a small point from our perspective, but it may be important to keep in mind when evaluating

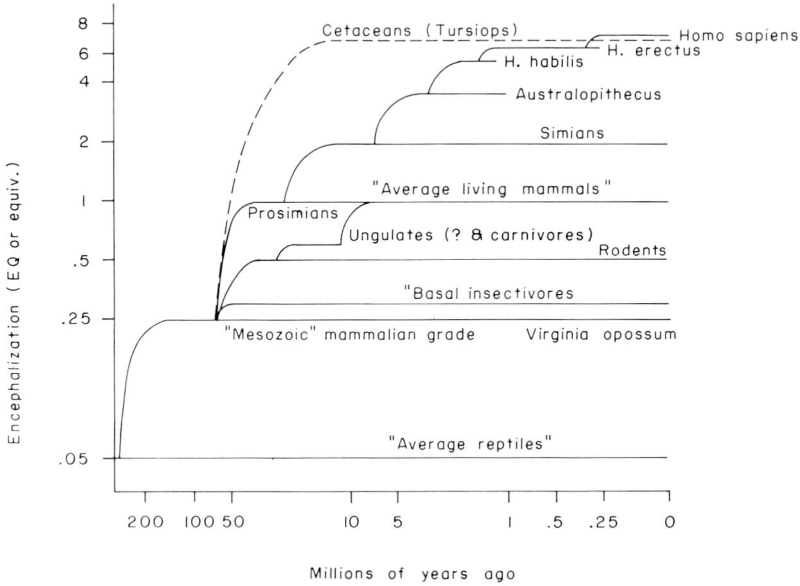

FIG. 15. Rates of mammalian brain evolution. Note log-scale for both time and encephalization quotient. Whales scaled on "extra neurons"—A in Eq. 9—measure of encephalization.

data on evolutionary rates. It is not always clear just what was evolving into what; taxonomists have not solved all of their problems.

Rate of Evolution of Processing Capacity

To sum up what we know about the evolution of brain size and encephalization in mammals, which is the core of this chapter, I have prepared Fig. 15. The picture is a bit different from one that I drafted earlier (Jerison, 1976). In the earlier picture, periods of rapid evolution alternated with periods of stasis. This alternation was shown for most groups, but I had suggested a continuous advance in encephalization in the carnivores and ungulates. My intention was to emphasize what appeared to be an evolutionary interaction corresponding to conventional ideas about predator-prey interaction, that groups of species maintained an ecological balance in which both progressed in behavior and encephalization. With a few more years to mull this over it is clear to me that no evidence, no matter how it is interpreted, would imply a continuous advance of the kind shown by the diagonal in Fig. 14. Figure 15 is, therefore, presented as a series of "punctuated equilibria" in the evolution of encephalization in all of the various groups of mammals.

Among the special features of Fig. 15, two are especially noteworthy. First, there has been a stability of encephalization, clustered about several grades, with an underlying conservatism throughout. Second, the rapid evolution of hominid encephalization during the past 3 or 4 million years (shown in detail in Fig. 14) contrasts sharply with the absence of evidence for advances as recently in any other group. A word of caution, however—there is no fossil evidence on the time course of cetacean encephalization to the grade of bottlenose dolphin, *Tursiops truncatus*. The conjectural line to *Tursiops* could represent a much more recent development in cetacean evolution than the one in Fig. 15, which shows this grade as having been reached almost 20 million years ago. The line is drawn in this way because of the evidence of *Aulophysiter morrisei* (B in Fig. 12, above) which is about 18 million years old and which was as encephalized as the living killer whale (*Orca*), a large, but behaviorally comparable, relative of *Tursiops*. A unique feature of hominid encephalization appears to be its recency, and this is emphasized in Fig. 15.

Let us keep in mind that the emphasis on encephalization in this section is appropriate for a chapter on processing capacity. Although total processing capacity, which is reflected in total brain size, evolved primarily in allometric conjunction with the evolution of larger bodies, the part of processing capacity that really interests us is the part independent of body size (see Lande, 1979). It is only in the hominid sequence, however, where body size has been fairly stable, that useful evidence (Fig. 14) can be taken from *total* processing capacity. In tracing the evolution of processing capacity in general, the adjustment for body size must be made, and this is done by analyzing allometry as well as encephalization.

Epilogue

This chapter has been concerned with what happened in evolution and not with why it happened. It is almost impossible to ignore the why of it, however, and I devote the final paragraphs to this issue. The important point to remember is that very advanced behavior can be governed by very small amounts of brain tissue. The behavioral adaptations of lower vertebrates are as remarkable as those of mammals in many ways. Encephalization in mammals, in primates, and in the human species is not easy to explain as a correlate of the refinement of behavior.

Information-processing of a kind that could be done only by very large amounts of neural tissue must have been evolving in the mammals. In discussing such processing the conventional language of neurophysiology that deals with sensory, motor, and even associative functions is inadequate, because these are categories that work as well for small as for large amounts of neural tissue. They are best understood at a spinal level. The evolution of encephalization must be understood in other terms. The key words may be consciousness (or awareness), intelligence, and other unusual species-typical adaptations (such as human language) that require enormous amounts of neural tissue for their control.

The evolution of processing capacity will be properly understood only when we understand the nature of information-processing in the most encephalized living species. There have been remarkable advances in that understanding for the human species in recent years (see Baddeley, 1976; Estes, 1978) and the beginnings of understanding of unusual processing ("language," etc.) in other highly encephalized species (Herman, 1980; Humphrey, 1976; Mason, 1976; Savage-Rumbaugh et al., 1980; Terrace et al., 1979). It is appropriate to end our discussion by noting but not including these issues, which are beset by controversies, perhaps necessary correlates of the developing understanding. But we can only postpone and cannot really avoid further discussion, which should be a natural sequel to this chapter.

Literature Cited

BADDELEY, A. D. 1976. The psychology of Memory. Harper & Row, New York, 430 pp.

BAILLARGER, J. 1853. De l'étendue de la surface du cerveau et de ses rapports avec le developpement de l'intelligence. Ann. Med.-Psych., ser. 2, 5:1–9.

BLUMENBERG, B. 1978. Hominid ECV versus time: available data does not permit a choice of a model. J. Human. Evol., 7:425–436.

BRODMANN, K. 1913. Neue Forschungsergebnisse der Grosshirnrindenanatomie mit besonderer Berucksichtung anthropologischer Fragen. Verh. Vers. Deutsch. Naturf. und Aertze in Wien, 85:200–240.

CHOW, K. L., AND A. L. LEIMAN. 1970. The structural and functional organization of the cortex. Neuroscience Res. Prog. Bull., 8:153–220.

CONNOLLY, C. J. 1950. External morphology of the primate brain. Thomas, Springfield, Illinois, 378 pp.

CRILE, G., AND D. P. QUIRING. 1940. A record of the body weight and certain organ and gland weights of 3690 animals. Ohio J. Sci., 40:219–259.

DIAMOND, I. T. 1979. The subdivisions of neocortex: a proposal to revise the traditional view of sensory, motor, and association areas. Prog. Psychobiol. Physiol. Psychol., 8:1–43.

EBBESON, S. O. E., AND R. G. NORTHCUTT. 1976. Neurology of anamniotic vertebrates. Pp. 115–146, *in* Evolution of brain and behavior in vertebrates (R. B. Masterton et al., eds.). Erlbaum/Wiley, Hillsdale, New Jersey, 482 pp.

ECCLES, J. C. 1979. The human mystery. (Gifford Lectures 1977–1978). Springer-Verlag, New York, 255 pp.

ELIAS, H., AND D. SCHWARTZ. 1971. Cerebo-cortical surface areas, volumes, lengths of gyri and their interdependence in mammals, including man. Z. Säugetierk., 36:147–163.

ESTES, W. K. (ed.). 1978. Handbook of learning and cognitive processes. Vol. 5: Human information processing. Erlbaum, Hillside, New Jersey, 337 pp.

FALK, D. 1978. Brain evolution in Old World monkeys. Amer. J. Phys. Anthrop., 48:315–320.

———. 1982. Mapping fossil endocasts. Pp. 216–217, *in* Primate brain evolution: methods and concepts (E. Armstrong and D. Falk, eds.). Plenum Press, New York, 332 pp.

HOLLOWAY, R. L. 1974. The casts of fossil hominid brains. Scientific American, 231:106–114.

———. 1976. Paleoneurological evidence for language origins. Ann. New York Acad. Sci., 280:330–348.

HOPSON, J. A. 1977. Relative brain size and behavior in archosaurian reptiles. Ann. Rev. Ecol. Syst., 8:429–448.

———. 1980. Paleoneurology. Pp. 39–146, *in* Biology of the Reptilia. Vol. 9: Neurology (A. C. Gans, R. G. Northcutt, and P. Ulinski, eds.). Academic Press, London, 462 pp.

HUMPHREY, N. K. 1976. The social function of intellect. Pp. 303–317, *in* Growing points in ethology (P. P. G. Bateson and R. A. Hinde, eds.). Cambridge Univ. Press, Cambridge, England, 548 pp.

JERISON, H. J. 1973. Evolution of the brain and intelligence. Academic Press, New York, 482 pp.

———. 1976. Paleoneurology and the evolution of mind. Sci. Amer., 234:90–101.

———. 1977. The theory of encephalization. Ann. New York Acad. Sci., 299: 146–160.

———. 1978. Brain and intelligence in whales. Pp. 159–197, *in* Whales and whaling (S. Frost, ed.). Australian Gov. Publ. Serv., Canberra, 2:1–220.

———. 1979*a*. The evolution of diversity in brain size. Pp. 29–57, *in* Development and evolution of brain size: behavioral implications (M. Hahn, C. Jensen, and B. Dudek, eds.). Academic Press, New York, 393 pp.

———. 1979*b*. Brain, body, and encephalization in early primates. J. Human Evol., 8:615–635.

———. 1982. Allometry, brain size, cortical surface, and convolutedness. Pp. 77–

84, *in* Primate brain evolution: methods and concepts (E. Armstrong and D. Falk, eds.). Plenum Press, New York, 332 pp.

KAAS, J. H. 1980. A comparative survey of visual cortex organization in mammals. Pp. 483–502, *in* Comparative neurology of the telencephalon (S. O. E. Ebbesson, ed.). Plenum, New York, 506 pp.

KENNEDY, G. E. 1980. Paleoanthropology. McGraw-Hill, New York, 479 pp.

LANDE, R. 1979. Quantitative genetic analysis of multivariate evolution, applied to brain:body size allometry. Evolution, 33:402–416.

MARTIN, R. D. 1979. Phylogenetic aspects of prosimian behavior. Pp. 45–77, *in* The study of prosimian behavior (R. D. Martin and P. D. Szalay, eds.). Academic Press, New York, 696 pp.

MASON, W. A. 1976. Environmental models and mental modes: representational processes in the great apes and man. Amer. Psychol., 31:284–294.

MCHENRY, H. M. 1975. Fossils and the mosaic nature of human evolution. Science, 190:425–431.

MERZENICH, M. M., AND J. H. KAAS. 1980. Principles of organization of sensory-perceptual systems in mammals. Prog. Psychobiol. Physiol. Psychol., 9:1–42.

MOUNTCASTLE, V. B. 1978. An organizing principle for cerebral function: the unit module and the distributed system. Pp. 7–50, *in* The mindful brain (G. M. Edelman and V. B. Mountcastle, eds.). MIT Press, Cambridge, Massachusetts, 100 pp.

RADINSKY, L. B. 1970. The fossil evidence of prosimian brain evolution. Pp. 209–224, *in* The primate brain (C. R. Noback and W. Montagna, eds.). Appleton, New York, 320 pp.

———. 1974. The fossil evidence of anthropoid brain evolution. Amer. J. Phys. Anthrop., 41:15–28.

———. 1978. Evolution of brain size in carnivores and ungulates. Amer. Nat., 112:815–831.

———. 1979. The fossil record of primate brain evolution. James Arthur Lecture on the Evolution of the Human Brain. Amer. Mus. Nat. Hist., New York, 27 pp.

———. 1980. Endocasts of amphicyonid carnivorans. Amer. Mus. Novitates, 2694: 1–11.

ROCKEL, A. J., R. W. HIORNS, AND T. P. S. POWELL. 1980. The basic uniformity in structure of the neocortex. Brain, 103:221–224.

SAVAGE-RUMBAUGH, E. S., D. M. RUMBAUGH, AND S. BOYSON. 1980. Do apes use language? Amer. Sci., 68:49–61.

SCHEIBEL, M. E., AND A. B. SCHEIBEL. 1970. Elementary processes in selected thalamic and cortical subsystems: the structural substrates. Pp. 443–457, *in* The neurosciences—second study program (F. O. Schmitt et al., eds.). Rockefeller Univ. Press, New York, 1069 pp.

SUGA, N., AND P. H. S. JEN. 1976. Disproportionate tonotopic representation for processing CF-FM sonar signals in the mustache bat auditory cortex. Science, 194:542–544.

SZENTÁGOTHAI, J. 1978. The neuron network of the cerebral cortex: a functional interpretation. (The Ferrier Lecture, 1977). Proc. Royal Soc. London, ser. B, 201:219–248.

TERRACE, H. S., ET AL. 1979. Can an ape create a sentence? Science, 206:891–900.

WELKER, W. I. (ed.). 1976. Neocortical mapping studies. Brain Behav. Evol., 13: 241–343.

WELKER, W. I., J. I. JOHNSON, JR., AND B. H. PUBOLS, JR. 1964. Some morphological and physiological characteristics of the somatic sensory system in raccoons. Amer. Zool., 4:75–94.

YOUNG, R. M. 1970. Mind, brain and adaptation in the nineteenth century. Clarendon Press, Oxford, 278 pp.

ZEKI, S. M. 1978. Functional specialization in the visual cortex of the rhesus monkey. Nature, 274:423–428.

PART II. MECHANISMS OF COMMUNICATION

INTRODUCTION

An organism interacts with its environment, both animate and inanimate. Sense organs deliver messages to the central nervous system to coordinate the activities of the animal with respect to objects perceived in the environment, as well as differences in temperature, lighting, and other less tangible stimuli. Both intra- and interspecific relations involve the transfer of information among individuals, often with selection having acted to increase the specificity of the message(s). For example, predator-prey relationships and their coevolution depend on perception and action by both the predator and the prey, mediated by the sense organs and motor system. It is, however, in the arena of intraspecific social behavior that the exchange of discrete signals among the interactants has relevance to the interpretation of the form and function of species-specific social structure. Classically, the communication channels of organisms are ordered according to the major sense organ involved in the reception of the message. Among mammals, with their nocturnal origin, olfaction and audition are prominent in communication systems, as in all life's activities. The importance of olfaction in mammalian phylogenetic history is demonstrated by the continued subtle reliance on chemical cues throughout the class, even the visually-dominated primates.

Müller-Schwarze presents an exhaustive review of scent glands and their functions in mammals. Clearly, scent glands have evolved in a parallel fashion among most of the mammalian orders, and functions overlap because of limitations and benefits associated with the use of chemical cues. In recent years, the exact chemical constituents of scent glands have been analyzed to the point where specific compounds can be associated with a specific response system. The importance of this recent advance can only be appreciated when one considers that visual and vocal signals for years have been definable and analyzeable through photography and the sound spectrograph. Studies of olfactory communication have been limited to an analysis of the effects of chemical cues on respondents and the behavior associated with the deposition of chemical cues. Analogizing to the study of bird song, researchers interested in the olfactory mode have until recently had to concentrate on the singing bird instead of the structure of the song.

Scent glands are not the only means of exchanging information among conspecifics. In the chapter by Bronson, the importance of chemical compounds in urine for demography, social organization, and reproduction is analyzed in great depth through a comparison of *Mus musculus* and *Peromyscus maniculatus*. In both species, compounds excreted in the urine have profound influences on sexual development and the integration of sexual activity between males and females. Both species of mice use a two-way feedback system with signals from the male stimulating the female's readiness to mate and signals from the female stimulating the male's spermatogenic activity. By the same token, chemical signals from males may inhibit sexual development in younger males and correspondingly maturation is retarded in young females by chemical cues from adult females. There are, however, interesting differences when these two common rodents are compared. Bronson rightly points out that the life history characteristics of members of the genus *Peromyscus* (for example, small litter sizes and a slower maturation rate), when compared with species of *Mus*, have profound consequences on their population biology and spatial organization which is reflected in their chemical communication systems.

In recent years, less attention has been devoted to the detailed analysis of visual signals, because the descriptive ethogram has been found to be too simplistic (see Part I). Especially, the functions of certain classical visual signals which pervade the Mammalia have been ignored because of difficulties of adequately documenting function. Wemmer and Wilson critically examine the adaptive advantage of crests in African carnivores, by correlating relative crest size and markings with ecological and behavioral characteristics. The function of piloerection in social encounters even received speculative comments by Darwin whose suggestion is still generally accepted that hair erection would increase the apparent bulk of the body, with the possible result that the actor might inhibit attack. By examining a series of carnivores, the authors conclude that small or medium-sized mammals foraging terrestrially in open habitats are vulnerable to predation and competition, and hence natural selection has favored intimidation displays through the evolution of crests. This interesting piece of correlative research suggests that fruitful avenues exist for exploration of visual display in mammals, and implies, at least in this group, that crest erection is more important as an interspecific display than in intraspecific communication.

Auditory communication is of paramount importance in mammals, especially in nocturnal forms. In the chapter by Gould, mammalian auditory communication is reviewed in depth. Gould reviews modes of sound production in mammals, but then concentrates on one class of sound production, namely, vocalizations. Gould's review emphasizes the functional classification of vocalization forms through an analysis of the ontogeny of vocalizations. He shows the importance of the maternal-neonatal communication system and how infantile vocalizations persist into adult life with altered function, but in contexts analogous to the maternal-neonatal interactive system.

In the final chapter of this section, Peter Marler examines some exciting new developments in the area of primate vocal communication, including one research area which suggests that species-specific vocalizations may be encoded in the brain such that perception is also species-specific. There is also strong evidence for the lateralization of brain function in the perception of vocalizations. Marler looks at the semantics of primate alarm calls and their ontogeny, through the analysis of the vocalizations produced by vervet monkeys (*Cercopithecus aethiops*) in the context of antipredator responses. Is there a semantic component to animal signal systems and if so how can we begin to understand it? Among vervet monkeys, there are discrete vocalizations produced when an animal perceives a snake, an aerial predator, or a ground predator. Conspecifics respond as if they clearly understand the vocal signal to mean snake, aerial predator, or ground predator. The ontogeny of the vocalizations and the specificity of responsiveness suggests a genetic determination of the system with some experiential components molding responsiveness as the infant develops. How these specific signal systems evolve in nature and develop during ontogeny will be the subject of continuing investigation and may provide exciting new insights into the evolution of mammalian communication.

D.G.K. and J.F.E.

SCENT GLANDS IN MAMMALS AND THEIR FUNCTIONS

DIETLAND MÜLLER-SCHWARZE

Abstract

THE occurrence of scent glands and their use in the behavior of mammals are reviewed, following a taxonomic approach. Classical information, such as from Schaffer's (1940) book on scent-gland anatomy and histology, is incorporated whenever possible, as that book is out of print and not available in English. Chemical communication using urine as a vehicle is not covered here.

Also reviewed are functions—individual and social—of skin-gland secretions, the chemical nature of active odor constituents that have been identified, some environmental factors and ecological correlates, and possible evolutionary steps in the use of skin glands.

As with other animal signals, scent-gland secretions may have multiple functions. The experimental analysis of chemical communication in mammals should allow for this possibility, and also for variability due to learning processes.

Introduction

In this review of mammalian scent glands and their functions I first present an overview of the types of skin glands found in mammals using a taxonomic approach with a description of the glands, behaviors associated with their use, and adaptive functions. In the second part, the demonstrated and assumed functions of skin glands are discussed and generalizations attempted.

Mammalian skin glands and their roles in behavior have been periodically reviewed. The standard source is still Schaffer's (1940) monograph on the structure and distribution of skin glands in mammals, which incorporates earlier studies by Brinkmann (1912), Pocock (1910), and Schiefferdecker (1922). Gabe (1967) wrote a shorter review of skin-gland structure. Since Schaffer (1940) an enormous amount of new material has accumulated. As Quay (1977) stated "any attempt at a sequal to Schaffer's (1940) still unique and useful compendium would require several volumes alone."

Reviews of scent-marking behavior include Ewer (1968*a*), Eisenberg and Kleiman (1972), Johnson (1973), Ralls (1971), and Thiessen and Rice (1976). Mykytowycz (1970) reviewed the role of skin glands in mammalian communication. Other reviews of mammalian scent-communication may be found in Sebeok (1978) and Doty (1976).

Mammalian skin glands that often are the source of odoriferous secretions consist of two glandular elements—sebaceous (alveolar) and sudoriferous (tubular) glands. There are two types of sudoriferous glands—merocrine ("sweat") and apocrine glands. Sebaceous glands were first described by Eichorn in 1826 and Weber in 1827 and sweat glands by Purkinje in 1833 (Schaffer, 1940).

Taxonomic Review of Occurrence of Skin Glands and Their Use

Phylogenetic Relationships of Gland Types

Schaffer (1940) considered the sebaceous glands (alveolar, holocrine, polyptychous glands, glandulae immusculatae) as the more original type of skin glands, because they are more widely distributed among mammals and have precursors with similar structure and secretory type among the Sauropsida. Thus, sebaceous glands are assumed to be phylogenetically older than hair, although in mammals they are associated with hair.

Of the two types of sudoriferous glands (tubular glands, glandulae musculatae), "sweat glands" (merocrine glands) seem to be a new acquisition of primates (Schaffer, 1940), whereas apocrine sudoriferous glands occur among all mammals and may be derived from secretory alveoli in reptiles (Quay, 1977). The following taxonomic review surveys the variety of glands and contexts in which they are used in the Mammalia.

Monotremata

The most conspicuous gland in this group is the femoral gland associated with the spur on the ankle. It occurs in both families of this order—in platypuses (*Ornithorhynchus anatinus*; Ornithorhynchidae) and spiny anteaters or echidnas (*Tachyglossus*, Tachyglossidae). The gland is present in males and degenerates during the female's development; the secretion is poisonous to conspecifics and even to people, who may lose power in spurred muscles for months (Walker et al., 1975). A platypus can kill a dog with its venom (Temple-Smith, *in* Griffiths, 1978). Temple-Smith assumed fighting and spacing between males to be the main function of the venom and spur. This apparatus is most active from August to October.

The male platypus also has glands at the base of the neck that emit a "musky" odor. In 1888, Klastsch described "sweat glands" in foot pads of fore- and hindlegs (Schaffer, 1940).

Spiny anteaters (*Tachyglossus aculeatus*) have been observed to extrude their cloaca and rub it over the substrate leaving an odor (Dobroruka, 1960; Hediger and Kummer, 1956).

Marsupials

Marsupials have cloacal ("anal") glands; frontal and sternal glands are present in some families.

Didelphidae.—*Didelphis* has a sternal gland, anal sacs, foot-pad glands, and proctodeal glands (Schaffer, 1940).

Dasyuridae.—Fat-tailed sminthopsis (*Sminthopsis crassicaudata*) have two pairs of cloacal glands and both sexes mark with these glands on the substrate in their territory. Sex and individual recognition is achieved by sniffing the mouth area, and saliva is applied to objects during "mouthing" (Ewer, 1968*b*).

The yellow-footed marsupial mouse (*Antechinus flavipes*) and the crest-tailed marsupial mouse (*Dasycercus cristicauda*) mark their territories with saliva by mouthing or chewing tree branches. *A. flavipes* drags its cloacal glands over the substrate and rubs branches with its sternal gland, which is larger in males (Ewer, 1968*b*).

Myrmecobiidae.—*Myrmecobius fasciatus* has an elaborate sternal gland (Beddart, 1887, *in* Schaffer, 1940).

Phalangeridae.—In this family there are frontal and sternal glands over the frontal end of the sternum, one pair of anal (paracloacal) glands, and sudoriferous glands in the foot pads.

The anal glands of *Trichosurus vulpecula* (brush-tailed phalanger) were described by Bolliger and Whitten (1948). Bolliger (1944) performed the first study of the hormonal control of a mammalian skin gland with the "sternal integument" of *Trichosurus vulpecula*. Castration retarded the gland's development, and testosterone facilitated it. In *T. vulpecula* and *T. caninus*, males respond with threat and agonistic vocalizations to secretions from male anal glands of both species (Thomson and Pears, 1962).

The anal glands of Queensland ring-tailed phalangers (*Pseudocheirus peregrinus*), along with the tail and skin flaps from the thighs, are rejected by owls (*Ninox strenua*) preying on this species (Thomson and Owen, 1964).

Schultze-Westrum (1965) analyzed scent-marking in a colony of captive flying phalangers (or sugar gliders), *Petaurus breviceps*. The scents of this New Guinea marsupial originate in frontal, sternal, and anal glands, as well as in saliva and urine. Males and females rub their frontal glands against the sternal glands of group members of the opposite sex, thus producing a clan-specific odor which is easily recognized by "insiders" and strangers. Male koalas (*Phascolarctos cinereus*) mark trees with the sternal gland.

Phascolomyidae.—Wombats have rectal (or anal) glands (Schaffer, 1940).

Macropodidae.—Only one of two pairs of embryonic rectal glands in this family develops. In the red kangaroo, *Megaleia rufa*, males have patches of tubular glands on the chest and abdomen ("pectoral glands"). Males lick or bite this "pre-sternal gland" when threatening another male (Sharman and Cataby, 1964). Neck glands have been described for the desert rat kangaroo, *Caloprymnus campestris* (Walker et al., 1975).

In the musky "rat" kangaroo (*Hypsiprymnodon moschatus*) both sexes emit a musky scent (Walker et al., 1975) from a pair of holocrine rectal glands (Schaffer, 1940).

Mykytowycz and Nay (1964) found apocrine glands in the breast and axillary regions of males and in the cloaca-pouch areas of females in the red kangaroo (*Megaleia rufa*), gray kangaroo (*Macropus canguru*), euro (*Macropus robustus*), and the swamp wallaby (*Wallabia bicolor*).

Insectivores

Solenodontidae.—Glands in the armpit and the groin produce a secretion with a goat-like odor. Solenodons mark with ventral glands and also have flank glands (Walker et al., 1975).

Tenrecidae.—Tenrecs deposit chemicals in the environment by a perineal drag and *Setifer, Echinops,* and *Microgale* have eyelid glands that produce a white secretion (Poduschka, 1977). There are also glands in the axillary, inguinal, head, ear, and caudal areas (Eisenberg and Gould, 1970).

In the small Madagascar "hedgehog" (*Echinops telfairi*), males and females salivate on environmental features, then alternate in scratching this licked spot and their head or side (Eibl-Eibesfeldt, 1965). This has been interpreted as a reaction to new or strong odor stimuli, and not as part of communication (Poduschka, 1977).

Erinaceidae.—Hedgehogs (*Erinaceus europaeus*) have eye, oral angle, proctodeal ("anal"), circumanal, and foot-pad glands (Poduschka, 1977; Schaffer, 1940).

During self-anointing, the hedgehog licks a substance until frothy saliva forms in the mouth. Then it places the froth with its tongue on the spines. This behavior has been interpreted as removal of chemical stimuli from the mouth after they have stimulated the vomeronasal organ during the Flehmen response (Poduschka, 1977). Male hedgehogs mark with a secretion from their penis area during courtship, and the females encounter these marks many times during the hours of their nightly "hedgehog roundabouts" (Poduschka, 1977).

The Malayan gymnure or moon rat (*Echinosorex gymnurus*) has anal glands (Weber, *in* Schaffer, 1940) that lend the animal a strong odor resembling "rotten onion", stale sweat, and even "Irish stew gone bad" (Walker et al., 1975). A similar strong odor is typical of the lesser gymnure (*Hylomys suillus*) of Southeast Asia.

Soricidae.—Large lateral glands are found in *Sorex, Myosorex, Neomys,* and *Crocidura* (white-toothed shrews, also "musk shrews"). They consist of sebaceous glands surrounded by apocrine glands. In males the sebaceous portion is enlarged during the reproductive season. Sebaceous and apocrine elements in the side glands of male *Suncus murinus* (musk shrew) atrophy after castration and can be restored by testosterone or progesterone (Dryden and Conaway, 1967). Estradiol has no effect. Similar endocrine responses are reported for the apocrine glands in the postauricular fields of *Suncus* (Ebling, 1977).

The Old World water shrew (*Neomys fodiens*) uses a poisonous secretion from its submaxillary glands to weaken its prey, such as fish, frogs, and aquatic invertebrates (Walker et al., 1975). The same is true for the greater North American short-tailed shrew (*Blarina brevicauda*) (Walker et al., 1975). The latter species has a ventral gland, consisting primarily of sebaceous glands and flank glands (Pearson, 1946).

Macroscelididae.—Among elephant shrews, *Rhynchocyon chrysopygus* has a subcaudal gland (Lang, 1923) that is periodically rubbed on the substrate while the animal walks. This may aid males in finding females (Rathbun, 1979). *Elephantulus rufescens* has "prominent" sternal glands in both sexes which are rubbed on the ground as the animal moves about. Rathbun (1979) described all three members of one family marking the same point within 9 min and suggests that the perineal drag, and localized defecation and urination, may also serve in chemical communication among elephant shrews.

Talpidae.—Moles of the genera *Parascalops, Condylura,* and *Scalopus* have skin glands on the dorsal snout, chin, breast and abdomen, wrists, and in the perineal region (Eadie, 1954).

Talpa europaea possesses sebaceous and apocrine glands on the snout, large single sebaceous glands (associated with hair) on the prepuce, large sebaceous gland pads ventral to the auditory duct, and anal and rectal glands. (Hoffman, *in* Schaffer, 1940). Schaffer did not find apocrine glands in the foot pads.

The Russian desman (*Desmana moschata*) has a sebaceous subcaudal gland and anal glands (Schaffer, 1940). The Spanish desman (*Galemys pyrenaicus rufulus*) has a similar subcaudal gland whose odor saturates the nest and the area around it (Richard, 1973).

Chrysochloridae.—Golden moles have paired anal glands (Schaffer, 1940).

Chiroptera

Bats have scent glands in the face, on the neck, chest, nape, shoulders and wings, as well as in the anal area.

Pteropodidae.—*Cynopterus* and *Pteropus* have shoulder glands with conspicuous hair tufts (Schaffer, 1940). *Eidolon* has glands on the foreneck and sides of the neck, and *Eonycteris* has anal glands (Schaffer, 1940).

Neuweiler (1962) described *Pteropus giganteus* (Indian flying foxes) sniffing each other at the genital and neck regions and suggested that they recognize one another by individual odors. Males produce a strong odor from their neck glands and females recognize their young by odor. The Australian flying fox (*Pteropus poliocephalus*) uses the scapular gland to mark territories (Nelson, 1965).

Nycteridae.—*Nycteris* has facial glands (Schaffer, 1940).

Rhinolophidae.—Circumanal and facial glands occur in *Rhinolophus* (Schaffer, 1940).

Phyllostomidae.—In *Phyllostomus hastatus,* the male has a chest gland which is rudimentary in females. One sex of *P. discolor* sniffs the chest gland of the other and males open their glands while flapping in front of the female.

Emballonuridae.—In *Saccopteryx bilineata,* only the male has a wing gland. Males display in front of females and males with a "gland shake" (Bradbury, 1975).

Vespertilionidae.—Large facial glands are characteristic of *Myotis, Pipistrellus, Vespertilio,* and *Nyctalus* (Schaffer, 1940).

Molossidae.—Throat glands have been described for *Cheiromeles, Molossus,* and *Tadarida.* Herreid (1960) characterized the odor of males of the Mexican free-tailed bat, *Tadarida brasiliensis,* as "sweet," and that of the females as dull and musk-like.

Primates

Tupaiidae.—The common tree shrew (*Tupaia glis*) has a jugulosternal gland similar to that in insectivores, and quite different from glands in other primates (Sprankel, 1962). Scent-marking takes two forms—"chinning," and "sledging," a forward movement of the posterior abdominal region (Martin, 1968). Sternal scent marks identify sex and individual (von Holst and Lesk, 1975). Chinning marks familiarize a tree shrew with new surroundings, warn off rivals, and help in formation and maintenance of a pair bond (von Holst and Buergel-Goodwin, 1975).

Lemuridae.—The ring-tailed lemur (*Lemur catta*) marks with its perineal gland, and with "palmar" scent of the antebrachial gland. This species shows a complicated "stink display" in agonistic encounters. Males rub their brachial glands (on the chest) against their antebrachial glands (on the inner forearm) and then in turn, rub the antebrachial gland openings on the tail. The tail is then held upright, quivers, and is pointed toward an opponent (Evans and Goy, 1968; Jolly, 1966). Male ring-tailed lemurs respond with Flehmen to genital secretions from both sexes. They do not discriminate between the marks of the sexes (Evans, 1980).

Lorisidae.—The nocturnal slender loris (*Loris tardigradus*), slow loris (*Nycticebus coucang*), and thick-tailed bushbaby (*Galago crassicaudatus*) use scent trails for orientation from urine rubbed on the soles and palms (Eibl-Eibesfeldt, 1953*a* for *Galago*; Ilse, 1955 for *Loris*; Seitz, 1969, for *Nycticebus*). *Nycticebus* also has a brachial gland that is used in defense (Seitz, 1969).

Cebidae.—The spider monkey (*Ateles geoffroyi*) has a sternal (pectoral) gland (Wislocki and Schultz, 1925). Sniffing the pectoral gland is coupled with a ritualized embrace and occurs most often between high-ranking males and females. Axillary glands are also present, and anogenital investigation occurs in social encounters (Eisenberg and Kuehn, 1966).

Species of *Alouatta* (howler monkeys) have a gular gland, *Callicebus* (titi monkeys) and *Saimiri* (squirrel monkeys) gular and sternal glands, and *Cebus* (capuchins) sternal and epigastric glandular areas (Epple and Lorenz, 1967). *Alouatta seniculus* marks branches with the gular gland, especially during agonistic episodes (Sekulic and Eisenberg, in press).

Callitrichidae.—The tamarin (*Saguinus nigricollis*) has sternal, periinguinal, and ulnar glands (Perkins, 1966). Both sexes of saddleback tamarins (*S. fuscicollis*) mark with their sternal and circumgenital-suprapubic gland pads, but females mark more often (Epple, 1980). In marmosets (*Callithrix jacchus*), both sexes mark with their circumgenital gland area; dominant individuals mark more than subordinate ones (Epple, 1970). In the golden lion tamarin (*Leontopithecus rosalia*), both sexes mark with sternal and circumgenital (suprapubic and anogenital) glands. Marking is inhibited by the presence of a conspecific of the same sex, especially sexually active males. Females show a marking peak in mid-pregnancy; both sexes mark less often when transporting infants (Kleiman and Mack, 1980).

Cercopithecidae.—Hill (1944) described a tuft of specialized hair on the middle of the chest in a male drill (*Mandrillus leucophaeus*). A captive drill at puberty rubbed twigs and bark of the mango tree on mouth, chin and sternal tuft, bringing it then back to the mouth, whereupon he intensely salivated. The female has no tuft or gland on the chest.

The role of the rhesus monkey's (*Macaca mulatta*) vaginal secretion in sexual behavior has been experimentally investigated by Goldfoot et al. (1976) and Michael et al. (1971).

Pongidae.—Sternal glands have been described for the gibbon *Hylobates leuciscus* (= *H. moloch*) (Pocock, *in* Schaffer, 1940) and the orangutan, *Pongo pygmaeus* (Schultz, 1921). The gland consists of sebaceous and apocrine glands. Axillary organs are known in chimpanzee (*Pan troglodytes*) and gorilla (*Gorilla gorilla*) (Brinkmann, 1909). The axillary organ consists of sebaceous, apocrine, and eccrine sweat glands. Parakkal et al. (1963) did not find an axillary organ in *Hylobates.*

Hominidae.—The most elaborate skin gland in man is the axillary gland. It consists of sebaceous, apocrine and eccrine glands. Emotional sweating occurs at the more distal part of the axillary organ and thermal sweating at the more proximal part (Rebell and Kirk, 1962).

There are more diffuse glandular areas around the anal and genital areas, prepuce, mammae, and mouth. Cerumen glands in the ears and glands on the eyelids are probably less significant for chemical communication.

Edentata

Schaffer (1940) found apocrine glands in the dorsal part of the snout of the giant anteater (*Myrmecophaga tridactyla*), but anal sacs seem to be absent. During social encounters, the anogenital region and urine are sniffed (Tembrock, 1968).

In *Dasypus* (armadillos) both sexes have dorsal glands that secrete a strong-smelling fluid (Schaffer, 1940). There are also anal glands. *Euphractus* has a pelvic gland (Tembrock, 1968).

The three-toed sloth (*Bradypus tridactylus*) has a mid-dorsal gland which it rubs against the substrate.

Pholidota

Pangolins (*Manis tricuspis* and *M. longicaudata*) have anal glands which they rub on the ground or trees. Dominant pangolins mark more often than subordinates (Pagés, 1968).

Lagomorpha

Ochotonidae.—The pika (*Ochotona princeps*) has cheek glands which are rubbed on rocks and sniffed by the opposite sex during courtship. Males mark more than females, and individual cheek-gland odors are discriminated (Meaney, in press). In addition, the territory is marked by localized urine and feces (Harvey and Rosenberg, 1960; Kilham, 1958). There are also paraproctodeal glands lateral to the penis glands; there are no inguinal glands (Tullberg, *in* Schaffer, 1940).

Leporidae.—The skin glands of the European rabbit, *Oryctolagus cuniculus*, and their behavioral functions have been studied extensively by Mykytowycz (1966*a*, 1966*b*, 1968). This species has chin, inguinal, and anal glands. The chin gland is rubbed on objects ("chinning"), and the anal glands add secretion to the fecal pellets. Both serve in territorial marking (Mykytowycz, 1962, 1968). The inguinal glands function in individual recognition and anal-gland secretion in giving "confidence" during encounters and fights (Mykytowycz and Nay, 1964). The European rabbit, like other leporids, also has a "pigment gland" on the tip of the snout (Schaffer, 1940). Hares (*Lepus*) also have inguinal, chin, and pigment glands. Swamp rabbits (*Sylvilagus aquaticus*) also rub the chin on objects (Marsden and Holler, 1964).

Rodents

Many rodents lack pronounced skin glands and use urine for communication, although some possess enormously developed scent glands, such as beavers (*Castor*) and the muskrat (*Ondatra zibethicus*). For priming phenomena and related chemical communication in *Mus* and *Peromyscus* see Bronson (this volume) and Vandenbergh (this volume).

A study of 79 species from 47 genera and 10 families showed that sebaceous glands in the oral lips and angle are ubiquitous in rodents, although sudoriferous glands in these body areas were not found in bathyergomorphs, hystricomorphs, *Aplodontia rufa*, Heteromyidae, Gerbillinae, Muridae, Zapodidae, and Dipodidae (Quay 1965*b*).

Anatomy, growth, metabolism, and functions of the preputial glands in myomorph rodents and the chemistry of its secretion have been reviewed by Clevedon Brown and Williams (1972).

Sciuridae.—The marmot (*Marmota marmota*) scent-marks with cheek glands; it is done more often by the male than the female. The anal-gland secretion serves in defense (Koenig, 1957).

Gray squirrels (*Sciurus carolinensis*) produce scent marks by gnawing bark and squirting urine on the spot, which results in a "dark, strong-smelling stain" (Taylor, 1968). In the oil-palm squirrel (*Protoxerus stangeri*), both sexes mark with the corners of the mouth (sebaceous oral glands), rub the anal region, and mark with urine (L. Emmons, in litt.).

Cynomys (prairie dogs) and *Spermophilus* (ground squirrels) possess three anal glands in the shape of papillae (Schaffer, 1940). These papillae are extended in agonistic encounters in Uinta ground squirrels (*Spermophilus armatus*) (Balph and Stokes, 1963). Kivett et al. (1976) compared oral, dorsal, and anal glands and associated marking behaviors in six species of ground squirrels (*Spermophilus*) and found some correlations between glandular endowment, frequency of marking, and mutual sniffing and sociality. Steiner (1974) also described marking behaviors of ground squirrels.

Heteromyidae.—Quay (1965*a*) described caudal glands and assumed that they are used in marking in the "perineal drag", described by Eisenberg (1963). *Perognathus* have preputial glands, while *Heteromys* are said to lack them (Tullberg, *in* Schaffer, 1940).

Castoridae.—Beavers (*Castor fiber* and *C. canadensis*) have enormous "castor" glands (up to 12 cm long) and also a pair of anal ("oil") glands, the secretion of which has been studied by perfume chemists (Lederer, 1950; Maurer and Ohloff, 1976). Beavers scent-mark with a touch of the cloaca on mud piles which they build themselves ("scent mounds") (Hogdon, 1978; Müller-Schwarze and Heckman, 1980; Svendsen, 1980). Resident beavers obliterate strange castor marks near their lodge (Müller-Schwarze et al., in press).

Cricetidae.—*Sigmodon* have preputial glands (Tullberg, *in* Schaffer, 1940). A mid-ventral sebaceous gland occurs in male woodrats (*Neotoma*). The odor of ventral gland secretion of males attracts females in the southern plains woodrat, *Neotoma micropus* (August, 1978). In the desert woodrat (*N. lepida*), ventral rubbing is under androgen control and "rolling" is under estrogen control (Fleming and Tambasso, 1980).

The black-bellied hamster (*Cricetus cricetus*) has flank, anal, preputial, and umbilical glands. There are also sebaceous glands on the snout and "sweat" glands on the foot pads (Vorontsov and

Gurtovoi, 1959; Vrtis, 1932). Male golden hamsters (*Mesocricetus auratus*) rub their flank glands on objects, particularly after winning a fight, and females are attracted to the flank gland. Females produce an odor in their clitoris glands that attracts conspecifics (Dieterlen, 1959). The vaginal secretion has attractant and excitant functions (Johnston, 1972). The attractant has been identified as dimethyl disulfide (Singer et al., 1976). The activity of the Harderian gland of female golden hamsters changes with the estrous cycle—gland weight and porphyrin levels are highest on the day of estrous (Payne et al., 1977).

Arvicola and *Microtus* possess a pair of lateral glands (Stoddart, 1970). In scent-marking in *A. terrestris,* the hindfoot scratches first the lateral gland, then the hindfeet drum the ground in alternating movements while the tail beats horizontally. These three behavioral elements are also known in *Clethrionomys* and in many species of the Muridae (Frank, 1956). *Microtus* have very large anal glands (Vrtis, *in* Schaffer, 1940). In *M. xanthognathus,* males and dominant individuals mark most often (Wolff and Johnson, 1979).

Both sexes of the muskrat (*Ondatra zibethicus*) have preputial glands which are larger in the male and most active during the main mating season in spring. Marking with this musk gland and with fecal piles (at a high population density) has been described by Akkermann (1975). Stevens and Erickson (1942) identified cyclopentadecanol, cycloheptadecanol, and the corresponding ketones that are odoriferous. Van Dorp et al. (1973) found five more macrocyclic ketones.

Many microtine rodents have paired flank glands, and some have paired hip glands (Quay, 1968). Quay suggests that the flank glands may represent the connection to the Old World Cricetinae.

The Mongolian gerbil's (*Meriones unguiculatus*) ventral (sebaceous) gland has been studied extensively. The gland is rubbed over the substrate (Baran and Glickman, 1970; Thiessen et al., 1970) and is under androgen control (Glenn and Gray, 1965; Thiessen et al., 1968). Phenylacetic acid has been identified as the constituent of the ventral gland secretion most important for attraction and investigation (Thiessen et al., 1974). Harderian glands, located behind the eyeball, produce an attractive secretion during grooming (Thiessen et al., 1976). *Peromyscus maniculatus* also possesses a midventral gland (Richmond and Roslund, 1952).

Male lemmings (*Lemmus lemmus*) have a large preputial gland

and a sebaceous juxtauricular gland (Schaffer, 1940). Lemmings also maintain latrines at their shelter entrances (Koch et al., 1969) and often use one of their subterranean chambers exclusively for their droppings. *Lemmus* also possess a single rump gland, and *Dicrostonyx* a single caudal gland (Quay, 1968).

Muridae.—In this large family (over 100 genera) are found preputial, anal, and foot-pad glands. In chemical communication, extensive use is made of urine. Only a few strictly gland-related studies in laboratory rats (*Rattus norvegicus*) and house mice (*Mus musculus*) are mentioned here. Preputial glands in these two species were described by Schaffer (1933), and further investigated by Beaver (1960), among others, and the foot-pad glands and the tracks they produce by Kemper (1950) and Ortmann (1956). Rats also mark with fecal pellets, which are particularly noticeable on traps (Steiniger, 1950). The preputial secretion of male mice is attractive to females (Bronson and Caroom, 1971). Female mice reared by intact or preputialectomized mothers later prefer the type of female (and male) that they were reared by (Hayashi, 1979). The attractiveness for conspecifics of chemical extracts from the preputial glands of laboratory rats has been examined by Gawienowski (1977).

A mid-ventral sebaceous gland in *Rattus exulans* has been described by Quay and Tomich (1963). Rudd (1966) found it in 10 of 12 species of *Rattus*. The typical body odor of mice (*Mus musculus*) is attributed to the preputial glands of the male (Lane-Petter, 1967). Preputialectomy did not directly influence individual aggressiveness but rendered the animal more sensitive to preputial-gland secretions. Preputial secretion contains an attack-eliciting stimulus (Mugford and Nowell, 1971). The size of the preputial gland is heavier in dominant male mice (Davis and Christian, 1957), and its weight is inversely related to population density (Christian, 1955).

An alarm odor emanates from the urine of *Mus,* and females are more sensitive to it than males (Müller-Velten, 1966).

Cricetomys gambianus, the African giant rat, marks with its cheeks and deposits feces in a handstand posture (Ewer, 1967).

Dipodidae.—Grosz (*in* Schaffer, 1940) described preputial glands in a species of jerboa (*Jaculus jaculus*). Sandbathing in *Salpingotus crassicauda* (three-toed dwarf jerboa) has been interpreted as marking behavior (Rogovin, 1979).

Gliridae.—Dormice (*Glis glis*) have anal, preputial, lip, and foot-pad glands (Schaffer, 1940). They mark with anal glands (Koenig, 1960).

Hystricidae.—*Hystrix cristata* has large anal glands in both sexes, and "sweat" glands on the foot pads (Schaffer, 1940).

Erethizontidae.—The North American porcupine (*Erethizon dorsatum*) lacks anal glands, but perineal glands are present (Tullberg, *in* Schaffer, 1940). Urine is used extensively in social and sexual behavior. During courtship, the male sprays the female with urine from his erected penis over distances of up to 2 m (Shadle et al., 1946).

Caviidae.—Both sexes of the domestic guinea pig (*Cavia porcellus*) have a (sebaceous) supracaudal gland, large anal glands, perineal glands, and sweat glands on foot pads (Schaffer, 1940). Guinea pigs mark with feces and urine and males spray urine at females during courtship, as do females in defense against males (Frühling, 1955). The supracaudal gland and genital region (perineal glands) are rubbed on the ground. Supracaudal and perineal glands are better developed in the male. The winner marks during and after a fight (Kunkel and Kunkel, 1964).

Maras (*Dolichotis patagonum*) have a supraanal (subcaudal) gland (Pocock, *in* Schaffer, 1940). They squirt urine at females during courtship, and in competitive situations (Kirchshofer, 1960).

Cavies (*Cavea aperea*) distinguish perineal, supracaudal, and urine odors (Martin and Beauchamp, 1982).

Hydrochoeridae.—A large (4 cm thick) sebaceous gland (larger in males) is found on the nose of the capybara (*Hydrochoerus hydrochoeris*). Males mark vegetation extensively with their nose gland (Ojasti, pers. comm.). There are also perineal glands (Schaffer, 1940).

Dasyproctidae.—Agoutis (*Dasyprocta aguti*) and acouchis (*Myoprocta pratti*) have anal glands (Schaffer, 1940) which they drag over branches or wood. They also spray each other with urine (Kleiman, 1972).

Chinchillidae.—*Chinchilla laniger* has a single anal gland (Tullberg, *in* Schaffer, 1940).

Capromyidae.—In the coypu, *Myocastor coypu,* a single anal gland is found in both sexes (Wagner, *in* Schaffer, 1940).

Octodontidae.—The degu (*Octodon degus*) has a single ventral anal gland, which Schaffer (1940) prefers to call a subanal gland.

Ctenomyidae.—Tullberg (*in* Schaffer, 1940) found a single, ventral anal gland in the tuco-tuco (*Ctenomys*).

Ctenodactylidae.—The gundi (*Ctenodactylus gundi*) of Northwest Africa has a pair of anal glands (Tullberg, *in* Schaffer, 1940).

Carnivora

Canidae.—Among the canids, the most prominent glands are anal sacs, first mentioned by Tyson in 1683 (Schaffer, 1940), and caudal glands.

The red fox (*Vulpes vulpes*) has a "violet gland" (after its odor) on the dorsal side of the tail. It consists of several types of sebaceous glands and is also found in other canids (*Vulpes* and *Canis*), with the possible exception of *Lycaon pictus* (Schaffer, 1940). The violet gland and circumoral glands in the red fox experience cyclical changes (Friedrich, 1959). Anal gland secretion from estrous bitches applied to other dogs stimulates sexual interest in males (Donovan, 1969). The secretory epithelium of the anal sac of the red fox, *Vulpes vulpes,* is most active in March–April and wanes in August–September and November (Spannhof, 1969).

The anal gland secretion of the red fox contains at least 12 volatile components, among them trimethylamine, and several fatty acids (Albone and Fox, 1971). In the vaginal secretion of the domestic dog bitch (*Canis familiaris*) in heat, Goodwin et al. (1979) identified para-methylhydroxybenzoate as stimulating sexual behavior in the male.

In wolves (*Canis lupus*) and most other canids, urine-marking provides the most important chemical signal (Kleiman, 1966; Peters and Mech, 1975). In the hunting dog (*Lycaon pictus*), both males and females mark with urine when a female is in heat (van Lawick and van Lawick Goodall, 1971). Raccoon dogs (*Nyctereutes procyonoides*), arctic foxes (*Alopex lagopus*), and golden jackals (*Canis aureus*) urine-mark, particularly during the breeding season (Heimburger, 1959).

Inguinal presentation in wolf, coyote (*Canis latrans*), fox, and gray fox (*Urocyon cinereoargenteus*) is a submissive or appeasement gesture (Fox, 1968).

Ursidae.—*Ursus arctos* (the brown bear) has anal glands and tubular glands in the soles (Schaffer, 1940). Bears rub their shoulders on trees after wallowing in urine. Males rub most often during the rut in May; females mark most in July when they are shedding. Seven of nine cubs were startled by rubbing marks of adults (Tschanz et al., 1970). A male sloth bear (*Melursus ursinus*) in Nepal rubbed several trees with its shoulders, leaving a caramel-like odor. At each tree it deposited feces (Müller-Schwarze, pers. observ.). In the giant panda (*Ailuropoda melanoleuca*), both sexes mark with an anogenital rub and with urine (Kleiman and Collins, 1972).

Procyonidae.—The raccoon (*Procyon lotor*) has anal glands which it rubs on the ground for marking (pers. observ.). Coatis (*Nasua nasua*) have anal glands. Males mark with urine when they threaten each other. During the breeding season, the male sniffs the female at the perineal area (Kaufmann, 1962). In kinkajous (*Potos flavus*) both sexes have mandibular, throat, and abdominal glands. Captive specimens marked environmental features with the mandibular gland and rubbed the throat and abdominal glands on familiar persons. The male licks all three glands of the female (Poglayen-Neuwall, 1966).

Anal-marking occurs in the lesser panda (*Ailurus fulgens*), coati (*Nasua nasua*), kinkajou (*Potos flavus*), ring-tailed cat (*Bassariscus astutus*) (Fiedler, 1957), and in olingos (*Bassaricyon*) (Poglayen-Neuwall and Poglayen-Neuwall, 1966).

Mustelidae.—Except .or the sea otter (*Enhydra*), mustelids have large anal glands that are used for marking and defense. The anal gland of the skunk (*Mephitis mephitis*) is used in defense, as in the Eurasian badger (*Meles meles*) (Hodgson, 1952), the zorilla (*Ictonyx striatus*), polecats (*Putorius*), and others. Marking with the anal glands occurs in wolverines (*Gulo gulo*) (Krott, 1961), pine marten (*Martes martes*), and Eurasian badger (*Meles meles*) (Östborn, 1976), and others.

Viverridae.—The civet of *Civettictis, Viverra,* and *Viverricula* has been one of the most important animal perfumes. It is secreted in perineal glands and owes its odor to civettone, a macrocyclic ketone with 17 carbon atoms (Ruzicka, 1926*a*). There are also anal glands.

Marking has been described for the Indian gray mongoose (*Herpestes edwardsi*) (Fiedler, 1957), slender-tailed meerkat (*Suricata suricatta*) (Ewer 1963), the marsh mongoose (*Atilax paludinosus*) (Hediger, 1949), *Civettictis civetta* (Ewer and Wemmer, 1974), *Genetta, Viverra, Nandinia binotata,* and *Paradoxurus* (Wemmer, 1977). In *Genetta genetta* anogenital marking is correlated with aggression. Males use flank marks to distinguish pregnant from non-pregnant females, and anogenital marks of females to detect estrous (Roeder, 1980).

The African dwarf mongoose (*Helogale parvula*) marks with both cheek and anal glands. Cheek-gland marks convey threat, and anal-gland secretion identifies individuals and carries information about the age of the mark. Allomarking is important in social bonding, especially in juveniles (Rasa, 1973). In the wild, perineal scent marks of *Civettictis* are more frequent near "civetries" (defecation sites) (Randall, 1979).

In the binturong, *Arctictis binturong,* both sexes have a perineal gland which may be rubbed on branches while the animal hangs upside down. Vertical objects are marked in an "upright quadrupedal" position (Kleiman, 1974).

Hyaenidae.—*Hyaena* and *Crocuta* have a supraanal (subcaudal) gland (Daubenton, *in* Schaffer, 1940). The gland is rubbed over tufts of grass at territory boundaries and near dens.

Felidae.—Cats have anal and foot-pad glands, and in the domestic cat (*Felis catus*) there is a dorsal tail gland (Schaffer, 1940). Anal-gland secretion together with urine is sprayed for marking in lions (*Panthera leo*), tigers (*Panthera tigris*), leopards (*Panthera pardus*), and cheetahs (*Acinonyx jubatus*).

Pinnipedia

Skin glands are poorly developed in seals, walruses, and sea lions.

Otariidae.—Fur seals (*Arctocephalus australis*) have a "musky and sweetish" odor during the rut. Their breath is pungent to humans, and the threat lunges of bulls are accompanied by violent expiration. Mothers recognize their young by smell (Bonner, 1968). *A. australis* males smell like an excited civet (Hamilton, 1956).

Phocidae.—Olfactory recognition of pups by the mother occurs in seals such as harp seals (*Pagophilus groenlandicus*) (Terhune et al., 1979) and southern elephant seals (*Mirounga leonina*) (pers. observ.).

Tubulidentata

The aardvark (*Orycteropus afer*) has large preputial glands in both sexes, but no anal glands (Pocock, *in* Schaffer, 1940).

Proboscidea

Both species of elephants (*Loxodonta africana* and *Elephas maximus*) have temporal glands between eye and ear. They consist of modified and complicated sweat glands (Eggeling, 1901), undergo drastic activity changes with the seasons, and are most active at mating time. Fighting elephants touch their own temporal gland, and a dominant one will touch the gland of others (Kühme, 1961). The temporal gland of male *Elephas* secretes during "musth," a period of aggressive and sexual arousal that may last from 2 weeks to 5 months. Only males show "musth," first between the ages of 10 and 20 years; "musth" was observed in 75% of 26 to 30 year-olds (Jainudeen et al., 1972), and is thought to be a remnant of a seasonal rut. Estes and Buss (1976) studied microanatomy and development of the gland. Stress and individual recognition were examined by Buss et al. (1976).

Hyracoidea

Hyraxes have a dorsal gland that is delineated on the fur by light-colored hair. Schaffer (1940) attributes a social attraction function to the gland. There are also "sweat" glands in the foot pads (Dobson, *in* Schaffer, 1940). Localized defecation is probably a form of scent-marking (Coe, 1962).

Sirenia

Dugongs and manatees have only simple scattered sebaceous glands (Leydig, *in* Schaffer, 1940).

Perissodactyla

Equidae.—The horse (*Equus caballus*) has been thoroughly searched for skin glands. Schaffer (1940) mentions circumanal, perineal, preputial, axillary, eyelid, circumoral, and foot-pad glands. Foot-pad glands are tubular; all other glands have tubular and holocrine elements (Schaffer, 1940). The so-called "chestnuts"—thickened epidermis on the forelegs and often also on the hindlegs of horses, donkeys, and zebras—are assumed to be derived from skin glands (metatarsal, carpal, or interdigital) (Schaffer, 1940).

Zebras (*Equus zebra hartmannae* and *E. z. zebra*) maintain group dung piles, and males paw urination sites of females (Klingel, 1968).

Tapiridae.—Schaffer (1940) designates the glands in the muzzle of *Tapirus terrestris* as related to apocrine glands. There are also foot-pad glands on all four legs on the posterior side of the hoof (Schaffer, 1940).

Rhinocerotidae.—Species of *Rhinoceros* have a pedal gland on the posterior side of each foot, about 7 cm above the sole (Owen, *in* Schaffer, 1940; Cave, 1962). The odor of rhino tracks is probably due to several sources, as dung is pawed and kicked with the hind-feet, which scatters the feces (Goddard, 1967). In the great Indian rhinoceros, *Rhinoceros unicornis,* the dung piles are located at points on communally used main trails where secondary trails branch off that lead to individually defended sleeping and grazing sites (Ullrich, 1967).

Urine is sprayed horizontally on plants in two to five bursts. This is done more often by the bull. Such spray-urination is often preceded by scraping with the legs and wiping of the horn over low brush or the ground (Schenkel, 1966 for the black rhinoceros, *Diceros bicornis,* and Owen-Smith, 1971, for the white rhinoceros, *Ceratotherium simum*). Backhaus (1964) described the flight of a male white rhinoceros from a bunch of grass that had been sprayed before by another male.

Artiodactyla

Suidae.—In pigs, antorbital, preputial, salivary, digital, proctodeal, perineal, preputial, teat, ear, eyelid, snout, and carpal glands have been described (Schaffer, 1940). In the domestic pig, *Sus scrofa,* the submaxillary gland of the boar is the source of 3-α-hydroxy-5α-androst-16-ene which stimulates sexual behavior, specifically the mating stance of the sow (Melrose et al., 1971; Patterson, 1968; Perry et al., 1973). The steroid is commercially available as an aerosol under the trade name "Jeyes Boar Mate" and is used in swine artificial insemination programs in Europe.

The preputial gland has also been implicated as a source of "boar odor" (Dutt el al., 1959).

Tayassuidae.—Peccaries have a dorsal gland about 20 cm in front of the tail. It consists of apocrine and sebaceous glands (Epling, 1956). Between mounts, a male collared peccary (*Tayassu tajacu*) rubs the dorsal gland of the female with his nose. In other social encounters between members of the same herd, the lower jaw and cheek are rubbed on the dorsal gland of the partner. Musk secretion can be ejected as far as 30 cm. When a group is alarmed, "a wave of musk may be left behind" (Neal, 1959). In captivity peccaries mark the walls of their enclosure with their dorsal gland (Hediger, 1956).

Hippopotamidae.—Hippopotamus amphibius has circumoral and snout glands that produce a mucous, red secretion ("bloody sweat"). There are also foot-pad glands. Male hippos distribute feces and urine by a circular movement of the tail, causing rivals to withdraw. In the pygmy hippopotamus, *Choeropsis liberiensis,* the tail of the male is more muscular and has more bristles than that of the female (Hediger, 1949). *H. amphibius* deposits dung on trails, usually at conspicuous places (Verheyen, 1954).

Camelidae.—Both species of *Camelus* have a pair of occipital glands that are dominated by apocrine glands, but also contain sebaceous glands. There are also interdigital glands (Schaffer, 1940). The occipital glands have been further investigated by Mimran (1962) and Singh and Baradwaj (1978). During the mating season, these glands swell by three to four times, and the animals rub them against objects and their own backs (Gauthier-Pilters, 1960; Wemmer and Murtaugh, 1980). In addition, there is a tail-urine flapping and defecation display that applies urine to the posterior hump and tail (Wemmer and Murtaugh, 1980).

Vicugna vicugna and *Lama peruana* have metatarsal glands; *L. guanacoe* has two glandular areas on the metatarsus (Schaffer, 1940). Most olfactory communication in *V. vicugna* and *L. guanacoe* involves dung piles (Franklin, 1980; this volume).

Tragulidae.—Tragulus javanicus has an intermandibular gland in both sexes. It consists primarily of tubular glands (Schaffer, 1940).

Cervidae.—The musk deer (*Moschus moschiferus*) is best known for its musk from the large abdominal gland, which is the size "of a small orange" and can contain up to 120 g of secretion. Muscone, a macrocylic ketone, is the compound largely responsible for the typical odor (Ruzicka, 1926*b*). *M. moschiferus* also possesses supracaudal (in front of tail base), circumcaudal (covering the whole tail), circumanal, pregenital (in female), crural (on caudolateral side of thigh, in male), and sebaceous glands between the forehooves (Schaffer, 1940). Musk deer back up to a branch and rub their tail on it. This results in a shiny, smelling scent mark (Frädrich, 1966).

Muntjacs (*Muntiacus*) have frontal, antorbital, and interdigital glands, the latter particularly well developed on the hindfeet (Schaffer, 1940). For scent-marking the frontal glands are pushed against the ground, grass, or the rump of females.

Interdigital glands are commonplace among cervids.

Of the two main branches of the cervids, the Plesiometacarpalia (with *Cervus* and *Dama*) have (in various combinations and degrees of development) antorbital, frontal, snout, infracaudal, circumcaudal, metatarsal, and interdigital glands. The Telemetacarpalia (with *Odocoileus, Rangifer, Alces, Capreolus*) have tarsal glands in addition. Sebaceous and apocrine glands in the velvet of the growing antler have also been reported (Lojda, 1955; Vacek, 1955).

In *Odocoileus hemionus columbianus,* the black-tailed deer, the forehead glands are used in marking, the tarsal glands (after accepting volatiles from the urine) function in social attraction and recognition, and the metatarsal glands emit an odor during alarm (Müller-Schwarze, 1971). The latter odor alerts conspecifics (Müller-Schwarze, 1980). The histology of scent glands in *Odocoileus hemionus* has been investigated by Quay and Müller-Schwarze (1970, 1971) and in *O. virginianus* by Quay (1959).

The large and behaviorally more important skin glands of *Rangifer tarandus* are the caudal and hindfoot interdigital glands (Müller-Schwarze et al., 1977*a,* 1978, and 1979). The histology of the glands in *R. tarandus* has been investigated by Quay (1955), Müller-Schwarze et al. (1977*a*), Källquist and Mossing (1977), and Mossing and Källquist (1981).

Hydropotes inermis is said to be the only cervid with inguinal glands (Walker, 1975). The male roe deer (*Capreolus capreolus*) has a large frontal (intercornual) gland whose activity changes with the seasons (Schumacher, 1936) and is used for scent marking on low-hanging twigs (Kurt, 1968).

Giraffidae.—The giraffe (*Giraffa camelopardalis*) and the okapi (*Okapia johnstoni*) are poorly endowed with skin glands. Schaffer (1940) mentions eyelid glands, consisting of sebaceous and apocrine elements in giraffes.

Antilocapridae.—The male pronghorn, *Antilocapra americana,* has a pair of subauricular glands and a mid-dorsal gland; both sexes possess ischiadic glands under the white rump patches and interdigital glands. The subauricular gland is rubbed on vegetation for territorial marking, and also "shown" to the female during courtship. The ischiadic gland emits an odor during alarm. The dorsal gland is activated when the buck is excited, and it is also licked by females during the mating season (Müller-Schwarze and Müller-Schwarze, 1972).

Bovidae.—The various species of cattle have lower lip, nasolabial, preputial, and circumanal glands (Schaffer, 1940). Much of the chemical communication in bovids utilizes urine and vaginal secretion (Donovan, 1967; Sambraus, 1971).

Duikers (*Cephalophus, Sylvicapra*) have antorbital, inguinal, and interdigital glands, and mark with antorbital glands (Rahm, 1960; Ralls, 1969).

Antelopes and gazelles have antorbital, inguinal, and interdigital glands. Many, such as the blackbuck (*Antilope cervicapra*) (Hediger, 1949), Thomson's gazelle (*Gazella thomsoni*) (Estes, 1967), *Gazella subgutturosa,* dik-diks (*Madoqua*), grysbok (*Raphicerus melanotis*), gerenuk (*Litocranius walleri*) (Gosling, 1981), oribi (*Ourebia ourebia*) (Gosling, 1972), and wildebeest (*Connochaetes taurinus*) (Fiedler, 1957) mark vegetation with their antorbital glands. Gerenuk and dibatag (*Ammodorcas clarkei*) males also rub their antorbital glands on females (Walther, 1966).

The behavioral role of the inguinal glands has remained obscure. Males of the Uganda kob (*Kobus kob thomasi*) nuzzle the inguinal glands of the female after copulation (Buechner and Schloeth, 1965).

Some antelopes and gazelles have special glands such as the dorsal gland in the springbok, *Antidorcas marsupialis,* and the metatarsal gland under the black hair tuft on the hindleg of the impala, *Aepyceros melampus,* and the large preputial gland of *Antilope gutturosa* (Schaffer, 1940).

Wildebeest (*Connochaetes taurinus*) rub their antorbital glands on the ground, paw these spots, and urinate and defecate on the site (Schaller, 1973).

Among the goats, the domestic goat (*Capra*) is best investigated. Here we find a sebaceous postcornual gland in both sexes, a pair of subcaudal glands, less pronounced preputial glands, and planonasal glands.

In the chamois, *Rupicapra rupicapra,* the most elaborate gland is the occipital gland (Schaffer, 1940). During the rut, this gland is rubbed on grass, conifers, rocks, and fence posts (Krämer, 1969). Interdigital glands are poorly developed. Mountain goats (*Oreamnos americanus*) mark tufts of grass with their occipital gland (Geist, 1965), as do chamois. Sewor (*Capricornis crispus*) and goral (*Naemorhedus goral*) lack occipital (postcornual) glands.

The musk ox, *Ovibos moschatus,* has antorbital glands and moderately-developed interdigital glands (Schaffer, 1940). The antorbital gland is rubbed on the foreleg during encounters with opponents or in danger (Tener, 1965). Cinnamaldehyde has been identified in the preputial secretion of the musk ox (Tener, 1965).

The tahr (*Hemitragus jemlahicus*) has subcaudal glands and traces of interdigital glands (Schaffer, 1940).

The domestic sheep, *Ovis aries*, has antorbital, planonasal, nasal, infracaudal, circumanal, inguinal, preputial, and interdigital glands. Mufflon females (*Ovis musimon*) recognize their young by smelling the anal region of the lamb. Infracaudal secretion of the own lamb applied to the rump of a strange lamb renders the latter acceptable (Tschanz, 1962).

Functions

Chemical signals, like stimuli in other sensory modalities, modulate physiology (see Vandenbergh, this volume, and Bronson, this volume), behavior, and ultimately regulate spacing between individuals and populations and thus contribute to a species' adjustments to its resources and general environment.

There are many possible functions of odors from scent glands. Odors can be produced directly by the glands or indirectly by urine constituents, dietary products, or bacterial fermentation. As deduced from the responses of conspecifics, responses to social odors vary according to circumstances: releasing responses; enhancing or inhibiting behavior; and priming the endocrine system (see Bronson, this volume; Vandenbergh, this volume). Functions may also be inferred from the odor donor's behavior and its context.

Just as vocalizations or visual signals, chemical signals can convey information in the following functional contexts: maternal (mutual attraction and attachment between mother and offspring); agonistic (threat, maintaining and increasing interindividual distance, dominance and subordination, territorial defense); social (group or family coherence by means of odor trails, group or nest odors); recognition of individuals and their physiological status or short-term mood; recognition of sex, group, or species; sexual (attraction and stimulation); and alarm.

Within each behavioral context there are numerous possible specific effects that an odor may have. For instance, in sexual behavior one can potentially discriminate between self-identification, mate attraction, stimulation of certain steps of sexual behavior (such as courtship, mounting, intromission, inhibition of aggression, or assuming the mating stance by the female as has been demonstrated by the response to 3α-hydroxy-5α-androst-16-ene in domestic pigs [Hafez and Signoret, 1969]). In the golden hamster, sexual approach by the male is stimulated by dimethyldisulfide; mounting is affected by a compound of higher molecular weight (Singer et al., 1980).

Similarly, in alarm behavior emitted odors may potentially express a physiological state of stress or alarm, alert conspecifics, cause conspecifics to retreat, flee, or disperse, or identify group members, notably mother and offspring, in the confusion of a retreat or escape (Müller-Schwarze, 1980). Moreover, an alarm odor may possibly function as an interspecific deterrent or even a decoy for a predator.

There are four major pathways for theriogenic (=of animal origin) odors that require different gland and secretion types as well as behaviors. 1) They can be released directly *into the air,* as in the metatarsal gland of black-tailed deer (*Odocoileus hemionus columbianus*) or the ischiadic gland of pronghorns (*Antilocapra americana*). Conspecifics may receive such odors over long distances (at least several meters) as in the above mentioned cases, or over short distances (centimeter range) during social interactions. In this pathway apocrine glands usually dominate, with sebaceous glands constituting a minor component. 2) A secretion may be deposited on the substrate, as in the numerous examples of scent-marking. A durable secretion usually originates in sebaceous glands. 3) Odoriferous secretions can be distributed to other parts of the body as in the stink display of *Lemur catta,* or in the rubbing of interdigital secretion of the hindfoot on the growing antler in reindeer (*Rangifer tarandus*) (Espmark, 1964; Müller-Schwarze et al., 1979). Urine is often applied to body hair on head, neck, or abdomen, as in goats (Coblentz, 1976) and elk (*Cervus canadensis*) (McCullough, 1969), on the back, as in camels (Wemmer and Murtaugh, 1980), and on the hindlegs, as in black-tailed deer (Müller-Schwarze, 1971). 4) Secretions or excretions can be applied to other members of the group. Females are marked by males in grysbok, *Raphiceros melanotis* (pers. observ.) and *Cricetomys gambianus* (Ewer, 1967). Smell-exchange and thus odor-sharing are achieved by the frontal sternal rub in the sugar glider, *Petaurus breviceps* (Schultze-Westrum, 1965), and occur in muskrats (*Ondatra zibethicus*), *Suricata suricatta* (Ewer, 1968a), Norway rats (*Rattus norvegicus*) (during crawling under and over, Steiniger, 1950), and bovids, as in the facial rub in mountain sheep (*Ovis canadensis*) (Geist, 1971). Male blue sheep (*Pseudois nayura*) rub the face on the rump of other males (Schaller, 1977). Peccaries rub the throat and shoulders on the dorsal gland of a conspecific (Neal, 1959).

The young are marked with the chin gland in rabbits (*Oryctolagus cunniculus*) (Mykytowycz and Dudzinski, 1972), the sternal gland in *Tupaia belangeri* (Martin, 1968), and with cheek glands in *Helogale undulata* (Zannier, 1965). Defeated opponents are marked in rats and mice, and also in hamsters (Dieterlen, 1959).

A possible mutual benefit of such allomarking is that the marker "secures" the social situation for an individual ("makes it safe"), analogous to marking an unfamiliar environment. The marked animal in turn is safe from attack; it "speaks the same odor dialect" ("wears the same uniform") as the potentially dangerous individual that marked it.

Individual Uses of Odors

The main effect may occur in the signalling individual itself. The uses of one's own odor can be familiarization with a strange area or object(s) (Kleiman, 1966), or reassurance (Johnson, 1973); the animal may "feel better" or "secure" (Ewer, 1968*a*). Gaining "confidence" in encounters with strangers as provided by the rabbit's chin gland secretion (Mykytowycz et al., 1976), or "increasing confidence and reducing anxiety" (Svendsen, 1980, for beavers), or "self-advertisement" by means of odors (Jolly, 1966), are feedbacks on the odor emitter in social encounters. They form a functional transition to truly social effects of odors. Marking unfamiliar objects or areas can also mean "checking off" ("settled," "needs no more attention"). This applies also to familiar situations, as for example the "bookkeeping system" in the red fox, *Vulpes vulpes* (Henry, 1977). Buried food items are repeatedly marked with urine and accumulate urine odor as the food becomes depleted. Strongly smelling sites are no longer investigated for food.

Related to the familiarization function is *orientation* in space. In his classical paper on territorial marking, Hediger (1949) suggested additional or alternative orientation functions of scent marks, and later studies have supplied strong evidence for this function in the black rhinoceros (Goddard, 1967), Thomson's gazelle (Walther,

1978), vicuña (Franklin, 1980), and beaver (Müller-Schwarze and Heckman, 1980).

Social Uses of Odors

Odors may signal the presence and identity of conspecifics, and thus provide specific information on species, sex, age, individual identity, and reproductive status of the odor donor. A specific response is not necessarily elicited. The instructions for particular action are not necessarily coded in the chemical signal, but rather in the relationship of the state of the receiver to the particular state of the donor for which the scent mark stands as *pars pro toto*. Thus, a scent mark often is not the animal equivalent of a sign saying "No trespassing," or "One-way street" (that is, instructions for prescribed actions), but rather it represents a sign announcing a community of a certain population size, information on what exists or has occurred in a given locale ("To whom it may concern," Walther, 1979). Such is the case for "bear trees" with rub marks (Tschanz et al., 1970), the lynx's (*Lynx lynx*) urine and feces at conspicuous spots, or the castor scent mounds of beavers. Individual recognition by odor has been demonstrated in virtually all species that have been investigated. The first example was the sugar glider, *Petaurus breviceps* (Schultze-Westrum, 1965). Selective habituation experiments have shown that *Lemur fulvus* can discriminate between odors (mixtures of urine, feces and glandular secretions) of conspecifics (Harrington, 1974), and *Lemur catta* discriminates the brachial gland secretions of different conspecifics (Mertl, 1975).

Given a particular recipient in a given physiological state, chemical signals from conspecifics can assume the following functions. 1) Threat (distance-increasing deterrent), meaning "proceed with caution," as in the roe deer's forehead marks (Kurt, 1968), or even spatial or social exclusion of the addressee, including signalling of dominance and territorial exclusiveness. 2) Attraction of conspecifics, especially of the opposite sex, and of young. The same effect can also be attained by the absence of a mark, a "to let" sign (Ewer, 1968*a*). 3) Alarm—alerting, retreat, flight, avoidance, group cohesion and individual recognition during flight (Müller-Schwarze, 1980). Inhibition of ongoing behavior can be the first step. Metatarsal secretion of conspecifics inhibits feeding in black-tailed deer (Müller-Schwarze, 1971).

Group odors may involve more than two animals. Communal use of scent piles has been described for the hippopotamus (Verheyen, 1954), and laboratory rats maintain odor trails that serve as communal guide marks for colony members and strangers (Barnett, 1963).

Species specificity.—In simultaneous choice tests, males prefer the odor of females of their own species in *Peromyscus* (Moore, 1965) and of their own subspecies in *Clethrionomys glareolus* (Godfrey, 1958) and in mule and black-tailed deer, *Odocoileus hemionus* (Müller-Schwarze and Müller-Schwarze, 1975). Wild and domestic male guinea pigs (*Cavia aperea* and *C. porcellus*) prefer the perineal secretion from males of the same species (Wellington et al., 1979).

Ontogeny of responses to social odors.—Odors from sources other than specialized skin glands, urine, or feces are saliva and amniotic fluid, both of which stimulate attachment to the mother's nipples in newborn laboratory rats (Teicher and Blass, 1977). Newborn mammals respond to odors of the mother, notably that of her nipples (Hofer et al., 1976; Teicher and Blass, 1977), and to odors from the nest (Rosenblatt et al., 1969). The sniffing response of male golden hamsters to vaginal secretions appears at puberty (40 to 43 days of age), and testosterone treatment brings about this response in males and females at 25 days of age (Johnston and Coplin, 1979).

Early experience of social odors determines later responses (Carter and Marr, 1970; Mainardi et al., 1965; Müller-Schwarze and Müller-Schwarze, 1971). Change of function may occur during individual development. Young European rabbits are attracted to anal gland secretion of adult rabbits; the same secretion is used by adults in aggressive and territorial behavior (Mykytowycz and Ward, 1971). Black-tailed deer fawns rub-urinate to attract the mother; the same behavior in mature bucks functions as a distance-maintaining (or distance-increasing) threat (Müller-Schwarze, 1971).

Chemistry of Mammalian Scents

There are seven species of mammals for which a pheromone or pheromone component has been identified. They are listed in Table 1.

TABLE 1
CHEMICALLY IDENTIFIED PHEROMONE COMPOUNDS IN MAMMALS

Species	Sex	Origin (secretion)	Chemical compound	Function	Reference
Domestic boar (*Sus scrofa*)	♂	Submaxillary gland	3α-hydroxy-5α-androst-16-ene	Stimulates female's sexual behavior	Patterson (1968)
Black-tailed deer (*Odocoileus hemionus columbianus*)	♂, ♀	Urine, in tarsal scent	*Cis*-4-hydroxy dodec-6-enoic acid lactone	Attracts conspecifics, especially females; part of recognition odor	Brownlee et al. (1968)
Rhesus monkey (*Macaca mulatta*)	♀	Vaginal secretion	Five fatty acids constitute "copulin": acetic, propionic, isobutyric, n-butyric isocaproic acid	Stimulates sexual behavior of male	Michael et al. (1971)
Pronghorn (*Antilocapra americana*)	♂	Subauricular gland	Isovaleric acid	Releases scent marking; attracts females	Müller-Schwarze et al. (1974)
Mongolian gerbil (*Meriones unguiculatus*)	♂, ♀	Ventral gland	Phenylacetic acid	Stimulates investigation and marking	Thiessen et al. (1974)
Golden hamster (*Mesocricetus auratus*)	♀	Vaginal secretion	Dimethyldisulfide	Stimulates sexual approach by male	Singer et al. (1976)
Dog (*Canis familiaris*)	♀	Vaginal secretion	Para-methyl hydroxybenzoate	Stimulates mounting in males	Goodwin et al. (1979)

Early interest in "animal perfumes" resulted in considerable chemical knowledge of glandular secretions. Because no biological function is known in these cases, they are not reviewed in detail here. Rather, a few milestones of that line of work follow.

Aldrich (1896) found a higher mercaptan (most likely iso-amyl mercaptan) to be the main component of the anal sac secretion of the skunk (*Mephitis mephitis*) and Beckmann (1896) found n-butylmercaptane and sulfides in the anal glands of the Phillipine badger (*Mydans marchei*). Ruzicka (1926*a*, 1926*b*, 1926*c*) found macrocyclic ketones in the preputial gland of the musk deer (*Moschus moschiferus*) and the anogenital glands of the civet cat (*Viverra zibethica*) and named them "muscone" and "civetone," respectively. Lederer (1950) identified 45 components of the "castor" secretion of the European beaver (*Castor fiber*).

The composition of glandular secretions and signal-carrying urine may vary with the diet. This has behavioral consequences for conspecifics. Pre- and post-weanling Mongolian gerbils (*Meriones unguiculatus*) respond more to odors from adult gerbils that are on the same diet as the pup's parents than to those on a different diet (Skeen and Thiessen, 1977). The production of phenylacetic acid in the ventral gland of the gerbil may be specific to Purina laboratory diet (Thiessen, 1977).

Male domestic guinea pigs (*Cavia porcellus*) discriminate urine from donors on different commercial diets (Beauchamp, 1976). Other diet-dependent social odors have been described from laboratory rat milk (Galef and Clarke, 1972) and rat fecal material (Leon, 1974).

Design Features of Mammalian Scent Production and Release Apparatus

Anatomy.—Glands specialized for release of odor into the air have primarily apocrine elements. Examples are the human axillary organ, the metatarsal gland of the black-tailed deer (*Odocoileus hemionus columbianus*), the ischiadic gland of the pronghorn (*Antilocapra americana*), and the dorsal gland of the springbok (*Antidorcas marsupialis*).

Glands for scent-marking are predominantly or exclusively sebaceous (holocrine) glands. Examples are the ventral gland of the Mongolian gerbil (*Meriones unguiculatus*), and the subauricular gland of the pronghorn.

Accessory structures may aid in the secretion's storage, social presentation, or application to features of the environment. The black-tailed deer's tarsal organ not only has a hair tuft resembling a paint-brush which can be opened and closed with arrector pili muscles, but each stiff bristle-like hair in the center of the tuft is covered with large chambers formed by the cuticular scales. In addition, the scales have surface sculpture that retains sebum very well (Müller-Schwarze et al., 1977*b*). The hair on the ventral gland of the Mongolian gerbil is C-shaped in cross-section; it is a "spatula" for holding sebum and applying it to objects during ventral rubbing (Müller-Schwarze et al., 1977*b*).

The regulation of emitted odor intensity employs many different, and often complementary, mechanisms. Such control of odor release can be hormonal, neural, postural, behavioral, or temporal.

Hormonal control of odor release.—Glands are often present only, or at least are larger, in one sex (usually the male). There are seasonal variations in gland size and odor production, as in the subauricular gland of the pronghorn (Moy, 1970) and the antorbital gland of reindeer, *Rangifer tarandus* (Mossing and Källquist, 1981). Castration diminishes or eliminates odor production and scent-marking, and testosterone stimulates these functions. Apocrine glands are generally stimulated by testosterone, and are inhibited by estrogen. Progesterone has no observable effect. The holocrine (sebaceous) glands are also stimulated by testosterone, and progresterone has similar, but weaker effects, whereas estrogen inhibits sebum production (Ebling, 1977). Adrenaline can stimulate odor discharge, as in the metatarsal gland in black-tailed deer (Müller-Schwarze, unpubl.).

Neural mechanisms.—Muscles can effect the opening of a gland sphincter or pouch, the discharge of secretion from glandular tissue, or the erection of hair, thus draining the pilosebaceous canals. Spreading of hair tufts also enlarges the odoriferous surface.

Neural control of scent-marking.—In the Mongolian gerbil, *Meriones unguiculatus,* testosterone implants in the preoptic area reinstate ventral-gland scent-marking in castrated males, whereas cholesterol, implanted at the same site, has no effect. Testosterone implants in cortex, caudate nucleus, amygdala, hippocampus, reticular formation, and septum also have no effect on scent-marking. The anterior aspect of the medial preoptic area is most androgen-sensitive for stimulating scent-marking in male Mongolian gerbils (Yahr, 1977). Ovariectomized female Mongolian gerbils show scent-marking behavior when estradiol benzoate is implanted in the anterior hypothalamus, preoptic area, and septum, but implants in hippocampus, amygdala, thalamus, or olfactory nucleus are not effective (Owen et al., 1974).

Postural modulation permits "unfurling the odor flag" or, conversely, covering odor-carrying surfaces. A young deer fawn, for instance, rests so that its body covers all possible odor sources (however weak)—muzzle, anogenital region, hooves with interdigital glands. Raising the tail and spreading tail hair (and often wagging the tail) increases odor release into the environment, as in bovids, cervids, and lagomorphs; tucking the tail between the legs achieves the opposite. In caribou, *Rangifer tarandus*, a large gland covers the ventral half of the tail and can be exposed to varying degrees. In canids with supracaudal glands, a gland (anal) can be covered with another gland (supracaudal). Humans possess a "scent box" (Brody, 1975) in their axilla. Sweat and apocrine gland secretions are fermented by bacteria, and the product is kept on a hair tuft in the confined cavity of the axilla; it can be released to the outside by varying degrees of arm-lifting.

Behavioral modulation.—Elaborate behavior patterns can serve in optimizing odor release and its impact on conspecifics. *Lemur catta* rubs the brachial glands on the tail and then carries the tail upright with the tail hair spread (Jolly, 1966). When startled, caribou create deep footprints by rearing up and deposit interdigital gland secretions. The site of such an "excitation jump" is sometimes avoided by other caribou (Lent, 1966).

Temporal patterns.—Many species of mammals produce or discharge scents only at certain times of the year, which is usually, but not always, the breeding season. The odor of red fox (*Vulpes vulpes*) urine is strongest during the prebreeding period (Henry, 1980), and both the development of the subauricular gland (Moy, 1970) and scent-marking (Gilbert, 1973) in pronghorns reach a peak in July and August, prior to breeding which occurs in September. The muskrat's preputial gland and odor are better developed during breeding time. Heightened development of glands prior to and during breeding is characteristic of many other species.

On the other hand, *Castor canadensis* scent-marks most often during May and June when the young have been born and the 2-year-olds disperse (Müller-Schwarze and Heckman, 1980; Svendsen, 1980).

Some glands, such as the ischiadic gland of pronghorns (Moy, 1970) and the metatarsal gland of black-tailed deer (Müller-Schwarze, pers. observ.), are active throughout the year.

Location of scent marks.—Many animals mark as high as possible. Pronghorn males, taken from a mountain area in Idaho with sagebrush only 30 cm high, marked in captivity on fence posts with the head extended upward as far as possible (pers. observ.). Black-tailed deer rub their foreheads on thin, terminal over-hanging twigs to which they sometimes have to rear up (pers. observ.). Viverrids often anal-mark in "handstand" posture (Wemmer, 1971, 1977). Musk oxen, lacking tall vegetation or objects in their native tundra, rub their antorbital glands on fenceposts in captivity (pers. observ.).

Wolves, red foxes, and domestic dogs often mark at conspicuous objects that are especially obvious in featureless terrain, as in the tundra, where arctic foxes (*Alopex lagopus*) mark at rocks, on eskers, or on cast caribou antlers (pers. observ.).

Scent marks are not confined to the periphery of a territory or home range, although they may be denser there, as in the case of wolves (*Canis lupus*) (Peters and Mech, 1975). Many species place scent marks throughout their territories, sometimes at a higher density near more frequently used trails, dens, lodges, or sleeping sites (for example, European rabbits (Mykytowycz and Gambale, 1969); American beavers, *Castor canadensis* (Müller-Schwarze and Heckman, 1980); hamsters, *Cricetus cricetus* (Eibl-Eibesfeldt, 1953*b*); water voles, *Arvicola terrestris* (Stoddart, 1970); black rhinoceroses (Goddard, 1967), vicuñas (Franklin, 1980), and Thomson's gazelles (Walther, 1966). In the brown hyena (*Hyaena brunnea*), the scent marks are densest in the center of the group territory (Gorman, pers. comm.).

Odor release from scent marks.—Regnier and Goodwin (1977) found that phenylacetic acid, a behaviorally active component of the secretion of the ventral gland of the Mongolian gerbil (Thiessen et al., 1974) evaporates faster from smooth surfaces such as platinum or stainless steel than from wood or glass. Lack of sebum and high (100%) humidity facilitate evaporation. A fourth factor is polarity; non-polar compounds evaporate faster. However, polar odorants will compete with water vapor for polar sites on surfaces and sebum, resulting in increased evaporation of polar compounds in high humidity. These facts have important ecological implications. Use of scent marks would be favored in arid climates where scent marks last longer and smelling animals can still release molecules from the mark "on demand" by the moisture of their breath.

It has been noted above that scent marks permit communication with conspecifics (or oneself) long after the signal has been emitted, thus resembling human handwriting. Additional information can be coded in changes over time. Odor intensity may decrease by evaporation, and odor quality may change by differential evaporation of odor mixture constituents, or by chemical decomposition. An example of differential evaporation is the interdigital secretion of reindeer (*Rangifer tarandus*), which has a vinegar odor when fresh because it contains acetic acid (Brundin et al., 1978); it is altered after several minutes to a cheesy smell because of fatty acids (Müller-Schwarze et al., 1978). Male hamsters (*Cricetus cricetus*) still respond to flank marks of conspecifics after 45 days and to vaginal marks after 100 days. Flank and vaginal marks 15- to 30-min old can be distinguished from those 24-h old (Johnston and Schmidt, 1979). Urine of female domestic guinea pigs loses its attractiveness after 48 h (Beauchamp and Berüter, 1973).

Whether or not tracks of rodents, lagomorphs, and artiodactyls provide clues to carnivores as to their direction is still unknown. Contrary to claims, dogs have not been shown conclusively to be able to determine the direction of a human track (Brisbin, in litt.).

Ecological Correlates

Aquatic mammals such as cetaceans and pinnipeds usually lack skin glands (Gabe, 1967). Instead of skin glands, aquatic rodents have large internal scent glands, such as the castor glands in beavers and preputial glands in muskrats. Aquatic species in otherwise terrestrial taxa may provide fascinating insights. Examples are the otter shrew (*Potamogale velox*) and water shrew (*Neomys fodiens*) among the insectivores. Among mustelids, the sea otter is the only species without anal sacs.

Arboreal mammals tend to have sternal glands, such as the sugar glider (*Petaurus breviceps*) among the marsupials and tree shrews and the orangutan among the primates. Tunneling small mammals have ventral glands (shrews and gerbils) or lateral glands (shrews, voles, hamsters). Mid-sized animals using dens may have supracaudal glands, as in the red fox.

Generally, in crepuscular and nocturnal forms scent glands are better developed than in diurnal forms (for example, murids versus sciurids).

Scent glands appear to be emphasized in species inhabiting arid areas. Suids and deer provide two examples of differences between arid and humid climate forms. The collared peccary (*Tayassu tayaca*) has a dorsal gland, whereas the European wild boar (*Sus scrofa*) does not. The mule deer (*Odocoileus hemionus*) has a large metatarsal gland, whereas this gland is very small in white-tailed deer (*O. virginianus*).

No clear correlation seems to exist between the complexity of scent communication and the degree to which a species leads a subterranean life, although, for instance, the European rabbit (*Oryctolagus cuniculus*) has a more elaborate chemical communication system than hares or cottontails.

In comparing vegetation types, open habitat would be expected to favor visual communication. One would have to compare forest and mountain reindeer, woodland and barren-ground caribou, woodchucks and alpine marmots, forest and plains antelopes.

Migratory species and populations are expected to scent-mark less than non-migratory ones (Graf, 1950). Indeed, less subauricular-marking occurred in a migratory Idaho population of the pronghorn (*Antilocapra americana*) than in those occupying territories year-round in the National Bison Range in Montana (pers. observ.).

The social system influences olfactory communication, and especially scent-marking. The swamp rabbit (*Sylvilagus aquaticus*) "chins" more frequently than the cottontail rabbit (*S. floridanus*). The former is highly territorial and has large chin glands; the latter is not territorial, and its chin glands are small (Marsden and Holler, 1964). A similar difference exists between *Lepus* and *Oryctolagus* (Mykytowycz, 1970). The more social species of ground squirrels (*Spermophilus*) have larger dorsal glands and more pronounced marking and social sniffing ("greeting") than less social ones (Kivett et al., 1976).

Evolution of Scent Gland Function

General physiological processes provide odoriferous end products that reach both the air and the substrate. Thus, odors of feces, urine, saliva, and breath probably were first used for chemical communication, with no energy required for synthesis of additional compounds. Some of these odors can be used by conspecifics to extract information on identity, status, and whereabouts of the emitting individual. Specialized behavior patterns would then be selected for, such as depositing excrement at certain sites at certain times, leg-lifting, urine-spraying, and so forth. Wemmer (1971) has noted the morphological similarities between anal-dragging, a cleaning movement, and anogenital marking with squatted hindquarters; the latter is the postulated common causal factor between scent-marking and elimination.

Further, glands associated with excretion, such as preputial or anal glands, add material to the excretion. Glands on other parts of the body originally in the service of integumentary maintenance may have also assummed communicatory functions. Such glands may be rubbed on the substrate, leading to new stereotyped motor patterns. Glandular odors may then have assumed meanings different from excretions, leading to further differentiation of function. Glands on certain areas of the body will be favored by the nature of the environment and the habits and size of the species, such as lateral glands in voles or sternal glands in arboreal forms.

Among bovids, the larger genera such as *Bos, Bubalus, Bison, Tragelaphus,* and *Taurotragus* lack head glands (antorbital, subauricular, and postcornual glands) and associated marking behavior, whereas smaller forms such as *Cephalophus, Madoqua, Oreotragus, Raphicerus, Ourebia, Antilope, Gazella,* and *Antilocapra* have head glands and mark frequently (Walther, 1979). Both groups contain forest and grassland forms, although the smaller bovids tend to utilize more brush and other cover. It appears that the better an animal can be seen by conspecifics, the less it relies on scent-marking.

A particular odor may have an individual and a social function at the same time. For example, scent mounds of beavers are often located at the beginning of trails at the water's edge so that resident beavers can orient themselves during their nocturnal landfalls, and at the same time strangers can obtain information that the area is used by resident beavers (Müller-Schwarze and Heckman, 1980).

Another experimentally better investigated example is the saddleback tamarin (*Saguinus fuscicollis*), whose scent marks communicate species, subspecies, gender, hormonal condition, individual identity, and perhaps rank. Scent marks can function as a substitute for injurious aggression, maintain the hierarchy within groups, enforce the pair bond, and perhaps function in territorial defense (Epple, 1980).

Therefore, one experimentally demonstrated function of a pheromone may often be only the beginning of an involved search for possible other functions. Even failure to demonstrate a behavioral function of an odor may not mean that there is none. The function(s) may be different from the one hypothesized, or more likely, other stimuli in the chemical or other sensory modalities (simultaneously or stepwise) may be required to influence behavior measurably, since mammalian behavior abounds with redundancy and multisensory control. These facts and learning processes render behavioral pheromone bioassays involving mammals very difficult.

Acknowledgments

This review and most of the cited studies by the author were supported by several grants from the National Science Foundation. Mrs. Christine Müller-Schwarze helped with proof-reading and assembling the references.

Literature Cited

AKKERMANN, R. 1975. Untersuchungen zur Ökologie und Populationsdynamik des Bisams (*Ondatra zibethicus* L.). III. Verhalten und Populations-dynamik. Z. Angew. Zool., 62:281–338.

ALBONE, E. S., AND M. W. FOX. 1971. Anal gland secretion of the red fox. Nature, 233:569–570.

ALDRICH, T. B. 1896. A chemical study of the secretion of the anal glands of *Mephitis mephitica* (Common skunk) with remarks on the physiological properties of this secretion. J. Expl. Med., 1:323.

AUGUST, P. V. 1978. Scent communication in the southern plains woodrat, *Neotoma micropus*. Amer. Midland Nat., 99:206–218.

BACKHAUS, D. 1964. Zum Verhalten des nördlichen Breitmaulnashorns (*Diceros simus cottoni* Lydekker 1880). Zool. Garten N.F., 29:93–107.

BALPH, D. F., AND A. W. STOKES. 1963. On the ethology of a population of Uinta ground squirrels. Amer. Midland Nat., 69:106–126.

BARAN, D., AND S. E. GLICKMAN. 1970. "Territory marking" in the Mongolian gerbil: a study of sensory control and function. J. Comp. Physiol. Psychol., 71:237–245.

BARNETT, S. A. 1963. A study in behaviour. Methuen, London, 288 pp.

BEAUCHAMP, G. K. 1976. Diet influences attractiveness of urine in guinea pigs. Nature, 263:587–588.

BEAUCHAMP, G. K., AND J. BERÜTER. 1973. Source and stability of attractive components in guinea pig (*Cavia porcellus*) urine. Behav. Biol., 9:43–47.

BEAVER, D. L. 1960. A re-evaluation of the rat preputial gland as a "dicrine" organ from the standpoint of its morphology, histochemistry, and physiology. J. Exp. Zool., 143:153–173.

BECKMAN, E. 1896. Über das Drüsensekret des Stinkdaches. Pharmaz. Zentralhalle Dtschlds., 37:557–558.

BOLLIGER, A. 1944. The response of the sternal integument of *Trichosurus vulpecula* to castration and to sex hormones. J. Proc. Royal Soc. New South Wales, 78:234–238.

BOLLIGER, A., AND W. K. WHITTEN. 1948. The paracloacal (anal) glands of *Trichosurus vulpecula.* J. Proc. Royal Soc. New South Wales, 82:36–43.

BONNER, W. N. 1968. The fur seal of South Georgia. British Antarctic Surv. Sci. Rep., London, 56:1–81.

BRADBURY, J. 1975. Social organization and communication. Pp. 1–72, *in* Biology of bats (W. Wimsatt, ed.). Academic Press, New York, 3:1–651.

BRINKMANN, A. 1909. Über das Vorkommen von Hautdrüsenorganen bei den anthropomorphen Affen. Anat. Anz., 34:513–520.

———. 1912. Die Hautdrüsen der Säugetiere (Bau und Sekretverhältnisse). Erbegn. Anat. Entw., 20:1173–1231.

BRODY, B. 1975. The sexual significance of the axillae. Psychiatry, 38:278–289.

BRONSON, F. H., AND D. CAROOM. 1971. Preputial gland of the male mouse: attractant function. J. Reprod. Fert., 25:279–282.

BROWNLEE, R. G., ET AL. 1968. Isolation, identification and function of the chief component of the male tarsal scent in black-tailed deer. Nature, 221:284–285.

BRUNDIN, A., ET AL. 1978. Short-chain aliphatic acids in the interdigital gland secretions of reindeer (*Rangifer tarandus* L.) and their discrimination by reindeer. J. Chem. Ecol., 4:613–622.

BUECHNER, H. K., AND R. SCHLOETH. 1965. Ceremonial mating behavior in Uganda kob (*Adenota kob thomasi* Neumann). Z. Tierpsychol., 22:209–225.

BUSS, I. O., L. E. RASMUSSEN, AND G. L. SMUTS. 1976. Role of stress and individual recognition in the function of African elephant's temporal gland. Mammalia, 40:437–451.

CARTER, C. S., AND J. N. MARR. 1970. Olfactory imprinting and age variables in the guinea pig, *Cavia porcellus.* Anim. Behav., 18:238–244.

CAVE, A. J. E. 1962. The pedal scent gland in *Rhinoceros*. Proc. Zool. Soc. London, 139:685–690.

CHRISTIAN, J. J. 1955. Effect of population size on the adrenal glands and reproductive organs of male mice in populations of fixed size. Amer. J. Physiol., 182:292–300.

CLEVEDON BROWN, J., AND J. D. WILLIAMS. 1972. The rodent preputial gland. Mammal Rev., 2:105–147.

COBLENTZ, B. 1976. Functions of scent urination in ungulates with special reference to feral goats. Amer. Nat., 110:549–557.

COE, M. J. 1962. Notes on the habits of the Mount Kenya Hyrax (*Procavia johnstoni mackinderi* (Thomas). Proc. Zool. Soc. London, 138:639–644.

DAVIS, D. E., AND J. J. CHRISTIAN. 1957. Nav. Med. Res. Inst. Res. Papers, 15: 311.

DIETERLEN, F. 1959. Das Verhalten des syrischen Goldhamsters (*Mesocricetus auratus* Waterhouse). Z. Tierpsychol., 16:47–103.

DOBRORUKA, L. J. 1960. Einige Beobachtungen an Ameisenigeln, *Echidna aculeata* Shaw (1792). Z. Tierpsychol., 17:178–181.

DONOVAN, C. A. 1967. Some clinical observations on sexual attraction and deterrence in dogs and cattle. Vet. Med. Small An. Clinician, 62:1047–1048.

———. 1969. Canine anal glands and chemical signals (pheromones). J. Amer. Vet. Med. Assn., 155:1995–1996.

DOTY, R. L. (ED.). 1976. Mammalian olfaction; reproductive processes; and behavior. Academic Press, New York, 344 pp.

DRYDEN, G. L., AND C. H. CONAWAY. 1967. The origin and hormonal control of scent production in *Suncus murinus*. J. Mamm., 48:420–428.

DUTT, R. H., ET AL. 1959. Identification of preputial glands as the site of production of sexual odor in the boar. J. Amer. Sci., 18:1557–1558.

EADIE, W. R. 1954. Skin gland activity and pelage descriptions in moles. J. Mamm., 35:186–196.

EBLING, F. J. 1977. Hormonal control of mammalian skin glands. Pp. 17–33, *in* Chemical signals in vertebrates (D. Müller-Schwarze and M. M. Mozell, eds.). Plenum Publishing Corp., New York, 609 pp.

EGGELING, V. 1901. Über die Schläfendrüse des Elefanten. Biol. Zentralbl., 21: 443–452.

EIBL-EIBESFELDT, I. 1953*a*. Eine besondere Form des Duftmarkierens beim Riesengalago, *Galago crassicaudatus* E. Geoffroy. Säugetierk. Mitt., 1:171–173.

———. 1953b. Zur Ethologie des Hamsters (*Cricetus cricetus* L.) Z. Tierpsychol., 10:204–254.

———. 1965. Das Duftmarkieren des Igeltenrec (*Echinops telfairi* Martin). Z. Tierpsychol., 22:810–812.

EISENBERG, J. F. 1963. The behavior of heteromyid rodents. Univ. California Publ. Zool., 69:1–100.

EISENBERG, J. F., AND E. GOULD. 1970. The tenrecs: a study in mammalian behavior and evolution. Smithsonian Contrib. Zool., 27:1–137.

EISENBERG, J. F., AND D. G. Kleiman. 1972. Olfactory communication in mammals. Ann. Rev. Ecol. Syst., 3:1–32.

EISENBERG, J. F., AND R. E. KUEHN. 1966. The behavior of *Ateles geoffroyi* and related species. Smithsonian Misc. Coll., 151:1–63.

EPLING, G. P. 1956. Morphology of the scent gland of the javelina. J. Mamm., 37:246–248.

EPPLE, G. 1970. Quantitative studies on scent marking in the marmoset (*Callithrix jacchus*). Folia Primatol., 13:48–62.

———. 1980. Relationships between aggression, scent marking and gonadal state in a primate, the tamarin (*Sagunius fuscicollis*). Pp. 87–105, *in* Chemical signals: vertebrates and aquatic invertebrates (D. Müller-Schwarze and R. M. Silverstein, eds.). Plenum Publ. Corp., New York, 445 pp.

EPPLE, G., AND R. LORENZ. 1967. Vorkommen, Morphologie and Funktion der Sternaldrüse bei den Platyrrhini. Folia Primatol. 7:98–126.

ESPMARK, Y. 1964. Rutting behaviour in reindeer (*Rangifer tarandus* L.). Anim. Behav., 12:159–163.

ESTES, J. A., AND I. O. BUSS. 1976. Microanatomical structure and development of the African elephant's temporal gland. Mammalia, 40:429–436.

ESTES, R. D. 1967. The comparative behavior of Grant's and Thomson's gazelles. J. Mamm., 48:189–209.

EVANS, C. S. 1980. Diosmic responses to scent-signals in *Lemur catta*. Pp. 417–420, *in* Chemical signals: vertebrates and aquatic invertebrates. (D. Müller-Schwarze and R. M. Silverstein, eds.). Plenum Publ. Corp., New York, 445 pp.

EVANS, C. S., AND R. W. GOY. 1968. Social behaviour and reproductive cycles in captive ring-tailed lemurs (*Lemur catta*). J. Zool. 156:181–197.

EWER, R. F. 1963. The behaviour of the meerkat, *Suricata suricatta* (Schreber). Z. Tierpsychol., 20:570–607.

———. 1967. The behaviour of the African giant rat (*Cricetomys gambianus* Waterhouse). Z. Tierpsychol., 24:6–79.

———. 1968*a*. Ethology of mammals. Plenum Publ. Corp., New York, 418 pp.

———. 1968*b*. A preliminary survey of the behaviour in captivity of the dasyurid marsupial, *Sminthopsis crassicaudata* (Gould). Z. Tierpsychol., 25:319–365.

EWER, R. F., AND C. WEMMER. 1974. The behavior in captivity of the African civet, *Civettictis civetta* (Schreber). Z. Tierpsychol., 34:359–394.

FLEMING, A. S., AND L. TAMBASSO. 1980. Hormonal and sensory control of scent-marking in the desert woodrat (*Neotoma l. lepida*). J. Comp. Physiol. Psychol., 94:564–578.

FOX, M. W. 1968. Ontogeny of inguinal approach and presentation in canids and foxes. Amer. Zool., 8:746.

FRÄDRICH, H. 1966. Einige Verhaltensbeobachtungen am Moschustier (*Moschus moschiferus* L.). Zool. Garten, 33:65–78.

FRANK, F. 1956. Das Duftmarkieren der Grossen Wühlmaus, *Arvicola terrestris* (L.) Z. Säugetierk., 21:172–175.

FRANKLIN, W. L. 1980. Territorial marking behavior by the South American vicuña. Pp. 53–66, *in* Chemical signals: vertebrates and aquatic invertebrates (D. Müller-Schwarze and R. M. Silverstein, eds.). Plenum Publ. Corp., New York, 445pp.

FRIEDRICH, H. 1959. Untersuchungen über zyklische Veränderungen der Violdrüse und der Zirkumoraldrüsen, sowie über die Haut des Rotfuches (*Vulpes vulpes* L.). Unpubl. dissert., Humboldt University, Berlin.

FRÜHLING, A. 1955. Über das "Harnspritzen" und ähnliche Verhaltensweisen beim Meerschweinchen, *Cavia porcellus* Linné, 1758. Säugetierk. Mitt., 3: 168–171.

GABE, M. 1967. Le tegument et ses annexes. Pp. 1–233, *in* Traité de Zoologie, Anatomie Systématique (P. P. Grassé, ed.). Masson et Cie, Paris, 16:1–1162.

GALEF, B. G., AND M. M. CLARKE. 1972. Mother's milk and adult presence: two factors determining initial dietary selection by weaning rats. J. Comp. Physiol. Psychol., 78:220–225.

GAUTHIER-PILTERS, H. 1960. Ergänzende Beobachtungen zum aggressiven Verhalten und Kampfspiel bei Tylopoden. Georg v. Opel Freigehege f. Tierforsch. Jahrb. 1959/60, pp. 104–109.

GAWIENOWSKI, A. M. 1977. Chemical attractants of the rat preputial gland. Pp. 45–59, *in* Chemical signals in vertebrates (D. Müller-Schwarze and M. M. Mozell, eds.). Plenum Publ. Corp., New York, 609 pp.

GEIST, V. 1965. On the rutting behavior of the mountain goat. J. Mamm., 45: 551–568.

———. 1971. Mountain sheep. Univ. Chicago Press, 383 pp.

GILBERT, B. K. 1973. Scent marking and territoriality in pronghorn (*Antilocapra americana*) in Yellowstone Park. Mammalia, 37:25–33.

Glenn, E. M., and J. Gray. 1965. Effect of various hormones on the growth and histology of the gerbil (*Meriones unguiculatus*) abdominal sebaceous gland pad. Endocrinology, 76:1115–1123.

Goddard, J. 1967. Home range, behaviour, and recruitment of two black rhinoceros populations. East African Wildl. J., 5:133–150.

Godfrey, J. 1958. The origin of sexual isolation between bank voles. Proc. Roy. Phys. Soc. Edinburgh, 27:47–55.

Goldfoot, D. A., et al. 1976. Lack of effect of vaginal lavages and aliphatic acids on ejaculatory responses in rhesus monkeys: behavioral and chemical analyses. Horm. Behav., 7:1–27.

Goodwin, M., K. M. Gooding, and F. Regnier. 1979. Sex pheromone in the dog. Science, 203:559–561.

Gosling, L. M. 1972. The construction of antorbital gland marking sites by male oribi (*Ourebia ourebia* Zimmermann 1783). Z. Tierpsychol., 30:271–276.

———. 1981. Demarkation in a gerenuk territory: an economic approach. Z. Tierpsychol., 56:305–322.

Graf, W. 1950. Territorialism in deer. J. Mamm., 37:165–170.

Griffiths, M. 1978. The biology of the monotremes. Academic Press, New York, 367 pp.

Hafez, E. S. E., and J. P. Signoret. 1969. The behaviour of swine. Pp. 349–390, *in* The behavior of domestic animals (E. S. E. Hafez, ed.). Williams and Wilkins, Baltimore, 647 pp.

Hamilton, J. E. 1956. Scent of otariids. Nature, 177:900.

Harrington, J. 1974. Olfactory communication in *Lemur fulvus*. Pp. 331–346, *in* Prosimian biology (G. A. Doyle, R. D. Martin, and A. Walker, eds.). Duckworth, London, 983 pp.

Harvey, E. B., and Rosenberg, L. E. 1960. An apocrine gland complex of the pika. J. Mamm., 41:213–219.

Hayashi, S. 1979. A role of female preputial glands in social behavior of mice. Physiol. Behav., 23:967–969.

Hediger, H. 1949. Säugetier-Territorien und ihre Markierung. Bijdr. Dierk., 28: 172–184.

———. 1956. Instinkt und Territorium. Pp. 521–541, *in* L'instinct dans le comportement des animaux et de l'homme. Masson et Cie, Paris, 796 pp.

Hediger, H., and H. Kummer. 1956. Das Verhalten der Schnabeligel (Tachyglossidae). Handb. Zool., 8(10):1–8.

Heimburger, N. 1959. Das Markierungsverhalten einiger Caniden. Z. Tierpsychol., 16:104–113.

Henry, J. D. 1977. The use of urine marking in the scavenging behavior of the red fox (*Vulpes vulpes*). Behaviour, 61:82–106.

———. 1980. The urine marking behavior and movement patterns of red foxes (*Vulpes vulpes*) during a breeding and post-breeding period. Pp. 11–27, *in* Chemical signals: vertebrates and aquatic invertebrates (D. Müller-Schwarze and R. M. Silverstein, eds.). Plenum Publ. Corp., New York, 445 pp.

Herreid, C. F., II. 1960. Comments on the odors of bats. J. Mamm., 41:396.

Hill, W. C. O. 1944. An undescribed feature in the drill (*Mandrillus leucophaeus*). Nature, 153:199.

Hofer, M. A., H. Shair, and P. Singh. 1976. Evidence that maternal ventral skin substances promote suckling in infant rats. Physiol. Behav., 17:131–136.

Ilse, D. 1955. Olfactory marking of territory in two young male loris, *Loris tardigradus lydekkerianus*, kept in captivity in Poona. British J. Anim. Behav., 3:118–120.

JAINUDEEN, M. R., G. M. MCKAY, AND J. F. EISENBERG. 1972. Observations on musth in the domesticated Asiatic elephant (*Elephas maximus*). Mammalia, 36:247–261.

JOHNSON, R. P. 1973. Scent marking in mammals. Anim. Behav., 21:521–535.

JOHNSTON, R. E. 1972. Sex pheromones of the golden hamster. Amer. Zool., 12. 662.

JOHNSTON, R. E., AND B. COPLIN. 1979. Development of responses to vaginal secretion and other substances in golden hamsters. Behav. Neur. Biol., 25: 473–489.

JOHNSTON, R. E., AND T. SCHMIDT. 1979. Responses of hamsters to scent marks of different ages. Behav. Neur. Biol., 26:64–75.

JOLLY, A. 1966. Lemur behavior: a Madagascar field study. Univ. Chicago Press, Chicago, 187 pp.

KÄLLQUIST, L., AND T. MOSSING. 1977. The distribution of sudoriferous glands in the hairy skin of reindeer (*Rangifer tarandus* L.). Acta Zool., 58:65–68.

KAUFMANN, J. H. 1962. Ecology and social behavior of the coati *Nasua narica* on Barro Colorado Island, Panama. Univ. California Publ. Zool., 60:1–222.

KEMPER, H. 1950. Über die Fusspuren von Ratten und Mäusen. Z. Hygien., 8: 225–245.

KIRCHSHOFER, R. 1960. Über das "Harnspritzen" der grossenmara. Z. Säugetierk., 25:112–127.

KIVETT, V. K., J. O. MURIE, AND A. L. STEINER. 1976. A comparative study of scent-gland location and related behavior in some northwestern nearctic ground squirrel species (Sciuridae): an evolutionary approach. Canadian J. Zool., 54:1294–1306.

KLEIMAN, D. 1966. Scent marking in the Canidae. Symp. Zool. Soc. London, 18: 167–177.

———. 1971. The Courtship and copulatory behavior of the green acouchi, *Myoprocta pratti*. Z. Tierpsychol., 29:259–278.

———. 1974. Scent-marking in the binturong, *Arctictis binturong*. J. Mamm., 55:224–227.

KLEIMAN, D., AND L. R. COLLINS. 1972. Preliminary observations on scent marking, social behavior and play in the juvenile giant panda, *Ailuropoda melanoleuca*. Amer. Zool., 12:644 (Abstr.).

KLEIMAN, D., AND D. S. MACK. 1980. Effects of age, sex, and reproductive status on scent marking frequencies in the golden lion tamarin, *Leontopithecus rosalia*. Folia Primatol., 33:1–14.

KLINGEL, H. 1968. Soziale Organisation und Verhaltensweisen von Hartmann- und Bergzebras *Equus zebra hartmannae* and *E. z. zebra*). Z. Tierpsychol., 25: 76–88.

KOCH, L. L., D. M. STODDART, AND H. KACHER. 1969. Notes on behavior and food supply of lemmings (*Lemmus lemmus* L.) during a peak density in southern Norway 1966/67. Z. Tierpsychol., 26:609–622.

KOENIG, L. 1957. Beobachtungen zur Reviermarkierung sowie Droh-, Kampf-, und Abwehrverhalten des Murmeltieres (*M. marmota*). Z. Tierpsychol., 14:510–521.

———. 1960. Das Aktionssytem des Siebenschläfers (*Glis glis* L.). Z. Tierpsychol., 14:427–505.

KRÄMER, A. 1969. Soziale Organisation und Socialverhalten einer Gemspopulation (*Rupicapra rupicapra* L.) der Alpen. Z. Tierpsychol., 26:889–964.

KROTT, P. 1960. Tupu-tupu, das seltenste Raubwild Nordeuropa. Paul Parey Verlag, Berlin und Hamburg, 201 pp.

KÜHME, W. 1961. Beobachtungen am afrikanischen Elefanten (*Loxodonta africana* Blumenbach) in Gefangenschaft. Z. Tierpsychol., 18:285–296.

KUNKEL, P., AND I. KUNKEL. 1964. Beiträge zur Ethologie des Hausmeerschweinchens *Cavia aperea* f. *porcellus*. Z. Tierpsychol., 21:602–641.

KURT, F. 1968. Das Sozialverhalten des Rehes *Capreolus capreolus* L. Eine Feldstudie. Mammalia Depicta, 102 pp.

LANE-PETTER, W. 1967. Odour in mice. Nature, 216:794.

LANG, H. 1923. Caudal and pectoral glands of African elephant shrews (*Elephantulus*). J. Mamm., 4:261–263.

LEDERER, E. 1950. Odeurs et parfums des animaux. Fortschr. Chem. Org. Naturstoffe, 6:87–153.

LENT, P. C. 1966. Calving and related social behavior in the barren-ground caribou. Z. Tierpsychol., 23:701–756.

LEON, M. 1974. Maternal pheromone. Physiol. Behav., 13:441–453.

LOJDA, Z. 1956. Histogenesis of the antlers of our *Cervidae* and its histochemical picture. Csl. Morfologie IV, 1:43–62.

MAINARDI, D., M. MARSAN, AND A. PASQUALI. 1965. Causation of sexual preferences in the house mouse. The behavior of mice reared by parents whose odor was artifically altered. Atti della Societa di Scienza Naturali e dei Museo covico di Storia Naturale di Milano, 104:325–338.

MARSDEN, H. M., AND N. R. HOLLER. 1964. Social behavior in confined populations of the cottontail and swamp rabbit. Wildl. Monogr., 13:1–39.

MARTIN, I. G., AND G. K. BEAUCHAMP. 1982. Olfactory recognition of individuals by male cavies (*Cavia aperea*). J. Chem. Ecol., 8:1241–1249.

MARTIN, R. D. 1968. Reproduction and ontogeny in tree shrews (*Tupaia belangeri*) with reference to their general behavior and taxonomic relationships. Z. Tierpsychol., 25:409–495, 505–532.

MAURER, B., AND G. OHLOFF. 1976. Zur Kenntnis der stickstoffhaltigen Inhaltsstoffe von Castoreum. Helv. Chim. Acta, 59:1169–1185.

MCCULLOCH, D. R. 1969. The Tule elk. Its history, behavior and ecology. Univ. California Press, Berkeley, 209 pp.

MEANEY, C. In press. Rate and location of scent marking by pikas during the breeding season. *In,* Chemical signals in vertebrates, III (D. Müller-Schwartze and R. M. Silverstein, eds.). Plenum Publ. Corp., New York.

MERTL, A. S. 1975. Discrimination of individuals by scent in a primate. Behav. Biol., 14:505–509.

MICHAEL, R. P., E. B. Keverne, AND R. W. BONSALL. 1971. Pheromones: isolation of male sex attractants from a female primate. Science, 172:964–966.

MIMRAM, R. 1962. Les glandes occipitales du dromedaire. Cahiers de la Faculte des Sciences, Rabat Ser. Biol. Animaux, 1:7–42.

MONTAGNA, W., AND C. R. NOBACK. 1946. The histology of the preputial gland of the rat. Anat. Rec., 96:41–54.

MOORE, R. E. 1965. Olfactory discrimination as an isolating mechanism between *Peromyscus maniculatus* and *Peromyscus polionotus*. Amer. Midland Nat., 73:85–100.

MOSSING, T., AND L. KÄLLQUIST. 1981. Variation in cutaneous glandular structures in reindeer (*Rangifer tarandus*). J. Mamm., 62:606–612.

MOY, R. F. 1970. Histology of the subauricular and rump glands of the pronghorn (*Antilocapra americana* Ord.). Amer. J. Anat., 129:65–88.

MUGFORD, R. A., AND N. W. NOWELL. 1971. The preputial glands as a source of aggression-promoting odors in mice. Physiol. Behav., 6:247–249.

MÜLLER-SCHWARZE, D. 1971. Pheromones in black-tailed deer (*Odocoileus hemionus columbianus*) Anim. Behav., 19:141–152.

———. 1980. Chemical signals in alarm behavior in deer. Pp. 39–51, *in* Chemical signals: vertebrates and aquatic invertebrates (D. Müller-Schwarze, and R. M. Silverstein, eds.), Plenum Publ. Corp., New York, 445 pp

MÜLLER-SCHWARZE, D., AND S. HECKMAN. 1980. The social role of scent marking in beaver (*Castor canadensis*). J. Chem. Ecol., 6:81–95.

MÜLLER-SCHWARZE, D., AND C. MÜLLER-SCHWARZE. 1971. Olfactory imprinting in a precocial mammal. Nature, 229:55–56.

———. 1972. Social scents in hand-reared pronghorn (*Antilocapra americana*). Zool. Africana, 7:257–271.

———. 1975. Subspecies specificity of response to a mammalian social odor. J. Chem. Ecol., 1:125–131.

MÜLLER-SCHWARTZE, D., S. HECKMAN, AND B. STAGGE. In press. Behavior of free-ranging beavers (*Castor canadensis*) at experimental scent marks. First Worldwide Beaver Symp., Helsinki.

MÜLLER-SCHWARZE, D., L. KÄLLQUIST, AND T. MOSSING. 1979. Social behavior and chemical communication in reindeer (*Rangifer t. tarandus* L.). J. Chem. Ecol., 5:483–517.

MÜLLER-SCHWARZE, D., W. B. QUAY, AND A. BRUNDIN. 1977*a*. The caudal gland in reindeer (*Rangifer tarandus* L.): its behavioral role, histology, and chemistry. J. Chem. Ecol., 3:591–601.

MÜLLER-SCHWARZE, D., N. J. Volkman, AND K. ZEMANEK. 1977*b*. Osmetrichia: specialized scent hairs in black-tailed deer. J. Ultrastructure Res., 59:223–230.

MÜLLER-SCHWARZE, D., ET AL. 1974. Mammalian pheromones: identification of active component in subauricular scent of pronghorn (*Antilocapra americana*). Science, 183:860–862.

MÜLLER-SCHWARZE, D., ET AL. 1978. Responses of reindeer to interdigital secretions of conspecifics. J. Chem. Ecol., 4:325–335.

MÜLLER-VELTEN, H. 1966. Über den Angstgeruch bei der Hausmaus (*Mus musculus*). Z. Vergl. Physiol., 52:401–429.

MYKYTOWYCZ, R. 1962. Territorial function of chin gland secretion in the rabbit, *Oryctolagus cuniculus* (L.). Nature, 193:799.

———. 1966*a*. Observations on odoriferous and other glands in the Australian wild rabbit, *Oryctolagus cuniculus* (L.) and the hare, *Lepus europaeus*. Part 1. The anal gland. CSIRO Wildl. Res., 11:11–29.

———. 1966*b*. Observations on odoriferous and other glands in the Australian wild rabbit, *Oryctolagus cuniculus* (L.) and the hare, *Lepus europaeus*. Part 2. The inguinal glands. CSIRO Wildl. Res., 11:49–64.

———. 1968. Territorial marking by rabbits. Sci. Amer., 218:116–126.

———. 1970. The role of skin glands in mammalian communication. Pp. 327–360, *in* Communication by chemical signals (J. W. Johnston, Jr., D. G. Moulton, and A. Turk, eds.). Appleton-Century-Crofts, New York, 412 pp.

MYKYTOWYCZ, R., AND M. L. DUDZINSKI. 1972. Aggressive and protective behaviour of adult rabbits *Oryctolagus cuniculus* (L) towards juveniles. Behaviour, 43:97–120.

MYKYTOWYCZ, R., AND S. GAMBALE. 1969. The distribution of dung-hills and the behaviour of free-living rabbits, *Oryctolagus cuniculus* (L.) on them. Forma et Functio, 2:1–12.

MYKYTOWYCZ, R., AND T. NAY. 1964. Studies of the cutaneous glands and hair follicles of some species of Macropodidae. CSIRO Wildl. Res., 9:200–217.

MYKYTOWYCZ, R., AND M. WARD. 1971. Some reactions of nestlings of the wild rabbit, *Oryctolagus cuniculus* (L.), when exposed to natural rabbit odours. Forma et Functio, 4:137–148.

MYKYTOWYCZ, R., ET AL. 1976. A comparison of the effectiveness of the odors of rabbits, *Oryctolagus cuniculus*, in enhancing territorial confidence. J. Chem. Ecol., 2:13–24.

NEAL, B. J. 1959. A contribution on the life history of the collared peccary in Arizona. Amer. Midland Nat., 61:177–190.

NELSON, J. E. 1965. Behavior of Australian Pteropodidae (*Megachiroptera*) Anim. Behav., 13:544–557.

NEUWEILER, G. 1962. Das Verhalten indischer Flughunde (*Pteropus giganteus gig.* Brunn). Naturwiss., 49:614–615.

ORTMANN, R. 1956. Uber die Musterbildung von Duftdrüsen in der Sohlenhaut der weissen Hausmaus (*Mus musculus albinus*). Z. Säugetierk., 21:138–141.

ÖSTBORN, H. 1976. Doftmarkering hos grävling. Zool. Revy, 38:103–112.

OWEN, K., P. WALLACE, AND D. D. THIESSEN. 1974. Effects of intracerebral implants of steroid hormones on scent marking in the ovariectomized female gerbil (*Meriones unguiculatus*). Physiol. Behav., 12:755–760.

OWEN-SMITH, N. 1971. Territoriality in the white rhinoceros (*Cerathotherium simum* Burchell). Nature, 231:294–296.

PAGÉS, E. 1968. Les glandes odorantes des pangolines arboricoles (*M. tricuspis* et *M. longicaudata*): morphologie, development et rôles. Biol. Gabon, 4:354–400.

PARAKKAL, P., W. MONTAGNA, AND R. A. ELLIS. 1963. The skin of primates. XI. The skin of the white-browed gibbon (*Hylobates hoolock*). Anat. Rec., 143: 169–177.

PATTERSON, R. L. S. 1968. Identification of 3 α-hydroxy-5-α-androst-16-ene as the musk odour component of boar submaxillary gland and its relationship to the sex odour taint in pork meat. J. Sci. Fd. Agric., 19:434–438.

PAYNE, A. P., ET AL. 1977. Cyclic and seasonal changes in Harderian gland activity in the female golden hamster. J. Endocrinol., 75:41P.

PEARSON, O. P. 1946. Scent glands of the short-tailed shrew. Anat. Rec., 94:615–630.

PERKINS, E. 1966. The skin of primates. XXXI. The skin of the black-collared tamarin (*Tamarinus nigricollis*). Amer. J. Phys. Anthropol., 25:41–50.

PERRY, G. C., R. L. S. Patterson, AND G. E. STINSON. 1973. Submaxillary salivary gland involvement in porcine mating behaviour. Proc. Int. Congr. Anim. Reprod. A. I., Munich, 7:386–389.

PETERS, R. P., AND L. D. MECH. 1975. Scent-marking in wolves. Amer. Sci., 63: 628–637.

POCOCK, R. J. 1910. On the specialised cutaneous glands of ruminants. Proc. Zool. Soc. London, pp. 840–986.

PODUSCHKA, W. 1977. Das Paarungsvorspiel des Osteuropäischen Igels (*Erinaceus e. roumanicus*) und theoretische Überlegungen zum Problem männlicher Sexualpheromone. Zool. Anz., 199:187–208.

POGLAYEN-NEUWALL, I. 1966. On the marking behaviour of the kinkajou (*Potos flavus* Schreber). Zoologica, 51:137–141.

POGLAYEN-NEUWALL, I., AND I. POGLAYEN-NEUWALL. 1966. Gefangenschaftsbeobachtungen an Makibären (*Bassaricyon* Allen, 1876). Z. Säugetierk., 30: 321–366.

QUAY, W. B. 1955. Histology and cytochemistry of skin gland areas in the caribou *Rangifer*. J. Mamm., 36:187–201.

———. 1959. Microscopic structure and variation in the cutaneous glands of the deer, *Odocoileus virginianus*. J. Mamm., 40:114–128.

———. 1965*a*. Variation and taxonomic significance in the sebaceous caudal glands of pocket mice (Rodentia: Heteromyidae). Southwestern Nat., 10:282–287.

———. 1965*b*. Comparative survey of the sebaceous and sudoriferous glands of the oral lips and angle in rodents. J. Mamm., 46:23–37.

———. 1968. The specialized posterolateral sebaceous glandular regions in microtine rodents. J. Mamm., 49:427–445.

———. 1977. Structure and function of skin glands. Pp. 1–17, *in* Chemical signals in vertebrates (D. Müller-Schwarze and M. M. Mozell eds.). Plenum Publishing Corp., New York, 609 pp.

QUAY, W. B., AND D. MÜLLER-SCHWARZE. 1970. Functional histology of integumentary glandular regions in black-tailed deer (*Odocoileus hemionus columbianus*). J. Mamm., 51:675–694.

———. 1971. Relations of age and sex to integumentary glandular regions in Rocky Mountain mule deer (*Odocoileus hemionus hemionus*) J. Mamm., 52:670–685.

QUAY, W. B., AND P. Q. Tomich. 1963. A specialized mid-ventral sebaceous glandular area in *Rattus exulans*. J. Mamm., 44:537–542.

RAHM, U. 1960. Territoriumsmarkierung mit der Voraugendrüse beim Maxwell-Ducker (*Philantomba maxwelli*). Säugetierk. Mitt., 8:140–142.

RALLS, K. 1969. Scent-marking in Maxwell's duiker, *Cephalophus maxwelli*. Amer. Zoo., 9:1072.

———. 1971. Mammalian scent marking. Science, 171:443–449.

RANDALL, R. M. 1979. Perineal gland marking by free-ranging African civets, *Civettictis civetta*. J. Mamm., 60:622–627.

RASA, O. A. E. 1973. Marking behaviour and its social significance in the African dwarf mongoose, *Helogale undulata rufula*. Z. Tierpsychol., 31:293–318.

RATHBUN, G. B. 1979. The social structure and ecology of elephant-shrews. Adv. Ethol. Suppl., 20:1–77.

REBELL, G., AND D. KIRK. 1962. Patterns of eccrine sweating in the human axilla. Adv. Biol. Skin, 3:108–126.

REGNIER, F. E., AND M. GOODWIN. 1977. On the chemical and environmental modulation of pheromone release from vertebrate scent marks. Pp. 115–133, *in* Chemical signals in vertebrates (D. Müller-Schwarze and M. M. Mozell, eds.). Plenum Publ. Corp., New York, 609 pp.

RICHARD, P. B. 1973. Le desman de Pyrenées (*Galemys pyrenaicus*): mode de vie, univers sensoriel. Mammalia, 37:1–16.

RICHMOND, N. D., AND H. R. ROSLUND. 1952. A mid-ventral gland in *Peromyscus maniculatus*. J. Mamm., 33:103–104.

ROEDER, J. J. 1980. Marking behavior and olfactory recognition in genets (*Genetta genetta* L. Carnivora-Viverridae). Behaviour, 72:200–210.

ROGOVIN, K. A. 1979. On the marker function of sand bathing in *Salpingotus crassicauda*. Zool. Zh., 58:770–772. (In Russian.)

ROSENBLATT, J. S., G. TURKEWITZ, AND T. C. SCHNEIRLA. 1969. Development of home orientation in newly born kittens. Trans. New York Acad. Sci. II, 31:231–250.

RUDD, R. L. 1966. The midventral gland in Malaysian murid rodents. J. Mamm., 47:331–332.

RUZICKA, L. 1926*a*. Zur Kenntnis des Kohlenstoffringes I. Über die Konstitution des Zibetons. Helv. Chim. Acta, 9:230.

———. 1926*b*. Zur Kenntnis des Kohlenstoffringes VII. Über die Konstitution des Muscons. Helv. Chim. Acta, 9:715.

———. 1926*c*. Zur Kenntnis des Kohlenstoffringes VIII. Weitere Beiträge zur Konstitution des Muscons. Helv. Chim. Acta, 9:1008.

SAMBRAUS, H. H. 1971. Das Sexualverhalten des Hausrindes, speziell des Stieres. Z. Tierpsychol., Suppl., 6:1–55.

SCHAFFER, J. 1933. Die Vorhautdrüsen vou Maus und Ratte. Z. Mikr. Anat. Forsch., 34:1–22.

———. 1940. Die Hautdrüsenorgane der Säugetiere. Urban and Schwarzenberg, Berlin and Wien, 464 pp.

SCHALLER, G. B. 1973. Golden shadows, flying hooves. A. A. Knopf, New York, 287 pp.

———. 1977. Mountain monarchs: wild sheep and goats of the Himalaya. Univ. Chicago Press, Chicago, 425 pp.

SCHENKEL, R. 1966. Zum Problem der Territorialität und des Markierens bei Säugern, am Beispiel des Schwarzen Nashorns und des Löwen. Z. Tierpsychol., 23:593–626.

SCHIEFFERDECKER, P. 1922. Die Hautdrüsen des Menschen und der Säugetiere, ihre biologische und rassenanatomische Bedeutung, sowie die Muscularis sexualis. Zoologica, 27:1–154.

SCHULTZ, A. H. 1921. The occurrence of a sternal gland in orangutan. J. Mamm., 2:194–196.

SCHULTZE-WESTRUM, T. 1965. Innerartliche Verständigung durch Düfte beim Gleitbeutler *Petaurus breviceps papuanus* Thomas (Marsupialia, Phalangeridae). Z. Vergl. Physiol., 50:151–220.

SCHUMACHER, S. 1936. Das Stirnorgan des Rehbockes (*Capreolus c. capreolus* L.), ein bisher unbekanntes Duftorgan. Z. Mikr. Anat. Forsch., 39:215–230.

SEBEOK, T. (ed.). 1978. How animals communicate. Indiana Univ. Press, Bloomington, Indiana, 1128 pp.

SEITZ, E. 1969. Die Bedeutung geruchlicher Orientierung beim Plumplori *Nycticebus coucang* Boddart 1785 (Prosimii, Lorisidae). Z. Tierpsychol., 26: 73–103.

SEKULIC, R., AND J. F. EISENBERG. In press. Throat-rubbing in red howler monkeys. *In,* Chemical signals in vertebrates, III (D. Müller-Schwarze and R. M. Silverstein, eds.). Plenum Publ. Corp., New York.

SHADLE, A. R., M. SMELZER, AND M. METZ. 1946. The sex reactions of porcupines (*Erethizon d. dorsatum*) before and after copulation. J. Mamm., 27: 116–121.

SHARMAN, G. B., AND J. H. CALABY. 1964. Reproductive behavior in the red kangaroo, *Megaleia rufa,* in captivity. CSIRO Wildl. Res., 9:58–85.

SINGER, A. G., ET AL. 1976. Dimethyl disulfide: an attractant pheromone in hamster vaginal secretion. Science, 191:948–950.

SINGER, A. G., F. MACRIDES, AND W. C. AGOSTA. 1980. Chemical studies of hamster reproductive pheromones. Pp. 365–375, *in* Chemical signals: vertebrates and aquatic invertebrates (D. Müller-Schwarze and R. M. Silverstein, eds.). New York: Plenum Publ. Corp., New York, 445 pp.

SINGH, U. B., AND M. B. BARAWAJ. 1978. Anatomical, histological and histochemical observations and changes in the poll glands of the camel (*Camelus dromedarius*). Part 3. Acta Anatomica, 102:74–83.

SKEEN, J. T., AND D. D. THIESSEN. 1977. Scent of gerbil cuisine. Physiol. Behav., 19:11–14.

SMITH, M. 1980. Behaviour of the koala, *Phascolarctos cinereus* (Goldfuss). IV. Scent-marking. Australian Wildl. Res., 7:35–40.

SPANNHOF, I. 1969. The histophysiology and function of the anal sac of the red fox (*Vulpes vulpes* L.). Forma et Functio, 1:26–45.

SPRANKEL, H. 1962. Histologie und biologische Bedeutung eines jugulosternalen Duftdrüsenfeldes bei *Tupaia glis* Diard 1820. Verh. Dtsch. Zool. Ges., 55: 198–206.

STEINER, A. L. 1974. Body-rubbing, marking, and other scent-related behavior in some ground squirrels, a descriptive study. Canadian J. Zool., 52:889–906.

STEINIGER, F. 1950. Über Duftmarkierung bei der Wanderratte. Z. Hygien. Zool. Schädlingsbek., 38:357–361.

STEVENS, P. G., AND J. L. E. ERICKSON. 1942. American musk: I. The chemical constitution of the musk of the Louisiana musk rat. J. Amer. Chem. Soc., 64:144–147.

STODDART, D. M. 1970. Individual range, dispersion and dispersal in a population of water voles (*Arvicola terrestris* L.) J. Anim. Ecol., 39:403–425,

SVENDSEN, G. E. 1980. Patterns of scent-mounding in a population of beaver (*Castor canadensis*). J. Chem. Ecol., 6:133–148.

TAYLOR, J. C. 1968. The use of marking points by gray squirrels. J. Zool., 155: 246–247.

TEICHER, M. H., AND E. M. BLASS. 1977. First suckling response of the newborn albino rat: the roles of olfaction and amniotic fluid. Science, 198:635–636.

TEMBROCK, G. 1968. Land mammals. Pp. 338–404, *in* Animal communication (T. A. Sebeok, ed.). Indiana Univ. Press, Bloomington, Indiana, 686 pp.

TENER, J. S. 1965. Musk oxen in Canada. Dept. N. Aff. Nation, Resources, Canadian Wildl. Serv.

TERHUNE, J. M., M. E. TERHUNE, AND K. RONALD. 1979. Location and recognition of pups by adult female harp seals. Appl. Anim. Ethol., 5:375–380.

THIESSEN, D. D. 1977. Methodology and strategies in the laboratory. Pp. 391–412, *in* Chemical signals in vertebrates (D. Müller-Schwarze and M. M. Mozell, eds.). Plenum Publ. Corp., New York, 609 pp.

THIESSEN, D. D., AND M. RICE. 1976. Mammalian scent gland marking and social behavior. Psychol. Bull., 83:505–539.

THIESSEN, D. D., S. L. BLUM, AND G. LINDZEY. 1970. A scent marking response associated with the ventral sebaceous gland of the Mongolian gerbil (*Meriones unguiculatus*). Anim. Behav., 18:26–30.

THIESSEN, D. D., A. CLANCY, AND M. GOODWIN. 1976. Harderian gland pheromone in the Mongolian gerbil, *Meriones unguiculatus*. J. Chem. Ecol., 2: 231–238.

THIESSEN D. D., H. C. FRIEND, AND G. LINDZEY. 1968. Androgen control of territorial marking in the Mongolian gerbil. Science, 160:432–433.

THIESSEN, D. D. ET AL. 1974. Identification of a ventral scent marking pheromone in the male Mongolian gerbil (*Meriones unguiculatus*). Science, 184:83–85.

THOMSON, J. A., AND F. N. PEARS. 1962. The functions of the anal glands of the brush-tailed possum. Victoria Nat., 78:306–308.

THOMSON, J. A., AND W. H. Owen. 1964. A field study of the Australian ringtail possum *Pseudocheirus peregrinus*. Ecol. Monogr., 34:37–42.

TSCHANZ, B. 1962. Über die Beziehungen zwischen Muttertier und Jungen beim Mufflon. Experientia, 18:187–190.

TSCHANZ, B., M. MEYER-HOLZAPFEL, AND S. BACHMANN. 1970. Das Informations-system bei Braunbären. Z. Tierpsychol., 27:47–72.

ULLRICH, W. 1967. Nashornstrassen in Assam. Pp. 56–67, *in* Die Strassen der Tiere (H. Hediger, ed.). Vieweg, Braunschweig, 313 pp.

VACEK, Z. 1955. The innervation of the velvet of the growing antlers of the Cervidae. Cs. Morfologie, 3:249–264.

VAN DORP, D. A., R. KLOK, AND D. H. NUGTEREN. 1973. New macrocyclic compounds from the secretions of the civet cat and the musk rat. Recl. Trav. Chim. Pays.-Bas, 92:915–928.

VAN LAWICK, H., AND J. VAN LAWICK-GOODALL. 1971. The innocent killers. Houghton-Mifflin, Boston, 222 pp.

VERHEYEN, R. 1954. Monographie ethologique de l'hippopotame (*Hippopotamus amphibius* L.). Explor. Parc. National Albert. Inst. parc. nat. Congo Belge, Brussels, 91 pp.

VON HOLST, D., AND U. BUERGEL-GOODWIN. 1975. Chinning by male *Tupaia belangeri*. The effects of scent marks on conspecifics and of other species. J. Comp. Physiol., 103:153–171.

VON HOLST, D., AND S. LESK. 1975. Über den Informationsgehalt des Sternaldrüsensekretes männlicher und weiblicher *Tupaia belangeri*. J. Comp. Physiol. 103A:173–188.

VORONTSOV, N. N., AND N. N. GURTOVOI. 1959. Structure of the midabdominal gland of the true hamster (Cricetini, Cricetinae, Rodentia, Mammalia). (Transl. of) Doklady Biol. Sci. Sect. Akad. Nauk, SSSR, 125:285–288.

WALKER, E. P., ET AL. 1975. Mammals of the world. 2 vols. Johns Hopkins Univ. Press, Baltimore, 1500 pp.

WALTHER, F. 1966. Mit Horn und Huf. Vom Verhalten der Horntiere. Parey Verlag, Hamburg, 171 pp.

———. 1978. Mapping the structure and the marking system of a territory of the Thomson's gazelle. East African Wildl. J., 16:167–176.

———. 1979. Das Verhalten der Hornträger (Bovidae). Handb. Zool., 8(54):1–84.

WELLINGTON, J. L. ET AL. 1979. Perineal scent gland of wild and domestic guinea pigs. J. Chem. Ecol., 5:737–751.

WEMMER, C. M. 1971. Scent marking and anointing: behavioral parallelism in mammals. Amer. Zool., 11:623 (abstr.).

———. 1977. Comparative ethology of the large-spotted genet (*Genetta tigrina*) and some related viverrids. Smithsonian Contrib. Zool., 239:1–93.

WEMMER, C. M., AND J. MURTAUGH. 1980. Olfactory aspects of rutting behavior in the Bactrian camel (*Camelus bactrianus ferus*). Pp. 107–124, *in* Chemical signals: vertebrates and aquatic invertebrates (D. Müller-Schwarze and R. M. Silverstein, eds.). Plenum Publ. Corp., New York, 445 pp.

WISLOCKI, G. B., AND A. H. SCHULTZ. 1925. On the nature of modifications of the skin in the sternal region of certain primates. J. Mamm., 6:236–243.

WOLFF, J. D., AND M. F. JOHNSON. 1979. Scent marking in Taiga voles, *Microtus xanthognatus*. J. Mamm., 60:400–404.

YAHR, P. 1977. Central control of scent marking. Pp. 547–562, *in* Chemical signals in vertebrates (D. Müller-Schwarze and M. M. Mozell, eds.). Plenum Publ. Corp., New York, 609 pp.

ZANNIER, F. 1965. Verhaltensuntersuchungen an der Zwergmanguste, *Helogale undulata rufula*, im Zoologische Garten Frankfurt am Main. Z. Tierpsychol., 22:672–695.

CHEMICAL COMMUNICATION IN HOUSE MICE AND DEER MICE: FUNCTIONAL ROLES IN REPRODUCTION OF WILD POPULATIONS

F. H. Bronson

Abstract

The ecological, physiological, and behavioral literature dealing with the social organization and reproduction of house mice (*Mus musculus*) and deer mice (*Peromyscus maniculatus*) are surveyed as a background for conceptualizing the uses of chemical cues in these two species. Three general categories of pheromone use are apparent: 1) broadcasting of time-delayed cues that advertise occupancy of a particular locale; 2) support for the acute, short-range identifications that precede sexual and aggressive behavior; and 3) priming of several endocrine events related to reproduction. The broadcasting function would seem to be the most important function for most deer mice populations whereas endocrine priming, particularly in relation to the pubertal processes, is of obvious importance in both species. Acute identification probably is the least important dimension of pheromonal communication in either species despite the fact that this is the function most concentrated upon by laboratory scientists. The present contribution also attempts to enumerate some of our most obvious areas of ignorance regarding chemical communication in small, nocturnal rodents in general.

Introduction

The liberation of a chemical cue by one animal, and its perception by way of primary or accessory olfactory pathways of another animal, probably constitutes the most commonly-used mode of communcation that exists in mammals. Typically the sources of these cues are either urinary or glandular and their transmission can be either passive or active; that is, marking gestures may or may not be involved. The responses elicited by mammalian chemical cues are varied but usually are associated in some way with the process of reproduction. More especially, chemical communication seems to serve three broad functions in support of the breeding activity of mammalian populations. First, chemical cues may be important in ensuring an appropriate dispersion of the social units that compose the breeding population. Thus, one of the most common behaviors associated with the daily foraging activities of most mammals is the routine deposition of time-delayed chemical signals that advertise the occupancy of a particular locale by an individual or a social group. Second, chemical cues often play important roles supporting the acute, short-range identifications that precede aggressive or sexual interactions. Third, chemical cues may evoke stereotyped endocrine responses, thereby timing important physiological events such as ovulation.

Mammalian communication probably involves multisensory input, at least under most social conditions. Intuitively, however, it seems as though chemical communication should attain a degree of dominance in any ecological situation involving any of three factors: 1) nocturnality, which would largely eliminate the use of visual cues; 2) existence on a physically complex substrate, a situation that would impede the use of both visual and high-frequency auditory cues; and 3) a geographically wide dispersion of individuals or social groups, which would necessitate the use of either long-distance or long-enduring signals. Many rodents exist in milieus that encompass all three of these environmental characteristics and, hence, the communicatory needs of these species seem to be dominated by the chemical modality. The purpose of this paper is to consider broadly the potentially diverse roles that chemical cues might play in the reproductive effort of two species of rodents, the house mouse (*Mus musculus*) and the deer mouse (*Peromyscus maniculatus*). A more general objective, however, is to use the state of our knowledge about chemical communication in these two species as a basis for assessing what we know and what we do not know about chemical communication in small, nocturnal rodents in general.

The limitation of this assessment to just deer mice and house mice stems from the nature of the literature dealing with this subject. Chemical communication per se has been studied only in the artificial confines of the laboratory. Because of the cryptic life styles of most rodents, we have almost no direct knowledge of the use of chemical cues in natural populations. Thus my effort will be largely an integrative attempt to couple the literature dealing with the social organization and reproduction of wild populations, as developed by ecologists and geneticists, with our laboratory-derived knowledge as revealed by physiologists and behaviorists. Field biologists and laboratory researchers interested in chemical communication seldom find themselves working on the same species of rodent, the house mouse and the deer mouse being the rare exceptions.

In this review I develop a broad view of chemical communication as it enhances the reproductive effort of a population. Thus, the two species first are considered briefly from the standpoint of some general ecological similarities and dissimilarities. Second, the ecological and genetical literature is used as a basis for relating our laboratory-derived knowledge about behaviorally-active chemical signals to the potential communicatory needs of breeding populations of these two species. Third, our knowledge about the specific ambient regulation of reproduction in natural populations of house mice and deer mice is used for assessing the roles that could be played by endocrinologically-active chemical substances. Finally, the relationships developed above are used to develop generalities about rodents as a group.

As a final introductory comment, a chemical cue that conveys information between animals has been referred to as a "pheromone" in both insects and mammals (Karlson and Luscher, 1959; Whitten, 1966). Although there are reasonable arguments against using this term in reference to the highly labile, experience-oriented behavior of mammals (Beauchamp et al., 1976), the term has proven useful in many regards and it is incorporated in this review. Chemical cues that may (or may not) elicit behavioral responses are referred to as signalling pheromones; those that evoke endocrine responses are referred to as priming pheromones (Bronson, 1971).

Ecological Generalities

House mice and deer mice exhibit many obvious similarities associated with the fact that both are small, nocturnal, omnivorous, prey species with a high reproductive potential. However, there are several relevant differences, as follows. Evolutionarily, the genus *Peromyscus* has proven to be quite sedentary. *Peromyscus* populations typically are highly viscous, each exhibiting the results of considerable selection for local environmental conditions. Deer mice probably are the least sedentary members of the genus *Peromyscus* but they are still markedly different than the house mouse, which has an almost unexcelled record of worldwide colonization; populations of this latter species are markedly fluid and much less geographical selection is evident today in the species as a whole. In this same vein, house mice are obvious opportunists in their choice of habitat whereas deer mice are not. As a gross example, house mice often live either commensally with man or in strictly feral populations while deer mice typically live only ferally. Finally, feral populations of house mice are highly irruptive while the densities maintained by deer mouse populations are remarkably stable from year to year.

The range of the deer mouse extends from 20° N in Mexico to the Northwest Territories and throughout Newfoundland in the northeast. Within this range, selection for local conditions has yielded some 40 subspecies (Hooper, 1968). Deer mice occur in a wide variety of wetlands, woodlands, grasslands, or deserts (Baker, 1968); each subspecies, however, seems to exhibit relatively little opportunism in its choice of habitat (Kirkland, 1976). One of the more interesting characteristics of deer mouse populations is their temporal stability. In his review of the records of the North American Census of Small Mammals, Terman (1966) found the year-to-year variation in the trapping success of deer mice to be among the smallest of all mammals recorded in this census. *P. maniculatus* may show a greater degree of annual variation in density at the higher latitudes (>60°, Fuller, 1969*a*) and occasional high densities of a local nature have been reported elsewhere (for example, see Hoffmann, 1955). Nevertheless, in this species over most of its range, each population stays in situ in a relatively unique and relatively stable habitat, adapts to that habitat and to other local conditions, and then remains more or less stable in numbers for long periods of time at or below carrying capacity.

House mice evolved on the dry steppes of southeastern Asia, but today they occur on all of the major land masses and even in the circumpolar islands (Berry, 1970; Schwarz and Schwarz, 1943). In part because it is metabolically adapted to dry environments, this species long ago adopted a commensal existence with man and has been transported throughout the world with our cereal crops. Being able to reproduce under exceedingly diverse conditions, the house mouse today is found in more places and in more habitats, feral or commensal, than almost any other mammal (Bronson, 1979*a*). Although genetic divergence has occurred because of the global breadth of the conditions colonized by this species, it still maintains marked genetic similarity over such broad areas as the entire continent of North America (Bruell, 1970; Schwarz, 1945). Commensal populations that occur in man's homes and outbuildings are stable both in occurrence and in density over long periods of time. Feral populations, on the other hand, are extremely opportunistic, changing habitats frequently, and they are extremely irruptive (Bronson, 1979*a*). In many harsher climates feral house-mouse populations cease to exist during the winter months, being reseeded each spring from commensal populations (for example, see Zejda, 1975). In marked contrast to deer mouse stability, then, the house mouse seems to exhibit a remarkable flexibility in most of its strategies, physiological and ecological.

Demography and Social Organization of Wild Populations

The potential uses of signalling pheromones in any mammal will be dependent ultimately upon the demographic aspects of its populations and, in particular, upon the nature of its basic social unit.

Wild Populations of House Mice

Most studies of house mice derive from one of two population types: relatively stable, high density, commensal populations found in barns, chicken coops or granaries, or created in the laboratory; and relatively unstable, low density, feral populations found in grasslands or cultivated areas. A third type of population, one occurring seasonally in stacked grain, can be considered as a variant of the low density, feral type because its founders typically come from feral habitats. The most common technique for studying house-mouse populations is either live- or dead-trapping. Blood enzyme studies, from which gene flow can be inferred, also have been conducted. Fortunately, such indirect methods also have been supplemented by actual observation of freely-growing populations in outdoor enclosures and indoor pens.

When house mice inhabit man-made structures where food is plentiful and stable, and where the substrate is of medium complexity, the result is a population that may endure at high density for many months to a few years. Indeed, densities approaching 10,000 mice per ha have been reported (reviewed by Anderson, 1961). This type of mouse population occurs commonly throughout the tropical and temperate zones wherever man has created the appropriate conditions. A host of workers have concerned themselves with the social organization of commensal populations, both natural and laboratory-confined (Adamczyk and Walkowa, 1971; Anderson and Hill, 1965; Andrzejewski et al., 1963; Brown, 1953; Christian, 1970, 1980*a*; Christian and Davis, 1964; Christian et al., 1965; Crowcroft, 1966; Crowcroft and Rowe, 1963; Davis, 1958; DeFries and McClearn, 1970; Eibl-Eibesfeldt, 1950; Harrington, 1976; Lidicker, 1965, 1976; Lloyd, 1975, 1980; Lloyd and Christian, 1969; Mackintosh, 1970, 1973; Nowak, 1971; Oakeshott, 1974; Petrusewicz, 1957; Petrusewicz and Andrzejewski, 1962; Poole and Morgan, 1976; Reimer and Petras, 1967; Reimov et al., 1968; Selander, 1970; Southwick, 1955*a*, 1955*b*; Strecker, 1954; Strecker and Emlen, 1953; Stueck and Barrett, 1978; Walkowa, 1964, 1971; Young et al., 1950).

The social organization of dense commensal populations of house mice typically includes a division of the living area into territories, each of which is defended by a single male. Thus, the basic social unit usually includes several breeding females, some of their offspring, and a few subordinate males, all in turn dominated by the one territorial male. Indirect evidence—for example, freedom of movement and reproductive success—suggests that there is a loose hierarchical ranking among the adult females within a territory, but the aggressive reinforcement of such ranking is not overt (Lloyd and Christian, 1969; see also Crowcroft, 1966). Nevertheless, females, particularly if pregnant, may aid the dominant male in territorial defense. Home-range movements are restricted to the territory and thus typically they total only a few square meters (for example, see Young et al., 1950).

Gene flow between territories is restricted because of aggressive defense; hence, each insular division of the mouse's living space has been conceptualized as a deme as well as a territory (Anderson,

1970; Selander, 1970). Given chronological stability of the ecological milieu, several generations may be born, but the number of territories apparently remains more or less constant (Lidicker, 1976). Two obvious consequences of such rigid territoriality are a lack of potential for irruptive population growth in situ and a high rate of dispersal. Dispersing females, usually young, may occasionally be successful in entering a dense and well-established territory; the mortality of dispersing young males and of adult subordinate males, however, seems to be routinely high.

House mice often occur as strictly feral populations in grasslands, old-fields and cultivated areas (Berry, 1968; Berry and Jakobson, 1974; Breakey, 1963; Caldwell, 1964; DeLong, 1967; Justice, 1962; Lidicker, 1966; Myers, 1974; Newsome, 1969*a,* 1969*b*; Pearson, 1963). The carrying capacity of such areas is much lower than it is for most commensal situations; hence, there is an expectation of both a low density and, correlatively, a larger home range. In the studies noted above, for example, non-irruptive densities typically have been reported in the range of one to 100 individuals per ha while individual home ranges may be as large as one hectare (for example, see DeLong, 1967; Justice, 1961; Quadagno, 1968). However, feral house mice are extremely irruptive and densities approaching 100,000 per ha have been recorded (see Anderson, 1961).

Unfortunately, we know little about the type, or even the existence of social structures in widely dispersed, feral populations of house mice. Intuitively the deme-territory mode of organization would seem impossible to maintain under these conditions. The constant patrolling and harassment necessary both for the maintenance of dominance within a territory, and for the defense of the territory itself, would seem incompatible with the size of the house mouse and a home range measured in thousands of square meters. The use of enduring olfactory cues for boundary marking would not necessarily modify this conclusion because, to be effective, such communication should require prior conditioning (subordination) and considerable reinforcement. Furthermore, one of the most obvious features of feral populations of house mice is their spatial instability. Seasonal or aperiodic changes in habitat quality, such as those related to cultivation practices and climatic patterns, result in considerable home-range shifting (for example, see Berry and Jakobson, 1974; Justice, 1962; Newsome, 1969*a,* 1969*b*; Newsome and Corbett, 1975). Indeed, Caldwell (1964) has considered feral

M. musculus to be semi-nomadic and questions whether adult house mice living in feral conditions have sufficient permanence of living locale even to justify the employment of a home-range concept (see also Gentry, 1966). Another characteristic of feral populations is their high turnover rate, relative to commensal populations. Mortality may run as high as 30% per month (Berry et al., 1973; Lidicker, 1966) and overwintering losses may be as high as 90% (Berry, 1968). Finally, these populations typically (but not always) have seasonal reproduction and dispersal. All of these factors yield marked spatial and temporal instability which, in turn, argues against the existence of the stable, deme-territory substructures that exist in barns and chicken coops.

The existence of aggression is not at question in feral populations of house mice, only the degree to which such aggression is locale-oriented. DeLong (1967), for example, reported that 42% of the adult males in one of his grassland samples were carrying fresh wounds; the comparable figure for females was less than 10%. These figures are not much different from those often seen in high-density, experimental populations. Thus, aggression between adult males seems to be a constant in house mouse populations, regardless of location.

As noted previously, seasonal populations occurring in stacked grain have been intensively studied in Britain and Czechoslovakia (Pelikan, 1974; Rowe et al., 1963; Southern and Laurie, 1946). Here grain is harvested in an unshucked form and then stacked in late summer, to remain in this condition until spring. Feral mice that live in fields and hedgerows during the rest of the year rapidly find and invade these stacks. Breeding often occurs throughout the winter, and spring densities sometimes exceed 15 mice per m^3 (Southern and Laurie, 1946). Populations in individual stacks range up to 1,600 mice (Rowe et al., 1963, 1964; Southwick, 1958). Loosely stacked (unshucked) grain could offer the structural complexity necessary for the development of many small territories. Whether or not this type of social organization actually occurs, however, is unknown. The larger than expected home ranges hinted at in such stacks (Southern and Laurie, 1946) argue against the capacity to defend territories in this structurally complex situation and the high rates of wounding that have been found in such populations probably indicate unstable social conditions (Rowe et al., 1964). Likewise, the movement out of these ricks by both adult males and females (Rowe et al., 1963) suggests that more social chaos occurs in ricks than in barns, where the dispersers are usually young.

All things considered, the social system of the house mouse probably is best characterized as being markedly flexible. Stable deme territories do occur where resources are plentiful and stable. Where resources are sparse, however, the basic social unit would seem most often to be simply the individual and its nearest neighbors, and where environmental conditions are in a state of flux there probably is an overriding dimension of social instability due to the species' more or less continuous, opportunistic shifting of locales.

Populations of Deer Mice

The social behavior and population dynamics of the deer mouse also have been studied extensively in the laboratory and in large outside pens. Data from natural populations consist largely of trapping records, but nest boxes sometimes have been added to a population's substrate to aid the collection of data relevant to social structuring. This section attempts to sort out the various sets of data and to arrive at some generalities. The following references form the basis for this discussion: Abramsky et al., 1979; Andrews et al., 1975; Beer and MacLeod, 1966; Blair, 1940, 1953, 1958; Burt, 1940; Eisenberg, 1962, 1968; Fairbairn, 1977*a*, 1977*b*, 1978*a*, 1978*b*; Fuller, 1969*a*, 1969*b*; Halpin and Sullivan, 1978; Hansen and Fleharty, 1974; Healey, 1967; Hill, 1977; Hoffmann, 1955; Holbrook, 1979; Howard, 1949; Metzgar, 1978; Mihok, 1979; O'Farrell, 1978; Petticrew and Sadleir, 1974; Sadleir, 1965, 1970*a*, 1970*b*, 1974; Savidge, 1974; Sheppe, 1963; Stickel, 1968; Sullivan, 1977, 1979; Terman, 1961, 1966, 1968, 1973*a*, 1973*b*, 1974, 1979, 1980; Vestal and Hellack, 1978.

The density of a breeding population of deer mice can vary from less than one to over 40 deer mice per ha, depending upon habitat quality (see Abramsky et al., 1979; Fairbairn, 1977*a*; Sullivan, 1979). Individual home-range size also varies with habitat quality, having been reported anywhere from 0.1 to over 2 ha (Stickel, 1968). However, deer mice seem to maintain a home range that is somewhat larger than might be expected for an animal of this size (O'Farrell, 1978). Also, in contrast to house mice, adult deer mice seldom shift their home ranges during the breeding season (Hansen and Fleharty, 1974; Howard, 1949; Terman, 1961).

The potential for social structuring of the widely-dispersed members of a deer-mouse population has intrigued many workers. Generally, a basic breeding unit consists of an adult male, one or more breeding females and a prepubertal litter. Extensive use of multiple-capture traps and underground nest boxes in both natural populations and outdoor enclosures have consistently verified the fact that some degree of pair-bonding may occur in this species (Blair, 1958; Howard, 1949; Metzgar, 1978; Mihok, 1979; Terman, 1961). Howard (1949) even provides convincing evidence of a sometimes quite close relationship between a father and his offspring; the attractive aspects of this relationship have been verified in the laboratory (Savidge, 1974). Thus the most typical pattern for a deer-mouse population probably involves over-dispersion of daily activities within sexes and attraction between sexes (see Metzgar, 1978). Because mortality can be high even for resident deer mice, however, perturbations in this pattern should be common and, depending upon food reserves, population density, and other factors, the basic pattern described above could easily be obscured in live-trap records (see Mihok, 1979). However, the basic social unit noted above must typically be organized somewhat loosely; nest-box studies often have documented the spatial separation of nearest neighbor males and females during the breeding season (for example, see Howard, 1949; Terman, 1961).

The question of whether or not either individual deer mice or the breeding units noted above show territoriality also has intrigued researchers. Based on a lack of overlap between home ranges, several authors have inferred territorial behavior (Burt, 1940; Hansen and Fleharty, 1974); others have observed considerable overlap and, consequently, have inferred an absence of territorial behavior (Blair, 1940; Howard, 1949). A degree of conflict exists even in laboratory studies designed to answer this question (see Eisenberg, 1962; Hill, 1977). Most authors agree, however, that there is a "negative force" of some kind operating to separate spatially the daily activities of same-sexed residents in a breeding population, and that this force is simple avoidance rather than the active defense of an area. The same arguments about small size and an inability to defend a large area that were applied previously to feral populations of house mice probably apply here also. In support of this argument, Metzgar (1978) reported less activity at the periphery of an individual deer mouse's home range, suggesting avoidance rather than territorial behavior. Additionally, Howard (1949) reported several instances of two females jointly bearing litters in one nest box in his field populations, and he also has reported some instances of multiple residence of adult males in a nest box in a breeding season. Finally, Terman (1961) observed mutual avoidance when deer mice encountered each other while foraging, and resident males avoiding their own nest box when an alien male was tethered near its entrance.

The daily social interactions of deer mice probably involve much less overt aggression than those of house mice (King, 1957; Terman, 1974). Nevertheless, overt aggression can be a potent force in the sense that large-scale dispersal of young is common in its breeding populations and this is associated with adult aggression. A seasonal cycle in male aggressiveness, undoubtedly caused by changes in circulating steroid hormones, has been elegantly documented in wild deer mice (Healey, 1967; Sadleir, 1965). Lactating females may show even more aggression toward unfamiliar young than do breeding males (Ayer and Whitsett, 1980; Savidge, 1974; Whitsett et al., 1979). In one of the most well-studied subspecies, *P. m. austerus* in British Columbia, several workers have defined a recurring pattern of seasonal changes in density because of changes in aggression and dispersal. Few young are recruited into the resident populations during the early part of the breeding season, and most become transient. Resident population density therefore remains relatively low and quite stable throughout the breeding season. The last litter of the season, however, is recruited, thereby yielding a dramatic increase in density during the fall months, a peak which then declines slowly throughout the winter until a period of social reorganization and dispersal of subordinate males prior to the onset of breeding in the spring (Fairbairn, 1977*a*, 1977*b*, 1978*a*, 1978*b*; Healey, 1967; Sadleir, 1965). This is not the only seasonal pattern shown by this species, however (for example, see Fuller, 1969*a*; Mihok, 1979; Sullivan, 1977). As discussed by Fairbairn (1978*a*), dispersal is a complicated process in deer mice; it is related both to the social pressures in a stable breeding population and to annual social disorganization.

Communication by Signalling Pheromones in Wild Populations of House Mice and Deer Mice

As already noted, chemical communication probably serves three broad functions in breeding populations of mammals—occupancy notification, short-range identification, and endocrine priming. This section considers the first two functions. In regard to occupancy notification, the available laboratory evidence suggests that both house mice and deer mice routinely broadcast chemical information by way of urine marks that are deposited on the ground during daily foraging and exploratory activity. Urine-marking has been examined extensively in the laboratory with both wild and domestic stocks of house mice, less so in the case of deer mice. In the laboratory, male house mice may deposit up to 2,000 marks in an overnight test (Desjardins et al., 1973). This behavior is generally an androgen-dependent response to any kind of environmental novelty and it is largely absent in subordinate males (Desjardins et al., 1973; Maruniak et al., 1974; Van Abeelen and deVries, 1978). Female house mice mark much less frequently than do males; their marking is independent of gonadal status, however, and an elevated marking frequency by a female seems to occur only in response to the presence of males or male urine (Maruniak et al., 1975*a*). Male deer mice are prodigious urine markers provided they are sexually experienced (Maruniak et al., 1975*b*). Because sexual experience is without such an effect in house mice, there are at least some differences in the modulation of this behavior in the two species, but in all probability there are more similarities than dissimilarities. We know little about possible sources of broadcast-type chemicals in addition to urine in either species. Some evidence suggests a communicatory role for the undifferentiated glands in the foot pads of house mice (Ropartz, 1967); deer mice possess a ventral gland (Doty, 1973).

Despite the fact that urine-marking has not been studied extensively in deer mice, we probably have sufficient information to state that this behavior and its associated broadcasting function probably constitutes the most important use of chemical cues by this species. The daily activity of a deer mouse is not spread equally about its home range. Blair (1940) estimated that a single deer mouse may use an average of 20 different nest sites, often varying the part of the home range that is used in any one day and the nest site in which it resides, depending upon seasonal availability of food and other factors (see also Howard, 1949; Metzgar, 1978; Terman, 1961). Thus, a deer mouse's home range should be visualized as a series of well-used trails connecting multiple nest sites and a complex of periodically traveled trails leading to various food sources (see Fuller, 1969*b*). Because a deer mouse probably uses only a small fraction of this complex of trails and nests during any one day's foraging, this argues strongly against the maintenance of within-sex spacing by overt agonistic behavior. Encounters with nearest neighbors of the same sex, given only a few individuals per ha, must be relatively rare. The real advantage for using a chemical cue in this situation is simply that the sender of information need not be present when it is perceived by another animal. Thus, the enforcement of spacing by avoidance of long-enduring urinary cues deposited along trails and around nest sites must be one of the most important characteristics of deer-mouse populations. Avoidance of such cues by transients as well as neighbors certainly would lead to, or at least strongly support, the stable densities that are characteristic of breeding deer-mouse populations.

The marking function might or might not be of equal importance in feral populations of house mice either in spite of, or because of, their tendency to shift home ranges frequently. We know little about either possibility at this time. The enforcement of territorial boundaries by urinary cues has been documented in high density, laboratory populations of house mice (for example, see Harrington, 1976). Nevertheless, the typically small size of these territories (often only a few meters square), and the continuous close proximity of animals in adjacent territories, would seem to suggest a somewhat lower importance for broadcasting occupancy via chemical cues.

The short-range identifications that precede sexual and aggressive behavior probably are not dependent ultimately on any one sensory modality and, hence, the olfactory aspects of such communication must be considered somewhat less important. From a theoretical standpoint, two animals meeting in the dark must accomplish a series of identifications through auditory (Nyby and Whitney, 1978) or chemical cues before the tactile aspects of sexual or agonistic behavior can ensue. Sex and relative age must be determined first, then sexual state and finally, possibly, individual identification. This suggests a sequence of information transfers starting when the animals are a few centimeters apart or, at least, when they are standing head-to-head, and ending when naso-anal orientation has been achieved. As already shown, all information necessary to identify sex, relative age and sexual state (and, therefore, to allow a fight to erupt or not) is contained in the urine of both sexes of the house mouse (Dixon and Mackintosh, 1971, 1976). Although urinary information may be extremely important for the broadcasting function, however, it is difficult to see how it could always function in a fool-proof way for the head-to-head, last-minute identifications necessary to organize sexual or aggressive behavior. First, the volatility requirements for long-enduring broadcast-type urinary cues must be markedly dissimilar to those used in acute short-range identification, and second, naso-anal orientation is not always achieved in male-male encounters before a fight erupts, at least not in house mice. Therefore it is probable that the chemicals comprising the information noted above are general throughout the body and occur in different chemical forms as well as being localized in particular components such as urine.

We know nothing about head-to-head communication in house mice except that strange males spend considerable time sniffing each other's muzzle and ear region. The possibility of a breath-pheromone emanating from the androgen-dependent submaxillary gland seems possible here (see Claus and Alsing, 1976). Vaginal secretions are known to contain information relevant to stages of the ovulatory cycle in house mice (Hayashi and Kimura, 1974) and the male's preputial glands contain lipids that arouse sexually-experienced but not sexually-inexperienced females (Caroom and Bronson, 1971). In this latter regard, preputial secretions undoubtedly are emitted during copulatory activity, thereby conditioning the females to associate this odor with sexual activity. Thus, the strong response of sexually-experienced females to preputial secretion is less a verification of the presence of a "pheromone" than it is a verification that copulatory activity has reinforcing properties in female mice.

We know little about the sequences of identifications necessary for deer mice to react sexually or aggressively except that male deer mice do not have preputial glands. Nevertheless, both species are nocturnal and the importance of chemical cues in such identifications is obvious, even if it is less important than the occupancy-broadcasting function.

Ambient Regulation of Reproduction

This section considers the ambient regulation of reproduction in the two species as background for viewing the functional roles that might be served by their endocrinologically-active primer pheromones. Physiologically, both house mice and deer mice are classified as short cycling, spontaneous ovulators whose corpora lutea become functional because of copulatory activity (Bradley and Terman, 1979; Bronson et al., 1966; Conaway, 1971; Dewsbury, 1979). To consider the ambient regulation of reproduction, each species can be classified according to three classes of environmental factors that influence the brain-pituitary-gonadal axis and, hence, the reproductive processes of any mammal. The first class encompasses those aspects of the physical and biological environment that vary seasonally and that affect gonadotropin secretion directly or indirectly to elicit seasonal breeding in a population: photoperiod, nutrition, available calories, or ambient temperature. Second are those factors that may depress the reproductive effort of particular individuals in a population during the breeding season; these include agonistic stimuli and caloric and nutritional availability in the sense that the freedom to forage may be suppressed in certain individuals by aggressive activity of other individuals. Finally, there are factors which operate along genetically-fixed pathways to time particular reproductive events during an individual's normal reproductive cycle. Of importance here are the day-night cycle and two types of social stimuli—tactile cues and priming pheromones.

Factors Influencing Seasonal Breeding

Both deer mice and house mice breed more or less continuously under constant laboratory conditions (for example, see Drickamer and Vestal, 1973; Lidicker, 1965). However, breeding throughout the year may or may not be characteristic of each species in its natural environs. Thus, our interest is with factors that might or might not cause a seasonal interruption in an inherent tendency for continuous breeding. Photoperiod, available calories, nutritional factors, and ambient temperature all vary seasonally and correlatively in the field environments normally occupied by these two species (Sadleir, 1969), and separation of one factor from another in regard to the production of seasonal breeding is impossible under field conditions. Fortunately this problem has been of some concern in the laboratory.

The annual patterns of reproduction that are exhibited by natural populations of house mice are not complicated. This subject has been reviewed previously (Bronson, 1979*a*). House mice in commensal populations appear to breed throughout the year. However, except for some maritime areas, feral populations always show seasonal breeding (Table 1). This suggests that house mice breed continuously in microclimates that show little seasonal variation (a house or a grain elevator) while breeding seasonally where seasonal variation exists. Thus, reproduction in the house mouse probably operates independently of photoperiodic control. This has been confirmed in the laboratory with both wild and domestic stocks; indeed, house mice reproduce nicely in continual darkness (Bronson, 1979*b*; Stoddart, 1970). Although not experimentally verified, seasonal breeding in house mice must reflect seasonal changes in a specific nutrient, in available calories, in ambient temperature, or some combination of these three factors.

Cold interferes with the reproduction of house mice in manifold ways (reviewed by Barnett, 1965, 1973; other relevant references can be found in Bronson, 1979*a*). However, when subjected to chronic temperatures even as low as −3°C, some house mice still breed more or less normally (Barnett, 1965) and, indeed, natural populations have been observed to breed in cold storage lockers maintained at −10°C (Laurie, 1946). Also, there always is a large discrepancy between a low ambient temperature and that of the nest occupied by a mouse. Thus, evidence indicates that seasonal breeding in house mice is not a simple response to temperature variation. Unfortunately we know little about the dependence of the reproductive processes of house mice upon either specific nutrients or available calories. However, DeLong (1967) and Newsome (1970) both were successful in producing winter breeding with food supplements in normally seasonal breeding populations of house mice in the field. In summary, seasonal breeding in house mice may or may not occur depending upon the constancy of the food supply; the proximal factor(s) underlying seasonal breeding must be dietary, with or without secondary interaction with physical effort and cold (Bronson, 1979*a*).

Deer mouse populations with their broad distribution also show considerable variation in their annual patterns of reproduction (see Table 2 for a summary). Although the probable reliability of the different data presented in Table 2 varies (this is true for Table 1 also), several patterns are obvious. First, even though deer mice usually occur only in feral environments, some populations still show continuous breeding throughout the year. For example, Schef-

TABLE 1
SEASONALITY OF BREEDING OF HOUSE MICE

Latitude	Location	Habitat	Breeding	Reference
57°N	Isle of Skye	Flour warehouses	Continuous	Laurie (1946)
54°S	Macquarie Island	Grassland	Continuous	Berry & Peters (1975)
51°N	Isle of Skokolm	Grassland	Seasonal, 6 months, spring/summer	Batten & Berry (1967)
51°N	Southern England	Urban buildings	Continuous	Laurie (1946)
51°N	Southern England	Cropland, hedgerows	Seasonal, 8 months, summer/fall	Southern & Laurie (1946)
51°N	Southern Alberta	Granery	Continuous	Anderson et al. (1977)
49°N	Central Czechoslovakia	Urban buildings	Continuous	Pelikan (1974)
49°N	Central Czechoslovakia	Cropland	Seasonal, 7 months, spring/fall	Pelikan (1974)
37°N	West-central California	Grassland	Seasonal, 7–10 months, spring/fall	Breakey (1963); DeLong (1967); Lidicker (1966); Pearson (1963)
34°S	South-central Australia	Wheatland, marshland	Seasonal, 6–8 months, spring/fall	Newsome (1969*a*)
30–35°N	State of Mississippi	Urban and farm buildings	Continuous	Smith (1954)

fer (1924) reported 11 to 66% of his trapped females as pregnant each month throughout the year in eastern Washington; there was no obvious seasonal pattern in this variation. Likewise, Svendsen (1964) reported continuous breeding year-round in Kansas (see also Brown, 1945, for evidence of winter breeding in Kansas), even though one relatively sharp 8-month breeding season was reported a few hundred miles northeast in Nebraska (Andrews et al., 1975). Indeed, occasional winter breeding has been reported even in Michigan in January among deer mice residing in corn shocks (Linduska, 1942).

Despite the occasional occurrence of continuous breeding, the data in Table 2 suggest a short, sharp breeding season that occurs in the spring and summer at high latitudes, grading into more prolonged breeding seasons starting in the spring and ending in the autumn in middle latitudes, with breeding strictly limited to winter months in Texas. Latitude is not always an overriding consideration. Some winter breeding is correlated with winter rains in deserts in eastern California, whereas deer mice show a short summer breeding season at 12,000 feet in the same area (Dunmire, 1960). Indeed workers who have studied *P. m. austerus* in British Columbia have collectively shown considerable year-to-year variation in the length of the breeding season and even variation between localities within the same year (Fairbairn, 1977*b*; Fordham, 1971; Sadleir, 1974; Sullivan, 1977, 1979).

Thus, each population's breeding season (or lack of it) appears tailored to the particular environment inhabited by that population. This also suggests that the reproductive efforts of the various populations probably are regulated by different combinations of environmental factors, a conclusion supported by laboratory studies documenting subspecies variation in the degree to which reproduction is responsive to photoperiodic variation. Laboratory stocks of *P. m. bairdii,* which originated in Michigan, show normal spermatogenic activity on 14 h of light per day but a dramatic, four-fold drop in such activity on a 10:14 L:D cycle; *P. m. palescens* from Texas, on the other hand, reproduce well on either long or short daylengths, as the experimental daylengths were defined by the annual photoperiodic cycle normal for that area (Desjardins and Lopez, 1980; see also Price, 1966). The best ambient correlate of the onset and cessation of breeding in natural populations of deer mice in British Columbia was temperature variation, with daylength being of lesser

TABLE 2
SEASONALITY OF BREEDING OF DEER MICE

Latitude	Location	Habitat	Breeding	Reference
61°N	Great Slave Lake, Northwest Territories	Coniferous forest	Seasonal, 2–3 months, spring/summer	Fuller (1969*a*, 1969*b*); Mihok (1979)
56°N	Manitoba	Coniferous forest	Seasonal, 4 months, spring/summer	Millar and Wille (1979)
49°N	Southwestern British Columbia	Coniferous forest	Seasonal, 3–8 months, spring/summer	Fairbairn (1977*a*); Fordham (1971); Sadleir (1974); Sullivan (1977, 1979)
48°N	Southeast Minnesota	—	Seasonal, 8 months, spring/fall	Beer et al. (1957); Beer and MacLeod (1966)
48°N	East-central Washington	Arid wheatland	Continuous	Scheffer (1924)
45°N	Western Oregon	Coniferous mixed forest	Continuous but most in 8 months, spring/fall	Gashwiler (1979)
43°N	Western Massachusetts	Hardwood forest	Seasonal, 5 months, spring/summer	Drickamer (1978)

TABLE 2
CONTINUED

Latitude	Location	Habitat	Breeding	Reference
43°N	Southern Michigan	Grassland	Seasonal, 7–8 months, spring/fall	Blair (1940); Howard (1949)
42°N	Southeastern Wyoming	Semi-arid grassland	Seasonal, 4 months, spring/summer	Brown (1966)
42°N	Eastern Nebraska	Cultivated areas, shelters	Seasonal, 8 months, spring/fall	Andrews et al. (1975)
40°N	Northern California	Woods and brush	Seasonal, 5 months	Jameson (1953)
40°N	Southern Pennsylvania	Grassland	Seasonal, 6 months, spring/summer	Christian (1980*a*, 1980*b*)
39°N	Kansas	Grassland	Continuous	Svendsen (1964)
37°N	West-central California	Chaparral	Seasonal, 7 months, spring/fall	McCabe and Blanchard (1950)
31°N	Central Texas	Fencelines, cultivated areas	Seasonal, 6 months, fall/winter	Blair (1958)

importance and precipitation least important (Sadleir, 1974). Andrews et al. (1975), on the other hand, found hard freezes uncorrelated with the onset of breeding activity in Nebraska. Both Sadleir et al. (1973) and Stebbins (1977) calculated that deer mice could not obtain sufficient calories in the winter at northern latitudes to support lactation. Importantly, Fordham (1971) was able both to accelerate the onset of breeding and to increase breeding intensity by giving food supplements to one natural population.

In summary, photoperiodic control of reproduction has never been seen in house mice and it may or may not occur in deer mice, depending upon the population under consideration. In all likelihood, dietary variation of some kind is supremely important in both species as a first-order cueing system. Whether there is interaction of physical effort and cold temperature with either the dietary or the photoperiodic pathways of control remains an open question.

Factors Depressing Reproductive Activity of Individuals During a Breeding Season

My concern here is with the action of agonistic stimuli on some members of a breeding population, or with dietary inadequacies that occur in some individuals owing to aggressive activity of the other members of a population. Because almost nothing is known about the latter problem, the present discussion focuses on the depressive effects of agonistic stimuli on reproduction.

Non-specific stimuli associated with fighting and/or subordination may act in various ways on the endocrine physiology of a house mouse to depress reproductive activity (reviewed by Christian, 1970, 1980*a*, 1980*b*; Christian and Davis, 1964; Christian et al., 1965; Lloyd, 1980). The most obvious hormonal correlates of this response include those of the adrenal and the reproductive glands. The nature of the relevant cue in this case is internal and emotional, a fact best shown by two studies. First, the release of corticosterone was shown to be a conditioned response to defeat at the hands of a trained fighting mouse (Bronson and Eleftheriou, 1965). More recently, Lloyd (1980) demonstrated a four-fold decrease in litter survival and a marked decrease in mammary gland weight when lactating female house mice and their litters were housed adjacent to groups of fighting males, from which they were separated by a wire mesh barrier.

The reproductive effects of this type of non-specific stimulation are best observed in freely-growing but confined populations of house mice where agonistic behavior is common. Adult males show smaller testes and sex accessories, while females show increased intrauterine and postnatal loss of offspring. Importantly, not all members of an experimental population show these effects. Dominant and/or territory-holding males are little affected, unless they are actively engaged in establishing dominance, whereas subordinated males exhibit these effects to a much greater degree. Some females seem unaffected but the relationship between these individuals and the female social structure of a population remains confusing. One of the more obvious characteristics of dense experimental populations of house mice, however, is a prolonged inhibition of sexual maturation in both sexes, again apparently caused by an emotional reaction to agonistic behavior.

With some variation, the general phenomenon described above for house mice also occurs in caged populations of deer mice. As shown in a series of papers by Terman and associates, freely-growing but confined populations of deer mice also show a massive failure to reproduce and/or a failure of any young to survive until weaning (Gardner and Terman, 1970; Terman, 1968, 1973*a*, 1973*b*, 1974, 1979). However, overt aggression is not an obvious correlate of this action in deer mice. The most dramatic effect of crowding in deer mice is an almost complete failure of sexual maturation. Indeed, in some of Terman's freely-growing populations, as many as 95% of the young that were alive at asymptote still had failed to show any reproductive activity by 90 days of age, a phenomenon that was related in some way to a critical period prior to 40 days of age (Gardner and Terman, 1970). Reproductive activity was eventually recoverable in 75% of such animals by removing them from their asymptotic population and housing them as bisexual pairs.

Specific Cues Timing Particular Reproductive Events

Certain physiological events in a normal reproductive cycle may depend upon external cueing for their temporal organization (see Vandenbergh, this volume). The coupling of day-night perception to circadian oscillators in order to time all of the physiological changes underlying locomotion, feeding, sexual behavior, and ovulation in a nocturnal rodent is a case in point. Social cues also are important in this regard. Two examples are reflex ovulation and the milk letdown response to suckling, both of which involve a tactile cue which evokes an endocrine response and times the reproductive event. Mammals routinely use tactile cues and priming pheromones in such cases. Indeed, I know of no examples in mammals in which other sensory modalities are so used, and often, perhaps always, tactile cues and priming pheromones work in concert to perform such tasks. Thus, in a sense, a priming pheromone can be considered as a long-range or a long-enduring "tactile cue," where long-range may be a matter of a few centimeters at most.

The urine of both house mice and deer mice contains chemicals that can act on the hypothalamo-hypophyseal unit of conspecific individuals to alter temporally certain reproductive processes. The evidence in house mice has been reviewed several times (Bronson, 1971, 1979*a*; Vandenbergh, 1975; Whitten and Champlin, 1973); thus, most of the following relevant citations are for less publicized deer-mouse research.

Generally the urine of one sex tends to accelerate certain reproductive events or processes in members of the opposite sex while decelerating the same events or processes in members of the same sex. Urine of the opposite sex affects ovulation in both the adult and peripubertal female, implantation, the release of reproductive hormones in adult males and maturation of young males (Table 3). First, in regard to the house mouse, female urine decelerates the estrous cycle of other females, often sending them into pseudopregnancy if they are adults (Kimura, 1971; Ryan and Schwartz, 1977; Whitten, 1959), and greatly delaying the pubertal ovulation of prepubertal females (Drickamer, 1974*a*; Vandenbergh et al., 1972). Exposure to a male or his urine accelerates the ovulatory cycle of both adult and immature females (Marsden and Bronson, 1964; Vandenbergh, 1969; Whitten, 1956). However, female cues override those of the male if the females are immature (Vandenbergh et al., 1972), but the male's cues override those of other females if the female is an adult (Whitten, 1956). The presence of a strange male (that is, any male other than the stud) will prevent implantation in a recently-inseminated female and return her to an estrous condition (Bruce, 1959, 1966). Pooled male urine also elicits this effect. Finally, urine of any genetic female house mouse releases luteinizing hormone and testosterone in adult male house mice

TABLE 3
SEXUAL COMBINATIONS IN WHICH URINE HAS OR HAS NOT BEEN SHOWN TO INFLUENCE A REPRODUCTIVE EVENT OR PROCESS

Urine Source	Recipient	Effect	Documented: Deer mouse	Documented: House mouse
Male	Peripubertal or adult female	Accelerates ovulation in adult and peripubertal females; can prevent implantation	Yes	Yes
Male	Immature male	Decelerates sexual maturation	Yes	No*
Female	Peripubertal or adult male	Releases LH, thereby increasing pheromonal potency (?)	No**	Yes
Female	Peripubertal or adult female	Decelerates ovulatory cycle; overrides male effect in the pubertal female	No**	Yes

* Sexual maturation is inhibited by the presence of an adult male in young male house mice but the potential role of urinary pheromones in this phenomenon has not been demonstrated.

** A potential action of female urine on adult male deer mice has not been investigated; grouping of adult deer mouse females, on the other hand, is known to delay their estrous cyclicity but the role of urine in this phenomenon has not been investigated.

(Macrides et al., 1975; Maruniak and Bronson, 1976), but the reproductive function served by this neuroendocrine reflex is obscure. The results of studies designed to determine whether or not female urine can accelerate the sexual maturation of young male house mice are somewhat contradictory (see Fox, 1968; Maruniak et al., 1978; Svare et al., 1978; Vandenbergh, 1971).

All the phenomena noted in house mice seem to occur in a similar manner in laboratory stocks of *P. m. bairdii*. Grouping of adult females suppresses their estrous cycles and exposure to a male or his urine accelerates these cycles (Bronson, 1971; Bronson and Dezell, 1968; Bronson and Marsden, 1964). Grouping of immature female deer mice delays the onset of puberty (Bronson, pers. observ.), whereas exposure to male urine accelerates sexual maturation in young females (Teague and Bradley, 1978). The pregnancy-block phenomenon occurs in deer mice and pooled urine will evoke this response (Bronson and Eleftheriou, 1963; Bronson et al., 1969; Eleftheriou et al., 1962; Terman, 1969). Deer mice may show one additional phenomenon, namely the suppression of sexual maturation in young males either by the presence of an adult male or exposure to his urine (Bediz and Whitsett, 1979; Lawton and Whitsett, 1979). Adult males suppress sexual maturation in young house mice but the role of pheromones, urinary or otherwise, has not been demonstrated and, in all probability, this is a non-specific reaction to agonistic stimuli.

Finally, the hormonal responses to priming pheromones often and possibly always involve interactions with another sensory modality. The induction of the pubertal ovulation by male-exposure in house mice involves a true synergism of action between a urinary pheromone and an as yet unidentified tactile component of the male's behavior (Bronson and Maruniak, 1975; Drickamer, 1974*b*). Likewise, the release of LH in a male house mouse in response to the presence of a female involves a urinary primer and an unknown cue operating via another sensory modality; in this case, however, the interaction between the two cues is redundant rather than synergistic (Coquelin and Bronson, 1980).

Regarding the potential utility of priming pheromones to natural populations, much previous concern has been with second-order adaptive advantages in relation to particular ecological strategies. However, a mammalian primer has the principal duty of regulating a normal reproductive event or process in the same sense that tactile cues are important in the induction of certain key reproductive events and processes. Thus, the most basic utility of a primer to a population of either deer mice or house mice is simply to promote normal reproduction.

In addition to this basic function are the possible interactions of urinary primers with important ecological strategies. The temporal modulation of the pubertal ovulation would seem to be a particularly important phenomenon. The endocrine response to other females, mother or sibling, both decelerates the attainment of the pubertal ovulation in a young female and blocks any accelerating action of the father. Populations of both deer mice and house mice are characterized by a high rate of dispersal, most of which occurs just prior to puberty. Importantly, the priming phenomenon can prevent conception before dispersal and, hence, the distance and time required for the successful termination of a dispersal movement would not be limited by pregnancy. As argued previously (Bronson, 1979*a*), the pheromonal modulation of puberty in relation to dispersal may have played an important role in the house mouse's successful history of colonization. Likewise, the phenomenon would seem important in deer mice for maintaining their very characteristic stable breeding densities; any breeding female that is lost to predators or other factors could be replaced immediately by a transient peripubertal female ready to ovulate after only two or three days of residency.

Laboratory evidence concerning the physiological importance of the transient episodic hormonal releases that are experienced by adult males in response to female urine has not been forthcoming. Ecologically, this reaction in males would seem best visualized simply as part of a system of mutual stimulation between the sexes. "Breeding" male deer mice and house mice, as well as subordinate and transient young males, may not be in top reproductive condition at all times. If true, stimulation of a male by a female to stimulate her own ovulation would seem to constitute a highly adaptive system. The fact that urine marking and pheromone synthesis are gonadally-dependent in male house mice and gonadally-independent in females, argues strongly for such a system of mutual stimulation (Bronson, 1979*a*).

The ecological significance of the temporal modulation of ovulation by pheromones in adult females, particularly when coupled with an implantation block, is much more puzzling. The mutual system of pheromonal stimulation that was envisioned above for pubertal females and either adult or young males would seem advantageous to adult females undergoing the home-range shifts that are such a normal dimension of most feral populations of house mice. How this would work is unknown and the arguments that have been presented so far for the secondary adaptive advantages of such modulation in either species seem unconvincing (Bronson and Coquelin, 1980).

One other priming phenomenon seems worth a comment, namely, the previously noted observation that adult male urine impedes sexual maturation in young male deer mice (Lawton and Whitsett, 1979; Whitsett et al., 1979). An inhibition of sexual maturation in both sexes is one of the most dramatic correlates of asymptote in freely-growing confined populations of deer mice. However, Terman and his associates have convincingly ruled out the potential role of urinary pheromones in this phenomenon, suggesting that non-specific reactions to agonistic stimuli are a better possibility

(Kipps and Terman, 1977; Terman, 1968, 1979). Because markedly different techniques were used in Terman's and Whitsett's experiments, the question still seems open, as does the ecological significance of any urinary effects on young males.

Conclusions, Speculations, and Areas of Ignorance

The original intent of this review was to sequentially 1) examine the literature dealing with populations of house mice and deer mice in terms of their ecological similarities and dissimilarities, 2) relate the resulting generalities to the ways in which the two species use pheromones in a laboratory setting and, in turn, 3) use these latter relationships as a basis for considering the functional roles that might be played by chemical communication in all rodent populations. The nature of the available literature has imposed severe limits on the success of this venture. Nevertheless, a few conclusions have emerged. Most obviously, the evolution of chemical communication, at least in these two species, seems to have been shaped far more by overriding ecological considerations than it has been by any particularly unique demand placed upon any population of either species. By and large, house mice and deer mice do about the same kinds of things with pheromones and they seem to accomplish these goals in about the same kinds of ways. Wild populations of the two species do differ in several ecologically fundamental ways but such differences do not seem to be reflected in the way the two species use pheromones in the laboratory. In the wild, both species routinely live in dark, physically complex environments, often in populations where the individuals are widely dispersed. Signalling and priming pheromones provide an efficient means of communication under such conditions and these general considerations seem to have provided the major force for promoting chemical communication in both species.

Additionally, it remains of interest to speculate about the more subtle relationships that must exist between a population's use of pheromones and its other social and ecological characteristics. However, this is not a simple question involving simple species comparisons. When one looks at the demographic aspects of the known populations of deer mice and house mice, one is struck by the large amount of variation that exists within as well as between species. Table 4 emphasizes this fact by summarizing the characteristics of both commensal and feral populations of house mice as well as those of typical deer-mouse populations. One type of house-mouse population may share characteristics with deer-mouse populations without sharing these same characteristics with the other type of *Mus musculus* population. For example, both deer-mouse and commensal house-mouse populations maintain markedly stable densities over time whereas feral house-mouse populations are highly irruptive. Conversely, a large home range, which certainly should be a factor of importance in pheromonal communication, is shared by feral house mice and deer mice whereas commensal house mice typically have exceptionally small home ranges. Thus population characteristics rather than species differences seem to provide a more meaningful framework for speculations about subtle aspects of pheromonal communication.

Within this context then, of the three general functions served by pheromones in breeding mammals—occupancy notification, short-range identification, and endocrine priming—the first would seem to be of most importance to most rodent populations. The routine broadcasting of chemical information relevant to an individual's occupancy of a geographical area should increase in its importance to the population's well-being as the distance between individual centers of activity increases. Furthermore, coupling this particular pheromonal function with temporal stability of home-range occupancy, and to avoidance or territorial behavior, should promote a stable population density. The typical deer-mouse population seems to provide a good example of the cumulation of all three factors, whereas the high-density commensal populations of house mice lack the long-distance characteristic and, hence, chemical broadcasting probably is less important to them. Individuals in a feral *Mus musculus* population, on the other hand, usually are widely spaced but the rapid home-range shifts in these populations make the importance of the broadcasting function to such populations somewhat enigmatic. Because most populations of small nocturnal rodents probably combine at least two or more of the factors noted above, it seems reasonable to assume that the occupancy-identification function is probably of wide importance in such animals.

The pheromonal function of next importance to most rodent populations probably is that of endocrine priming. The flexible control

TABLE 4
SOME CHARACTERISTICS OF HOUSE-MOUSE AND DEER-MOUSE POPULATIONS

Characteristics	House mouse		Deer mouse
	Commensal	Feral	
Food resources	Stable, plentiful	Seasonally sparse	Seasonally sparse
Reproduction	Non-seasonal	Usually seasonal	Usually seasonal
Breeding density (non-irruptive)	Up to 10,000/ha	Up to 100/ha	Up to 40/ha
Home range	Often <.001 ha	Up to 1 ha	Up to 2 ha
Social unit, breeding season	Dominant male, one or more breeding females, few subordinate males	The individual	Individual and/or one adult male with one or more breeding females
Social organization of breeding population	Stable; deme-territory with internal hierarchial organization	Unstable; probably nearest neighbor dominance for resident males and lactating females if possible	Stable; overdispersion of daily activities due to mutual avoidance within sexes
Potential for irruption	None	High	None
Potential for dispersal	High	High	High

of pubertal ovulation by multiple social cues and its correlation with dispersal would seem particularly relevant because high productivity, high turnover, and dispersal of young are constants of most rodent populations. This particular adaptation may have reached its greatest importance in supplementing the other physiological and behavioral adaptations that have supported the well-known global colonization by the house mouse (Bronson, 1979*a*). The timing of an adult's ovulation in widely dispersed populations would seem of equal importance but little is known about such timing, even if it exists, in natural populations. Unfortunately, the ecological literature simply will not support more elegant theories about the adaptive significance of primer use at this time.

The short-range identification function probably is the least important of the three reproductively-related functions of pheromones in rodents because such identification typically involves multiple sensory input. Short-range identification, however, is the function most commonly studied in rodents by laboratory scientists.

Finally, this review should serve a purpose beyond simply comparing in depth the use of chemical cues in two species; it should provide a vehicle for enumerating general areas of ignorance about chemical communication in rodents. If deer mice and house mice are typical of small, nocturnal rodents, then three gaps in our knowledge seem notably apparent at this time. First, we actually have very little information even in deer mice and house mice about normal social relations and normal reproductive characteristics in over-dispersed field populations, the types that are typical of most rodents. The transient release of reproductive hormones in males when exposed to female urine, and the pheromonal timing of ovulation in the adult female, both might make considerable sense if the real state of readiness of "breeding" residents in wild populations of house mice and deer mice was better known. How much pheromonal and hormonal fine-tuning by social cues is normally needed in the field to yield a sexually-ready male and a pregnant female? Likewise, the potential utility of broadcast-occupancy cues to an animal that routinely shifts its home range, like the opportunistic feral *Mus musculus,* is a totally unexplored subject. Unfortunately, there have been few attempts to experimentally approach any of these questions in the field. Studies of nearest-neighbor recognition would seem particularly rewarding in this regard (see Healey, 1967; Vestal and Hellack, 1978), as would the use of scented traps (for example, see Daly et al., 1978).

Secondly, we know little about the types and amounts of information that can be encoded in chemical cues. In addition to communication of sex, relative age, and estrous state of a female, there is reasonably good laboratory evidence that house mice can detect alarm in other mice, differentiate between victors and losers in fights and, possibly, identify particular individuals, all by odor (Bowers and Alexander, 1967; Carr et al., 1970; Kalkowski, 1967; Müller-Velten, 1966; Rottman and Snowdon, 1972). How much such information actually is communicated between resident neighbors in a natural population? How much is in urine and what other sources of such cues exist? These are difficult questions. For example, at least theoretically, a message that says "male" actually needs to be sent in several different ways by a free-roaming individual. For a broadcasting function, this message should be combined with a carrier so that it becomes long-enduring. If the cue is water-soluble, as most urinary signals must be, then it should be tailored chemically in some way to the humidity of the microclimate inhabited by the deme. A water-soluble message left on the surface of a desert should be long-enduring, for example, whereas the same chemical deposited in a wetland might exhibit a duration too short for efficient communication. Similarly, a male probably needs to emit notification of his sex as a highly volatile, short-range signal when standing head-to-head with another individual in the dark. Male primers likewise should have their own chemical identities because sustained actions must be maintained. Finally, we know nothing of the numbers of discrete signals and primers that occur in any species nor do we know about the potential for the gradation of such cues. The chemical diversity of pheromones in rodents must be immense and our lack of knowledge about this diversity is only matched by our ignorance of their reception and the relative importance of the various specific functions thereby served in the different species.

The third area of deep ignorance concerns the more subtle functions that might be served by chemical cues in rodents, that is, the type of phenomena that would not be readily apparent in either the field or the laboratory. Our limited knowledge about the life styles of cryptic rodents limits our imagination but two possibilities are apparent. First, are there population, deme or subspecies differences in chemical cues that are important? Subspecies diversity has been noted repeatedly in other characteristics of deer mice, such as food preferences (Gray, 1979), environmental control of hoarding behavior (Barry, 1976), circadian periodicity (King, 1975), aggression and dispersal characteristics (Halpin and Sullivan, 1978); even r- versus K-selection differences have been postulated (Redfield, 1976). Can either signalling or priming pheromones act as isolating mechanisms, as has been suggested in a comparison of *P. polionotus* with *P. maniculatus* (Moore, 1965)? Additionally, small rodents may compete for resources with other species of small rodents. *Mus musculus*, for example, may compete with species of *Peromyscus*, and species of *Microtus* may compete with both. The exact nature of such direct competition, if it occurs, is not well understood but certainly a genetically-wired reaction to another species' chemical signals is possible, although it has not been explored.

In summary, deer mice and house mice use the same general kinds of chemical signals and primers to accomplish the same general behavioral and physiological ends. Overall, the use of chemical cues seems to be tailored more to the grossly similar environments occupied by these species than it is to the smaller differences in life styles adopted by each. This probably is true for all rodents. The present comparison also has attempted to enumerate some of our more obvious areas of ignorance regarding chemical communication in small, nocturnal rodents in general.

Acknowledgments

This work was supported by U.S. Public Health Grant HD13155 from the National Institute of Child Health and Human Development.

Literature Cited

Abramsky, Z., M. I. Dyer, and P. D. Harrison. 1979. Competition among small mammals in experimentally perturbed areas of the short-grass prairie. Ecology, 60:530–536.

Adamczyk, K., and W. Walkowa. 1971. Compensation of numbers and production in a *Mus musculus* population as a result of partial removal. Ann. Zool. Fennici, 8:145–153.

Anderson, P. K. 1961. Density, social structure, and nonsocial environment of house-mouse populations and the implications for regulation of numbers. Trans. New York Acad. Sci., ser. 2, 23:447–451.

———. 1970. Ecological structure and gene flow in small animals. Symp. Zool. Soc. London, 26:299–325.

Anderson, P. K., and J. L. Hill. 1965. *Mus musculus*: experimental induction of territory formation. Science, 148:1753–1755.

ANDERSON, P. K., ET AL. 1977. *Mus musculus* and *Peromyscus maniculatus*: homing in relation to habitat utilization. Canadian J. Zool., 55:169–182.

ANDREWS, R. V., ET AL. 1975. Physiological consequences of experimentally altering the population structure of *Peromyscus maniculatus* in the field. Comp. Biochem. Physiol., 51A:785–792.

ANDRZEJEWSKI, R., K. PETRUSEWICZ, AND W. WALKOWA. 1963. Absorption of newcomers by a population of white mice. Ekol. Pol., ser. A, 11:223–240.

AYER, M. L., AND J. M. WHITSETT. 1980. Aggressive behaviour of female prairie deer mice in laboratory populations. Anim. Behav., 28:763–771.

BAKER, R. H. 1968. Habitats and distribution. Pp. 98–126, *in* Biology of *Peromyscus* (Rodentia) (J. A. King, ed.). Spec. Publ., Amer. Soc. Mamm., 2: 1–593.

BARNETT, S. A. 1965. Adaptation of mice to cold. Biol. Rev., 40:5–51.

———. 1973. Maternal processes in cold-adaptation of mice. Biol. Rev., 48:477–508.

BARRY, W. J. 1976. Environmental effects on food hoarding in deermice (*Peromyscus*). J. Mamm., 57:731–746.

BATTEN, C. A., AND R. J. BERRY. 1967. Prenatal mortality in wild-caught house mice. J. Anim. Ecol., 36:453–463.

BEAUCHAMP, G. K., ET AL. 1976. The pheromone concept in mammalian chemical communication: a critique. Pp. 143–160, *in* Mammalian olfaction, reproductive processes, and behavior (R. L. Doty, ed.). Academic Press, New York, 344 pp.

BEDIZ, G. M., AND J. M. WHITSETT. 1979. Social inhibition of sexual maturation in male prairie deer mice. J. Comp. Physiol. Psychol., 93:493–500.

BEER, J. R., AND C. F. MACLEOD. 1966. Seasonal population changes in the prairie deer mouse. Amer. Midland Nat., 76:277–289.

BEER, J. R., C. F. MACLEOD, AND L. D. FRENZEL. 1957. Prenatal survival and loss in some cricetid rodents. J. Mamm., 38:392–402.

BERRY, R. J. 1968. The ecology of an island population of the house mouse. J. Anim. Ecol., 37:445–470.

———. 1970. The natural history of the house mouse. Field Stud., 3:219–262.

BERRY, R. J., AND M. E. JAKOBSON. 1974. Vagility in an island population of the house mouse. J. Zool., 173:341–354.

BERRY, R. J., AND J. PETERS. 1975. Macquarie Island house mice: a genetical isolate on a sub-Antarctic island. J. Zool., 176:375–389.

BERRY, R. J., M. E. JAKOBSON, AND G. S. TRIGGS. 1973. Survival in wild-living mice. Mammal Rev., 3:46–57.

BLAIR, W. F. 1940. A study of prairie deermouse populations in southern Michigan. Amer. Midland Nat., 24:273–305.

———. 1953. Population dynamics of rodents and other small mammals. Adv. Genet., 5:1–41.

———. 1958. Effects of x-irradiation of a natural population of the deer-mouse (*Peromyscus maniculatus*). Ecology, 39:113–118.

BOWERS, J. M., AND B. K. ALEXANDER. 1967. Mice: individual recognition by olfactory cues. Science, 158:1208–1210.

BRADLEY, E. L., AND C. R. TERMAN. 1979. Ovulation in *Peromyscus maniculatus bairdii* under laboratory conditions. J. Mamm., 60:543–549.

BREAKEY, D. R. 1963. The breeding season and age structure of feral house mouse populations near San Francisco Bay, California. J. Mamm., 44:153–168.

BRONSON, F. H. 1971. Rodent pheromones. Biol. Reprod., 1:344–357.

———. 1979*a*. The reproductive ecology of the house mouse. Quart. Rev. Biol., 54:265–299.

———. 1979*b*. Light intensity and reproduction in wild and domestic house mice. Biol. Reprod., 21:235–239.

BRONSON, F. H., AND A. COQUELIN. 1980. The modulation of reproduction by priming pheromones in house mice: speculations on adaptive function. Pp. 243–265, *in* Chemical signals (D. Müller-Schwarze and R. M. Silverstein, eds.). Plenum Publishing Corp., New York, 445 pp.

BRONSON, F. H., AND H. E. DEZELL. 1968. Studies on the estrus-inducing (pheromonal) action of male deermouse urine. Gen. Comp. Endocrinol., 10: 339–343.

BRONSON, F. H., AND B. E. ELEFTHERIOU. 1963. Influence of strange males on implantation in the deermouse. Gen. Comp. Endocrinol., 3:515–518.

———. 1965. Adrenal response to fighting in mice: separation of physical and psychological causes. Science, 147:627–628.

BRONSON, F. H., AND H. M. MARSDEN. 1964. Male-induced synchrony of estrus in deermice. Gen. Comp. Endocrinol., 4:634–637.

BRONSON, F. H., AND J. A. MARUNIAK. 1975. Male-induced puberty in female mice: evidence for a synergistic action of social cues. Biol. Reprod., 13:94–98.

BRONSON, F. H., C. P. DAGG, AND G. D. SNELL. 1966. Reproduction. Pp. 187–204, *in* Biology of the laboratory mouse (E. L. Green, ed.). McGraw-Hill, New York, 706 pp.

BRONSON, F. H., B. E. ELEFTHERIOU, AND H. E. DEZELL. 1969. Strange male pregnancy block in deermice: prolactin and adrenocortical hormones. Biol. Reprod., 1:302–306.

BROWN, H. L. 1945. Evidence of winter breeding of *Peromyscus*. Ecology, 26: 308–309.

BROWN, L. N. 1966. Reproduction of *Peromyscus maniculatus* in the Laramie Basin of Wyoming. Amer. Midland Nat., 76:183–189.

BROWN, R. Z. 1953. Social behavior, reproduction and population changes in the house mouse (*Mus musculus* L.). Ecol. Monogr., 23:217–240.

BRUCE, H. M. 1959. An exteroceptive block to pregnancy in the mouse. Nature, 184:105.

———. 1966. Smell as an exteroceptive factor. J. Anim. Sci., 25:83–89.

BRUELL, J. H. 1970. Behavioral population genetics and wild *Mus musculus*. Pp. 261–291, *in* Contributions to behavior—genetic analysis (G. Lindzey and D. D. Thiessen, eds.). Meredith Corp., New York, 336 pp.

BURT, W. H. 1940. Territorial behavior and populations of some small mammals in southern Michigan. Misc. Publ. Mus. Zool., Univ. Michigan, 45:1–58.

CALDWELL, L. D. 1964. An investigation of competition in natural populations of mice. J. Mamm., 45:12–30.

CAROOM, D., AND F. H. BRONSON. 1971. Responsiveness of female mice to preputial attractant: effects of sexual experience and ovarian hormones. Physiol. Behav., 7:659–662.

CARR, W. J., R. D. MARTORANO, AND L. KRAMES. 1970. Responses of mice to odors associated with stress. J. Comp. Physiol. Psychol., 71:223–228.

CHRISTIAN, J. J. 1970. Social subordination, population density, and mammalian evolution. Science, 168:84–90.

———. 1980*a*. Regulation of annual rhythms of reproduction in temperate small rodents. Pp. 367–380, *in* Testicular development, structure, and function (A. Steinberger and E. Steinberger, eds.). Raven Press, New York, 483 pp.

———. 1980*b*. Endocrine factors in population regulation. Pp. 55–116, *in* Biosocial mechanisms of population regulation (M. Cohen, R. Malpass, and H. Klein, eds.). Yale Univ. Press, New Haven, Connecticut, 406 pp.

CHRISTIAN, J. J., AND D. E. DAVIS. 1964. Endocrines, behavior and population. Science, 145:1550–1560.

CHRISTIAN, J. J., J. A. LLOYD, AND D. E. DAVIS. 1965. The role of endocrines in the self-regulation of mammalian populations. Recent Prog. Horm. Res., 21:501–578.

CLAUS, R., AND W. ALSING. 1976. Occurrence of 5α-Androst-16-en-3-one, a boar pheromone, in man and its relationship to testosterone. J. Endocrinol., 68: 483–484.

CONAWAY, C. H. 1971. Ecological adaptation and mammalian reproduction. Biol. Reprod., 4:239–247.

COQUELIN, A., AND F. H. BRONSON. 1980. Secretion of luteinizing hormone in male mice: factors that influence release during sexual encounters. Endocrinology, 106:1224–1229.

CROWCROFT, P. 1966. Mice all over. G. T. Foulis and Company, London, 275 pp.

CROWCROFT, P., AND F. P. ROWE. 1963. Social organization and territorial behavior in the wild house mouse (*Mus musculus* L.). Proc. Zool. Soc. London, 140:517–531.

DALY, M., M. I. WILSON, AND S. F. FAUX. 1978. Seasonally variable effects of conspecific odors upon capture of deer mice (*Peromyscus maniculatus gambelii*). Behav. Biol., 23:254–259.

DAVIS, D. E. 1958. The role of density in aggressive behavior of house mice. British J. Anim. Behav., 6:207–210.

DEFRIES, J. C., AND G. E. MCCLEARN. 1970. Social dominance and Darwinian fitness in the laboratory mouse. Amer. Nat., 104:408–411.

DELONG, K. T. 1967. Population ecology of feral house mice. Ecology, 48:611–634.

DESJARDINS, C., AND M. J. LOPEZ. 1980. Sensory and nonsensory modulation of testis function. Pp. 381–388, *in* Testicular development, structure, and function (A. Steinberger and E. Steinberger, eds.). Raven Press, New York, 483 pp.

DESJARDINS, C., J. A. MARUNIAK, AND F. H. BRONSON. 1973. Social rank in house mice: differentiation by ultra-violet visualization of urinary marking patterns. Science, 182:939–941.

DEWSBURY, D. A. 1979. Copulatory behavior of deer mice (*Peromyscus maniculatus*). III. Effects on pregnancy initiation. J. Comp. Physiol. Psychol., 93: 178–188.

DIXON, A. K., AND J. H. MACKINTOSH. 1971. Effect of female urine on the social behavior of adult male mice. Anim. Behav., 19:138–140.

———. 1976. Olfactory mechanisms affording protection from attack to juvenile mice. Z. Tierpsychol., 41:225–234.

DOTY, R. L. 1973. Reactions of deer mice (*Peromyscus maniculatus*) and white-footed mice (*Peromyscus leucopus*) to homospecific and heterospecific urine odors. J. Comp. Physiol. Psychol., 84:296–303.

DRICKAMER, L. C. 1974*a*. Contact stimulation, androgenized females and accelerated sexual maturation in female mice. Behav. Biol., 12:101–110.

———. 1974*b*. Sexual maturation of female house mice: social inhibition. Dev. Psychobiol., 7:257–265.

———. 1978. Annual reproduction patterns in populations of two sympatric species of *Peromyscus*. Behav. Biol., 23:405–408.

DRICKAMER, L. C., AND B. M. VESTAL. 1973. Patterns of reproduction in a laboratory colony of *Peromyscus*. J. Mamm., 54:523–528.

DUNMIRE, W. W. 1960. An altitudinal survey of reproduction in *Peromyscus maniculatus*. Ecology, 41:174–182.

EIBL-EIBESFELDT, I. 1950. Beitrage zur Biologie der Haus- und der Ahrenmaus nebst einigen Beobachtungen an anderen Nagern. Z. Tierpsychol., 7:558–587.

EISENBERG, J. F. 1962. Studies on the behavior of *Peromyscus maniculatus gambelii* and *Peromyscus californicus parasiticus*. Behaviour, 19:177–207.

———. 1968. Behavior patterns. Pp. 451–495, *in* Biology of *Peromyscus* (Rodentia) (J. A. King, ed.). Spec. Publ., Amer. Soc. Mamm., 2:1–593.

ELEFTHERIOU, B. E., F. H. BRONSON, AND M. X. ZARROW. 1962. Interaction of olfactory and other environmental stimuli on implantation in the deer mouse. Science, 137:764.

FAIRBAIRN, D. J. 1977*a*. The spring decline in deer mice: death or dispersal? Canadian J. Zool., 55:84–92.

———. 1977*b*. Why breed early? A study of reproductive tactics in *Peromyscus*. Canadian J. Zool., 55:862–871.

———. 1978*a*. Dispersal of deer mice, *Peromyscus maniculatus*. Oecologia, 32: 171–193.

———. 1978*b*. Behaviour of dispersing deer mice (*Peromyscus maniculatus*). Behav. Ecol. Sociobiol., 3:265–282.

FORDHAM, R. A. 1971. Field populations of deermice with supplemental food. Ecology, 52:138–146.

FOX, K. A. 1968. Effects of prepubertal habitation conditions on the reproductive physiology of the male house mouse. J. Reprod. Fert., 17:75–85.

FULLER, W. A. 1969*a*. Changes in numbers of three species of small rodent near Great Slave Lake, N.W.T. Canada, 1964–1967, and their significance for general populations theory. Ann. Zool. Fennici, 6:113–144.

———. 1969*b*. An unusual winter movement of *Peromyscus maniculatus*. Canadian Field-Nat., 83:275.

GARDNER, R. H., JR., AND C. R. TERMAN. 1970. The relationship between age of grouping and weight of selected organs of prairie deermice. Res. Popul. Ecol., 12:1–18.

GASHWILER, J. S. 1979. Deer mouse reproduction and its relationship to the tree seed crop. Amer. Midland Nat., 102:95–104.

GENTRY, J. B. 1966. Invasion of a one-year abandoned field by *Peromyscus polionotus* and *Mus musculus*. J. Mamm., 47:431–439.

GRAY, L. 1979. Feeding diversity in deer mice. J. Comp. Physiol. Psychol., 93: 1118–1126.

HALPIN, Z. T., AND T. P. SULLIVAN. 1978. Social interactions in island and mainland populations of the deer mouse *Peromyscus maniculatus*. J. Mamm., 59:395–401.

HANSEN, C. M., AND E. D. FLEHARTY. 1974. Structural ecological parameters of a population of *Peromyscus maniculatus* in west-central Kansas. Southwestern Nat., 19:293–303.

HARRINGTON, J. E. 1976. Recognition of the territorial boundaries by olfactory cues in mice. Z. Tierpsychol., 41:295–306.

HAYASHI, S., AND T. KIMURA. 1974. Sex-attractant emitted by female mice. Physiol. Behav., 13:563–567.

HEALEY, M. C. 1967. Aggression and self-regulation of population size in deermice. Ecology, 48:377–392.

HILL, J. L. 1977. Space utilization of *Peromyscus*: social and spatial factors. Anim. Behav., 25:373–389.

HOFFMANN, R. S. 1955. A population-high for *Peromyscus maniculatus*. J. Mamm., 36:571–572.

HOLBROOK, S. J. 1979. Habitat utilization, competitive interactions, and coexistence of three species of cricetine rodents in east-central Arizona. Ecology, 60:758–769.

HOOPER, E. T. 1968. Classification. Pp. 27–74, *in* Biology of *Peromyscus* (Rodentia) (J. A. King, ed.). Spec. Publ., Amer. Soc. Mamm., 2:1–593.

HOWARD, W. E. 1949. Dispersal, amount of inbreeding, and longevity in a local population of prairie deermice on the George reserve, Southern Michigan. Contrib. Lab. Vert. Biol., Univ. Michigan, 43:1–50.

JAMESON, E. W., JR. 1953. Reproduction of deer mice (*Peromyscus maniculatus* and *P. boylei*) in Sierra Nevada, Calif. J. Mamm., 34:44–58.

JUSTICE, K. E. 1961. A new method for measuring home ranges of small mammals. J. Mamm., 42:462–470.

———. 1962. Ecological and genetical studies of evolutionary forces acting on desert populations of *Mus musculus*. Arizona-Sonora Desert Museum, Inc., Tucson, 47 pp.

KALKOWSKI, W. 1967. Olfactory bases of social orientation in the white mouse. Folia Biol. (Cracow), 16:69–87.

KARLSON, P., AND M. LUSCHER. 1959. "Pheromones": a new term for a class of biological active substances. Nature, 183:55.

KIMURA, T. 1971. Modifications of the estrous cycles in mice housed together with normal or ovariectomized females. Sci. Papers Coll. Gen. Educ., Univ. Tokyo, 21:161–166.

KING, J. A. 1957. Intra- and interspecific conflict of *Mus* and *Peromyscus*. Ecology, 38:355–357.

———. 1975. Stimulus selection (light) in four subspecies of deermice (*Peromyscus maniculatus*). J. Interdiscipl. Cycle Res., 6:267–278.

KIPPS, P. L., AND C. R. TERMAN. 1977. The influence of pheromones produced by freely growing laboratory populations on the reproductive maturation of prairie deermice following prenatal and postnatal exposure. Res. Popul. Ecol., 18:261–266.

KIRKLAND, G. L., JR. 1976. Small mammals of a mine waste situation in the central Adirondacks, New York: a case of opportunism by *Peromyscus maniculatus*. Amer. Midland Nat., 95:103–110.

LAURIE, E. M. O. 1946. The reproduction of the house-mouse (*Mus musculus*) living in different environments. Proc. Roy. Soc. London, 133:248–281.

LAWTON, A. D., AND J. M. WHITSETT. 1979. Inhibition of sexual maturation by a urinary pheromone in male prairie deer mice. Horm. Behav., 13:128–138.

LIDICKER, W. Z. 1965. Comparative study of density regulation in confined populations of four species of rodents. Res. Popul. Ecol., 7:57–72.

———. 1966. Ecological observations on a feral house mouse population declining to extinction. Ecol. Monogr., 36:27–50.

———. 1976. Social behavior and density regulation in house mice living in large enclosures. J. Anim. Ecol., 45:677–697.

LINDUSKA, J. P. 1942. Winter rodent populations in field-shocked corn. J. Wildl. Mgmt., 6:353–363.

LLOYD, J. A. 1975. Social structure and reproduction in two freely-growing populations of house mice (*Mus musculus*). Anim. Behav., 23:413–424.

———. 1980. Interaction of social structure and reproduction in populations of mice. Pp. 3–22, *in* Biosocial mechanisms of population regulation (M. Cohen, R. Malpass, and H. Klein, eds.). Yale Univ. Press, New Haven, Connecticut, 406 pp.

LLOYD, J., AND J. C. CHRISTIAN. 1969. Reproductive activity of individual females in three experimental freely growing populations of house mice (*Mus musculus*). J. Mamm., 50:49–59.

MACKINTOSH, J. H. 1970. Territory formation by laboratory mice. Anim. Behav., 18:177–183.

———. 1973. Factors affecting the recognition of territory boundaries by mice (*Mus musculus*). Anim. Behav., 21:464–470.

MACRIDES, F., A. BARTKE, AND S. DALTERIO. 1975. Strange females increase plasma testosterone levels in male mice. Science, 189:1104–1105.

MARSDEN, H. M., AND F. H. BRONSON. 1964. Estrous synchrony in mice: alteration by exposure to male urine. Science, 144:3625.

MARUNIAK, J. A., AND F. H. BRONSON. 1976. Gonadotropic responses of male mice to female urine. Endocrinology, 99:963–969.

MARUNIAK, J. A., A. COQUELIN, AND F. H. BRONSON. 1978. The release of LH in male mice in response to female urinary odors: characteristics of the response in young males. Biol. Reprod., 18:251–255.

MARUNIAK, J. A., C. DESJARDINS, AND F. H. BRONSON. 1975*a*. Adaptations for urine marking in rodents: prepuce length and morphology. J. Reprod. Fert., 44:567–570.

MARUNIAK, J. A., ET AL. 1974. Urinary marking in male house mice: responses to novel environmental and social stimuli. Physiol. Behav., 12:1035–1039.

———. 1975*b*. Urinary marking in female house mice: effects of ovarian steroids, sex experiences and type of stimulus. Behav. Biol., 13:211–217.

MCCABE, T. T., AND B. BLANCHARD. 1950. Three species of *Peromyscus*. Rood Association, Santa Barbara, 136 pp.

METZGAR, L. H. 1978. Dispersion patterns in a *Peromyscus* population. J. Mamm., 60:129–145.

MIHOK, S. 1979. Behavioral structure and demography of subarctic *Clethrionomys gapperi* and *Peromyscus maniculatus*. Canadian J. Zool., 57:1520–1535.

MILLAR, J. S., AND F. B. WILLE. 1979. Breeding by *Peromyscus* in seasonal environments. Canadian J. Zool., 57:719–727.

MOORE, R. E. 1965. Olfactory discrimination as an isolating mechanism between *Peromyscus maniculatus* and *Peromyscus polionotus*. Amer. Midland Nat., 73:85–100.

MÜLLER-VELTEN, H. 1966. Uber der Angstgeruch bei der Hausmaus (*Mus musculus* L.). Z. Vergleich Physiol., 52:401–429.

MYERS, J. H. 1974. Genetic and social structure of feral house mouse populations on Grizzly Island, California. Ecology, 55:747–759.

NEWSOME, A. E. 1969*a*. A population study of house-mice permanently inhabiting a reedbed in South Australia. J. Anim. Ecol., 38:361–377.

———. 1969*b*. A population study of house-mice temporarily inhabiting a South Australian wheatfield. J. Anim. Ecol., 38:341–359.

———. 1970. An experimental attempt to produce a mouse plague. J. Anim. Ecol., 39:299–311.

NEWSOME, A. E., AND L. K. CORBETT. 1975. Outbreaks of rodents in semi-arid and arid Australia: causes, preventions and evolutionary considerations. Pp. 117–153, *in* Rodents in desert environments (I. Prakash and P. K. Gosh, eds.). Dr. W. Junk, The Hague, 214 pp.

NOWAK, Z. 1971. The effect of removing a dominant on the social organization of laboratory mice populations. Acta Theriol., 16:61–71.

NYBY, J., AND G. WHITNEY. 1978. Ultrasonic communication of adult myomorph rodents. Neurosci. Biobehav. Rev., 2:1–14.

OAKESHOTT, J. G. 1974. Social dominance, aggressiveness and mating success among male house mice (*Mus musculus*). Oecologia, 15:143–158.

O'FARRELL, M. J. 1978. Home range dynamics of rodents in a sagebrush community. J. Mamm., 59:657–668.

PEARSON, O. P. 1963. History of two local outbreaks of feral house mice. Ecology, 44:540–549.

PELIKAN, J. 1974. On the reproduction of *Mus musculus* L. in Czechoslovakia. Acta Sc. Nat. Brno, 8(12):1–42.

PETRUSEWICZ, K. 1957. Investigation of experimentally induced population growth. Ekol. Pol., ser. A, 5:281–301.

PETRUSEWICZ, K., AND R. ANDRZEJEWSKI. 1962. Natural history of a free-living population of house mice (*Mus musculus* L.), with particular reference to groupings within the population. Ekol. Pol., ser. A, 10:85–122.

PETTICREW, B. G., AND R. M. F. S. SADLEIR. 1974. The ecology of the deer mouse *Peromyscus maniculatus* in a coastal coniferous forest. I. Population dynamics. Canadian J. Zool., 52:107–118.

POOLE, T. B., AND H. D. R. MORGAN. 1976. Social and territorial behaviour of laboratory mice (*Mus musculus* L.) in small complex areas. Anim. Behav., 24:476–480.

PRICE, E. O. 1966. Influence of light on reproduction in *Peromyscus maniculatus gracilis*. J. Mamm., 47:343–344.

QUADAGNO, D. M. 1968. Home range size in feral house mice. J. Mamm., 49: 149–151.

REDFIELD, J. A. 1976. Distribution, abundance, size and genetic variation of *Peromyscus maniculatus* on the Gulf islands of British Columbia. Canadian J. Zool., 54:463–474.

REIMER, J. D., AND M. L. PETRAS. 1967. Breeding structure of the house mouse *Mus musculus,* in a population cage. J. Mamm., 48:88–99.

REIMOV, R., K. ADAMCZYK, AND R. ANDRZEJEWSKI. 1968. Some indices of the behaviour of wild and laboratory house mice in a mixed population. Acta Theriol., 13:129–150.

ROPARTZ, P. 1967. Rôle des communications olfactives dans le comportment social des souris males. Colloq. Int. de Centre Nat. de la Res. Scient., 173.

ROTTMAN, S. J., AND C. T. SNOWDON. 1972. Demonstrations and analysis of an alarm pheromone in mice. J. Comp. Physiol. Psychol., 81:483–490.

ROWE, F. P., E. J. TAYLOR, AND A. H. J. CHUDLEY. 1963. The numbers and movements of house mice (*Mus musculus* L.) in the vicinity of four cornricks. J. Anim. Ecol., 32:87–97.

———. 1964. The effect of crowding on the reproduction of the house mouse (*Mus musculus* L.) living in corn-ricks. J. Anim. Ecol., 33:477–483.

RYAN, K. D., AND N. B. SCHWARTZ. 1977. Grouped female mice: demonstration of pseudopregnancy. Biol. Reprod., 17:578.

SADLEIR, R. M. F. S. 1965. The relationship between agonistic behaviour and population changes in the deermouse, *Peromyscus maniculatus* (Wagner). J. Anim. Ecol., 34:331–352.

———. 1969. The ecology of reproduction in wild and domestic mammals. Methuen and Company, London, 310 pp.

———. 1970*a*. The establishment of a dominance rank order in male *Peromyscus maniculatus* and its stability with time. Anim. Behav., 18:55–59.

———. 1970*b*. Population dynamics and breeding of the deermouse (*Peromyscus maniculatus*) on Burnaby Mountain, British Columbia. Syesis, 3:67–74.

———. 1974. The ecology of the deermouse *Peromyscus maniculatus* in a coastal coniferous forest. II. Reproduction. Canadian J. Zool., 52:119–131.

SADLEIR, R. M. F. S., K. D. CASPERSON, AND J. HARLING. 1973. Intake and requirements of energy and protein for the breeding of wild deermice, *Peromyscus maniculatus*. Suppl., J. Reprod. Fert., 19:237–252.

SAVIDGE, I. R. 1974. Social factors in dispersal of deermice (*Peromyscus maniculatus*) from their natal site. Amer. Midland Nat., 91:395–405.

SCHEFFER, T. H. 1924. Notes on the breeding of *Peromyscus*. J. Mamm., 5:258–260.

SCHWARZ, E. 1945. On North American house mice. J. Mamm., 26:315–316.

SCHWARZ, E., AND H. K. SCHWARZ. 1943. The wild and commensal stocks of the house mouse, *Mus musculus* L. J. Mamm., 24:59–72.

SELANDER, R. K. 1970. Behavior and genetic variation in natural populations. Amer. Zool., 10:53–66.

SHEPPE, W. 1963. Population structure of the deer mouse, *Peromyscus,* in the Pacific Northwest. J. Mamm., 44:180–185.

SMITH, W. W. 1954. Reproduction in the house mouse, *Mus musculus* L., in Mississippi. J. Mamm., 35:509–515.

SOUTHERN, H. N., AND E. M. O. LAURIE. 1946. The house-mouse (*Mus musculus*) in corn ricks. J. Anim. Ecol., 15:134–149.

SOUTHWICK, C. H. 1955*a*. Regulatory mechanisms of house mouse populations: social behavior affecting litter survival. Ecology, 36:627–634.

———. 1955*b*. The population dynamics of confined house mice supplied with unlimited food. Ecology, 36:212–225.

———. 1958. Population characteristics of house mice living in English corn ricks: density relationships. Proc. Zool. Soc. London, 131:163–175.

STEBBINS, L. L. 1977. Energy requirements during reproduction of *Peromyscus maniculatus*. Canadian J. Zool., 55:1701–1704.

STICKEL, L. F. 1968. Home range and travels. Pp. 373–411, *in* Biology of *Peromyscus* (Rodentia) (J. A. King, ed.). Spec. Publ., Amer. Soc. Mamm., 2: 1–593.

STODDART, R. C. 1970. Breeding and growth of laboratory mice in darkness. Lab. Anim., 4:13–16.

STRECKER, R. L. 1954. Regulatory mechanisms in house-mouse populations: the effect of limited food supply on an unconfined population. Ecology, 35: 249–253.

STRECKER, R. L., AND J. T. EMLEN. 1953. Regulatory mechanisms in house-mouse populations: the effect of limited food supply on a confined population. Ecology, 49:1169–1172.

STUECK, K. L., AND G. W. BARRETT. 1978. Effects of resource partitioning on the population dynamics and energy utilization strategies of feral house mice (*Mus musculus*) populations under experimental field conditions. Ecology, 59:539–551.

SULLIVAN, T. P. 1977. Demography and dispersal in island and mainland populations of the deer mouse, *Peromyscus maniculatus*. Ecology, 58:964–978.

———. 1979. Demography of populations of deer mice in coastal forest and clear-cut (logged) habitats. Canadian J. Zool., 57:1636–1648.

SVARE, B., A. BARTKE, AND F. MACRIDES. 1978. Juvenile male mice: an attempt to accelerate testis function by exposure to adult female stimuli. Physiol. Behav., 21:1009–1013.

SVENDSEN, G. 1964. Comparative reproduction and development in two species of mice in the genus *Peromyscus*. Trans. Kansas Acad. Sci., 67:527–538.

TEAGUE, L. G., AND E. L. BRADLEY. 1978. The existence of a puberty accelerating pheromone in the urine of the male prairie deermouse (*Peromyscus maniculatus bairdii*). Biol. Reprod., 13:314.

TERMAN, C. R. 1961. Some dynamics of spatial distribution within semi-natural populations of prairie deermice. Ecology, 42:288–302.

———. 1966. Population fluctuations of *Peromyscus maniculatus* and other small mammals as revealed by the North American Census of Small Mammals. Amer. Midland Nat., 76:419–426.

———. 1968. Inhibition of reproductive maturation and function in laboratory populations of prairie deermice: a test of pheromone influence. Ecology, 49:1169–1171.

———. 1969. Pregnancy failure in female prairie deermice related to parity and social environment. Anim. Behav., 17:104–108.

———. 1973*a*. Recovery of reproductive function by prairie deer-mice (*Peromyscus maniculatus bairdii*) from asymptotic populations. Anim. Behav., 21: 443–448.

———. 1973*b*. Reproductive inhibition in asymptotic populations of prairie deermice. Suppl., J. Reprod. Fert., 19:457–463.

———. 1974. Behavioral factors associated with cessation of growth of laboratory populations of prairie deermice. Res. Popul. Ecol., 15:138–147.

———. 1979. Inhibition of reproductive development in laboratory populations of prairie deermice (*Peromyscus maniculatus bairdii*): influence of tactile cues. Ecology, 60:455–458.

———. 1980. Behavior and regulation of growth in laboratory populations of prairie deermice. Pp. 23–36, *in* Biosocial mechanisms of population regulation (M. Cohen, R. Malpass, and H. G. Klein, eds.). Yale Univ. Press, New Haven, 406 pp.

Van Abeelen, J. H. F., and A. E. de Vries. 1978. Genetic control of sniffing and marking responses in mice. Behav. Genet., 8:219–222.

Vandenbergh, J. G. 1969. Male odor accelerates female sexual maturation in mice. Endocrinology, 84:658–660.

———. 1971. The influence of the social environment on sexual maturation in male mice. J. Reprod. Fert., 24:383–390.

———. 1975. Hormones, pheromones and behavior. Pp. 551–584, *in* Hormonal correlates of behavior, Vol. 2 (B. E. Eleftheriou and R. L. Sprott, eds.). Plenum Press, New York, 527 pp.

Vandenbergh, J. G., L. C. Drickamer, and D. R. Colby. 1972. Social and dietary factors in the sexual maturation of female mice. J. Reprod. Fert., 28:397–405.

Vestal, B. M., and J. J. Hellack. 1978. Comparison of neighbor recognition in two species of deer mice (*Peromyscus*). J. Mamm., 59:339–346.

Walkowa, W. 1964. Rate of absorption of newcomers by a confined white mouse population. Ekol. Pol., ser. A, 12:326–335.

———. 1971. The effect of exploitation on the productivity of laboratory mouse populations. Acta Theriol., 16:295–328.

Whitsett, J. M., L. E. Gray, Jr., and G. M. Bediz. 1979. Gonadal hormones and aggression toward juvenile conspecifics in prairie deer mice. Behav. Ecol. Sociobiol., 6:165–168.

Whitten, W. K. 1956. Modifications of the oestrus cycle of the mouse by external stimuli associated with the male. J. Endocrinol., 13:399–404.

———. 1959. Occurrence of anoestrus in mice caged in groups. J. Endocrinol., 18:102–107.

———. 1966. Pheromones and mammalian reproduction. Pp. 155–177, *in* Advances in reproductive physiology, Vol. 1 (A. McLaren, ed.). Academic Press, New York, 437 pp.

Whitten, W. K., and A. K. Champlin. 1973. The role of olfaction in mammalian reproduction. Pp. 109–123, *in* Handbook of physiology II(7) (R. O. Greep, ed.). American Physiological Society, Washington, D.C., 323 pp.

Young, H., R. L. Strecker, and J. T. Emlen, Jr. 1950. Localizations of activity in two indoor populations of house mice (*Mus musculus*). J. Mamm., 31: 403–410.

Zejda, J. 1975. Habitat selection in two feral house mouse (*Mus musculus* L.) lowland populations. Zool. Listy, 23:99–111.

STRUCTURE AND FUNCTION OF HAIR CRESTS AND CAPES IN AFRICAN CARNIVORA

CHRISTEN WEMMER AND DON E. WILSON

Abstract

THIS study examines ecological correlates of specialized hair tracts in 55 of the 64 species of African Carnivora. Seventy-five percent of the species have torso hair of more or less uniform length. The remainder have hair that is elongated as narrow middorsal crests or broad dorsal capes. Threat displays that increase apparent body size are widespread in carnivores, but piloerection of crests and capes produces disproportionate increases in body size. The occurrence of crests and capes among taxa refutes the hypothesis that these structures have evolved primarily for intimidation in species capable of inflicting fatal wounds. Crests and capes are attributes of many relatively defenseless species, as well as those specialized for killing large vertebrates. More evidence supports the hypothesis that long dorsal hair is primarily an antipredator adaptation. A significant number of long-haired species are poor climbers unable to seek arboreal refuge from predators. Among viverrids long-haired species occur in open habitats and short-haired species occur in forests and woodlands. Herpestids are generally poor climbers and exhibit no correlation between hair length and habitat. However, larger and perhaps more conspicuous species tend to have long dorsal hair. The relative dorsal hair length of the torso in canids and herpestids was significantly correlated with number of sympatric carnivores not exceeding by 50% the head and body length of each species being evaluated. Inability to climb trees, intermediate body size, and occurrence in open habitats may increase vulnerability to predators, and place a selective premium on hair patterns that maximize the illusion of a larger body.

"Hardly any expressive movement is so general as the involuntary erection of the hairs, feathers and other dermal appendages; .... The action serves to make the animal appear larger and more frightful to its enemies or rivals, and is generally accompanied by various voluntary movements adapted for the same purpose, and by the utterance of savage sounds."

Charles Darwin, 1872

Introduction

Elements of the mammalian integument have been important in the evolution of visual, olfactory, and auditory modes of communication. Social functions of the skin emerged by modifications of the widespread and probably primitive association of follicle, hair(s), arrector pili muscle, and apocrine gland. The diversity of dermal scent glands (Schaffer, 1940), richness of hair color and length patterns (Cott, 1940; Miller, 1931), and bizarre adaptations such as sound-producing quills (Darwin, 1872; Ewer, 1968; Gould, 1965), attest to the evolutionary significance of the integument.

Color, length, and density characteristics of the pelage, collectively termed hair patterning (Miller, 1931), have been the focus of numerous taxonomic studies, but have stimulated fewer investigations of their adaptive significance. Thermal properties of insulation and color are well known (Herrington, 1951; Hutchinson and Brown, 1969; Scholander, 1950; Tregear, 1965). Social and antipredator functions were summarized in 1940 by Cott, but little research was done on these topics for the next three decades. Renewed interest and controversy have been stimulated by recent work on mimicry (Gingerich, 1975; Goodhart, 1975; Greene, 1977) and pursuit invitation signals (Coblentz, 1980; Hirth and McCullough, 1977; Smythe, 1970). The present study deals with visual communication mediated by erection of specialized zones of hair, in Darwin's words, "generalized expressive movement." Elongation of dorsal pelage to enhance the illusion of larger body size is a recurrent adaptation of terrestrial mammalian predators and omnivores in Africa, Asia, South America, and probably Australia. The objectives of this study are to describe some design features of specialized pilomotor tracts within sub-Saharan carnivores, and to examine ecological correlates of such characteristics.

Methods

Hair-length measurements were taken from flat or stuffed museum specimens in adult pelage. A middorsal and lateral series of guard hair measurements was taken at shoulder, midback, and rump. Dorsal and lateral hair length was then calculated by averaging the three measures. The length of hair of the tail was measured at two places, one third the total length from both the base and the tip. A ruler was placed perpendicular to the skin, and the length of the majority of guard hairs was measured. With one exception (*Civettictis civetta*), no consistent or significant difference could be found between the length of dorsal and lateral tail hair. We averaged tail hair length, therefore, using only the two dorsal measures. Although these are somewhat crude estimates, they give a fair appraisal of species differences. Specimens in molting condition or missing hair were avoided or the hair in the molting section of the skin was not measured if samples were small. Fifty-five species were examined, and an average of 10 specimens per species was measured (range 1 to 24). Head and body length measurements were obtained for all species. We calculated nearly all indices using body measurements taken from the tags of those specimens used for hair-length measurements. Dorst and Dandelot (1970) was used as the source for head and body length of five additional species (Fig. 1). Taxonomic usage follows Ellerman et al. (1953) and Meester and Setzer (1971). Cross-sectional profiles are based on seven hair-length measures taken in the midbody region. Hair length was measured on the spine, side, and ventrum on one half of the skin. Two other measurements were made at points equidistant between the lateral and dorsal, and lateral and ventral measurement points. Diagrams show the measured half of the skin and its mirror image. Circumferences were normalized to facilitate comparisons of hair-length profiles.

Information on threat displays was gathered from the sources listed in Table 1, including photographs of species for which descriptions are lacking or incomplete. Photographs by field workers and from the National Zoological Park photofiles were additional sources of information on displays. A polar planimeter was used to measure body surface areas from photographs of animals in normal and piloerected postures.

We excluded otters from analyses because of their aquatic habits,

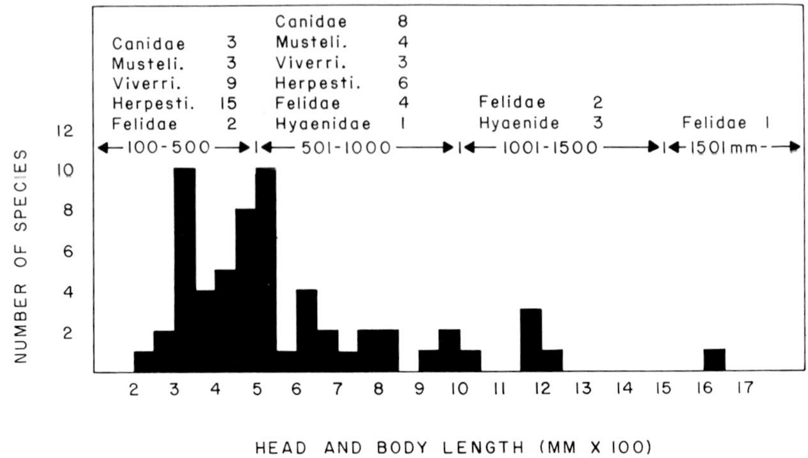

FIG. 1. Frequency histogram of body size (head and body length) in the sub-Saharan carnivore fauna (60 species).

but included the aquatic genet (*Osbornictis piscivora*) because its specializations for aquatic life are less developed. The mane of the male African lion differs from the hair structures we are concerned with; therefore, only measurements of a female lion were used.

Results

Sub-Saharan carnivore fauna.—There are about 64 species of living carnivores representing six families south of the Tropic of Cancer. These species range in size from the dwarf mongoose (*Helogale parvula*) (0.3 kg) to the lion (*Panthera leo*) (172 kg). One-half of the 64 species are less than 500 mm in head and body length, and 90% are less than 1 m long (Fig. 1). Two families, the Felidae and Hyaenidae, constitute the largest carnivores, but all families have representatives less than 1 m in body length.

Hair-length patterning.—We were able to examine 55 (90%) of the non-aquatic species (n = 61). Most short-haired species (lateral

---

→

FIG. 2. Scatter diagram of lateral hair length plotted against dorsal hair length in 55 species of sub-Saharan carnivores.

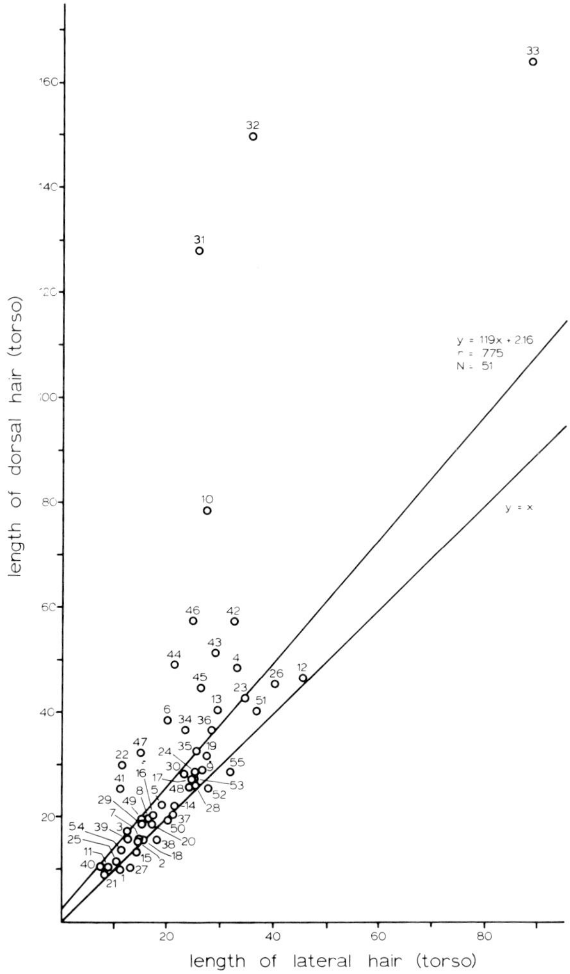

VIVERRIDAE
1 Poiana
2 Genetta thierryi
3 G. pardina
4 G. angolensis
5 G. tigrina
6 G. genetta
7 G. servalina
8 G. victoriae
9 Osbornictis
10 Civettictis
11 Nandinia

HERPESTIDAE
12 Herpestes ichneumon
13 H. naso
14 H. pulverulentus
15 H. sanguineus
16 Crossarchus ansorgei
17 C. obscurus
18 Cynictis
19 Atilax paludinosus
20 Dologale
21 Helogale
22 Liberiictis
23 Ichneumia
24 Mungos mungo
25 M. gambianus
26 Rhynchogale
27 Suricata
28 Paracynictis
29 Bdeogale crassicauda
30 B. nigripes jacksoni

HYAENIDAE
31 Proteles
32 Hyaena hyaena
33 H. brunnea
34 Crocuta

FELIDAE
35 Felis libyca
36 F. nigripes
37 F. serval
38 F. caracal
39 Panthera pardus
40 P. leo
41 Acinonyx

CANIDAE
42 Canis aureus
43 C. adustus
44 C. mesomelas
45 C. simensis
46 Lycaon
47 Otocyon
48 Vulpes ruppelli
49 V. chama
50 V. pallida
51 V. vulpes
52 Fennecus

MUSTELIDAE
53 Ictonyx
54 Poecilictis
55 Mellivora

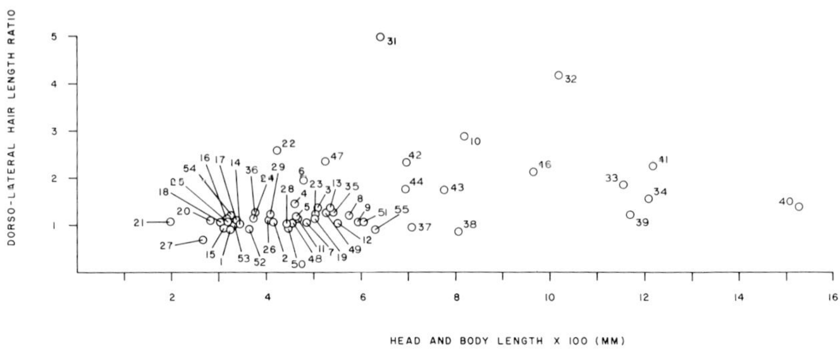

FIG. 3. Dorso-lateral length ratios of torso hair versus head and body length in sub-Saharan carnivores. Refer to Fig. 2 for species corresponding to numbers.

hair < 25 mm) show a one-to-one relationship between length of lateral and dorsal body hair (Fig. 2). However, many species have dorsal hair longer than lateral hair. To predict dorsal from lateral hair length, we calculated the formula y = 1.19x + 2.16 (r = 0.775, n = 51). We excluded the four most deviant species from this calculation (*Civettictis civetta, Proteles cristatus, Hyaena hyaena, H. brunnea*). When these species are included, r = 0.74 (n = 55) and the slope increases (y = 1.854x + 7.75). Most species show relatively slight differences in length of dorsal and lateral hair, but 25% of 55 species are exceptions (Fig. 3). *Proteles cristatus* and *Hyaena hyaena* exhibit extreme ratios (>4.0); six species show moderate ratios (2.0 to 3.0), and six other species show ratios between 1.5 and 2.0. Large ratios exist in both long- and short-haired species, but nearly all species with the greatest ratios have disproportionately long dorsal hair.

Length of lateral and dorsal hair increases as a function of head and body length (Fig. 4). However, relative length of dorsal body hair varies considerably. Thirty-five species (63%) have hair less than 7% of the head and body length, and 17 species (31%) have hair between 7 and 14% of the head and body length. Only three

→

FIG. 4. Scatter diagrams of head and body length versus lateral hair length (A), and dorsal hair length (B) in 55 sub-Saharan carnivores. Refer to Fig. 2 for species corresponding to numbers.

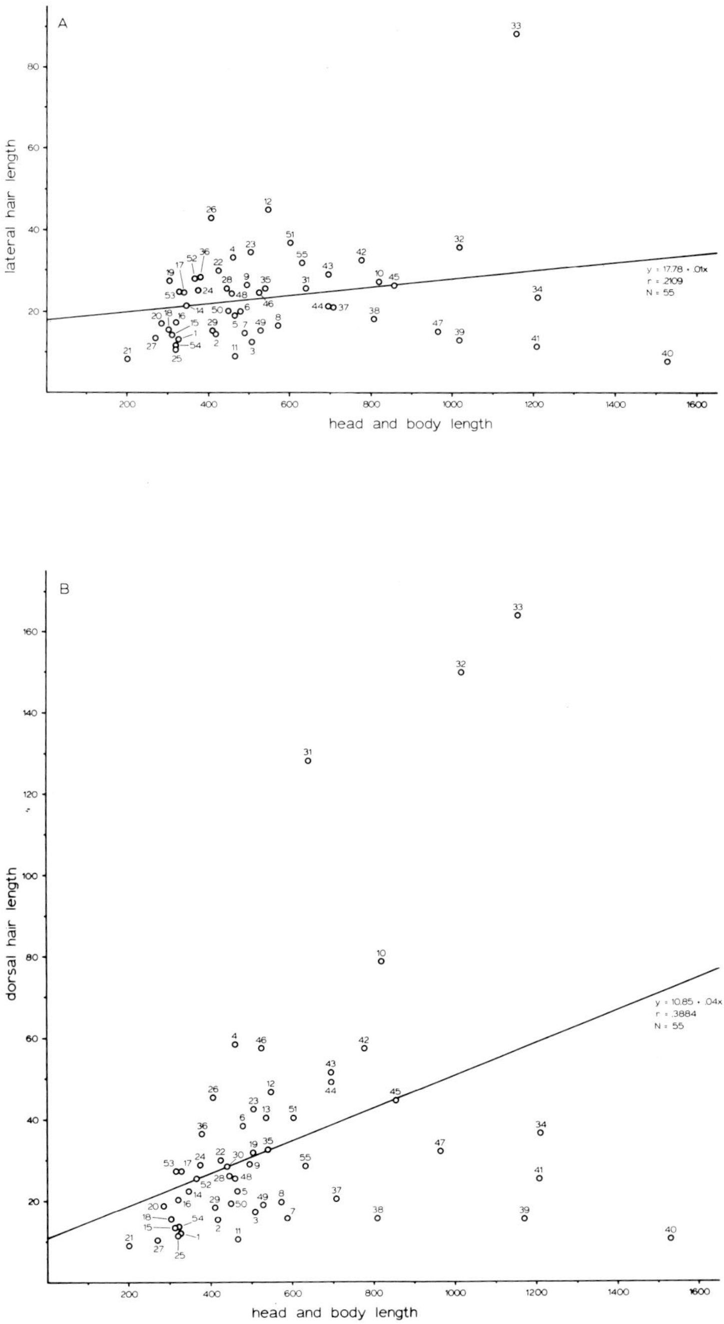
A
lateral hair length
head and body length
y = 17.78 + .01x
r = .2109
N = 55
B
dorsal hair length
head and body length
y = 10.85 + .04x
r = .3884
N = 55

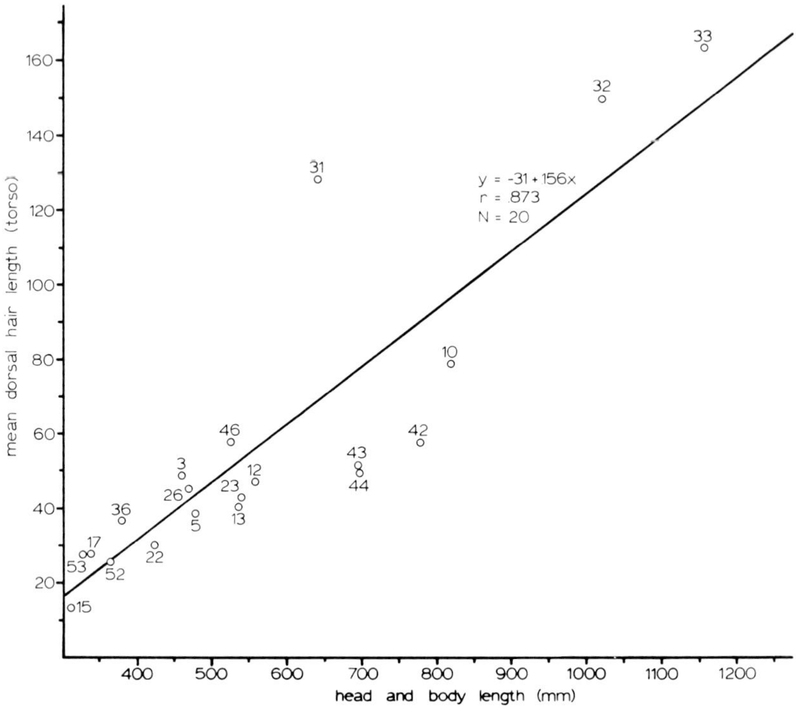

FIG. 5. Scatter diagram of head and body length versus dorsal hair length of the torso in 20 species of sub-Saharan carnivores having dorsal body hair exceeding 7% of head and body length. Refer to Fig. 2 for species corresponding to numbers.

species (5%) have hair exceeding 14% of the head and body length. When those 20 species having hair exceeding 7% of the head and body length are considered as a group, 76% of the variance in mean absolute dorsal hair length is predicted by length of head and body (Fig. 5). The correlation is still significant when the four largest species are deleted ($y = 4.96 + 0.72x$, $r = 0.806$, $n = 16$). Tail length and length of dorsal hair of the tail are not significantly correlated ($y = 45.65 + 0.02x$, $r = -0.144$, $n = 54$), but species with long dorsal pelage also have long-haired tails. Bushy tails occur in both short- and long-tailed species.

Variations in crest form.—Three categories of hair form are recognizable by viewing the torso in cross-section (Fig. 6). Long hair may be confined to a narrow middorsal zone forming a crest (*Prot-*

eles cristatus, Civettictis civetta, Genetta tigrina), or long hair may form a broad cape across the back (*Genetta genetta, Canis mesomelas*). The third category includes species in which there is no zone of pronounced elongation (*Herpestes sanguineus, Atilax paludinosus*).

Threat displays.—Intraspecific and interspecific threat behavior has been described in 25 sub-Saharan species representing all carnivore families (Table 1). The completeness of behavioral descriptions, however, is far from satisfactory. Although some differences are real, others undoubtedly are due to incomplete observation. Nearly all species show fundamental similarities. The shared elements include leg extension, spinal convexity, piloerection, and tail elevation. Differences between species in the expression of these components are probably a function of anatomy. Pronounced arching of the back occurs in species with long torsos and supple spines (for example, in species of *Felis* and *Genetta*). Species of *Canis* extend the hindlegs strongly, but often retain flexed forelimbs; both the forelegs and hindlegs are extended in *Proteles cristatus* and *Civettictis civetta*. The lion, leopard (*Panthera pardus*), and cheetah (*Acinonyx jubatus*) lack display elements that increase stature (leg extension, back-arching). The serval (*Felis serval*) feigns increased size like most other species, but also threatens like the larger felids.

Scant information on body orientation toward rivals during threat suggests this element of threat has been overlooked. The findings are also tentative because some species (*Civettictis civetta*) require very strong stimulation to elicit lateral orientation. Lateral orientation is absent in some small forms (*Herpestes* spp.), but present in another (*Ictonyx striatus*). Many medium-sized carnivores display laterally, but incomplete information prevents conclusions. We predict that lateral displays will be most common in species having strongly developed crests. To summarize, stature-enhancing displays are widespread in African carnivores, regardless of degree of crest development.

Piloerection.—Piloerection in social contexts is assumed to be an autonomic response to exciting stimuli, motivated by aggression, fear, or a combination thereof. It is also a non-specific reaction to many other circumstances, and can be regarded as an excitation indicator because of apparent correspondence between the intensity of stimulation and the degree and duration of the pilomotor response (Gould, 1971). Besides the illusion of increased body size, piloerec-

TABLE 1
AGGRESSIVE DISPLAYS IN SUB-SAHARAN CARNIVORA

Species	Display elements												References
						Depressed head and shoulders	Tail position			Body orientation			
	Legs spread	Leg extension	Stiff-legged gait	Spinal convexity	Spinal cavity		Down	Horizontal	Up	Lateral, tangential	Frontal	Piloerection	
Hyaenidae													
Hyaena hyaena				−	−	−		+		+	+	+	Kruuk (1976); Rieger (1979)
Hyaena brunnea				−	−			+			+	+	Mills (1978); Owens and Owens (1978)
Proteles cristatus		+		−	−	+		+			+	+	Dobroruka (1967); von Ketelhodt (1966)
Crocuta crocuta				−	−	−		+	+	+	+	+	Kruuk (1972)
Felidae													
Felis aurata		+		+	−	+	+	−	−	+	+	+	Leyhausen (1979)
Felis margarita		+		+	−	+	+	−	−	+	+	+	Hemmer (1974, 1977)
Felis libyca				+	−							+	Armstrong (1977)
Felis caracal	+	+	−	−	+	+				−	+	−	Kingdon (1977); York (1973)
Felis serval				+					+	+	+	+	Leyhausen (1979)
Acinonyx jubatus		−	−	−	+	+	+	−	−		+		Leyhausen (1979)
Panthera pardus		−	−		+	+	+	−	−		+		Leyhausen (1979)

TABLE 1
CONTINUED

Species	Display elements: Legs spread	Leg extension	Stiff-legged gait	Spinal convexity	Spinal concavity	Depressed head and shoulders	Tail position: Down	Horizontal	Up	Body orientation: Lateral, tangential	Frontal	Piloerection	References
Panthera leo		+	+	−	−	−	−	−	+	+	+	−	Schaller (1972)
Panthera leo	+	+	−	−	+	+				−	+	−	Leyhausen (1979); Schaller (1972)
Viverridae													
Genetta tigrina	−	+	+	+	−	±	+	+	−	+	+	+	Wemmer (1977)
Genetta genetta	−	+	+	+	−	±	+	+	−	+	+	+	Gangloff and Ropartz (1972)
Civettictis civetta	−	−	−	−	−	+	+	+	−	+	+	+	Ewer and Wemmer (1974); Wemmer (1977)
Nandinia binotata	−	+	+	+	−	−	−	+	+	+	+	+	Wemmer (1977, pers. observ.)
Herpestidae													
Suricata suricatta	−	+	+?	+	−	−	−	+	+	−	+	+	Ewer (1963); pers. observ.
Crossarchus obscurus	−	+	+	+	−		+	−	−	−	+	+	Ewer, *in* Kingdon (1977)
Herpestes ichneuman				+	−	+	−	+	−	−	+	+	Leyhausen (1979)

TABLE 1
CONTINUED

Species	Display elements												References
						Depressed head and shoulders	Tail position			Body orientation			
	Legs spread	Leg extension	Stiff-legged gait	Spinal convexity	Spinal concavity		Down	Horizontal	Up	Lateral, tangential	Frontal	Piloerection	
Ichneumia albicauda	−	+	−			±	−	+	+	+	+	+	Leyhausen (1979); Waser (pers. comm.)
Canidae													
Canis aureus	−	+	?	+	−	±	−	+	+	+	+	+	Estes and Goddard (1967); Fox (1971); Golani and Keller (1971); Golani and Mendelssohn (1975)
Canis mesomelas		+	−	+	−	+		+			+	+	Moehlman (pers. comm.)
Vulpes vulpes		+	+	+	−	±	±	+	±	+	+	+	Fox (1969); Tembrock (1957)
Otocyon megalotis	−	+	−	+	−	±	−	−	+	±	±	+	Kleiman (1967); Lamprecht (1979)
Mustelidae													
Ictonyx striatus	+	+	+		−	+	−	−	+	±	±	+	Rowe-Rowe (1978)
Poecilogale albinucha	+	+	+		−	+	−	−	+	±	±	+	Rowe-Rowe (1972, 1978)

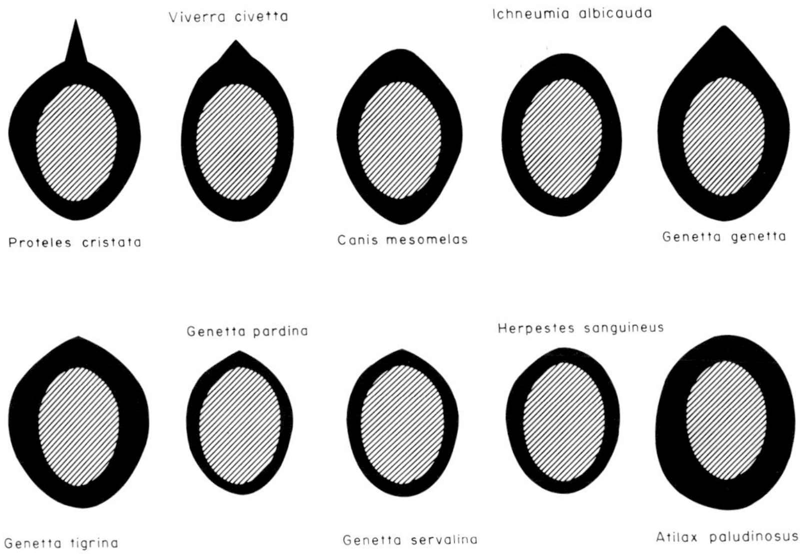

FIG. 6. Cross-sectional profiles of guard hair at mid-body in 12 species of carnivores. *Viverra civetta* = *Civettictus civetta.*

tion almost always modifies body shape or profile, and sometimes is accompanied by changes in color pattern (Wemmer, 1977).

The illusion of increased body size during piloerection depends on the hair pattern of the displaying animal and its body orientation to the rival (Fig. 7). The erectile crest of *Proteles cristatus* can increase its lateral aspect by as much as 74%, but it is obvious that a frontal perspective is much less enhanced. Piloerection in *Civettictis civetta* increases lateral aspect at least 30%, and the white spinal stripe formed by the base of the crest undergoes a six-fold increase in size. The presence of this white area appears to vary geographically, and is more prominent in forested than in savanna populations. *Hyaena hyaena* increases lateral aspect by 38%. These measures, based on photographs of single animals, confirm the magnitude of the illusion of increased body size.

Evolution of Long Dorsal Hair

Probably several factors have influenced the evolution of long dorsal hair, but likely only one played a primary role. In searching

FIG. 7. Lateral views of *Proteles cristatus* (left) and *Civettictis civetta* (right) with dorsal hair relaxed (bottom) and erected (above).

for relations we divided the 55 species into long-haired and short-haired categories. This was an arbitrary choice because the relative lengths of dorsal body hair approximate a normal distribution (Fig. 8). Species exhibiting moderate to long dorsal crests and capes (relative hair lengths > 7.0%) were classified as long-haired. No food habit predominates in this group, and no differences were found in mean head and body length ($t = 0.008$, d.f. = 53, $P = 0.9$), or in nocturnal and diurnal activity ($\chi^2 = 0.514$, d.f. = 1, $P = 0.48$) between long- and short-haired groups. Hypotheses are discussed below.

Thermoregulation hypothesis.—Hair has important insulative properties. Therefore, could long dorsal body pelage be an adaptation to conserve heat? Measures of hair density and insulation quality are necessary to answer this question conclusively. The undercoat of the pelage is specialized to create dead air space, and guard hairs, the components of crests and capes, contribute little to

→

FIG. 8. Frequency distribution of dorsal hair length in species of sub-Saharan carnivores.

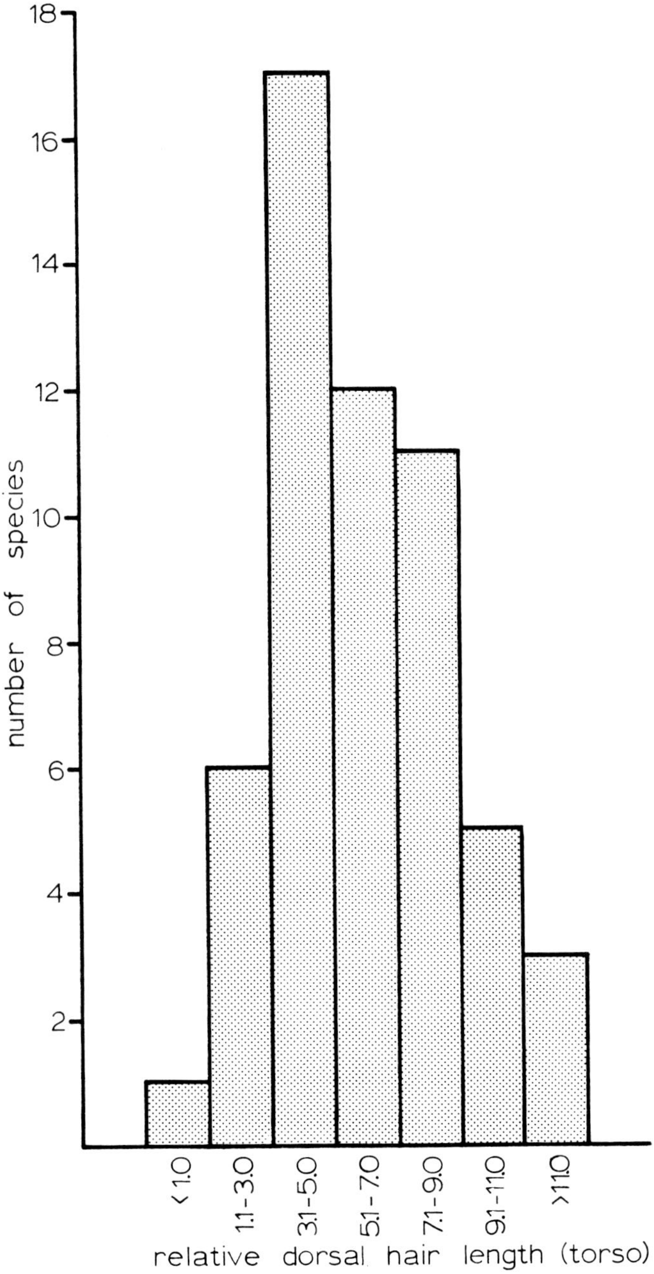
18
16
14
12
10
8
6
4
2
number of species
< 1.0
1.1-3.0
3.1-5.0
5.1-7.0
7.1-9.0
9.1-11.0
>11.0
relative dorsal hair length (torso)

insulation (Herrington, 1951; Scholander, 1955). A range of thermoregulatory adaptations may be inferred from differences in the undercoat and the microenvironments of tropical carnivores. Diurnal *Mungos* spp., for example, lack undercoats (Rosevear, 1974), and this probably is true for *Suricata suricatta.* On the other hand, nocturnal desert foxes (*Vulpes pallida, V. ruppelli,* and *Fennecus zerda*) have well-developed insulation layers. In general, however, long dorsal hair consists of bristles and awns dense enough to create an illusion of substance when erect, but are probably insufficiently packed to impede convection and aid thermal resistance. We doubt that differences in dorsal pelage length and conformation are related to thermoregulation patterns.

Intraspecific conflict hypothesis.—Effecting larger body size may serve primarily to resolve intraspecific conflict. Interspecific similarities in pilomotor threat displays may have resulted from convergent or parallel evolution, and differences in length of dorsal hair may be related to social organization. We tested the prediction that the number of short- and long-haired species would differ when sorted into solitary versus group-living categories. The latter category included all monogamous and clan-living species, and species reported in anecdotal sources to forage, den, or travel communally (Kingdon, 1977; Roberts, 1951). There was no significant difference between groups ($\chi^2 = 0.189$, d.f. = 1, $P = 0.667$).

There are two corollaries of this hypothesis. Intimidation structures (that is, crests or capes) should be largest in species able to inflict fatal wounds. Piloerection almost always accompanies fighting. Owens and Owens (1978) reported that *Hyaena brunnea* raised their hackles during competitive interactions around food and during fights involving neck-biting and chasing. Rieger (1978) made similar observations of *Hyaena hyaena. Proteles cristatus* has weak jaws and reduced dentition, but has the largest crest of all carnivores. Cats are among the most effective killers, but only one of the six African species examined has dorsal hair exceeding 7% of head and body length.

The second corollary is that polygynous carnivores should exhibit sexual dimorphism in crest size. No species examined except the African lion shows such differences. Schaller (1972) described the mane of *Panthera leo* as a secondary sexual characteristic permitting recognition at a distance of sex and possibly of individuals. The mane's ancillary function is probably to decrease chance of injury during fights by shielding the head and neck. Schaller gave no evidence that the mane is erected; thus the mane of the lion differs in several ways from carnivore crests and capes.

There is insufficient evidence to support the intraspecific conflict hypothesis. Many relatively defenseless species have long dorsal hair, and crests and capes are not a common attribute of taxa with strong jaws and teeth or other specializations for killing large vertebrate prey. On the contrary, there is now evidence that dermal shields evolved in some species to protect body parts frequently bitten during intraspecific conflict (Kruuk, 1972; Owens and Owens, 1979).

Interspecific conflict hypothesis.—Is long dorsal body hair primarily an antipredator adaptation? Long-haired species share one feature which might predispose them to predation. Eighty-five percent of the 20 species lack adept climbing ability; only 31% are nonclimbers in the short-haired group. The difference is significant ($\chi^2 = 13.72$, d.f. = 1, $P < 0.001$). Many species depend on the ability to climb trees to escape ground predators. If this is true of African carnivores, climbing species living in open habitats should have longer dorsal hair than forest species. This prediction is based on the assumption that forest species have ready access to an arboreal refuge, and that the chance of being close to a tree is much less in savanna.

In the Viverridae the prediction generally holds. *Civettictis civetta* is the only African species that cannot climb. *Osbornictis piscivora,* a genet with incipient aquatic adaptations, is probably a less skillful climber than other genets. Forest viverrids have short to medium-length hair, whereas desert and savanna species have longer dorsal hair (Fig. 9).

Most African mongooses are terrestrial and are poor climbers. *Herpestes sanguineus* is the only known exception; it forages in bushes and small trees (Rood, pers. comm.; Taylor, 1975). Fifteen of the 20 herpestid species are diurnal, and an equal number occur in savanna or desert habitats. Five species are restricted to primary or secondary forest. There is no trend within the family for longer-haired species to occur in more open habitats (Fig. 10). On the contrary, *Suricata suricatta* and *Cynictis penicillata* from arid southern Africa have very short pelage.

The 18 species of African herpestids average 279 mm in head and body length (range 200 to 547), 25% smaller than the average

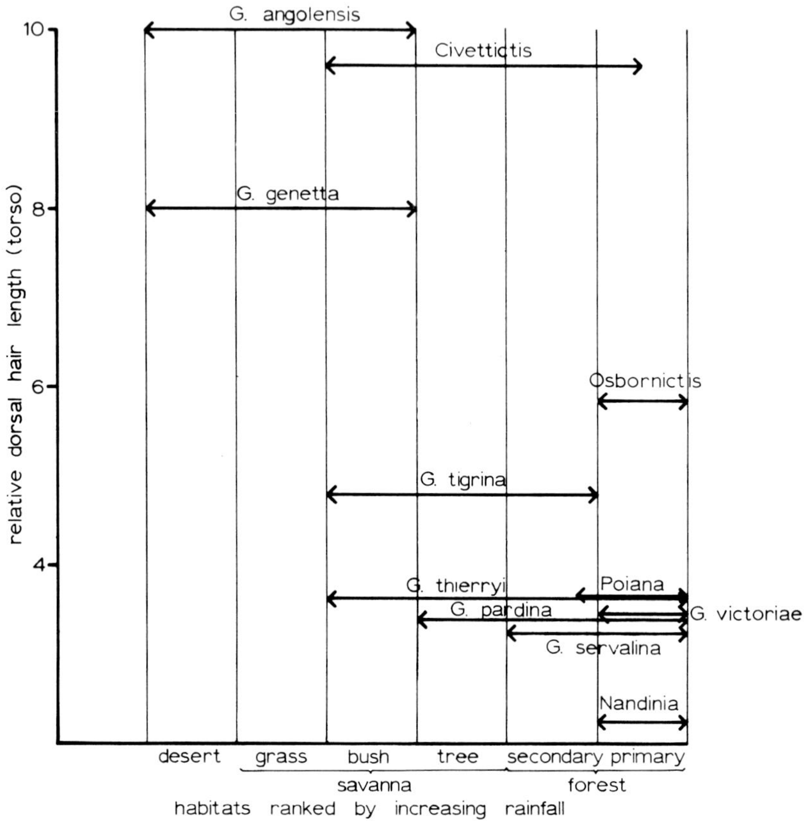

FIG. 9. The relationship of habitat selection and dorsal hair length of the torso in sub-Saharan Viverridae.

viverrid ($\bar{X} = 499$ mm, range 326 to 819). It could be argued that smaller, open-habitat herpestids evade predators more effectively than larger species by hiding in ground vegetation. The "tweed" coat pattern of most species is clearly cryptic in well-lit grassy and scrub habitats, and grassland can be regarded as a closed or three-dimensional habitat for small-bodied mongooses. Indeed, the seven long-haired mongooses have significantly larger bodies ($\bar{X} = 437.8$ mm, range 311 to 547) than short-haired species ($\bar{X} = 350.4$ mm, range 200 to 504, $t = 1.98$, d.f. $= 15$, $P < 0.05$, one sided). How-

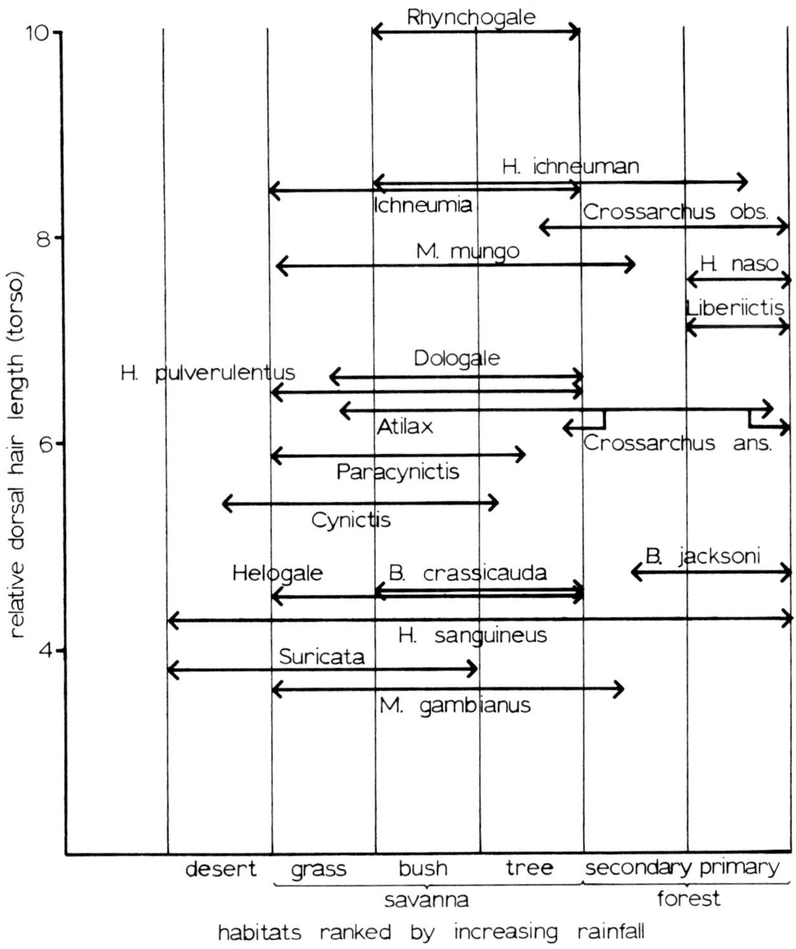

FIG. 10. The relationship of habitat selection and dorsal hair length of the torso in sub-Saharan Herpestidae.

ever, we question whether this slight size difference between the two groups appreciably affects vulnerability.

Vulnerability to interspecific conflict.—If hair length has anti-predator function, a correlation might exist between dorsal hair length and number of sympatric carnivores. This hypothesis was tested by plotting relative dorsal hair length against the number of carnivores that 1) overlapped the range of the species by >25%, 2) were equal to or larger than the species, and 3) had similar activity periods. No significant relationship was found (y = 4.9x + 0.12, r = 0.139, n = 51). Two further refinements were made on the sympatric carnivores variable. The potential to visually intimidate a larger rival probably decreases with increasing size difference between two contenders. Therefore, two size ranges of sympatric carnivores were specified. The smaller range included all sympatric species no smaller than 20%, nor larger than 30% of the head and body length of each species being evaluated. The larger range included sympatric carnivores with the same lower limits in body size, but not exceeding 50% of the head and body length of each species being evaluated. A small percentage of variance in relative dorsal hair length is predicted by number of sympatric species (y = 4.18 + 0.11x, r = 0.204, n = 51 for sympatric carnivores <50% larger; y = 3.41 + 0.129x, r = 0.191, n = 49 for sympatric carnivores <30% larger). The results are not surprising because there are fewer large than small carnivores. However, when the effects of large size were removed by considering only carnivores between 300 and 600 mm in head and body length, there was no improvement in fit (y = 3.76 + 0.25x, r = 0.184, n = 34 for sympatric carnivores <50% larger; y = 3.87 + 0.02x, r = 0.016, n = 31 for sympatric carnivores <30% larger).

Finally, each family having an adequate number of species was treated separately. In the Canidae and Herpestidae, relative dorsal hair length of the torso was significantly correlated with number of sympatric carnivores not exceeding head and body length by 50% ($P < 0.05$) (Fig. 11). When the four hyaenids were included with the canids on the basis of similar body form and ecology, the correlation was not significant (y = 4.51 + 0.177x, r = 0.288, n = 15, $P > 0.05$).

Discussion

Crests and capes of long dorsal hair occur in both sexes in a range of medium-sized African carnivores. Guard hair length is

---

→

FIG. 11. Correlations between relative dorsal hair length of the torso and number of sympatric carnivores not exceeding 50% head and body length in the Herpestidae and Canidae. Refer to Fig. 2 for species corresponding to numbers.

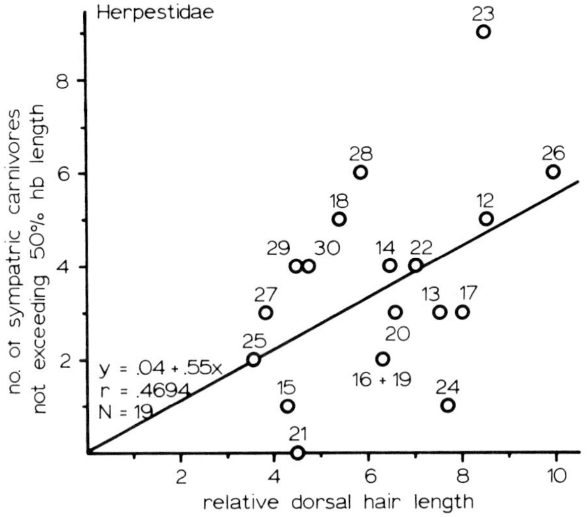
Herpestidae
y = .04 + .55x
r = .4694
N = 19
no. of sympatric carnivores not exceeding 50% hb length
relative dorsal hair length

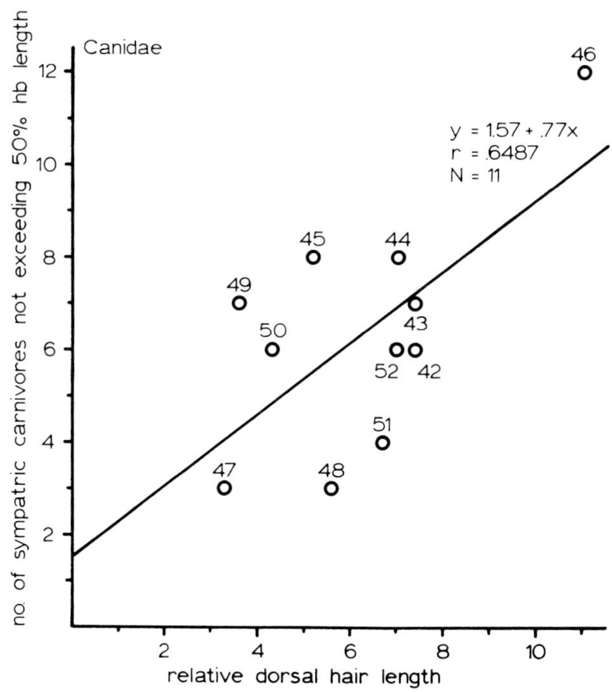
Canidae
y = 1.57 + .77x
r = .6487
N = 11
no. of sympatric carnivores not exceeding 50% hb length
relative dorsal hair length

probably related to underhair length, but the long guard hair probably lacks significant insulative properties. Erection of body hair commonly occurs in social contexts, and the illusion of increased stature and girth is appreciable in species having crests and capes. Estes and Goddard (1967:66) remarked that "A little 15 pound jackal, coat fluffed, head down, and snarling would stand its ground and snap ferociously if the hunting dog continued to advance . . . it proved a surprisingly effective intimidation display." Intraspecific rivalry for food and space undoubtedly selected for the intimidation value of long dorsal hair. However, long-haired species do not share behavioral or ecological features that reveal the adaptive character of these specialized hairs. Better understanding of natural history should clarify intraspecific factors favoring evolution of capes and crests.

We hypothesize that interspecific aggression also has selected for long dorsal hair. Widespread lack of climbing ability, intermediate body size, and occurrence in open habitats may increase an animal's vulnerability or reduce its escape options during interspecific conflict. These characteristics are often shared by species having well-developed crests and capes (that is, >7% head and body length). Carnivores usually avoid one another unless there are benefits of association. However, competitive associations often precipitate hostility, which sometimes leads to the death of the smaller or outnumbered species. Though flight is a prevalent first reaction in many species, it may also invite pursuit under certain circumstances (Smythe, 1970). Conduct diminishing the probability of attack probably is a more fit alternative if the defending species lacks speed.

The safest action for a weaker animal confronted at close range by an aggressor is probably threat rather than attack or flight. Subsequent escape is most likely to be successful when the attention of the aggressor is diverted. The ability to intimidate by feigning larger body size is undoubtedly a function of the size difference between rivals. In the preceding analyses, sympatric carnivores were considered to be potentially responsive to intimidation if they did not exceed the head and body length of the species by 50%. Only mongooses and canids, families of primarily terrestrial and open habitat species, showed a significant correlation between relative length of dorsal hair and number of sympatric carnivores as we have defined them. A refined analysis should focus on type and intensity of interspecific relationship, but this information is scant.

Dorsal hair length of hyaenids and viverrids could not be predicted by number of sympatric carnivores according to our definition. Three of the four hyaenids have exceptionally long dorsal body hair. All forage long distances for dispersed foods, but only *Crocuta crocuta* achieves high population densities and cooperatively kills large-bodied prey (Kruuk, 1972, 1976; Kruuk and Sands, 1972; Mills, 1978; Owens and Owens, 1978). Cooperative group behavior may have replaced the need of the individual for intimidation by piloerection. This reasoning, however, does not explain the existence of poorly developed capes in some social mongooses (for example, *Crossarchus obscurus* and *Mungos mungo*).

Neither viverrids nor felids show correlations between relative dorsal hair length and number of sympatric carnivores no larger than 50% of head and body length. However, nearly all of these species are fleet-footed, adept climbers, and have relatively short dorsal body hair (<7% of head and body length). The exceptions include the long-crested *Civettictis civetta,* which cannot climb, and the long-haired *Genetta genetta* and *G. angolensis*. Like all genets they are good climbers, but *G. genetta* lives in arid open habitats, and *G. angolensis* is a woodland savanna species. *Felis nigripes* is the only felid with dorsal body hair exceeding 7% of its head and body length. It is an able climber, but occupies barren wastelands of southern Africa. These species support the contention that climbing forms living in open habitats tend to have longer dorsal hair than those living in forests. However, several exceptions to the general trends require explanation, and underscore the fact that there is no universal set of characters predicting the occurrence of long crests and capes. Vulnerability to interspecific conflict for food and space is hypothesized to be a critical factor. Intermediate body size, poor climbing ability, or occurrence in open habitats are features that may increase vulnerability. Additional factors probably also play a role. There is limited evidence that the importance of individual intimidation incorporating crests and capes is reduced in group-living species. Explanation of exceptions to the general trends should clarify additional factors that have been critical in the evolution of crests and capes.

Acknowledgments

Special thanks for permission to examine specimens are extended to Drs. C. O. Handley, Jr. and H. W. Setzer of the National

Museum of Natural History (Washington, D.C.), L. de la Torre of the Field Museum of Natural History (Chicago), R. G. Van Gelder of the American Museum of Natural History (New York), and J. E. Hill of the British Museum (London). Thanks are due to S. Bruch, P. Mehlhop, L. Morland, T. Portillo, and L. Wemmer for assistance in measuring specimens, data compilation and analysis, and preparation of figures. Drs. G. Mace and M. Biben gave timely statistical help. Photos of *Hyaena hyaena* and *Proteles cristatus* were kindly provided by Dr. I. Rieger and E. Maliniak. Dr. J. F. Eisenberg gave time and stimulating feedback during the long incubation of this study.

Literature Cited

Armstrong, J. 1977. The development and hand-rearing of black-footed cats. Pp. 71–80, *in* The world's cats, contributions to breeding biology, behavior and husbandry (R. L. Eaton, ed.). Univ. Washington, Seattle, 3:1–144.

Coblentz, B. E. 1980. On the improbability of pursuit invitation signals in mammals. Amer. Nat., 115:438–442.

Cott, H. B. 1940. Adaptive coloration in animals. Methuen and Co., Ltd., London, 508 pp.

Darwin, C. 1872. The expression of the emotions of man and animals. Murray, London, 372 pp.

Dobroruka, L. J. 1967. Zur Haltung des Erdwolfen, *Proteles cristatus*. Zool. Garten, 6:307–314.

Dorst, J., and P. Dandelot. 1970. A field guide to the larger mammals of Africa. Collins, London, 287 pp.

Ellerman, J. R., R. C. S. Morrison-Scott, and R. W. Hayman. 1953. Southern African mammals, 1758 to 1951: a reclassification. London, 363 pp.

Estes, R. F., and J. Goddard. 1967. Prey selection and hunting behavior of the African wild dog. J. Wildl. Mgmt., 31:52–70.

Ewer, R. F. 1963. The behavior of the meerkat, *Suricata suricatta* (Schreber). Z. Tierpsychol., 20:570–607.

———. 1968. A preliminary survey of the behaviour in captivity of the dasyurid marsupial, *Sminthopsis crassicaudata* (Gould). Z. Tierpsychol., 25:319–365.

Ewer, R. F., and C. Wemmer. 1974. Behavior of the African civet (*Civettictis civetta*) in captivity. Z. Tierpsychol., 34:359–394.

Fox, M. W. 1969. The anatomy of aggression and its ritualization in the Canidae: a developmental and comparative study. Behaviour, 35:242–258.

———. 1971. The behavior of wolves, dogs and related canids. Harper and Row, New York, 220 pp.

Gangloff, B., and P. Ropartz. 1972. Le repertoire comportemental de la genetta (*Genetta genetta*) (Linne). La Terre et la Vie, 4:489–560.

Gingerich, P. D. 1975. Is the aardwolf a mimic of the hyena? Nature, 253:349–356.

Golani, I., and A. Keller. 1975. A longitudinal field study of the behavior of a pair of golden jackals. Pp. 303–335, *in* The wild canids, their systematics, behavioral ecology and evolution (M. W. Fox, ed.). Van Nostrand Reinhold Co., New York, 508 pp.

GOLANI, I., AND H. MENDELSSOHN. 1971. Sequences of precopulatory behavior of the jackal (*Canis aureus* L.). Behaviour, 38:169–192.

GOODHART, C. G. 1975. Does the aardwolf mimic a hyena? Zool. J. Linnean Soc., 57:349–357.

GOULD, E. 1965. Evidence for echolocation in the Tenrecidae of Madagascar. Amer. Phil. Soc., 109:352–360.

———. 1971. Studies of maternal infant communication and development of vocalizations in the bats *Myotis* and *Eptesicus*. Commun. Behav. Biol., 5:263–313.

GREENE, H. W. 1977. The aardwolf as hyaena mimic: an open question. Anim. Behav., 25:245.

HEMMER, H. 1974. Studien zur Systematik und Biologie der Sandkatze (*Felis margarita* Locke, 1858). Z. Kölner Zoo, 17:11–20.

———. 1977. Biology and breeding of the sand cat. Pp. 13–21, *in* The world's cats, contributions to breeding biology, behavior and husbandry (R. L. Eaton, ed.). Univ. Washington, Seattle, 3:1–144.

HERRINGTON, L. P. 1951. The role of the piliary system in mammals and its relation to thermal environment. Ann. New York Acad. Sci., 53:600–607.

HIRTH, D. H., AND D. R. MCCULLOUGH. 1977. Alarm signals in ungulates with special reference to white-tailed deer. Amer. Nat., 111:31–42.

HUTCHINSON, J. C. D., AND G. D. BROWN. 1969. Penetrance of cattle coats by radiation. J. Appl. Physiol., 20:454–464.

KETELHODT, H. F. VON. 1966. Der Erdwolf, *Proteles cristatus* (Sparrman, 1783). Z. Säugetierk., 31:300–308.

KINGDON, J. 1977. East African mammals. An atlas of evolution in Africa. Vol. 3, Part A (Carnivores). Academic Press, New York, 476 pp.

KLEIMAN, D. G. 1967. Some aspects of social behavior in the Canidae. Amer. Zool., 7:365–372.

KRUUK, H. 1972. The spotted hyena, a study of predation and social behavior. Univ. Chicago Press, Chicago, 355 pp.

———. 1976. Feeding and social behavior of the striped hyaena (*Hyaena vulgaris* Desmarest). E. African Wildl. J., 14:91–111.

KRUUK, H., AND W. A. SANDS. 1972. The aardwolf (*Proteles cristatus* Sparrman, 1783) as predator of termites. E. African Wildl. J., 10:211–227.

LAMPRECHT, J. 1979. Field observations on the behavior and social system of the bat-eared fox (*Otocyon megalotis* Desmarest). Z. Tierpsychol., 49:260–284.

LEYHAUSEN, P. 1979. Cat behavior, the predatory and social behavior of domestic and wild cats. Garland STPM Press, New York, 340 pp.

MEESTER, J., AND H. SETZER. 1971. The mammals of Africa: an identification manual. Smithsonian Institution Press, Washington, D.C.

MILLER, G. S. 1931. Human hair and primate patterning. Smithsonian Misc. Coll., 85:1–13.

MILLS, M. G. L. 1978. Foraging behavior of the brown hyena (*Hyaena brunnea* Thunberg, 1820) in the southern Kalahari. Z. Tierpsychol., 48:113–141.

OWENS, M. J., AND D. D. OWENS. 1978. Feeding ecology and its influence on social organization in brown hyenas (*Hyaena brunnea,* Thunberg) of the Central Kalahari desert. E. African Wildl. J., 16:113–135.

———. 1979. Notes on social organization and behavior of brown hyenas (*Hyaena brunnea*). J. Mamm., 60:405–408.

RIEGER, I. 1978. Social behavior of striped hyenas at the Zürich Zoo. Carnivore, 1:49–60.

———. 1979. A review of the biology of striped hyenas, *Hyaena hyaena* (Linne, 1758). Säugetierk. Mitt., 40:81–95.

Roberts, A. 1951. Mammals of South Africa. Trustees of the Mammals of South Africa, Johannesburg, 700 pp.

Rosevear, D. R. 1974. The carnivores of West Africa. British Mus. Nat. Hist., 723:1–548.

Rowe-Rowe, D. T. 1972. The African weasel *Poecilogale albinucha* Gray. Observation on behavior and general biology. Lammergeyer, 15:39–58.

———. 1978. Reproduction and post-natal development of South African mustelines (Carnivora: Mustelidae). Zool. Africana, 13:103–114.

Schaffer, J. 1940. Die Hautdrüsenorgane der Säugetiere. Urban und Schwarzenberg, Berlin, 367 pp.

Schaller, G. B. 1972. The Serengeti lion. A study of predator-prey relations. Univ. Chicago Press, Chicago, 480 pp.

Scholander, P. F. 1950. Adaptation to cold in arctic and tropical mammals and birds in relation to body temperature, insulation, and basal metabolic rate. Biol. Bull., 99:259–271.

———. 1955. Evolution of climatic adaptation in homeotherms. Evolution, 9: 15–26.

Smythe, N. 1970. On the existence of "pursuit invitation" signals in mammals. Amer. Nat., 104:491–494.

Taylor, M. E. 1975. Herpestes sanguineus. Mammalian Species, 65:1–5.

Tembrock, G. 1957. Zur Ethologie des Rotfuchses (*Vulpes vulpes* L.) unter besonderer Berücksichtigung der Fortpflanzung. Zool. Garten, 23:289–532.

Tregear, R. T. 1965. Hair density, wind speed, and heat loss in mammals. J. Appl. Physiol., 20:796–801.

Wemmer, C. 1977. Comparative ethology of the large-spotted genet (*Genetta tigrina*) and some related viverrids. Smithsonian Contrib. Zool., 239:1–93.

York, W. 1973. A study of serval melanism in the Aberdares and some general behavioral information. Pp. 191–197, *in* The world's cats, ecology and conservation (R. L. Eaton, ed.). World Wildlife Safari, Winston, Oregon, 1:1–349.

MECHANISMS OF MAMMALIAN AUDITORY COMMUNICATION

EDWIN GOULD

Abstract

PARSIMONY characterizes mammalian auditory communication. Most vocalizations are derived from precursor mechanisms of the respiratory system. In the neonate they occur in the context of a rather uniform mother-infant configuration. Some calls that originate in infancy and appear in the mother-infant context reappear later in life in similar contexts but serve different functions. A small repertoire of sounds serves many functions in many situations. Auditory communication like other systems almost always operates in combination with other modalities and contextual cues. Even a call with a powerful effect, such as the playback of the male hamster's ultrasonic pulse that induces lordosis in an estrous female, requires a specific endocrine state in the receiver.

Most functions of auditory communication have their counterpart in olfactory communication; these similarities would have permitted a "copying" of integrative processes (circuitry) during the evolution of one of the systems. The neurophysiological findings of Jurgens (1979) and the motivational structural rules governing vocal structure of Morton (1977) corroborate the subjective statements of other workers that most vocal systems are graded and related to motivational state. Vocal redundancy in the infant, reactivation during subadult or adult life and the shared, partially overlapping schemata of infant and subadult greeting and adult courtship behaviors may be the mechanism by which vocal behaviors are recalled in the mature animal.

Introduction

Reviews of the literature on animal communication have been extensive and wide-ranging (Eisenberg and Kleiman, 1972; Petrovich and Hess, 1978). A number of recent reviews have been organized by taxonomic categories in "How Animals Communicate" (Sebeok, 1977); these included separate treatments of the Metatheria, Insectivora, Lagomorpha and Rodentia, Artiodactyla, Perissodactyla, Canidae, Felidae, Mustelidac, Procyonidae, Ursidae, Cetacea, Sirenia, Pinnipedia, Prosimia, New and Old World monkeys and apes; all modes of communication are discussed. Special topics of vocal communication have also been detailed in papers that deal with one or a few species in relation to other mammals: vocal analogues and homologues in marsupials (Eisenberg et al., 1975); adaptations of calls to different habitats in hystricomorph rodents (Eisenberg, 1974); motivational basis for vocalization in ungulates (Kiley, 1972), bats and insectivores (Gould, 1971), and rodents (Eisenberg, 1974); and dialects in macaques (Green, 1975*b*). Whereas some biologists have proudly hailed the application of animal studies to human behavior, Green and Marler (1979) reverse the trend by examining animal communication in light of our knowledge of human communication. Still other investigators consider specific functions such as alarm signals (Sherman, 1977) or mechanisms of spacing (Waser and Wiley, 1979). Behavior and physiology of rodent ultrasound are reviewed by Floody (1979). The purpose of this paper is to review mammalian vocal communication from a functional viewpoint and to consider the ontogeny of auditory communication in light of recent investigations in other areas of developmental behavior.

Sound Production

Once information is transmitted from one animal to another and the second animal has responded to that information, communication has occurred. A first step in communication studies is describing the signal source. Most mammalian vocalizations derive from the controlled emission of air through the respiratory system. The larynx, tongue, nasal passages, and mouth (including the pharynx) are four major sources or contributors to the quality of mammalian vocalizations. The primary function of the larynx is to protect the respiratory system. The larynx acts as a valve by preventing foreign material from entering the lungs and by expelling objects that might enter the trachea. During the early stages of mammalian evolution, the main functions of the sound-producing organs (the larynx, pharynx, mouth, and nose) were digestion and respiration.

Today the motor system for the mouth and pharynx coordinates feeding, vocalization, and breathing (Bosma, 1971). Separate sensory guiding mechanisms and separate central representation control the system (Bosma, 1971).

Because its primary function is protective, the larynx has not been associated with any set of typical sounds (Kiley, 1972; Negus, 1949). Kiley (1972) concluded ". . . that the individual structure of the calls and their interrelations depend on the social behavior and the habit of the animal and its general motivational state, to a greater degree than its laryngeal structure." It is possible that other factors such as the nasal, oral, and pharyngeal cavities are more relevant to vocal qualities. Negus' main thesis is that laryngeal form is an essential accommodation to locomotion or activity. But Negus emphasizes the great versatility of the larynx: "A very simple vocal apparatus is sufficient for powers of speech and this is present in most mammals. . . ."

Nevertheless, a few mammals show extreme anatomical modification. Social behavior seems to have been a powerful selective force in the evolution of the larynx of the howler monkey (*Alouatta palliata*; Southwick, 1963), elephant seal (*Mirounga angustirostris*), siamang (*Symphalangus syndactylus*), and epomophorine bats (Fig. 1; Wickler and Seibt, 1976). But Negus (1949) warned that large laryngeal air sacs are not necessarily indicative of a specialized vocal apparatus. Some very vocal species of gibbons have no laryngeal sacs. However, a specialized larynx is associated with echolocation in the Microchiroptera (Griffin, 1958; Matsumura, 1979).

Though many mammalian sounds may result from modification of the respiratory system, the larynx is not always central. Nasal sounds are prominent in the repertoire of viverrids (Wemmer, 1977) and artiodactyls (Walther, 1977). Lengthening and narrowing of the skull for feeding specializations, such as we find in species of *Hemicentetes* and *Manis,* may have imposed limits on vocal abilities; vocalizations of both taxa are mainly limited to nasal sounds (Eisenberg and Gould, 1970; Kingdon, 1971).

The pharynx is central in its location among the phonation structures, but published functional anatomy is lacking (Schoen, pers. comm.). Very little is known about where phonation occurs and, as Eisenberg et al. (1975) have argued, this void is the caveat in establishing homologues. Sales and Pye (1974) indicated that rodent ultrasonics derive from a whistle mechanism, whereas bat ultrasonics derive from vocal folds.

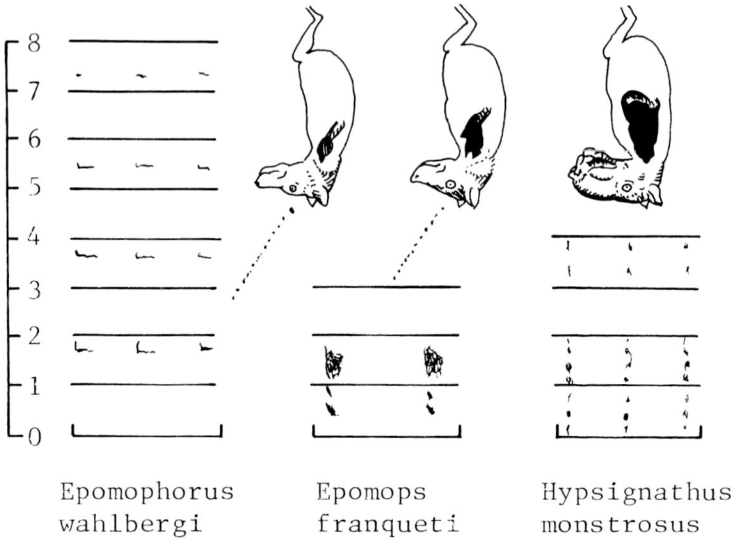

FIG. 1. Increase in size of larynx and differences between male calls in related genera of epomophorine bats, and spectrograms of *Epomophorus wahlbergi, Epomops franqueti,* and *Hysignathus monstrosus.* Increased laryngeal size in the three species is associated with increased audible range, decreased fundamental frequency, and decreased duration of calls. From left to right the male of each species is roughly 120, 145, and 300 g. Vertical scale = kHz. Horizontal brackets = 1 s. (Adapted from Wickler and Seibt, 1970.)

The tongue also plays a role in sound production. From radiocinematographic studies, Busnel (1976) found that human whistles are produced by modifications of the point of the tongue. The difficulty of establishing homologues can be appreciated by considering the diverse derivation of human clicks. Thirteen African bushman tribes, each with a different language, use a total of five distinct clicks. Four are produced with the tongue (against the teeth, front palate, and backward or forward movement against the back palate) and a fifth is produced by the lips only (Bleek, 1929). From the standpoint of ontogeny and the establishment of homologues, it would be interesting to determine if some close contact sounds—low chirps and whistles—depend on tongue movements that are involved in sucking. There are numerous other mechanisms for producing sounds that convey messages, such as foot-drumming and quill stridulation, but these classes of sounds will not be considered in this review.

Clicks

Clicks may be produced by tooth-chattering, the tongue, or the lips. The latter two organs are involved in suckling by the neonate. Sucking in mammals is universal. The occurrence of clicks in the context of nursing suggests that this sound is an old one in the phylogenetic sense. Clicks are brief, discrete, unstructured, broadband signals with a rapid onset (Broughton, 1963; Buchler and Mitz, 1980). Clicks, easy to produce, hear, and locate, may be the most widespread sounds among mammalian species. Andrew (1964) suggested two sources of clicks—one derived from the motor controls for sucking and the other from glottis closure associated with a protective response. Purring in cats may be related to the latter. Purring results from active glottal closure controlled by a highly regular, alternating activation 20 to 30 times per second, of the diaphragm and laryngeal muscles; it may derive from some oscillatory mechanism within the central nervous system (Remmers and Gautier, 1972). Purring also occurs in *Genetta tigrina* and *Nandinia binotata* (Wemmer, 1977), in many felids (Ewer, 1973), *Myoprocta pratti* (Eisenberg, 1974; Kleiman, 1972), and *Cavia porcellus* (Berryman, 1976).

In the human infant, the tongue fills a small space. The soft palate, the tongue and the pharyngeal walls move in order to maintain patency of the airway. These respiratory movements adjust to reflexes, such as rooting, swallowing, and sucking, and give rise to brief clicks and other noises, both expiratory and inspiratory (Stark, 1979).

Lip-smacking, which produces a click sound in the infant stumptail macaque (*Macaca speciosa*), is indistinguishable in form from sucking except that the lips contact something (the mother's nipple, or an object) in "sucking" and do not contact anything in "lip smacking" (Chevalier-Skolnikoff, 1974). Early lip-smacking occurred in response to tactile stimulation of the oral region, and non-social and social visual stimuli. Later lip-smacking occurs in increasingly specific social-visual contexts and less in association with tactile stimuli. Lip-smacking may have evolved from sucking behavior through a process of ritualization (Chevalier-Skolnikoff, 1974; Hinde, 1966). Sucking behavior is probably related ontogenetically to fetal movements as occurs in the human fetus (Humphrey, 1964, 1969).

All clicks are unlikely to have a common function. Clicks can function in both affiliative and agonistic contexts. For example, during territorial defense against intruders, the colobus monkey (*Colobus guereza*) opens the mouth wide and "the tongue is clicked from roof to floor of the mouth making a soft smack" (Marler, 1969). Various nocturnal primates use clicks when establishing contact (Charles-Dominique, 1977). Infant shrews and opossums also emit clicks, and female *Blarina brevicauda* (Gould, 1969) and *Marmosa mitis* (Thrasher et al., 1971) retrieve displaced clicking infants. In rodents, clicking is closely associated with suckling (Zippelius and Schleidt, 1956). Clicks are produced at the same time as ultrasounds in infant laboratory rats (*Rattus norvegicus*) and by some cricetids (Gould, pers. observ.; Sales and Pye, 1974). The number and intensity of clicks are comparable to those of ultrasonic whistles (Gould, pers. observ.), although investigators have concentrated mainly on the ultrasounds of rodent infants. The indiscriminate retrieval behavior of rodents (Smith, 1975) could mean that a common cue, repetitive clicking, is significant.

As yet the functional aspects of each basic sound type have not been explored. Zippelius (1974) found that the pulse type was not as important as the sound-pulse repetition rate. Gish (pers. comm.) suggests that simultaneous emission of click-bursts and whistles by *Tursiops truncatus* ensures signal transmission in the midst of background noise. When whistles "disappear," click-bursts continue to retain the whistle rhythm. This same hypothesis might apply to the function of click-whistle combination of infant rodents.

Families of Sounds

The same sound producing structures can account for different classes of vocalizations emitted by a species. Thus, a number of investigators have noted the similarity among a series of calls from animals of different ages. Bartholomew and Collias (1962) detected structural similarities among the sounds of the pup protest, the yearling threat, the female threat, and the roar of the male elephant seal (*Mirounga angustirostris*). Infant and adult calls of bats (Gould, 1971; Woolf, 1973) and primates (Chevalier-Skolnikoff, 1974) show a similar continuum. Woolf (1973) referred to these as a "sonar family." The fundamental frequencies of a class of sounds are so related as to be considered a "family," the members of which are

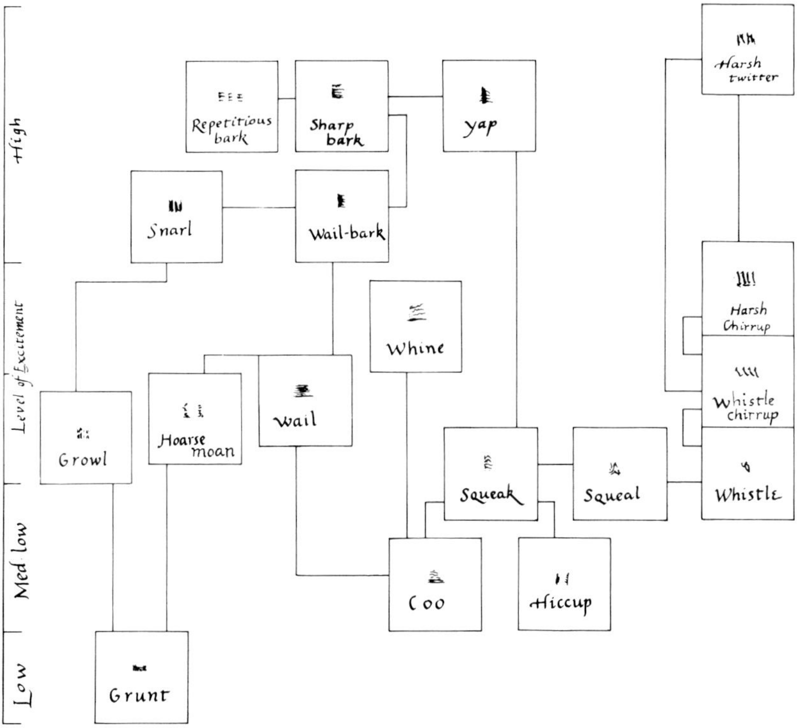

FIG. 2. Schematic illustration of the interrelationships of vocal sounds of the adult rock hyrax, *Procavia capensis,* based on their physical characteristics. Intermediate sounds, not included in the diagram, are represented by the interconnecting lines, which do not imply pathways that have to be followed in situations where, for example, the level of excitement rises or decreases. (From Fourie, 1977.)

produced by the same physiological mechanism (Klatt, 1973; Levitt and Rabiner, 1971; Woolf, 1973). This concept may be applied to sounds which change with age or adult sounds with similar frequency profiles.

The 26 types of howler monkey (*Alouatta palliata*) sounds can be subdivided into five or six families of vocalizations based on similarities of structure and usage; such grouping emphasizes the relatedness of many of the calls and demonstrates which calls form graded series with each other (Baldwin and Baldwin, 1976). Similar classifications exist for *Procavia capensis* (Fourie, 1977; Fig.

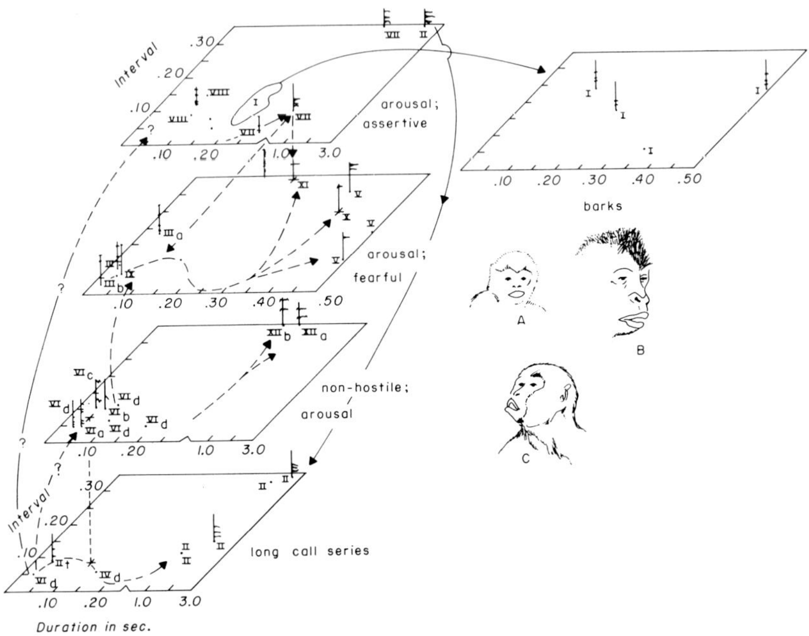

FIG. 3. Call transformations and syllabic similarity in the spider monkey (*Ateles fusciceps*), a composite diagram derived from four sonographic records that typify temporal patterns of syllables. The discrete call forms (Roman numerals) are shown with transitional syllables. Calls believed to be indicative of the same general mood are portrayed on the same plane. Broken lines indicate pathways of syllabic transformation. The three faces include: A, expression during a growl; B, expression during a squeal; C, expression during a long call. (From Eisenberg, 1976.)

2), *Ateles fusciceps* (Eisenberg, 1976; Fig. 3), and *Saimiri sciureus* (Schott, 1975). Electrode implant studies demonstrating 47 calls of five basic forms corroborate the "family" concept in squirrel monkeys (Jurgens, 1979; Fig. 4). Vocal families are particularly relevant to ontogenetic studies that attempt to explain the expanding number of referents to which a call becomes associated.

Ultrasonics

The tendency to treat ultrasonics as a special case in mammal communication in part reflects our own sensory limitations. Functionally, ultrasonics are not very different from audible sounds in

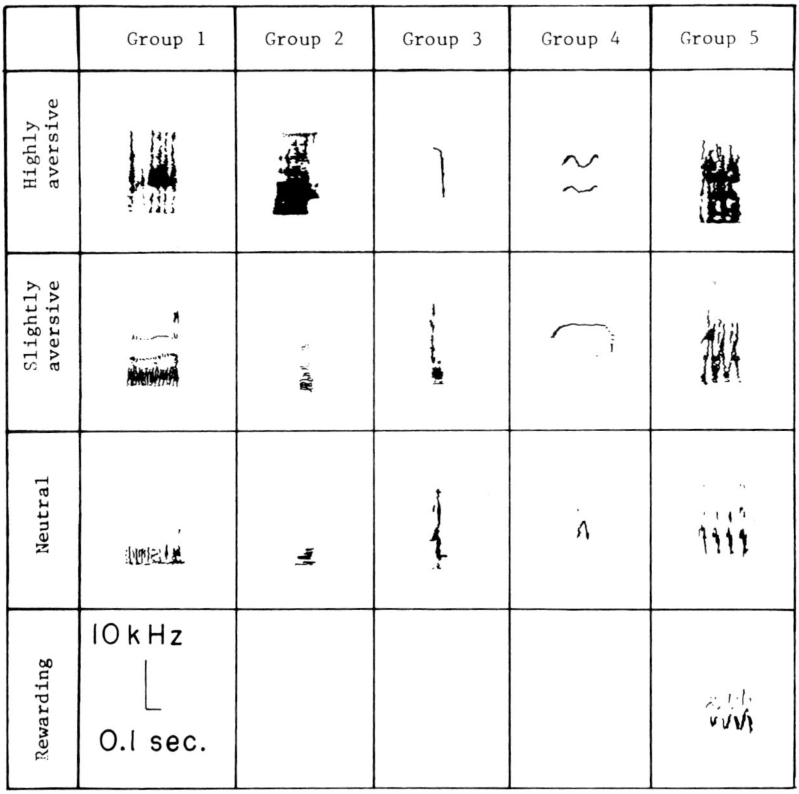

FIG. 4. Schematic summary of the relationships between call type and emotional state in squirrel monkeys (*Saimiri scuireus*) as determined by the self-stimulation procedure. Each group represents a different type of call: Group 1, spitting; 2, shrieking; 3, alarm peep; 4, squealing; 5, shriek-cackling. (From Jurgens, 1979.)

that they encompass nearly all the functions of signals occurring in lower frequencies, including maternal, sexual (Fig. 5), and agonistic behaviors. Thermal and tactile stress evoke ultrasonics from infant rodents (Okon, 1972) and bats (P. Brown, 1976; Gould, 1971). Following ejaculation or defeat in an agonistic interaction, male laboratory rats emit a 22- to 30-kHz call. A 50-kHz pulse is emitted by aggressive males and lactating female rats when interacting with other males.

Ultrasonics attenuate rapidly in air (Pye, 1979). Why rodents

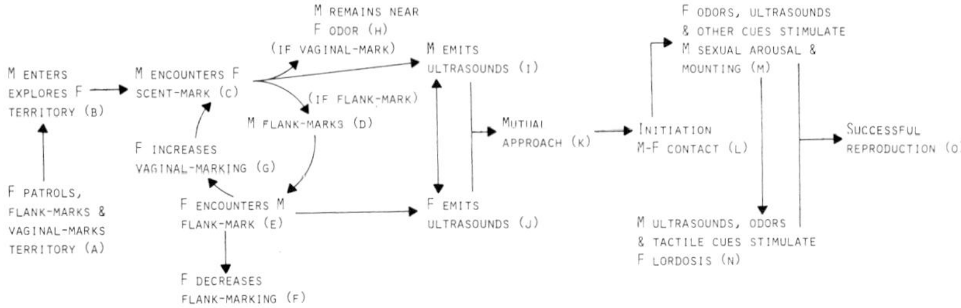

FIG. 5. A model of communication signals and responses that facilitate reproduction in golden hamsters (*Mesocricetus auratus*). Male's behavior (M) above; female's (F) below. Lower-case letters in parentheses code behavior. Many of the "links" in this chain of social behaviors and signals depend on gonadal hormones. In particular, testosterone facilitates male responses at least at points d, h, m, and o. Estrogen and progesterone facilitate female responses at least at points a, g, j, k, n, and o. (From Floody and Pfaff, 1977*b*.)

use ultrasonics rather than lower frequencies remains obscure (Brown and Pye, 1975). However, their rapid attenuation may confer special antipredator advantages for communication over short distances (Sales and Pye, 1974). Thus, "cryptic" sounds may protect many small mammals from predators.

Probably all Microchiroptera, most altricial rodents, many insectivores, cetaceans, infant mustelids, and a few small primates communicate with ultrasonics (Brown and Pye, 1975). This implies that about 50% of mammalian species produce ultrasounds and that high-frequency sounds probably occurred early in the history of mammalian evolution. Although no specialized anatomical adaptations have been detected in rodents (A. Brown, 1976), the microchiropteran larynx has vocal cords that cover vocal pits in the side of the larynx (Griffin, 1958).

Properties and Transmission of Mammalian Vocalizations

Mammalian vocalizations range in frequency from the horse's roar of 50 to 100 Hz (Kiley, 1972) up to 150 kHz in some bats (Novick, 1977). High-frequency sounds are extremely directional

(Pye, 1979) and attenuate quickly with distance (Morton, 1975). Low-frequency sounds attenuate slowly with distance and are relatively omnidirectional. The transmission properties of a vocalization in nature are by no means uniform (Marten and Marler, 1977; Marten et al., 1977; Wiley and Richards, 1978). Temperature, humidity, landscape, and vegetation are all important considerations (Wiley and Richards, 1978). Depending on the vertical height of the sound source, curved shadows of silence due to heat radiation are generated near the ground or at heights; sometimes channels of transmission are produced as a result of temperature gradients (Beranek, 1949; Saby and Nyborg, 1946).

The range of a vocal signal is influenced by intensity at the source, the level of background noise, the rates of signal degradation, and the perceptual abilities of the receiver (Waser and Wiley, 1979). This implies that the spread of a signal may be unequally dispersed in the environment depending on local conditions and how closely a species patterns its vocal activity with optimal transmission conditions. All things being equal, low-frequency sounds carry farther in forests than do high-frequency sounds of the same amplitude. A complex distribution of vegetation in a forest and the effects of micrometeorology in open environments influence sound transmission so variably that no consistent difference can be discerned when sounds of equivalent frequency are tested in open and closed habitats (Marten and Marler, 1977). Indeed, Waser and Waser (1977) demonstrated that four species of monkeys living in the Uganda rainforest maximize the audible range of their spacing signals by calling from heights of 15 to 20 m before dawn when the forest is relatively quiet and when attenuation (owing to temperature gradients) of the calls is least. Furthermore, the frequency spectrum of the signals, 500 to 2,000 Hz, is particularly efficient for transmission through the forest canopy. The low frequency and greater duration of rainforest scuirid calls, compared to temperate species, probably reflects adaptations to dense vegetation and high background noise of the rainforest (Emmons, 1978).

The long-distance vocalizations of varied taxa have several frequency peaks, but the similarity of calls of some unrelated taxa inhabiting similar environments is remarkable; for example, the calls of the pika (*Ochotona princeps*) and those of the rodent (*Lagidium peruanum*), both of which live on mountain slopes or rock exposures, are about 5 kHz (Eisenberg, 1974; Somers, 1973). Cheetahs (*Acinonyx jubatus*), lions (*Panthera leo*), and wild dogs (*Lycaon pictus*) inhabit similar open country and exhibit a similar frequency range of calls (Schaller, 1972*b*).

But animal vocalizations have probably been selected for other important features. Because frequencies of a variable-frequency call degrade in a predictable pattern, a listener with experience could estimate the signaler's distance. Features of a vocalization that provide species identification information might propagate further than information providing signature information (Wiley and Richards, 1978).

Though investigations have centered on the range of relatively high-intensity calls, many sounds of rodents are either of relatively low intensity or, if ultrasonic, attenuate rapidly. Morton (1975) speculates that in grassland habitats temporal aspects of signals are critical for information transfer because pulse-timing is less distorted in temperature- and windspeed-stratified environments. The same relationships of sounds and microclimate may apply to crepuscular mammals that are active when greatest temperature stratification and air movements occur (Geiger, 1957).

Perception of Auditory Signals

Most of the information on hearing in mammals derives from audiograms which are based on the animal's sensitivity to long-duration pure tones, sounds that are relatively rare in nature. Increasingly, auditory neurophysiologists are interested in animal sound communication (Capranica, 1972), and hearing experiments are being tried in the field. For example, Waser (1977*b*) determined that accurate localization of conspecifics several hundred meters away in tropical forest is possible within 6° for mangabeys (*Cercocebus albigena*).

Two major selection forces have undoubtedly been active in the evolution of hearing in mammals—predation and conspecific communication. However, body size imposes primary constraints. For example, high-frequency hearing is directly correlated with the functional distance between the two ears, where functional distance is defined as the distance between the ears divided by the speed of sound. Small mammals with close-set ears can hear high-frequency sounds better than species with large heads and wide-set ears. Asiatic elephants (*Elephas maximus*) hear from 17 Hz to 10.5 kHz (Heff-

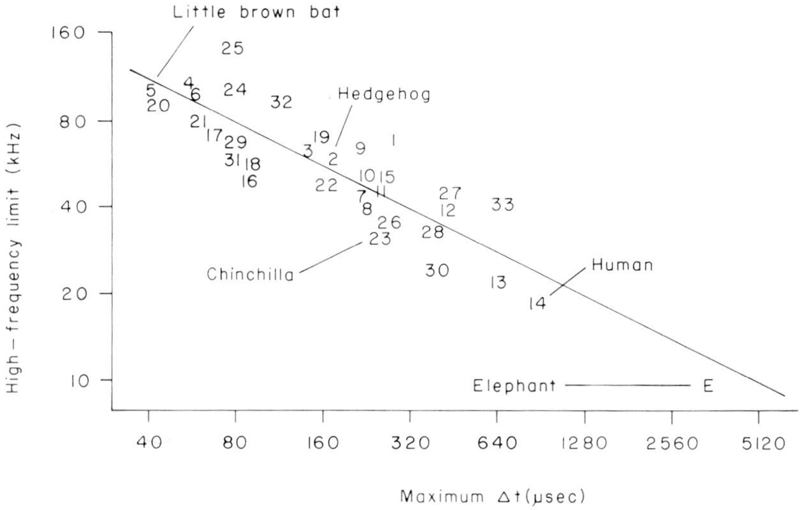

FIG. 6. Relationship between maximum Δt (maximum interaural distance divided by the speed of sound) and high-frequency hearing limit (highest frequency audible at 60-dB sound pressure level). Numbers and letter represent points for individual species. All high-frequency limits were determined in air except as noted. Key: E, elephant (*Elephas maximus*); 1, opossum (*Didelphis virginiana*); 2, hedgehog (*Hemiechinus auritus*); 3, tree shrew (*Tupaia glis*); 4, horseshoe bat (*Rhinolophus ferrumequinum*); 5, little brown bat (*Myotis lucifugus*); 6, big brown bat (*Eptesicus fuscus*); 7, slow loris (*Nycticebus coucang*); 8, potto (*Perodicticus potto*); 9, bushbaby (*Galago senegalensis*); 10, owl monkey (*Aotus trivirgatus*); 11, squirrel monkey (*Saimiri sciureus*); 12, macaque (*Macaca mulatta*); 13, chimpanzee (*Pan troglodytes*); 14, human (*Homo sapiens*); 15, rabbit (*Oryctolagus cuniculus*); 16, kangaroo rat (*Dipodomys merriami*); 17, cotton rat (*Sigmodon hispidus*); 18, gerbil (*Meriones unguiculatus*); 19, laboratory rat (*Rattus norvegicus*); 20, feral house mouse (*Mus musculus*); 21, laboratory mouse (*Mus musculus*); 22, guinea pig (*Cavia porcellus*); 23, chinchilla (*Chinchilla laniger*); 24, dolphin (*Inia geoffrensis*) (underwater); 25, porpoise (*Tursiops truncatus*) (underwater); 26, killer whale (*Orcinus orca*) (underwater); 27, dog (*Canis familiaris*); 28, sea lion (*Zalophus californianus*) (in air); 29, harbor seal (*Phoca vitulina*) (underwater); 30, harbor seal (*Phoca vitulina*) (in air); 31, ringed seal (*Pusa hispida*) (underwater); 32, harp seal (*Pagophilus groenlandicus*) (underwater); and 33, domestic sheep (*Ovis aries*).

ner and Heffner, 1980; Fig. 6). Larger animals can not only hear low frequencies better, but they can also better detect brief sounds from a longer distance than small mammals. This finding is also consistent with the physical limits on the frequency of vocalizations that a large or small animal can produce.

Selective pressure to detect brief sounds results in selective pressure to detect high frequencies (Masterton and Diamond, 1973). Brief sound detection—that is, protection from predators or facilitation of prey location—has probably been a source of selective pressure to hear sounds as high as 8 to 10 kHZ. Detection of sounds up to 50 kHz probably results from selective pressure for more accurate sound localization (Masterton and Diamond, 1973). The advantage to small mammals of using ultrasonic communication signals that attenuate rapidly but function effectively at short range may have been a complementary factor in the evolution of mammalian hearing in the higher frequency range. A bimodal pattern of frequency sensitivity in some rodents (Brown and Pye, 1975) suggests the existence of different selective forces which are related to predation in the lower range and to communication in the higher range.

Low-frequency sensitivity in *Dipodomys* is adaptive in predator avoidance in open, desert habitats (Webster and Webster, 1980). In other closely related heteromyids, there is less specialization going from desert to mesic habitats. The xeric genera (*Dipodomys* and *Microdipodops*) have greatly inflated middle ears, a very efficient tympano-ossicular system, modified organ of Corti, and acute sensitivity to low frequencies (Webster and Webster, 1980). Similar modifications of the bullae occur in other desert rodents, carnivores, and fossil South American marsupials (Prakash, 1959; Webster and Webster, 1980). Other rodents, such as deer mice (*Peromyscus*), pack rats (*Neotoma*), ground squirrels (*Spermophilus*), and grasshopper mice (*Onychomys*), which have similar predator problems but occur in habitats with more cover, lack most of the heteromyid syndrome and emphasize higher pitched vocalizations (e.g., *Peromyscus* [Smith, 1972] and *Onychomys* [Hafner and Hafner, 1979]) and hearing (A. Brown, 1976; Brown and Pye, 1975).

Generally, vertebrates can perceive the sounds they emit (Stevens and Warshofsky, 1971). In nine species of rodents, the peak sensitivity to ultrasonic frequencies measured from the inferior colliculus corresponds with the frequency of vocalizations emitted (Brown, 1973). Though extremely vocal at birth, *Antrozous pallidus,* the most altricial of bats studied to date, cannot hear during the first week of life (Brown and Grinnell, 1980). Dogs too are deaf until about two weeks of age (Fox, 1971), despite being able to vocalize (Cohen and Fox, 1976; Fox, 1971).

Encoding

The scheme by which complex sounds, such as vocalizations, are encoded in the auditory nervous system may be fundamentally similar throughout mammalian species (Capranica, 1972). Neuronal specialization is one mode of encoding. For example, neurons in the auditory cortex of the squirrel monkey (*Saimiri sciureus*) respond only to certain components of species-specific vocalizations (Winter and Funkenstein, 1971; Wollberg and Newman, 1972). Through operant conditioning, it was found that Japanese macaques (*Macaca fuscata*) performed best when responding to position of the peak frequency inflection in a species-specific call, as opposed to initial pitch, whereas other monkey species perform best when responding to initial pitch. These special strategies for perceptual processing of vocal sounds resemble attributes of human speech perception (Zoloth et al., 1979; see also Marler, this volume).

Neurophysiological studies of bat echolocation suggest other modes of encoding auditory information. In *Rhinolophus ferrumequinum*, there is an expansion of the region on the basilar membrane sensitive to 82 to 86 kHz that is analogous to the foveal representation in the visual central nervous system (Neuweiler, 1980; Schuller and Pollak, 1979). "Acoustic foveation" probably enhances the resolving power for fine target features "by devoting a disproportionately large number of neurons to the processing of the echo CF (constant frequency) components" (Pollak, 1980).

In *Pteronotus parnelli*, the cerebral auditory cortex has two expanded regions: one devoted to processing FM (frequency modulation) signals (50 to 60 kHz) and another devoted to the CF component (61 to 63 kHz) of the echo (Suga and O'Neill, 1980). Descriptions of neural maps of auditory space and of target range as well as amplitopic and tonotopic representation in the cortex (Suga and O'Neill, 1980) indicate the bat's capacity to encode and segregate sonic information in a systematic way. Some of these mechanisms undoubtedly function in the processing of microchiropteran vocal communication signals.

As bat neurophysiologists have focused on specific sensitivity to echolocation sounds and their echoes, future investigators will emphasize the encoding of complex auditory signals that are used in communication (Capranica, 1972).

Context of Auditory Communication

Increasingly, the study of auditory communication is the study of context (Chevalier-Skolnikoff, 1974; Fentress et al., 1978; Green, 1975*a*; Marler, 1976*a*; Shalter et al., 1977). Context is considered all that accompanies the signal including the past and present state of the signaler/receiver (Shalter et al., 1977; Smith, 1977), and provides information on both function and motivation.

Vocalizations, like visual, tactile, or olfactory signals, may evoke a broad spectrum of responses. The context of the sender and/or receiver may have considerable influence over the outcome. Responses may be immediate or delayed; they may be vocal, postural, locomotor, or physiological. Some are discrete; others can only be described as a "general shift in arousal level" shown by an increase in exploratory or vocal activity. Their functions may have immediate consequences, such as fleeing in response to an alarm call, or may be more long term, perhaps by promoting spacing in primate troops (Waser, 1975).

Playback experiments demonstrate that vocalizations unsupported by visual, olfactory, or tactile cues elicit approach and investigation of the sound source by mother cats (Haskins, 1977), dogs (Bleicher, 1963), and big brown bats, *Eptesicus fuscus* (Gould, pers. observ.). Mother elephant seals emit a specific call in response to their infants' calls (Petrinovich, 1974). Estrous female hamsters (*Mesocricetus auratus*) increase their rate of vocalizing and assume lordosis during playback of artificial or natural ultrasounds simulating male hamster calls (Floody and Pfaff, 1977*b*). Playback studies of rhesus monkeys (*Macaca mulatta*) suggest that infant calls may evoke motor elements of retrieval and support postures as well as gazing (at their breasts) by the mothers (Erwin, 1974).

Combinations of stimuli are sometimes necessary to evoke a response. Composite signals are probably much more a part of mammalian communication than the exclusive function of any one modality. In Japanese monkeys (*Macaca fuscata*) "vocal behavior is inextricably tied to simultaneous olfactory, tactile, and visual signals . . ." (Green, 1975*a*). These other signals are contextual elements of vocal behavior. Particularly during high levels of arousal, virtually all the signaling systems may be mobilized to repel an adversary. Agonistic displays involving grunting and rhythmic tail-wagging occur in *Lemur fulvus* during intergroup encounters in trees (Sussman and Richard, 1974). The jump-yip display of the prairie dog, *Cynomys ludovicianus* (Smith et al., 1976), as its name implies, is clearly a mixture of signals. An animal jumps upright and simultaneously vocalizes. This display identifies the communicator as a black-tailed prairie dog, indicates its location (acoustically), and may communicate that the performer is less likely to flee.

Ground squirrels (*Spermophilus beecheyi*) emit the same call in two entirely different contexts—aggressive interactions and predator detection. Leger and Owings (1978) hypothesized that contextual features, such as the number of ground squirrels vocalizing, could indicate to silent squirrels the type of elicitor (one or two calls from aggressively interacting squirrels versus many calls from squirrels responding to a predator). Leger et al. (1979) experimentally demonstrated that flight responses of ground squirrels were stronger to multiple than to single whistles. "Thus an earlier whistle acts as a contextual determinant of reaction to a subsequent whistle." ". . . An event that alarms many individuals is more likely to be dangerous to the receiver than an event that alarms only one of many individuals." In the titi monkey (*Callicebus moloch*), different sequences of vocalizations occur in different contexts. Silent monkeys can discriminate different sequences on the basis of order of calls in the sequence (Robinson, 1979).

Ultrasounds emitted by male hamsters (*Mesocricetus auratus*) in the presence of estrous females differ from those emitted by solitary males in regard to frequency profile and duration (Floody and Pfaff, 1977*a*). Thus, the context may affect or modify the internal structure of the signal. Additional examples indicate the importance of other sensory modalities for the elicitation of vocalizations (DeGhett, 1978; Nyby et al., 1977*b*; Okon, 1970; Oswalt and Meier, 1975; Whitney and Nyby, 1979) and emphasize the degree to which several sensory modalities have been integrated in mammals.

Context could be determined by altered sensitivity. Sensory thresholds in relation to endocrine condition are relatively unexplored. Hormonal conditions at birth might alter a mother's sensitivity to her infant by affecting peripheral sensitivity (Klopfer, 1976; Komisaruk et al., 1973). Angular localization of sounds varies significantly in women depending on the stage of their menstrual cycle, perhaps because progesterone levels are affecting fluid pressure in the inner ear (Haggard, 1979). Considerations of context punctuate the enormous variability and the number of contingencies involved in analyzing auditory communication.

Functional Aspects of Auditory Communication

Spacing

Marler (1968) identified regulation of distance as a basic function of vocalization. A receiver may increase, retain, or decrease the distance to the caller. Green and Marler (1979) emphasize temporal pattern and structure as important determinants of whether vocalizations are classified as discrete or graded. Auditory signals are discrete when the receiver perceives each sound distinctly. Graded signals are so similar to each other that two sounds emitted close together are indistinguishable to the receiver. However, in a series of signals the receiver may detect a change. Most primates communicating at night or calling in the day over long distances in the forest use discrete calls. "Conversely, graded signals are more common in close-range signalling, often within-group, where rich supplementation by visual and tactile signals is the rule" (Green and Marler, 1979). Thus graded signals seem to typify the calls of most mammals, including insectivores, rodents, bats, and primates. Discrete signals have been described in *Saimiri sciureus* (Winter et al., 1966), *Callicebus moloch* (Robinson, 1977), *Cercopithecus aethiops, C. mitis,* and *C. ascanius* (Marler, 1973; Struhsaker, 1967), and *Leontopithecus rosalia* (Green, 1979).

Long-distance calls, whose main function is spacing, are prominent in forest-living primates, and in open country carnivores and rodents, where the absence of visual contact with other individuals or groups is common. Most loud calls range between approximately 200 and 2,000 Hz (see Properties and Transmission of Vocalizations).

Waser's (1975, 1976) study gives us some insight into functional mechanisms. Free-ranging mangabeys (*Cercocebus albigena*) emit whoopgobbles that are clearly audible over a distance equal to the straight-line distance in which a group normally moves each day. Whoopgobbles are used to maintain an avoidance-based system in harmony with an irregular temporal distribution of major food resources. Responses to playbacks vary with distance but not with location. If the call is nearby the group retreats; if distant (>600 m), there is no response (Waser, 1975).

No single hypothesis has adequately explained the relationship between the loud calls of primate species and distance between conspecific troops (Robinson, 1979). In the greatest number of species observed, troops usually approach one another if they were close when calling began. But there are exceptions. Robinson (1979) argues in favor of fundamental differences in the "aggressive field" (Wiley, 1973) of different primate species as occurs in ungulates (Owen-Smith, 1977). Some primates focus their greatest territorial defense near the boundary; others concentrate their efforts near the center of the home range.

The distance at which high-intensity vocalizations can be heard by humans has been plotted in relation to the home range of 15 species of mammals (Fig. 7). In most instances I used the largest published home-range estimate. I assumed that sound carried equally in all directions and I used the audible distance as a radius to compute the area of each animal's potential communication space. Most of the 15 calls function in group spacing. Bushbabies (*Galago demidovii*), however, use their loud vocalizations as dawn gathering calls before reuniting in tree dens. At least two trends seem to emerge from a plot of home range and audible distance of loud calls. Some calls (for example, on the "grasshopper mouse–wolf curve") seem to correlate well with home range; nearest neighbors presumably hear these calls. An explanation for the clustering of points representing small home ranges of species with large communication spaces (as in indri, colobus, ringtailed lemurs, siamang, and howler) must be speculative at best. Vocalizations that reach far beyond the caller's home range could accommodate spacing over a period of many years; drastic population fluctuations and small shifts of home ranges have been documented (see Waser, 1977*a*).

Perhaps all of these explanations are incomplete. Many primate troops live near one another for years. There may be familiar relatives in nearby or distant but audible troops. Some troops such as howlers may intermingle peacefully (Baldwin and Baldwin, 1976). Perhaps demic or interdemic "cohesion" is another long term function that ensures periodic flow of genes throughout the extensive distribution of some species that occupy the vast and contiguous tropical forests. Interdemic communication by long calls might be especially functional in species with calls so intense they greatly exceed size of the home range (Fig. 7).

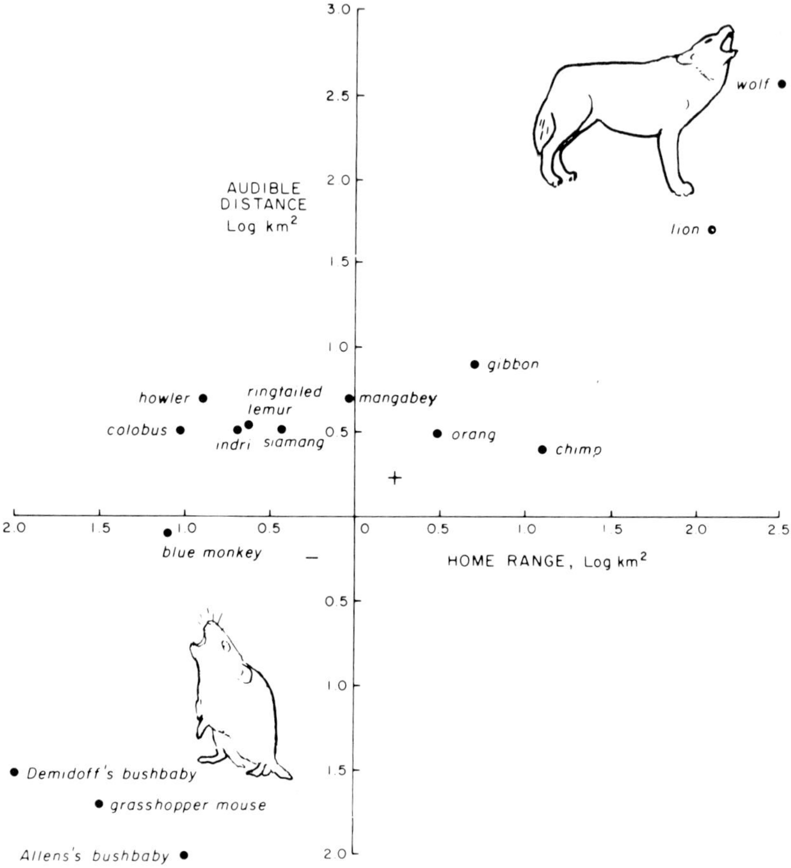

FIG. 7. Audible distance in relation to home range of 15 species of mammals. Sources: grasshopper mouse (*Onychomys leucogaster*; Ruffer, 1968); Demidoff's and Allen's bushbaby (*Galago demidovii* and *G. alleni*; Charles-Dominique, 1977); blue monkey (*Cercopithecus mitis*; Aldrich-Blake, 1970; Waser and Waser, 1977); colobus (*Colobus guereza*; Aldrich-Blake, 1970; Marler, 1972; Struhsaker, 1975; Waser and Waser, 1977); howler (*Alouatta palliata*; Baldwin and Baldwin, 1976; Chivers, 1969); ringtailed lemur (*Lemur catta*; Budnitz and Dainis, 1975; Jolly, 1966); indri (*Indri indri*; Petter and Charles-Dominique, 1979; Pollock, 1975); siamang (*Symphalangus syndactylus*; Chivers, 1974; Gould, pers. observ.); mangabey (*Cercocebus albigena*; Waser and Waser, 1977); orang (*Pongo pygmaeus*; MacKinnon, 1974); chimpanzee (*Pan troglodytes*; Reynolds and Reynolds, 1965; Goodall, 1965; Waser and Waser, 1977); gibbon (*Hylobates lar*; Ellefson, 1968); lion (*Panthera leo*; Schaller, 1972*b*); wolf (*Canis lupus*; Harrington and Mech, 1978; Mech, 1970; Theberge and Falls, 1967).

Silence

To avoid detection silence is usually a safe and advantageous strategy of prey species. The silent predator is also legendary, but all mammals produce some sounds by way of the respiratory system. Minimizing neonatal vocalizations is achieved in some species by extended periods of infant attachment to the nipple, as in megachiropterans (Nelson, 1964), or close body contact, as in primates.

In some species, vocalizations occur in discrete and rather brief stages during the life history. Vocal behavior is so seasonal that these species seem to be totally silent. Sounds of raccoons, *Procyon lotor* (Pruitt and Burghardt, 1977), and reindeer, *Rangifer tarandus* (Espmark, 1971), are emitted mainly during the breeding season and during infancy.

A mother's response to silent infants may depend on the flexibility or openness (Mason, 1968) of a species' maternal care repertoire. If the newborn seal pup is motionless and quiet, the mother nips at it and calls are usually provoked (Peterson, 1965). If an infant bat (*Eptesicus fuscus*) is silent or emits abnormal sounds, the mother will ignore it (Gould, 1971, 1975*b*). Young calves of some cervids spend long periods silently crouched in protective vegetation where they are occasionally visited by their mothers, at which time vocal exchanges take place (Altmann, 1952; Bubenik, 1965). In addition, all vocalizations have an associated silent period before and after each signal.

Most mammals are silent most of the time. Even bats, which when active emit an average of about 10 pulses/s, are silent more than 90% of the time. Thus, the temporal patterning of the silent period becomes potentially significant as information (see Repetitive Calls and Excitation Indicators).

Primates, such as baboons (*Papio ursinus*) and patas monkeys (*Erythrocebus patas*), live in open country and tend to have large undefendable home ranges (Marler, 1969). Hall (1968) remarked on the almost complete lack of any vocalizations in patas monkeys. "Their general vocal pattern is one 'of adaptive silence in which muted calls have occasional function.'" Some animals with infrequent needs to reinforce territorial limits also remain relatively silent; despite their notable "roars," howlers are among this group (Klein, 1974; Richard, 1970). Orangutans (*Pongo pygmaeus*) move slowly and as a result can predict with only a few cues the location of each individual in the forest. One call per 3 days is quite sufficient for a male to advertise his whereabouts (MacKinnon, 1974).

Recognition

Generally, the juvenile stages of most mammals are distinguishable from the adult stage on the basis of voice. Adults have lower-pitched voices than juveniles. The reverse is true of microchiropteran ultrasounds, which function in both echolocation and communication (Matsumura, 1981).

Some large animals give us some insight into the effect of body size on vocalizations. In elephant seals (*Mirounga angustirostris*), pitch of vocalizations decreases with increasing body size; pulse interval and sound intensity tend to increase directly with body size (Bartholomew and Collias, 1962).

Sexually dimorphic vocalizations are more likely if body size differs between the sexes because respiratory structures are probably larger in larger animals. Defense of females and resources is also a likely correlate. Bull elephant seals (Bartholomew and Collias, 1962) and bull bison, *Bison bison* (Gunderson and Mahan, 1980), have lower-pitched roars than those of cows. When played back at reduced speeds, the purer sounds of males are distinguishable from the noisy calls of female *Saccopteryx bilineata* (Bradbury and Emmons, 1974). In various primates, the pitch with which a call is emitted is often higher in one sex or the other; but differences from one study to another suggest that age, context, and ontogeny may be significant determinants (Marler, 1976*b*; Mitchell, 1979). In both the siamang (*Symphalangus syndactylus*) and the gibbon (*Hylobates lar*), temporal differences characterize sex-specific patterns of calling which derive from similar vocal elements (Chivers, 1974). Songs of Bornean (*Hylobates muelleri*) and pileated gibbon (*H. pileatus*) females are nearly identical, whereas males differ considerably (Marler and Tenaza, 1977). The song of male gibbons probably functions in mate attraction as well as in territorial interaction. Yet even in species with minimal size dimorphism, sexually-dimorphic ultrasounds are demonstrable, such as in laboratory mice (Nyby et al., 1977*a*) and hamsters (Floody and Pfaff, 1977*a*).

Species-Typical Vocalizations

Since the physical structure of most, if not all, mammalian vocalizations seems to be genetically fixed, some calls might provide good taxonomic characters to measure species differences. The greatest species distinctiveness occurs in vocalizations that function to maintain interspecific segregation or promote intraspecific cohesion—the loud vocalizations of *Cercopithecus ascanius* and *C. mitis* (Marler, 1973), those of various subspecies of *Colobus badius* (Struhsaker, 1975), and the calls of eight species of gibbons and the siamang (Marler and Tenaza, 1977; Marshall and Marshall, 1976; Marshall et al., 1972). Calls that do not relate to species separation or group cohesion are more likely to be similar across species. For example, a detailed computer analysis of vocal signals in four species and one hybrid of *Spermophilus* revealed no conclusive systematic relationships (Koeppl et al., 1978). Although the effect of body size was not examined in *Spermophilus,* larger body size in ten species of *Eutamias* correlates positively with sound frequency and negatively with chipping rate (Brand, 1976).

The sonar calls of many bat species are species-typical and capable of discrimination with the aid of electronic equipment (Simmons et al., 1979). Infant bat calls are also species-typical. Since roosting sites are critical limiting factors to bat populations, species-typical vocalization would facilitate spacing within roosts and reuniting of mothers and their infants in multiple-species roosts (Gould, 1979, 1980).

Dialects

Local dialects are characterized by consistent differences in the vocalizations of animals from different populations of the same species over a period of many years. Genetic determinants probably account for the vocal differences in some races and subspecies of mammals (Green, 1975*b*)—populations of deer (Tembrock, 1965), grey langurs, *Presbytis entellus* (Vogel, 1973), squirrel monkeys (Winter, 1969) and possibly pikas, *Ochotona princeps* (Somers, 1973).

Discrete agonistic vocalizations exist in four island populations of male elephant seals (*Mirounga angustirostris*). Differing pulse rates typify each population but a change in mean pulse rate has occurred over a period of five years (LeBoeuf and Petrinovich, 1974). The vocalizations were believed to represent dialects that were learned (LeBoeuf and Peterson, 1969) and arose as a consequence of population expansion and isolation (LeBoeuf and Petrinovich, 1974), but the recent shift in pulse rate weakens the argument for genetically-determined dialects. Furthermore, the broad range of variation in pulse rate within and among individuals minimizes the importance of differences in mean pulse rate because pulse rates on various islands probably overlap extensively (Shipley et al., in press). Dialects have been reported in humpback whales, *Megaptera novaeangliae* (Hafner et al., 1979); killer whales, *Orcinus orca* (Ford, 1980); and in porpoises, *Tursiops truncatus* (Caldwell and Caldwell, 1979).

At three sites, Japanese macaques emit different vocalizations associated with food provisioning. The calls arose independently at each site "yet they reflect modifications of a shared theme," suggesting that they are determined in part by the genetic heritage of the animals (Green, 1975*a*, 1975*b*). Green (1975*a*, 1975*b*) combines several effects to explain the genesis of new calls, including the behavioral founder effect and socially-facilitated mimicry enhanced by cultural propagation with feeding as a reinforcer.

The apparent absence of dialects in most mammals could support one or both of two notions: genetic controls strongly influence auditory communication, or the context of vocal ontogeny is so conservative that vocal behavior within species is very uniform.

Individual Recognition

Recognition of a familiar versus a strange conspecific is the next level of complexity but most studies have concentrated on the individual rather than groups. The presence of vocal signatures in neonatal primates, such as talapoin monkeys, *Miopithecus talapoin* (Gautier, 1974) and squirrel monkeys, *Saimiri sciureus* (Lieblich et al., 1980), suggests that individual recognition is important throughout the life history.

The study of individual recognition on the basis of voice as the exclusive cue is usually approached in two phases. First, one can document the distinctiveness of an individual's vocalizations (signatures) (Table 1). A second approach to the study of signatures is by examining the specific responses to calls. The mother-infant relationship is the predominant context in which individual recognition has been documented; it occurs mainly in species in which mother and young need a means of identification prior to physical contact, either because of the social mileu in which young are reared or the frequent separation of mother and young. Approach and retrieval are two stereotyped, easily observed measures of the receiv-

TABLE 1
VOCAL SIGNATURES OF MAMMALS

Taxon	Signature information*	References
Chiroptera		
Antrozous pallidus	i L	P. Brown (1976)
Rhinolophus ferrumequinum	i L	Matsumura (1981)
Primates		
Saimiri sciureus	i L	Kaplan et al. (1978); Lieblich et al. (1980)
Papio ursinus	A S F	Hamilton and Arrowood (1978)
Cercocebus albigena	A S F	Waser (1976)
Macaca mulatta		Rowell and Hinde (1962)
Macaca nemestrina		Grimm (1967)
Gorilla gorilla		Fossey (1972)
Pan troglodytes	A S F	Marler and Hobbett (1975)
Aotus trivirgatus	A F	Moynihan (1964)
Callicebus moloch	A F	Robinson (1977)
Homo sapiens	m	DeCasper and Fifer (1980)
Ungulates		
Equus caballus	A F	Kiley (1972)
Rangifer tarandus	i m F	Espmark (1975)
Equus zebra	A F	Klingel (1968)
Pinnipedia		
Mirounga angustirostris	i F A S	Petrinovich (1974)
Halichoerus grypus	i F	Fogden (1971)
Cetacea		
Megaptera novaeangliae	A F	Hafner et al. (1979)
Tursiops truncatus	i A L	Caldwell and Caldwell (1979)

* Signature information: A, adult; F, field; i, infant; L, lab; m, mother; S, sex.

er's (mother's) response that have been studied. After vocal recognition successfully brings female and offspring together, the mother sniffs and then repels or accepts the infant bat (P. Brown, 1976; Nelson, 1965), seal (Bartholomew, 1959; Fogden, 1971), or reindeer (Espmark, 1971), which implies that odor is a more reliable or confirming clue. Experiments on adult-adult recognition are not so easily accomplished.

The apparent absence of signatures and individual recognition in mother-infant interactions of most carnivores, insectivores, and rodents may be a concomitant of larger litters and/or communal nesting (for example, as in lions, *Panthera leo*; Ewer, 1973). The entire kin group may take on a distinct odor in which case vocal signatures would be unnecessary. Individual recognition in many ungulates relies on a combination of visual, olfactory and acoustic cues (Altmann, 1963; Espmark, 1975).

The significance of individual recognition in bats is documented only for the young. However, the stability of harem members in some bat species (Bradbury, 1977*b*) implies the presence of individual recognition in adults. In a maternity ward of 10 to 200 *Eptesicus fuscus,* females usually nurse and care for their infants in a warm building. Infant *Myotis lucifugus* and *E. fuscus* vocalize immediately after parturition, that is, just as they clear the birth canal. As *E. fuscus* neonates vocalize the mother licks them and guides them (usually twins in northeastern United States) to the inguinal nipples. Without vocalizations, the mother will not begin normal maternal care. Premature infant *Eptesicus* vocalize abnormally: 1) calls are 10 to 15 db lower than those calls emitted by normal infants; and 2) many calls that are normally one-syllable are two-syllable. Such infants are usually ignored by their mothers (Gould, 1971, 1975*b*).

Isolation calls of *Antrozous pallidus* possess a vocal signature by which mothers identify and then reunite with their own infants, even though during the first 12 days the upper frequency range of infant *A. pallidus* calls changes dramatically by 20,000 Hz (P. Brown, 1976). The frequency profile retains some idiosyncratic features which endow the calls with a signature (P. Brown, 1976).

For more than three months, each mother and infant flying fox (*Pteropus poliocephalus*) reunite in the midst of many other conspecifics. When the female returns from foraging, she calls to her young and the young replies. The female lands near the infant, smells its chest, and only then accepts the infant (Nelson, 1965), suggesting that tentative recognition occurs by vocal signature and confirmation by olfaction.

Although the frequency profiles of bat sonar calls are relatively similar within broad taxonomic categories, this is not the case with infant isolation ("i") calls, particularly in the *Hipposideros* complex (Gould, 1979, 1980) and phyllostomids. As many as seven species of bats raise their young in the same cave in Malaysia. The dominant frequency of sonar sounds differs for different species of *Hip-*

posideros. However, during development, the dominant frequencies overlap. Thus, the species-specific "i" calls are probably essential to the efficient reunion of mother and infant. In *H. diadema* and *H. armiger,* two species that nurse infants in the same cave, *H. diadema* emits "i" calls with descending FM pulses and *H. armiger* with ascending FM pulses, respectively (Gould, 1979, 1980).

When a female *Rhinolophus ferrumequinum* approaches or flies away from her infant (1 to 4 weeks old), the calls of the infant become high-pitched and irregular, particularly when the adult flies toward the infant. Mother and infant *R. ferrumequinum* call antiphonally after a preliminary stage in which they coordinate their vocal intervals (Matsumura, 1981). The timing of mother and infant calls is at first asynchronous, then antiphonal, and finally simultaneous. The precision of overlap in the last stage resembles single sounds (Matsumura, 1981). Matsumura suggests that in *R. ferrumequinum* each mother-infant pair may have a fixed range of emission timing of high-pitched phrases which is the basis of recognition. Although the vocal exchange is accomplished with the typical sonar signals, the antiphonal character of exchange makes active echolocation unlikely. Matsumura's (1981) observations corroborate the double function of ultrasonic signals for communication and echolocation (Bradbury, 1977*b*; P. Brown, 1976; Gould, 1971; Mohres, 1966).

Mother squirrel monkeys (*Saimiri sciureus*) discriminate their infants by the isolation peep, a distinctive tonal call emitted by young squirrel monkeys when leaving their mothers (Fig. 8) (Kaplan et al., 1978). On the basis of a computer analysis of eight factors related to pitch profile, calls were found to be distinctive in two subspecies. Vocal individuality is present by one week and retains acoustic distinctiveness until the changing features stabilize at six weeks (Lieblich et al., 1980). Isolation peeps are "pre-wired," genetically determined and unaffected by the rearing environment, according to Lieblich et al. (1980) and Winter et al. (1973).

From a detailed analysis of calls emitted by ten different 17- to 21-day-old reindeer calves (*Rangifer tarandus*), Espmark (1975) concluded that no single factor is responsible for the uniqueness of a call but that the "Gestalt" of the call (that is, the subtle combinations and interactions of different features) accounted for the signature.

During a communicative interaction of reuniting, the mother and

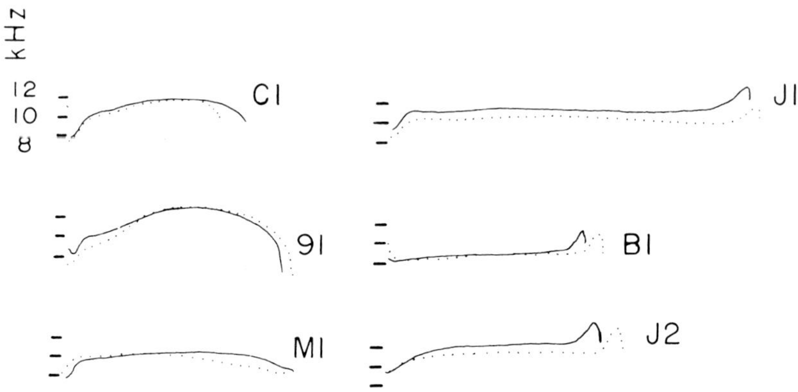

FIG. 8. Superimposed average pitch profiles at two ages in six squirrel monkeys (*Saimiri sciureus*), illustrating stability of shape in laboratory-related animals. Solid line is one year, dotted line is two years for C1, J1, and M1. Solid line is 36 weeks, dotted line is one year for 91, B1, and J2. (From Lieblich et al., 1980.)

infant may each use different sensory modalities to recognize one another. For example, Merino lambs (*Ovis aries*) deprived of auditory cues sought out their dams but failed to recognize them. Lambs deprived of visual cues failed to seek their dams at all (Arnold et al., 1975). In contrast, ewes recognized their lambs entirely by smell. Hearing and sight, although playing no part in recognition per se, helped the ewe to locate her lamb (Morgan et al., 1975).

Vocal exchange between mother and pup California sea lions (*Zalophus californianus*) begins at birth and continues for several days. When the mother returns from feeding, she raises her head and trumpets the penetrating "pup attraction call" (PAC). As she approaches a pod of pups, the mother repeats the call; her pup often responds to the call with a "bleat-like mother response call" and moves toward her (Peterson and Bartholomew, 1967). The two then vocalize alternately while moving toward each other. At night ". . . individual pups suddenly rouse, reply to a distant pup attraction call by uttering the mother-response call and then move away to join the vocalizing female, while other pups in the pod remain sound asleep" (Peterson and Bartholomew, 1967).

In a similar fashion, a female elephant seal (*Mirounga angustirostris*) recognizes her own infant and responds to its call with a PAC and with approach. The PAC is emitted by the mother shortly after the pup is born and usually elicits a vocal response from the pup. Selective recognition in fur seals (*Callorhinus ursinus*) appears to depend exclusively on the cow (Bartholomew, 1959); PACs also occur in *Arctocephalus pusillus, A. doriferus,* and *A. forsteri* (Stirling and Warneke, 1971).

Studies of adult individual recognition are restricted to only a few species (Table 1). Despite annual changes in song format, individual humpback whales (*Megaptera novaeangliae*) can be distinguished on the basis of their cries alone (Hafner et al., 1979). Adult mangabeys (*Cercocebus albigena*) emit a loud call, the "whoopgobble" that varies from one monkey to the next; in playback experiments, mangabeys responded differently to calls of different groups and different individuals depending on their distance from the loudspeaker (Waser, 1976). We know little about how long kin or acquaintances remember each other and more importantly how that memory contributes to long term reproductive relationships.

Antipredator Behavior

Alarm calls.—The use of silence as an antipredator response has already been mentioned. Alarm calls are vocalizations that alert other animals to impending danger. Although the same call may occur during conspecific aggression and threat of a predator, the response depends on the context (Leger and Owings, 1978). For example, roars, snorts, and squeals of *Colobus guereza* double as an intergroup spacing call and predator alarm call (Marler, 1972). Greater specificity occurs in African vervet monkeys (*Cercopithecus aethiops*), which emit distinct alarm calls that announce danger from a venomous snake crossing the territory, an eagle overhead, and an approaching leopard, *Panthera pardus* (Struhsaker, 1967; see also Marler, this volume).

Vocal responses to a mobile predator have evolved convergently in ungulates, lagomorphs, primates, bats, and some small carnivores (Eisenberg and Kleiman, 1977). The specific form and pitch of the call derive from a complex of factors, including the vocal apparatus and physical features of the habitat that affect sound propagation (Eisenberg, 1974). The significance of the close timing between an alarm call and the appearance of a predator might be learned after a few experiences; for example, young barasinghas (*Cervus duvauceli*) and chital deer (*Axis axis*) respond to alarm calls of birds, langurs, and various ungulates (Schaller, 1967).

Sherman (1977) studied a population of individually marked Belding's ground squirrels (*Spermophilus beldingi*) of known age, sex, and familial relationship. By noting the reproductive condition and the familial relationship of squirrels that emitted alarm calls, Sherman showed that alarm calls probably function to alert relatives. Diversion of predators, discouraging predator pursuit, or reducing the likelihood of later attacks by the same predators were hypotheses eliminated by field observations. A similar interpretation emerged from studies of alarm calls in California ground squirrels (*Spermophilus beecheyi*) and in chipmunks (*Eutamias sonomae*); Leger and Owings (1978) concluded that kin selection accounted for the greater responsiveness of parous versus non-parous females to alarm calls. In *Eutamias sonomae,* 2-year-old or older lactating females and previously lactating females, the individuals which probably have close relatives nearby (their adult daughters or their newest litter, respectively), are most prone to give alarms (Smith, 1978). Alternatively, parental care, rather than nepotism, may explain sciurid alarm calls (Shields, 1980).

Mobbing.—Bats trapped in a net emit a call that attracts multiple species of bats which swoop at the sound source (reviewed in August, 1979). The mobbing calls in *Artibeus jamaicensis* and *Phyllostomus hastatus* are acoustically similar to avian mobbing calls (August, pers. comm.). The repetition rate of these bat calls is 4 to 7 per second, which is in the range of infant bat calling rates; in both situations the calls attract adult bats. *Saguinus oedipus* and other callitrichids also exhibit threat behaviors (Green, 1979; Moynihan, 1970; Neyman, 1977), including single units of whines, trills, and clucks in response to predators, while constantly exposing themselves and disappearing from view. In both phyllostomid bats and callitrichids, the intense activity in response to a disturbance may serve to thwart or confuse a potential predator.

Fenton et al. (1976) found that in a wild *Myotis lucifugus* colony bats responded to a playback of distress calls by swooping flights near the loudspeaker, by increasing their level of activity, and by emitting low-frequency vocalizations. The presence of the distress-call mobbing response in vespertilionids and phyllostomids attests to the early appearance of the behavior in ancestral groups (August, 1979). Loud piercing calls may function to repel predators, as in howlers, *Alouatta palliata* (Carpenter, 1934), or in torpid bats as they arouse (Martin and Fenton, 1978), or to induce a predator to release its grip, as in *Ptilocercus lowi* (Gould, 1978).

Sounds Associated with Reproductive Behavior

Beach's (1976) classification of reproductive states in female mammals is useful for associating vocalizations with specific functions: "Attractivity refers to the female's stimulus value in evoking sexual responses by the male. Proceptivity connotes various reactions by the female toward the male which constitute her assumption of initiative in establishing or maintaining sexual interaction. Receptivity is defined in terms of female responses necessary and sufficient for the male's success in achieving intravaginal ejaculation." A fourth category relevant to auditory communication is post-ejaculatory activity. Table 2 indicates that investigators have noted certain vocalizations to be associated with each of the four phases of courtship. I have included under proception those sounds of males that are associated with the male's readiness to copulate since these sounds may prime or facilitate the proceptive phase. Some sounds occur during more than one phase; usually I assigned those to the phase in which the sound occurred most often. The absence of many examples during attractivity probably reflects the greater initial involvement of olfactory communication.

Coordination of Suckling

Survival of each infant in a litter should be maximized if signals coordinate and synchronize suckling. If familiar sounds are more attractive to infant mammals and if some sounds are audible to the fetus before birth (as in domestic sheep; Armitage et al., 1980) then the purrs of acouchis (*Myoprocta pratti*), grunts of pigs (*Sus scrofa*), and purrs of cats (*Felis catus*) could be particularly effective contact stimuli. Purring by the mother appears before parturition and during the first two weeks postpartum in acouchis. "Purring seems to be a contact call designed to keep the young informed of the mother's whereabouts as well as to induce them to remain near her . . . ." Young acouchis followed the mother more when she purred (Kleiman, 1972).

Moelk (1979) discriminated two types of low-intensity closed-mouth murmurs from domestic cats. The purr is a continuous vibra-

TABLE 2

SOUNDS ASSOCIATED WITH FOUR PHASES OF MAMMALIAN REPRODUCTIVE BEHAVIOR

Phase and species	Vocalizations	Sex of emitter	Function and comments	References
Attractivity				
Mesocricetus auratus	Ultrasounds	♀	Attracts ♂	Floody (1979)
Rattus norvegicus	Ultrasounds	♂ and ♀	Solicitation	Barfield et al. (1979)
Dicrostonyx groenlandicus	Rhythmic mating call	♀	Calms ♂ and stimulates ♂ to mate	Brooks and Banks (1973)
Ailuropoda melanoleuca	Rhythmic "bleat" call	♂ and ♀	Attraction	Kleiman et al. (1979)
Proceptivity				
Sus scrofa	Chant de coeur	♂	Stimulates ♀	Signoret et al. (1960)
Mesocricetus auratus	Ultrasounds	♂	Attracts ♀ and facilitates estrous	Floody (1979)
Rattus norvegicus	Ultrasound	♂	Attracts ♂ and facilitates estrous	Barfield et al. (1979)
Cavia porcellus	Purr	♂	♂ ready to copulate	Eisenberg (1974)
Myoprocta pratti	Squeak series	♂	♂ ready to copulate	Eisenberg (1974)
Proechimys semispinosus	Whimper	♂	♂ ready to copulate	Eisenberg (1974)
Dinomys branickii	Twitter	♂	♂ ready to copulate	Eisenberg (1974)
Sciurus aberti	Squeal	♂	During mating	Farentinos (1974)
Hypsignathus monstrosus	Repetitive metallic click	♂	Lek behavior	Bradbury (1977*a*)
Epomophorus wahlbergi	Repetitive	♂	Attracts ♀	Wickler and Seibt (1976)
Epomops franqueti	Repetitive	♂	Attracts ♀	Kingdon (1974)
Myotis lucifugus	Variable low frequency	♂	Emitted while ♀ struggles, less when she ceases struggling	Barclay and Thomas (1979); Barclay et al. (1979)

TABLE 2
CONTINUED

Phase and species	Vocalizations	Sex of emitter	Function and comments	References
Receptivity				
Microtus pinetorum	Ultrasound		During prolonged intromission	Geyer (1979)
Mesocricetus auratus	Ultrasound	♂	Induces lordosis	Floody and Pfaff (1977*a*)
Proechimys semispinosus	Duet whimper	♂ and ♀	During 5-min mount	Eisenberg (1974)
Dicrostonyx groenlandicus	Low frequency squeaks	♀	Before & after ejaculation; during & after lordosis	Brooks and Banks (1973)
Suncus murinus	Chirp	♀	Associated with tail wagging	Dryden (1968); Gould (1969)
Papio ursinus	Variable repetitive	♂ and ♀	Mutual stimulation	Hamilton and Arrowood (1978)
Hylobates hoolock	Repetitive	♂ and ♀	Incite ♂ competition	Hamilton and Arrowood (1978)
Ailuropoda melanoleuca	Chirp, whine	♀	During receptivity	Kleiman et al. (1979)
Post-ejaculatory				
Rattus norvegicus	22-kHz whistle	♂	Refractory	Barfield et al. (1979)
Myoprocta pratti	Cough, snort			Kleiman and Eisenberg (1974)
Octodon degus	Squeak		Repels animal and/or allows second ejaculation	Kleiman and Eisenberg (1974)
Octodontomys gliroides	Squeak			Kleiman and Eisenberg (1974)
Papio ursinus		♂ and ♀		Hamilton and Arrowood (1978)
Adenota kob	Whistle	♂		Buechner and Buechner (1971)

tion occurring during exhalation and inhalation. "The independent mhrn murmur is like an expanded single stroke of purring. It is usually uttered in one to four distinctive rolls . . . ." that can be counted (Moelk, 1979). The house cat mother greets and summons her infants with mhrns. When she returns to the nest, "she greets them with mhrns, calls them with mhrns, and licks them before lying down to permit them to suckle" (Moelk, 1979).

Ewer (1973) described feline purring as "an all's well signal" between nursing mother and her young that facilitates the formation of a bond. As the kittens mature, the first to suckle will purr loudly; the others then come to suckle probably in response to the purring. At least in domestic cats, the mother at first coordinates nursing; as kittens develop, their vocalizations contribute to the synchrony (Moelk, 1979).

While nursing, a sow (*Sus scrofa*) utters rhythmic vocalizations, partly because of stimulation of the udder by piglets (Fraser, 1973). As soon as the grunting rate increases, piglets stop nosing the udder and begin slow sucking. The rate of grunting increases substantially about 25 to 35 s before piglets begin rapid sucking (lasting 10 to 20 s), which might signal the piglets that milk-flow will soon occur (Whittemore and Fraser, 1974). This interval is about the same duration as that between the injection of oxytocin into the blood and the ejection of milk (15 s) (Whittlestone, 1953).

Closed-mouth purring or humming occurs in squirrel monkeys (Jurgens, 1979), mouse lemurs, *Microcebus murinus* (Charles-Dominique and Martin, 1972; Petter, 1962), bears (Tembrock, 1968), and guinea pigs (*Cavia porcellus*) in association with nursing. Young pikas (*Ochotona princeps*) chirp repetitively while nursing (M. Whitworth, pers. comm.). Because ultrasonics are difficult to study in the "noisy nest" situation of rodents and other nesters, we know nothing about acoustic synchrony of suckling in these species. However, the early function of purring for close contact and coordination of young (Ewer, 1973; Kleiman, 1972) probably later serves as a social signal for groups of sleeping females and paired males (Charles-Dominique and Martin, 1972).

Daily and Seasonal Factors

Daily and seasonal factors affect the frequency of vocal behaviors. Playbacks of a lion's (*Panthera leo*) roar during the day evoked no response from lions, whereas the same sounds played back at night evoked an immediate reply (Schaller, 1972*b*). Night-time playback of distress calls of little brown bats (*Myotis lucifugus*) in a barn colony of 700 bats evoked frequent swooping at the loudspeaker, whereas no response occurred during the day when bats were usually resting (Fenton et al., 1976). Daily patterns of calling probably typify species that use vocalizations for spacing. In the Malayan forest, the siamang (*Symphalangus syndactylus*), gibbon (*Hylobates lar*), dusky leaf monkey (*Presbytis obscurus*), and banded leaf monkey (*P. melalophus*) have distinctive temporal patterns of calling, with differences in time of first calling or peak periods of calling, probably corresponding to the timing and type of group behaviors (Chivers, 1973).

The most predictable trend in vocalization behavior may be the seasonal peaks that occur in diverse species, which are often related to reproduction. In wolves (*Canis lupus*), late winter peaks may relate to the breeding season and seasonal reorganization of packs with pup-howling occurring in late summer and fall (Field, 1978; Harrington and Mech, 1978). Male hammer-headed bats (*Hypsignathus monstrosus*) exhibit lek behavior and call incessantly at night between June and August (Bradbury, 1977*a*). Numerous cervids (Lincoln, *in* Kiley, 1972; Walther, 1977), pikas, *Ochotona princeps* (Somers, 1973), sciurids (Farentinos, 1974; Horwich, 1972; Layne, 1954), and spotted seals, *Phoca largha* (Beier and Wartzok, 1979) also call more during the breeding season.

Seasonal fluctuations in vocal behavior probably derive from changes in photoperiod, nutrition, endocrine condition, and reproductive processes. For example, whereas lactating female laboratory rats leave the nest and accurately localize the sound source during a playback of rat-pup ultrasounds, males and virgin females merely orient to the sound (Allin and Banks, 1972). Endocrine levels also influence the amount of calling in hamsters, *Mesocricetus auratus* (Floody, 1979), laboratory rats, *Rattus norvegicus* (Allin and Banks, 1972), and probably red deer, *Cervus elaphus* (Lincoln, 1970).

Duetting

Duetting has been noted in *Proechimys semispinosus* during copulation (Maliniak and Eisenberg, 1971), in adult *Leontopithecus rosalia* (Green, 1979), and in *Callicebus moloch* when adult pairs are within 0.5 m of one another (Robinson, 1979). It also occurs in *Rhinolophus ferrumequinum* when mothers reunite with infants (Matsumura, 1981). Duetting is common in all known monogamous hylobatids (Marshall and Marshall, 1976; Marshall et al., 1972). Antiphonal calling is a special form of duetting in which two animals use different or similar notes and sing alternately; a marked precision in timing may prevail (Hooker and Hooker, 1969). In communal bats we see highly-developed antiphonal calling between mother and infant, perhaps to accommodate the severe problems of localization among so many conspecifics in a roost.

Duetting probably functions to maintain the pair bond for a brief duration in the case of copulating rodents or an extended period as in primates. Perhaps duetting is as close as auditory communication can come to achieving "acoustic touch" without contact. Wickler (1980) hypothesized that the energetic commitment of vocal duetting prevents a partner from deserting. The investment reduces the likelihood that a partner would later change mates because he would have to invest time and energy in adjusting to a new mate (Wickler, 1980).

Motivational Basis for Vocalization

The study of vocal communication has been approached from different viewpoints by investigators interested in causal, ontogenetic, phylogenetic, or sociobiologic aspects of ethology. Descriptive as opposed to experimental approaches have probably influenced interpretations. Kiley (1972) focused on the stimuli that evoke vocalizations; she noted few calls that were associated with specific contexts. In studies of shrews (Gould, 1969), bats (Gould, 1971), and tenrecs (Eisenberg and Gould, 1970), arousal state and location of the sender, rather than discrete information, were considered the primary messages. In his analysis of infant rodent ultrasounds, Bell (1974) stressed the capacity of vocalization to induce arousal changes in the receiver. Others have emphasized the utility of functional classes of sounds occurring across taxa and associated with typical contexts (Eisenberg et al., 1975; Morton, 1977). In a field study of the Japanese monkey (*Macaca fuscata*), Green (1975*a*) characterized the acoustic structure of seven variants of the graded "coo" vocal repertoire. He then showed how each variant was associated with two to five situations, but with one modal situation in particular. He hypothesized that all sound patterns are linked to the internal state of the vocalizing animal.

Repetitive Calls and Excitation Indicators

Arousal state or level, motivational state, frustration level, and state of tension have all been used to describe the ever-shifting state of the organism above and below some baseline level. Some vocal emissions seem to be coupled to changing arousal state. A special class of auditory signal, sonic excitation indicators (SEI) (Gould, 1971) are often continuously emitted within an activity period, are graded in expression, and vary in repetition rate (and sometimes intensity and duration) commensurate with subtle changes in excitation as evinced by locomotor transitions and hesitations (Gould, 1971). The coupling of sonic emission with changing excitatory state and locomotor activity occurs in a diversity of distantly related taxa.

Redundancy is a predominant feature of these animal vocalizations and may protect a signal from background noise and increase intelligibility (Busnel, 1968). The continuous battery of calls could persuade or manipulate the receiver because the call often ceases once an "end" has been achieved (Dawkins and Krebs, 1978). For example, once a galago meets with nesting mates, the gathering call ceases (Charles-Dominique, 1977). Once an infant mouse is retrieved to the nest, the ultrasonic calls cease (Gould, pers. observ.).

A few descriptions will illustrate how widespread is the phenomenon: 1) a repetitive "put" mediates body-contact avoidance at close range in *Suncus murinus* (Gould, 1969); 2) repetitive calls coordinate social activities in primates (Marler, 1965); 3) sounders of pigs grunt and facilitate group cohesion (Kiley, 1972); 4) repetitive sonar calls are modified by mother and infant bats to produce antiphonal calling during reunion (Matsumura, 1981); 5) the repetitive calls of infant rodents evoke approach and retrieval by the mother (Sales and Pye, 1974). The rhythm, rather than the unit structure, of repetitive calls uttered by young foxes and young rodents is believed to be the critical feature to the responding mother (Tembrock, 1958; Zippelius, 1974).

Using two entirely different operational approaches, Haskins (1979) and Moelk (1979) came to similar conclusions. Haskins (1979) eschewed terms such as "distress" vocalizations or assumptions about the "internal state" of kittens (*Felis catus*). He showed how kitten vocalizations (all types of calls were lumped) are elicited or reduced by normal conditions in the litter situation. Moelk (1979) identified each vocal type and recorded the associated behavior, context, and "tension level" of the kitten. Both Moelk (1979) and Haskins (1979) concluded that kittens vocalize as a result of environmental factors and that their calls act as stimuli to evoke maternal behaviors. Through a process of reinforcement during normally occurring social contexts, calls become increasingly confined to a limited number of situations.

Floody (1979) conducted a physiological analysis of ultrasound production in hamsters (*Mesocricetus auratus*). Estrous females produced high rates of ultrasounds that attracted males. The female odor then seemed to evoke ultrasound calling in the males, resulting in a hormone-dependent communication chain that regulated hamster reproduction (Floody and Pfaff, 1977*a*). Floody (1979) argued convincingly that "concepts such as arousal suffer losses for explanatory power from ambiguous definitions, for example, how is general arousal to be measured or controlled?" Nevertheless, he acknowledged that "ultrasound production is characteristic of one sort of aroused state but not another, . . ." (sexual but not agonistic interactions) (Floody, 1979).

Bear in mind that causal studies are but one level of ethological investigation. By using another level of ethological analysis, Morton (1977) reemphasized Darwin's principle of antithesis by showing the universal relationship between sound structures of birds and mammals and their motivation. Sounds that express a sender's rapidly changing motivational state are associated with close contact when attack or escape are likely; these transitional sounds usually attenuate rapidly. In birds and mammals low, harsh sounds are consistently associated with hostile motivation, whereas higher puretoned sounds are associated with friendly motivation (Morton, 1977). These generalizations about vocal signals enabled Morton to predict the adaptive features of communicative signals in the context of different environments and social interactions. Morton's methodology is as acceptable in the field of sociobiology as Floody's explanation is in the field of behavioral physiology.

Morton's (1977) classification is corroborated by electrophysiological studies on the squirrel monkey, *Saimiri sciureus* (Jurgens et al., 1978). Squirrel monkeys have two different voice registers; each register is controlled by different nerves, one for the lower-pitched call (220 to 700 Hz), and the other for higher pitched calls (2,000 to 17,000 Hz). Jurgens (1979) calculated for each electrode position the percentage of the total session time during which the animal self-stimulated and thus categorized these positions as highly rewarding or highly aversive. He showed how each of five groups of calls lies on a continuum of motivational states. Furthermore, within each group, a call type exhibited a graded series. Most of his classifications are consistent with interpretations of squirrel monkey behavior (Schott, 1975; Winter et al., 1966). Eisenberg's (1976) multilevel illustration (Fig. 3) of five call transformations in spider monkeys (*Ateles fusciceps*) is added support for Jurgens' thesis.

Morton's thesis permits prediction about sound structure in association with major behavioral categories and in certain habitats, but it does not predict the receiver's response because that is dependent on the receiver's state. However, some vocalizations are particularly noteworthy because of their contagious effect on other conspecifics. At least one interpretation is that the vocalization changes the arousal level of other individuals; a group may become synchronized in its expectation of a common event, such as feeding. Vocalizations spread like a wave through various groups of primates in a forest (Chivers, 1969; Ellefson, 1968; Jurgens, 1979; Nashida, 1970). Contagious calling also occurs in cows and pigs (Kiley, 1972); prairie dogs, *Cynomys ludovicianus* (Smith, 1958); tree hyraxes, *Dendrohyrax arboreus* (Kingdon, 1971); wolves, *Canis lupus* (Harrington and Mech, 1978; Murie, 1944); howler monkeys, *Alouatta palliata* (Carpenter, 1965); and lions, *Panthera leo* (Schaller, 1972*b*); but see also Walther (1977) for discussion of contagious behavior.

Pause Patterns

Speech psychologists have taken special interest in the pattern of speech pauses which constitute messages of their own (Hayes and Bouma, 1975). The phrases "loud mouth," "smooth talker" and "fast talker" all denote pejorative implications about non-categorical features of speech (Feldstein and Crown, in litt.). Humans process separately information about content and information about the pattern of vocal participation (Hayes and Bouma, 1975). In fact, attending to the acoustic and message aspects of speech are incompatible operations (Martin and Strange, 1968). By synchronizing the pattern of flashing lights to the pattern of speech and speech pauses, one can present to human observers a stimulus that avoids the complications of sound frequency and context variables inherent in many speech experiments. Human observers make the same judgment of the flashing-light "speakers" as they do of audio presentation. Harmoniously interacting speakers tend toward a temporal congruence of conversational pattern (Feldstein and Welkowitz, 1978).

The heuristic value of the speech studies could be applied to the study of mammalian communication. One might ask: 1) Does each animal have its own pause pattern? 2) Does the pause pattern of one individual affect the pause pattern of another animal? 3) Do two interacting animals develop congruent pause patterns?

We tend to conceptualize the intervals between bouts of olfactory and auditory communication in two extremes but along a continuum. For example, an aggressive animal marks intensively (Ralls, 1971) and/or vocalizes with a "greater global muscular effort" (Jurgens, 1979), whereas a less aroused animal, when alone, may self-mark or mark the substrate at a low frequency (Eisenberg and Kleiman, 1972) or emit a continuous battery of low-repetition-rate, low-intensity vocalizations. Continuously emitted excitation indicators provide an auditory field which may be comparable to the olfactory field proposed by Eisenberg and Kleiman (1972). The auditory field, like the olfactory field, is persistent and may change in rate, intensity, or pitch over time. For example, as a male may anoint a female by rubbing both her and the substrate with glandular secretions, he may also saturate an area with his sounds at different distances, at first with "put-puts" in the case of *Suncus murinus* (Gould, 1969), and later with continuous clicks as the male settles down next to the female (at first she is very aggressive). Such similarities in the temporal patterning and changes in intensity of two modes of communication are consistent with Jerison's (1973) concept of one sensory system using the design features of another during the evolution of neural integration of the central nervous system.

A Unifying Concept for Motivational State and Vocal Emission of Sonic Excitation Indicators

Natural selection has exaggerated normal respiratory patterns in numerous ways (Brown, 1975). For example, shrews and certain tenrecs emit a repetitive sound ("put-put") whose rate changes in relation to respiratory and sniffing rates, body extension, and nasal movements, suggesting that "put" is an acoustic consequence of more active olfactory exploration (Gould, 1969, 1977). Sniffing in rats, including moving the vibrissae back and forth rhythmically in synchrony with inhalation sniffs and saltatory fixations of the head (Welker, 1964), is accompanied by the so-called theta rhythm of the brain 97% of the time (Cooper, 1973). Rhythmical sniffing likely generates rhythmical tactile, olfactory, proprioceptive, visual, and vestibular sensory input. A one-to-one correlation between individual theta waves and exploratory sniffing movements in laboratory rats (Komisaruk, 1970) and hamsters (Komisaruk, 1977; Macrides, 1975) has been seen (Table 3).

"Each theta cycle may reflect the activity of a neural mechanism which organizes the individual's reactions to the environment into such a pattern of sequential processes. Each sequence may be repeated rhythmically, approximately 7 times per second (that is, the frequency of the theta rhythm). By having a rhythmical mechanism (for example, at 7 Hz), each process system could be increased by increasing the sampling rate. The finding that the theta rate increases just before a laboratory rat makes a jump (Vanderwolf, 1975) is consistent with this hypothesis. The hypothetical mechanism therefore is a common pacemaker for sniffing, chewing, and licking. Located in the lower brainstem, it directly drives or modulates the cranial motor nuclei and can then recruit the forebrain limbic system to its frequency when the individual explores the environment, thereby facilitating the integration between sensory and motor activity" (Komisaruk, 1977). If an SEI was originally coupled to olfaction and tactile investigation, the simplest way for a receiver to process the incoming auditory signals might have been a rhythmic system comparable to that suggested by Komisaruk (1977). An evolving system that combined common rhythmic patterns with signal perception and production provides "a sense of economy" in design (Green and Marler, 1979) (Table 3). This

TABLE 3

REPETITION RATES OF VOCALIZATIONS, MOTOR ELEMENTS, AND THETA RHYTHM. MOST VOCALIZATIONS HAVE BEEN ASSOCIATED WITH STATE OF AROUSAL BY THE RESPECTIVE AUTHORS; THE SOUNDS ARE PRESUMED TO BE SEI

Species	Age*	Sound or behavior	Repetition rate/sec (range)	References
Rodentia				
Rattus norvegicus	a	Licking	3–7	Walker et al. (1964)
	a	Sniffing	5–11	Komisaruk (1977)
	a	Theta rhythm	3–7	Komisaruk (1977)
Mus musculus	y	Ultrasounds	4	Gould (1971); Zipellius 1974)
Apodemus favicollis	y	Ultrasounds	20	Zipellius (1974)
Clethrionomys sp.	y	Ultrasounds	5	Zipellius (1974)
Micromys minutus	y	Ultrasounds	2	Zipellius (1974)
Lagidium peruanum	a	Twitter†	12	Eisenberg (1974)
Insectivora				
Suncus murinus	a	Put	1–13	Gould (1969, 1971)
Hemicentetes semispinosus	a	Put		Eisenberg and Gould (1970)
Hemicentetes semispinosus and *H. nigriceps*	a	Stridulation	1–18	Eisenberg and Gould (1970); Gould (1971)

TABLE 3
CONTINUED

Species	Age*	Sound or behavior	Repetition rate/sec (range)	References
Sorex minutus	a	Whisper†	5	Hutterer (1976)
Sorex minutus	y	Cries†	4–5	Hutterer (1976)
Carnivora				
Thalarctos maritimus	a	Chuf†	1–4	Wemmer et al. (1976)
Viverricula sp.	a	Coughing†	12	Wemmer (1977)
Genetta tigrina	a	Coughing†	3	Wemmer (1977)
Mustela putorius	y	Ultrasounds†	8	Solmsen and Apfelbach (1979)
Artiodactyla				
Sus scrofa	y	Grunt†	1	Kiley (1972)
Primates				
Galago alleni	y	Tsic†	9–15	Charles-Dominique (1977)
Saimiri sciureus	y	Twitter	11–17	Jurgens (1979)
(36 species, 7 orders)**	y	Sucking	1–4.6	Wolff (1968)

* Ages: a, adult; y, young.
† From sonographs.
** Rodentia, Pinnipedia, Carnivora, Primates, Artiodactyla, Perissodactyla, Proboscidea.

theoretical framework is consistent with the hypothesis that bat sonar evolved from continuous, graded communication signals of early insectivores (Gould, 1970, 1971).

Auditory Communication in Maternal-Infant Configurations

Background

One may conceptualize the fundamental social unit of mammals in at least two forms of parental care: 1) a "mother-family," and 2) sets (demes) of parents with offspring (Anderson, 1970; Eisenberg, 1977). These forms are not only consistent with most sociobiological discussions, but they are also convenient references for these discussions of vocal communication. The first, the "mother-family," sets the stage for the private and intimate interaction among mother and infants that occurs shortly after birth. The second encompasses the interaction from nearby individuals. An understanding of the vocalizations exchanged within the "mother-family" unit could be a key to understanding the ontogeny of at least one category of signals that persists in adulthood—that is, those sounds which function in social contact behaviors such as greeting and courtship.

What was the context of the early mother-family unit? Early mammals were probably small, densely-furred, nocturnal insectivores (Hopson, 1973). Judging from most extant conservative placentals, we might surmise that the early mammals were nursed in a protected, dark, and relatively quiet nest site. Tactile, olfactory, and auditory communication would have been predominant while visual signals would have been of minor importance. The clicking and wheezing sounds between nursing bouts (caused by lip-smacking and mouth-opening) permitted the mother to acoustically orient toward the infants. As the infants detached from the nipples and rolled away from the nest site, they must have clicked and wheezed in their first efforts to swallow properly, encouraging the flow of fluids toward the pharynx and esophagus while protecting the respiratory system from blockage. Slight uncoordinated movements caused coughing and associated sounds in the mouth (judging from observations of human infants [Stark, 1979] and of rat pups [Gould, pers. observ.]).

Any analysis of parent-offspring communication in mammals must take into account the social organization of a species, the concomitant ecological adaptations, and the degree of development at birth (Eisenberg, 1977). As mammalian adaptive radiation occurred, these multiple factors impinged on the parent-infant communicative interaction. However, the context of the interaction remained relatively conservative. Two fundamental configurations influenced the extent of vocal communication between parent and offspring: whether infants were carried by the parent for a protracted period or whether they were left at a protected site. The carried infant does not need vocal signals at a distance, whereas a rodent pup displaced from the nest and an infant bat reuniting with its mother in the midst of other infants do. We can classify the configurations into four basic profiles and suggest the major trends within different mammalian groups.

1. Nest or den
   - Insectivora
   - Edentata
   - Rodentia
   - Carnivora
2. Follow (or hide and follow)
   - Rodentia
   - Cetacea
   - Sirenia
   - Ungulates
3. Carry
   - Marsupialia
   - Primates
   - Edentata (some)
4. Refuge
   - Chiroptera
   - Pinnipedia

Subcategories, such as "parking," occur across profiles in certain ungulates, bats, and nocturnal prosimians. In some species or groups, there is a change of category from one stage of rearing to the next (for example, marsupials). The classification is useful not only because the profiles seem to represent normal contextual circumstances but because each profile relates to some obvious consequences in regard to auditory communication. From an evolutionary standpoint, these profiles are the boundaries within which natural selection favored certain auditory signals.

Nesting and Denning

Most rodents, insectivores, and carnivores restrict early infant care to a confined nest or den. Quiet, darkness, and insulation against heat loss are usually characteristic of the site. Infants of this category emit more than one type of contact-promoting call, including repetitive calls that evoke approach and sometimes retrieval by the mother. The calls of many infant rodents and at least one carnivore, *Mustela putorius* (Solmsen and Apfelbach, 1979) are ultrasonic. Three factors in particular increase the rate of calling in rodents: 1) a drop in core body temperature of the infant (A. Brown, 1976; DeGhett, 1978); 2) tactile stimulation (Okon, 1970); and 3) the absence or presence of nest odors (Geyer, 1979; Oswalt and Meier, 1975). The lack of conspicuous auditory signals in neonatal elephant shrews, *Elephantulus rufescens* and *Rhynchocyon chrysopygus* (Rathbun, 1979) and rabbits, *Sylvilagus floridanus* and *S. palustris* (Marsden and Holler, 1964), is associated with precociality, reduced parental care, and high risk of predation in the absence of protected burrow systems.

Refuging and Nursery Creches

Refuging is the rhythmical dispersal of groups of animals from and their return to a fixed point in space (Hamilton and Watt, 1970). Pinnipeds use islands and bats use caves, buildings, or trees as refuges. In both groups the situation is cacophonous and offspring are suckled at the refuge. Females usually have but one infant, nurse none but their own, and periodically leave that infant among many others. Bats and pinnipeds living in colonies have the same basic problem—reuniting with their infants in the midst of many same-aged animals.

In both pinnipeds and bats, mother-infant vocal activity is intense during a brief period just after birth. Infant bat calls increase with a drop in body temperature (P. Brown, 1976). Mother and infant pinnipeds call continuously; mother and infant bats reunite several times after short separations. These periods of vocal exchange, which occur in unison with other sensory modalities, probably reinstate important social elements of the mother-infant unit that persist in adult life. A vocal signature seems to be a prominent feature of both groups. Furthermore, in both groups individual recognition by vocal signals is often confirmed by olfactory cues.

Carrying

Most non-human primates are relatively quiet during the first few weeks of life. Carrying is associated with a complex set of selective forces that have led to reduced neonatal vocalizing. Emphasis on tactile stimulation, neoteny, and social behaviors is significant. Tactile and kinesthetic modes account for 99% of the communication in newborn stumptail macaques, *Macaca arctoides* (Chevalier-Skolnikoff, 1974). By the third week, there is slight increase in vocal communication but the tactile and visual mode remain dominant. There is a marked increase in vocalizations from the infant when weaning begins at about six months of age (Chevalier-Skolnikoff, 1974). This is the reverse of the nest-den configuration; rodents tend to reduce vocal emission after weaning. Marsupials are born in such an undeveloped state that the protracted close contact and transport by the mother is obligatory. Despite this dependence, a more advanced infant that falls from the pouch emits a low-intensity chirp or squeak. The mother approaches the calling infant and the infant climbs into the pouch or onto her back. Infant calls are produced by *Marmosa cinerea* (Beach, 1939; Thrasher et al., 1971), *Sminthopsis crassicaudata* (Ewer, 1970), and in the kangaroos, *Megaleia rufa* and *Macropus giganteus* (Frith and Calaby, 1969). In some small species, infants are left in the nest when they exceed the size of the pouch.

Following

Predation has probably been a strong selective force to reduce ungulate vocalizations and restrict auditory communication to maternal care and courtship. Young ungulates either follow the mother or hide between nursing bouts. Mothers usually call the infant just before nursing, as in pigs, *Sus scrofa* (Whittemore and Fraser, 1974), in blackbucks, *Antilope cervicapra,* and in dorcas gazelles, *Gazella dorcas* (Walther, 1977). Species with infants that follow the mother usually possess vocal signals that inform the mother and/or infant of each other's location, as in caviomorph rodents, artiodactyls (Kleiman, 1972), and certain tenrecs (Eisenberg and Gould, 1970).

The four mother-young configurations show a graded trend in the use of auditory communication. Refuging and nesting species are highly vocal; diurnal carrying and following (as well as parking) profiles are all associated with reduced vocal activity.

An important element of maternal care that relates to the prominence or reduction of auditory signals is the reunion after a period of separation. Frequent reunions become periodic rehearsals of contact signals throughout the preweaning period in species exhibiting nesting, refuging, and following (parking) categories. In primates, the mother is the "nest" (Mason, 1968); reunions become more frequent after weaning.

In nearly all species of mammals, isolated or stressed neonates vocalize and their calls attract the mother. Infant calls may be repetitive and high-pitched (rodents, bats, shrews, and polecats) or less repetitive but rasping and irritating (for example, primates). In either case the effect may be to "persuade" the mother to make contact, which turns off the sound (Dawkins and Krebs, 1978; Mason, 1968).

Ontogeny of Vocalizations

Introduction

The number and sequence of infant vocalizations follow a similar pattern of development in many species of mammals. Shortly after birth, most mammals emit sounds that evoke approach or contact by the mother. As the infant matures, sounds that elicit withdrawal are produced (Cohen and Fox, 1976). These sounds are often called agonistic vocalizations. The number of different vocalizations tends to increase from infant to adult in bats (Gould, 1975*a*), rock hyraxes, *Procavia capensis* (Fourie, 1977), gorillas, *Gorilla gorilla* (Fossey, 1979), and stumptail macaques, *Macaca arctoides* (Chevalier-Skolnikoff, 1974), but there is a marked decrease in vocal activity late in juvenile life. Calls that attract the mother are less and less frequent in bats that are near the age of weaning (Gould, 1977). A reduction in the number of ultrasounds emitted by infant rodents is concomitant with eye-opening, greater locomotor activity and an ability to thermoregulate (Geyer, 1979). Audible squeals are first emitted by laboratory mice at about the time ultrasonics cease, 13 to 14 days of age (Whitney, 1970); ultrasonics reappear in adults.

While neonatal rodents are vocally active, neonatal primates are relatively quiet, as already discussed. In the stumptail macaque (*Macaca arctoides*), vocalizations change in form and function during development (Chevalier-Skolnikoff, 1974). Calls initially tend to concentrate their energy in only one harmonic band. There is a gradual decline in frequency with a greater emphasis on the higher of two bands. The repertoire begins with three calls (chirp, gecker, and scream) in the first week; there are 12 vocalizations by the sixth month and 17 vocalizations in the adult.

The sequential appearance of vocalizations in different species of bats varies according to how developed the infant is at birth; for example, *Antrozous pallidus* (P. Brown, 1976) and *Eptesicus fuscus* (Gould, 1971, 1975*a*) are among the most altricial bats. They emit one type of call at birth; echolocation signals are detectable by 9 and 3 days, respectively. In contrast, *Macrotus californicus* and *Carollia perspicillata* are precocial. In these species, the isolation call has not been detected and three calls that occur after 9 days in *A. pallidus* and *E. fuscus* are detectable on the day of birth.

A single call type may develop in different ways depending on the species. A long isolation call may fragment into combinations of long-short or short-long syllables, as in *Eptesicus fuscus* (Gould et al., 1973; Woolf, 1973), while concomitantly the frequency increases (Woolf, 1973). Vocal changes from noisy to pure-tone sounds may parallel a gradual separation of pathways for food and air in the nasopharyngeal region of *Rhinolophus ferrumequinum* (Matsumura, 1979).

Sequelae of Behavioral Elements in Maternal-Infant Relationship and in Courtship

In a recent analysis of animal communication, Green and Marler (1979) stress the lack of ethological studies on the phylogeny or ontogeny of encoding and decoding procedures. "While the development of many signal actions is largely innate, it seems probable that, at least in higher animals, learning plays an important role in the acquisition and development of functionally appropriate 'assessments' of situations or signals. Thus, one may argue that the focus on the morphology of signaling behaviors in past animal research has led to overemphasis on the inherited components of the communicative performance." This discussion focuses on the relationship of two compatible ontogenetic concepts: specification—validation (Jacobson, 1974), and Piaget's schemata (Vernon, 1955; Flavell, 1963)—to the similarities between vocalization occurring during maternal care and courtship.

Vocal repertoires are not the only behaviors that are repeated during different life-history stages of a species. Leyhausen (1956*a*, 1956*b*) suggested that in felids the female pattern of sexual presentation has evolved into a social greeting ceremony for both sexes similar to presentation in primates. Wickler (1967) enumerated comparable similarities in other mammalian species and concluded that "sexual behavior is necessarily bound to social behavior; a socio-sexual link is present *a priori*." Moelk (1979) argues that all feline friendly-approach behavior derives from the initial nursing-suckling touch and sound relationship between mother and kittens. "Because all the sensory and stimulus elements of greeting are present during the maternal infant context and because sexual behavior contains a number of greeting elements it is more logical to regard sexual behavior as a special form of friendly approach behavior rather than vice versa (Leyhausen, 1956*a*, 1956*b*; Schaller, 1972*b*)" (Moelk, 1979). The examples listed in Table 4 support Moelk's argument. Young males and females develop with a common set of social contact experiences because of the conservative set of contexts related to the four mother-young configurations (see above). This phenomenon of ontogenetic sequelae of auditory, tactile, and olfactory signals is widespread and may contribute to the synchronization of reproductive behavior in male and female.

The organization of mammalian behavior patterns has been conceptualized in the general terms applied to development and remembering in humans. "Schemata are considered to be inherent species-characteristic modes of organizing experience and of acting on information received (Vernon, 1955; Flavell, 1963). They play a major role in the performance of biologically essential tasks, such as mating, caring for young, forming filial bonds, reacting to predators, and capturing prey. These performances are not based on a single schema, however, but on multiple schemata, usually organized hierarchically in ascending levels of complexity and control. In this form, they provide the basic ground plan within which behavioral development proceeds" (Mason, 1979).

During courtship adult mammals repeat some behaviors that occurred earlier in life. Reinstatement is achieved by multiple experiences that are similar but not identical to the first learning experience. Reinstatement of a behavior can alleviate forgetfulness in young and adult laboratory rats (Mactutus et al., 1979) and in 3-month-old human infants after a retention interval as long as 4 weeks (Rovee-Collier et al., 1980). "An infant's reencounters with contextual aspects of prior training or an earlier experience can prime or recycle the remaining memory attributes and enhance access to them, alleviating forgetting which otherwise appeared complete weeks earlier" (Rovee-Collier et al., 1980). "Reinstatement is a potent mechanism through which experiences of early

TABLE 4
VOCALIZATIONS THAT ARE SIMILAR IN ADULT AND MATERNAL-INFANT CONTEXTS

Order and species	Vocalization	Context during infancy (preweaning)	Context during adult life	References
Marsupialia				
Didelphis virginianus	Click	Mother calls infant; isolated infants call	♂ during courtship	Reynolds (1952)
Marmosa mitis	Click	Isolated from mother	♂ during mating	Thrasher et al. (1971)
Chiroptera				
Carollia perspicillata	Double-note	Emitted by infant during reunion; rate increased as mother flies	♂/♀ pairs chasing in flight; just prior or subsequent to a chase; chases that precede mating	Porter (1979)
Antrozous pallidus	Infant isolation call; directive of adult third and fourth harmonic drop out	Emitted by infant in response to isolation, decreasing body temperature, tactile stimulation, and by 1-week-old in response to mother's directive	Single bat isolated from main roosting group emits directive until answered by others	P. Brown (1976)
Insectivora				
Suncus murinus	Low intensity chirp	Caravaning and before suckling	Courtship, receptive	Dryden (1968); Gould (1969)
	Click	About to suckle during caravan formation; separation from female	Exploration, increased rates in courting ♂	Gould (1969)
Cryptotis parva	Click	Displaced from mother	Courtship	Gould (1969)

TABLE 4
CONTINUED

Order and species	Vocalization	Context during infancy (preweaning)	Context during adult life	References
Rodentia				
Tamiasciurus hudsonicus	Coughing or buzzing note	1 month, exploration	Mating chase	Layne (1954)
Sciurus aberti	Repetitive squeal	Displaced from nest	Mating chase	Farentinos (1974)
Sciurus carolinensis	Sneeze	Search from human feeder to nurse	Mating chase	Horwich (1972)
Aethosciurus poensis	C calls	Mother-infant greeting before contact	♂ solicits ♀	Emmons (1978)
Rattus norvegicus (lab rat)	50 kHz	Isolation	Solicitation mounting	Barfield et al. (1979)
Microtus pinetorum	Ultrasound	Isolation	Copulation	Geyer (1979)
Mesocricetus auritus	Ultrasound	When not nursing; drop in body temperature	♂ odor ♀ calls; ♀ odor ♂ calls	Floody and Pfaff (1977*a*); Sales and Pye (1974)
Mus musculus	Ultrasound	Away from nest; drop in body temperature; strange nest odors	♂ exposure to ♀; odor of ♀; mounting (mediated by androgen)	Nyby et al. (1977*b*); Sales and Pye (1974)
Cavia	Repetitive chirp	Suckling	Courtship	Berryman (1976)
Dicrostonyx groenlandicus	Ultrasonic chirp and twitter	Decrease in body temperature; maternal retrieval	Courtship	Eibl-Eibesfeldt (1958)

TABLE 4
CONTINUED

Order and species	Vocalization	Context during infancy (preweaning)	Context during adult life	References
Primates				
Galago alleni	Short clicks 2.5 ms	When arousing; mother comes to sound	Adult seeking to establish contact (e.g., ♀♀ regrouping to sleep in tree hollow)	Charles-Dominique (1977)
Rhesus monkey and other macaques	Coo	Isolation	Separated ♀ in estrous	Mitchell (1979)
Saimiri sciureus	Purr	During and before suckling	Huddling and between copulatory activity	Jurgens (1979); Schott (1975); Winter et al. (1966)
Miopithecus talapoin	Type 4	Infants surprised but not frightened by a new stimulus	During copulation	Gautier (1974)
Carnivora				
Panthera leo	Miaowing	Small cubs moving around and behind group; approach an adult; being licked roughly	Lost, unsure; prevented from reaching goal, i.e., conditions of light distress; ♂ during copulation	Schaller (1972*b*)
Acinonyx jubatus	Chipr	Mother summons	♂ during courtship	Schaller (1972*b*)
Felis catus	Mhrn	Excitation indicator	Greeting, courtship	Moelk (1979)

infancy can continue to influence behavior" (Rovee-Collier et al., 1980).

Recall of a past experience is enhanced if the motivational state of the animal is the same at the time of the first and subsequent experiences (Haroutunian and Riccio, 1979; Spear, 1978). In addition, the more contextual elements related to the initial experience that occur during subsequent reinstatement the greater the fidelity of recall (Spear, 1978). The redundancy of infant vocalizations associated with various body contact behaviors that occur during infancy and courtship seems to fulfill these criteria for recall (*Suncus murinus*; Table 5, Fig. 9). However, the argument depends on the assumption that motivation during infancy is the same as that of adult courtship (or components of courtship). At least two similarities prevail. First, many of the same behaviors and contextual circumstances occur. Second, infancy and courtship both culminate in behaviors that are endpoints in themselves—nipple attachment, attaining a state of homeostasis in the case of vocal behaviors of neonates, and intromission in the case of courtship behaviors of adults.

Thus, a case can be made for "similarities," but how do vocalizations become connected to schemata? Jacobson (1974) distinguished between two classes of neurons. Class I neurons constitute the hard wiring of the brain; their structure and function are genetically constrained or specified. Class II neurons exhibit greater variability, perform integrative functions, exist abundantly and are reduced in numbers by a seemingly "Darwinian struggle for survival . . . . The surviving neurons will become more and more closely matched with the functional requirements of the organism but will have a progressively reduced reserve of potential functions." Survival of neurons depends on sensory or hormonal stimulation. For example, separation of domestic kittens from the mother for the first 6 or 7 days after birth does not affect nipple attachment in kittens (Kovach and Kling, 1967). Separation of pups from the mother laboratory rat severely retards nipple attachment despite an initial period of normal nursing (Stoloff, 1980). In Jacobson's (1974) conceptual framework the cat's suckling system is relatively more "specified" than the rat's; the sucking system requires less practice for its "validation" (Stoloff, 1980).

Some features of infant calls must be highly specified at birth because the calls are emitted on day 1 and because they result in specific maternal responses. Bats (Gould, pers. observ.), domestic cats, and dromedary camels, *Camelus dromedarius* (Gauthier-Pilters, 1959), vocalize just as the head emerges from the vulva or clears its birth sac (Moelk, 1979). Calls of premature infant walruses (*Obodenus rosmarus*) are very similar to those emitted by 4-month-old walruses (Kibalchich and Lisitzyna, 1979). Alaskan fur seals (*Callorhinus ursinus*) often begin vocalizing within 15 to 45 s of delivery (Petrinovich, 1974). No deprivation experiments have resulted in the permanent alteration of development of the fundamental structure of vocalizations (for example, deprivation studies with squirrel monkeys [Winter et al., 1973] and big brown bats [Gould, 1975*b*; Woolf, 1973]). The context, rate of emission and intensity of calls may be modified but the frequency (pitch) profile remains relatively unmodified. In some birds, the vocalization itself is subject to modification (Kroodsma, 1978; Marler and Peters, 1977; Nottebohm, 1969), whereas in mammals it is probably the situations with which the vocalization becomes associated that are subject to modification. Deprivation experiments with bats did not change the physical structure of vocalizations. However, manipulations (for example, hand-raised versus normal animals) did alter the emission rate of two graded bat communication calls—short-long and long-short double notes. Hand-raised bats seemed less excited and less aggressive than those raised by their mothers. Accordingly one would expect a change in the occurrence of call types that fall on a graded continuum and that couple to excitation levels (Gould, 1975*b*; Gould, in press; see also deprivation studies of monkeys, Hinde and Spencer-Booth, 1971).

There are no deprivation experiments demonstrating sensitive periods of early vocal experience in mammals. However, there is abundant evidence in diverse species that sounds occurring in infancy are identical or similar to those occurring during courtship (Table 4). In many mammals juvenile habits reappear in the adult (Beach and Jaynes, 1954). During subadult life some social calls may not be emitted in solitary species. A fundamental problem in performing behaviors that occur only occasionally throughout an animal's lifetime is the requirement for recall. Some species are solitary between weaning and their first courtship experience; other species continue to associate with conspecifics. Memory in both sets of species may be enhanced by frequent repetition and reinstatement during infancy as well as by a short period of solitary life prior to first courtship. In the case of continuously social species, rehearsals may be facilitated by various displays of greeting, play, and presentation.

Social vocalizations occurring during infancy may show up in a number of adult contexts: 1) a call may occur in the same context but be emitted by the adult when communicating with her young (for example, in cervids; Schaller, 1972*a*); 2) a call may be repeated during one of the 4 phases of reproductive behavior (Table 2); and 3) a call may occur during a friendly greeting, as in domestic cats (Moelk, 1979).

Tests of the hypothesis that infant greeting and courtship behaviors are related could take several forms. Deprivation experiments might disrupt the entire schemata rather than alter a specific vocalization. Playback of adult courtship calls to nursing mothers with variable age infants could be informative. A third approach, the sort taken by Moelk (1979), involves a longitudinal study of an animal from birth. If variation occurs in the adult repertoire (Klopfer and Boskoff, 1979), one should be able to trace that variable from the early stages of parental care.

The following examples illustrate the proposed sequelae of auditory signals.

Opossum (Didelphis virginianus).—The female oppossum's clicks stimulate the young to leave the den with her (Reynolds, 1952). Males click during courtship and infants click when separated from the mother; the mother usually approaches the clicking young.

House shrew (Suncus murinus).—In both caravan formation and in courtship behavior of house shrews, *Suncus murinus,* we can identify similar behavioral elements including 1) body orientation, 2) body contact points, 3) unified odor fields, 4) vocalizations, and 5) body and tail movements (Table 5). A house shrew's maternal association may terminate after three weeks (at weaning); sexual behaviors activate 10 days later in the female and 30 days later in the male, with a phase of solitary life between weaning and courtship (Dryden, 1968, 1969). However, for two weeks during nursing, intermittent caravaning experiences occur (Stine, 1976). During that period multiple components of courtship may be rehearsed and committed to memory. Although learning decrements occur (in laboratory rats) after about one week depending on age and number of trials (Mactutus, pers. comm.), emission of a particular call in the adult context might be facilitated if the call were

TABLE 5

PARTIALLY SHARED BEHAVIORS OF OVERLAPPING SCHEMATA: MATERNAL-INFANT INTERACTION AND COURTSHIP IN HOUSE SHREWS, *Suncus murinus* (STINE, 1976; AND STINE'S VIDEO RECORDS)

Behavior	Maternal-infant behaviors	Courtship	Comments
Tactile			
Lip-licking	+	–	♂ and ♀ avoid naso-naso contact
Tail-bite	+ (caravan)	+	♂ holds ♀'s tail or rump but does not follow
Rump-bite	+ (caravan)	+	
Belly-to-back contact	+ drape	+ ♂ rubs ♀	
Anogenital lick	+ ♀ licks infant on back	+ ♂ lifts ♀ off substrate as ♂ licks anogenital region	
Tail-wag	+ ♀ before caravan	+ ♀ before mount	♀ wags tail during courtship
Auditory			
Clicks	+ ♀	+ ♂	
Put	+ ♀	+ ♂	
Chirp	+ infant	+ receptive ♀	Unified auditory field
Olfactory	♀ odor	♂ glandular rub	Unified odor field (see Eisenberg and Kleimen, 1972)
Locomotor	+ follows mother	+ ♂ follows ♀	
Intromission	–	+	
Sequence of courtship	–	+	

Abbreviations: +, present; –, absent.

associated with a constellation of other sensory and motor elements. There is a correlation in mammals between the sequential emergence of some behavioral elements during ontogeny and the subsequent temporal sequencing of the same elements in adult life (Horwich, 1972).

Rodents (Cricetidae and Muridae).—Eibl-Eibesfeldt (1958) eobserved that when female lemmings (*Dicrostonyx groenlandicus*) hear

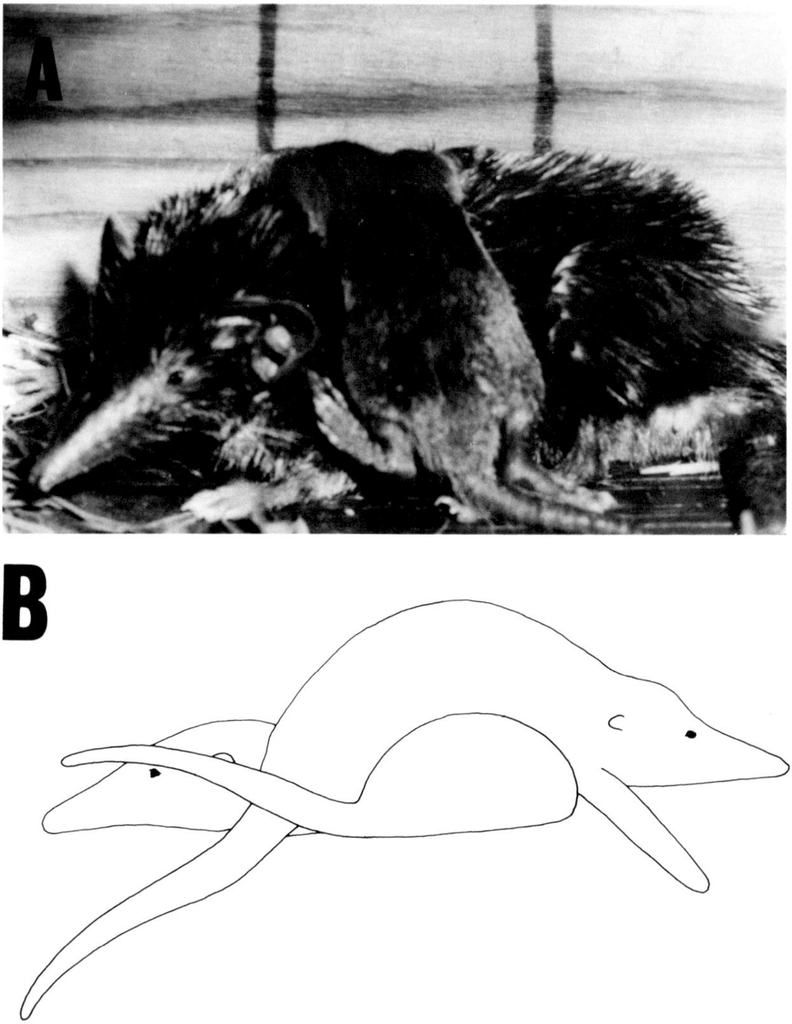

FIG. 9. The body contact points of a 10-day-old (check) infant shrew (*Suncus murinus*) draping over the mother's back (A) and the body contact of adults during courtship (B) are similar. Photograph (A) and video tape recording (B) courtesy of Dr. Charles Stine.

the twitter courting call, they displayed care-giving behavior. Of particular relevance to the sequelae hypothesis is that adult courtship and infant calls are similar and even more similar when older pups (11 to 12 days) were used for comparison (Brooks and Banks, 1973). Infant calls might become increasingly specified to the adult condition toward the end of weaning (see Fentress' discussion of three stages of grooming ontogeny, this volume). Infant and adult calls are both given in rapid bursts of ultrasonic units of short duration and with clear tonal structure and similar frequency profiles (Brooks and Banks, 1973). A consideration of recent studies of rodent ultrasounds in hamsters (Floody, 1979; Sales and Pye, 1974: 157), laboratory rats (Adler and Anisko, 1979; Barfield et al., 1979; Barfield and Geyer, 1972; Sales and Smith, 1978), mice (Sales and Pye, 1974; Whitney et al., 1973; Whitney and Nyby, 1979), and lemmings (Brooks and Banks, 1973) bear out ideas that infant and adult vocalizations are ontogenetically related and that sensory and hormonal factors contribute to the ontogeny of ultrasounds. Descriptions of adult and infant ultrasounds indicate great variability in the frequency of infant calls as compared to those of adults.

An infant rodent may emit ultrasonic whistles during a 2 to 3 week period of nursing while an adult may emit (or receive) similar ultrasounds for only a few hours of each mating bout. Thus, the redundancy of infant calls that is so effective in evoking approach by the mother may also allow infants to reinstate a call that promotes contact. During infancy, calls are emitted under different circumstances and in response to tactile, olfactory, and temperature stimuli; the same or similar calls emitted by adults may depend on endocrine, social and environmental stimuli.

Ungulates.—In most ungulates, vocal signals have been reduced and visual cues emphasized in association with herd mobility in open country and antipredator strategies. It is significant that vocalizations in reindeer (*Rangifer tarandus*) are manifested primarily during the rut and the mother-infant association. However, there is little documentation of similar auditory repertoires during infancy and courtship (except in *Tragulus napu*; Ralls et al., 1975).

Primates.—Overlapping schemata may include or exclude auditory signals depending on the multiplicity of combinations that have been emphasized because of normal selection. Auditory signals in early infancy are reduced in many primate species in favor of continuous tactile, thermal, and olfactory stimuli (for example, see Chevalier-Skolnikoff, 1974). Nevertheless, some infant calls do occur in courtship. In Table 4 we can see that the situations with which vocalizations are associated during infancy and adulthood in four species of primates are different, yet the aroused state and some stimuli may be quite similar. In the ontogeny of human sexual behavior visual, tactile, and olfactory "images" predominate; auditory components are rare (Money, 1980, pers. comm.; Money and Musaph, 1977). "As a general rule, courtship corresponds to the establishment of bonds between male and female" (Charles-Dominique, 1977). Likewise, infants form a temporary bond with the mother (and sometimes peers). Thus the continuity between the elements of these two behavioral repertoires is not surprising. Overlapping schemata with partially shared behavioral elements constitute one hypothetical framework for the ontogeny of courtship and the welding of contact signals into social behavior patterns. This hypothesis does not demand that all vocalizations have their genesis during infancy. For example, female copulatory vocalizations in *Macaca fascicularis* and male loud calls in *Cercopithecus cephus* occur suddenly at the onset of puberty (Deputte and Goustard, 1980; Gautier, 1978).

Vocal and Suckling Behavior Compared

The study of suckling ontogeny in the laboratory rat may help us to conceptualize the development of mammalian auditory communication. Suckling is important in this discussion, not only because it is fundamental to all mammals, but because some vocalizations are so closely associated with it. Suckling mechanisms, like other developing systems, are not completely specified at birth, but become specified later, as shown by behavioral modification experiments (Bruno et al., 1980; Pedersen and Blass, in press). Suckling must be practiced to be maintained (Blass et al., 1979; Stoloff et al., in press). If suckling can be modified, why is it so stereotyped and "successful" in its development? A somewhat analogous question about vocalization is: How do sounds become fixed to specific contexts (see Green, 1975*a*)?

Laboratory rats (*Rattus norvegicus*) are born with five neuromotor elements (searching, rooting, probing, mouthing, and nipple attachment) related to suckling (Pedersen and Blass, in press).

Moreover, 1-day-old rat pups (Johanson and Hall, 1979), as well as 2- or 3-day-old human infants (DeCasper and Fifer, 1980), have a great capacity to learn new skills. Results of experiments in which the nipples of virgin female rats (painted with amniotic fluid and saliva of the parturient and virgin females) were offered to pups suggest that prenatal experiences influence postnatal behaviors (Blass et al., 1979). A rat may come to associate a specific olfactory stimulus with suckling; the stimulus can then gain control over the five basic motor elements.

There is a great likelihood that "prenatal" uterine fluid will be on the nipple to which a neonatal rat attaches. Probably just as great is the likelihood that an infant mammal will sometimes not attach to the nipple when it is so motivated; at the moment of distress, the infant will emit a particular species-specific call associated with a particular motivational state. Repetition of the call usually culminates in approach and contact by the mother.

Could the association of infant vocalizations with the inability of an infant to gain access to the mother's nipple be an homologous mammalian context? Note the following: 1) elephant seals (*Mirounga angustirostris*) give a shrill cry "when ready to suckle and the mother's nipple is unavailable" (Petrinovich, 1974); 2) lion cubs (*Panthera leo*) emit harsh, snarly miaows, lasting as long as two to three seconds when they want to suckle but are not permitted to do so by the lioness (Schaller, 1972*b*); 3) young puppies whine when hurt or when they cannot attach to a nipple (Bleicher, 1963); 4) infant macaques (*Macaca arctoides*) emit geckers in distressing situations such as in loss of nipple contact (Chevalier-Skolnikoff, 1974); galagos (*Galago alleni*) behave similarly (Charles-Dominique, 1977). To be sure, many other contexts in which these vocalizations are given exist, but these contexts are maternal-care imperatives.

Both suckling and vocal behavior are structurally stereotyped and occur soon after birth. Termination of neonatal vocalizations occurs upon return of baseline thermal, olfactory, or tactile (including nipple attachment) conditions (when calling stops and/or purring begins in felids). The vocalization and suckling paradigms differ in that suckling in the very young mammal appears to be an endpoint in itself, whereas vocalizations are emitted in the absence of optimal endpoints (thermo-tactile-olfactory conditions) (Fig. 10). In both cases the endpoint changes with age.

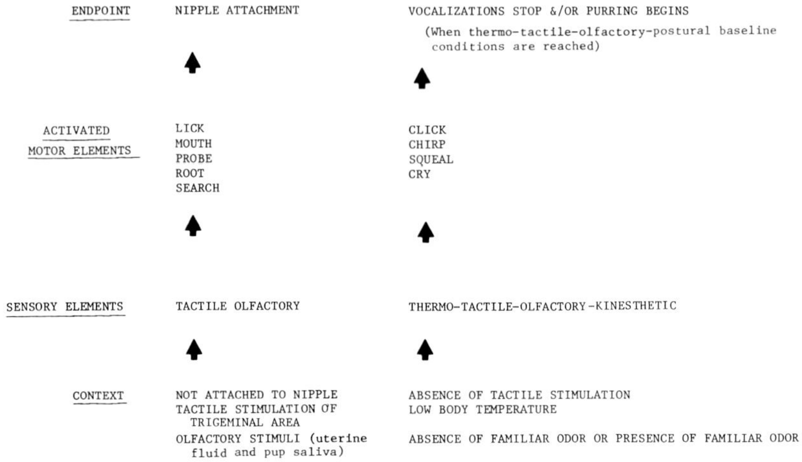

FIG. 10. Comparison of neonatal suckling and vocal behaviors.

Conservative Behavior

Eisenberg (1967) commented on the conservative characteristics of mammalian behavior patterns. "Relative differences rather than absolute differences in behavior" are the rule. How extraordinary is the similarity of mammalian vocal behaviors! How have so many species "protected" these patterns from the massive onslaught of natural selection? Roeder (1967) remarked on how the encoding of organic evolution, social evolution, and the integration of many cells for unified action was accomplished through rather changeless coding systems of 1) gene arrangement, 2) language and writing, and 3) nerve impulses and hormones, respectively. What then is the encoding system that has retained the uniformity of such stereotyped behaviors as vocalizations and suckling?

First, a neonate is born with activated motor systems—the ability to click, bleat, chirp, purr, whistle, and cry. Food activates not only the feeding system but the entire body. "Such behavior activation may be related to developing neural arousal and reward systems associated with consummatory behaviors and may prove useful in studying the development and organization of motivation systems" (Hall, 1979).

Second, sensory systems are responsive to thermo-tactile-olfactory stimuli (that is, cold, touch, or new odors) that divert the neonate from a species-specific baseline motivational state. The sensory systems regulate early vocalization by sensing changes in the baseline and somehow evoking vocal responses. Warmth and nipple attachment turn off distress calls and in some species may turn on purring.

Third, the neonate has an enormous capacity for associative learning of new tasks. "Suckling is an adequate incentive to support instrumental learning in rats as young as seven days of age" (Blass et al., 1979). The neonatal rat also displays long-term retention of early aversive odor conditioning (Rudy and Cheatle, 1977), inhibition of learned responses (Amsel et al., 1976; Brake, 1978), and at day 1, the ability to learn an appetitive response and use this response in making (and later reversing) two-choice discriminations (Johanson and Hall, 1979). The infant probably learns quickly what sounds result in the return of its body to a motivational baseline.

Fourth, quantitative differences in behaviors may derive from the differential specializations of the sensory systems (for example, thresholds of hearing). Fifth, the conservative, rather uniform context of the mother and infant association tends to confine the total array of stimuli. Finally, behavior redundancy during infancy, reactivation of these behaviors during subadult life and their incorporation into the shared, partially overlapping schemata of infancy, greeting, and courtship, has ensured the protection of these behaviors from the "continuous energy flow and transformation that is characteristic of life" (Roeder, 1967).

Acknowledgments

I deeply appreciate the fruitful discussion with J. Eisenberg, L. Gould, J. Heinz, R. Stark-Seitz, C. Stine, and D. Wartzok. I am particularly indebted to S. Petrovich, E. Blass, and J. Childs for their time, correction, and stimulating suggestions. Thanks are extended to L. Gould for improvements in style and clarity and to E. Poplovski, who typed the manuscript, and D. Stayman, who cross-checked the citations and conducted literature searches.

Literature Cited

ADLER, N., AND J. ANISKO. 1979. The behavior of communicating: an analysis of the 22 kHz call of rats (*Rattus norvegicus*). Amer. Zool., 19:493-508.

ALDRICH-BLAKE, F. P. G. 1970. Problems of social structure in forest monkeys. Pp. 79-99, *in* Social behaviour in birds and mammals (J. H. Crook, ed.). Academic Press, New York, 492 pp.

ALLIN, J. T., AND E. M. BANKS. 1972. Functional aspects of ultrasound production by infant albino rats (*Rattus norvegicus*). Anim. Behav., 20:175-185.

ALTMANN, M. 1952. Social behavior of elk, *Cervus canadensis nelsoni,* in the Jackson Hole area of Wyoming. Behaviour, 4:116-143.

———. 1963. Naturalistic studies of maternal care in moose and elk. Pp. 233-253, *in* Maternal behavior in mammals (H. L. Rheingold, ed.). John Wiley and Sons, Inc., New York, 349 pp.

AMSEL, A., D. R. BURDETTE, AND R. LETZ. 1976. Appetitive learning, patterned alternation, and extinction in 10-d-old rats with non-lactating suckling as reward. Nature, 262:816-818.

ANDERSON, P. K. 1970. Ecological structure and gene flow in small mammals. Symp. Zool. Soc. London, 26:299-352.

ANDREW, R. J. 1964. The displays of the primates. Pp. 227-309, *in* Evolutionary and genetic biology of primates (J. Buettner-Janusch, ed.). Academic Press, New York, 2:1-330.

ARMITAGE, S. E., B. A. BALDWIN, AND M. A. VINCE. 1980. The fetal sound environment of sheep. Science, 208:1173-1174.

ARNOLD, G. W., ET AL. 1975. The roles of sight and hearing in the lamb in the location and discrimination between ewes. Applied Anim. Ethol., 1:167-176.

AUGUST, P. V. 1979. Distress calls in *Artibeus jamaicensis*: ecology and evolutionary implications. Pp. 151-160, *in* Vertebrate ecology in the northern neotropics (J. F. Eisenberg, ed.). Smithsonian Inst., Washington, D.C., 271 pp.

BALDWIN, J. D., AND J. I. BALDWIN. 1976. Vocalizations of howler monkeys (*Alouatta palliata*) in southwestern Panama. Folia Primatol., 26:81-108.

BARCLAY, R. M. R., AND D. W. THOMAS. 1979. Copulation call of the bat *Myotis lucifugus*: a discrete situation-specific communication signal. J. Mamm., 60:632-634.

BARCLAY, R. M. R., M. B. FENTON, AND D. W. THOMAS. 1979. Social behavior of the little brown bat, *Myotis lucifugus*. II. Vocal communication. Behav. Ecol. Sociobiol., 6:137-146.

BARFIELD, R. J., AND L. A. GEYER. 1972. Sexual behavior: Ultrasonic postejaculatory song of the male rat. Science, 176:1349-1350.

BARFIELD, R. J., ET AL. 1979. Ultrasonic vocalizations in rat sexual behavior. Amer. Zool., 19:469-480.

BARTHOLOMEW, G. 1959. Mother-young relations and the maturation of pup behavior in the Alaska fur seal. Anim. Behav., 7:163-171.

BARTHOLOMEW, G. A., AND N. E. COLLIAS. 1962. The role of vocalization in the social behavior of the northern elephant seal. Anim. Behav., 10:7-14.

BEACH, F. A. 1939. Maternal behavior of the pouchless marsupial, *Marmosa cinerea,* J. Mamm., 20:315-322.

———. 1976. Sexual attractivity, proceptivity, and receptivity in female mammals. Horm. Behav., 7:105-133.

BEACH, F. A., AND J. JAYNES. 1954. Effects of early experience upon the behavior of animals. Psychol. Bull., 51:239-263.

BEIER, J. C., AND D. WARTZOK. 1979. Mating behaviour of captive spotted seals (*Phoca largha*). Anim. Behav., 27:772–781.

BELL, R. W. 1974. Ultrasounds in small rodents: arousal produced and arousal-producing. Dev. Psychobiol., 7:39–42.

BERANEK, L. L. 1949. Acoustic measurements. John Wiley and Sons, New York, 914 pp.

BERRYMAN, J. C. 1976. Guinea-pig vocalizations: their structure, causation and function. Z. Tierpsychol., 41:80–106.

BLASS, E. M., W. G. HALL, AND M. H. TEICHER. 1979. The ontogeny of sucking and ingestive behaviors. Prog. Psych. Physiol. Psychol., 8:243–299.

BLEEK, D. F. 1929. Comparative vocabularies of bushman languages. Cambridge Univ. Press, London, 94 pp.

BLEICHER, N. 1963. Physical and behavioral analysis of dog vocalizations. Amer. J. Vet. Res., 24:415–427.

BOSMA, J. F. 1971. Progressive motor disabilities of the mouth. Postgrad. Med., 49:162–166.

BRADBURY, J. W. 1977*a*. Lek mating behavior in the hammer-headed bat. Z. Tierpsychol., 45:225–255.

———. 1977*b*. Social organization and communication. Pp. 1–72, *in* Biology of bats (W. A. Wimsatt, ed.). Academic Press, New York, 3:1–651.

BRADBURY, J. W., AND L. H. EMMONS. 1974. Social organization of some Trinidad bats. I. Emballonuridae. Z. Tierpsychol., 36:137–183.

BRAKE, S. C. 1978. Discrimination training in infant, preweanling, and weanling rats: effects of prior learning experiences with the discriminanda. Anim. Learning Behav., 6:435–443.

BRAND, L. R. 1976. The vocal repertoire of chipmunks (genus *Eutamias*) in California. Anim. Behav., 24:319–335.

BROOKS, R. J., AND E. M. BANKS. 1973. Behavioural biology of the collared lemming [*Dicrostonyx groenlandicus* (Traill)]: an analysis of acoustic communication. Anim. Behav. Monogr., 6:1–83.

BROUGHTON, W. B. 1963. Method in bio-acoustics terminology. Pp. 83–124, *in* Acoustic behavior of animals (R. G. Busnel, ed.). Elsevier Publishing Company, New York, 933 pp.

BROWN, A. M. 1973. High levels of responsiveness from the inferior colliculus of rodents at ultrasonic frequencies. J. Comp. Physiol., 83:393–406.

———. 1976. Ultrasound and communication in rodents. Comp. Biochem. Physiol. Comp. Physiol., 53:313–317.

BROWN, A. M., AND PYE, J. D. 1975. Auditory sensitivity at high frequencies in mammals. Adv. Comp. Physiol. Biochem., 6:1–73.

BROWN, J. L. 1975. The evolution of behavior. W. W. Norton Company, New York, 761 pp.

BROWN, P. 1976. Vocal communication in the pallid bat, *Antrozous pallidus*. Z. Tierpsychol., 41:34–54.

BROWN, P. E., AND A. D. GRINNELL. 1980. Echolocation ontogeny in bats. Pp. 355–377, *in* Animal sonar systems (R. G. Busnel and J. F. Fish, eds.). Plenum Press, New York, 1135 pp.

BRUNO, J. P., M. H. TEICHER, AND E. M. BLASS. 1980. Sensory determinants of suckling behavior in weanling rats. J. Comp. Physiol. Psychol., 94:115–127.

BUBENIK, A. B. 1965. Beitrag zur Geburtskunde und zu den Mutter-Kind-Beziehungen des Reh- (*Capreolus capreolus L.*) und Rotwildes (*Cervus elaphus L.*). Z. Säugetierk., 30:65–128.

BUCHLER, E. R., AND A. R. MITZ. 1980. Similarities in design features of orientation sounds used by simpler, non-aquatic echolocators. Pp. 871–874, *in*

Animal sonar systems (R. G. Busnel and J. F. Fish, eds.). Plenum Press, New York, 1135 pp.
BUDNITZ, N., AND K. DAINIS. 1975. *Lemur catta*: Ecology and behavior. Pp. 219–235, *in* Lemur biology (I. Tattersall and R. W. Sussman, eds.). Plenum Press, New York, 365 pp.
BUECHNER, H. K., AND J. H. BUECHNER. 1971. The Uganda kob. Territoriality and ceremonial mating behavior. Color film (length: 20 min.). The Pennsylvania State University, Audio Visual Services, University Park, Pennsylvania, distributor.
BUSNEL, R-G. 1968. Acoustic communication. Pp. 127–153, *in* Animal communication (T. A. Sebeok, ed.). Indiana Univ. Press, Bloomington, 686 pp.
———. 1976. Information in the human whistled language and sea mammal whistling. Pp. 914–938, *in* Speech and surrogates: drum and whistle systems, Part 2 (T. A. Sebeok and D. J. Umiker-Sebeok, eds.). Mouton, The Hague, 1456 pp.
CALDWELL, M. C., AND D. K. CALDWELL. 1979. The whistle of the Atlantic bottlenosed dolphin (*Tursiops truncatus*)—Ontogeny. Pp. 369–401, *in* Behavior of marine animals. Vol. 3. Cetaceans (H. E. Winn and B. L. Olla, eds.). Plenum Press, New York, 438 pp.
CAPRANICA, R. R. 1972. Why auditory neurophysiologists should be more interested in animal sound communication. Physiologist, 15:55–60.
CARPENTER, C. R. 1934. A field of study of the behavior and social relations of howling monkeys. Comp. Psychol. Monogr., 10:1–168.
———. 1965. The howlers of Barro Colorado Island. Pp. 250–291, *in* Primate behavior (I. DeVore, ed.). Holt, Rinehart and Winston, New York, 654 pp.
CHARLES-DOMINIQUE, P. 1977. Ecology and behavior of nocturnal primates. Prosimians of equatorial West Africa. Columbia Univ. Press, New York, 277 pp.
CHARLES-DOMINIQUE, P., AND R. D. MARTIN. 1972. Behaviour and ecology of nocturnal prosimians. Parey, Berlin, 89 pp.
CHEVALIER-SKOLNIKOFF, S. 1974. The ontogeny of communication in the stumptail macaque (*Macaca arctoides*). Contrib. Primatol., 2:1–174.
CHIVERS, D. J. 1969. On the daily behaviour and spacing of howling monkey groups. Folia Primatol., 10:48–102.
———. 1973. An introduction to the socio-ecology of Malayan forest primates. Pp. 102–146, *in* Comparative ecology and behaviour of primates (R. P. Michael and J. H. Crook, eds.). Academic Press, London, 847 pp.
———. 1974. The siamang in Malaya. A field study of a primate in tropical rain forest. Contrib. Primatol., 4:1–335.
COHEN, J. A., AND M. W. FOX. 1976. Vocalizations of wild canids and possible effects of domestication. Behav. Processes, 1:77–92.
COOPER, R. L. 1973. Relationship between rhythmical behavioral patterns and limbic system EEG activity in the rat. Unpubl. Ph.D. dissert., Rutgers University, New Jersey, 228 pp.
DAWKINS, R., AND J. R. KREBS. 1978. Animal signals: information or manipulation. Pp. 282–312, *in* Behavioural ecology (J. R. Krebs and N. B. Davies, eds.). Sinauer Associates, Sunderland, Massachusetts 494 pp.
DECASPER, A. J., AND W. P. FIFER. 1980. The fetal sound environment of sheep. Science, 208:1173–1176.
DEGHETT, V. J. 1978. The ontogeny of ultrasound production in rodents. Pp. 343–365, *in* The development of behavior: comparative and evolutionary aspects (G. M. Burghardt and M. Bekoff, eds.). Garland STPM Press, New York, 429 pp.

DEPUTTE, B. L., AND M. GOUSTARD. 1980. Copulatory vocalizations of female macaques, *Macaca fascicularis*. Primates, 21:83–99.

DRYDEN, G. L. 1968. Growth and development of *Suncus murinus* in captivity on Guam. J. Mamm., 49:51.

———. 1969. Reproduction in *Suncus murinus*. J. Reprod. Fert., Suppl., 6:377–396.

EIBL-EIBESFELDT, I. 1958. Das verhalten der nagetiere. Handb. Zool., 10:1–88.

EISENBERG, J. F. 1967. A comparative study in rodent ethology with emphasis on evolution of social behavior. Proc. U.S. Natl. Mus., 122(3597):1–51.

———. 1974. The function and motivational basis of hystricomorph vocalization. Symp. Zool. Soc. London, 34:211–247.

———. 1976. Communication mechanisms and social integration in the black spider monkey, *Ateles fusciceps robustus,* and related species. Smithsonian Contrib. Zool., 213:1–108.

———. 1977. The evolution of the reproductive unit in the class mammalia. Pp. 39–71, *in* Reproductive behavior and evolution (J. S. Rosenblatt and B. R. Komisaruk, eds.). Plenum Press, New York, 170 pp.

EISENBERG, J. F., AND E. GOULD. 1970. The tenrecs: a study in mammalian behavior and evolution. Smithsonian Contrib. Zool., 27:1–138.

EISENBERG, J. F., AND D. G. KLEIMAN. 1972. Olfactory communication in mammals. Ann. Rev. Ecol. Syst., 3:1–32.

———. 1977. Communication in lagomorphys and rodents. Pp. 634–654, *in* How animals communicate (T. A. Sebeok, ed.). Indiana Univ. Press, Bloomington, 1128 pp.

EISENBERG, J. F., L. R. COLLINS, AND C. WEMMER. 1975. Communication in the Tasmanian devil (*Sarcophilus harrisii*) and a survey of auditory communication in the marsupialia. Z. Tierpsychol., 37:379–399.

ELLEFSON, J. O. 1968. Territorial behavior in the common white-handed gibbon, *Hylobates lar* Linn. Pp. 180–199, *in* Primates: studies in adaptation and variability (P. C. Jay, ed.). Holt, New York, 529 pp.

EMMONS, L. H. 1978. Sound communication among African rainforest squirrels. Z. Tierpsychol., 47:1–49.

ERWIN, J. 1974. Responses of rhesus monkeys to separation vocalizations of a conspecific infant. Percept. Motor Skills, 39:185–197.

ESPMARK, Y. 1971. Mother-young relationship and ontogeny of behavior in reindeer (*Rangifer tarandus L.*). Z. Tierpsychol., 29:42–81.

———. 1975. Individual characteristics in the calls of reindeer calves. Behaviour, 54:50–59.

EWER, R. F. 1970. The ethology of mammals. Logos Press, London, 418 pp.

———. 1973. The carnivores. Cornell Univ. Press, Ithaca, New York, 494 pp.

FARENTINOS, R. C. 1974. Social communication of the tassel-eared squirrel (*Sciurus aberti*): a descriptive analysis. Z. Tierpsychol., 34:441–458.

FELDSTEIN, S., AND J. WELKOWITZ. 1978. A chronography of conversation: in defense of an objective approach. Pp. 329–378, *in* Nonverbal behavior and communication (A. W. Siegman and S. Feldstein, eds.). Lawrence Erlbaum Associates, Hillsdale, New Jersey, 400 pp.

FENTON, M. G., ET AL. 1976. Responses of *Myotis lucifugus* (Chiroptera: Vespertilionidae) to calls of conspecifics and to other sounds. Canadian J. Zool., 54:1443–1448.

FENTRESS, J. C., R. FIELD, AND H. PARR. 1978. Social dynamics and communication. Pp. 67–106, *in* Behavior of captive wild animals (H. Markowitz and V. J. Stevens, eds.). Nelson-Hall, Chicago, 314 pp.

FIELD, R. 1978. Vocal behavior of wolves (*Canis lupus*): variability in structure, context, annual/diurnal patterns, and ontogeny. Unpubl. Sc.D. dissert., Johns Hopkins Univ., Baltimore, Maryland, 243 pp.

FLAVELL, J. H. 1963. The developmental psychology of Jean Piaget. Van Nostrand, Princeton, 472 pp.

FLOODY, O. R. 1979. Behavioral and physiological analyses of ultrasound production by female hamsters (*Mesocricetus auratus*). Amer. Zool., 19:443–455.

FLOODY, O. R., AND D. W. PFAFF. 1977*a*. Communication among hamsters by high-frequency acoustic signals: I. Physical characteristics of hamster calls. J. Comp. Physiol. Psychol., 91:794–806.

———, AND ———. 1977*b*. Communication among hamsters by high-frequency acoustic signals: III. Responses evoked by natural and synthetic ultrasounds. J. Comp. Physiol. Psychol., 91:820–829.

FOGDEN, S. C. L. 1971. Mother-young behavior at grey seal breeding beaches. J. Zool., 164:61–92.

FORD, J. K. B. 1980. Group-specific vocalizations of the killer whales. Abstr. 146th meet. Amer. Assoc. Adv. Sci., symp. on "Communication and behavior of whales."

FOSSEY, D. 1972. Vocalizations of the mountain gorilla (*Gorilla gorilla beringei*). Anim. Behav., 20:36–53.

———. 1979. Development of the mountain gorilla (*Gorilla gorilla beringei*): the first thirty-six months. Pp. 139–186, *in* The great apes (D. A. Hamburg and E. R. McCown, eds.). Benjamin Cummings, Menlo Park, California, 554 pp.

FOURIE, P. B. 1977. Acoustic communication in the rock hyrax, *Procavia capensis*. Z. Tierpsychol., 44:194–219.

FOX, M. W. 1971. Integrative development of brain and behavior in the dog. Univ. Chicago Press, Chicago, 348 pp.

FRASER, D. 1973. Nursing and suckling behavior of pigs. I. Importance of stimulation of anterior teats. British Vet. J., 129:424–436.

FRITH, H., AND J. CALABY. 1969. Kangaroos. C. Hurst, London, 209 pp.

GAUTHIER-PILTERS, H. 1959. Einige Beobachtungen zum Droh-, Angriffs- und Kampf-verhalten des Dromedarhengstes, sowie uber Geburt und Verhaltensentwicklung des Jungtieres, in der nortwestlichen Sahara. Z. Tierpsychol., 16:593–604.

GAUTIER, J-P. 1974. Field and laboratory studies of the vocalizations of talapoin monkeys (*Miopithecus talapoin*). Behaviour, 51:209–272.

———. 1978. Repertoire sonore de *Cercopithecus cephus*. Z. Tierpsychol., 46: 113–169.

GEIGER, R. 1957. The climate near the ground. Harvard Univ. Press, Cambridge, Massachusetts, 494 pp.

GEYER, L. A. 1979. Olfactory and thermal influences on ultrasonic vocalization during development in rodents. Amer. Zool., 19:419–430.

GOODALL, J. 1965. Chimpanzees of the Gombe Stream Reserve. Pp. 425–473, *in* Primate behavior (I. DeVore, ed.). Holt, Rinehart and Winston, New York, 654 pp.

GOULD, E. 1969. Communication in three genera of shrews (Soricidae) *Suncus, Blarina,* and *Cryptotis*. Comm. Behav. Biol., 3:11–31.

———. 1971. Studies of maternal-infant communication and development of vocalizations in the bats *Myotis* and *Eptesicus*. Comm. Behav. Biol., 5: 263–313.

———. 1975*a*. Neonatal vocalization in bats of eight genera. J. Mamm., 56:15–29.

———. 1975*b*. Experimental studies of the ontogeny of ultrasonic vocalizations in bats. Dev. Psychobiol., 8:333–346.

———. 1977. Echolocation and communication. Pp. 247–280, *in* Biology of bats of the new world family Phyllostomatidae, Part 2 (R. J. Baker, J. K. Jones, Jr., and D. C. Carter, eds.). Spec. Publ. Mus., Texas Tech Univ., 364 pp.

———. 1978. The behavior of the moonrat, *Echinosorex gymnurus* (Erinaceidae) and the pentail shrew, *Ptilocercus lowi* (Tupaiidae) with comments on the behavior of other insectivora. Z. Tierpsychol., 48:1–27.

———. 1979. Neonatal vocalizations of ten species of Malaysian bats (Megachiroptera and Microchiroptera). Amer. Zool., 19:481–491.

———. 1980. Vocalizations of Malaysian bats (Microchiroptera and Megachiroptera). Pp. 901–904, *in* Animal sonar systems (R. G. Busnel and J. F. Fish, eds.). Plenum Press, New York, 1135 pp.

———. In press. Echolocation by young bats on their initial and subsequent flights. Dev. Psychobiol.

GOULD, E., N. K. WOOLF, AND D. C. TURNER. 1973. Double-note communication calls in bats: occurrence in three families. J. Mamm., 54:998–1001.

GREEN, K. M. 1979. Vocalizations, behavior, and ontogeny of the golden lion tamarin *Leontopithecus rosalia rosalia*. Unpubl. Sc.D. dissert., Johns Hopkins Univ., Baltimore, Maryland, 208 pp.

GREEN, S. 1975*a*. Variation of vocal pattern with social situation in the Japanese monkey (*Macaca fuscata*): A field study. Pp. 1–102, *in* Primate behavior: developments in field and laboratory research (L. A. Rosenblum, ed.). Academic Press, New York, 4:1–407.

———. 1975*b*. Dialects in Japanese monkeys, vocal learning and cultural transmission of local-specific vocal behavior. Z. Tierpsychol., 38:304–314.

GREEN, S., AND P. MARLER. 1979. The analysis of animal communication. Pp. 73–158, *in* Handbook of behavioral neurobiology. Vol. 3. Social behavior and communication (P. Marler and J. G. Vandenbergh, eds.). Plenum Press, New York, 321 pp.

GRIFFIN, D. R. 1958. Listening in the dark. Yale Univ. Press, New Haven, 413 pp.

GRIMM, R. J. 1967. Catalogue of sounds of the pigtailed macaque. J. Zool., 152: 361–373.

GUNDERSON, H. L., AND B. R. MAHAN. 1980. Analysis of sonograms of American bison (*Bison bison*). J. Mamm., 61:379–381.

HAFNER, M. S., AND D. J. HAFNER. 1979. Vocalizations of grasshopper mice (genus *Onychomys*). J. Mamm., 60:85–94.

HAFNER, G., ET AL. 1979. Evidence for signature information in the song of the humpback whale. Proc. Conf. Biol. Marine Mammals, 3.

HAGGARD, M. P. 1979. Menstrual fluctuations in hearing and their relevance to hearing disorders. Clin. Otolaryngol., 4:148.

HALL, K. R. L. 1968. Behaviour and ecology of the wild patas monkey, *Erythrocebus patas,* in Uganda. Pp. 32–119, *in* Primates (P. C. Jay, ed.). Holt, Rinehart and Winston, New York, 529 pp.

HALL, W. G. 1979. Feeding and behavioral activation in infant rats. Science, 205: 206–209.

HAMILTON, W. J., AND P. C. ARROWOOD. 1978. Copulatory vocalizations of chacma baboons (*Papio ursinus*), gibbons (*Hylobates hoolock*), and humans. Science, 200:1405–1409.

HAMILTON, W. J., AND K. E. F. WATT. 1970. Refuging. Ann. Rev. Ecol. Syst., 1:1–406.

HAROUTUNIAN, V., AND D. C. RICCIO. 1979. Drug-induced "arousal" and the effectiveness of CS exposure in the reinstatement of memory. Behav. Neural Biol., 26:115-120.

HARRINGTON, F. H., AND L. D. MECH. 1978. Wolf vocalizations. Pp. 109-132, *in* Wolf and man: evolution in parallel (R. L. Hall and H. S. Sharp, eds.). Academic Press, New York, 210 pp.

HASKINS, R. 1977. Effect of kitten vocalizations on maternal behavior. J. Comp. Physiol. Psychol., 91:830-838.

———. 1979. A causal analysis of kitten vocalization: an observational and experimental study. Anim. Behav., 26:726-736.

HAYES, D. P., AND G. D. BOUMA. 1975. Patterns of vocalization and impression formation. Semiotica, 13:112-129.

HEFFNER R., AND H. HEFFNER. 1980. Hearing in the elephant (*Elephas maximus*). Science, 208:518-520.

HINDE, R. A. 1966. Ritualization and social communication in rhesus monkeys. Phil. Trans. Royal Soc. London, 251:285-294.

HINDE, R. A., AND Y. SPENCER-BOOTH. 1971. Effects of brief separation from mother on rhesus monkeys. Science, 173:111-118.

HOOKER, T., AND B. I. HOOKER. 1969. Duetting. Pp. 185-205, *in* Bird vocalizations (R. A. Hinde, ed.). Cambridge Univ. Press, Cambridge, 394 pp.

HOPSON, J. A. 1973. Endothermy, small size, and the origin of mammalian reproduction. Amer. Nat., 107:446-452.

HORWICH, R. H. 1972. The ontogeny of social behavior in the gray squirrel (*Sciurus carolinensis*). Forts. Verhaltenforsch., 8:1-103.

HUMPHREY, T. 1964. Some correlations between the appearance of human fetal reflexes and the development of the nervous system. Prog. Brain Res., 4: 93-135.

———. 1969. Postnatal repetition of human prenatal activity sequences with some suggestions of their neuroanatomical basis. Pp. 43-84, *in* Brain and early behavior (R. J. Robinson, ed.). Academic Press, London, 205 pp.

HUTTERER, V. R. 1976. Observations on the delivery-behavior and postnatal development of pygmy shrews, *Sorex minutus L.* Z. Säugetierk., 41:1-22.

JACOBSON, M. 1974. A plentitude of neurons. Pp. 151-166, *in* Aspects of neurogenesis (G. Gottlieb, ed.). Academic Press, New York, 287 pp.

JERISON, H. J. 1973. Evolution of the brain and intelligence. Academic Press, New York, 482 pp.

JOHANSON, I. G., AND W. G. HALL. 1979. Appetitive learning in 1-day-old rat pups. Science, 205:419-421.

JOLLY, A. 1966. Lemur behavior: a Madagascar field study. Univ. Chicago Press, Chicago, 187 pp.

JURGENS, U. 1979. Vocalization as an emotional indicator. A neuroethological study in the squirrel monkey. Behaviour, 69:88-117.

JURGENS, U., M. HAST, AND R. PRATT. 1978. Effects of laryngeal nerve transection on squirrel monkey calls. J. Comp. Physiol. Psychol., 123:23-29.

KAPLAN, J. N., A. WINSHIP-BALL, and L. SIM. 1978. Maternal discrimination of infant vocalizations in squirrel monkeys. Primates, 19:187-193.

KIBALCHICH, A. A., AND T. Y. LISITZYNA. 1979. Some acoustic signals of the young in the Pacific walruses. Zool. Zhurnal, 58:1247-1249.

KILEY, M. 1972. The vocalizations of ungulates, their causation and function. Z. Tierpsychol., 31:171-222.

KINGDON, J. 1971. East African mammals. An atlas of evolution in Africa. Academic Press, London, 1:1-446.

———. 1974. East African mammals. Insectivores and bats. Academic Press, London, New York 2(A):1-341.

KLATT, D. H. 1973. Discrimination of fundamental frequency contours in synthetic speech: implications for models of pitch perception. J. Acoust. Soc. Amer., 53:8–15.

KLEIMAN, D. G. 1972. Maternal behavior of the green acouchi (*Myoprocta pratti,* Pocock), a South American caviomorph rodent. Behaviour, 43:48–84.

KLEIMAN, D. G., AND J. F. EISENBERG. 1974. Discussion following article by J. F. Eisenberg, The function and motivational basis of hystricomorph vocalizations. Symp. Zool. Soc. London, 4:211–247.

KLEIMAN, D. G., W. B. KARESH, AND P. R. CHU. 1979. Behavioural changes associated with oestrus in the giant panda, *Ailuropoda melanoleuca,* with comments on female proceptive behaviour. Internatl. Zoo Yearb., 19:217–223.

KLEIN, L. L. 1974. Agonistic behavior in Neotropical primates. Pp. 77–122, *in* Primate aggression, territoriality and xenophobia (R. L. Holloway, ed.). Academic Press, New York, 513 pp.

KLINGEL, H. 1968. Soziale organisation und Verhaltensweisen von Hartmann- und Bergzebras (*Equus zebra hartmannae* und *E. z. zebra*). Z. Tierpsychol., 25:76–88.

KLOPFER, P. H., AND K. J. BOSKOFF. 1979. Maternal behavior in prosimians. Pp. 123–156, *in* The study of prosimian behavior (G. A. Doyle and R. D. Martin, eds.). Academic Press, New York, 696 pp.

KOEPPL, J. W., R. S. HOFFMANN, AND C. F. NADLER. 1978. Pattern analysis of acoustical behavior in four species of ground squirrel. J. Mamm., 59:677–696.

KOMISARUK, B. R. 1970. Synchrony between limbic system theta activity and rhythmical behavior in rats. J. Comp. Physiol. Psychol., 70:482–492.

———. 1977. The role of rhythmical brain activity in sensory-motor integration. Pp. 55–90, *in* Progress in psychobiology and physiological psychology (J. M. Sprague and A. N. Epstein, eds.). Academic Press, New York, 7:1–375.

KOMISARUK, B. R., N. T. ADLER, AND J. HUTCHINSON. 1973. Genital sensory field: enlargement by estrogen treatment in female rats. Science, 178: 1295–1298.

KOVACH, J. A., AND A. KLING. 1967. Mechanisms of neonate suckling behavior in the kitten. Anim. Behav., 15:91–101.

KROODSMA, D. E. 1978. Aspect of learning in the ontogeny of bird song: where, from whom, how many, which, and how accurately? Pp. 215–230, *in* The development of behavior: comparative and evolutionary aspects (G. M. Burghardt and M. Bekoff, eds.). Garland Press, New York, 429 pp.

LAYNE, J. N. 1954. The biology of the red squirrel, *Tamiasciurus hudsonicus loquax* (Bangs), in central New York. Ecol. Monogr., 24:227–267.

LEBOEUF, B. J., AND R. S. PETERSON. 1969. Dialects in elephant seals. Science, 166:1654–1656.

LEBOEUF, B. J., AND L. F. PETRINOVICH. 1974. Dialects of northern elephant seals *Mirounga angustirostris*: origin and reliability. Anim. Behav., 22:656–663.

LEGER, D. W., AND D. H. OWINGS. 1978. Responses to alarm calls by California ground squirrels: effects of call structure and maternal status. Behav. Ecol. Sociobiol., 3:177–186.

LEGER, D. W., D. H. OWINGS, AND L. M. BOAL. 1979. Contextual information and differential responses to alarm whistles in California ground squirrels. Z. Tierpsychol., 49:142–155.

LEVITT, H., AND L. RABINER. 1971. Analysis of fundamental frequency contours in speech. J. Acoust. Soc. Amer., 49:569–582.

LEYHAUSEN, P. 1956*a*. Das Verhalten der Katzen. Handb. Zool., 10:1–34.

———. 1956*b*. Verhaltensstudien an Katzen. Parey, Berlin, 232 pp.

LIEBLICH, A. K., ET AL. 1980. Development of the isolation peep in laboratory-bred squirrel monkeys. Anim. Behav., 28:1–9.

MACKINNON, J. 1974. The behavior and ecology of wild orang-utans (*Pongo pygmaeus*). Anim. Behav., 22:3–74.

MACRIDES, F. 1975. Temporal relations between hippocampal slow waves and exploratory sniffing in hamsters. Behav. Biol., 14:295–308.

MACTUTUS, C. F., D. C. RICCIO, AND J. M. FEREK. 1979. Retrograde amnesia for old (reactivated) memory: some anomalous characteristics. Science, 204:1319–1320.

MALINIAK, E., AND J. F. EISENBERG. 1971. Breeding spiny rats (*Proechimys semispinosus*) in captivity. Internatl. Zoo Yearb., 11:93–98.

MARLER, P. 1965. Communication in monkeys and apes. Pp. 544–584, *in* Primate behavior (I. DeVore, ed.). Holt, Rinehart and Winston, New York, 654 pp.

———. 1968. Aggregation and dispersal: two functions in primate communication. Pp. 420–438, *in* Primates (P. C. Jay, ed.). Holt, Rinehart and Winston, New York, 529 pp.

———. 1969. *Colobus guereza*: territoriality and group composition. Science, 163: 93–95.

———. 1972. Vocalizations of East African monkey. II. Black and white colobus. Behaviour, 42:175–197.

———. 1973. A comparison of vocalizations of red-tailed monkeys and blue monkeys, *Cercopithecus ascanius* and *C. mitis,* in Uganda. Z. Tierpsychol., 33: 223–247.

———. 1976*a*. Social organization, communication and graded signals: the chimpanzee and the gorilla. Pp. 239–280, *in* Growing points in ethology (P. P. G. Bateson and R. A. Hinde, eds.). Cambridge Univ. Press, New York, 548 pp.

———. 1976*b*. Ethological theory of vocal learning. Ann. New York Acad. Sci., 280:386–395.

MARLER, P., AND L. HOBBETT. 1975. Individuality in a long-range vocalization of wild chimpanzees. Z. Tierpsychol., 38:97–109.

MARLER, P., AND S. PETERS. 1977. Selective vocal learning in a sparrow. Science, 198:519–521.

MARLER, P., AND T. TENAZA. 1977. Signalling behavior of apes with special reference to vocalization. Pp. 965–1033, *in* How animals communicate (T. A. Sebeok, ed.). Indiana Univ. Press, Bloomington, 1128 pp.

MARSDEN, H. M., AND N. R. HOLLER. 1964. Social behavior in confined populations of the cottontail and the swamp rabbit. Wildl. Monogr. 13:1–39.

MARSHALL, J. T., AND E. R. MARSHALL. 1976. Gibbons and their territorial songs. Science, 193:235–237.

MARSHALL, J. T., B. A. ROSS, AND S. CHANTHAROJVONG. 1972. The species of gibbons in Thailand. J. Mamm., 53:479–486.

MARTEN, K., AND P. MARLER. 1977. Sound transmission and its significance for animal vocalization. I. Temperate habitats. Behav. Ecol. Sociobiol., 2:271–290.

MARTEN, K., D. QUINE, AND P. MARLER. 1977. Sound transmission and its significance for animal vocalization. II. Tropical forest habitats. Behav. Ecol. Sociobiol., 2:291–302.

MARTIN, J. G., AND W. STRANGE. 1968. The perception of hesitation in spontaneous speech. Perception and Psychophysics, 3:427–438.

MARTIN, K. A., AND M. B. FENTON. 1978. A possible defensive function for calls given by bats (*Myotis lucifugus*) arousing from torpor. Canadian J. Zool., 56:1430–1432.

Mason, W. A. 1968. Early social deprivation in the nonhuman primates: implications for human behavior. Pp. 70–101, *in* Environmental influences (D. C. Glass, ed.). Rockefeller Univ. Press, New York, 304 pp.

———. 1979. Ontogeny of social behavior. Pp. 1–28, *in* Handbook of behavioral neurobiology. Vol. 3. Social behavior and communication (P. Marler and J. G. Vandenbergh, eds.). Plenum Press, New York, 441 pp.

Masterton, B., and I. T. Diamond. 1973. Hearing: Central neural mechanisms. Pp. 407–448, *in* Handbook of perception. Vol. 3. Biology of perceptual systems (E. C. Carterette and M. P. Friedman, eds.). Academic Press, New York, 521 pp.

Matsumura, S. 1979. Mother-infant communication in horseshoe bats (*Rhinolophus ferrumequinum nippon*): development of vocalization. J. Mamm., 60:76–84.

———. 1981. Mother-infant communication in horseshoe bat (*Rhinolophus ferrumequinum nippon*): Vocal communication in three-week-old infants. J. Mamm., 62:20–28.

Mech, L. D. 1970. The wolf. Natural History Press, Garden City, 384 pp.

Mitchell, G. 1979. Behavioral sex differences in non-human primates. Van Nostrand, New York, 515 pp.

Moelk, M. 1979. The development of friendly approach behavior in the cat: a study of kitten-mother relations and the cognitive development of the kitten from birth to eight weeks. Adv. Stud. Behav., 10:163–225.

Mohres, F. P. 1966. Communicative character of sonar signals in bats. Pp. 939–945, *in* Animal sonar systems (R. G. Busnel, ed.). Lab. Physio. Acoust., Jouy-en-Josas, 2:714–1233.

Money, J. 1980. Love and love sickness: the science of sex, gender difference and pair-bonding. Johns Hopkins Press, Baltimore, 256 pp.

Money, J., and H. Musaph (eds.). 1977. Handbook of sexology. Elsevier/North Holland Biomedical Press, New York, 1402 pp.

Morgan, P. D., et al. 1975. The roles played by the senses of the ewe in the location and recognition of lambs. Applied Anim. Ethol., 1:139–150.

Morton, E. S. 1975. Ecological sources of selection on avian sounds. Amer. Nat., 109:17–34.

———. 1977. On the occurrence and significance of motivation-structural rules in some bird and mammal sounds. Amer. Nat., 111:855–869.

Moynihan, M. 1964. Some behavior patterns of platyrrhine monkeys. I. The night monkey (*Aotus trivirgatus*). Smithsonian Misc. Coll., 146:1–84.

———. 1970. Some behavior patterns of platyrrhine monkeys. II. *Saguinus geoffroyi* and some other tamarins. Smithsonian Contrib. Zool., 28:1–77.

Murie, A. 1944. The wolves of Mount McKinley. National Park Service Fauna Ser., U.S. Govt. Printing Office, Washington, D.C., 5:1–238.

Nashida, T. 1970. Social behavior and relationships among wild chimpanzees of the Mahali mountains. Primates, 11:47–87.

Negus, V. E. 1949. The comparative anatomy and physiology of the larynx. Hafner, New York, 230 pp.

Nelson, J. E. 1964. Vocal communication in Australian flying foxes (Pteropodidae: Megachiroptera). Z. Tierpsychol., 21:857–870.

———. 1965. Behaviour of Australian Pteropodidae (Megachiroptera). Anim. Behav., 13:544–557.

Neuweiler, G. 1980. Auditory processing of echoes: peripheral processing. Pp.

519–548, *in* Animal sonar systems (R. G. Busnel and J. F. Fish, eds.). Plenum Press, New York, 1135 pp.

NEYMAN, P. F. 1977. Some aspects of the biology of free-ranging cotton-top tamarins (*Saguinus o. oedipus*) and conservation status of the species. Pp. 39–71, *in* The biology and conservation of the Callitrichidae (D. G. Kleiman, ed.). Smithsonian Inst. Press, Washington, D.C., 354 pp.

NOTTEBOHM, F. 1969. The "critical period" for song learning in birds. Ibis, 111: 386–387.

NOVICK, A. 1977. Acoustic orientation. Pp. 73–287, *in* Biology of bats (W. A. Wimsatt, ed.). Academic Press, New York, 3:1–651.

NYBY, J., G. DIZINNO, AND G. WHITNEY. 1977*a*. Sexual dimorphism in ultrasonic vocalizations of mice (*Mus musculus*): gonadal hormone regulation. J. Comp. Physiol. Psychol., 91:1424–1431.

NYBY, J., ET AL. 1977*b*. Pheromonal regulation of male mouse ultrasonic courtship (*Mus musculus*). Anim. Behav., 25:333–341.

OKON, E. K. 1970. The ultrasonic responses of albino mouse pups to tactile stimuli. J. Zool., 162:485–492.

———. 1972. Factors affecting ultrasound production in infant rodents. J. Zool., 168:139–148.

OSWALT, G. L., AND G. W. MEIER. 1975. Olfactory, thermal and tactual influences on infantile ultrasonic vocalizations in rats. Dev. Psychobiol., 8:129–135.

OWEN-SMITH, N. 1977. On territoriality in ungulates and an evolutionary model. Quart. Rev. Biol., 52:1–38.

PEDERSEN, P., AND E. M. BLASS. In press. Olfactory control over suckling. *In* Development of sensory systems (R. Aslin and J. Alberts, eds.). Erlbaum Publ., New Jersey.

PETERSON, R. S. 1965. Behavior of the northern fur seal. Unpubl. Sc.D. dissert., Johns Hopkins Univ., Baltimore, Maryland, 214 pp.

PETERSON, R. S., AND G. A. BARTHOLOMEW. 1967. The natural history and behavior of the California sea lion. Spec. Publ., Amer. Soc. Mamm., 1:1–67.

PETRINOVICH, L. 1974. Individual recognition of pup vocalization by northern elephant seal mothers. Z. Tierpsychol., 34:308–312.

PETROVICH, S. B., AND E. H. HESS. 1978. An introduction to animal communication. Pp. 17–53, *in* Nonverbal behavior and communication (A. W. Siegman and S. Feldstrin, eds.). Lawrence Erlbaum Assoc., Hillsdale, 400 pp.

PETTER, J. J. 1962. Recherches sur l'écologie et l'éthologie des lémuriens malagaches. Mém. Mus. Hist. Nat. (nouv. sér.), Sér. A: Zool., 27:1–146.

PETTER, J. J., AND P. CHARLES-DOMINIQUE. 1979. Vocal communication in prosimians. Pp. 247–305, *in* The study of prosimian behavior (G. A. Doyle and R. D. Martin, eds.). Academic Press, New York, 696 pp.

POLLAK, G. D. 1980. Organizational and encoding features of single neurons in the inferior colliculus of bats. Pp. 549–587, *in* Animal sonar systems (R- G. Busnel and J. F. Fish, eds.). Plenum Press, New York, 1135 pp.

POLLOCK, J. I. 1975. Field observations on *Indri indri*: a preliminary report. Pp. 287–311, *in* Lemur biology (I. Tattersall and R. W. Sussman, eds.). Plenum Press, New York, 365 pp.

PORTER, F. L. 1979. Social behavior in the leaf-nosed bat, *Carollia perspicillata*. II. Social communication. Z. Tierpsychol., 50:1–8.

PRAKASH, I. 1959. Hypertrophy of the bullae tympanicae in the desert mammals. Science and Culture, 24:580–582.

PRUITT, C. H., AND G. M. BURGHARDT. 1977. Communication in terrestrial carnivores: Mustelidae, Procyonidae, and Ursidae. Pp. 767–793, *in* How animals communicate (T. A. Sebeok, ed.). Indiana Univ. Press, Bloomington, 1128 pp.

PYE, J. D. 1979. Why ultrasound? Endeavor, 3:57–62.

RALLS, K. 1971. Mammalian scent marking. Science, 171:443–449.

RALLS, K., C. BARASCH, AND K. MINKOWSKI. 1975. Behavior of captive mouse deer, *Tragulus napu*. Z. Tierpsychol., 37:356–378.

RATHBUN, G. B. 1979. The social structure and ecology of elephant-shrews. Adv. Ethol., Verlag Paul Parey, Berlin, 20:1–74.

REMMERS, J. E., AND H. GAUTIER. 1972. Neural and mechanical mechanisms of feline purring. Resp. Physiol., 16:351–361.

REYNOLDS, H. C. 1952. Studies on reproduction in the opossum (*Didelphis virginiana virginiana*). Univ. California Publ. Zool., 52:223–275.

REYNOLDS, V., AND F. REYNOLDS. 1965. Chimpanzees of the Budongo Forest. Pp. 368–424, *in* Primate behavior (I. DeVore, ed.). Holt, Rinehart and Winston, New York, 654 pp.

RICHARD, A. 1970. A comparative study of the activity patterns and behavior of *Alouatta villosa* and *Ateles geoffroyi*. Folia Primatol., 12:241–263.

ROBINSON, J. G. 1977. Vocal regulation of spacing in the titi monkey, *Callicebus moloch*. Unpubl. Sc.D. dissert., Univ. North Carolina, 195 pp.

———. 1979. Vocal regulation of use of space by groups of titi monkeys *Callicebus moloch*. Behav. Ecol. Sociobiol., 5:1–15.

ROEDER, K. D. 1967. Nerve cells and insect behavior. Harvard Univ. Press, Cambridge, Massachusetts, 238 pp.

ROVEE-COLLIER, C. K., ET AL. 1980. Reactivation of infant memory. Science, 208: 1159–1161.

ROWELL, T. E., AND R. A. HINDE. 1962. Vocal communications by the rhesus monkey (*Macaca mulatta*). Proc. Zool. Soc. London, 138:279–294.

RUDY, J. W., AND M. D. CHEATLE. 1977. Odor-aversion learning in neonatal rats. Science, 198:845–846.

RUFFER, D. G. 1968. Agonistic behavior of the northern grasshopper mouse (*Onychomys leucogaster fuscogriseus*). J. Mamm., 49:481–487.

SABY, J. S., AND W. L. NYBORG. 1946. Ray computation for non-uniform fields. J. Acoust. Soc. Amer., 18:316–322.

SALES, G., AND D. PYE. 1974. Ultrasonic communication by animals. Chapman and Hall, London, 275 pp.

SALES, G. D., AND J. C. SMITH. 1978. Comparative studies of the ultrasonic calls of infant murid rodents. Devel. Psychobiol., 11:595–619.

SCHALLER, G. B. 1967. The deer and the tiger. Univ. Chicago Press, Chicago, 370 pp.

———. 1972*a*. The deer and the tiger. Second ed. Univ. Chicago Press, Chicago, 369 pp.

———. 1972*b*. The Serengeti lion. Univ. Chicago Press, Chicago, 480 pp.

SCHOTT, D. 1975. Quantitative analysis of the vocal repertoire of squirrel monkeys (*Saimiri sciureus*). Z. Tierpsychol., 38:225–250.

SCHULLER, G. AND G. D. POLLAK. 1979. Disproportionate frequency presentation in the inferior colliculus of the Doppler-compensating horseshoe bats: evidence for an acoustic fovea. J. Comp. Physiol., 132:47–54.

SEBEOK, T. A. 1977. How animals communicate. Indiana Univ. Press, Bloomington, 1128 pp.

SHALTER, M. D., J. C. FENTRESS, AND G. W. YOUNG. 1977. Determinants of response of wolf pups to auditory signals. Behavior, 60:98–114.

SHERMAN, P. W. 1977. Nepotism and the evolution of alarm calls. Science, 197: 1246–1253.

SHIELDS, W. M. 1980. Ground squirrel alarm calls: nepotism or parental care. Amer. Nat., 116:599–603.

SHIPLEY, C., M. HINES, AND J. S. BUCHWALD. In press. Individual differences in threat calls of northern elephant seal bulls. Anim. Behav.

SIGNORET, J. P., F. DU MESNIL DU BUISSON, AND R-G. BUSNEL. 1960. Role d'un signal acoustique du verrat dans le comportement reactionnel de la truie en oestrus. C. R. Sci. Acad. Sci., 250:1355–1357.

SIMMONS, J. A., ET AL. 1979. Apparatus for research on animal ultrasonic signals. Life Sci. Misc. Publ., Royal Ontario Mus., 31 pp.

SMITH, J. C. 1972. Sound production by infant *Peromyscus maniculatus.* (Rodentia: Myomorpha). J. Zool., 168:369–379.

———. 1975. Sound communication in rodents. Pp. 317–330, *in* Sound reception in mammals (R. J. Bench, A. Pye, and J. D. Pye, eds.). Symp. Zool. Soc. London, 37:1–357.

———. 1977. Factors affecting the transmission of rodent ultrasounds in natural environments. Amer. Zool., 19:431–441.

SMITH, S. F. 1978. Alarm calls, their origin and use in *Eutamias sonomae.* J. Mamm., 59:888–893.

SMITH, W. J. 1958. Natural history of the prairie dog in Kansas. Misc. Publ. Mus. Nat. Hist., Univ. Kansas, 16:1–36.

———. 1977. The behavior of communication: an ethological approach. Harvard University Press, Cambridge, 497 pp.

———, ET AL. 1976. The jump-yip displays of the black-tailed prairie dog *Cynomys ludovicianus.* Anim. Behav., 24:609–621.

SOLMSEN, E. AND R. APFELBACH. 1979. Behavioral effect of high frequency components in young ferret distress calls. Z. Tierpsychol., 50:337–347.

SOMERS, P. 1973. Dialects in southern rocky mountain pikas, *Ochotona princeps* (Lagomorpha). Anim. Behav., 21:124–137.

SOUTHWICK, C. H. 1963. Primate social behavior. Van Nostrand, Princeton, 191 pp.

SPEAR, N. E. 1978. The processing of memories: forgetting and retention. Lawrence Erlbaum Associates, Hillsdale, New Jersey, 553 pp.

STARK, R. E. 1979. Prespeech segmental feature development. Pp. 15–32, *in* Language acquisition: studies in first language development (P. Fletcher and M. Garman, eds.). Cambridge Univ. Press, Cambridge, 507 pp.

STEVENS, S. S., AND F. WARSHOFSKY. 1971. Sound and hearing. Life Science Library, Time-Life Books, New York, 194 pp.

STIRLING, I., AND R. M. WARNEKE. Implications of a comparison of the airborne vocalizations and some aspects of the behavior of the two Australian fur seals, *Arctocephalus* spp., on the evolution and present taxonomy of the genus. Australian J. Zool., 19:227–241.

STINE, C. J. 1976. Behavioral ontogeny and mother-offspring interaction of the house shrew, *Suncus murinus.* Unpubl. Sc.D. thesis, Johns Hopkins Univ. Press, Baltimore, Maryland, 176 pp.

STOLOFF, M. L. 1980. From suckling to feeding: transitions in ingestive behavior in albino rats. Unpubl. Ph.D. dissert., Johns Hopkins Univ. Press, Baltimore, Maryland, 165 pp.

STOLOFF, M. L., ET AL. In press. The role of experience in suckling maintenance in albino rats. J. Comp. Physiol. Psychol.

STRUHSAKER, T. 1967. Auditory communication among vervet monkeys (*Cercopithecus aethiops*). Pp. 281–324, *in* Social communication among primates (S. Altmann, ed.). Univ. Chicago Press, Chicago, 392 pp.

———. 1975. The red colobus monkey. Univ. Chicago Press, Chicago, 311 pp.

SUGA, N., AND W. E. O'NEILL. 1980. Auditory processing of echoes: Representation of acoustic information from the environment in the bat cerebral cortex. Pp. 589–611, *in* Animal sonar systems (R-G. Busnel and J. F. Fish, eds.). Plenum Press, New York, 1135 pp.

SUSSMAN, R. W., AND A. RICHARD. 1974. The role of aggression among diurnal prosimians. Pp. 49–76, *in* Primate aggression, territoriality and xenophobia (R. L. Holloway, ed.). Academic Press, New York, 513 pp.

TEMBROCK, G. 1958. Lautentwicklung beim Fuchs: sichtbar gemacht. Umschau, 18:566–568.

———. 1965. Untersuchungen zur intraspezifischen Variabilität von Lautäusserungen bei Säugetieren. Z. Säugetierk., 30:257–273.

———. 1968. Land mammals. Pp. 338–404, *in* How animals communicate (T. A. Sebeok, ed.). Indiana Univ. Press, Bloomington, 1128 pp.

THEBERGE, J. B., AND J. B. FALLS. 1967. Howling as a means of communication in timber wolves. Amer. Zool., 7:331–338.

THRASHER, J. D., ET AL. 1971. The colony management of *Marmosa mitis,* the pouchless opossum. Lab. Anim. Sci., 21:526–536.

VANDERWOLF, C. H. 1975. Neocortical and hippocampal activation in relation to behavior: effects of atropine, eserine, phenothiazines, and amphetamines. J. Comp. Physiol. Psychol., 88:300–323.

VERNON, M. D. 1955. The functions of schemata in perceiving. Psychol. Rev., 62: 180–192.

VOGEL, C. 1973. Acoustical communication among free-ranging common Indian langurs (*Presbytis entellus*) in two different habitats of North India. Amer. J. Phys. Anthropol., 38:469–479.

WALKER, E. P., ET AL. 1964. Mammals of the world. The Johns Hopkins Press, Baltimore, 1500 pp.

WALTHER, F. 1977. Artiodactyla. Pp. 655–714, *in* How animals communicate (T. A. Sebeok, ed.). Indiana Univ. Press, Bloomington, 1128 pp.

WASER, P. M. 1975. Experimental playbacks show vocal mediation of intergroup avoidance in a forest monkey. Nature, 255:56–58.

———. 1976. *Cercocebus albigena*: site attachment, avoidance, and intergroup spacing. Amer. Nat., 110:911–935.

———. 1977*a*. Individual recognition, intragroup cohesion and intergroup spacing: evidence from sound playback to forest monkeys. Behaviour 60:28–74.

———. 1977*b*. Sound localization by monkeys: a field experiment. Behav. Ecol. Sociobiol., 2:427–431.

WASER, P. M., AND M. S. WASER. 1977. Experimental studies of primate vocalization: specializations for long-distance propagation. Z. Tierpsychol., 43: 239–263.

WASER, P. M., AND R. H. WILEY. 1979. Mechanisms and evolution of spacing in animals. Pp. 159–223, *in* Handbook of behavioral neurology. Vol. 3. Social behavior and communication (P. Marler and J. G. Vandenbergh, eds.). Plenum Press, New York, 409 pp.

WEBSTER, D. B., AND M. WEBSTER. 1980. Morphological adaptations of the ear in the rodent family Heteromyidae. Amer. Zool., 20:247–254.

WELKER, W. I. 1964. Analysis of sniffing of the albino rat. Behavior, 22:223–244.

WEMMER, C. M. 1977. Comparative ethology of the large-spotted genet (*Genetta tigrina*) and some related viverrids. Smithsonian Contrib. Zool., 239:1–93.

WEMMER, C. M., M. VON EBERS, AND K. SCOW. 1976. An analysis of the chuffing vocalization in the polar bear (*Ursus maritimus*). J. Zool., 180:425–439.

WHITNEY, G. 1970. Ontogeny of sonic vocalizations of laboratory mice. Behav. Genetics, 1:269–273.

WHITNEY, G., AND J. NYBB. 1979. Cues that elicit ultrasounds from adult male mice. Amer. Zool., 19:457–463.

WHITNEY, G., ET AL. 1973. Ultrasonic emissions: do they facilitate courtship of mice? J. Comp. Physiol. Psychol., 84:445–452.

WHITTEMORE, C. T., AND D. FRASER. 1974. The nursing and suckling behaviour of pigs. II. Vocalization of the sow in relation to sucking behaviour and milk ejection. British Vet. J., 130:346–356.

WHITTLESTONE, W. G. 1953. The milk-ejection response of the sow to standard doses of oxytocic hormone. J. Dairy Res., 20:13–15.

WICKLER, W. 1967. Socio-sexual signals and their intra-specific imitation among primates. Pp. 69–147, *in* Primate ethology (D. Morris, ed.). Aldine Publishing Co., Chicago, 374 pp.

———. 1980. Vocal duetting and the pair-bond: I. Coyness and partner commitment, a hypothesis. Z. Tierpsychol., 52:201–209.

WICKLER, W., AND U. SEIBT. 1976. Field studies on the African fruit bat *Epomophorus wahlbergi* (Sundevall) with special reference to male calling. Z. Tierpsychol., 40:345–376.

WILEY, R. H. 1973. Territoriality and non-random mating in sage grouse, *Centrocercus urophasianus*. Anim. Behav. Monogr., 6:87–169.

WILEY, R. H., AND D. G. RICHARDS. 1978. Physical constraints on acoustic communication in the atmosphere: implications for the evolution of animal vocalizations. Behav. Ecol. Sociobiol., 3:69–94.

WINTER, P. 1969. Dialects in squirrel monkeys: vocalizations of the roman-arch type. Folia Primatol., 10:216–229.

WINTER, P., AND H. H. FUNKENSTEIN. 1971. The auditory cortex of the squirrel monkey: neuronal discharge patterns to auditory stimuli. Proc. Third Internatl. Congr. Primatol., 2:24–28.

WINTER, P., D. PLOOG AND J. LATTA. 1966. Vocal repertoire of the squirrel monkey (*Saimiri sciureus*), its analysis and significance. Exp. Brain Res., 1:359–384.

WINTER, P., ET AL. 1973. Ontogeny of squirrel monkey calls under normal conditions and under acoustic isolation. Behaviour, 47:230–239.

WOLFF, P. H. 1968. Suckling patterns of infant mammals. Brain Behav. Evol., 1: 354–367.

WOLLBERG, Z., AND J. D. NEWMAN. 1972. Auditory cortex of the squirrel monkey: response patterns of single cells to species-specific vocalizations. Science, 175:212–214.

WOOLF, N. K. 1973. The ontogeny of bat sonar sounds: with special emphasis on sensory deprivation. Unpubl. Sc.D. thesis, Johns Hopkins Univ. Press, Baltimore, Maryland, 140 pp.

ZIPPELIUS, H-M. 1974. Ultraschall-laute nestjunger mause. Behaviour, 49:197–204.

ZIPPELIUS, H-M., AND W. M. SCHLEIDT. 1956. Ultraschall-laute bei jungen mausen. Naturwissenschaften, 43:502.

ZOLOTH, S. R., ET AL. 1979. Species-specific perceptual processing of vocal sounds by monkeys. Science, 204:870–873.

MONKEY CALLS: HOW ARE THEY PERCEIVED AND WHAT DO THEY MEAN?

PETER MARLER

Abstract

EVIDENCE is presented that monkeys use a special perceptual mode for processing vocal signals of their species as compared with sounds that lack natural meanings for them. Monkey calls recorded in the field were employed as stimuli in perceptual learning tests in the laboratory. Animals learned faster in discrimination tests when the task required a classification of conspecific calls that mapped readily onto patterns of natural usage. They learned more slowly when there was a mismatch with patterns of natural usage or in a complex discrimination with calls of another species. In effect, monkeys behave as though species-specific features of calls possess a natural salience for them, even in artificial laboratory situations. With regard to the semantics of monkey calls, field experiments with playback of alarm calls specific to different predators suggest that "eagle calls" or "snake calls" actually represent different things to the monkeys. Monkey calls may actually function symbolically and not just as manifestations of emotional arousal.

Introduction

Research on animal communication has made almost no progress in understanding the semantics of natural signalling behavior because it is an exceedingly difficult problem in both practical and theoretical terms. Despite the much vaunted intuitive ability of ethologists to empathize with their subjects, even they are prone to slip into a subtle anthropomorphism when thinking about animal semantics (Griffin, 1976). There is an irrestible tendency to use language as a model, either for comparison or contrast. This would be more appropriate if we really understood human semantics and the processes by which language acquires meaning in the course of our own early development. In many respects our picture of how words acquire meaning in human infancy is hardly any clearer than our understanding of the meanings of signals for animals, and the ways in which they develop. In both human and animal studies, for example, there are assertions of the overwhelming importance of contextual cues in understanding meaning (Smith, 1977), but precisely how the context influences meaning in particular cases has hardly been explored.

Evidence for Special Processing

Aside from questions about what animal signals mean, there is even uncertainty about how they are perceived. In particular, there is debate about whether there are special modes of perceptual processing for signals distinct from those employed for stimuli in general. The issue of special processing of signals is really a new form of the classic ethological question about innate release mechanisms. Many young animals are innately responsive to particular, simple "sign stimuli," usually emanating from other species members and serving as social signals for communication (Tinbergen, 1951). This has often been cited as evidence that animals are instinctual machines. I believe that this view is mistaken, and that many examples of innate responsiveness in young animals must be reinterpreted as indicating genetic guidelines for learning how best to process biologically important stimuli, especially stimuli relating to communication (Marler et al., 1980). It is in this sense that I believe special processing of monkey calls may occur.

In developing sensitivity to the many nuances of meaning in animal communication, and in our own speech behavior, there would be considerable benefit in providing genetic instructions for accomplishing this difficult task quickly and efficiently. Further advantages accrue if species members follow similar courses of learning in unravelling the many levels of signal complexity so that all individuals employ similar rules, and improve their chances of communicating without misunderstanding. We have obtained what is, I believe, the first evidence of such species-specific rule-sharing in the processing of vocal signals by monkeys.

Green (1975), in a detailed study of sounds used by the Japanese macaque (*Macaca fuscata*) in social communication, found that the fine acoustic details of monkey calls were much more important than anyone had guessed. Calls differing in quite subtle ways were given in distinct situations, at least potentially encoding different kinds of information for another listening monkey. The features involved were subtle enough for us to question whether they could even be detected by the monkeys themselves, and this skepticism led to a collaboration between Steven Green and others of us at The Rockefeller University, and a group of psychologists at the University of Michigan in the laboratory of William Stebbins, a leading student of hearing in non-human primates (Beecher et al., 1976; Beecher et al., 1979; Zoloth et al., 1979; Zoloth and Green, 1979). Stebbin's team of researchers trained Japanese macaques in the laboratory in a sophisticated test of their ability to discriminate between subtly different sounds.

Attention focused on a group of sounds known as "coos." Green had suggested that the position of a frequency inflection in the call, sometimes occurring early, sometimes late, seemed to encode meaningful information for the monkeys. Good quality field recordings from Green's work were presented to the monkeys in a test situation. One of the calls, perhaps with an early peak, would be presented as a repeating background stimulus, with a late-peak sound occasionally being inserted in the series. The monkey was trained to respond when it heard this different pattern of sound by lifting its hand from a lever; it received a food reward for a correct response. The monkeys quickly learned the task. Once trained with first examples of a "smooth early high," and a "smooth late high," they quickly generalized to other examples of these calls, recorded from different individuals on different occasions.

In addition, these calls varied in other features, such as tonal quality and pitch, much as human voices do. In particular, the fundamental frequency of the sounds also varied quite strikingly, just as the voice pitch of different human speakers varies according to age and gender (Green, 1981).

The experiment was designed to determine whether there is special processing of Japanese monkey calls, that is, whether a Japanese macaque is more attentive to variations in peak position than to variations in pitch. We expected any monkey to be sensitive to the pitch of a sound, and ready to employ this feature in its perceptual analysis of virtually any important sounds from the environment. By contrast with sound pitch, the feature of frequency-peak position is a more obscure and subtle cue, which Green's field research had suggested might in some communicative situations be a more reliable cue to meaning to the Japanese macaque than call pitch.

We added a comparative dimension to the experiment by including other species. Monkeys with a different vocal system might well lack any special predisposition to focus on this odd feature of peak position unlike the Japanese monkey. In this event, sound pitch might be more important than peak position as a basis for general perceptual processing of sounds. Thus, we designed the following experiment, knowing that the monkeys could discriminate both peak-position variation and variations in the pitch of the fundamental.

A set of coo calls from field recordings was arranged to vary both in fundamental frequency and in position of the frequency peak. In one test, calls were arranged in two groups according to whether the peak was early or late, with the pitch of the fundamental as an irrelevant, random variable. We then rearranged the same set of calls according to whether the pitch of the fundamental was high or low, with the position of the frequency peak as the random, irrelevant variable. Both Japanese macaques and several other monkey species were tested for the discriminability of these groups of stimuli. In particular, we took careful account of the rates at which each monkey became fully proficient in the different discriminations. The results provided evidence of special processing in the Japanese macaque.

The rate at which a Japanese macaque learned to generalize from the first two examples of an early-peak and a late-peak coo call to other exemplars of the same class is presented in the top left of Fig. 1. These are cumulative plots of the number of trials necessary to achieve a certain level of performance so that a gradual slope means rapid learning. Clearly this Japanese macaque had no trouble in learning quickly and accurately to classify a variety of coo calls on the basis of the peak position, while ignoring pitch variations.

When the Japanese monkey was asked to classify them, not on the basis of peak position but rather on the basis of the pitch of the fundamental, he accomplished the task, but it took more trials than in the former case. The monkeys seemed more reluctant to classify them on this basis, as though peak position is a more salient or

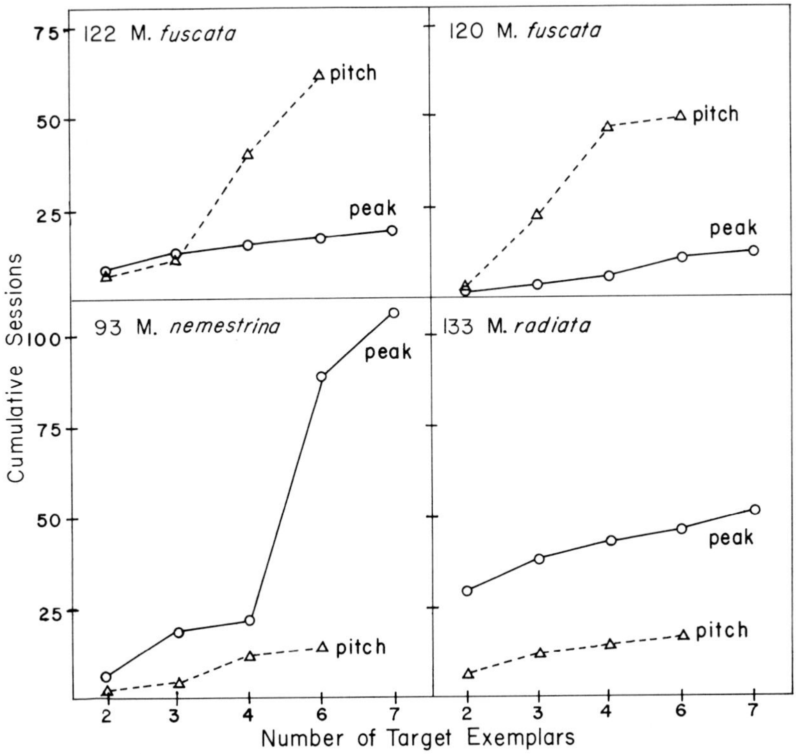

FIG. 1. Rates at which Japanese macaque (*Macaca fuscata*) and pigtail and bonnet macaques (*M. nemestrina* and *M. radiata*) learned to classify different types of field-recorded "coo" calls of Japanese monkeys. They were required to do this on the basis either of position of a frequency inflection ("peak" task, with early or late peaks) or frequency of the fundamental ("pitch" task, with high- or low-pitched calls). Cumulative learning curves from the "pitch" and "peak" discrimination tasks are shown for four monkeys. Each point represents the cumulative number of sessions necessary to meet the performance criteria and to advance beyond that stage of the problem. The total number of calls of one type discriminated at each stage is shown; they were discriminated from an equal number of the other call type at each stage. The subjects on the left were tested on the pitch task first, those on the right, on the peak task first. (From Zoloth et al., 1979.)

potent feature of the vocal sounds of their species. Another Japanese macaque with the same two tasks but in a reverse order, gave essentially the same result (Fig. 1, upper right). In repetitions with other Japanese macaques, the outcome has been consistent. They are more ready to classify coo calls on the basis of frequency-peak position than on the basis of pitch, though they can do both.

We then presented the same problems to other monkey species, in which we did not expect particular predispositions for perceptual processing of these calls. Remember that these were calls of an alien species. The results for a pigtail macaque (*Macaca nemestrina*) were almost mirror images for those obtained with Japanese macaques (Fig. 1, bottom left). The fastest solution to the problem was achieved when the monkey was required to classify the calls on the basis of the pitch of the fundamental rather than on the basis of frequency peak position. This result has been replicated for the bonnet macaque (*Macaca radiata*) (Fig. 1), and also for the vervet monkey (*Cercopithecus aethiops*). It is a predicted outcome if the fundamental frequency of a tonal sound is a generally salient feature, likely to be employed when monkeys are discriminating between calls of another species.

One basis for invoking special processing for human speech is the preferential processing of speech sounds by the left hemisphere of the brain. This preference is most readily shown by presenting speech and non-speech sounds in competition to the two ears, revealing a right-ear, left-hemisphere advantage for detecting sounds of speech. In our monkey experiments, the calls were played to the animals through headphones, and we presented sounds to one ear at a time, sometimes the right, sometimes the left. As a result, it was later possible to compare the numbers of errors made with right-ear and left-ear presentations. In this way Petersen et al. (1978) made the remarkable discovery that Japanese macaques show a right-ear advantage for the processing of their sounds with significantly fewer errors in right-ear presentations than in those to the left. This seems to be another indication of special processing of conspecific vocal sounds by Japanese macaques because the control species showed no asymmetry in their perceptual processing of Japanese macaque coo calls, except inexplicably for one vervet monkey. We believe this to be the first indication of an animal analogy with the human hemispheric dominance involved in the perceptual processing of communicative sounds.

The Semantics of Alarm Calling

Although important, the issue of special brain mechanisms in signal processing tells us nothing about what different calls mean.

One would like to know what kind of mental images they call forth, and how monkeys act upon the assessments (Green and Marler, 1979) that the brain makes of their significance. Some years ago, Struhsaker (1967) discovered a remarkable array of alarm calls in African vervet monkeys that seemed to represent different classes of predators to the monkeys.

The investigation of the semantic content of any signal is enormously complicated, but it is at least simplified if the signal has a clear, distinct external referent which the signal might symbolize. It is rare to find animal signals with such clear differentiated referents as these vervet calls, and the range of different calls for different predators seemed to offer a unique opportunity to see whether anything approaching symbolism is involved.

The problem was recently addressed again by Cheney and Seyfarth (Seyfarth and Cheney, 1981; Seyfarth et al., 1980*a*, 1980*b*). They determined to put the issue to a critical test, by conducting playback experiments with recorded calls. Because some researchers give overwhelming importance to *context* as a factor in meaning, the playbacks were conducted in a variety of natural situations.

First they repeated Struhsaker's (1967) descriptive analyses in the field confirming the array of quite distinct, discrete calls, three of which seemed to be given to different classes of predators, at least by adult monkeys. One was given to ground predators, most typically in response to leopards (*Panthera pardus*). Another was given to two species of eagle that prey on the monkeys. A third was given to venomous snakes and to certain large non-venomous species, such as python, that may prey on vervets. Again each call type seemed to evoke responses from other monkeys appropriate to the situation—rushing into dense cover and out of treetops in response to the eagle alarm call, rushing out of ground cover and up into trees in response to the leopard call, and approaching and mobbing in response to snake alarm calls. Without experiments, however, one could not be sure that responding animals did not detect the predator directly, or alternatively, indirectly, by looking at the vocalizer, following its line of view and after locating and identifying the cause of danger, reacting appropriately. To ensure a full record of the monkeys' reactions, playbacks were conducted on camera. Subsequent film analyses showed clearly and unequivocally that the animals do indeed respond distinctly to the three classes of calls. Furthermore, the crucial differences in response persisted in the face of radical changes in context such as when responding animals were on the ground or up in a tree. It is in this sense that each call type may be viewed as representing or perhaps even symbolizing a particular class of predator.

When we speak of symbolic semantic representation, there is not only a requirement for evidence of representation but also a presumption of particular, mental connotations developed in response to variable referents with shared meaning. Griffin (1976) was surely correct in insisting that such mental constructs are no less appropriate with animals than with humans. The problem is how to characterize the mental processes that are involved. Are *cognitive procedures* playing a role, or is it an *emotion* that mediates between the referent and the signal? I believe that various lines of evidence point to a cognitive interpretation, though with emotion or arousal unquestionably involved.

First, you have to consider how the monkeys know which type of predator is being encountered, and how that taxonomic or categorical judgment comes to be linked with the production of the appropriate alarm calls? The first hint of the developmental basis of this linkage arises from another discovery by Seyfarth and Cheney (1981). Whereas adult vervet monkeys rarely give the three alarm calls to anything other than leopards, martial eagles, or pythons, young vervets often seem to make mistakes. In infant monkeys, a snake alarm may be given to a mouse, an eagle alarm may be given to a harmless dove, a leopard alarm to an impala coming to graze under the tree in which the monkeys are feeding.

When the data on such errors are assembled they create an interesting picture (Fig. 2). Infant monkeys do indeed give these alarm calls to a much wider array of other animal species than do adults, suggesting strongly that specific responsiveness must be learned. As infants mature, the relation between referents and alarm calls gradually becomes more highly focused. Beginning with a large and imprecisely-definable set, each call finally comes to focus on particular predator species—those which are major, real-life representatives of that class at Amboseli in Kenya where the studies were conducted.

The kinds of experience guiding the sharpening of the specificity of the relationship between referent and signal are unknown. Perhaps there is something equivalent to tutelage by adults, or maybe maturation is involved. However, while infants display confusion

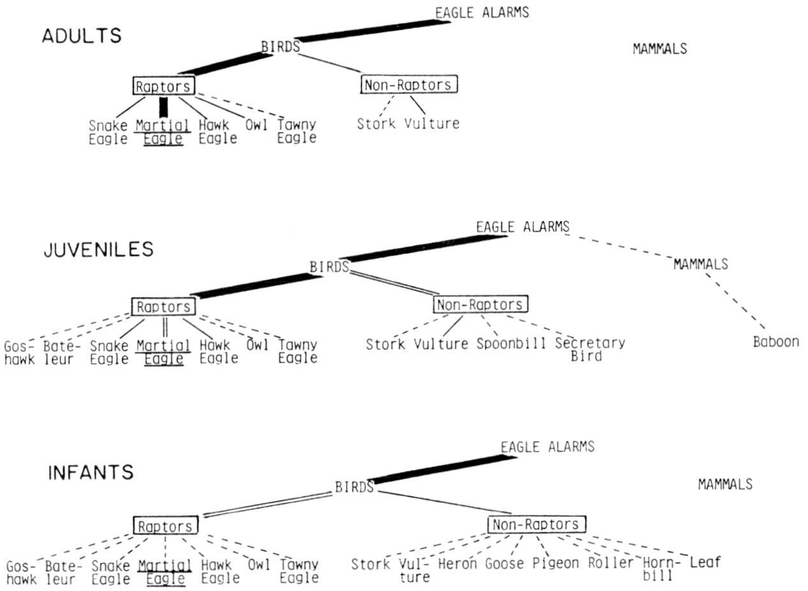

FIG. 2. Stimuli for "eagle alarm" calls in wild vervet monkeys. Results for adult, juvenile and infant monkeys are given separately. Only the species connected by lines were effective. The width of the lines provides a rough index of the relative frequency of elicitation of alarm calling. Martial eagles were the major referent for eagle alarm calling in adults. (From Seyfarth et al., 1980.)

with regard to which "name" to use for which predator, they do not behave as though they were beginning with a *tabula rasa.*

The many different stimulus situations that evoke eagle calling in infants in fact have something in common—an object moving above them in free space, a feature even of a floating leaf. Correspondingly, they do not give eagle calls to a leopard, though a leopard call may be given by an infant to lions (*Panthera leo*), hyenas (*Crocuta crocuta*), or even an antelope. Similarly there is a certain specificity in the large class of stimuli evoking snake calls so that the young are behaving as though they already possess a set of rules for the perceptual analysis and classification of predatory animals. Although the rules are sufficient to divide them into three lumped classes, they are still inadequate for the diagnostic classification of different predator species. While the developmental work remains to be done, the general findings are nevertheless consonant with those that ethologists have discovered in other perceptual studies. More and more cases are being discovered of unspecific, generalized innate release mechanisms that set the path of perceptual development of the young on certain species-specific trajectories. These are shared with other members of the species, to be subject to adjustment and refinement in the face of individual experience (Marler et al., 1980).

So the monkeys behave as though they start with generalized responsiveness to, say, moving aerial stimuli, which becomes sharpened from experience to a very small subset (for example, just two eagle species out of many). This is reminiscent of the process by which word meaning becomes more sharply focused in the course of semantic development in children. One might expect the monkeys to perform a perceptual analysis of stimulus objects and then classify them as a consequence of what the experiences mean to them. However, even if there is, as a first step, cognitive processing of varying stimuli into a series of perceptual categories with particular connotations, the next stages still could be emotional, with call types corresponding to different degrees of arousal. Anticipating this argument, Seyfarth et al. (1980*a*, 1980*b*) incorporated in their playback tapes both brief and long versions of the various alarm calls so that, if the difference was one of intensity of an emotion, the response to long and thus intense stimulation by one call type might overlap with the response to a brief and thus weaker experience of another call. Although responses did vary with the duration of the call sequence, the qualitative difference between responses to the three classes of calls remained.

Approaching animal signalling from a different but not unrelated viewpoint, one may ask whether vervet alarm calls show any sign of being under voluntary control, as Sutton and his colleagues have recently demonstrated with some rhesus macaque (*Macaca mulatta*) calls (Sutton, 1979). Alternatively, are the alarm calls involuntarily and inextricably bound together with emotional behaviors concerned with fearfulness and escape? Of relevance here is that the animals responded vigorously to the playbacks but only once did animals ever vocalize in response to the call playback. Thus, alarm calling cannot be viewed as an obligatory component in a cluster of emotionally-based escape behaviors. The calls seem to be under voluntary control, and they do in fact serve as symbols, reflecting a cognitive appraisal of the nature of different causes of danger, and implying an assessment of the most appropriate responses in the circumstances. If this is a correct interpretation, it helps us to understand the otherwise astonishing behavior of captive chimpanzees (*Pan troglodytes*) and gorillas (*Gorilla gorilla*) trained with language-like systems of signalling and thus susceptible to interrogation by the experimenter (Gardner and Gardner, 1971; Premack, 1976; Rumbaugh, 1977). These animals clearly exhibit a capacity for symbolic signalling, and it would be remarkable if this were never manifest under natural conditions, where signalling behavior is often a matter of life or death. It is my conviction that once we learn how to ask the right questions, symbols will prove to be widespread in animal signalling. The evidence accumulating on the meaning of alarm calls of California ground squirrels (*Spermophilus beecheyi*) is consistent with the vervet findings (Coss and Owings, 1978; Leger and Owings, 1978; Leger et al., 1979, 1980). As in humans, I expect that symbolic and "emotional" or "affective" components will often be mingled together in animals, sometimes serving rather different communicative functions, sometimes similar ones.

It may still be true that few, if any, monkey calls are free of emotional content. In fact Eisenberg (1976) and Green (1975) have presented cogent evidence that variations in the quality and degree of arousal associated with different monkey calls correlate well with certain acoustic features. They make a strong case that entire repertoires can be ordered into a meaningful classification on the basis of such arousal-related properties. The burden of this paper is not to counter such interpretations but rather to complement them. It seems clear that affective components are there, and that they are important. However, I would urge serious consideration of the likelihood that there is a symbolic component in monkey calls as well, designed to work in harmony with the affective component, as in our own speech behavior.

On the question of special, species-specific processing both of signals and of some environmental stimuli, I take a nativist position. The ontogeny of complex behavior with all of its cognitive underpinnings will not, I believe, be intelligible until we have diagnosed the innate predispositions that organisms bring to bear on their perceptions of the world and their social signals. In this guise, studies of the innate release mechanisms of Lorenz (1950) and

Tinbergen (1951) have a significant role to play, opening a new window in the ancient and refractory problem of innate knowledge.

The empiricist view of the development of perception and of knowledge, at least in pure form, has never made any sense. On the other hand, it is equally absurd to insist on treating animals as though they were instinctual automata. The knowledge that monkeys have of companions and their surroundings is probably as intricate and complex as that which we have of our own worlds, though with an emphasis that is unique to each species. Understanding how we ourselves come by such knowledge is a problem of long standing. Philosophers have tried to solve it by invoking "innate knowledge" that the child brings to bear on the task of developing understanding. These animal studies suggest that we can bring the problem into the experimental domain, and perhaps illuminate this vitally important yet notoriously difficult issue of innate knowledge and the role that it plays in the development of social communication.

Acknowledgments

The field work on vocal communication in vervet monkeys, conducted in collaboration with Robert Seyfarth and Dorothy Cheney Seyfarth, was supported by research grant NIMH MH07446, a grant from the Wenner-Gren Foundation, an NSF National Needs postdoctoral fellowship, National Geographic Society Grant No. 1767, and NSF BNS 16894. I am indebted to them for permission to reprint Fig. 1.

For the Japanese monkey data, the author is indebted to Michael Beecher, Steven Green, David Moody, Michael Petersen, William Stebbins, and Stephen Zoloth for all aspects of the collaboration supported by NSF research grants BNS 77-19524 and BNS 75-19431, P.H.S. grants MH 31386-01 and MH 24269, P.H.S. training grant GM 1789-08, NIH grant NS 05785, NICHD grant N01-HD-1-2420, PHS predoctoral fellowship MH 05993, a Sigma Xi grant-in-aid-of-research, and an H. H. Packham dissertation grant.

Literature Cited

BEECHER, M. D., ET AL. 1976. Perception of conspecific communication signals by Japanese macaques (*Macaca fuscata*). J. Acoust. Soc. Amer., 36:26–58.

———. 1979. Perception of conspecific vocalization by Japanese macaques: evidence for selective attention and neural lateralization. Brain Behav. Evol. 16:443–460.

Coss, R. G., and D. H. Owings. 1978. Snake-directed behavior by snake-naive and experienced California ground squirrels in a simulated burrow. Z. Tierpsychol., 48:421–435.

Eisenberg, J. F. 1976. Communication mechanisms and social integration in the black spider monkey, *Ateles fusciceps robustus*. Smithsonian Contrib. Zool., 213:1–108.

Gardner, B. T., and R. A. Gardner. 1971. Two-way communcations with an infant chimpanzee. Pp. 117–184, in Behavior of non-human primates. Vol. 4 (A. M. Schrier and F. Stollnitz, eds.). Academic Press, New York, 239 pp.

Green, S. 1975. Communication by a graded vocal system in Japanese monkeys. Pp. 1–102, *in* Primate behavior (L. A. Rosenblum, ed.). Academic Press, New York, 407 pp.

———. 1981. Sex differences and age gradations in vocalizations of Japanese and lion-tailed monkeys (*Macaca fuscata* and *Macaca silenus*). Amer. Zool., 21:165–183.

Green, S., and P. Marler. 1979. The analysis of animal communication. Pp. 73–158, *in* Handbook of behavioral neurobiology. Vol. 3. Social behavior and communication (P. Marler and J. G. Vandenbergh, eds.). Plenum Press, New York, 411 pp.

Griffin, D. R. 1981. The question of animal awareness. Rockefeller University Press, New York, 209 pp.

Leger, D. W., and D. H. Owings. 1978. Responses to alarm calls by California ground squirrels. Behav. Ecol. Sociobiol., 3:177–186.

Leger, D. W., D. H. Owings, and L. M. Boal. 1979. Contextual information and differential responses to alarm whistles in California ground squirrels. Z. Tierpsychol., 49:142–155.

Leger, D. W., D. H. Owings and D. L. Gelfand. 1980. Single-note vocalizations of California ground squirrels: graded signals and situation-specificity of predator and socially-evoked calls. Z. Tierpsychol., 52:227–246.

Lorenz, K. Z. 1950. The comparative method in studying innate behavior patterns. Symp. Soc. Exp. Biol., 4:221–268.

Marler, P., R. Dooling, and S. Zoloth. 1980. Comparative perspectives on ethology and behavioral development. Pp. 189–230, *in* The comparative method in psychology (M. Bornstein, ed.). Sinauer Associates, New York, 302 pp.

Petersen, M. R., et al. 1978. Neural lateralization of species-specific vocalizations by Japanese macaques (*Macaca fuscata*). Science, 202:324–327.

Premack, D. 1976. Intelligence in ape and man. Wiley, New York, 370 pp.

Rumbaugh, D. M. 1977. Language learning by a chimpanzee: The Lana project. Academic Press, New York, 312 pp.

Seyfarth, R. M., and D. L. Cheney. 1981. The ontogeny of vervet alarm calling behavior: A preliminary report. Z. Tierpsychol., 54:37–56.

Seyfarth, R. M., D. L. Cheney, and P. Marler. 1980*a*. Vervet monkey alarm calls: semantic communication in a free-ranging primate. Anim. Behav., 28:1070–1094.

———. 1980*b*. Monkey responses to three different alarm calls: evidence of predator classification and semantic communication. Science, 210:801–803.

Smith, W. J. 1977. The behavior of communicating. Harvard Univ. Press, Cambridge, Massachusetts, 545 pp.

Struhsaker, T. T. 1967. Auditory communication among vervet monkeys (*Cerco-*

pithecus aethiops). Pp. 281–324, *in* Social communication among primates (S. A. Altmann, ed.). Univ. Chicago Press, Chicago, 392 pp.

SUTTON, D. 1979. Mechanisms underlying vocal control in nonhuman primates. Pp. 45–67, *in* Neurobiology of social communication in primates (H. D. Steklis and M. J. Raleigh, eds.). Academic Press, New York, 325 pp.

TINBERGEN, N. 1951. The study of instinct. Oxford Univ. Press, Oxford, 228 pp.

ZOLOTH, S. R., AND S. GREEN. 1979. Monkey vocalizations and human speech: parallels in perception? Brain, Behav. Evol., 16:430–442.

ZOLOTH, S. R., ET AL. 1979. Species-specific perceptual processing of vocal sounds by Old World monkeys. Science, 204:870–873.

PART III. CASE STUDIES OF MAMMALIAN BEHAVIOR: A PERSPECTIVE

INTRODUCTION

IN this section, we present case studies of the behavioral ecology and sociobiology of several representatives of the major subclasses of the Mammalia. The Class Mammalia contains at least 17 living orders (Simpson, 1945), among which are representatives that carry forward into the present time attributes which we believe are characteristic of mammals at an earlier stage of evolution. The three major subdivisions of the Class Mammalia include: the Prototheria, including the order Monotremata; the infraclass Metatheria, which includes the order Marsupialia; and the infraclass Eutheria, which includes the remaining 15 living orders. This tripartite classification reflects basic anatomical differences, many of which are tied up in the processes of reproduction. Although all mammals provide early nutrition to the young in the form of milk secreted from the mother's mammary glands, the monotremes lay eggs and incubate them and the marsupials give birth to extremely underdeveloped young which attach to a teat (usually in a pouch) and complete their postuterine development externally. By contrast, the eutherians give birth to young in a much more advanced state and have tended to reduce the length of the lactation phase relative to the marsupials.

T. R. Grant deals with the behavioral ecology of monotremes, a poorly studied and little understood mammalian group. Grant's data emphasize what highly specialized and beautifully adapted organisms the monotremes are and how they carry forward into the present their rather unique and archaic method of reproduction.

Pierre Charles-Dominique presents data from the first comparative study of South American marsupial socioecology. The South American marsupials of the family Didelphidae are characterized by large litters, rapid population turnover, and a simplified social structure. Indeed, only the genus *Caluromys* appears to have deviated from the basic didelphid pattern in that it has smaller litters with a longer maturation and a greater potential longevity for an adult. At the level of *Caluromys,* the didelphid marsupials may approximate the ecology and social structure so distinctive of their eutherian ecological counterparts, the prosimians (lorises and galagos) of West Africa. Charles-Dominique's essay stands as a benchmark in the study of small, conservative, nocturnal mammals and demonstrates the utility of comparative ecological studies.

The remaining chapters deal with eutherians and are designed to compare and contrast socioecology through the class. In Patricia Moehlman's study of two jackal species, she demonstrates that at least in one species pup survivorship is significantly increased if some siblings from the previous litter remain with the parents and assist in the rearing of the following year's young. Both species are monogamous and pairs typically have older offspring assisting in parental care. The costs and benefits of remaining with the parents or dispersing and attempting to find a mate and territory are discussed. Clearly, kin selection could have favored the type of social system and the "helpers" observed in jackals.

Jon Rood presents similar observations on the dwarf mongoose, a small monogamous carnivore in which one female usually bears a litter which is cared for by other group members. Unlike the jackals, membership in a dwarf mongoose group does not depend on genetic relatedness. In fact, exchanges of unrelated individuals between groups is the rule and care-giving behavior is seen in all group members regardless of genetic relatedness. Rood suggests that care-giving behavior in the dwarf mongoose is a case of reciprocity in that individuals which care for unrelated young may eventually have their own offspring cared for by these same non-relatives. The evolution of such reciprocity depends on immigrants to groups eventually achieving breeding status, a condition which does exist.

In the chapter by Stirling, a series of polygynous species is considered. Stirling's paper evaluates the ecological and phylogenetic conditions which led to the variety of mating and social systems exhibited by pinnipeds. He demonstrates how the form and stability of the breeding substrate and the presence of terrestrial versus aquatic predators profoundly influence whether females will aggregate and how long they will remain in a terrestrial phase while rearing young. Polygyny will evolve when females are bound to a fixed substrate in sufficient numbers and in such proximity that a male can defend a set of females against intruders. Stirling's general conclusions expand on those of earlier workers because his research has led him to consider the subtle differences between pack-ice versus shelf-ice as substrates for breeding females. Finally, the limitations of phylogeny are evaluated through a comparison of seals, sea lions, and the walrus, even within similar habitats.

In the chapter by Michener, the evolution of sociality in ground squirrels in the genus *Spermophilus* is carefully reviewed from a consideration of specific habitat type, length of growing season, and the duration of time that adults and juveniles spend above ground. By limiting the discussion to hibernators, Michener has developed testable hypotheses concerning the correlation of sociality and the duration of time that socialization between adults and young can occur and the length of the growing season.

In the final chapter by William Franklin, we again see how the ecological comparison of related species in different habitats clarifies the evolution of social systems and behavior. This comparative study, contrasting the adaptations of the vicuña to high semi-arid plains and the guanaco to cool temperate meadows, clarifies the similarities and profound differences in social systems. Franklin's studies echo previous themes; the social organization of these ungulates, like tropical forest marsupials, reflect an adaptational syndrome based on ecological constraints.

D.G.K. and J.F.E.

BEHAVIORAL ECOLOGY OF MONOTREMES

T. R. GRANT

Abstract

CURRENT research on the behavioral ecology of the three extant species of the order Monotremata has been mainly concerned with answering fairly simple questions about the natural history of the platypus (*Ornithorhynchus anatinus*), the Echidna (*Tachyglossus aculeatus*), and the New Guinea long-beaked echidna (*Zaglossus bruijnii*). The distribution, food and feeding ecology, temperature regulation and thermoregulatory behavior, and movements of the three species are discussed. Although similarities are pointed out, the emphasis is on the nature of specialization associated with particular ways of life in the individual species. Some comment is made on the value of the current research and on the direction future research should take.

Introduction

The fact that Griffiths (1968, 1978) has written two monographs on the biology of monotremes would suggest that much is known about the biology of the three species belonging to this group of mammals. Such a suggestion is erroneous. Although their discovery by Europeans (in the late 18th century in the case of *Ornithorhynchus anatinus* and *Tachyglossus aculeatus,* and nearly a century later in *Zaglossus bruijnii*) sparked intense interest in the group, most information presented by Griffiths (1968, 1978) has only been collected in the last 15 to 20 years, primarily because it has been difficult to locate and capture echidnas and hard to maintain platypuses in captivity. To some extent these difficulties have been overcome by persistent research workers but they have emphasized the laboratory biology of the species, that is physiology (reproductive, thermal, respiratory, nervous and endocrine), genetics, taxonomy, and anatomy, all reviewed by Griffiths (1978).

The basic field biology of all three extant monotreme species has been almost totally ignored. Many early naturalists, including Burrell (1927), made valuable observations on their natural history but this information mainly consists of fragmentary observations because of the secretive behavior of the animals and the nature of the research. The "one-man-and-friends" approach to the study of the ecology of species which are shy, retiring, mainly fossorial, often nocturnal, fairly small, mobile, and, in the case of the platypus, amphibious is totally inadequate. I know of no systematic, intensive, well-manned and well-funded study on the ecology of any monotreme either in Australia or Papua New Guinea. As a result the information presented will be limited and open to conflicting interpretations.

The behavioral ecology of monotremes is discussed broadly along the following lines: where do they live, what do they eat, when do they breed, when are they active, and how far do they move? These questions are often asked by people outside the field of monotreme biology, and are of considerable importance. The research to date can yield at least some tentative answers.

The review covers each species in turn as it would be too confusing to consider species of such diverse and specialized ways of life together under specific subject headings. No apologies are made for the paucity of data presented in the New Guinea long-beaked echidna (*Z. bruijnii*) because little research has been published on this species. The platypus (*O. anatinus*) is the only monotreme that I have personally studied.

The Platypus, Ornithorhynchus anatinus

The exact distribution, especially the western extent, of the platypus is unknown, but Griffiths (1978) indicated that the species occurs over a large J-shaped area of eastern Australia, extending from Cooktown in north Queensland to Mildura in South Australia and including Victoria and Tasmania in the south (Fig. 1A). In this vast area, bounded by the sea to the east and probably the extent of river permanency in the west, are many types of freshwater creeks, rivers, lakes, and artificial water empoundments, many of which are occupied by platypuses (Grant and Carrick, pers. comm.; Griffiths, 1978; Temple-Smith, 1973).

O. anatinus depends on water for feeding. However, its flexibility

A.

B.

of habitat selection within the limits of available freshwater habitats is quite striking. Platypuses have been reported or seen in such varied habitats as abandoned orchard irrigation trenches, polluted streams in the Sydney metropolitan area, clear mountain streams, large rapidly flowing rivers, and a variety of lake situations, including water storage reservoirs. My studies have been conducted in a typical habitat in the upper reaches of the Shoalhaven River in New South Wales (Grant and Carrick, 1978) where animals were caught in gill nets in deep pools which occur in the river, separated by boulder and bedrock riffle areas. The study area is at 686 m elevation in the southern tableland area of New South Wales where summer water temperatures reach above 25°C and those in winter fall below 5°C (air temperature range from below −10°C to above 30°C). The Shoalhaven River, like most Australian rivers, is subject to periodic flooding and low water.

Since this study area experiences certain extremes of climate and river flow, the results permit inferences regarding platypus natural history in areas where they have been less intensively studied.

Temperature Regulation and Thermoregulatory Behavior

Despite earlier reports to the contrary (Smyth, 1973; Sutherland, 1897), Grant and Dawson (1978*a*, 1978*b*) showed that the platypus, although having a body temperature (T_b) lower than that of many marsupials and eutherians, is capable of maintaining this T_b (around 32°C) relatively constant over a range of environmental temperatures, including water temperatures as low as 5°C and as high as 30°C. The platypus has a higher metabolic rate than either *Z. bruijnii* or *T. aculeatus*; this may be an adaptation to the thermal stresses of its amphibious way of life (Dawson et al., 1979).

Recordings of T_b from animals taken in nets were close to those recorded in the laboratory at similar temperatures. More recently radio-telemetric studies of free-ranging platypuses indicate that a

---

←

FIG. 1. Distributions of the three monotreme species. A, *Zaglossus bruijnii* (black) and *Ornithorhynchus anatinus* (horizontal lines). Arrow identifies introduced population of *O. anatinus* on Kangaroo Island. B, *Tachyglossus aculeatus: T. a. acathino* (vertical lines); *T. a. aculeatus* (diagonal lines); *T. a. lawesii* (stipple); *T. a. multiaculeatus* (arrow); *T. a. setosus* (cross-hatching).

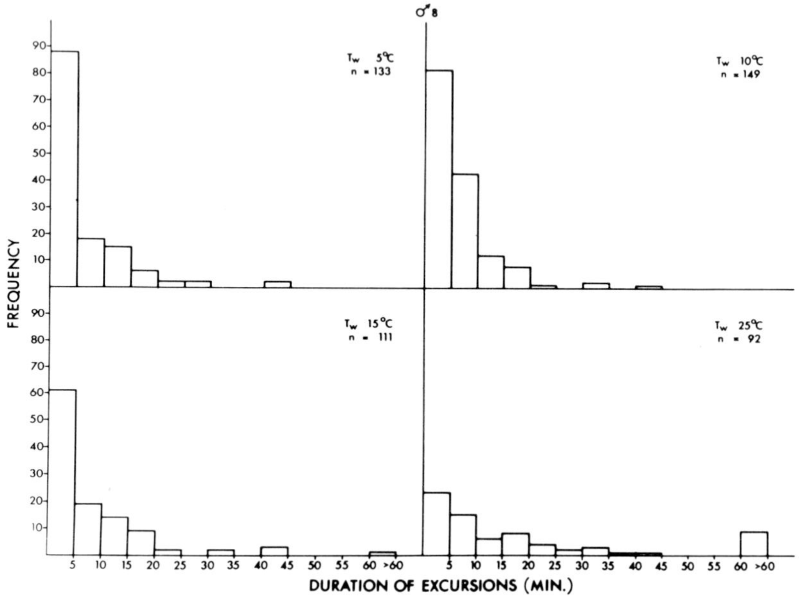

FIG. 2. The frequency of excursions of various duration made by one captive male platypus into water of different temperatures (5, 10, 15 and 20°C) during four 24-h periods. Sample size (n) identifies the total number of excursions made in each 96 h of observation. (From Grant, 1976.)

similar range of T_b is maintained in the wild (Grant, pers. observ.). Body temperatures of two free-swimming platypuses in winter (T_w 5°C) were found to be more variable than those recorded from swimming animals in laboratory studies (Grant and Dawson, 1978*a*). During immersion in water these two animals had a T_b which fluctuated between 30.0 and 33.5°C, a range well above the 21°C mentioned by Smyth (1973).

On the basis of limited data on captive platypuses, Smyth (1973) postulated that in the field these animals are poor thermoregulators and probably spend only brief periods in the water in winter, returning frequently to the bank or burrow to rewarm. Certainly, given a choice, captive platypuses will only enter the water at various temperatures for short periods when food is freely available; however, captive animals will spend greater than 60 min in water at 15°C at one excursion (Fig. 2), and a combined period of immer-

TABLE 1
TOTAL NUMBER AND MEAN DURATION OF EXCURSIONS MADE INTO WATER OF VARIOUS TEMPERATURES BY TWO CAPTIVE PLAYTPUSES OVER SEVERAL 24-HOUR OBSERVATION PERIODS (FROM GRANT, 1976)

	Total no. of excursions	Mean no. hours per day in water	Water temp. (T_w)
Male	133	3.9	5
	149	4.9	10
	111	4.5	15
	92	6.9	25
Female	100	8.1	5
	108	6.9	10
	100	7.1	15
	53	9.7	20

sion of up to 8 h over 24-h intervals has been recorded at a water temperature of 5°C (Table 1). Marked platypuses in the wild also immersed for short durations, but as Table 2 shows, during winter platypuses were observed in water for periods approaching 3 h (Grant, 1976). Two platypuses radio-tracked in winter maintained the T_b within the range mentioned above while spending between 6 and 12 h continuously in the river (Grant, pers. observ.). Such periods spent in water are in keeping with the species' mode of feeding which involves the collection of quite small food items from the benthos of its habitat (Faragher et al., 1979).

The added thermal stress of winter climatic conditions would of necessity force the platypus to spend at least the same amount of

TABLE 2
DURATION OF OBSERVED EXCURSIONS BY WILD PLATYPUSES IN WATER. THE NUMBER OF OBSERVATIONS GREATER THAN AND LESS THAN 15 MIN ARE SHOWN, ALONG WITH THE MEAN (±SD), TIMES, AND RANGES OF TIMES GREATER THAN 15 MIN (FROM GRANT, 1976)

	Summer (T_w21–24°C)	Winter (T_w6–9°C)
$\bar{X} \pm SD$	57.0 ± 47.4 min	51.5 ± 33.0 min
Range	15–90	15–155
>15 min	38.0	24.0
<15 min	57.0	35.0

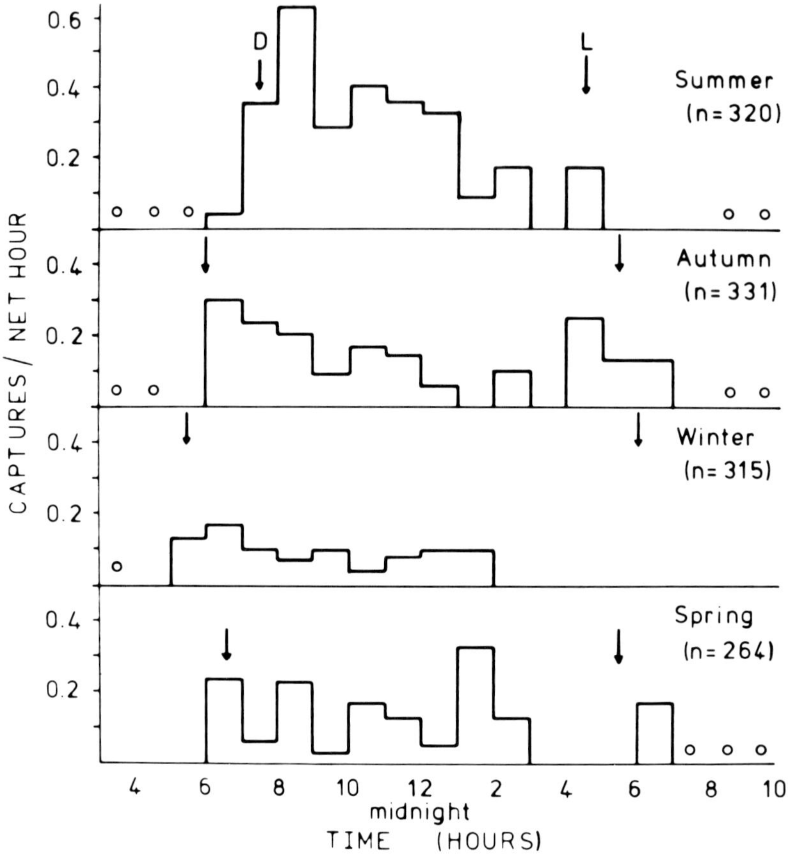

FIG. 3. Capture rates of platypuses in the upper Shoalhaven River, New South Wales Australia. Abbreviations are: n, numbers of net hours; D, darkness; L, daylight; o, no trapping. (Grant, pers. comm.)

time feeding in winter as at other seasons (Grant and Carrick, 1978) unless the animal was capable of becoming torpid as is its relative, the echidna, *Tachyglossus aculeatus* (Augee and Ealey, 1968). There is some evidence for torpor in platypuses in captivity (Eadie, 1935; Fleay, 1944), and although this is possible in the wild, in most areas platypuses are active and can be seen during daylight in the winter months (Figs. 3 and 4), although fewer animals are caught at that time. There may be a winter shift to diurnal

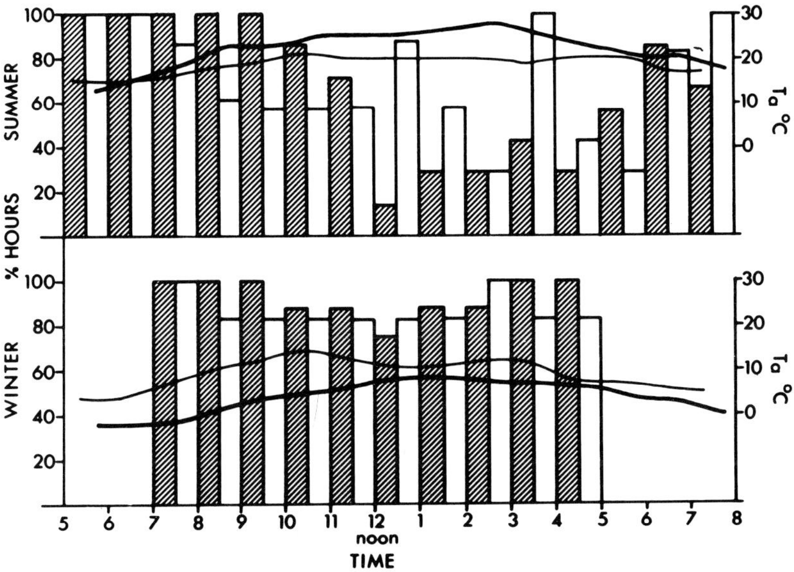

FIG. 4. The percentage of hours of observation at various times of the day in which platypuses were seen active in the upper Shoalhaven River, New South Wales, Australia. Mean hourly temperatures are also shown. Shaded bars, summer, 1974 and winter, 1973; open bars, summer, 1975 and winter, 1974. (From Grant, 1976.)

activity so animals are not foraging at night when air temperatures above the water surface fall below freezing. In winter there is little difference in water temperature from day to night but air temperatures are well above freezing in the day (Grant, 1976). Figure 4 shows more consistent observations of platypuses during the day in winter months in the Shoalhaven River.

When platypuses are not in water the microhabitat of their burrows is of great thermal importance. Even in an unoccupied artificial burrow the insulation of layers of earth was found to provide a significant buffering effect against outside ambient temperature changes both in winter and summer (Fig. 5). Data collected recently from temperature transmitters in burrows suggest a further modifying effect of the animal's presence on the microhabitat temperature, elevating it several degrees above that of an unoccupied burrow (Grant, pers. observ.).

Very little is known about platypuses in rivers and lakes in trop-

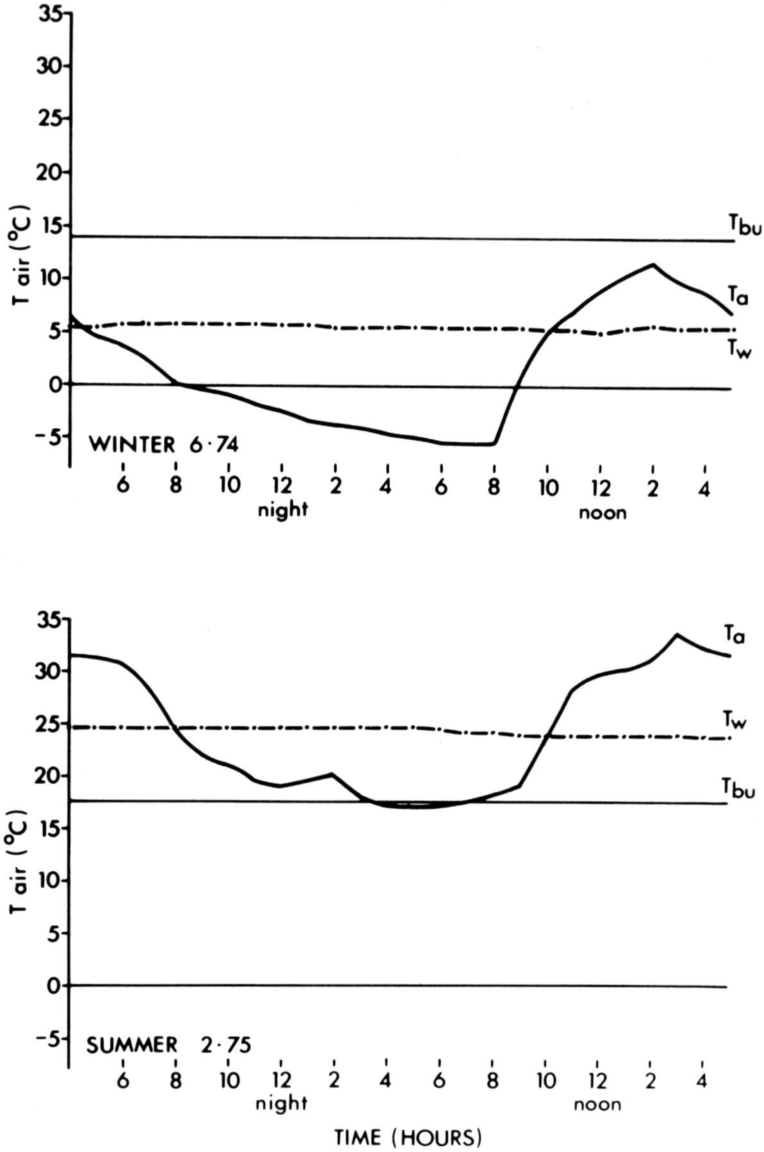

FIG. 5. Air (T_a), burrow (T_{bu}), and dee water (T_w) temperatures in the upper Shoalhaven River, New South Wales, Australia, taken over two 24-h periods. (From Grant, 1976.)

ical areas of its range. Grant and Carrick (pers. observ.) caught platypuses in Marooby Creek on the Atherton Tablelands in north Queensland during the night in summer and found that the water temperature of the first 2 m of the pool in which they were caught reached close to 30°C during the heat of the next day. Platypuses can regulate body temperature in water temperatures this close to their T_b but were found to be relatively inactive in such conditions (Grant, 1976). However, no platypus was observed during the day in this location and animals may avoid such high temperatures by occupying burrows during the day in tropical areas, emerging at night to feed when water temperatures are much lower (Grant and Carrick, pers. observ.). Initially I suspected that water temperatures above the body temperature of the platypus may have been responsible for the species' absence from the rivers of the Cape York area of Australia. However, water temperatures measured at night in several rivers on Cape York in November, 1976 were found to be well below platypus body temperature, ranging between 25 and 27°C (Grant and Carrick, pers. observ.). The truncation of the range in the northern distribution of *O. anatinus* around the Cooktown area remains a mystery (Griffiths, 1978), highlighting the need for more research in the tropical northern extension of platypus distribution.

Temple-Smith (1973) captured platypuses in the Crackenback River in the Snowy Mountains region of New South Wales where he recorded a water temperature of 0.6°C. Grant and Dawson (1978*b*) found that a low conductance (high insulation) of both tissue and fur, and possibly the lack of necessity to maintain the T_b elevated to the eutherian level, permitted *O. anatinus* to cope with its harsh winter environment. However, the experimental animals were not subjected to such extremes of temperature as those noted by Temple-Smith (1973). Although extrapolation of the Grant and Dawson (1978) metabolic data to a water temperature of 0°C indicates that the animals could tolerate such an extreme, it would be fruitful to investigate both the possibility of torpor in this situation and the aquatic activities of platypuses at these extremes of temperature. The productivity of cold mountain streams in Australia is unknown; however, Hynes (1970) noted that unlike terrestrial species, aquatic invertebrates can remain active at low temperatures. On the other hand Faragher (pers. comm.) has found that the dragonfly species *Hemicordulia tau* is present mainly as small

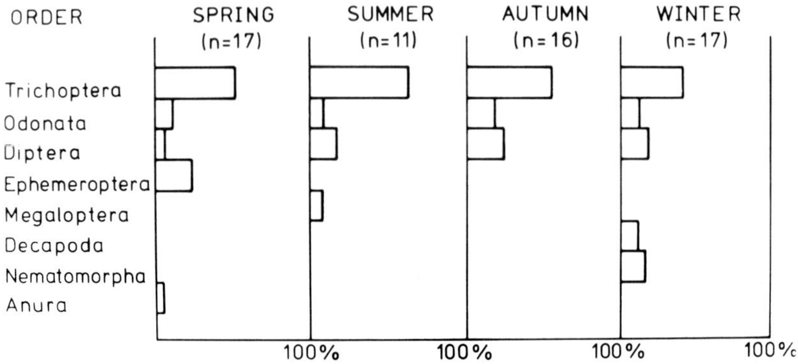

FIG. 6. Seasonal composition of dominant food items in the diet of the platypus. (Data from Faragher et al., 1979.)

nymphal instars in winter in Lake Eucumbene in the Snowy Mountain area of New South Wales, although trout do manage to find some larger stages at this time. If food organisms are more difficult to find, or are present mainly as smaller growth stages in high mountain habitats in winter, the activities of platypuses occupying these habitats might be food limited, a possibility which surely warrants investigation.

Food and Feeding Ecology

Faragher et al. (1979) found that the platypus is predominantly an opportunistic predator on benthic invertebrates, utilizing mainly larval stages of insect species in the pools and riffle areas of the upper Shoalhaven River (Fig. 6). The animals were unselective and mainly took benthic species in the relative proportions in which they occurred in the benthos of the river and in an adjacent creek (Ward, 1979) at all seasons. Drift organisms did not appear to be of importance in the platypus diet, although Griffiths (1978) reported platypuses taking cicadas from the water surface in the wild and captive animals have been observed taking moths from the water surface (Faragher et al., 1979).

Ward (1979) attempted to determine the feeding location of free living platypuses by analysis of food organisms and benthic samples from various substrates in both pool and riffle areas. Although the results were somewhat equivocal, this and earlier work (Faragher et al., 1979) suggest that both riffle and pool benthos are used by platypuses as a food resource. Although animals are most frequently seen and, by the nature of the trapping method, always caught in deep pools, it seems likely that a variety of potential food sources in rivers are exploited by *O. anatinus*.

In all of these studies platypus diet was determined from cheek pouch contents which consisted of food materials stored during underwater foraging or food parts ready to be rejected after being ground by the horny pads which replace the teeth in adult platypuses. Material from gut contents is totally unrecognizable and useless for identification as it is so finely ground before being swallowed. Both the whole animals and debris extracted from the cheek pouches are of very small organisms, presumably explaining the long time periods spent in the water by platypuses during foraging.

Some dietary overlap with other vertebrate species occurring over the range of platypuses probably occurs, including trout (*Salmo trutta* and *S. gairdneri*), the water rat (*Hydromys chrysogaster*), and various species of waterfowl. Dietary overlap with the water rat is probably not marked as *Hydromys* eats a wide variety of food, not all of which is obtained in the water. It also tends to select larger adults of aquatic insect species (Woollard et al., 1978) whereas the platypus takes mainly immature forms. Faragher et al. (1979) also found little diet overlap with brown trout (*S. trutta*), which consumed some midwater and drift species, including the freshwater decapod (*Paratya australiensis*), and took different proportions of the benthic organisms it consumed.

Many Australian species of ducks are omnivorous (Frith, 1967; Frith et al., 1969). Of the species likely to occur in the Shoalhaven River area, the black duck (*Anas superciliosa*), and the grey teal (*A. gibberifrons*) are dabbling ducks, the wood duck (*Chenonetta jubata*) grazes green herbage, and the hardhead (*Athya australis*), the musk duck (*Riziura lobata*), and the blue-billed duck (*Oxyura australis*) are diving ducks. Possible overlap in the diet of ducks and platypuses is probably small. The diets of the black duck and grey teal are quite similar and consist of up to 31% insect material (mainly adult Corixidae and Dystiscidae) (Frith, 1967). In diving ducks the amount of animal food ranges from 43.5 to 72.6%, with Mollusca, Trichoptera, Odonata, and Hemiptera being important sources. Faragher et al. (1979) noted that only the black duck, wood duck, and grey teal were seen in the Shoalhaven River study area during their study. Overlap of diet in other areas, especially in coastal waterfowl refuge areas during inland drought periods, could be much more marked and would be worth investigating.

Aspects of platypus feeding ecology still to be investigated (including the already mentioned study of feeding behavior in cold mountain streams) are the feeding selectivity and composition of cheek pouch contents of captive platypuses fed known diets. Detailed observations of actual platypus feeding behavior, including how the food is sorted and consumed, should be conducted. Also, food selection in different habitats occupied by the species should be studied. Dredge samples taken recently from a water storage dam in the Australian Capital Territory suggest a depauperate benthic fauna, and analysis of cheek pouch samples collected at the same time should prove to be interesting. Two platypuses taken from Lake Burleigh Griffin in Canberra city had cheek pouches containing mainly freshwater crustacean fragments (Grant, pers. observ.). Captive platypuses can catch and consume free-swimming mosquito fish (*Gambusia affinis*) and freshwater crayfish (*Cherax* sp.) (Grant et al., 1977). These isolated observations suggest that *O. anatinus* could have different feeding patterns and food selection techniques in the habitats it occupies. In fact, such differences might be an essential adaptation to different habitats and permit the avoidance of direct dietary competition with co-existing vertebrate species.

The fact that platypuses locate food at all seems amazing as they close both their ears and eyes underwater. It has always been surmised (Burrell, 1927) that the sensitive bill was used in searching for food and in orientation underwater through a sense of touch. Recent neurological investigations indicate that the tactile sensitivity of the platypus bill may be more highly developed than could have been anticipated by naturalists observing the animal in captivity or in the wild. Bohringer and Rowe (1977) found that the entire bill has densely packed arrays of mechanoreceptors on its outer surface, and particularly on the upper bill. Cortical mapping showed that a large proportion of the somatic sensory cortex was devoted to receiving precise sensory information from these areas, and that fine details of the speed and direction of moving objects might be perceived by patterns of sequential firing of columns of cortical neurones receiving information from the peripheral mechanoreceptors of the bill. Platypuses can locate moving prey species and navigate in unfamiliar surroundings in captivity without noticeably bumping into objects (Grant, pers. observ.). Behavioral studies are needed to complement the laboratory experiments of Bohringer and Rowe (1977) to determine how *O. anatinus* locates food and orients in its aquatic environment.

Movements

Studies of *O. anatinus* movements are difficult to interpret because of the different marking methods used, the loss of some marks, the variety of localities netted, and the irregular nature of the trapping effort after marking (Grant and Carrick, 1978; Griffiths, 1978; Temple-Smith, 1973). Temple-Smith (1973) marked over 100 animals in various locations on the tablelands of New South Wales, in the Australian Capital Territory and in the Snowy Mountains area. He recaptured 25 of these animals and found that only five were retaken outside the pool of initial capture. Griffiths and Elliot (Griffiths, 1978) recaptured 12 out of 34 marked platypuses in the same pools or areas as the original capture point over periods of time ranging from just under 2.5 months of 3.5 years in several localities near Canberra. From these limited data Griffiths (1978) agreed with Temple-Smith (1973) that "site attachment appears to be strong in the platypus." However, Grant and Carrick (pers. observ.) and Grant and McBlain (1976) netted platypuses in two areas where Temple-Smith released marked animals from 1973 to 1980, and only captured one of 22 marked. In spite of the release of 16 more marked individuals in an area on the Murrumbidgee River in New South Wales, no marked animals were caught between 1973 and 1976 (Grant and Carrick, pers. observ.). Grant and McBlain (1976) sighted another of Temple-Smith's marked platypuses in a second area on the upper Shoalhaven River, but neither this animal nor the other previously recaptured one were recaptured in subsequent intensive trapping over the next 7 years (Grant, pers. observ.).

Further doubt was cast on the interpretation of the recapture data of Griffiths (1978) when recent netting by Grant and Scott (pers. observ.) in an area where I had released marked individuals yielded seven platypuses (4 males and 3 females), none of which was marked.

Grant and Carrick (1978) followed visually marked individuals of *O. anatinus* over 8 to 10 day periods in both summer and winter

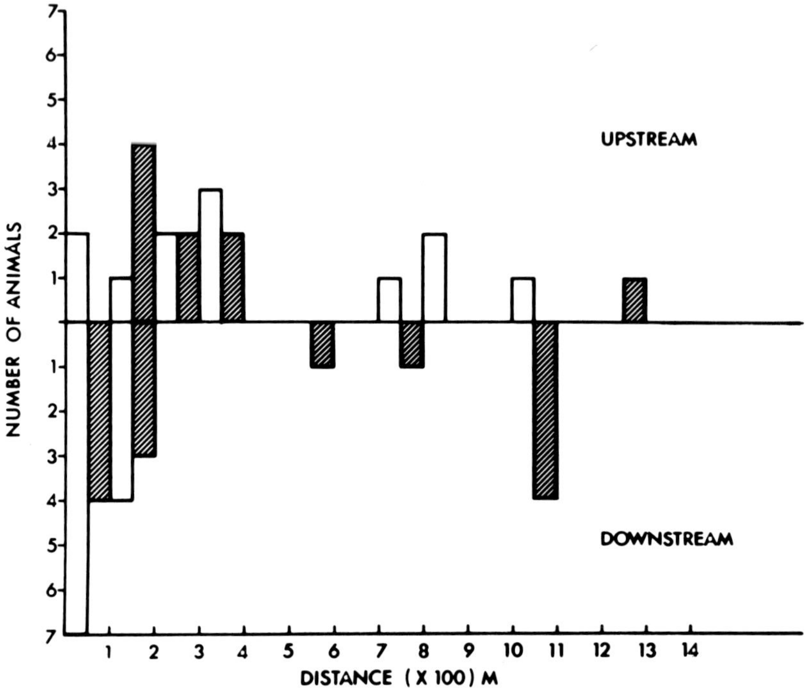

FIG. 7. Maximum distance moved by platypuses in the upper Shoalhaven River, New South Wales, Australia during four 10-day observation periods. Open bars, winters (1973, 1974); shaded bars, summers (1974, 1975). (From Grant and Carrick, 1978.)

and found these animals moving some distance within the original capture pools, but despite free movement of four animals between two adjoining pools in the Shoalhaven River study area, only one animal moved outside these two pools during four such observation periods. This platypus moved a distance of 1.3 km in 6 days, less than the 2 mi recorded by Temple-Smith (1973) for his most mobile animal. Most of the visually marked platypuses were resighted (60 to 80%) and had moved less than 400 m from the capture point (Fig. 7).

Trapping studies carried out in 1977 outside Grant and Carrick's (1978) main study area showed a similar limited movement of many platypuses. Very few animals marked before the end of 1977 in the two pools of the main study area were caught outside them,

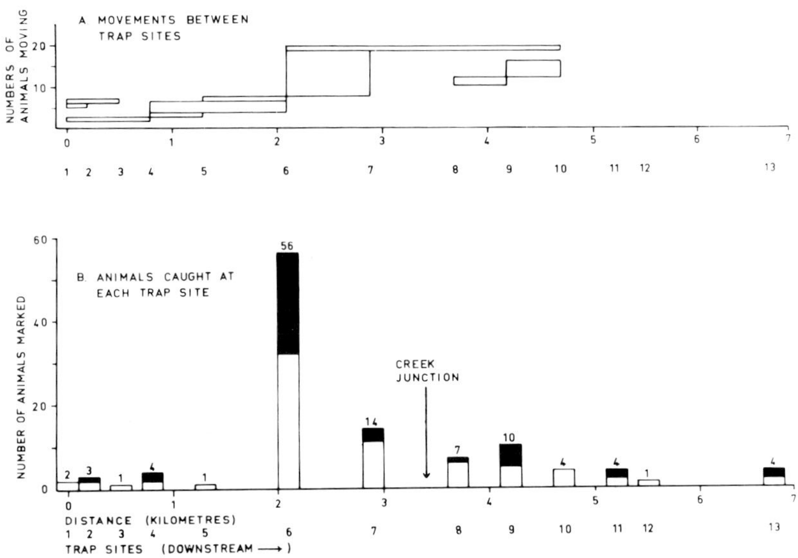

FIG. 8. Distances moved by marked animals between sites (A) and numbers of platypuses marked at each trap-site (B). In A, the thickness of the horizontal bars indicates the number of individuals moving between trap-sites (the site numbers and distances apart are shown on the horizontal scale of both A and B). In B, solid bars represent males and open bars represent females; the numbers at the top of each bar are the numbers caught at each site in the upper Shoalhaven River, New South Wales, Australia. (From Grant and Carrick, 1978.)

regardless of the extensive netting operations of May and December of that year (222 net h, defined as a net of 50 m length in water for 1 h). Four animals moved out of the area to pools upstream, one moved to a pool 2.6 km downstream and another was caught early in 1978 in the adjoining creek (some 2.4 km from its point of release 6 months earlier). In Fig. 8 the numbers of both male and female platypuses caught at each trap-site are shown, along with a graph indicating the numbers of animals moving to trap-sites other than those at which they were marked. Figure 8 suggests that movements were restricted, especially between the two pools of the main study area (trap-sites 6 and 7) and trap-sites upstream and downstream from them.

Platypuses recaptured outside the area where they were marked often moved reasonable distances in relatively short times. Recaptures of animals at sites over 1.0 km apart during netting operations separated by only a few days occurred several times. For example, one female was caught and marked at trap-site 6 on 30 March 1976. She was next caught at trap-site 4 on 1 December 1977, then recaught back at site 6, a distance of 1.3 km, only 5 days later. These data suggest that many platypuses in the area are relatively sedentary but that some may be more free-ranging. The sedentary habits of many animals was further indicated by the fact that 57% of all platypuses marked in the area were recaptured at least once during the study and 26% were recaptured more than once, with one female being caught 10 times after she was marked in July 1973.

Netting activities in 1978, 1979, and early 1980 (Grant, pers. observ.) substantiated the earlier findings of Grant and Carrick (1978); marked platypuses caught at these times moved distances similar to those shown in Figs. 7 and 8. The female which was caught 10 times since its capture in 1973 (Grant and Carrick, 1978) was captured three more times (each capture being within the two pools making up the main study area). However, when this female was tracked by radio-telemetry in June 1979, she moved 1.5 km from the main study area pool where she had occupied a burrow during the day after her release, to another pool just outside the main study area (between sites 7 and 8; see Fig. 8). She spent part of the night and all of the next day in a burrow in this latter pool at the junction of the river and adjoining creek until she was recaptured when she emerged. Two other female platypuses which had moved less than 400 m (based on three and four recaptures within the main study area pools between 1973 and 1979) were also captured in this Junction Pool in February 1980. Both were found to be lactating.

Thus, animals originally thought by Grant and Carrick (1978) to be relatively sedentary may be more mobile than anticipated. Greater mobility and possibly a larger total population of *O. anatinus* in the upper Shoalhaven River are also suggested because after marking 137 platypuses along a river and adjoining creek approximately 11 km long between 1973 and 1979, extensive netting in February 1980 still yielded 33 unmarked adult animals of 44 caught (Grant, pers. observ.).

Apparently, floods need not cause the displacement of animals. During Grant and Carrick's study (1978), three major floods (a rise of river level above the normal banks of the river) and four minor ones occurred. The largest, in April 1974, represented a ten-fold rise in normal river level (Fig. 9). However, platypuses marked before the floods were recaught in subsequent months. For example, in 1977 the main study area was trapped on the nights of 15 to 17 February before a 6.4-m flood (eight times the minimum level for that year). The area was netted again in March (3 weeks after the flood peak), and of eight platypuses caught, five were recaptures, and two of these were last caught immediately before the flood. In Fig. 9 the numbers of platypuses caught and recaptured in the two main pools at each sampling period are shown. The times and levels of floods passing through the area during this time are also shown.

How the platypuses cope with floods is not known. After recapturing a marked animal in the same area after a severe flood, Griffiths (1978) remarked that this animal "apparently 'rode-out' severe flooding or managed floods," suggesting that there is strong site attachment as do the data of Grant and Carrick (1978).

Floods in much of the distribution of *O. anatinus* are normally of quite short duration, the flood peak passing quickly. In northern Australia during the wet season, however, rivers may run high for extended periods. Indeed, the absence of the platypus from the rivers of the Cape York Peninsula (Carrick and Grant, pers. observ.) may be due to its inability to cope with prolonged flooding. Griffiths (1978) pointed out that predation by crocodiles does not explain the truncation of the platypus' range because the much slower-swimming water rat (*Hydromys chrysogaster*) inhabits these rivers. *Hydromys* is probably less affected by floods because it does not obtain all of its food in the water (Woollard et al., 1978), and so is more independent of river conditions than is *O. anatinus*.

Burrell (1927) attributed the reworking of the long and complex nesting burrows by female platypuses to the repair of flood damage. He also indicated that, at least in his time, platypuses could be captured in hollow logs on land. More recent anecdotal reports of platypuses being killed on roads and appearing in farm dams, only infrequently connected to a stream, suggest possible terrestrial mobility in the species in response to floods or drought (Burrell, 1927). Occupation by platypuses of habitats where little or no suitable banks are available for the construction of burrows near the water (for example, Bendora Dam, a water storage reservoir with very little soil covering rocky margins; Griffiths, 1978) may be facilitated by such terrestrial movement.

Platypuses breed in the spring, although the breeding season is

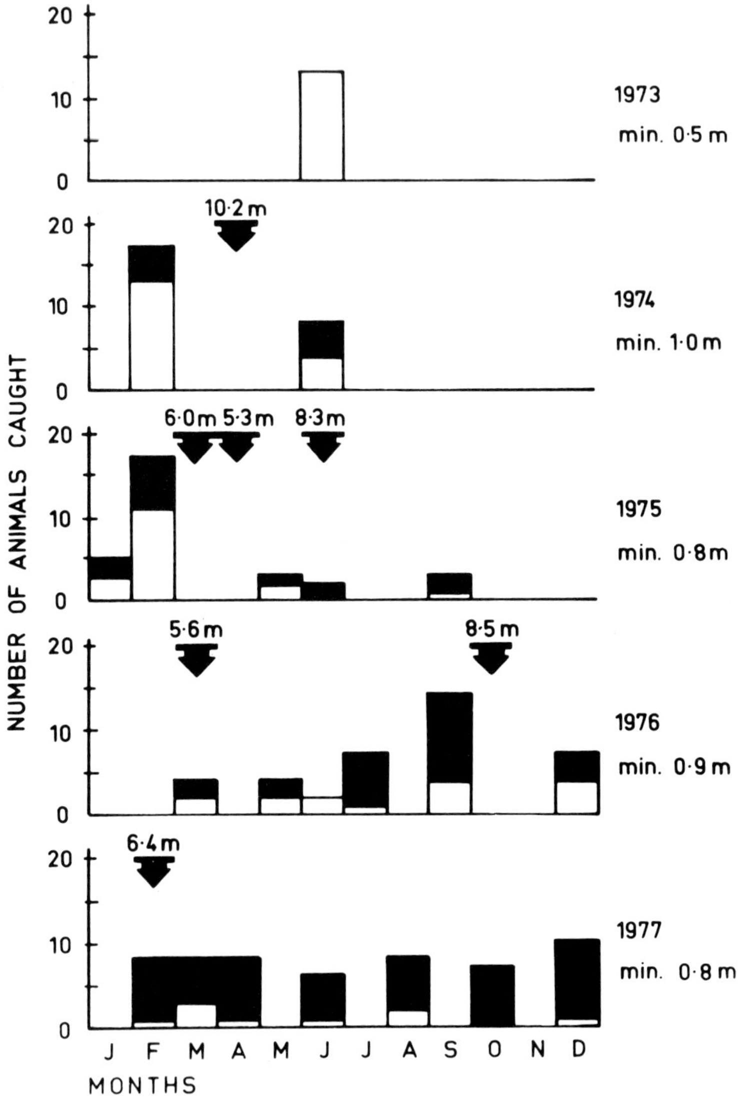

FIG. 9. Numbers of animals caught and recaptured in the upper Shoalhaven River, New South Wales, Australia. Flood peaks and their maximum heights are shown as arrows. Minimum water levels are shown for each year (m, meters). Histograms represent total captures, including recaptures which are shown as shaded portions. (From Grant and Carrick, 1978.)

earlier in the north (early August in Queensland) and later in the south (mid- to late-October in Victoria) (Griffiths, 1978). The platypus gestation period is unknown, although Hughes and Carrick (1978) have indicated that it is at least 9 days based on eggs laid in captivity after the capture of a female. Griffiths (1978) suggested that females incubate the eggs for a further 10 days after laying, and it is known that the young are suckled in burrows for in excess of 3.5 months. The young emerge from burrows in New South Wales in late February when they are already consuming the same food as adults (Ward, 1979). In the upper Shoalhaven River, 23 juvenile platypuses were marked in early 1978 and 1979 but only two have been recaptured during subsequent netting activities including February 1980 (when an additional 12 were marked). Juveniles are recognizable by spur morphology for periods from 9 to 12 months after hatching (Temple-Smith, 1973). However, between July 1973 and February 1980, only two animals of this age were caught unmarked in the upper Shoalhaven River (Grant, pers. observ.). What becomes of most of the young after each breeding season is unknown. Radio-tracking of one juvenile male platypus which moved 3.5 km in 5 days and occupied three different burrows suggests greater mobility in newly emerged young animals (Grant, pers. observ.). However, a few captures of even unmarked juveniles from outside the study area suggests that mobility is not the only answer.

There may be both resident and more mobile segments of the platypus population in the upper Shoalhaven River which can only be determined by a much more intensive netting operation over a wider area than that previously investigated by Grant and Carrick (1978). Such an approach, coupled with radio-tracking of a number of animals (males, females, juveniles) over a considerable time period, would yield information on movements as well as basic data on the population dynamics of *O. anatinus.* Data from a relatively undisturbed area could be valuably compared with disturbed areas such as those affected by flood mitigation work, dams or pollution.

The Echidna, Tachyglossus aculeatus

Griffiths (1978) recognized five subspecies of *Tachyglossus aculeatus* with hybridization in areas where the ranges are adjacent

(Fig. 1B). This classification is based mainly on pelage characteristics which seem related to the habitat types occupied by the various subspecies. The tropical subspecies *T. a. lawesii* has a pelage of fine long hairs with a few long slender spines, whereas *T. a. acanthion* from arid regions possesses sparse bristles rather than hair and has long stout spines, and the Tasmanian echidna (*T. a. setosus*) has dense wooly hair and short stout spines. The echidnas found in the southeast corner of Australia were called *T. a. aculeatus* by Griffiths (1978); these have an intermediate but variable pelage, with intermediates between *acanthion* and *aculeatus* being found on the northern edge of the *aculeatus* distribution. Kangaroo Island echidnas, according to Griffiths, are the most distinct of all echidnas with a pelage of long, fine hair and many long slender spines (*T. a. multiaculeatus*).

T. a. acanthion has the widest range, occupying most of central and western Australia. *T. a. aculeatus* and *T. a. lawesii* have distributions restricted to the east of the continent, with the latter extending into New Guinea. As their names suggest, both the Tasmanian and Kangaroo Island echidnas are confined to those two islands, although *T. a. setosus* does occur on the larger islands off Tasmanian in Bass Strait (Griffiths, 1978).

Temperature Regulation and Thermoregulatory Behavior

The relationship between pelage of the various subspecies of *T. aculeatus* and their respective thermal environments was investigated by Augee (1969, 1978). Individuals of *T. a. acanthion, T. a. aculeatus,* and *T. a. setosus* were exposed to temperatures from 20°C to 5°C over a period of 37 days. The Tasmanian echidna maintained the highest body temperature (28.2°C) and the Queensland species the lowest (23.5°C); *T. a. aculeatus* from Victoria exhibited an intermediate T_b (28.2°C). Oxygen consumption measurements closely paralleled the trends in T_b, with *T. a. acanthion* having a higher consumption of oxygen than the others at this temperature. Griffiths (1978) cited a personal communication from Hales and Germmell (C.S.I.R.O. Division of Animal Production) which indicated that the echidna has little capacity for peripheral vasomotor adjustment. These differences in thermoregulatory abilities in echidnas from differing thermal environments could be largely due to differences in the insulation of the pelage. Augee (1978) calculated thermal conductances and indeed found significant differences between subspecies. However, the differences in conductances disappeared after periods of acclimatization which produced no noticeable changes in pelage. Augee (1969) suggested a shift in the major site of thermogenesis from peripheral muscle masses to the deep musculature of the body in the initially poorer thermoregulators as a much more efficient means of heat production in response to cold, although he tendered no real explanation of the acclimation effect (Augee, 1978). Augee (pers. comm.) intends to further investigate conductance changes in the echidna in response to acclimation. The actual contribution of the coats of individuals of the three subspecies to overall thermal balance can only be determined by direct measurements of the insulative values of pelts of echidnas taken from different habitats (Hulbert and Dawson, 1974). Certainly, the pelage makes a significant contribution to temperature regulation of platypuses, accounting for between 37 and 94% of the total body insulation at air temperatures ranging from 5 to 30°C (Grant and Dawson, 1978*b*).

Under thermoneutral conditions *T. aculeatus* has a T_b of 31.3 ± 0.6°C (Dawson et al., 1978). However, Augee and Ealey (1968) showed that the T_b fluctuates daily in captive animals exposed to normal summer and winter temperatures in Victoria (below 5 to over 30°C). Active echidnas had a T_b of 31.1 ± 1.0°C while resting animals had a lower temperature of 27.2 ± 2.5°C. Echidnas are also heterothermic in that they lowered the T_b at successively lower air temperatures (Augee, 1976; Martin, 1902). For example, the mean T_b of seven individuals kept at 25°C for 7 days was 32.2 ± 0.7°C and 28.0 ± 1.3°C at 5°C (Augee and Ealey, 1968).

Augee et al. (1970) found that the activity patterns of echidnas were not governed by light or darkness, but rather by ambient temperatures. In winter their captive echidnas moved out of an artificial shelter containing sand for burrowing in the morning when air temperatures reached 18.1 ± 1.5°C, and returned when the temperature declined to 18.6 ± 20.0°C. There was a significant correlation between emergence from the sand and the time at which air temperature exceeded the sand in which the animals had burrowed.

In summer the echidnas studied by Augee and Ealey apparently retreated to shelter and burrowed into the sand when the air temperature exceeded 32°C. They also returned to the shelter to avoid rain. Augee (1976), Martin (1902), and Robinson (1954) found that *T. aculeatus* did not thermoregulate well at ambient temperatures above their body temperature. All of these workers indicated that echidnas became lethargic and fainted at a T_a of above 35°C, and would stop feeding and die if kept at this temperature (Augee, 1976). This poor thermoregulatory performance in response to heat stress is due to the almost complete lack of mechanisms for heat loss in echidnas. They do not pant, lick themselves, or sweat at high temperatures; in fact, they have few, if any, sweat glands (Augee, 1976; Griffiths, 1978; Martin, 1902; Robinson, 1954). Anesthetized echidnas studied by Augee (1976) did not secrete water from their skin either in response to body heating or to the injection of adrenaline subcutaneously.

Such poor temperature regulation would seem anomalous for a species inhabiting many hot and arid habitats. However, Griffiths (1978) suggested that echidnas adapt to this environment by simply avoiding the hottest parts of the day by retreating into caves and other shaded microhabitats. Davies (cited in Griffiths, 1968) studied *T. a. acanthion* in an area of Western Australia where maximum summer shade temperatures reach higher than 44°C, a temperature that would cause an explosive rise in body temperature and death in even the tropical subspecies of echidna (Augee, 1976). Echidnas in Davies' study area occupied small caves formed by wind and soil erosion under laterite formations. Davies measured temperatures in these caves throughout a year and found a buffering effect of as much as 12°C between them and the air outside. At the extreme air temperature of 44°C the cave temperature was 32°C, a temperature close to the limits of thermal tolerance of the species but far enough below it to allow survival (Fig. 10).

Behavioral avoidance of high temperature is also indicated by the data of Augee et al. (1970) presented earlier. Limited data collected from free-ranging echidnas by Augee et al. (1975) showed patterns of activity in wild echidnas similar to those of captive individuals. Daylight activities were related to rising and falling air temperatures and to rain. Some echidnas were active at shade temperatures in excess of 32°C; the authors suggested that echidnas use more shaded microenvironments at these times (although the study did not allow direct observation of this). Obviously further study is required to determine the behavioral repertoire and actual ranges of habitats occupied by the various subspecies of the ubiquitous *T. aculeatus*.

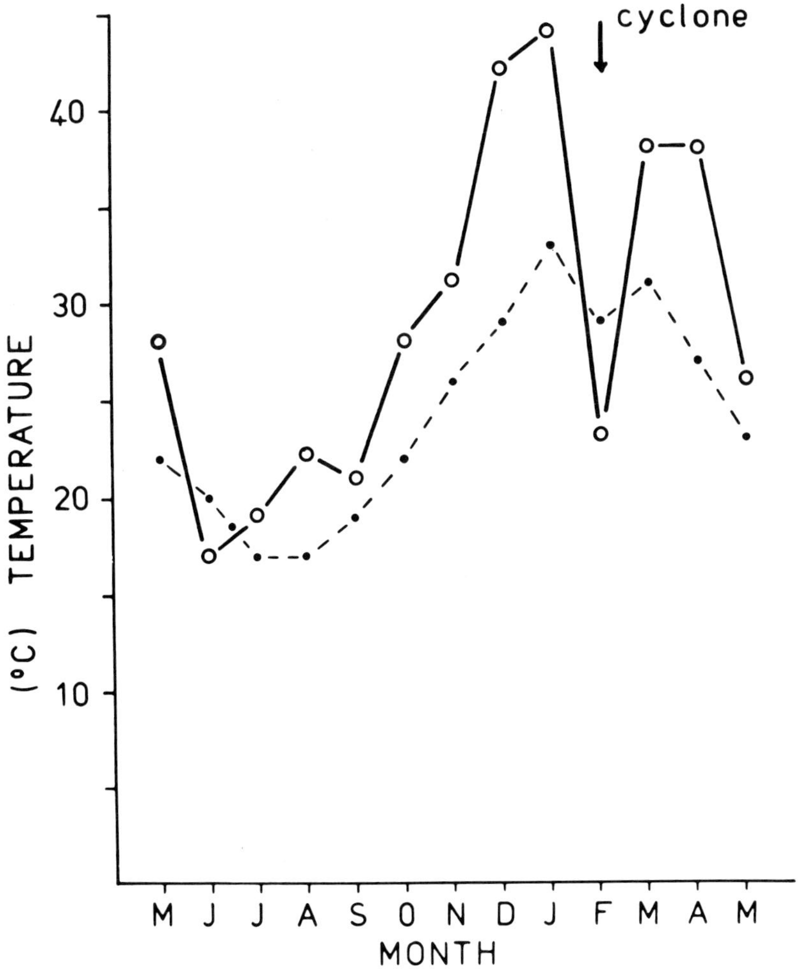

FIG. 10. Maximum air temperatures (open circles) and cave temperatures (solid circles) at noon on the 15th day of each month at Mileura, Western Australia, Australia. (Data from Davies, cited by Griffiths, 1978.)

Perhaps the most striking aspect of the thermal capabilities of the echidna is its ability to become torpid at low temperatures, with its T_b drifting close to ambient at these temperatures (5 or 10°C) (Augee and Ealey, 1968; Griffiths, 1965; Martin, 1902; Mikluoho-Maclay, 1884), and its oxygen consumption reduced to as low as 6% of its non-torpid level at an ambient temperature of 5°C (Augee and Ealey, 1968).

A young suckling echidna, dug from an abandoned rabbit stop and subsequently returned to the damaged stop, dug itself a burrow of 30 cm and exhibited periods of reduced T_b over 17 days, when its temperature fell close to the ambient soil temperatures. This temperature fluctuation suggests that young undergo torpor, which may be a mechanism in suckling echidnas for conserving energy between feeding visits by mothers after the young are evicted from the pouch (Griffiths, 1978).

Although Augee et al. (1970) substantiated the claims of natural torpor in adult echidnas, they could only induce torpor in captives by depriving them of food. Such a reduction in metabolism in response to both low temperatures and food shortage would obviously be beneficial to a species which occupies such a wide variety of habitats in mainland Australia, Tasmania, and Papua New Guinea, where cold conditions could reduce the availability of its food.

Food and Feeding Ecology

Tachyglossus aculeatus is an insectivore whose diet includes small beetles, beetle larvae, lepidopteran larvae (Coleman, 1938; Griffiths, 1968; Semon, 1899), and earthworms (Semon, 1899). The main diet, however, comprises ants and termites, which it gathers by probing its snout into the galleries of nests and mounds, and into rotting wood and other places occupied by these insects. In many instances, access of the snout is gained by an initial excavation with the forefeet, and the food organisms are then drawn into the buccal cavity by the sticky protrusable tongue.

Because insect exoskeletons are passed undigested through the alimentary system of the echidna, Griffiths (1978) was able to analyze scats from a variety of habitats. *T. aculeatus* consumed predominantly ants in cool wet areas (New South Wales tablelands, Snowy Mountains, and northeastern Tasmania), and mainly termite species in the hot, dry areas of Queensland, Northern Territory, and Western Australia (Table 3). Griffiths sought an ecophysiological explanation of this diet preference. Although he found species of termites to contain over 74% of their weight as water compared with only 64% in ants, a termite preference was still seen in the Queensland study area in spite of availability of free water and numerous species of ants. He found no difference in the salt concentrations in the bodies of the two groups of insects, and con-

TABLE 3
PERCENTAGES OF ANTS AND TERMITES IN DIETS OF *T. aculeatus* FROM DIFFERENT HABITATS (FROM GRIFFITHS, 1978)

Habitats	Ants	Termites
Hot, dry		
Queensland	18	82
Northern Territory	29	71
Western Australia	30	70
Cool, moist		
New South Wales tablelands	77	23
Snowy mountains	97	3
Northeast Tasmania	98	2

cluded, tentatively, that the choice may derive from the fact that termites do not bite. Earlier, Griffiths (1968) noted that echidnas showed considerable distress when ants climbed on them as they fed on mounds of the meat ant, *Iridomyrmex detectus*. Restricted availability of termites in some areas may force the selection of ants. Griffiths (1978) noted that in Tasmania termites are sparsely distributed in rotting logs and not in mounds, and that in the Snowy Mountain area mound-building species are not plentiful owing to the cold winter weather. In both of these areas *T. aculeatus* is predominantly an anteater. The occurrence of various types of termites in the scats of echidnas from the Canberra area was related either to the commonness of the particular species or to its accessibility to echidnas (Griffiths, 1968).

The feeding ecology of *T. aculeastus* has been only sparingly researched and detailed studies of the behavior of both this animal and its prey species in various habitats and at different seasons of the year would lead to a much better understanding of the nature of the present distribution of *T. aculeatus*.

Echidnas are responsive to aspects of the biology of prey species (Griffiths and Simpson, 1966). For example, in late winter and early spring, echidnas attack predominantly on the northern warm side of mounds of the meat ant (*I. detectus*) where winged males and virgin queens congregate after leaving the deep galleries of the mound prior to their nuptial flight. Virgin females have a large proportion of body weight made up of stored fats (47.2% of body weight). Griffiths (1968) suggested that echidnas use this food source during arousal in order to replace weight lost during the cold months. Food availability is probably also scarce in winter. In certain areas echidnas may depend on such high energy food sources for continued survival and reproduction.

The apparent restricted mobility in the species (see below) must also profoundly influence feeding behavior and food selection.

Movements

Griffiths (1978) marked 67 echidnas collected in gardens or roads in the Canberra area and later released them in two other areas, one already supporting an echidna population and the second in an area of cleared farmland which was unsatisfactory for echidnas. From his results, it appears that echidnas may remain relatively sedentary for quite long periods in suitable areas because most animals which were mobile were ones released in the farmland. These animals moved distances ranging from 0.8 to 17.6 km; six moved distances of over 1.6 km in periods ranging from 5 days to just under 4 years. Some of these echidnas moved quite rapidly (up to 0.79 km/day); however, others took much longer to cover the distances observed. One was recaptured 7.2 km from its release point, 1,388 days later; this is a rate of movement of only 0.005 km/day! Griffiths did not indicate whether this animal had finally found suitable habitat after its 7.2-km walk. Obviously, conclusions other than those made by Griffiths—that is, "if an echidna is released in a suitable habitat it will probably stay for a long time in a small area around its point of release,"—are unwarranted (Griffiths, 1978).

Augee et al. (1975) systematically studied the movements of radio-tracked echidnas on Kangaroo Island. Over a period of 2 years, radio-tracked animals occupied home ranges of between 635 and 1,067 m in diameter (mean 808 $\pm$ 161 m); several conspecifics (13 in one instance) overlapped and shared home ranges. Interestingly, Augee et al. (1978) found that seven echidnas in the Taronga Zoo in Sydney grouped together in preferred areas which were sheltered from view and from exposure to maximum winter sunlight. These echidnas were mutually tolerant of each other, which substantiates similar general conclusions of Griffiths (1978). Griffiths (1978; pers. comm.) indicated that interpretations of the results of Kangaroo Island study were difficult because the population of echidnas in the Flinders Chase on the island was continually replenished with animals from outside the area for 13 years. However, individuals of *T. aculeatus* exhibited fixed home ranges even at manipulated population densities. Obviously, only more intensive field studies can determine the nature of long-term movements, resource utilization, and the fate of recruits added to natural populations of an extremely sedentary species.

The Long-beaked Echidna, Zaglossus bruijnii

This species, or species closely related to it, once occurred on mainland Australia and Tasmania (Griffiths, 1978; Murray, 1978*a*, 1978*b*). The one extant species of the genus, *Zaglossus bruijnii* (Griffiths, 1978; Van Deusen and George, 1969), is now restricted to the upland areas of Papua New Guinea and Irian Jaya (Fig. 1A). Here it occupies humid and often cold habitats from open rain forest at 1,200 m, through beech forest (*Nothofagus*) at elevations of 3,000 m, to alpine grassland at heights of over 4,000 m (Griffiths, 1978; G. S. Hope, 1976; Hope and Hope, 1976; Mackay, 1976; Van Deusen and George, 1969). Because of its geographical isolation, this monotreme has been the subject of very little research.

Temperature Regulation and Thermoregulatory Behavior

Although Griffiths (1978) and Van Rijnberk (1913) both recorded some variation in the cloacal temperatures in the species, Griffiths suggested that because of its thick pelage with few spines, *Z. bruijnii* would be shown to be a good thermoregulator and have a low total body conductance. In fact, Dawson et al. (1978) showed that the species has a remarkably low total body conductance, which they attributed to its thick pelage and possible cardiovascular adjustments. Conductance is half that of the platypus (Grant and Dawson, 1978*b*), and *Z. bruijnii* has been found to be capable of maintaining its T_b over a restricted range of air temperatures of 15 to 30°C. In that study, the standard metabolic rates (S.M.R.) of the two zoo-kept *Z. bruijnii* were found to be only 25% of those predicted for eutherian mammals of the same body weights (Kleiber, 1961). The maintenance of constant T_b with such a low metabolic rate is not so anomalous because the level of T_b maintained is quite low (31.7°C), and the animal has a high body insulation and is quite large. Thus, the animal has a relatively small surface area for loss of body heat. Although these animals were not exposed to temperatures which would be experienced in the higher altitudes of the species' range, where summer minimum temperatures may fall close to 0°C and summer maxima are below 10°C (Allison and Bennett, 1976), the physiological adaptations found are consistent with the types of habitats occupied by the species. However, the response of *Z. bruijnii* to the very cold and wet conditions known to occur in parts of its distribution, where night temperatures fall below freezing, should be further investigated (Griffiths, 1978).

An apparent anomaly of *Z. bruijnii* is the possession of sweat glands (Griffiths, 1978) and the ability to elevate evaporative heat loss to account for 73% of the small heat production at ambient temperatures above 30°C (Dawson et al., 1978). Obviously, the range of habitats actually occupied by the long-beaked echidna, especially in the less elevated parts of its range, also need to be further investigated.

Food and Feeding Ecology

Popular literature indicates that *Z. bruijnii* feeds on ants and termites (Mackay, 1976). Most biologists, however, believe that the species feeds predominantly on earthworms by probing in the leaf litter and humus with its long curved beak. The tongue is grooved and has recurved keratinous spines for capturing and holding the prey. The prey is drawn into the buccal cavity by the retraction of the tongue and is masticated at its base (Griffiths, 1978). Worms are apparently abundant in both montane forest litter and in peaty soils of highland New Guinea (J. H. Hope, 1976; Van Deusen and George, 1969). Although worms are available in the present range of *Z. bruijnii* and the curved rostrum is apparently suited to vertical probing (Murray, 1978*b*), the long-beaked echidna may be capable of utilizing alternative prey in different habitats or in different seasons. Even Griffiths (1978), who strongly advocates an earthworm diet for *Z. bruijnii,* reported that individuals may also eat larvae of the beetle family Scarabidae. Van Deusen and George (1969) recorded conflicting evidence for insect-feeding by the species and suggested more observations of non-captive animals were needed before the real nature of food selection by the long-beaked echidna would be known.

According to Hawkins and Jewell (1962) earthworms have a caloric value on a wet-weight basis of only 53% that of a mixed insect diet. Such a low energy intake per prey organism may, however, be adequate to maintain *A. bruijnii* with such a low metabolism (Dawson et al., 1978). Considering both their food supply and the cold temperatures of the higher altitudes of parts of their range, it would not be surprising to find that *Z. bruijnii* is capable of entering torpor. Such an ability could be used either in times of food scarcity or perhaps when temperatures fall below freezing at night as they do in the Carstensz area of Irian Jaya (Allison and Bennett, 1976), where the species is known to occur (J. H. Hope, 1976).

Movements

The movements of *Z. bruijnii* within and along gradients of vegetation types with increasing and decreasing altitude in New Guinea have not been investigated. Such studies might answer those questions posed by the limited studies on thermal adaptations and feeding ecology in the species.

Conclusions

All three species of the order Montremata are highly specialized organisms, showing peculiar adaptations of anatomy, physiology, and behavior to their particular way of life. Recently, Dawson et al. (1979) pointed out marked differences in thermoregulatory capabilities of the three species of monotreme. All species maintain a very similar T_b by the utilization of different combinations of thermoregulatory adaptations in widely different thermal environments. Many of these attributions, as in other modern organisms, are more reflective of environmental adaptations than phylogenetic position. From the studies of Augee (1969, 1976,) Dawson et al. (1978), and Grant and Dawson (1978*a*, 1978*b*), it can be seen that the adaptations (anatomical, behavioral and physiological) involved in temperature regulation in *Tachyglossus aculeatus*, *Zaglossus bruijnii*, and *Ornithorhynchus anatinus* have been moulded by habitat rather than by phylogenetic relationships.

Z. bruijnii and *T. aculeatus* are restricted in selection of food organisms. Murray (1978*a*) suggested that the entomophagous *Tachyglossus* species competitively displaced the *Zaglossus* genus on the mainland of Australia after the climatic changes of the Pleistocene changed the availability of food organisms. Although the extinct mainland echidnas were insect-feeders, the smaller species of *Tachglossus* were at a competitive advantage after this change in food availability. In spite of its dietary specialization for ants and termites, *T. aculeatus* has been able to occupy a wide variety of habitats because of the ubiquity of its food source.

The platypus is largely an opportunistic insectivore, but it feeds aquatically and competes with other amphibious vertebrate species. Partitioning of resources, both temporally and spatially, seems to have permitted waterfowl, the native water rat, and various species of native and introduced fish to coexist with *O. anatinus*.

The reproductive physiology and behavior of both *T. aculeatus* and *O. anatinus* are unmistakably the product of K-selection (Pianka, 1970; Stearns, 1976). A maximum of one (*T. aculeatus*) or two (*O. anatinus*) offspring are normally produced each year and females of both species do not breed every year (Grant and Griffiths, pers. observ.). This low reproductive potential is largely compensated for by a considerable expenditure of time and energy (protracted lactation) for the successful rearing of young, including construction of complex breeding-burrow systems in *O. anatinus* (Burrell, 1927) and the use of simple burrows (pre-existing or constructed) in *T. aculeatus* (Griffiths, 1978). The conservative nature of this reproductive strategy is revealed by the dramatic decline in platypus numbers which occurred when it was hunted during the last century. However, the steady recovery made by the species since its protection in all states of eastern Australia suggests that the expenditure of such effort on a few young is a successful strategy.

Z. bruijnii has survived hunting in Papua New Guinea for 26,000 years (Murray, 1978*a*). However, recent depredations have resulted from use of dogs in hunting and from increasing habitat destruction, so that it is now considered a vulnerable species in Papua New Guinea (George, 1978). Little is known of the breeding biology of this species, except that it probably has a protracted lactation period like the other two species (Griffiths, 1978). Reductions of its populations from hunting pressure by indigenous people (Hope, cited in Griffiths, 1978) further suggest that *Z. bruijnii* also has a low potential reproductive rate.

It would seem that the successful adaptations of the monotremes, the result of millions of years of evolution (Archer et al., 1978; Murray, 1978*a*, 1978*b*; Woodburne and Tedford, 1975), have been responsible for their presence in the recent fauna of Australia and New Guinea. All three living species are quite specialized in anatomy and physiology. However, at least in *Tachyglossus aculeatus* and *Ornithorhynchus anatinus,* the generality or opportunism of behavior has permitted these species to occupy a range of habitats within their present distributions.

Acknowledgments

I would first like to thank Dr. Mervyn Griffiths for his encouragement of my own work and his unceasing interest in "things Monotreme." Thanks are also extended to him for reading and criticizing this manuscript, which was also read by Drs. A. J. Hulbert, R. C. Bohringer, M. L. Augee, and P. D. Temple-Smith. The opinions expressed are not necessarily held by these people, although I have accepted most of their criticisms and modified my writing accordingly. I must thank the many friends who assisted with my field work and others who have contributed to the success of the other studies mentioned in this review.

Literature Cited

Allison, I., and J. Bennett. 1976. Climate and micro climate. Pp. 61–80, *in* The equatorial glaciers of New Guinea (J. S. Hope et al., eds.). A. A. Balkema, Rotterdam, 244 pp.

Archer, M., M. D. Plane, and N. S. Pledge. 1978. Additional evidence for interpreting the Miocene *Obdurodon insignis* Woodburne and Tedford, 1975, to be a fossil platypus (Ornithorhynchidae: Monotremata) and a reconsideration of the status of *Ornithorhynchus agilis* De Vis, 1885. Australian Zool., 20:9–19.

Augee, M. L. 1969. Temperature regulation and adrenal function in the echidna. Unpubl. Ph.D. thesis, Monash Univ., Clayton, Australia, 240 pp.

———. 1976. Heat tolerances of monotremes. J. Thermal Biol., 1:181–184.

———. 1978. Metabolic consequences of subspecific pelage variations in the echidna. Australian Zool., 20:105–109.

Augee, M. L., and E. H. M. Ealey. 1968. Torpor in the echidna, *Tachyglossus aculeatus*. J. Mamm., 49:446–454.

Augee, M. L., T. J. Bergin, and C. Morris. 1978. Observations on the behaviour of echidnas at Taronga Zoo. Australian Zool., 20:121–199.

Augee, M. L., E. H. M. Ealey, and I. P. Price. 1975. Movements of echidnas, *Tachyglossus aculeatus,* determined by marking-recapture and radio tracking. Australian Wildl. Res., 2:93–101.

Augee, M. L., E. H. M. Ealey, and H. Spencer. 1970. Biotelemetric studies of temperatures regulation and torpor in the echidna, *Tachyglossus aculeatus.* J. Mamm., 51:561–570.

Bohringer, R. C., and M. J. Rowe. 1977. The organization of the sensory and motor areas of the cerebral cortex in the platypus (*Ornithorhynchus anatinus*). J. Comp. Neurol., 174:1–14.

Burrell, H. 1927. The platypus. Angus and Robertson, Sydney, 227 pp.

Coleman, E. 1938. Notes on the hibernation, ecdysis and sense of smell of the echidna under domestication, Victorian Nat., 55:105–7.

Dawson, T. J., D. Fanning, and T. J. Bergin. 1978. Metabolism and temperature regulation in the New Guinea monotreme, *Zaglossus bruijnii.* Australian Zool., 20:99–103.

Dawson, T. J., T. R. Grant, and D. Fanning. 1979. Standard metabolism of monotremes and the evolution of homeothermy. Australian J. Zool., 27: 511–515.

Eadie, R. 1935. Hibernation in the platypus. Victorian Nat., 52:71–72.

Faragher, R. A., T. R. Grant, and F. N. Carrick. 1979. Food of the platypus (*Ornithorhynchus anatinus*) with notes on the food of brown trout (*Salmo trutta*) in the Shoalhaven River, N.S.W. Australian J. Ecol., 4:171–179.

Fleay, D. 1944. Observations on the breeding of platypus in captivity. Victorian Nat., 61:8–14.

Frith, H. J. 1967. Waterfowl of Australia. Angus and Robertson, Sydney, 328 pp.

Frith, H. J., L. W. Braithwaite, and J. L. McKean. 1969. Waterfowl in an inland swamp in N.S.W. II. Food. C.S.I.R.O. Wildl. Res., 14:17–64.

George, G. G. 1978. The status of endangered Papua New Guinea mammals. Pp. 93–100, *in* The status of endangered Australasian wildlife (M. J. Tyler, ed.). Royal Zool. Soc. South Australia, Adelaide, 210 pp.

Grant, T. R. 1976. Thermoregulation of the platypus, *Ornithorhynchus anatinus.* Unpubl. Ph.D. thesis, School Zoology, Univ. New South Wales, Kensington, 161 pp.

Grant, T. R., and F. N. Carrick. 1978. Some aspects of the ecology of the platypus *Ornithorhynchus anatinus,* in the upper Shoalhaven River, New South Wales, Australia. Australian Zool., 20:181–199.

Grant, T. R., and T. J. Dawson. 1978*a*. Temperature regulation in the platypus, *Ornithorhynchus anatinus*: maintenance of body temperature in air and water. Physiol. Zool., 51:1–6.

———. 1978*b*. Temperature regulation in the platypus, *Ornithorhynchus anatinus*: production and loss of metabolic heat in air and water. Physiol. Zool., 51:315–332.

Grant, T. R., and R. B. McBlain. 1976. Observations on daily and seasonal movements, and activity in a population of platypus, *Ornithorhynchus anatinus.* Bull. Australian Mamm. Soc., 3:41–42.

Grant, T. R., R. Williams, and F. N. Carrick. 1977. Maintenance of the platypus (*Ornithorhynchus anatinus*) in captivity under laboratory conditions. Australian Zool., 19:117–124.

Griffiths, M. 1965. Digestion, growth and nitrogen balance in an egg-laying mammal, *Tachyglossus aculeatus* (Shaw). Comp. Biochem. Physiol., 14: 357–375.

———. 1968. Echidnas. Permagon Press, London, 282 pp.

———. 1978. The biology of the monotremes. Academic Press, New York, 367 pp.

Griffiths, M. E., and K. S. Simpson. 1966. A seasonal feeding habit of the spiny anteater. C.S.I.R.O. Wildl. Res., 11:137–143.

Hawkins, A. E., and P. A. Jewell. 1962. Food consumption requirements of captive British shrews and the mole. Proc. Zool. Soc. London, 138:137–157.

Hope, G. S. 1976. Vegetation. Pp. 113–172, *in* The equatorial glaciers of New Guinea (G. S. Hope et al., eds.). A. A. Balkema, Rotterdam, 244 pp.

Hope, G. S., and J. H. Hope. 1976. Man on Mt. Jaya. Pp. 225–240, *in* The equatorial glaciers of New Guinea (G. S. Hope et al., eds.). A. A. Balkema, Rotterdam, 244 pp.

Hope, J. H. 1976. Fauna. Pp. 207–244 *in* The equatorial glaciers of New Guinea (G. S. Hope et al., eds.). A. A. Balkema, Rotterdam, 244 pp.

Hughes, R. L., and Carrick, F. N. 1978. Reproduction in female monotremes. Australian Zool., 20:233–254.

Hulbert, A. J., and T. J. Dawson. 1974. Thermoregulation in perameloid marsupials from different environments. Comp. Biochem. Physiol., 47A:591–616.

Hynes, H. B. N. 1970. The ecology of running waters. Univ. Toronto Press, Toronto, 555 pp.

Kleiber, M. 1961. The fire of life. Wiley, New York, 454 pp.

Mackay, R. D. 1976. New Guinea. Time-Life International, Amsterdam, 184 pp.

Martin, C. J. 1902. Thermal adjustment and respiratory exchange in monotremes and marsupials. A study in the development of homeothermism. Philos. Trans. Roy. Soc., ser. B, 195:1–37.

Miklouho-Maclay, N. de. 1884. Temperature of the body of *Echidna hystrix* Cuv. Proc. Linnean Soc. New South Wales, 8:425–426.

Murray, P. F. 1978*a*. Late Cenozoic monotreme anteaters. Australian Zool., 20: 29–56.

———. 1978*b*. A Pleistocene spiny anteater from Tasmania (Monotremata: Tachyglossidae, Zaglossus). Proc. Roy. Soc. Tasmania, 112:39–68.

Pianka, E. R. 1970. On r- and K-selection. Amer. Nat., 104:592–597.

Robinson, K. W. 1954. Heat tolerance of Australian monotremes and marsupials. Australian J. Biol. Sci., 7:348–360.

Semon, R. 1889. In the Australian bush. Macmillan & Co. Ltd., London.

Smyth, D. M. 1973. Temperature regulation in the platypus, *Ornithorhynchus anatinus* (Shaw). Comp. Biochem. Physiol., 45A:705–716.

Stearns, S. C. 1976. Life-history tactics: a review of the ideas. Quart. Rev. Biol., 51:3–47.

Sutherland, A. 1897. Temperatures of reptiles, monotremes and marsupials. Proc. Roy. Soc. Victoria, 9:57–67.

Temple-Smith, P. D. 1973. Seasonal breeding biology of the platypus, *Ornithorhynchus anatinus* (Shaw, 1799), with special reference to the male. Unpubl. Ph.D. thesis, Dept. Zoology, Australia Natl. Univ., Canberra, 252 pp.

Van Deusen, H. M. and G. G. George. 1969. Results of the Archibald Expeditions. No. 90. Notes on the echidnas (Mammalia: Tachyglossidae). Amer. Mus. Novitates, 2383:1–23.

Van Rijnberk, G. 1913. Kleine Beitrage zur vergleichenden Physiologie. III. Einige Physiologische Beibachtungen und Versuche an zwei Proechidnidae. Bijd. Bierkd., 19:187–196.

WARD, A. 1979. Utilisation of benthic and drift organisms by adult and juvenile platypus, *Ornithorhynchus anatinus* (Shaw) in the Shoalhaven River and Jerrabattgulla Creek, N.S.W. Honours thesis, Univ. Wollongong, Wollongong, New South Wales.

WOODBURNE, M. O., AND R. H. TEDFORD. 1975. The first teritary monotreme from Australia. Amer. Mus. Novitates, 2588:1–11.

WOOLLARD, P., W. J. M. VESTJENS, AND L. MCLEAN. 1978. The ecology of the eastern water rat *Hydromys chrysogaster* at Griffith, N.S.W.; food and feeding habits. Australian Wildl. Res., 5:59–73.

ECOLOGY AND SOCIAL ADAPTATIONS IN DIDELPHID MARSUPIALS: COMPARISON WITH EUTHERIANS OF SIMILAR ECOLOGY

PIERRE CHARLES-DOMINIQUE

Introduction

TROPICAL forest ecosystems of different continents show many similarities in faunal composition. Equivalent ecological categories can be represented either by closely related species (Primates in Africa, Asia and America) or by species of different taxa (African Manidae and American Myrmecophagidae in anteater niches). Anatomical convergence correlated to the exploitation of equivalent food sources, have been discussed by many authors, but the problem of social evolution is still debated. It has been demonstrated that social structures are often correlated with ecological roles, but the different taxa on which these comparisons have been made are generally phyletically closely related (same family or same suborder). In these conditions, it is difficult to evaluate that which results from ecological convergence and that due to common ancestry. In addition, sociality has often been considered only in terms of gregariousness or non-gregariousness, without attention to the inherent mechanisms (for example, systems of social communication, territoriality, social structure).

In this paper, two groups are considered: didelphid marsupials in South America and lorisid primates in Africa. These medium-sized animals live in similar habitats and exploit the same categories of food. They are characterized by the following features: Arboreal-Nocturnal-Climbers and (or) Leapers-Animalivores-Frugivores-Gumivores (=A.N.C.L.A.F.G., Charles-Dominique, 1978*a*). However, the marsupials (metatherians) and placentals (eutherians) have been separated since the early Cretaceous (about 110 mybp) (Hoffstetter, 1976; Keast, 1977) and have followed two parallel lines of evolution. They, therefore, offer good examples for testing theories about the relationship between ecology and social structure.

Data on lorisid primates were previously obtained in Gabon (Charles-Dominique, 1971, 1977*a*, 1977*b*) and comparative information on didelphid marsupials was recently collected in French Guyana. During a 16-month study (August 1978 to December 1979), 670 individuals were marked (plus 1,988 recaptures) and 50 animals were equipped with radio transmitters. Direct observations with headlamps and the determination of animal locations with portable radio receivers were complemented by automatic goniometric recorders set in three stations. The team engaged in this program (see Acknowledgments) included five mammalogists, two botanists, and one entomologist, who cooperated to interpret ecological phenomena. Our results on marsupial social structure are presented and discussed in the present paper, with emphasis of *Caluromys philander* and *Philander opossum,* two sympatric species of similar size but markedly different ecology.

The relation between ecology and social structure shall first be examined for didelphids, and then didelphids will be compared to lorisid primates.

Partition of Food Resources between Sympatric Marsupials

Body Size and Spatial Distribution in Forest

In French Guyana, five species of marsupials correspond more or less to the A.N.C.L.A.F.G. characters (Arboreal-Nocturnal-Climber and (or) Leaper-Animalivore-Frugivore-Gumivore, Charles-Dominique, 1978*a*). Except for body size, percentage of total muscle, and relative size of the eye, their anatomy is very similar (for example, teeth, gut, feet, and hands). Competition for food is avoided mainly through differences in body size and the level of forest strata exploited (Charles-Dominique et al., 1981) (Table 1). This system of niche separation among phyletically related species, combining height of the strata used and size of the animal, is common among many tropical forest taxa, including the five lor-

TABLE 1
RELATIONSHIP BETWEEN BODY SIZE AND LEVEL OF FOREST STRATA EXPLOITED IN DIDELPHID MARSUPIALS AND PROSIMIAN PRIMATES

Species	Body weight (g)	Stratum
Didelphidae (Marsupialia)		
Marmosa murina	40	low (shrubs)
Marmosa cinerea	80	high
Caluromys philander	300	high
Philander opossum	400	low (essentially ground)
Didelphis marsupialis	1,000	low (may climb to collect fruits)
Lorisidae (Primates)		
Galago demidovii	60	all strata of dense structure (lianas)
Galago alleni	300	low
Galago elegantulus	300	high
Arctocebus calabarensis	250	low
Perodicticus potto	1,000	high

isid prosimian species studied in Gabon (Charles-Dominique, 1971, 1977*a*, 1977*b*) (Table 1).

The anatomy of didelphids is somewhat affected by the degree of arboreality. For example, the tail and claws are better developed in *Marmosa cinerea* and *Caluromys philander* (upper strata), and the legs are more elongated in *Philander opossum* (the more terrestrial). Muscular anatomy also reflects such adaptations (Grand, pers. comm.). However, in the didelphids, the degree of anatomical specialization correlated to ecological niche is not as evident as in the Lorisidae.

Diet

A combination of direct feeding observations (n = 120) and examination of stomach contents (n = 90) permitted a diet determination for the five marsupial species considered in this study (Table 2). The two large terrestrial species (*Didelphis marsupialis* and *Philander opossum*) are mainly opportunistic feeders, as shown by Fitch and Sandidge (1953), Hunsaker (1977), and Lay (1942) for the North American opossum. In French Guyana, between

TABLE 2
DIETS OF FIVE SYMPATRIC DIDELPHID MARSUPIALS OF FRENCH GUYANA
(N = NUMBER OF STOMACHS ANALYZED)

Food items	*Didelphis marsupialis* (N = 14)	*Philander opossum* (N = 31)	*Marmosa murina* (N = 5)	*Marmosa cinerea* (N = 3)	*Caluromys philander* (N = 37)
Fruit pulp (trees and lianas)	X	X	X	X	X
Fruit pulp (shrubs)	X	X	X		
Gums					X
Flowers and nectaries (trees)		X			X
Flowers and nectaries (herb)	X	X	X	X	X
Arthropods	X	X			
Earthworms	X	X	X		
Small vertebrates	X	X	X		
Carrion	X	X			

December and June, *D. marsupialis* and *P. opossum* feed largely upon insects and big earthworms that emerge from the damp soil after heavy rains. Carrion is also systematically exploited by these two species, as observed twice in the field (carcasses of the arboreal porcupine *Coendou prehensilis*) and on many other occasions on discarded remains of dissected animals near the laboratory (opossums, sloths, porcupines, rodents). During the peak of fruit production of the palm *Attalea regia*, these two species pass many hours chewing the fibrous husks of fallen fruit and licking the exudates of the pulp. Fruit and animal matter are systematically eaten according to availability, but *D. marsupialis* and *P. opossum* (Gerard, pers. comm.) are mainly opportunistic predators that occasionally eat fruit (see Tamar, 1961, for the same conclusions concerning the North American opossum).

The arboreal species are more frugivorous than the terrestrial ones, but they also prey upon small invertebrates (mainly insects). In our study area, *Caluromys philander* was observed eating fruits on 80 occasions (28 species), eating flower nectar 11 times (three species), and once eating the gums of a single species. Analysis of 37 stomachs indicated that 75% of the diet is composed of vegetable matter (mostly fruits; Atramentowicz, 1982).

Analysis of the different diets (presented in more detail elsewhere; Charles-Dominique et al., 1981) thus indicates opportunistic feeding strategies, especially for the species exploiting lower stratas. Compared to lorisid primates, which are specialized on some categories of food, didelphids appear to be generalists.

Seasonal Variation in Food Availability

To evaluate the seasonal variation in fruit production, systematic collections were made weekly along 1,300 m of transects in the study area. The fruits were identified and weighed (fresh weight) for each collection point. Mapping and phenological observations of fruiting tree species were made simultaneously. Minimum fruit production during the study period was recorded in June–July–August at the end of the rains and beginning of the dry season (Fig. 1). During this period of food scarcity, fruits are particularly sought and exploited by arboreal species and few fall to the ground. Consequently, the ground-living marsupials (*Didelphis marsupialis* and *Philander opossum*) are more affected by fruit shortages than are the arboreal species.

The variation in flying insect density was estimated from light traps, one at 1.5-m and a second at 15-m height, set monthly from three nights before to the second night following the new moon. The insects collected (belonging mainly to 10 orders) were weighed (dry weight). There were pronounced monthly variations in relative abundance. The minimum density occurred during April–May–June; the relative abundance in the undergrowth varied from 1 to 7.8 g (in terms of biomass) and from 1 to 2.75 in the canopy. Nonflying prey could not be evaluated by this method, but the frequency of earthworms emerging from the soil after heavy rains was positively correlated with the amount of precipitation.

The period of food scarcity determined both by records of fruit and insect abundance and also by body-weight changes in the marsupials begins in May, is most critical during June–July–August, and ends in September. Data collected in 1976 (pers. observ.) and from 1967 to 1972 by the Institut Pasteur de Cayenne (variation in weight and reproductive condition of marsupials) suggest that this period of food scarcity is a regular phenomenon, probably correlated with seasonal variations in rainfall.

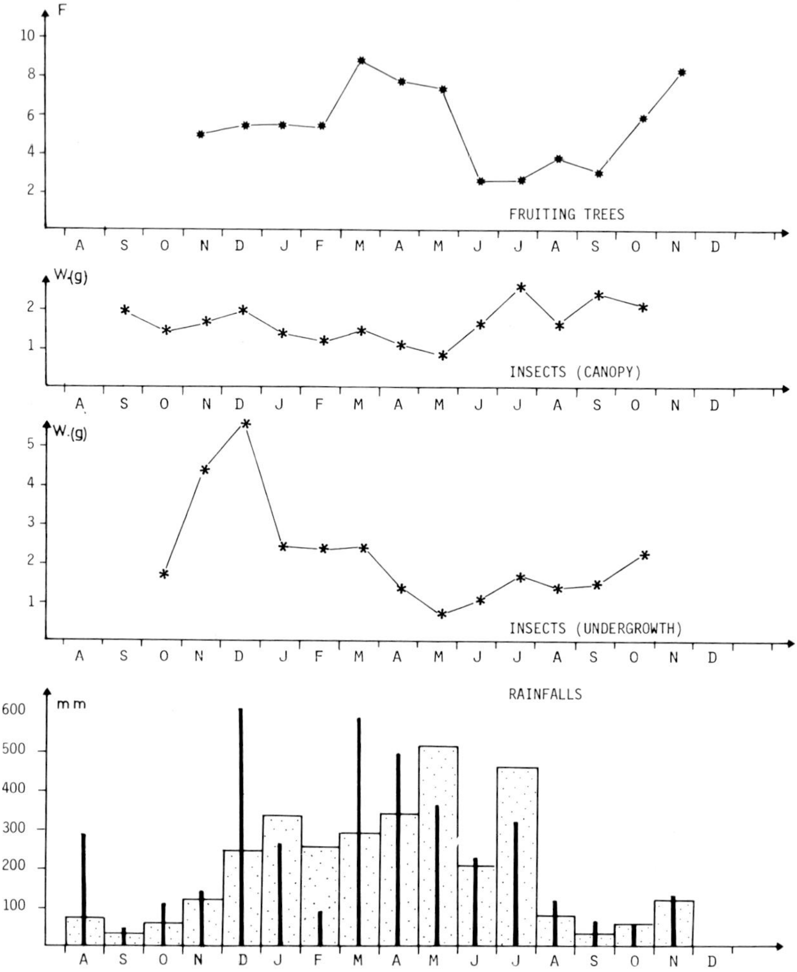

FIG. 1. Correlation between rainfall, monthly variation of relative insect density in the undergrowth and in the canopy, and number of fruiting species. (Rainfall: dotted areas identify monthly means established over a 20-year period; black columns identify amounts during the study period.)

Variation in food supply is particularly critical for marsupial species living in the lowest strata of the forest, where food sources undergo larger fluctuations than in the canopy. This is of great importance for interpreting eco-ethological differences between

Caluromys philander (canopy) and *Philander opossum* (undergrowth and ground).

Social Structures of Marsupials

As with their anatomy, the social structures of different didelphids exhibit small differences correlated to their ecological peculiarities. Before examining this problem it is necessary to review the general patterns of didelphid social structure.

General Pattern of Didelphid Social Structure

Home ranges.—No territories, in the sense of an exclusive defended area, have been observed in didelphid marsupials; each individual occupies a home range, sometimes for long periods (one year in *Caluromys philander*), but there is no exclusion of conspecifics from this zone, which overlaps with other home ranges (Figs. 2 and 3). In fact, if an artificial food source (bananas) is regularly provided, or if a large fruiting (or flowering) tree is producing during several weeks in an area poor in other food resources, the density of marsupials increases progressively, and under these conditions we observed high concentrations of animals.

In contrast to many eutherian mammals (prosimians for example), which regularly explore their entire territory every few days (Charles-Dominique 1971, 1977*a*), the movements of marsupials are essentially confined to feeding and going to and from the den. During the peak of sexual activity, however, males are more active than females (Atramentowicz, 1982) and on rare occasions make displacements that are not related to feeding (drift of home range; see below). Depending on the density and location of food sources, the daily home range and activity rhythm were very variable (Fig. 4)—in four female *C. philander* the daily home range varied from 3,000 to 10,000 m^2 (compared to about 50,000 m^2 in female *Galago alleni,* an African prosimian of equivalent size and ecology) and locomotor activity constituted 35% to 95% of the night, depending on the season (Atramentowicz, 1982). When fruit production decreased, *C. philander* explored the surroundings in an area of about 18,000 m^2. The animals seemed familiar with this area, in which they established five to 27 dens. These dens were used alter-

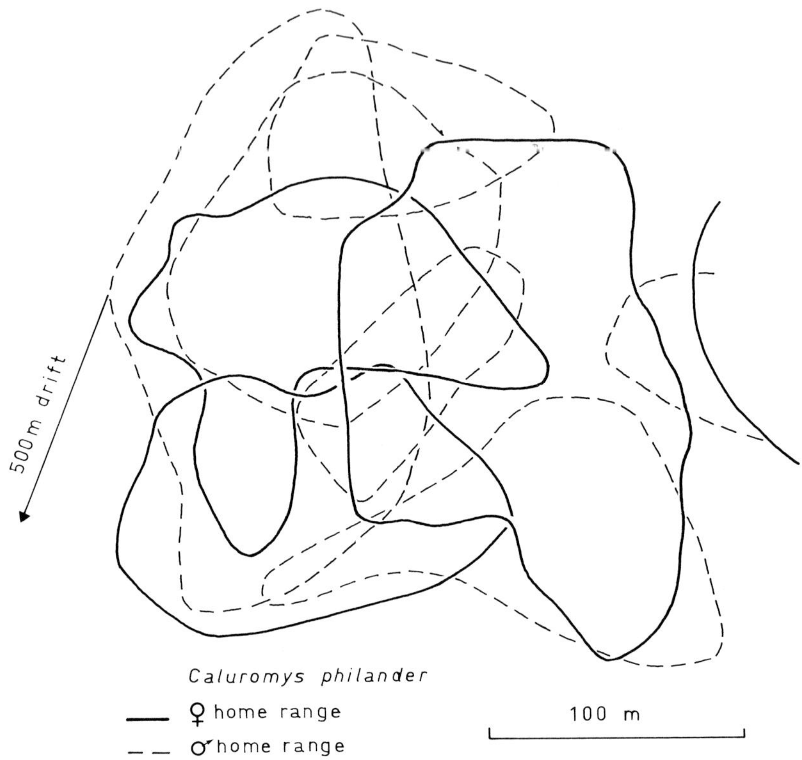

FIG. 2. *Caluromys philander* home ranges determined by radio-tracking. Note the territory of a male which drifted 500 m after 2 months of observations.

nately by conspecifics with overlapping home ranges. Compared to *C. philander, P. opossum* explores a larger area during the night (about 25,000 m^2), probably resulting from its intensive exploration of the milieu to find animal prey (Gerard, pers. comm.).

An interesting behavior is the urine-marking displayed by *C. philander*. By lowering the hindquarters this species deposits urine in a point or on a 10 to 20 cm trail along a branch. This is also typical of several prosimian species, such as *Microcebus, Perodicticus,* and *Nycticebus*. In the absence of any true territory, this behavior, frequently performed in fruiting trees, cannot be interpreted as a territorial signal as is suggested by many authors for

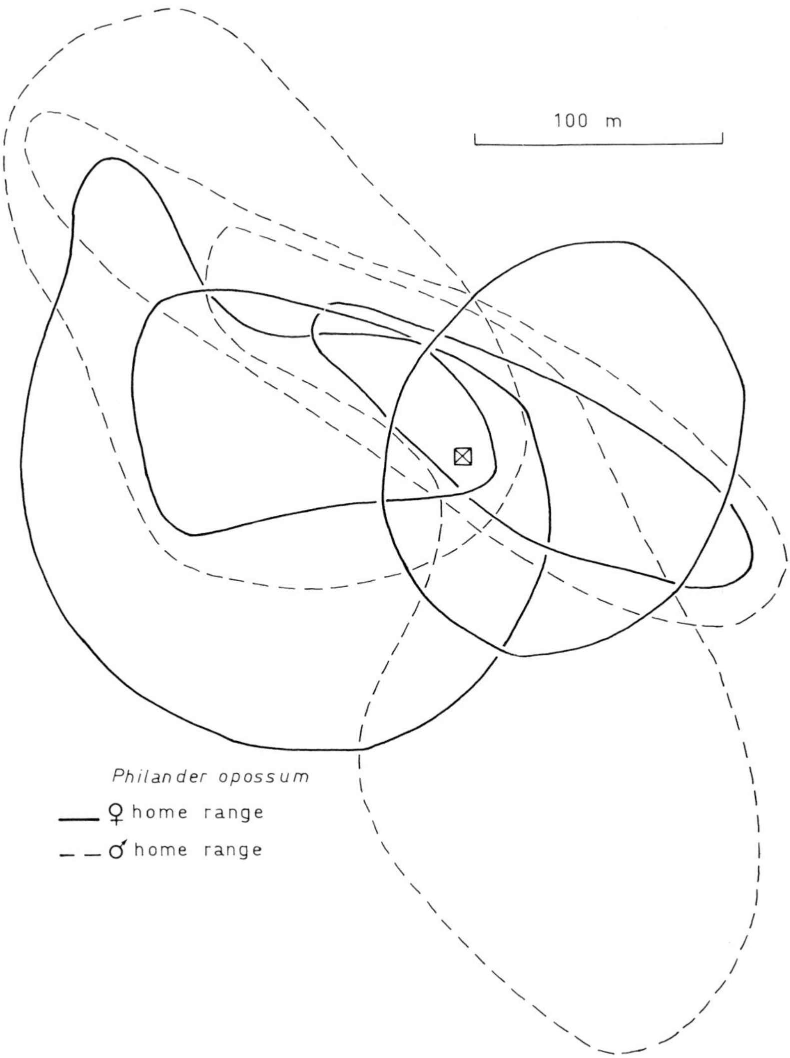

FIG. 3. *Philander opossum* home ranges determined by radio-tracking in the 1976 study area (2,000 m from the 1978–1979 study area). The cross identifies the baiting site where animals congregated to eat bananas.

eutherian mammals. The urine marks may play a role of individual familiarization (and recognition?) within the home range although experimental data are needed.

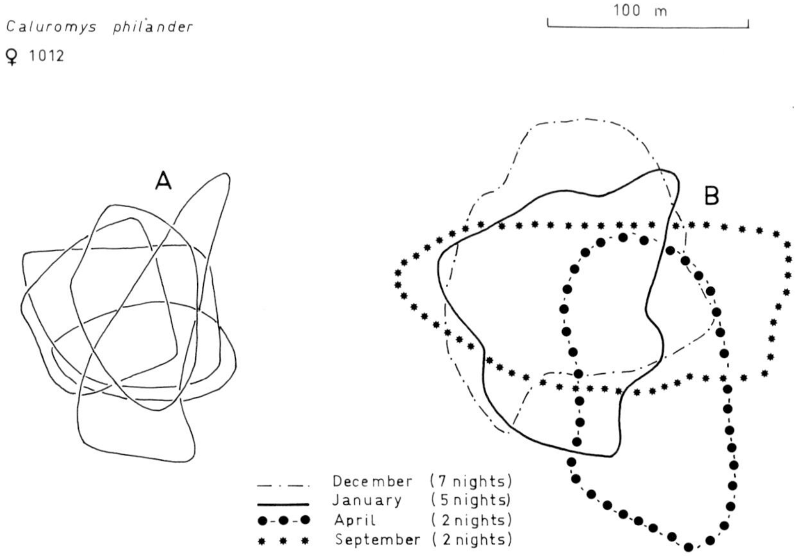

FIG. 4. Details of a female *Caluromys philander* home range determined by radio-tracking. A, areas exploited by the female during five different nights in January; B, areas exploited by the same female during four different months.

Home range drift.—By systematically trapping, marking, and releasing, we have observed a regular loss of individuals, compensated by immigrations from other areas (juveniles and adults of both sexes). (Trapping was conducted using Tomahawk and Sherman traps placed on the ground and special "home-made" traps set at 15 to 20 m in trees; traps were baited with bananas.) This nomadic tendency was noted by Fleming (1972) and Hunsaker (1977) for the genus *Didelphis* and by O'Connell (1979) for six sympatric didelphid species in Venezuela. In our study area, the following data concern new arrivals (juveniles and adults) from May to October 1979, after 8 months of regular trapping and marking (Table 3). This influx of new animals is general, but the phenomenon is very variable when individuals are considered separately. Some young born in the study area remained there. For example, 13 of 78 young *C. philander* marked in the pouch of the mother and raised until emergence from the pouch were recaptured, whereas 32 juveniles of the species immigrated from other areas. The same phenomenon was observed with adults; some of them

TABLE 3

Marsupials Immigrating into the 20 ha Study Area During a Regular Trap-Mark-Release Program

Species	May	June	July	August	September	October
Philander opossum	8	2	1	3	4	3
Caluromys philander	10	5	7	6	1	2
Didelphis marsupialis	2	1	3	5	0	4
Marmosa murina	4	1	3	0	5	5

stayed several months in one area (sometimes more than one year for *C. philander*), whereas others were observed only once or during a few weeks. Opossums readily enter traps, and do not avoid them even after repeated experience. During our nocturnal observations we verified that almost all individuals encountered on the study area had been caught (plastic ear tags). Migration seems to be more important in *P. opossum* than in *C. philander*—after one year, the entire population of *P. opossum* had been replaced by new individuals, but only one-third of the original *C. philander* population was replaced. Mortality rates seem higher in *P. opossum* and probably add to home-range drift, resulting in a more rapid replacement of the population. One male *C. philander,* equipped with a radio transmitter, drifted 500 m in one night and remained permanently in his new home range. Home range drift cannot be directly correlated to food location.

Solitary Animals

During their nocturnal activity, as well as during diurnal resting periods, didelphids are always solitary. However, during development, immediately after the young have released the nipples, they remain for a short period in a nest (8 to 15 days for *P. opossum* and 30 to 45 days for *C. philander*). During the first days (the first weeks for *C. philander*), the young maintain a certain cohesion between themselves (they often grasp each other) and explore the surroundings of the nest. This gregarious behavior indicates the persistence of a need for body contact, directed indiscriminately towards the littermates as well as towards the mother. Sometimes one or several young of the litter cling to the mother's fur during nocturnal activity. Isolated young *C. philander* (especially in situations of discomfort) can emit a high-pitched click which attracts the mother as well as other lactating females (observed experimentally and described for *Didelphis* by Eisenberg and Golani, 1977). On one occasion we trapped a lactating female *C. philander* carrying a young previously marked in the pouch of another neighboring female. This observation suggests that individual recognition may not exist between lactating females and young. In the middle of the night, and again at dawn, the mother returns to the nest for 2 to 3 h. This cohesive behavior disappears after weaning, mother and young becoming indifferent or aggressive when they meet. In captivity, many cases of cannibalism (young–young; young–mother) are observed at this period (pers. observ. of *C. philander, P. opossum,* and *Didelphis marsupialis*; Hunsaker, 1977, for *D. virginiana*).

Social Bonds

We observed no durable social bonds between adult individuals. When two animals met (for example, in a fruiting tree) they often threatened each other with an aggressive vocalization and then continued their individual activities. In a large flowering tree, *Inga ingoides,* twelve adult *C. philander,* feeding on nectar secretions (in September, a period of food scarcity), were indifferent to the presence of conspecifics, except at a distance of less than 2 to 3 m, which resulted in aggression. Aggressive behavior was seen on several occasions in large fruiting or flowering trees (*Ficus* spp. and *Hymenea courbaril*) where animals threatened, chased each other, and emitted aggressive calls. The four-eyed opossum (*P. opossum*), showed similar behavior when artificially attracted to a place baited with banana. To test for a dominance hierarchy among different individuals inhabitating the same area, we conducted a field experiment during the dry season when food was scarce. Nineteen adult four-eyed opossums (10 females; nine males) habitually fed on a 1 m high platform, baited daily with bananas. Animals were individually identified by collars (radio-transmitters and color bands) and only had access to the platform via a single branch. Observations under weak lighting were made from a distance of 3 m. Conflicts were observed principally during the first 2 h of the night when the four-eyed opossums came to take the first meal (12 evenings of observation). Usually, an animal which had not eaten chased one which was already eating on the platform. I observed direct fighting (bites) on only three occasions, but usually only threats were exchanged (hissing, open mouth, and start of chase). No rank order was observed and an individual (male or female) chased by another (male or female) one night could chase it in turn the following night depending on the situation (degree of hunger).

In captivity, a male and a female can be caged together; for the first few days they threaten each other, then they become indifferent and eat and sleep separately (*Marmosa murina, C. philander*). We have never observed any allogrooming between adults, even during mating periods. In eutherian mammals, allogrooming is regularly performed between male and female, even in crocidurine shrews (*Suncus etruscus*; Fons, 1974) which are known for intraspecific aggressivity. Once we saw a radio-tagged female *C. philander* in estrous being followed for two nights by three locally-tagged males. The female was probably not ready to accept copulation and she emitted aggressive calls and rejected the males. After this period the animals returned to a solitary mode of life without any exchange of social contacts. During sexual pursuit, the male emits a series of "clicks" (*C. philander*) or "buzzes" (*M. murina*). This special vocalization has been mentioned by Grassé (1955) and McManus (1970) for the Virginia opossum.

In didelphid marsupials only two or three types of vocalizations are known to be emitted by adults: a weak sexual call (by the male, very rarely), and an aggressive call (by both sexes). The sexual vocalization analyzed in *Marmosa murina* is composed of a series of three to four "clicks" (white noises) separated by 19 millisecond intervals (see Fig. 5); the series are emitted every 0.5 to 1 s. In *C. philander* the sexual vocalization (although not recorded) seems similar but of lower frequency. The aggressive vocalization emitted in situations of inter- and intraspecific conflict as well as in distress is a stereotyped hiss performed with an open mouth (Eisenberg et al., 1975; Hunsaker and Shupe, 1977). The frequency (pitch) of this "hiss" depends on the size of the species, and can be affected by the degree of excitation of the emitter. In *D. marsupialis* it develops towards a growl more or less modified by abrupt changes of intensity when the animal is extremely excited. Except for *C. philander,* which emits a more complex and intense aggressive call (see Fig. 6), didelphids seem to have no long distance calls and a very limited vocal repertoire. Emmons (pers. comm.) once heard *D. marsupialis* give a series of loud moaning sounds audible for more than

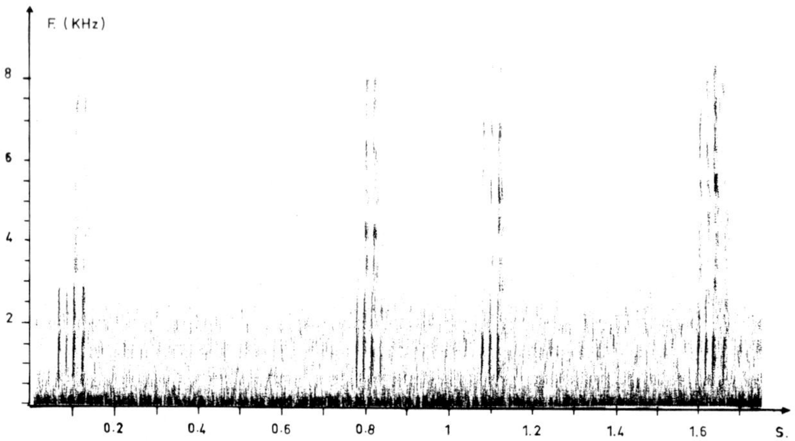

FIG. 5. Didelphid vocalizations: "buzz" emitted by an adult male *Marmosa murina* courting a female.

50 m. In comparison, the Galaginae have a repertoire that includes at least eight types of vocalizations, four of them involving long distance signals (Charles-Dominique, 1977*a*; Doyle, Bearder, and Anderson, pers. comm.).

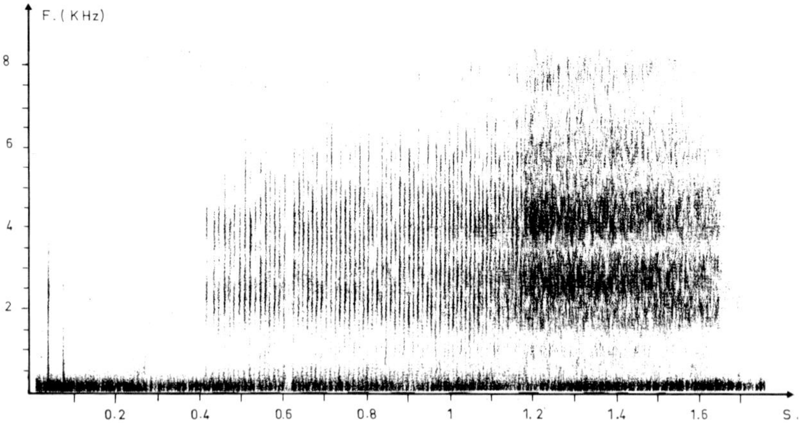

FIG. 6. Didelphid vocalizations: "growl" finishing in a "screech" emitted by an adult *Didelphis marsupialis* menacing a conspecific.

Comparison of Caluromys philander *and* Philander opossum *Social Structures*

These two equivalent-sized species occupy markedly different ecological niches. Variations in food abundance are very pronounced in the undergrowth where the four-eyed opossum lives, but less so in the upper strata used by *C. philander*. In addition, *C. philander* is more frugivorous than *P. opossum*. The two species respond to these situations by adjusting their reproductive rate accordingly. Our data, described in more detail elsewhere (Atramentowicz, 1982, for *C. philander*; and Gerard, pers. comm., for *P. opossum*), permit us to schematize the two reproductive systems (Figs. 7 and 8). The time devoted to raising one litter (birth to weaning) is short for the four-eyed opossum (68 to 75 days) and nearly twice as long for *C. philander* (110 to 125 days). However, at weaning, young of both species reach about the same size (50 to 75 g). A similar situation has been described for the swamp wallaby (*Wallabia bicolor*) and the red kangaroo (*Megaleia rufa*) by Merchant and Sharman (1966), who demonstrated by exchanging the litters that the difference in growth rates was largely due to the foster mother (probably specific milk composition). In addition, young female *P. opossum* can breed at 6 months, whereas the youngest female *C. philander* observed with newborn in the pouch was about 10 months old.

The ovarian cycle in marsupials is inhibited by lactation (Sharman, 1970) and by physiological weakness owing to food scarcity (Newsome, 1965, for the red kangaroo), but not by gestation (Sharmon, 1970; Tyndale-Biscoe, 1973). Because of this peculiarity and the very short gestation period (11 to 13 days for *Didelphis marsupialis,* Tyndale-Biscoe, 1973; 14 days for *Marmosa robinsoni,* Barnes, 1968; and probably similar values for other didelphids), a very short time separates consecutive generations (variations in *C. philander*). In a year, *P. opossum* can have three successive litters (G1, G2, G3) and *C. philander* two (G1, G2) (see Figs. 7 and 8). In fact, G3 of *P. opossum* generally occurs at the beginning of the period of food scarcity and in most cases the females lose weight and young die in the pouch at a more or less early stage. Exceptionally, another litter can follow and abort, but generally reproduction is interrupted until the beginning of good recrudescence. Among the Didelphidae, *P. opossum* behaves more like an *r*-strat-

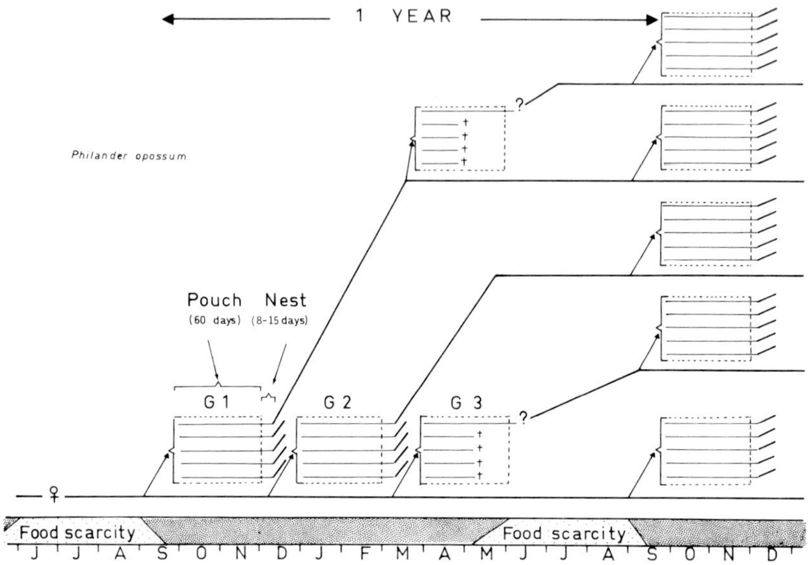

FIG. 7. Schema of reproduction of *Philander opossum*. The different generations (G1, G2, G3) are closely spaced compared to *Caluromys philander* (Fig. 8). Note the death of the young in the pouch and the interruption of reproduction during the period of food scarcity. The question mark symbolizes the possibility of a successful rearing (of the whole litter) observed on very rare occasions.

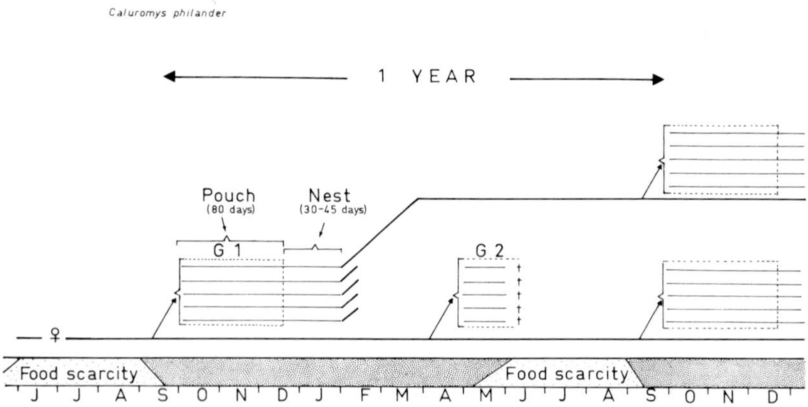

FIG. 8. Schema of reproduction in *Caluromys philander*. The different generations (G1, G2) are at longer intervals than in *Philander opossum* (same symbols as in Fig. 7).

TABLE 4
AVERAGE LITTER SIZE OF FIVE SPECIES OF DIDELPHID MARSUPIALS (N = NUMBER OF LITTERS EXAMINED IN THE POUCH AT AN EARLY STAGE; m = MINIMUM; M = MAXIMUM)

Species	Mean	Standard deviation	Variance	N (m–M)	Confidence interval (t = .05)
Caluromys philander	4.1	1.41	1.96	67 (1–7)	3.76 ≤ mean ≤ 4.44
Philander opossum	4.24	1.31	1.71	179 (1–7)	4.05 ≤ mean ≤ 4.43
Didelphis marsupialis	4.57	1.06	1.10	42 (2–7)	4.24 ≤ mean ≤ 4.9
Marmosa cinerea	6.25	2.75	5.69	4 (3–9)	1.88 ≤ mean ≤ 10.62
Marmosa murina	8.45	2.50	6.07	33 (1–11)	7.57 ≤ mean ≤ 9.33

egist and *C. philander* more like a *K*-strategist, adaptations that can be easily correlated with their niche specializations.

Eisenberg and Wilson (1981) compared different didelphid species and concluded that arboreality is associated with characteristics such as augmentation of longevity, metabolic rate, brain size, and decreased litter size. In French Guyana, we observed slight differences (not significant) between mean litter sizes of *C. philander* and *P. opossum* (Table 3); however, these two species differ greatly in their productivity because of differences in developmental time of young (especially during the nidicolous period: 8 to 15 days in *P. opossum*; 30 to 45 days in *C. philander*).

A comparative study conducted in our captive colony in Brunoy (Aubert, pers. comm.) has permitted us to observe the behavior of these two species in the nest. The young begin emitting very short high-pitched "clicks" (4 milliseconds) at a very early stage (about one month before they release the nipples), especially during bouts of allogrooming (the mother licks the young attached to the nipples in her pouch) or when young are experimentally pulled off the nipples (see Fig. 9). Sonographic analysis of both species shows very short white noise isolated or repeated in irregular series (Fig. 9). The clicks progressively organize into a more regular series with the energy mostly distributed between 1 and 2 kHz (fundamental frequency) and few harmonics (Fig. 10). When a young is very excited (when pulled off the nipple during experimental manipulation; and later, when it is naturally isolated from the mother) the clicks develop towards a tonal structure with a fundamental fre-

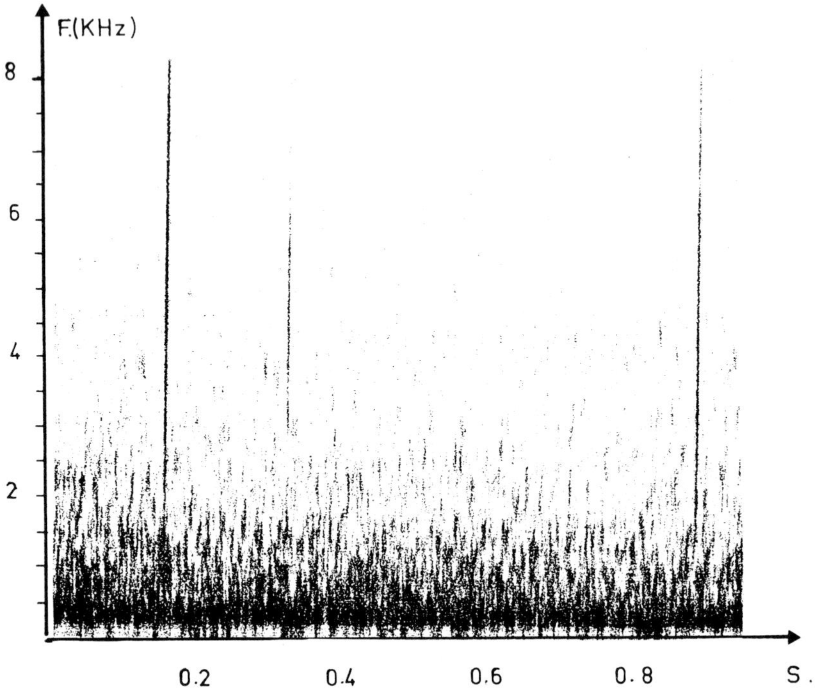

FIG. 9. Didelphid vocalizations: "clicks" emitted by a 1-month old *Caluromys philander* pulled off the nipple.

quency at 1.5 to 2.5 kHz and several harmonics modulated during 10 to 100 milliseconds (Fig. 11). This is about the period when the young leave the nipples, and vocalizations have a more or less structured or unstructured pattern according to the degree of excitation of the young (more variable in *C. philander*).

In *P. opossum* these "clicks" (Fig. 12) progressively disappear when the young release the nipple and the first "hiss"—associated with the typical posture (opened mouth)—is emitted very early in conditions of fear (for example, during manipulation by the experimenter) or at an early stage when it is pulled off the nipple artificially. The hiss develops rapidly towards the adult form although it is initially weak (Fig. 13). Similar observations have been made on a hand-reared *D. marsupialis*.

In *C. philander* the hiss appears later in comparison to *P. opossum* (30 days after the young release the nipple) and the clicks

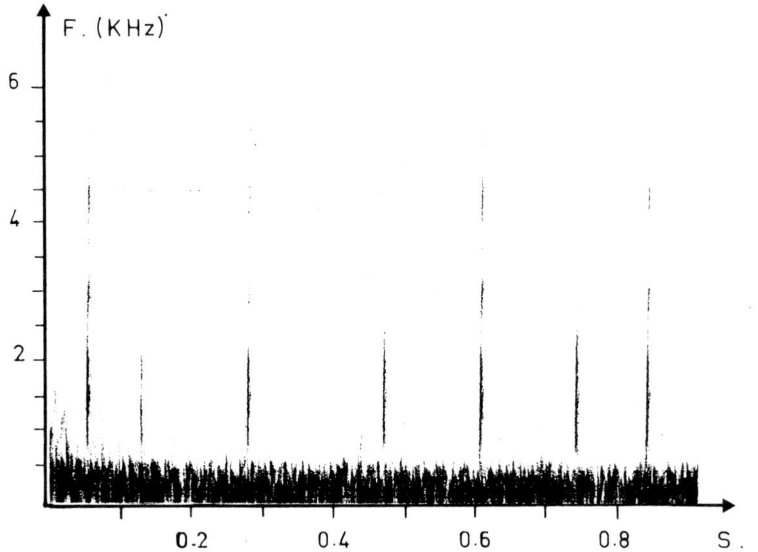

FIG. 10. Didelphid vocalizations: "clicks" of a young *Caluromys philander* at a later stage than the individual shown in Fig. 9. Note the more regular emission rhythm and the irregular energy distribution (outline of fundamental and overtones?).

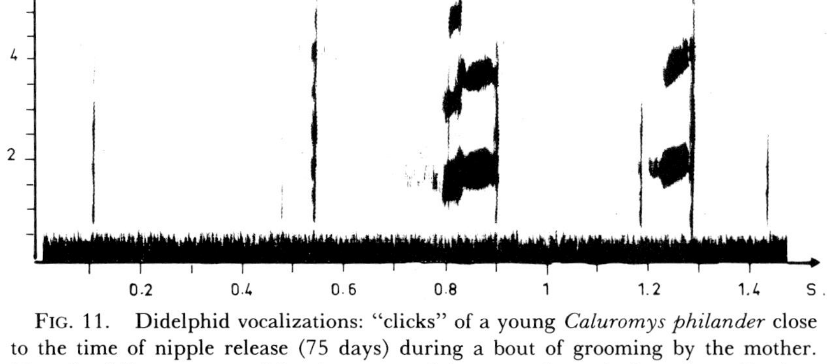

FIG. 11. Didelphid vocalizations: "clicks" of a young *Caluromys philander* close to the time of nipple release (75 days) during a bout of grooming by the mother. Note the different stages of the elements evolving toward a typical tonal vocalization.

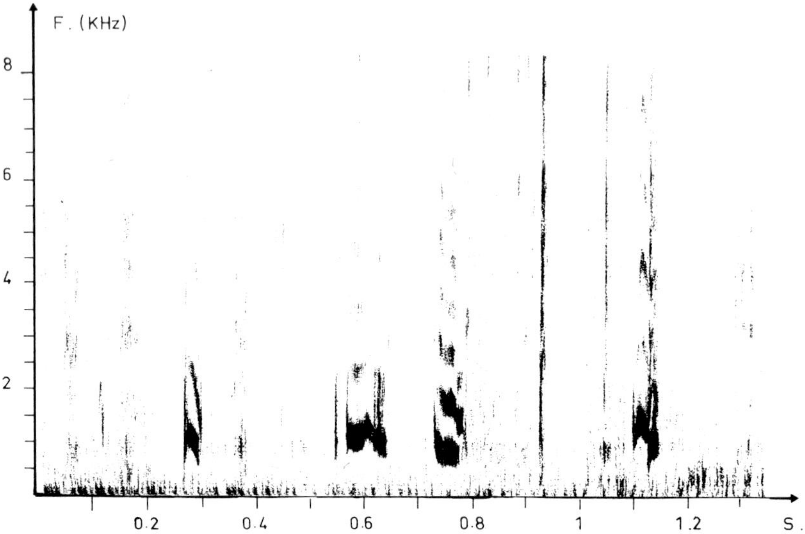

FIG. 12. Didelphid vocalizations: vocalization of a 100-day-old *Philander opossum*. Note the presence of clicks, tonal elements, and hiss.

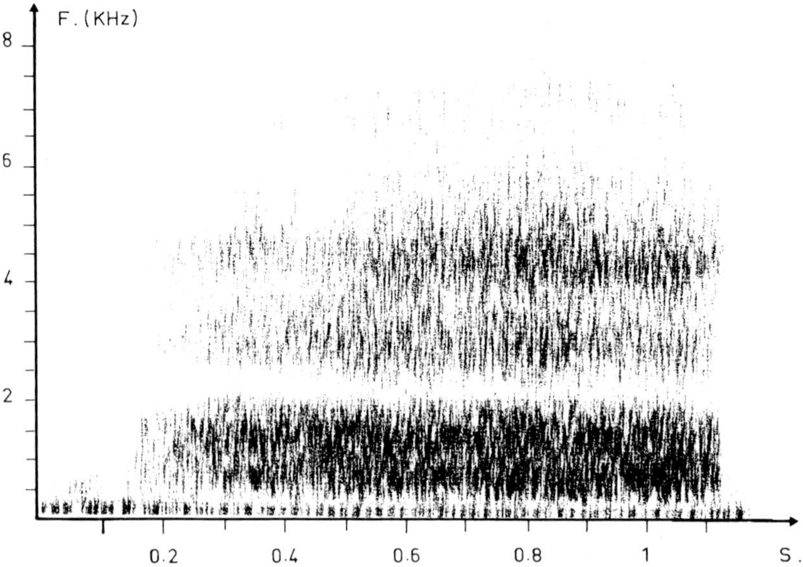

FIG. 13. Didelphid vocalizations: hiss of a 150-day-old *Philander opossum* in situation of discomfort.

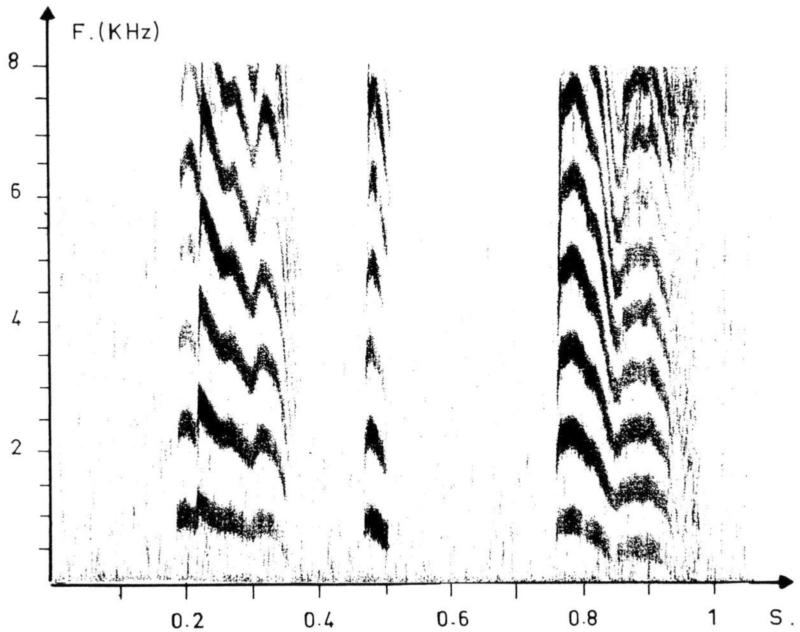

FIG. 14. Didelphid vocalizations: vocalization of a 110-day-old *Caluromys philander* (close to being weaned) in situation of discomfort. Note the appearance of the hiss following the tonal elements.

increase in strength and duration and play a role in social contacts with the mother during this period. This vocalization progressively develops into the species-typical aggressive distress call associated with the hiss. A loud aggressive distress call, audible in the forest at a distance of 100 to 200 m, is emitted in situations of pain (fighting, predation) or intense aggression. This call varies in form and in strength depending on the excitation level, from a weak hiss to a loud "ki-hein—ki-hein—ki-hein . . . ," in which the tonal element alternates with a non-tonal element (probably corresponding to alternations of expirations and inspirations; Figs. 14 and 15).

Calyromys philander is the only didelphid exhibiting such a long distance call (compared to the genera *Philander, Didelphis, Marmosa,* and *Monodelphis*). This characteristic can be interpreted in the light of its ecological features. Among the different marsupial species studied in French Guyana, *C. philander* was the most frugivorous-nectarivorous, and frequently (30 observations) several

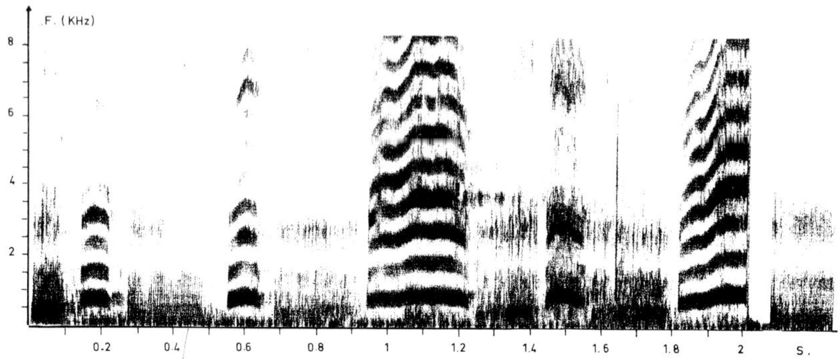

FIG. 15. Didelphid vocalizations: vocalization of an adult *Caluromys philander* in situation of distress.

individuals fed simultaneously in the same tree. No social interactions occurred between these animals, which were simply attracted by the same fruiting or flowering tree. The other more insectivorous species were always seen alone, although up to 19 individual *P. opossum* congregated in the same place after baiting with banana. Foraging behavior, which chiefly involved hunting animal prey, was a solitary activity, and when *P. opossum* fed on fruit, it was during a rapid visit to a tree.

During the night, large fruiting or flowering trees can attract not only several species of frugivorous mammals, but also their predators. On two occasions we saw a *C. philander* caught by a large owl in our study area (once near a group of fruiting palms, *Astrocaryon vulgare,* and another time in a flowering *Hymenea courbaril*). By tracing the source of the powerful distress call we were able to locate each animal dying in the talons of the raptor. When we approached, the bird flew about 200 m and the distress calls continued for 15 min.

Conspecifics feeding in the same tree can probably identify the predator and later flee at its approach. Distress calls are common in many forest mammals such as *Galago, Microcebus,* rodents, and bats, and probably inform conspecifics that are attracted by these vocalizations. The aggressive distress call of *C. philander* could have a similar function.

Comparison of Didelphid and Lorisid Social Structures

All Lorisidae have a common social scheme from which the specific adaptive social structures are derived (Charles-Dominique, 1977*a*, 1977*b*, 1978*a*) as follows:

During the active period at night the animals are solitary, but individuals of a social unit remain in contact by scent markings (= deferred social contacts) and by long-range vocalizations (depending on the species). In contrast with diurnal gregarious primates, individual displacements and activities are independent and unsynchronized. The male may rejoin his female(s) for a short period (sexual visits), even during the non-breeding season and sometimes there is direct contact. During the resting period (day) several individuals of a social unit (adults and juveniles) sleep together in a nest or in vegetation (contact groups). A special vocal signal at dawn permits gathering of the individuals which were dispersed. Social grooming is common during this period.

The territory, signalled to conspecifics by scent marks (urine) and vocalizations (variable), is sexually exclusive: adult females exclude adult females and adult males exclude adult males. However, adult females of the same family can form associations (matriarchies) and their territories may largely overlap. Male territory overlaps with one or several female territories; permanent social bonds are shared between this male and these females independent of reproduction. Subordinate males are generally excluded at the periphery of the territory but in some cases they are tolerated by the dominant male. Juveniles remain in their social group, especially young females; young males emigrate after puberty.

This social structure, based upon territorial organization, is similar to that found in other eutherian mammals, particularly the more primitive families without diurnal adaptations, such as the Viverridae (Charles-Dominique, 1978*b*), Tragulidae (Dubost, 1975), Manidae (Pages, 1975), Tenrecidae (Eisenberg and Gould, 1970), and Erinaceidae (Berthoud, 1978). These social structures probably represent the ancestral eutherian features retained in several primitive families of different orders. In these groups, only minor details of the social structure are affected by ecological adaptations.

A comparison of the nocturnal prosimians and didelphid marsupials reveals fundamental differences. In fact, the peculiarly elementary nature of didelphid social structure is more reminiscent of reptiles than of eutherians.

Didelphid marsupials share a complex of common characteristics that give the family a great overall homogeneity. Despite large differences in ecological specializations (for example between *C. philander* and *P. opossum*), there are relatively few differences between them in either morphology, physiology, or behavior.

Morphological differences related to arboreal locomotion.—These are not pronounced. *Caluromys* has a lower percentage of body muscle than *Philander,* a direct correlate of arboreality. Its eyes are relatively larger than in any other genus and its brain and claws are also relatively large (Eisenberg and Wilson, 1981; Grand, pers. comm.).

Longevity.—*Caluromys philander* has a longer expected life span than other didelphids (Collins, 1973, *in* Eisenberg and Wilson, 1981).

Metabolic rate.—Basal metabolic rate is higher in *Caluromys* than in other didelphids (Eisenberg and Wilson, 1981; McNab, 1978).

Growth rate.—*Caluromys philander* has a slower growth rate than *Philander opossum,* which results in a longer dependency period of the young and a decreased reproductive rate because of a longer generation time.

Vocal repertoire.—The vocal repertoire is small and uniform among the didelphids studied. In *C. philander* the call of juveniles develops into a loud call given in contexts of aggression and distress. This call probably serves to warn conspecifics of predators.

In their comparative study of relative brain sizes of didelphids, Eisenberg and Wilson (1981) pointed out a number of characters that are associated with arboreality (especially for the most arboreal genus, *Caluromys*), including increased brain volume, greater longevity, and ". . . perhaps an increased percentage of the juvenile life span spent in a social learning situation." Our behavioral data lead to a similar conclusion. The distinctiveness of *Caluromys,* which belongs to a separate subfamily (Microbiotheriinae), is based only on a small number of characters that do not affect its typical didelphid social structure: nocturnal, solitary activity; no defended territory; alternation of sedentariness and home-range drift without relation to age or sex; more or less complete home-range overlap between many males and females that avoid close contact; no social bonds between adults outside of a short period of copulation; no allogrooming.

This extremely simple type of social organization has no exact equivalent among eutherian mammals and is reminiscent of that of reptiles. The great age of the didelphid lineage (about 100 million years; Hoffstetter, 1976; Keast, 1977) and the large number of archaic morphological characters that they retain, suggests that this kind of social structure represents a conservative condition which, like their anatomy, has undergone little modification up to the present day.

Although didelphids and nocturnal prosimians occupy homologous ecological positions, these two groups have fundamentally different social structures. The diets of the two require no complex morphological or physiological adaptations, and each has retained many primitive characters (marsupial on the one hand and eutherian on the other) compatible with the kinds of foods that they exploit. Do two such different systems of social organization permit equally efficient exploitation of the environment? A description of diet and the foraging substrates used provides only a crude delineation of the ecological niche, but if the dynamics of the habitats are included, the social structures and the ecological adaptations of the two groups may be compared. Didelphids, in contrast to nocturnal prosimians, are much more numerous in secondary than in mature forest (Charles-Dominique, 1971). In French Guyana, the biomass of the five sympatric didelphids was estimated at 200 kg/km^2 in secondary forest (Charles-Dominique et al., 1981), whereas in primary forest densities are much lower. Trapping yielded one individual/night/150 traps in primary forest, but 30 to 50 individuals/night/150 traps in secondary forest. The frequency of opossums encountered is 10 to 15 times higher in secondary forest than in primary forest.

Secondary forests, especially during the early stages of regeneration, consist mainly of pioneer plant species with high productivities and short life spans. These are quickly succeeded by other species that in turn colonize the habitat until a homeostatic equilibrium is established in mature, or primary, forest (Halle et al., 1978). Drawing a parallel between the reproductive strategies of plants and animals, Halle et al. (1978) defined pioneer species as *r*-strategists, compared to the *K*-strategists that dominate mature forests. Mature tropical rain forest is thus a relatively stable environment, with a slow turnover occurring chiefly in limited areas of treefalls. Young secondary forest is a less stable habitat with rapid turnover. In Africa, nocturnal prosimians, with their low reproductive rate and large home ranges, are adapted to the stable habitat of mature forest. In South America, however, didelphid marsupials, with their high reproductive rate, rapid growth, and lack of territorality that permits high densities around abundant food sources, are adapted to the unstable environment of secondary forest. One might conjecture that in primary forest didelphids principally use areas undergoing regeneration (treefalls), which approach the condition of young secondary forest. The studies that we are continuing in French Guyana will permit us to test this hypothesis.

Acknowledgments

This research was part of a larger multidisciplinary program (nocturnal, arboreal, and frugivorous mammals; plant-animal interactions) undertaken with the cooperation of Martine Atramentowicz, Mireille Charles-Dominique, Nicolas Dégallier, Hugues Gérard, Annette Hladik, Marcel Hladik, Marie-Francoise Prévost, and myself. Additional publications are in progress. It was sponsored by the Centre National de la Recherche Scientific and the Délégation Générale de la Recherche Scientifique et Technique, Paris (A. T. P. no. 7970025). I particularly thank the Office de la Recherche Scientifique et Technique Outre Mer, at Cayenne, for their hospitality and material help in Guyana, and Howard Cooper and Louise Emmons for helping to translate the manuscript.

Literature Cited

Atramentowicz, M. 1982. Influence du milieu sur l'activité locomotrice et la reproduction de *Caluromys philander* (L.). Rev. Ecol., 36.

Barnes, R. D. 1968. Small marsupials as experimental animals. Lab. Anim. Care, 18:251–257.

Berthoud, G. 1978. The movements of the European hedgehog: a preliminary study. La Terre et la Vie, 22:73–82.

Charles-Dominique, P. 1971. Eco-éthologie des Prosimiens du Gabon. Biol. Gabon, 7:121–228.

———. 1977*a*. Ecology and behaviour of nocturnal primates. Duckworth, London, 277 pp.

———. 1977*b*. Urine marking and territoriality in *Galago alleni* (Waterhouse, 1837—Lorisoidea, Primates)—A field study by radio-telemetry. Z. Tierpsychol., 43:113–138.

———. 1978*a*. Ecological position of the family Lorisidae compared to other mammalian families. Bull. Carnegie Mus. Nat. Hist., 6:26–30.

———. 1978*b*. Ecologie et vie sociale de *Nandinia binotata* (Carnivores, Viverridés). Comparaison avec les Prosimiens sympatriques du Gabon. La Terre et la Vie, 32:477–528.

CHARLES-DOMINIQUE, P., ET AL. 1981. Les Mammifères frugivores arboricoles nocturnes de la forêt guyanaise: inter-relations plantes animaux. La Terre et la Vie, 35:341–435.

DUBOST, G. 1975. Le comportement du Chevrotain africain *Hyemoschus aquaticus* Ogilby (Artiodactyla, Ruminantia). Sa situation écologique et phylogénétique. Z. Tierpsychol., 37:403–501.

EISENBERG, J. F., AND I. GOLANI. 1977. Communication in Metatheria. Pp. 575–599, *in* How animals communicate (T. A. Sebeok, ed.). Indiana Univ. Press, Bloomington, 1128 pp.

EISENBERG, J. F., AND E. GOULD. 1970. The tenrecs: a study in mammalian behavior and evolution. Smithsonian Contrib. Zool., 27:1–137.

EISENBERG, J. F., AND D. WILSON. 1981. Relative brain size and demographic strategies in didelphid marsupials. Amer. Nat., 118:1–15.

EISENBERG, J. F., L. R. COLLINS, AND C. WEMMER. 1975. Communication in the Tasmanian devil (*Sarcophilus harrisii*) and a survey of auditory communication in the Marsupialia. Z. Tierpsychol., 37:379–399.

FITCH, H. S., AND L. SANDIDGE. 1953. Ecology of the opossum on a natural area in northeastern Texas. Publ. Mus. Nat. Hist., Univ. Kansas, 7:305–338.

FLEMING, T. H. 1972. Aspects of the population dynamics of three species of opossums in the Panama Canal Zone. J. Mamm., 53:439–455.

FONS, R. 1974. Le répertoire comportemental de la Pachyure etrusque *Suncus struscus* (Savi 1822). La Terre et la Vie, 1:131–197.

GRASSÉ, P. P. 1955. Marsupiaux. Pp. 93–142, *in* Traite de Zoologie (P. P. Grasse, ed.). Masson, Paris, 17.1:1–1170 pp.

HALLE, F., R. A. A. OLDEMAN, AND P. P. TOMLINSON. 1978. Tropical trees and forests, an architectural analysis. Springer-Verlag, Berlin, 441 pp.

HOFFSTETTER, R. 1976. Histoire des Mammifères et dérive des continents. La Recherche, 64:124–138.

HUNSAKER, D. 1977. Ecology of New World marsupials. Pp. 95–156, *in* The biology of marsupials (D. Hunsaker, ed.). Academic Press, New York, 537 pp.

HUNSAKER, D., AND D. V. SHUPE. 1977. Behavior of New World marsupials. Pp. 279–347, *in* The biology of marsupials (D. Hunsaker, ed.). Academic Press, New York, 537 pp.

KEAST, A. 1977. Historical biogeography of the marsupials. Pp. 69–95, *in* The biology of marsupials (B. Stonehouse, and D. Gilmore, eds.). University Park Press, Baltimore, 416 pp.

LAY, D. W. 1942. Ecology of the opossums in eastern Texas. J. Mamm., 23:147–159.

MACMANUS, J. J. 1970. Behavior of the captive opossum, *Didelphis marsupialis virginiana*. Amer. Midland Nat., 84:144–169.

MCNAB, B. K. 1978. The comparative energetics of neotropical marsupials. J. Comp. Physiol., ser. B, 215:115–128.

MERCHANT, J. C., AND G. B. SHARMAN. 1966. Observations on the attachment of marsupial pouch young to the teats and on the rearing of pouch young by foster-mothers of the same or different species. Australian J. Zool., 14: 593–609.

Newsome, A. E. 1965. Reproduction in natural populations of the red kangaroo (*Magaleia rufa*) in Central Australia. Australian J. Zool., 13:735–759.

O'Connell, M. A. 1979. Ecology of didelphid marsupials from northern Venezuela. Pp. 73–87, *in* Vertebrate ecology in the northern neotropics (J. F. Eisenberg, ed.). Smithsonian Institution Press, Washington, D.C., 271 pp.

Pages, E. 1975. Etude éco-éthologique du Pangolin *Manis tricuspis* par radio-tracking. Mammalia, 39:613–641.

Sharman, G. B. 1970. Reproductive physiology of marsupials. Science, 167:1221–1228.

Tamar, H. 1961. Taste reception in the opossum and the bat. Physiol. Zool., 34: 86–91.

Tyndale-Biscoe, H. 1973. Life of marsupials. American Elsevier, New York, 254 pp.

*SOCIOECOLOGY OF SILVERBACKED AND GOLDEN JACKALS (*CANIS MESOMELAS *AND* CANIS AUREUS*)*

PATRICIA D. MOEHLMAN

Abstract

THE behavior and ecology of silverbacked and golden jackals were studied for 34 months over a five-year period near Lake Ndutu, Serengeti Plain, Tanzania. This locale provided the opportunity to study two closely related species inhabiting adjoining but ecologically different habitat. Silverbacked jackals typically occur in the brush woodlands peripheral to Lake Ndutu. Their whelping season is concentrated from July to September. Golden jackals inhabit the adjacent shortgrass plains and typically whelp from December to March.

Both species have long-term pair bonds, cooperative hunting, food sharing, and year-round territories. Some offspring remain with the parents and "help" in the provisioning, guarding, and socialization of the subsequent litter.

In 17 litters of silverbacked jackals, families with helpers exhibited significantly greater survivorship of offspring. This correlated with a higher rate of provisioning and greater percent of time that at least one adult was guarding the pups at the den. By contrast, golden jackal litters with helpers did not have improved offspring survivorship. Presence of helpers in golden jackals did correlate with higher provisioning and guarding rates, but pups were exposed to environmentally dependent mortality from severe thunderstorms and den flooding.

Species differences in the temporal distribution of food resources result in differences in whelping season and dispersal pattern. There are also subtle differences in social behavior, with a higher rate of affiliative interactions seen in golden jackals. In golden jackals 100% of known surviving pups stayed and helped, whereas only 24% of surviving silverbacked jackal pups remained as helpers. The higher frequency of affiliative interactions in golden jackals may indicate tolerance for larger group sizes in this species under conditions where there are sufficient food resources.

Introduction

In recent years the relationship between social systems and ecological constraints has been the focus of much research and review both within species and among closely related taxa (Alexander, 1974; Clutton-Brock and Harvey, 1977; Crook, 1964; Crook and Gartlan, 1966; Crook and Goss-Custard, 1972; Eisenberg, 1966; Jarman, 1974; Kruuk, 1975; Lack, 1968; Wilson, 1975). Major factors influencing a population's social organization are its mating system, spacing system, and feeding ecology. The behavioral substrate for these factors is dynamic and thought to be responsive to such important ecological parameters as abundance and distribution of food resource, nesting site availability, and predation pressure.

In mammals, monogamy is a rare mating system (Eisenberg, 1966; Kleiman, 1977). This is attributed to the heavy physiological investment of the female in the gestation and nursing of the young. When monogamy does occur it tends to be correlated with 1) minimal sexual dimorphism, 2) a long period of dependency in offspring, 3) high paternal investment, 4) delayed sexual maturation of juveniles in the family group, and 5) parental investment by juveniles in younger siblings (Eisenberg, 1966; Kleiman, 1977). Monogamy, especially in the Carnivora, is also closely associated with the occurrence of social groups. The evolution of sociality and the tendency to live in social groups have been reviewed extensively (Alexander, 1974; Eisenberg, 1966; Wilson, 1975) and the selective pressures responsible potentially involve 1) reduced susceptibility to predation, 2) greater efficiency in acquiring food, and 3) extreme localization of a valuable resource such as nesting sites.

Within carnivores the evolution of sociality and increased group size has most often been attributed to increased foraging efficiency (Bowen, 1978; but see Rood, this volume; Kleiman and Eisenberg, 1973; Kruuk, 1975; Lamprecht, 1978; Macdonald, 1977, 1979*a*,

1979*c*). Recent theory (Dawkins, 1976; Hamilton, 1964; Trivers, 1972; Wilson, 1975) has focused additional attention on the potential action of kin selection in social groups of carnivores (Bertram, 1976; Macdonald, 1979*a*; Malcolm, 1979; Moehlman, 1979; Rood, 1978).

This study examines the behavioral interactions and social organization of two sympatric species of jackals (*Canis mesomelas* and *C. aureus*) in light of the theory of kin selection and attempts to relate the social system and pup survival to feeding ecology, predation pressure, and climate.

Materials and Methods

Research on the behavioral ecology of silverbacked jackals (often referred to as black-backed jackals) and golden jackals (*Canis mesomelas, C. aureus*), begun in July 1974, has involved 34 months of observations over a 5-year period. The silverbacked-jackal study site is an area of approximately 25 km^2 in the *Acacia* sp.–*Balanites aegyptiaca* brush woodlands peripheral to Lake Ndutu, Serengeti Plain, Tanzania; golden jackals were studied on the short grasslands adjacent to this area and approximately 15 km to the northeast (Fig. 1).

Individual jackals were identified by natural markings such as ear notches, scars, and color differences. Observations were made at dens and by following individuals as they foraged. Most jackals in this locale are tolerant of cars and I could stay within 20 to 30 m of them during their daily activities. Some individuals were "spooky" and only limited data were taken on them and their families. Most data sets involved 12-h focal samples on individuals or at den sites. Twelve-h watches were not made at dens in which one of the parents was intolerant of the observer so as not to interfere with the normal provisioning and guarding of the pups. Most observations were during the day (0630 to 1930 h) but some nocturnal data were taken using an image intensifier scope on moonlit nights. Jackals were active at night and their provisioning and guarding of pups did not appear to be appreciably different from daytime observations.

A catalogue of approximately 125 coded behaviors was developed and used during most of the study. All activity states and behavioral interactions were voice recorded on tape and later transcribed or

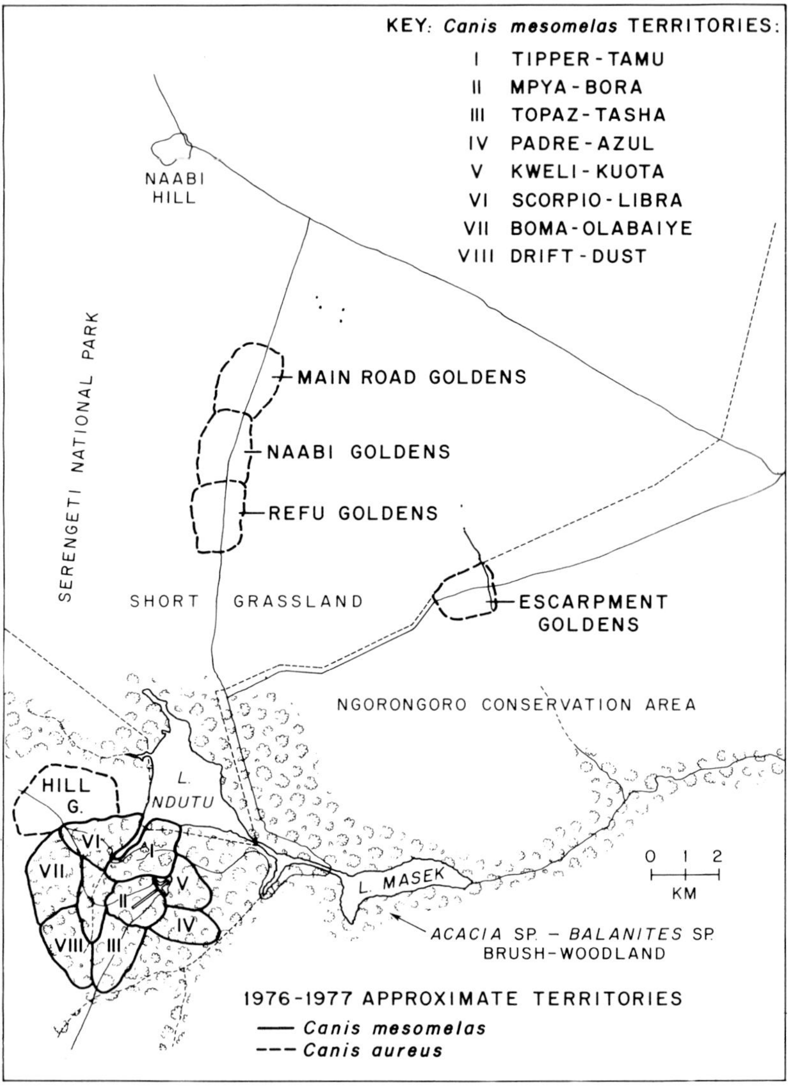

FIG. 1. Approximate jackal territories, 1976–1977.

were recorded directly on an SSR keyboard system (Stephenson and Roberts, 1977). This methodology resulted in frequency and duration data on the coded behaviors. Cine film, still photographs, and sound recordings were made to supplement these data.

Results and Discussion

Mating System

Silverbacked jackals.—Both silverbacked and golden jackals exhibited long-term pair bonds. Of five pairs of silverbacked jackals identified in 1974 (Fig. 2), two were still bonded in 1978. The other three pairs had disappeared and were presumed to be dead. One pair may have been bonded for as long as 6 years. This pair, Padre and Azul, had puppies and a helper in 1974. Since helpers are likely to be offspring from the previous litter, they had probably been mated for at least a year in 1974. They disappeared between October 1978 and January 1979, having been paired for approximately 6 years, perhaps a life-time bond. In each of seven pairs that were observed for 2 to 6 years, no mate changes occurred.

Silverbacked jackals are small omnivores (11 kg) that typically inhabit brush woodlands. In this study population, there was little sexual dimorphism, either physically or behaviorally. Although the female was responsible for gestation and lactation, the mate brought food to her during the nursing period. The female initiated all changes in the location of the den but was assisted by the mate. Pairs exhibited a high degree of synchrony of behavior and tended to scent-mark the territory in tandem, forage together, and rest within 5 m of each other. For example, in one pair providing for a litter of pups (Drift, male, and Dust, female, in 1978), the male and female spent 46% of 121 h within 100 m of each other (resting, 36%; foraging, 10%). When foraging together, this pair scent-marked (raised leg urination) in tandem in 76% of scent marks.

Tandem marking potentially advertises that both members of a pair are in residence. Individual recognition from urine-marks has not been demonstrated for canids, but its has been shown that domestic dogs (*Canis familiaris*) can detect a female's reproductive status (Beach and Gilmore, 1949) and that chemical compounds present in male red fox urine (*Vulpes vulpes*) are absent in female red fox urine (Jorgenson et al., 1978). Macdonald (1979*b*) found

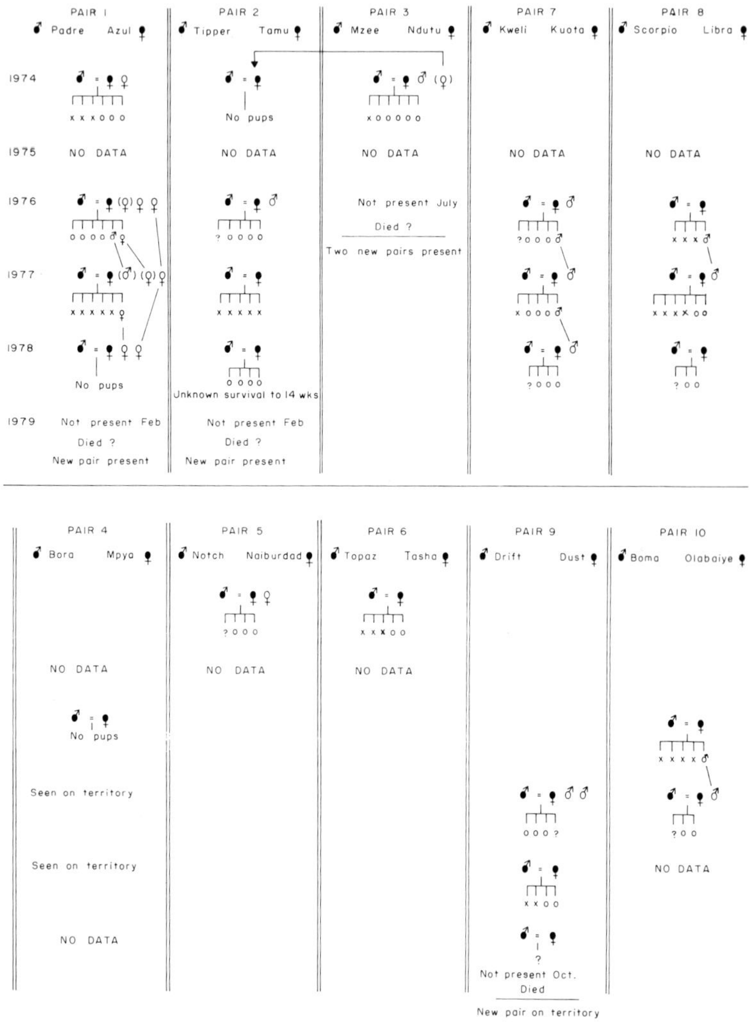

FIG. 2. History of *Canis mesomelas* pairs. Symbols are: ♂, male parent; ♀, female parent; ♂, male helper; ♀, female helper; (♂) and (♀), helpers that disappear before pups reach 14 weeks of age; ×, pup that dies prior to 14 weeks of age; 0, pup that survives to 14 weeks of age; ?, litter size at 3 weeks not known.

that a leashed tame red fox vixen could distinguish between her own and other urine.

Pairs exhibited affiliative behavior, consisting of food sharing, cooperative hunting, mutual grooming, and locating each other by contact calls. They shared killed food items ranging in size from hares (*Lepus* sp.) to adult Thomson gazelle (*Gazella thomsoni*). Small rodents such as *Arvicanthis niloticus* (unstriped grass rat, 60 g) were generally not shared when caught, but on occasion a female harassed the mate until he dropped it. Males provisioned mates during the lactation period. Females initiate the majority of allo-grooming, but are groomed when they solicit the mate.

If one member of a pair dies, the mate potentially could form another bond, but this was not observed in the present study, and single animals may not be able to retain a territory for very long. In late September 1979, Drift disappeared and was presumed dead. His mate (Dust) had recently had a litter of young pups (<3 weeks old). On 4 October a new pair trespassed on and scent-marked Dust's territory, and also ate (and possibly killed) the pups (Camenzind, 1978, observed two cases of possible cannibalism and pup-killing in coyotes, *Canis latrans*). Dust was unable to drive the new pair away and the new female attacked and harassed Dust for the next three days. By the evening of October 7, Dust was apparently exhausted and could barely walk. She was never seen thereafter, and the new pair became permanent residents on the territory.

Although acquiring a new mate is possible, this may be more difficult during the whelping season. However, Dust's inability to maintain the territory and pups by herself may have been due to increased vulnerability from illness (from 1977 to 1979 an undiagnosed disease was inflicting high mortality on silverbacked jackals throughout the Serengeti area).

In the 1980 whelping season, the resident male of a pair raising a 13-week-old cub disappeared. At this time a strange male trespassed on the territory and near the den and resident female. The next day, the resident female was seen for the last time, with serious wounds on her neck and back. On the third day, the unfamiliar male and a female were scent-marking near the den. Four days later the pup of the original pair, who had been staying near the den, disappeared. The unfamiliar pair then established residency on the territory.

Golden jackals.—The mating system of golden jackals in this study area was similar to that of silverbacked jackals. Data of my own and others indicate that golden jackals are monogamous and probably mate for life (Golani and Keller, 1975; Lamprecht, 1978; Schaller, 1967; van Lawick and van Lawick-Goodall, 1970; Wyman, 1967). I observed individuals that were paired for 6+ years (for

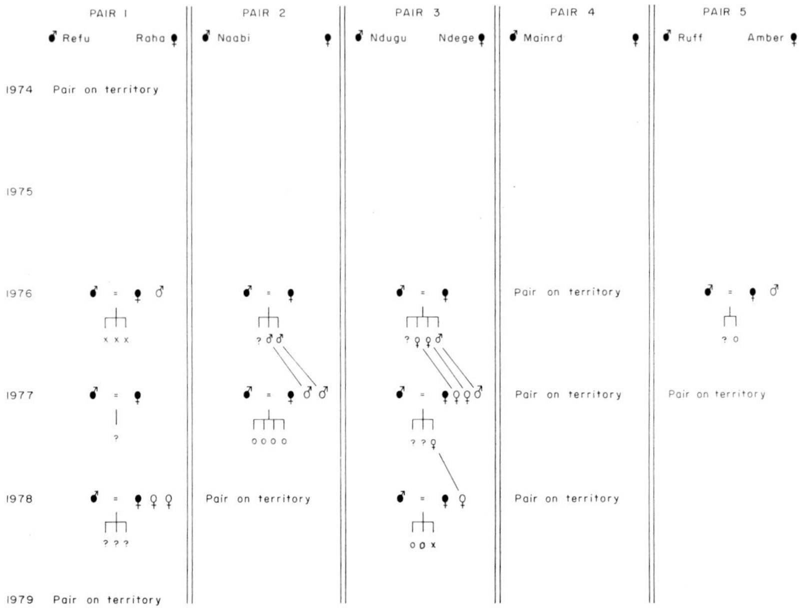

FIG. 3. History of *Canis aureus* pairs. For identity of symbols, see Fig. 2.

example, Refu and Raha, Fig. 3). However, Macdonald's (1979*c*) observations on golden jackals in Israel indicate that the species' social system can be quite flexible. He observed two stable territorial groups of 20 and 10 individuals each with several reproductive females (mating system unknown). He attributed this relatively large group size to the highly clumped and defensible nature of the food resource.

In this study, the golden jackals inhabited the short grasslands where they foraged successfully for items ranging in size from dung beetle larvae to adult Thomson's gazelles. They also scavenged on wildebeest (*Connochaetes taurinus*) afterbirth and herbivore carcasses. Unlike silverbacked jackals, males appeared to be 5 to 10% larger than females. Although females have the major reproductive responsibility, males invested heavily by regurgitating to the mother and pups and driving off potential predators. A golden-jackal male (Ruff) also fed his mate when she was an invalid and unable to forage for herself.

During parental care golden-jackal pairs without a helper exhibited a high degree of synchrony of behavior similar to that of silverbacked jackals. In 25.2-h of focal samples, members of one pair spent 43% of their time within 100 m of each other (resting, 28%; foraging, 15%). They also marked in tandem in 77% of scent-marks when foraging together.

Spacing System

In the Ndutu study area, silverbacked and golden jackals hold territories throughout the year. Data from 1974–1979 indicate that a pair may hold a territory for its lifetime with only slight changes in boundaries. All silverbacked-jackal pairs inhabited the *Acacia* sp.–*Balanites aegyptiaca* brush woodland peripheral to Lake Ndutu. There were differences in both vegetation cover and rodent densities from territory to territory.

Boundaries and the internal area of the territory were actively defended. All observed territorial conflicts involved aggression between animals of the same sex. Female trespassers were threatened, attacked, and driven off by the resident female. If the resident male was present he did not assist his mate. Similarly, if the intruder were male, the resident male was the aggressor.

Territories were maintained indirectly by scent-marking and vocalization. Recorded contact calls of Mzee's family were played once to Padre's family in Padre's territory. There was no response. Contact calls of Padre's family were then played to family members and they responded with contact calls. Though more experiments would be desirable, the results are consistent with the possibility that jackals can distinguish between family and non-family vocalizations. Both members of the pair marked by urination on grass tufts, bushes, and trees within and at the boundary of their territories. If members of a pair were foraging together then they tended to both mark the same spot at nose height.

Mated pairs leave their territories to drink water and scavenge from carcasses. When leaving its territory for water a resident will trot steadily in a fairly direct line and will not scent-mark. Jackals are sensitive to the flight patterns of vultures and raptors. If a jackal observes a large bird making a rapid descent and landing, it will swiftly run to that spot, and as a result, may trespass. Among silverbacked jackals, residents initially will threaten and drive away

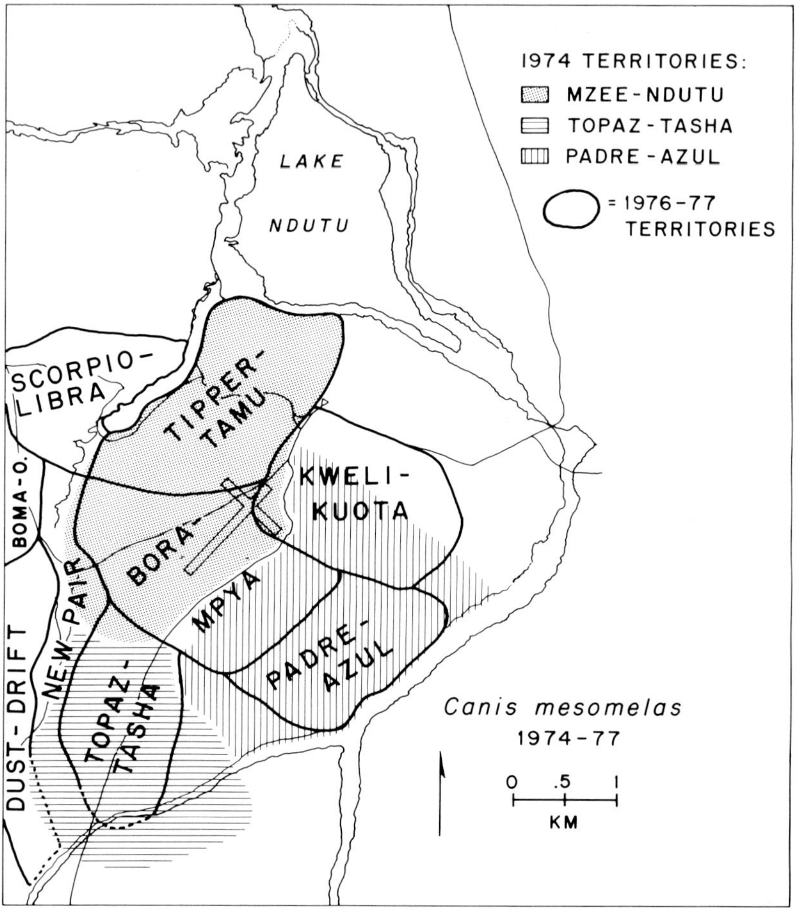

FIG. 4. Silverbacked jackal territories, 1974–1977.

intruders, but if individuals from several territories converge then it is impossible for residents to drive them all away. On occasion large carcasses attracted individuals from as many as six different known territories. Anecdotal accounts of jackals aggregated at large carcasses often have led to assumptions that they operate in large packs.

Golden jackals seldom allowed individuals other than family members to trespass and feed on a carcass. But this simply may reflect the greater abundance and visibility of carcasses on the short grass plains compared to the brush woodlands.

Among silverbacked jackals there was a substantial change in average territory size between 1974 and 1976 (Fig. 4). In 1974, three pairs of jackals (Mzee and Ndutu, Padre and Azul, Topaz and Tasha) held adjoining territories between two gullies leading into Lake Ndutu. When I returned in 1976 Mzee and Ndutu had disappeared and the old territory was split between two new pairs (Tipper and Tamu and Mpya and Bora). Padre and Azul were still in residence but were utilizing and defending a smaller portion of the original territory with Kweli and Kuota and Mpya and Bora in residence on part of the old territory. This change in territory size did not correlate with the number of adults in the family. In 1974 Padre and Azul had one helper, while in 1976 they had two to three helpers.

The change in territory size may have been related to improved food resources. In the 1973–1974 rainy seasons, precipitation was comparatively low (47 cm), but then increased in the 1974–1975 (56 cm), 1975–1976 (51 cm), and 1976–1977 (77 cm) rainy seasons. In the 1976 and 1977 dry season Senzota (pers. comm.) estimated rodent populations by a trap-recapture study in the silverbacked-jackal study area. At the end of the dry season in November 1976 he found densities varying from 3,125 (Scorpio-Libra territory with little ground cover) to 7,792 rats/km^2 (Tipper-Tamu territory). In July 1977 his estimates ranged from 13,125 (Scorpio-Libra territory) to 32,083 rats/km^2. The ratio of rat density in the two areas was 1:2.5. However, the lowest biomass estimate was extremely high (3,000 × 60 g = 180 kg/km^2). In Tipper-Tamu's territory, the biomass was approximately 470 kg/km^2. This relatively high rodent biomass may have enabled jackal pairs to defend a smaller feeding territory and still have sufficient resources.

Helpers

One of the most intriguing aspects of both jackal species in the Ndutu area was the tendency of some offspring to remain with their parents an extra year and "help" in the raising of the next litter. Helpers are usually non-breeding adults approximately 11 to 20 months old, although one female remained with her parents for 2 years (Fig. 2). Golden jackals (male and female) are sexually mature at 11 to 12 months (Kingdon, 1977; Taryannikov, 1976), and data for coyotes (*Canis latrans*), wolves (*C. lupus*), and domestic dogs (*C. domesticus*), indicate that males and females are physiologically capable of breeding at a similar age (Anderson and Simpson, 1973; Gier, 1975; Kennelly, 1977; Seal et al., 1979). Information on age of first reproduction is not available for silverbacked jackals, but because the *Canis* species are closely related, they probably closely resemble other species in the genus. Thus when year-old jackals remain with their parents they are presumably delaying reproduction for a year. With the one female exception mentioned above, all helpers emigrated and/or disappeared by two years of age.

Within the family, jackal helpers have a subordinate status to the parents and may experience suppression of endocrine function and/or reproductive behavior. Sexual behavior was not observed among helpers and male helpers had small testes when visually compared with their fathers. An effect of the social environment on pheromone production and gonadal function has been documented for mice, *Mus musculus* (Lombardi and Vandenburgh, 1977; Vandenburgh, 1971, 1973), collared lemmings, *Dicrostonyx groenlandicus* (Buhl et al., 1978; Hasler and Banks, 1975), and common marmosets, *Callithrix jacchus* (Abbott, 1979; Abbott and Hearn, 1978). Although female wolves (*Canis lupus*) do not reproduce as yearlings in the wild (Medjo and Mech, 1976), three females have done so in captivity. Seal et al. (1979) reported that a female wolf pup came into estrous after the death of the adult alpha female, while two similarly-sized sisters who were subordinate to her remained anestrous. Seal et al. concluded that wolves are physiologically capable of reproducing during their first year, but that the social environment can suppress the first seasonal estrus.

Silverbacked jackals.—From 1974 to 1979 observations were made on 19 litters of silverbacked jackals. In 12 of these litters helpers were present (63%). The sex ratio of known surviving pups (n = 49 at age 14 weeks) and helpers (n = 20) was 1:1 (Fig. 2); 24% of surviving pups remained to help the parents.

From 1976–1979 it was possible to observe successive litters of known pups. Within this sample, all helpers were pups from the previous year's litter (n = 7). Thus they were full siblings to the pups they were helping (r = ½). On average, the degree of relatedness of helpers to full siblings is the same as would be the relationship to offspring should they have reproduced.

Within the family the social status of helpers was always submissive with respect to the parents. Silverbacked-jackal parents rarely regurgitated to helpers but did share food with the older offspring and engage in such affiliative behavior as allogrooming. Same-sex helpers assisted the appropriate parent in defending the territory against same-sex intruders. However, helpers did no scent-marking. Unlike the parents, helpers trespassed on adjoining territories and foraged for rodents and fruit. Generally they encountered a resident within 1 to 2 h and were driven back to the home territory. This type of trespassing may provide important information to young jackals on the availability of unoccupied and suitable habitat.

Golden jackals.—Among golden jackals, a similar pattern to that of silverbacked jackals is seen, although there are interesting differences. Golden-jackal helpers are also subordinate to their parents, but they experience more grooming and food-sharing from them. Golden-jackal pups leave the natal territories for extended periods during the first year. In the dry season (July to November) of 1977, when Ndugu and Ndege's three pups were 6+ months old, they could not be located. Because the short grasslands are very open it seemed unlikely that I would not see the pups while spending 48-h periods on the territory (the observation conditions in brush woodlands do not allow the same certainty). The three pups reappeared in the rainy season (December) and helped to raise the parents' next litter. Again in the 1978 dry season a female offspring could not be located but returned to help in 1979. The dry season is a harsh time on the short grasslands and golden-jackal diets appear to be limited to lizards and insects. Nomadic individuals are often seen in the brush woodlands at this time. They may be young of the year whose natal territory has insufficient food resources. Thus offspring can disperse for several months and then return. Similar behavior has been observed in coyotes (Bekoff and Wells, 1980).

Observations were made on eight litters of golden jackals. Twelve pups (8 males; 4 females) were known to survive to 14 weeks. However, the sex ratio of helpers (n = 10) was equal (Fig. 3). In contrast to silverbacked jackals, 100% of *known* surviving pups stayed as helpers, and 75% of pairs with litters had helpers. These

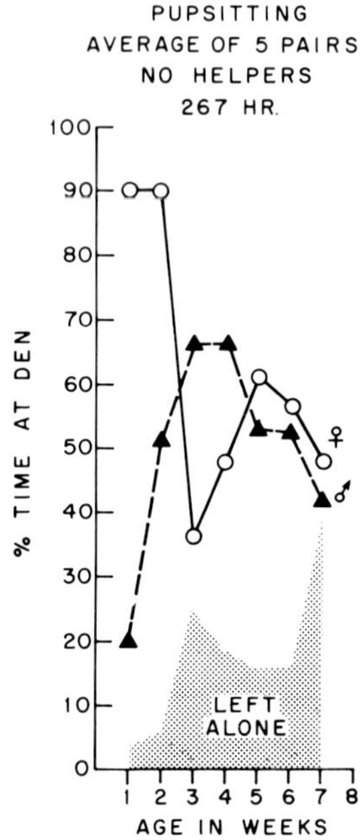

FIG. 5. Pupsitting by male and female silverbacked jackals with no helpers (n = 5 pairs; 267 h of observations).

observations may reflect significant differences or may reflect higher mortality in silverbacked jackals from age three and one-half to 11 months.

Patterns of Alloparental Care

Helpers are not just resident on the home territory, but actively assist the family by 1) guarding pups, 2) feeding pups, 3) socializing pups, 4) feeding the lactating female, and 5) defending the territory. The time budget of parents without helpers differs from parents with helpers. In five silverbacked pairs, each of which raised a litter without helpers, pups were left alone up to 40% of the time (267 h of observations). During the first 3 weeks mothers spent a majority of time in the den, which may have prevented hypothermia. When the pups were about three weeks old the mothers decreased the time at the den from 90% to approximately 35%. Concurrently the mates started spending more time within 100 m of the den.

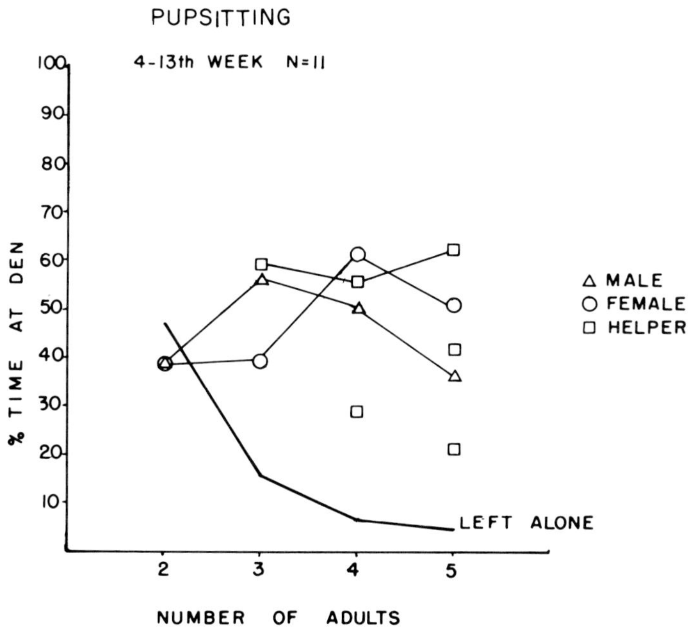

FIG. 6. Mean pupsitting time by silverbacked jackal families with varying number of helpers during weeks 4 to 13 (n = 11; 622 h of observations).

From weeks 4 to 13 both parents averaged about 40% of the time at the den and 60% foraging (Fig. 5).

By contrast, when a helper was present, the pups were left alone on average only 15% of the time (Fig. 6). Mothers spent slightly less time at the den and fathers slightly more, but helpers averaged 60% time at the den, ensuring that the pups were seldom left alone. With two helpers the pattern was more complicated. The mother spent more time at the den and presumably less time foraging. But, because the mother was fed by the mate and helpers during this period, her need to forage for the pups and for herself may have been reduced. The father spent less time at the den and presumably more time foraging. In 12-h tracking of parents, the majority of resting behavior was at the den, while most time away from the den was spent foraging. On average the two helpers continued to spend in total about 55% of the time at the den, and pups were left

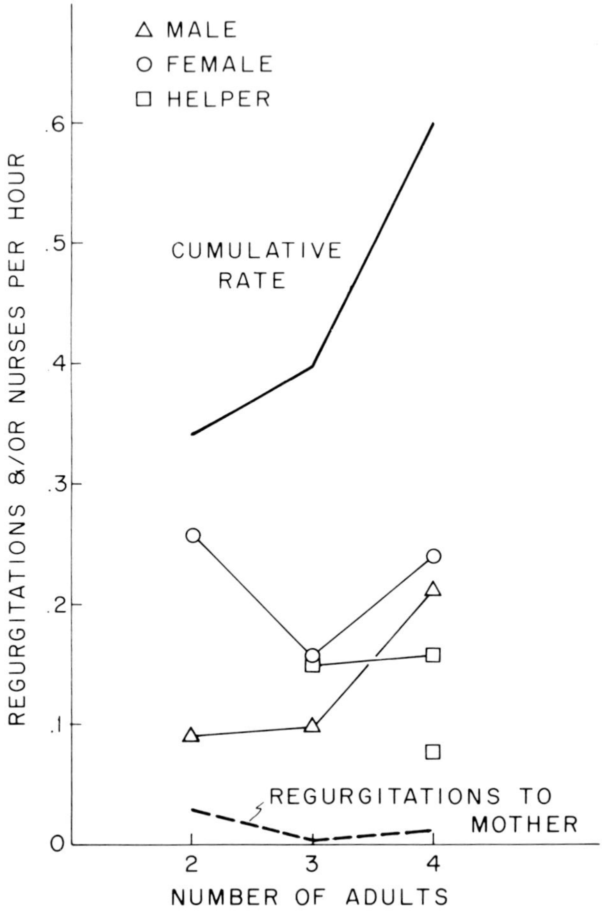

FIG. 7. Mean hourly rate of feeding pups (nursing or regurgitation) at ages up to 14 weeks in silverbacked jackal families (n = 5; 414 h of observations).

alone less than 10% of the time. This same general pattern persisted when a family had three helpers.

The presence of a single adult at the den apparently constitutes protection for the pups. Not only will an adult "rumble growl" or "predator bark," which warn the pups to take refuge and threaten predators, but an adult also can successfully drive large predators away from the den area. Small jackals (11 kg) will chase and bite the rump of a 55-kg hyena (*Crocuta crocuta*). Jackal parents and helpers are quick and can dart away before a hyena can turn and defend its hindquarters.

Silverbacked-jackal helpers regularly contributed food to young pups and the lactating mother. Careful observations of litters indicated significantly higher rates of regurgitations and/or nursing bouts per h in families with more adults (that is, those with helpers) (Fig. 7). Helpers varied in the amount of food they provided but on an average performed approximately 30% of all regurgitations to pups and/or the lactating mother. Helpers may also improve the provisioning of pups indirectly by allowing the parents to spend more time foraging alone or hunting as a pair. Wyman (1967) found that silverbacked- and golden-jackal pairs were more successful at hunting Thomson gazelle fawns than individuals and Lamprecht (1978) found that success rates at hunting gazelle fawns were 36.4% for an individual and 84.6% for jackals hunting in groups of two to three.

A jackal family in possession of a carcass (either killed or scavenged) can defend and exploit it more fully than a single individual. I have observed single jackals at carcasses spending most of the time threatening vultures rather than feeding, and occasionally being driven away by vultures. Thus the feeding ecology of both jackal species involves not only availability, distribution, and size of food items, but also intraspecific and interspecific competition to retain and consume food. Lamprecht (1978) has also suggested that hunting in pairs allowed faster exploitation of prey and reduced the probability that it would be lost to scavengers.

The presence of helpers in a silverbacked-jackal family correlated significantly with pup survival at 14 weeks of age (Fig. 8). For 17 litters the correlation coefficient for number of adults in a family and number of pups surviving at 14 weeks was $R_s = 0.85$ ($P < .01$). One litter did not follow this pattern, but the group was aberrant in that 1) the parents were old and did not reproduce at all the next year, and 2) only one helper of three was active during the critical period of pup development (this helper was the only individual who ever helped for more than one year). As a result, only one pup survived.

Initial litter size was not known although zoo records indicate that it may range from one to eight. Dens were not disturbed because of the potential detrimental effect on pup survival and subsequent

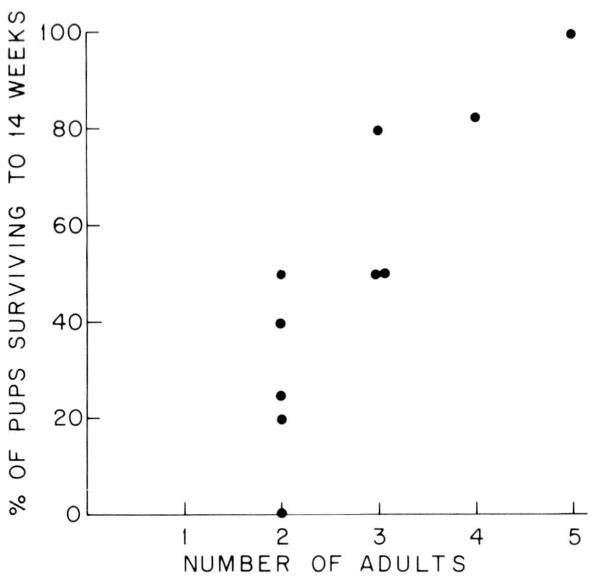

FIG. 8. Relationship between number of adults per family and pup survivorship (3–14 weeks) in *Canis mesomelas* (n = 10 litters). Spearman rank correlation coefficient: $R_s = 0.88$, $P < .01$.

den use (dens may be used again). Pups first emerged from the den at approximately 3 weeks of age, probably after some mortality had already occurred. In laboratory-reared domestic dogs 37% of mortality occurred in the first three weeks (Anderson, 1957). The number of adults and number of pups at three weeks of age (n = 10) correlated significantly ($R_s = 0.83$; $P < .01$) as did the number of adults and percentage of pups surviving from 3 to 14 weeks ($R_s = 0.88$; $P < .01$). The number of pups at 3 and 14 weeks also correlated significantly, but with a lower R_s value (n = 10; $R_s = 0.68$; $P < .05$). Thus it appears reasonable to conclude that helpers improve pup survival rates, possibly from a very early age, and they may even affect initial litter size.

However, there are at least two alternative explanations for the above correlations. Variation in pup survival may reflect variation in parental competence. For example, a primiparous female might be less efficient in caring for pups (she also would not have helpers). Long term observations on one pair indicate that helper presence may be critical. In 1976, Tipper and Tamu had one helper and four pups survived for six months. In 1977 this same pair had no helper and five three-week-old pups, none of which survived to 14 weeks. Thus in one case experienced parents without helpers did not successfully raise any pups.

Alternatively, variation in pup survival could reflect variation in territory quality. Lack of helpers or surviving pups might reflect poor habitat and food resources. In 1976 and 1977 rainfall was high at Ndutu and an important prey species, *Arvicanthis niloticus* (the unstriped grass rat), was abundant (see above). This 60-g rodent is diurnal and could often be observed running back and forth between thornbushes. Despite the abundance of prey, when Tipper and Tamu had no helper, no pups survived (five pups were present at 3 weeks in 1977). Similarly, in 1976, Scorpio and Libra, in a similar-sized territory, had no helper and only one surviving pup, whereas in 1977, with the surviving pup as a helper, they had three of six pups (at 3 weeks) surviving to 14 weeks. The density of *Arvicanthus niloticus* was lower on this pair's territory (November 1976: 3,125 rats/km^2; July 1977: 13,125 rats/km^2) than on Tipper and Tamu's territory, although the available biomass was still high (approximately 800 kg/km^2). Even with abundant food, adults may be limited in their ability to provide for themselves and the pups, thus making the presence of helpers a critical factor. In families where there were more pups than adults, most of the pups tended to look malnourished, rough-coated, weak-legged, and exhibited a lower rate of play activity.

Data on golden jackals are more difficult to interpret. Pup survival did not correlate significantly with the presence of helpers (Fig. 9), although golden jackal helpers appear to contribute substantially to pups. A pair raising a litter alone left it unguarded up to 29% of the time. With the addition of at least one helper the pups were left alone less than 6% of the time (see Fig. 10). However, having an adult member of the family within 100 m of the den may not be critical for pup survival; visibility is so good on the open plains that a golden jackal can be at the far end of its territory and still keep its den in sight.

Golden-jackal helpers also contribute food to the litter directly by regurgitating to pups, and indirectly by regurgitating to the mother. Female parents contribute significantly more food to the offspring than either the male parent or the helpers. There does

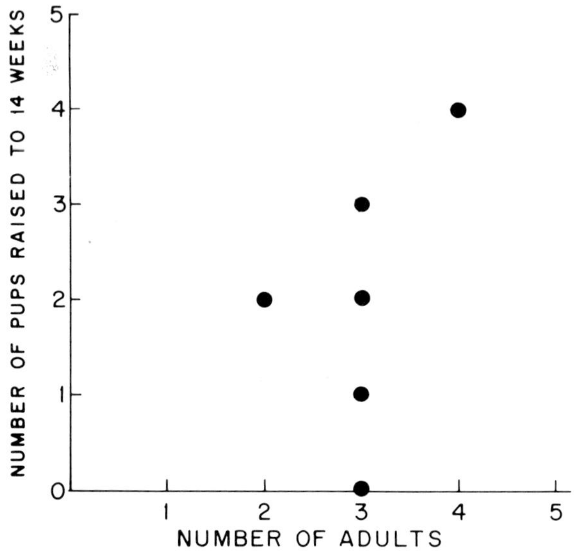

FIG. 9. Relationship between number of adults per family and number of pups raised to 14 weeks in *Canis aureus* (n = 6 families).

not appear to be a significant increase in the cumulative rate of feeding with the addition of helpers. However, the addition of helpers does lead to an increase in the number of regurgitations the mother receives. Comparison of feeding rates of golden (Fig. 11) and silverbacked (Fig. 7) jackals indicates that pairs of golden jackals with no helpers feed their pups at a rate close to that of silverbacked jackals with two helpers. Golden jackals whelp in December and January and are feeding pups during the period of greatest food abundance on the plains. Wildebeest are in the area and there are adult carcasses, afterbirths, and orphans. In addition, there is a peak in fawning of Thomson gazelles at this time. Hence the parents may be able to provide the pups with a more than adequate food supply; consequently, helpers mainly relieve the female's reproductive burden by contributing to her nutritional needs.

Golden-jackal pups are better provisioned and guarded than silverbacked-jackal pups. Despite this they suffer a much higher pup mortality. Golden jackals whelp during the rainy season, resulting in dens being repeatedly flooded and pups dying of exposure and illness, factors which appear to negate the apparent benefits of helpers.

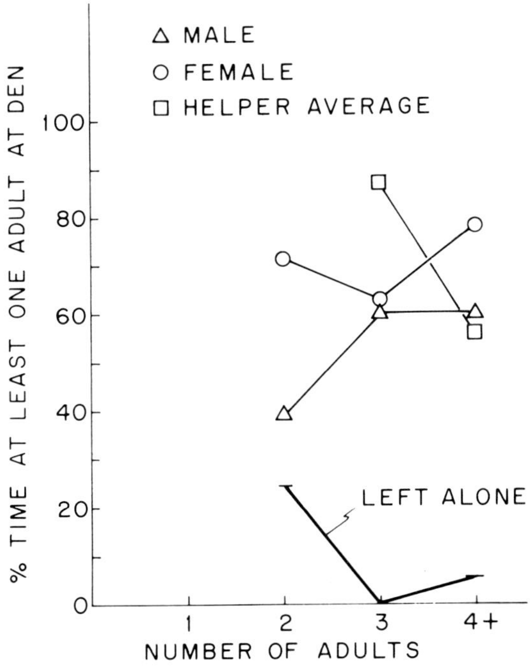

FIG. 10. The relationship in *Canis aureus* between number of adults per family and the percent time that pups up to 14 weeks of age are left unguarded (n = 6 families; 221 h of observations).

Selective Pressures for Monogamy

Jackals have a relatively large litter size and there is a long period of infant dependency. Litters of silverbacked jackals raised solely by a pair of adults produced only 1.3 young which survived to 14 weeks of age. If this sample is representative then pairs in this locale on average can raise only one pup. If paternal investment were lowered by a male dividing his care between several litters, then quite possibly no pups would survive. Thus paternal care is likely to be an important factor in pup survival.

Long-term pair bonds are perhaps reinforced by holding territories year-round. Pair bonds also ensure that subsequent litters of pups will be full siblings. The latter has important implications for the kind of selective pressure exerted on some pups for staying and helping to raise the next litter.

Selection for monogamy may reflect several different types of

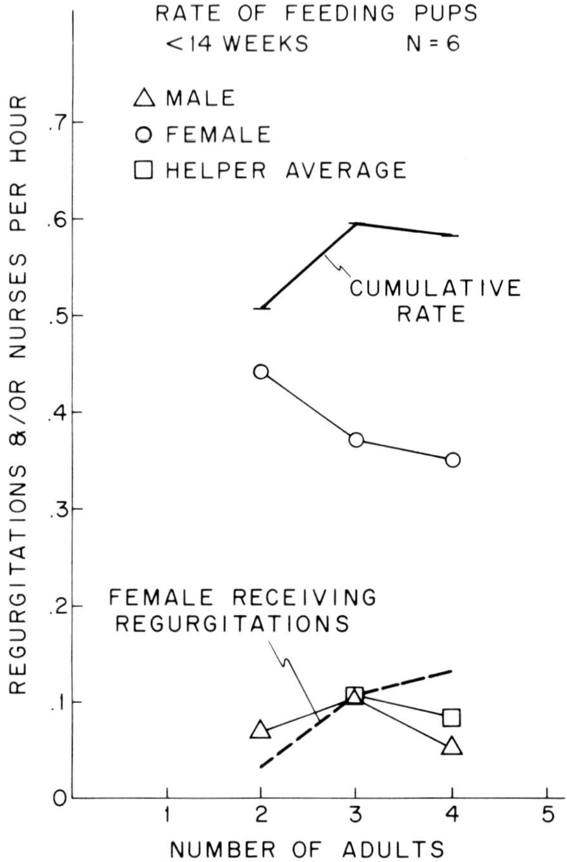

FIG. 11. The relationship in *Canis aureus* between number of adults per family and hourly rate of feeding pups (nursing and/or regurgitation) (n = 6 families; 221 h of observations).

physiological, social, and ecological pressures. Mating exclusivity may be reinforced by the fact that canids can produce litters with multiple sires (Beach, pers. comm.). Thus the monogamous male canid needs to ensure that he alone mates with the female and that he does not invest in pups that he did not sire. Conversely, if the female permits another female on the territory, she may have to share the male and his paternal investment.

TABLE 1
RELATIONSHIP BETWEEN PUP SURVIVAL AND NUMBER OF HELPERS IN SILVERBACKED JACKALS

Parents and helpers	Success (N)
Pair	Raised 1.3 pups (6)
Pair and 1 helper	Raised 3.3 pups (8)
Pair and 2 helpers	Raised 4.0 pups (2)
Pair and 3 helpers	Raised 6.0 pups (1)

Selection for Helping

Jackals that stay on the natal territory for an extra year and provide "help" derive important benefits. Extended experience on familiar terrain may improve fitness in helpers by increasing their survivorship and by improving the quality of parental care they give to their own pups in the future. Some helpers may benefit by inheriting a portion of the natal territory. The potential costs of helping are 1) delaying production for a year, 2) feeding the pups, and 3) guarding the pups. By contrast, year-old jackals that emigrate may not defer reproduction but may suffer greater mortality.

Jackal helpers are on average as closely related to their full siblings as they would be to their own offspring. Because their contributions as helpers appear to increase pup survivorship, their inclusive fitness is enhanced by kin selection until such time as breeding is attempted. In 17 of 18 litters of silverbacked jackals observed between 1974 and 1978 there was a direct relationship between average pup survival and the number of helpers present (see Table 1). The average number of pups alive at 14 weeks was greater by 1.7 pups with a helper present.

Attempting to determine whether inclusive fitness is increased via individual and/or kin selection is a difficult problem. Moreover, a year-old jackal can emigrate and still have its "inclusive fitness" improved as long as another sibling stays and helps. Data on success at emigrating, mating, and establishing a territory at 1 year of age versus 2 years of age are needed to fully answer questions concerning the evolution of delayed reproduction and helping behavior.

In this study the behavior of parents toward helpers was of an affiliative nature and ranged from grooming helpers to sharing food with them. Parents are consistently dominant to helpers, but there have been no observations of parents threatening or driving their young from the territory. Parents would tend to benefit more from the presence of helpers investing in their pups (r = ½) than by helpers producing their own pups (r = ¼—grandpups). Females may also improve their own future reproductive success by tolerating helpers because helpers may alleviate their nutritive stress during the lactation period. Thus it is in the parents' best interest to retain helpers and accure increased reproductive success. The benefits of retaining helpers are limited by available food resources and the upper limit set on efficiency by the limits of litter size (six to eight pups).

Once again, golden-jackal data are difficult to interpret. Helpers potentially benefit by staying on the home territory an extra year. The fact that some surviving pups emigrated during the dry season and later returned to their natal territory may indicate how difficult it is to set up new territories. However, considering the observed rate of pup survival, neither helpers nor parent appear to accrue direct reproductive success. The mother may accrue future reproductive success because helpers alleviate nutritive stress.

Dispersal Patterns

To date this study has not followed emigrating pups, and observations have been limited to events occurring within a circumscribed area. From this limited perspective the dispersal pattern for silver-backed jackals has been as follows: 1) lone surviving pups always stay as helpers (n = 3); 2) approximately 24% of all surviving pups stay as helpers; 3) in one family with three sisters as helpers, one female repeatedly attacked one of her sisters that disappeared within 10 days. In 1974, 1976, and 1977, 60 to 83% of the pairs with litters had helpers, but in late 1977 the jackals started to die in the study area of an undiagnosed disease, and by 1978 only 20% of pairs with litters had helpers. In 1979 most of the known pairs had disappeared and strange pairs were established on their home territories.

Dominance interactions among littermates may determine who remains on the home territory. Dominant pups tend to be more independent and better foragers; they might have a higher rate of survival and reproductive success when they emigrated, *if* one or more of their siblings stayed and helped. In fact pups do stray off the home territory; they may be testing the potential for emigration. However, if successful emigration were difficult (for example, if the jackal population was at high density) then a pup might improve its fitness by remaining with its family for an additional year. Thus if emigration was not feasible, dominant pups could remain on the home territory and possibly force siblings out. In the one observed case of an individual being driven out of the family, the aggression occurred between sisters. The most reasonable model for ensuring that helpers are always present would involve having dominant pups staying rather than emigrating (Dawkins, pers. comm.).

In summary, the choice of whether to emigrate or to remain and help may depend on: 1) abundance of food resources and the number of additional adults which could be sustained; 2) density of jackals and the availability of territories which could affect incidence of disease; 3) dominance status of an individual relative to its same-sex siblings; and 4) number of surviving litter mates (litter size may affect the development of exploratory behavior).

Van der Merwe (1953) suggests that mate preferences are established in silverbacked-jackal litters before weaning and that these pairs emigrate together, thereby increasing the probability of inbreeding. My data do not support this observation. Two known individuals formed pair bonds with strangers to the area. However, data on emigration, pair formation, and territory acquisition are needed to fully answer questions concerning dispersal.

In golden jackals, all surviving pups stayed as helpers and no severe aggression has been observed among same-sex siblings (but data are few). Between 1976 and 1979, 75% of pairs with litters had helpers. Young of the year left the natal territory during the dry season and returned to help in the rainy season.

Thus most golden-jackal offspring stay with the parents until almost two years of age. Affiliative behaviors are common, inter-individual distances are low, and parents tend to break up fights among offspring. The contrast with silverbacked jackals may reflect the flexible nature of social organization in golden jackals. Macdonald (1979*c*) observed a stable group of 20+ golden jackals in which several females appeared to be breeding in Israel. In his study area there were large, clumped, and defendable food resources. The Israeli jackal population was territorial; it provides the only known example of jackals using feces piles (middens). Macdonald suggested that large social groups were possible because the clumped nature of the food resources made it economically defensible (Bradbury and Vehrencamp, 1976; Emlen and Oring, 1977). The presence of golden-jackal helpers in this population does not correlate with pup survivorship. The presence of helpers may have been more directly related to availability of open territories and food resources. If helpers remain because food resources are available and there is an increased ability to utilize prey through cooperative hunting and carcass defense, then it appears odd that helpers should also regurgitate food to pups. Individuals who are not "outstanding" helpers are still tolerated by the parents. As Malcolm (1979) has stressed for wild dogs (*Lycaon pictus*), the benefits accrued by remaining with the pack do not seem sufficient reason for a helper to incur the costs, unless kin selection is included in the equation. The density independent nature of pup mortality of golden jackals during my study may distort the relationship between helping and the potential reproductive success of both parents and helpers.

Summary and Conclusions

Silverbacked and golden jackals (*Canis mesomelas* and *C. aureus*) are two of the few mammalian species to have long-term pair bonds and to exhibit a tendency for some offspring to help in the provisioning, guarding, and socialization of subsequent litters. Some pairs of both species were observed to form bonds for 6 years, a period which may approach the life expectancy of a jackal in the wild.

The silverbacked jackal occurs in brush woodland habitat and whelps mainly during the dry season (July to September). The golden jackal inhabits the adjacent short grasslands and typically whelps during the wet season (December to March).

Both species are small ($\sim$11 kg) omnivores and are opportunistic feeders. Their diet includes dung beetle larvae, fruit, and adult Thomson gazelles, which they kill. They hold year-round territories and both male and female scent-mark and defend the territory and its boundaries.

The monogamous bond and exclusive mating were maintained by a high degree of affiliative behavior, food sharing, synchrony of activity, year-round maintenance of the feeding territory, cooperative hunting, and intrasexual aggression. There is a relatively large litter size. The fitness of a female tolerating bigamy would potentially be lower than the reproductive success of monogamous females because of the division of the male's parental investment. In silverbacked jackals, pairs successfully raised a mean of 1.3 pups. In one case when the male parent disappeared, the whole litter died and the mother subsequently disappeared.

During the study period (1974, 1976 to 1979) 12 of 19 observed litters of silverbacked jackal had helpers (63%). Six of eight (75%) observed litters of golden jackals had helpers. From 1976 to 1979, four silverbacked jackal families and two golden-jackal families were observed with consecutive litters. In all cases where helpers were present (n = 13), they were pups from the previous year's litter. Thus with monogamous parents the helpers were full siblings to the litter of pups in which they were investing.

Helpers regurgitate to the pups and to the mother during the period of lactation. They also babysit, guard, defend, groom, and play with the pups. In silverbacked jackals the presence of helpers correlates positively with pup survivorship from three to 14 weeks (Spearman rank correlation coefficient, $R_s = 0.88$, $P < 0.01$, n = 10). The presence of golden-jackal helpers did not correlate with pup survival. Helpers did provide higher provisioning and babysitting rates, but pups were exposed to a density independent variable (severe thunderstorms and flooding) which resulted in high mortality and negated adult contributions. Helper provisioning of the mother may be important in terms of her future reproductive success, and thus indirectly increase the inclusive fitness of golden-jackal helpers.

Jackal helpers are on average as closely related to full siblings as they would be to their own offspring. If their contributions as helpers increase pup survivorship, their own inclusive fitness may be enhanced via kin selection until such time as breeding is attempted. In silverbacked jackals (n = 17), helpers delaying reproduction for one year potentially increased their inclusive fitness by on average 1.74+ × ½, assuming that helping incurred a minimal cost.

In silverbacked jackals it appears reasonable to conclude that helper presence positively influences pup survival rates. This may exert strong selection pressure on the parental pair to retain helpers and accrue increased reproductive success.

Factors affecting dispersal include parental behavior toward young, same-age, same-sex sibling interactions, jackal density, and ecological parameters such as food distribution and availability. Golden jackals show more affiliative behavior towards their pups and helpers and rarely threaten them. Of pups surviving to 14 weeks, 100% were helpers the next year. By contrast, only 24% of surviving silverback pups became helpers. In both species the sex ratio of helpers was equal and they typically delayed reproduction for one year.

There was more aggression among same-age, same-sex siblings in silverbacked jackals. In 1976, a female helper repeatedly attacked her sister and within 10 days the subordinate helper disappeared. Such aggression was never observed among golden-jackal helpers.

For young jackals to remain on the home territory there must be a sufficient food supply. Young golden jackals emigrate during the dry season when food resources are low, but then return in the wet season when food is abundant. Golden-jackal density remained approximately the same from 1976–1979.

In silverbacked jackals the option of helping versus emigrating appears to be more open. Between 1974 and 1976 the size of territories in the Ndutu area decreased by half. This probably reflected greater abundance of an important prey item, *Arvicanthis niloticus*. But jackal density doubled and it may have been difficult for young of the year to establish territories. From 1976 to 1977, 60 to 83% of pairs with litters had helpers. In late 1977 jackals with an undiagnosed disease started to die in the study area; and in 1978 only 20% of pairs with litters had helpers. Either young of the year in 1977 succumbed to disease and/or so many jackals died that potential helpers could successfully emigrate to open territories. In silverbacked jackals the age of dispersal may depend on food resources, jackal densities, and social development within a litter.

Acknowledgments

I am grateful to Jane Goodall, Derek Bryceson, James Malcolm, Leo Kunkel, Ramathani Senzota, Inayet Anajri, Chimen Patel, the staff of Ndutu Tented Camp, and Hugo van Lawick for assistance. I thank the Tanzanian Scientific Council, Tanzanian National Parks, the Ngorongoro Conservation Area Authority, and the Serengeti Research Institute for making this research possible. This is SRI publication no. 240.

Conversations with and comments by James Malcolm, David Macdonald, Peter Waser, Jon Rood, Lori Herbison Frame, George Frame, and Richard Dawkins contributed greatly to this manuscript. D. Macdonald, P. Waser, J. Rood, and J. Malcolm kindly criticized a draft of this paper.

Financial and moral support were provided by the National Geographic Society, the Harry Frank Guggenheim Foundation, and the University of Wisconsin.

Literature Cited

ABBOTT, D. H. 1979. The sexual development of the common marmoset monkey (*Callithrix jacchus*). Unpubl. Ph.D. dissert., Univ. Edinburgh, 383 pp.

ABBOTT, D. H., AND J. P. HEARN. 1978. Physical, hormonal, and behavioral aspects of sexual development in the marmoset monkey (*Callithrix jacchus*). J. Reprod. Fert., 53:155–166.

ALEXANDER, R. D. 1974. The evolution of social behavior. Ann. Rev. Ecol. Syst., 5:325–383.

ANDERSON, A. C. 1957. Puppy production to weaning age. J. Amer. Vet. Med. Assn., 130:151–158.

ANDERSON, A. C., AND M. E. SIMPSON. 1973. The ovary and reproductive cycle of the dog (beagle). Geron-X Inc., Los Altos, California, 290 pp.

BEACH, F., AND R. W. GILMORE. 1949. Response of male dogs to urine from females in heat. J. Mamm., 30:391–392.

BEKOFF, M., AND M. C. WELLS. 1980. The social ecology of coyotes. Sci. Amer., 242(4):130–151.

BERTRAM, B. C. R. 1976. Kin selection in lions and evolution. Pp. 281–301, *in* Growing points in ethology (P. P. G. Bateson and R. A. Hinde, eds.). Cambridge-London, Cambridge Univ., 548 pp.

BOWEN, D. 1978. Social organization of the coyote in relation to prey size. Unpubl. Ph.D. dissert., Univ. British Columbia, Vancouver, 230 pp.

BRADBURY, J. W., AND S. L. VEHRENCAMP. 1976. Social organization and foraging in Emballonurid bats. II. A model for the determination of group size. Behav. Ecol. Sociobiol., 1:383–404.

BUHL, A. E., ET AL. 1978. The effects of social rank on reproductive indices in groups of males collared lemmings (*Dicrostonyx groenlandicus*). Biol. Reprod., 18:317–324.

CAMENZIND, F. J. 1978. Behavioral ecology of coyotes (*Canis latrans*) on the National Elk Refuge, Jackson, Wyoming. Unpubl. Ph.D. dissert., Univ. Wyoming, Laramie, 97 pp.

CLUTTON-BROCK, T. H., AND P. H. HARVEY. 1977. Primate ecology and social organization. J. Zool., 183:1–39.

CROOK, J. H. 1964. The evolution of social organisation and visual communication in the weaver birds (Ploceinae). Behaviour, Suppl., 10:1–178.

CROOK, J. H., AND S. S. GARTLAN. 1966. Evolution of primate societies. Nature, 210:1200–1203.

CROOK, J. H., AND J. D. GOSS-CUSTARD. 1972. Social ethology. Ann. Rev. Psychol., 23:277–312.

DAWKINS, R. 1976. The selfish gene. Oxford Univ. Press, Oxford, 224 pp.

EISENBERG, J. F. 1966. The social organizations of mammals. Handbuch der Zoologie, VIII (10/7), Lieferung 39, De Gruyter, Berlin, 92 pp.

EMLEN, S. T., AND L. W. ORING. 1977. Ecology, sexual selection, and evolution of mating systems. Science, 197:215–223.

GIER, H. T. 1975. Ecology and social behavior of the coyote. Pp. 247–262, *in* The wild canids (M. W. Fox, ed.). Van Nostrand Reinhold, New York, 508 pp.

GOLANI, I., AND A. KELLER. 1975. A longitudinal field study of the behavior of a pair of golden jackals. Pp. 303–335, *in* The wild canids (M. W. Fox, ed.). Van Nostrand Reinhold, New York, 508 pp.

HAMILTON, W. D. 1964. The genetical evolution of social behavior. I. II. J. Theor. Biol., 7:1–52.

HASLER, J. F., AND E. M. BANKS. 1975. The influence of mature males on sexual maturation in female collared lemmings (*Dicrostonyx groenlandicus*). J. Reprod. Fert., 42:583–586.

JARMAN, P. J. 1974. The social organisation of antelope in relation to their ecology. Behaviour, 48:215–267.

JORGENSON, J. W., ET AL. 1978. Chemical scent constituents in the urine of the red fox (*Vulpes vulpes* L.) during the winter season. Science, 199:796–798.

KENNELLY, J. J. 1977. Coyote reproduction. Pp. 73–92, *in* Coyote biology: evolution, behavior, ecology and management (M. Bekoff, ed.). Academic Press, New York, 384 pp.

KINGDON, J. 1977. East African mammals. Carnivores. Academic Press, New York, 3A:1–476 pp.

KLEIMAN, D. 1977. Monogamy in mammals. Quart. Rev. Biol., 52:39–69.

KLEIMAN, D., AND J. F. EISENBERG. 1973. Comparisons of canid and felid social systems from an evolutionary perspective. Anim. Behav., 21:637–659.

KRUUK, H. 1975. Functional aspects of social hunting in carnivores. Pp. 119–141, *in* Function and evolution in behaviour (G. Baerends, A. Mannings, and C. Beer, eds.). Oxford Univ., Oxford, 393 pp.

LACK, D. 1968. Ecological adaptations for breeding in birds. Methuen & Co., London, 409 pp.

LAMPRECHT, J. 1978. The relationship between food competition and foraging group size in some larger carnivores: a hypothesis. Z. Tierpsychol., 46: 337–343.

VAN LAWICK, H., AND J. VAN LAWICK-GOODALL. 1970. The innocent killers. Collins, London, 221 pp.

LOMBARDI, J. R., AND J. G. VANDENBERGH. 1977. Pheromonally induced sexual maturation in females: regulation by the social environment of the male. Science, 196:545–546.

MACDONALD, D. W. 1977. The behavioural ecology of the red fox *Vulpes vulpes*. A study of social organization and resource exploitation. Unpubl. Ph.D. dissert., Oxford Univ., Oxford.

———. 1979*a*. "Helpers" in fox society. Nature, 282:69–71.

———. 1979*b*. Some observations and field experiments on the urine marking behavior of the red fox, *Vulpes vulpes* L. Z. Tierpsychol., 51:1–22.

———. 1979*c*. Flexibility of the social organization of the golden jackal, *Canis aureus*. Behav. Ecol. Sociobiol., 5:17–38.

MALCOLM, J. 1979. Social organization and communal rearing of pups in African wild dogs (*Lycaon pictus*). Unpubl. Ph.D. dissert., Harvard Univ., Cambridge, Massachusetts.

MEDJO, D. C., AND L. D. MECH. 1976. Reproductive activity in nine- and ten-month-old wolves. J. Mamm., 57:406–408.

VAN DER MERWE, N. J. 1953. The jackal. Fauna Flora, 5:1–80.
MOEHLMAN, P. D. 1979. Jackal helpers and pup survival. Nature, 277:382–383.
ROOD, J. 1978. Dwarf mongoose helpers at the den. Z. Tierpsychol., 48:277–287.
SCHALLER, G. B. 1967. The deer and the tiger. Univ. Chicago Press, Chicago, 370 pp.
SEAL, U. S., ET AL. 1979. Endocrine correlates of reproduction in the wolf. Biol. Reprod., 21:1057–1066.
STEPHENSON, G. R., AND T. W. ROBERTS. 1977. The SSR system 7: a general encoding system with computerized transcription. Behav. Res. Methods Instr., 9:434–441.
TARYANNIKOV, V. I. 1976. Reproduction of the jackal (*Canis aureus aureus* L.) in Central Asia. Ekologiya, 2:107.
TRIVERS, R. L. 1972. Parental investment and sexual selection. Pp. 136–179, *in* Sexual selection and the descent of man, 1871–1971 (B. Campbell, ed.). Aldine Publ. Co., Chicago, 378 pp.
VANDENBERGH, J. G. 1971. The influence of the social environment on sexual maturation in male mice. J. Reprod. Fert., 24:383–390.
———. 1973. Acceleration and inhibition of puberty in female mice by pheromones. J. Reprod. Fert., Suppl., 19:411–419.
WILSON, E. O. 1975. Sociobiology: the new synthesis. Harvard Univ. Press, Cambridge, Massachusetts, 697 pp.
WYMAN, J. 1967. The jackals of the Serengeti. Animals, 10:79–83.

THE SOCIAL SYSTEM OF THE DWARF MONGOOSE

JON P. ROOD

Abstract

OVER 300 individually-marked dwarf mongooses (*Helogale parvula*) were studied in the Serengeti National Park, Tanzania, from 1974 to 1979. Dwarf mongooses live in packs typically containing a dominant breeding pair, mongooses born in the pack, and immigrants. The mean size of 66 packs recorded just prior to the onset of the birth season was 8.4. Maximum recorded pack size was 27. Pack ranges usually overlapped and aggressive interactions with neighboring packs occurred in overlap areas. Some packs maintained stable home ranges throughout the study period. Small groups of transients, consisting mainly of males, were occasionally observed.

Dwarf mongooses are unusual among group-living mammals previously studied in that both males and females commonly transfer between groups. A few mongooses of both sexes attained breeding status by remaining in their natal packs, but most emigrated and either formed new packs by joining with opposite-sexed individuals in an uninhabited area, or joined existing packs. Most immigrants were young adults but interpack transfer occurred in animals as old as 5 years. Behavior occurring during immigration is described. The rate of immigration varied considerably between packs; those with at least two adult males and two adult females were the least open to immigration. Males attempting to join packs frequently faced considerable aggression from same-sexed residents, particularly the beta male. Two cases of male takeovers, with expulsion of resident males, are described, as well as one female takeover.

Intragroup relationships are characterized by a high degree of cooperation. Mutual grooming is common and primarily heterosexual. Alpha females were usually the main pack leaders whereas alpha males scored highest in vigilance behavior. Pack members cooperate in guarding, feeding, transporting, and grooming the young. Unrelated immigrants are active helpers. Some subordinate females lactated and nursed the young of the alpha female although they had not been visibly pregnant.

The social organization of the dwarf mongoose is compared with that of other species of social mongooses. Predation has probably been the most important selective pressure moulding sociality in these small carnivores.

Introduction

Cooperative hunting is the usual explanation for the evolution of group life in large social carnivores (for example, see Kruuk, 1972; Schaller, 1972). Small carnivores are primarily solitary, but several species of viverrids form multi-male social groups in which members find and kill food individually. The selective pressures producing group life in these forms must be different from those operating on the carnivores which hunt in cooperative groups. As Gorman (1979) points out, a knowledge of viverrid social organizations is important for an understanding of the evolution of social behavior among carnivores.

Ewer (1973) lists 36 genera and 72 species in the Viverridae, the largest and least known family of carnivores. Only in the Herpestinae (mongooses) are groups larger than a single family unit found. In four genera (*Helogale, Mungos, Crossarchus,* and *Suricata*), individuals live in cohesive packs which forage and den together. It is possible that the little known *Liberiictis kuhni* might form social groups (Ewer, 1973), but confirmatory field data are not yet available.

My research has been concentrated on the smallest member of the family Viverridae, the dwarf mongoose (*Helogale parvula*), in which adult weight averages about 320 g. Dwarf mongooses are diurnal and feed primarily upon invertebrates, particularly beetles. Rodents, especially *Arvicanthis niloticus,* and other small vertebrates are chased when encountered and are occasionally captured and eaten, but they form a small proportion of the diet.

This study was conducted in the Serengeti National Park, Tanzania, from February 1974 to December 1979, with absences from December 1977 to August 1978, and May 1979 to September 1979. Early in the study, I became impressed with the high degree of

FIG. 1. Social grooming among dwarf mongooses. Note the white mark produced by freeze-marking.

cooperation shown by the groups in predator defense, and in caring for the young. My objectives were to describe social structure and breeding strategies, determine the relationship between kinship and helper behavior, and assess the adaptive significance of group living in this small carnivore.

Methods

It was necessary to develop a method of permanently marking animals for individual identification from a distance. Nyanzol dye and ear tags were initially used, but the former wore off after about 4 months and the latter eventually pulled out. Freeze-marking with Arcton 12 was found to be an ideal method. The animals were live-trapped at their termite-mound dens and each was weighed, sexed, and its teeth examined. Marking involved spraying a small section of skin with Arcton 12 (CCl_2F_2) from an aerosol can. The hair grew in white within about 6 weeks (see Fig. 1), providing a permanent distinctive mark which could be readily identified at a distance (for additional information on the marking procedure, see Rood and Nellis, 1980). At the termination of observations in December 1979, 323 mongooses had been permanently freeze-marked. Using known age animals as a baseline, it was possible to estimate ages of mongooses of unknown birth date from the degree of toothwear.

Observations were made with 10× binoculars at termite-mound dens from a Land Rover parked at a distance of approximately 50 m. Regular counts were made of the marked packs and all social interactions recorded on data sheets or with a tape recorder. Twenty-four packs with all individuals identified were observed for periods of up to 5½ years and all changes in pack composition were recorded. The number of packs under observation increased as the study progressed and also varied because of the disappearance of some packs and the formation of new ones. The total monitoring time was 58 pack-years (1 pack year = one pack monitored for 1 year).

Most observations commenced when the mongooses first emerged from the den soon after sunrise and terminated when they left the den to forage. Observations were usually not possible during the day once a pack left its elevated termite mound because of the tall grass and other cover on the study areas. Because packs frequently change dens, it was necessary to locate the animals at den sites in the afternoon before they went in for the night. Increased observation time was possible at breeding dens because a pack with young centers its activities around the termite mound. Series of all day watches were conducted at breeding dens to record social interactions and helper behavior. Data recorded included the amount of time each individual spent in, on, and away from the den, the time spent atop the den, the order of emergence and leaving the den site, all responses to predators, and all social interactions.

The Study Area

In the Serengeti, the dwarf mongoose is the most abundant mammalian carnivore in the woodlands (see Herlocker, 1976, and Norton-Griffiths and Sinclair, 1979, for a description of the Serengeti woodlands). The main study area, 2.2 km^2 in open woodland is located approximately 4 km north of the Serengeti Research Institute. A central ridge runs east to west sloping to a seasonal river

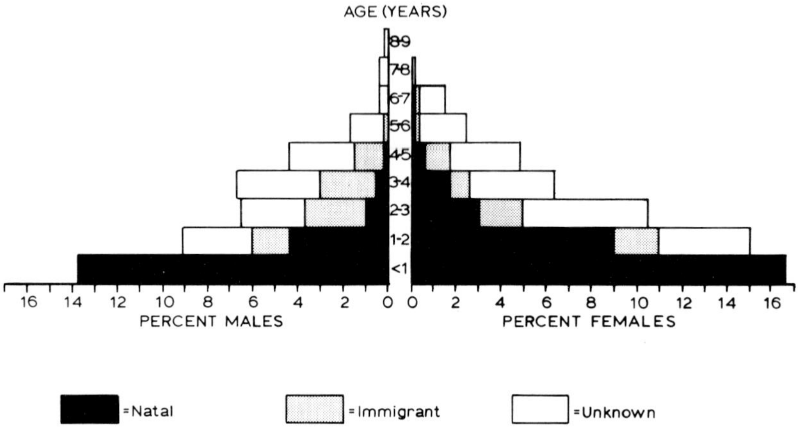

FIG. 2. Age pyramid showing dwarf mongoose pack compositions in October just prior to the onset of the birth season. All packs of known composition in October from 1974 to 1979 were used. The age pyramid is based on 66 packs containing 541 animals.

(the Sangere) and two drainage lines. A field which was not used by the mongooses lies to the east. Acacias, particularly *A. tortilis* and *A. clavigera,* are the dominant trees on the area, and *Commiphora trothae* is also common. Termite mounds (*Macrotermes subhyalinus*) are abundant on the study area, occurring at a density of about 70/km^2. The main area of mongoose occupation was along the ridge and coincided with the highest density of termite mounds. While the river and drainage lines could be crossed by mongooses, the long grass growing along them was avoided by the resident packs; consequently, approximately one-third of the study area was unused.

Mongoose packs denned in nearly all termite mounds within the home range, including numerous active mounds, and neighbors used the same dens successively in areas of range overlap. Termite mounds are an important resource to the mongooses both as den sites and by providing a food resource. After a rainfall, mongooses were frequently seen digging at fresh termite workings. Each pack had from 10 to 30 mounds within its range which were used as dens and entered via the termite ventilation shafts. Other less commonly used den sites were hollow trees and crevices in kopjes (granite outcrops).

During the study three packs resident near the Serengeti Research

Institute and most of the packs in woodland between this and the main study area were marked. In addition, a pack which had been monitored by A. Kreulen from 1970 to 1974 at Seronera was kept marked throughout the study and three other Seronera packs were marked in late 1979.

Three packs were also marked and observed at Kirawira in the western corridor of the Serengeti National Park. This area consists of short grassland with scattered *Acacia* and *Balanites* trees and thickets. Most of the area is grazed year round, largely by a resident herd of wildebeest (*Connochaetes taurinus*), and this short grass permitted observations of mongooses after they left a den site. The general aspect is much more open than the main study site along the Sangere River, but dwarf mongooses occur because of the cover and den sites provided by a high density of termite mounds, made by *Odontotermes tanganicus* and *O.* sp.

Pack Size and Composition

A dwarf mongoose pack typically consists of a dominant breeding pair (the alpha male and female) who are usually the oldest pack members, mongooses born in the pack, and immigrants. Mongooses less than one year are termed juveniles; thereafter, they are termed adults. Table 1 shows the number and size of packs of known composition, in October, from 1974 to 1979, just prior to the onset of the birth season when pack size is at its lowest. Mean size of the total of 66 packs was 8.4 (maximum size 18). The maximum pack size recorded in the study was 27.

Of 38 young trapped before the age of 3 months, 21 were males and 17 females. Table 1 suggests a bias toward adult females in the packs; comparison of sex ratios within age classes (Fig. 2) indicates that the main difference was in mongooses 1 to 3 years old. Significantly more females than male occurred only in the 1 to 2 ($\chi^2 = 7.88$; $P < 0.01$) and the 2- to 3-year age classes ($\chi^2 = 4.85$; $P < 0.05$), suggesting a greater tendency for young adult females to remain in the natal packs.

Figure 2 indicates that approximately half the mongooses living in packs at the start of the birth season were less than 2 years old and that immigrants were common among the adults. Among animals over 1 year old of known origin, 58% of 77 males and 30% of 112 females were immigrants. The oldest individual in the study pop-

TABLE 1

Helogale PACK COMPOSITION AND SIZE IN OCTOBER (RANGE GIVEN IN PARENTHESES)

Statistics	1974	1975	1976	1977	1978	1979	All packs
Number of packs	4	7	10	15	14	16	66
Mean size of pack	9.5 (6–15)	10.0 (3–12)	8.2 (3–15)	9.2 (2–18)	8.1 (2–16)	7.6 (2–18)	8.4 (2–18)
Mean no. adult males	1.8 (1–2)	2.0 (0–3)	2.4 (0–4)	2.5 (1–4)	3.0 (1–7)	2.3 (1–4)	2.4 (0–7)
Mean number of adult females	4.0 (3–5)	4.1 (1–9)	3.8 (1–7)	2.7 (1–7)	3.1 (1–8)	3.3 (0–8)	3.3 (0–9)
Mean number of juveniles	3.8 (0–10)	2.7 (0–5)	2.0 (0–5)	4.0 (0–9)	2.1 (0–6)	2.1 (0–7)	2.7 (0–10)

ulation was a male estimated to be 9 years old when observations were terminated.

Although the majority of dwarf mongooses on the study areas lived in resident packs, transient groups of one to three mongooses, consisting mainly of males, were occasionally observed. The sex composition of 13 transient groups was nine all-male groups, two all-female groups, and two groups of mixed sex.

Growth curves for males and females indicate that body weight reached asymptotic levels at about 14 to 18 months of age. Only mongooses over 2 years in age were used to compute adult weights. Mongooses living around the Serengeti Research Institute received supplemental feeding from research workers and averaged 50 to 100 g heavier than mongooses from other study areas. These were likewise excluded from the calculations. Mean body weight was 326.4 g (range 265 to 415 g) for 59 males, and 315.5 g (range 221 to 395 g) for 80 non-pregnant females, a non-significant difference ($t = 1.84$, $P > 0.05$). Greatest weights in both sexes were recorded in June at the end of the rains (mean for males = 364.5 g, n = 2; mean for females = 355 g, n = 4) and lowest weights were recorded at the end of the dry season and start of the rains (in October for males, mean = 310.1 g, n = 16, and November for females, mean = 289.9 g, n = 9).

TABLE 2
PACK SIZE AND DENSITY ON SANGERE RIVER STUDY AREA IN OCTOBER

Pack	Year 1975	1976	1977	1978	1979
E	12	15	18	16	6*
L	10	10	18	14	11
M	9	13	14	12	11
W	9	3	6		
S	5	6	5	5	
P	8	5			
B		6	7	6	4
J				5	6
Total	53	58	68	58	38
Number of packs	6	7	6	6	4
Number of mongooses/km^2	24.1	26.4	30.9	26.4	17.3

* Six male survivors of E-pack reappeared on study area in late October and subsequently took over the L-pack females.

Mortality

Juvenile mortality was high. Only 46% of 148 juveniles counted when less than two months old were still present in the natal pack at 1 year of age. Most (78%) of the young which disappeared did so before they were six months old. Causes of mortality were usually unknown but predation is probably a major factor. Birds of prey, slender mongooses (*Herpestes sanguineus*), banded mongooses (*Mungos mungo*), and silverbacked jackals (*Canis mesomelas*) are common on the study areas and elicit antipredator responses from dwarf mongooses. On one occasion a banded mongoose was observed eating a month old juvenile, and in another instance, a young silverbacked jackal caught an adult male which had become separated from its pack and was attempting to return to its den.

Population Density and Home Range

Population density on the main study area in October—just prior to the onset of the birth season—between 1975 and 1979 varied

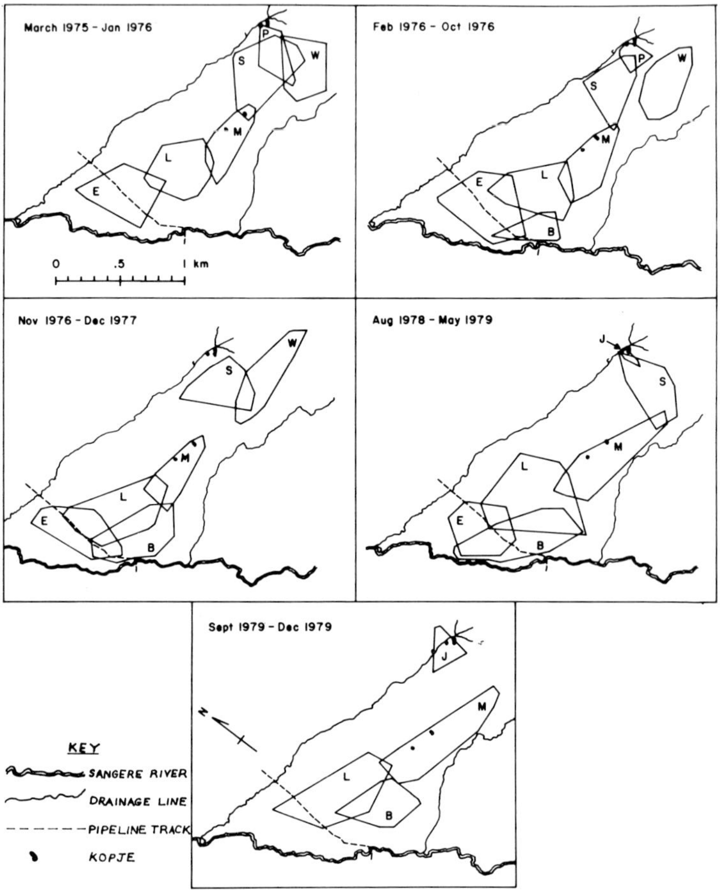

FIG. 3. Pack ranges on the main study area during five periods.

from 17.3 to 30.9 mongooses per km^2, and the number of packs using the area varied from four to seven (Table 2).

The packs had overlapping home ranges as shown in Fig. 3 for five time periods. Range boundaries were obtained by connecting the outermost sightings; the number of mornings and afternoons each pack was seen in each time period is shown in Table 3. The first map in Fig. 3 shows the ranges of the six packs using the study area from March 1975 to January 1976 and each subsequent map

TABLE 3
PACK RANGES AND NUMBER OF SIGHTINGS ON SANGERE RIVER STUDY AREA

Period	Packs								
	P	W	S	M	L	E	B	J	Total
March 1975 to January 1976	20	80	69	153	140	62			524
February 1976 to October 1976	24	103	85	225	202	71	30		740
November 1976 to December 1977		124	47	199	195	62	86		713
August 1978 to May 1979			36	129	140	64	116	5	490
September 1979 to December 1979				60	62		27	9	158
Total	44	307	237	766	739	259	259	14	2,625
Home range (ha) during entire study	10.2	28.1	36.6	33.6	47.6	28.1	33.6	8.1	

Maximum sightings = one morning and one afternoon per pack per day.

shows the ranges following the recording of a change in the number of packs.

In February 1976 a new pack (B pack) formed when a male and female from outside the study area moved separately onto a formerly unoccupied area bordering the Sangere River. The B pack grew as a result of natal recruitment and immigration, and the enlargement of its range and its encroachment onto the areas of neighboring packs is shown in the two subsequent maps. The P pack disappeared in October 1976 and, from November 1976 to December 1977, six packs again used the area. Some time during my absence from the study area between December 1977 and August 1978, the W pack disappeared and from August 1978 to May 1979 the westernmost part of their former range was used by both the M and S packs. The J pack was first seen during this time period using part of the area formerly occupied by the P pack. Between May and August 1979, both the S and E packs disappeared and thus, in the September to December 1979 period, only four packs used the study area. The M pack enlarged its range to the east and the L pack was seen farther west than usual. But much of the former ranges of the S and E packs were unused and appear to be likely sites for the formation of new packs.

Long-term monitoring of individual packs shows that some maintained stable home ranges for many years. The L and M packs were first identified in October 1974 and were still occupying the same home ranges and using most of the same termite-mound dens at the end of the study. When observations terminated in December 1979, there had been nearly a complete turnover in L- and M-pack membership; only one M-pack female present in October 1974 was still in the M pack.

The reasons for the disappearance of packs were usually unknown. The surrounding area was frequently searched and no evidence was obtained that entire packs migrated. The death of the alpha female may affect the stability of the pack, leading to disintegration. For example, the E-pack alpha female was very old (at least 7 years) prior to the pack's disappearance between April and August 1979, and she may have subsequently died. In the S pack, the alpha female died in March 1979; the alpha male later left the pack range and was found the following September in a pack off the study area. The remainder of the pack (one young adult male and two juveniles) was not located again.

The total area used by each pack throughout the study was estimated by connecting the outermost points of all sites where the pack was seen and computing the size of the resulting polygon. Estimated home ranges of the packs on the main study area are shown in Table 3, with the number of mornings and afternoons in which that pack was seen. Insufficient sightings were obtained for the P and J packs to show their entire ranges. Mean home-range size of the other six packs was 34.6 ha (range 28.1 to 47.6 ha). Home-range estimates for two Kirawira packs were much larger (160 ha for 119 sightings, and 84.6 ha for 26 sightings); the number of observations in the latter case is probably insufficient to show the full extent of the range. No range overlap occurred in the more open Kirawira habitat.

Dwarf mongoose home ranges can be considered territories, sensu Wilson (1975). Although there is often some overlap, each pack range includes an area of exclusive use. Spacing is mediated by frequent scent marking which occurs at dens, by avoidance when sighting a neighboring pack and sometimes by brief aggressive

TABLE 4
MONTHLY DISTRIBUTION OF LITTERS IN DWARF MONGOOSE PACKS

Months	Unprovisioned packs	Provisioned packs	Total
January	9	4	13
February	9	1	10
March	10	3	13
April	9	3	12
May	4	1	5
June	2	0	2
July	0	0	0
August	0	0	0
September	0	0	0
October	0	4	4
November	7	5	12
December	6	1	7
Total	56	22	78

interactions in areas of range overlap. Chasing, occasional fighting, anal marking of objects, and allomarking (scent marking of conspecifics) of pack members are common components of encounters which usually end with one or both packs moving toward the center of their ranges. If a smaller pack observed the approach of a larger one, it avoided contact. For example, the B pack (numbering six) became suddenly alert at the den, looked eastward and then ran off to the west. Two minutes later, the L pack (numbering 15) appeared from the tall grass to the east and climbed on the mound formerly occupied by the B pack. Large packs sometimes harrass small packs in dens, rushing at the holes with the tail hair bristling, and frequently lunging and attempting to bite the inhabitants within.

Reproduction

In the Serengeti most packs produce three litters per year in a birth season extending from November to June (see Table 4). The onset of the birth season depends on the food supply at the end of the dry season. If sufficient rainfall occurs in September or early October, food resources (particulary termites) increase in availability, the mongooses rapidly gain weight and mate, and the first births occur in November. When rainfall is slight in September and October, breeding is delayed and the first births may not occur until December. The packs living near the Serengeti Research Institute, receiving supplemental feeding, regularly breed earlier than unprovisioned packs. Four provisioned packs produced litters in October, whereas none from the other study areas did. Most births occurred in the November to April period when food resources were most abundant.

Mating relationships within the pack have been previously described (Rood, 1980; see also Rasa, 1977, for a description of mating postures). Usually only one female (the alpha individual) breeds in each pack. Other sexually mature females come into estrus in synchrony with the alpha female, but, in the few cases when these became pregnant, the resulting litter size was independent of the number of pregnant females, suggesting that the subordinate females may have lost their litters. The alpha male copulates exclusively with the alpha female during the first few days of her estrus, driving off any other males which approach her. Subordinate adult males often mate with subordinate adult females at this time. Later the alpha male copulates with other adult females and the beta male mates with the alpha female. The alpha pair are likely to be the parents of the young born in the pack (Rood, 1980).

The earliest observed pregnancy occurred in a female which mated at 15 months of age. The earliest recorded male copulation occurred in a year-old individual who mounted and achieved intromission (it was not possible to determine if he ejaculated). Most breeding females came into estrus and mated two to four weeks after giving birth. The earliest recorded post-partum mating period commenced seven days after parturition. The mean $\pm$ SD of 13 gestations (calculated from the first day copulation was observed until parturition) was 50.3 $\pm$ 0.85 days (range 49 to 52 days; Rood, 1980). Pregnant females were sometimes mounted with apparent intromission; mounts were observed when females were 12 to 34 days pregnant. A total of 68 litters was counted before the young were six weeks old. Mean litter size was 2.88 (range 1 to 6).

Of particular interest was the occurrence of lactation in eight subordinate females. These had not been visibly pregnant prior to the parturition of the alpha female, although they may have aborted in early pregnancy. In one instance the presence of a second lactating female (an aunt) was directly responsible for saving a litter whose mother had died (Rood, 1980). In all cases the lactating females were related to the young, but the sample is too small to exclude the possibility of immigrants nursing unrelated young.

Intergroup Transfer

To produce viable offspring dwarf mongooses ordinarily must achieve alpha status within a pack. A mongoose can remain in its natal pack and attain alpha status on the deaths of older same-sexed individuals, but the wait may be long. Alternatively, a mongoose can leave its natal pack and attempt to found a new one by joining with individuals of opposite sex on a previously unoccupied area or by transferring to an existing pack. Mongooses of both sexes attained alpha status in this study by adopting either strategy, but most became dominant breeders after emigration. Twenty-five (12 males and 13 females) attained alpha status after emigrating; six (2 males and 4 females) attained this position by remaining in the natal packs. Once alpha status was achieved, it was often maintained for several years. The L-pack alpha male in 1974 was still present in October 1979 when his pack was taken over by six invading E-pack males. The Y-pack alpha female in 1975 still survived and maintained her position when observations were terminated in spite of losing a hindleg to a predator three years previously.

Except in one case (to be discussed separately) in which six E-pack males took over an adjacent pack, immigration occurred either singly or in same-sex pairs. During the study 20 males and 17 females immigrated into monitored packs, giving a mean rate of immigration of approximately one mongoose per pack in 19 months. The immigration rate varied considerably among packs. The highest recorded was five immigrants to the F pack in two years and five months, whereas one large pack allowed none in four years and seven months. Most mongooses which joined packs or formed new ones were young adults. Age at immigration or new pack formation was known or estimated from tooth wear in 14 males and 20 females. Median age was 22 months (range 15 to 38 months) in males and 20 months (range 14 to 59 months) in females. Immigration and new pack formation most frequently occurred early in the breeding season, at the end of the dry season, and during the first part of the rainy season in the period from October to March (92% of males, n = 13; and 84% of females, n = 19).

Table 5 shows the mongooses which transferred between monitored groups during the study and also includes marked individuals which left known packs and joined mongooses of opposite sex. The greater number of females shown in the table is not indicative of a higher transfer rate in this sex, because most immigrant females came from packs within the study area, whereas most males came from outside it. This suggests that males may disperse farther before successful immigration. Mongooses frequently joined overlapping packs but some moved distances of up to 5 km. Frequent scent-marking probably allows mongooses in neighboring packs to assess pack composition and quickly fill a breeding opening.

Eight females transferred in same-sex pairs (Table 5). In two cases the pairs were known to be sisters and the other two pairs were of similar age and may also have been close relatives. Emigrating males were not always related. A young male which left the K pack was later found associating with two strange males, and the six E-pack males, which took over the L pack, consisted of the alpha male (EM11), three of his sons, and two presumably unrelated males which and had joined the E pack four years previously. Three male pairs of unknown origin joined packs on the study area.

Table 5 indicates that transferring mongooses moved from packs with greater numbers of older same-sex animals to packs with fewer, thereby improving reproductive potential. The exception (SM8) emigrated after the alpha female had died and no other female remained. He was relocated 5 months later as the beta male in another pack. The E-pack alpha male (EM11) emigrated with five other males, probably after the death of the alpha female, and became the alpha male in the adjacent L pack following a male takeover (see below). Two females were known to transfer more than once (Table 5). IF9 transferred five times from September to December between two neighboring packs, each containing a single older female. The following year, the alpha female of one of these packs had disappeared (presumed dead) and IF9 had replaced her. LF8 transferred with her sister (LF10) in September 1976 to the B pack (composition: an alpha male and female, an immigrant male, and a juvenile female) and LF10 was observed copulating with the alpha male. In November 1976 both females and the immigrant male disappeared and could not be relocated despite considerable searching. Three years later LF8 was found as the alpha female in a pack off the study area.

Packs containing less than two adult males and females were the most open to immigration. One young female (CF4) was seen being vigorously groomed, anogenitally licked and allomarked by the older alpha female and the alpha male within 2 h of contacting the pack, and two other females and one male were known to successfully join packs with only one female or male, respectively, in less than 2 days. Another young female (BF2) was solitary for a month during which she travelled across the F pack range, eventually encountering the I pack which had lost its alpha female in the previous month. The new female established friendly contacts with the adult males but was chased by a 9-month-old female (the only female in the pack) and slept in a separate den. When the pack was relocated 4 days later, BF2 was established as the new alpha female.

When aggression was directed toward potential immigrants, it came from animals of the same sex. Three instances of attacks and chases on solitary females by resident females were observed. CF9 trailed the T pack for 2 days but was repeatedly attacked by the alpha female although this individual had accepted a younger immigrant (CF4) four days previously. CF9 left the area and was found 2 years later in another pack (Table 5). In each of the two other observed cases, a transient female was chased by a subordinate resident female and soon left the area.

Most aggression toward immigrant males was by the resident beta male. Two males of unknown origin which eventually joined the L pack in January 1976 trailed it for a month before they were allowed to share the den. Once able to contact the females, the encounters were friendly, and the males often groomed, anogenitally licked, and allomarked the females. Throughout the month the intruders appeared extremely nervous, frequently giving alarm calls while trailing the pack. Of 26 observed chases, 21 were by the beta male, three by the alpha male, and two by both males together. Two males, which attempted to join the M pack in early June 1976, were vigorously chased by the beta male whenever he encountered them. The two intruders established friendly contacts with the females on the second day after arrival in the M pack range. After 6 days, one was found badly injured in the hindquarters and could not run, and the other male was missing.

Two cases of extrapack mating following temporary excursions from the pack range were recorded. In December 1979 the K-pack

TABLE 5
INTERGROUP TRANSFERS IN *Helogale parvula*

	Pack left			Pack joined						
Individuals	Pack	Size	No. older same-sex individuals	Pack	Size	No. older same-sex individuals	Age at immigration (months)	Month of immigration	Distance moved (km)	Comments
							Males			
EM12	E	15	3	L	13	1	34	Nov.	A	Became integrated after trailing new pack for 6 days.
EM27	E	6–17	5	M	11–19	3–4	13–17	May to Sept.	1.2	
SM8	S	0–3	0	K	8–9	1	30–35	Apr. to Sept.	1.3	Left pack after death of alpha female.
EM11, 29, 30, 18, 20, 28	E	0–12	0	L	11	1	81, ca. 66, ca. 66, 34, 31, 9	Oct.	A	Male takeover; see text.
							Females			
LF8, LF10	L	10	4	B	4	1	18	Sept.	A	Littermates. Reemigrated following Nov., LF8 found 3 years later as alpha female in a previously unmonitored pack.
MF1, MF3	M	13	1	B	7	0	59	Oct.	A	MF3 returned to M pack after 5 days. Transfer probably resulted in expulsion of only adult female (BF2) from B pack.

TABLE 5
CONTINUED

Individuals	Pack left			Pack joined			Age at immigration (months)	Month of immigration	Distance moved (km)	Comments
	Pack	Size	No. older same-sex individuals	Pack	Size	No. older same-sex individuals				
BF2	B	8	2	I	7	0	20	Oct.	1.0	Solitary for 30 days.
CF4	C	16	5	T	4	1	18	Oct.	A	Transfer took less than 1 day.
CF9	C	15	3	D					5.0	Emigrated in Oct. when 31 months old. Relocated 2 years later.
CF13	C	13	4	N	1	0	13–23	Nov.	4.1	Found with young adult male 10 months after emigration.
KF3	K	10	1	U	2	0	32	Nov.	A	Joined two males remaining after deaths of alpha pair.
MF5, MF8	M	11	3	W	3	0–1	22 & 20	Nov.	1.0	Sisters.
LF25	L	13–17	4–7	B	6–7	1	12–20	Dec. to Sept.	A	
YF8, YF1	Y	9–11	2	H	2	0	Both 23–25	Nov. to Dec.	2.1	Joined two adult males.
IF9	I	3	1	0	5	1	41	Sept.	A	Transferred five times in 4 months between two adjacent packs.

Individuals are referred to by the letter of their natal pack (or, if unknown, the pack in which they were first trapped) and their sex and pack number (for example, BF2 = female 2 from the B pack).

Pack sizes exclude transferees.

When time of transfer was not known, ranges of possible pack compositions, ages, and months are given.

Transfer distances are measured from center of old to new range.

A = joined adjacent pack with range overlap.

beta male was observed outside his range mating with a solitary female who had left the G pack several months earlier. On the following day, he was found back with the K pack. In November 1975 the A pack was reduced to three females through mortality and emigration. These temporarily left their natal range for a period of ten days. When they returned, the alpha female was pregnant and all three cooperated in raising her litter. By November 1976 only the alpha female remained. She stayed within the natal range and was joined by an immigrant male 2 months later.

Interpack transfer of two or more mongooses occasionally resulted in the expulsion of residents. Two M-pack females (MF1 and MF3) transferred to the B pack 7 weeks after its alpha female died (Table 5). MF3 returned to the M pack 5 days later, leaving her companion as the new B pack alpha female. The only resident female then in the B pack (BF2) left the same day the two older females joined. Two male takeovers with expulsion of the resident males were also recorded. The S-pack alpha male (the only adult male in the pack) was seen alone twice, shortly after two immigrant males joined the pack. Subsequently he disappeared. The E pack was last seen in its home area in May 1979. At this time it consisted of 18 mongooses. None of the females was subsequently found but on 31 October six of the males (the alpha male, two immigrants, two adult natal males, and one juvenile) reappeared on the study area and two were seen copulating with females in the neighboring L pack. The females rejoined the two resident L-pack males on 1 November, but 2 days later the E males were found near the L pack and the alpha male of the E group chased one of the L male residents from the seven L-pack females. The L females thereafter remained with the E males in the L-pack range. The two L males were subsequently seen together several times within their range, but disappeared in mid-November.

Intrapack Relationships and Care of Young

Dwarf mongooses live in a highly cooperative society with well-developed social roles. The pack forages and sleeps together and nearly all social interactions are amicable. Pack members frequently scan the environment from high points, such as termite mounds or rock formations. Vision is acute and raptors are often detected when

TABLE 6
NUMBER OF HEAD TURNS IN 50-MIN FOCAL SAMPLES IN THE M PACK

Individual		Age	Time atop mound (min:sec)	Head turns/min when atop	Head turns/min when not atop	Total head turns
Alpha	M2	ca. 6 years	32:40	7.3	1.8	270
	M10	ca. 3 years	12:10	3.8	1.6	107
	M9	1 year 7 months	10:40	4.8	2.3	141
	M12	9 months	15:38	2.8	1.3	88
Alpha	F6	ca. 6 years	4:47	7.9	1.7	113
	F1	ca. 4 years	0:51	7.1	2.9	146
	F7	1 year 10 months	—	—	2.0	99
	F13	9 months	5:19	5.5	2.3	132
Mean			10:05	5.5	2.0	136.4

they are mere specks in the sky. If a potential predator, such as a jackal or bird-of-prey is sighted, loud repeated alarm calls (a penetrating "tchee") warn other pack members in the vicinity. Frequently several animals give alarm calls at the same predator and these usually last until it leaves the area. Mongooses respond to the close approach of a predator with an alarm "chitter" as they run to cover, followed by repeated alarm calls when the cover is reached. Alarm responses are often given to bird calls.

Social roles in vigilance behavior were investigated by collecting focal animal data on each of eight members of the M pack in September and October of 1976. At this time the pack consisted of 12 individuals; those selected as focal animals were the alpha male and female, two each of subordinate adult males and females, and a juvenile male and female. Ages of these mongooses are given in Table 6. Sampling began when all individuals had emerged on the termite mound in the morning and ended when they began to leave the mound to forage. The number of head-turns made by each focal animal when atop and not atop the mound was recorded for 10 min using a continuously running tape recorder. If an individual was out of view, recording was stopped until it reappeared. Five samples were obtained for a total of 50 min on each animal. Table 6 shows that the alpha male (M2) made a much greater total number of head-turns and spent considerably more time atop the mound than any other individual. All pack members except F7 spent some time atop the mound and all showed higher rates of head-turning when atop than when lower down. Data from 15 all-day watches of the B pack at a breeding den also suggest that the alpha male is the most vigilant member of the pack. While at the den, the alpha male spent 15% of his time scanning the environment from atop the mound. Figures for the other five members varied from 7 to 10%.

Cooperation in dwarf mongoose packs extended to the rescue of trapped individuals. Early in the study, while attempting to capture a pack of nine mongooses with a large wire mesh trap, an alpha male rescued three pack members—the alpha female and two yearling females—which were trapped inside a termite mound. The trap (2 m × 1 m × ½ m high) with a hole cut in the bottom was placed over a den in an *Odontotermes* mound when the mongooses were inside. The wire mesh surrounding the hole was staked down and small live-traps placed inside to catch the mongooses when they emerged from the mound holes and began to search the trap for an escape route. Food and water were placed inside the large trap. One mongoose was caught on the second day and five more, including the alpha male, on the third. When these animals were released after marking, all but the alpha male scattered and ran into holes in nearby mounds. The alpha male ran to the mound covered by the trap and began to dig at the periphery. However, he later ran to a nearby mound, perhaps frightened by my presence. Three hours later, the male had dug back into the mound containing the trapped mongooses and had been recaptured in one of the small live-traps. When released, he immediately ran back to the mound and into the hole that he had dug previously. A few minutes later, he emerged inside the large trap and began to dig at the point where the wire mesh was staked to the ground. Eventually the male was able to tunnel under the trap to the outside thereby enabling the three trapped animals to escape. All four mongooses were seen in the evening with the remainder of the pack.

This alpha male managed to release his mate (which was not in estrous at the time) and two young females who were probably his daughters. Although mongooses in nature rarely encounter live-trapping scientists, they might experience a cave-in or other den damage from animals, such as elephants, which could result in part of a pack being stranded and in need of rescue from outside. This

TABLE 7
HELPER BEHAVIOR IN B PACK

Statistics	Alpha male M4	Alpha female F1	Immigrant male M5	Immigrant female F25	Natal male M6	Natal female F10
Age (months)	>48	ca. 72	ca. 47	23	23	23
r	.5	.5	0	0	.25	.25
Babysitting (hours)	10.5	5.4	16.3	21.8	32.2	21.8
Alone with young	2.8	3.2	8.0	13.6	17.1	11.3
Carrying food to young	0	0	6	18	11	6
Number of times groomed young per 10 h at den	6.5	7.0	16.6	19.0	19.8	20.5

r = Degree of relatedness to young

incident demonstrates that dwarf mongooses have the capacity and inclination to aid group members even at a risk to themselves.

The B pack provides an example of intrapack relationships occurring at a breeding den. This pack was selected for more intensive observation since it formed on the study area in February 1976 and all kinship relationships were known from regular monitoring. In November and most of December 1978, the grass within its range was unusually short because of previous burning and heavy grazing, greatly facilitating field observations. At this time the pack contained both an immigrant male and female whose behavior could be compared with that of animals born in the pack.

The alpha female gave birth to a litter on 21 November 1978, and detailed observations were begun the following day. The pack then consisted of six mongooses: the alpha male (M4), the alpha female (F1), the immigrant male (M5) and female (F25), and a natal male (M6) and female (F10) (offspring of the alpha male by a previous alpha female). Ages of these animals are given in Table 7. My wife and I conducted all-day watches at the breeding den at approximately 2-day intervals until the young were 32 days old, when they were starting to leave the den and tall grass precluded further detailed records. Observations commenced when the first mongoose emerged from a hole in the termite mound in the morning and terminated when the last one went in for the night. Fifteen all-day watches were conducted for a total observation time of 160 h.

In dwarf mongooses intrapack aggression is rare, except by males

TABLE 8
CROUCH GREETINGS IN B PACK

	Animal being greeted						
Greeter	M4	M5	M6	F1	F25	F10	Total
M4		0	0	0	0	0	0
M5	7		0	2	0	0	9
M6	1	4		0	2	0	7
F1	0	0	0		0	0	0
F25	0	1	0	1		0	2
F10	0	0	0	1	0		1
Total	8	5	0	4	2	0	19

in the presence of estrous females. Ritualized "crouch-greeting" sometimes occurs when two mongooses meet after separation. Crouch-greeting indicates submission and is not reciprocated. The subordinate approaches the dominant in a flattened crouch while darting the head from side to side near the mouth of the other animal and maintaining a high-pitched twittering call. Frequently the greeter lifts a forepaw during the interaction and it may roll on its side (see Rasa, 1977, for a more complete description). At the B-pack breeding den, this pattern was most frequently given by the beta male (M5) to the alpha male (M4) when the latter returned to the den (see Table 8). The beta male was also crouch-greeted four times by the younger natal male (M6). Table 8 indicates a dominance hierarchy among the males that was confirmed when the alpha female came into estrous 23 days after parturition. Both younger females crouch-greeted F1 and other social interactions indicated that between the two same-age females, the immigrant (F25) was dominant to the natal female (F10). On several occasions F25 reared in threat to F10, and F10 sometimes turned her head away from F25 when the two met.

In dwarf mongooses, allogrooming is an indicator of social bonds; it is primarily heterosexual and mutual. In the B pack, 200 grooming interactions were recorded (Fig. 4), of which 187 (94%) were heterosexual and 142 reciprocated (71%). The alpha female usually reciprocated grooming but initiated it much less frequently than other pack members (Fig. 4).

Subordinate females sought contact with the alpha male and frequently approached and rubbed along his side or across his scrotum.

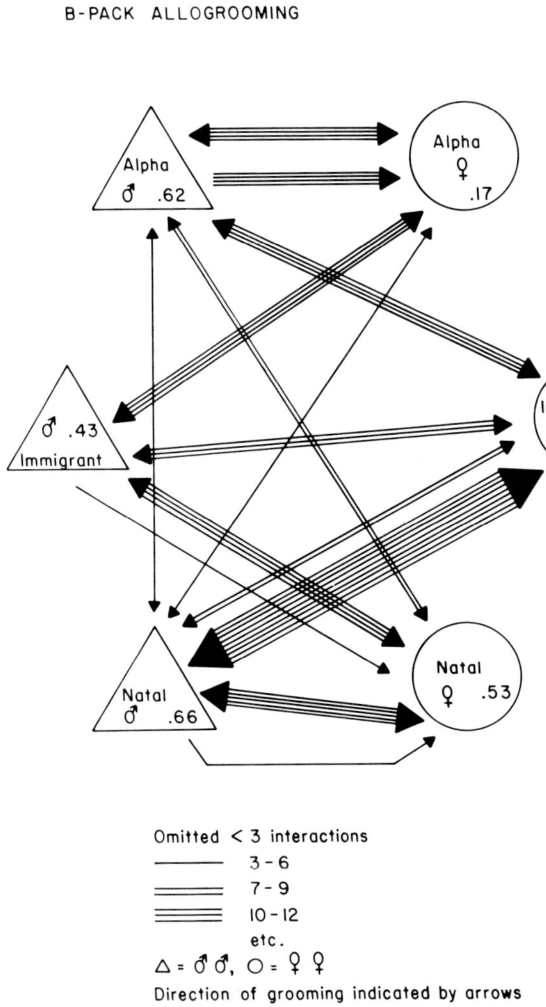

FIG. 4. Sociogram of allogrooming in the B pack. Numbers inside symbols represent proportion of mutual grooming which was initiated by each individual.

They also commonly allomarked the alpha male by rubbing the anal or cheek glands over him, and the alpha male sometimes reciprocated. Twelve of 18 incidences of body rubbing observed at the B-pack breeding den were directed by the immigrant female (F25) to the alpha male.

The alpha female appeared to take the main leadership role in most dwarf mongoose packs. Alpha females were frequently the first to emerge from the den in the morning and the first to leave the den site to forage. In the B pack, the alpha female was the first to emerge on 12 of the 15 mornings and the first to leave the den on 10 of the 13 mornings when there was a clear leader.

Cooperation in dwarf mongooses is particularly evident in the care of the young. While confined to the breeding den, young mongooses are typically guarded throughout the day by one or more pack members, but changeovers occur frequently allowing all mongooses some time to forage. Pack members also cooperate in feeding, grooming, and transporting the young (Rood, 1978).

In the B pack, babysitting was considered to take place when less than half the pack (that is, one or two mongooses) were at the den. This occurred during approximately half (51%) of the 160 h observation time (a single mongoose stayed with the young for 68% of the babysitting time). Three or more pack members were at the den for 37% of the observation time. Occasionally the young were left alone, sometimes for several hours. The B-pack young were alone for 12% of the observation time, but this figure is high compared to other observed packs. For example, in 49 h of observation at a T-pack breeding den, the young were left alone by the five pack members for only 6% of the time (Rood, 1978).

Potential ground predators which approached a dwarf mongoose breeding den were actively chased from it. The B pack chased a slender mongoose (*Herpestes sanguineus*) from the den site and, on three occasions, they attacked and held off a larger intruding pack of dwarf mongooses (the E pack, numbering 15) until it retreated. All B-pack members present at the den participated in chasing the intruders during these encounters.

The B pack used three termite mound dens during the 32-day study period. Straightline distances over which the young were moved were 310 m and 160 m, but the actual distances were much greater because the mongooses frequently changed course to take advantage of available cover. Den moves occurred when the young were 14 and 27 days old. When they were 23 days old, they were moved from the den in the morning and carried to three separate sites during the day before being returned to the same den in the late afternoon. All pack members assisted in transporting the young between dens. They were carried just behind the head. The two den moves were initiated by the natal and immigrant males, respectively, and the alpha female initiated the move in which the young were returned to the den later in the day.

No observable explanatory event preceded den changes. Most den moves appeared haphazard and confused, and the young were frequently exposed and more vulnerable to predators. One animal sometimes carried an infant to a new den only to have another carry it back to the old one. Frequently young were carried in relays. A pack member drops a young enroute and continues, leaving another to pick it up and complete the journey. Den moves involved general excitement within the pack and increased scent marking. The new den was frequently extensively cheek and anal marked by all pack members, particularly at grass stalks or branches close to den entrances.

Day-old young are sometimes carried to the surface of the mound where they are extensively groomed by pack members. The B-pack young were first observed to emerge from the den hole without assistance when 17 days old and from then on they spent much time above ground. Two of the four young disappeared at age 22 days; their cause of death was unknown. During the last all-day watch, the 32-day-old young persistently followed pack members. In the late morning, they followed two adults to a mound 50 m from the main den and remained there with babysitters throughout the day, returning with the pack in the evening. When next seen at 36 days, the young left the den with the pack in the morning to forage. Data from other packs show that young mongooses first regularly accompany the pack at 5 to 6 weeks.

The B-pack alpha female spent relatively little time with her young during the 15 all-day watches, averaging 67 min/day (range 8 to 217 min). Thirteen nursing periods were observed in the eight days after the young first emerged from the den at 17 days of age. Average nursing time was 4 min 38 s (range 15 s to 20 min). Data from the entire study indicate that young dwarf mongooses are usually weaned when between 6 and 7 weeks old. No young were observed suckling after 55 days.

Data on helper contributions (Table 7) indicate that the main helpers in the B pack were the immigrants and natal mongooses, with the alpha pair, the closest relatives of the young, spending less time in babysitting. The alpha female spent the most time away from the den, enabling her to fulfill the greater nutritional requirements of lactation. Pack members fed young mongooses by collecting insects and carrying them to the young. Again, the main helpers were the immigrants and natal mongooses, with the immigrant female making the greatest number of feeding visits. The alpha pair of the B pack was not observed to bring food to the young, although this was occasionally recorded in other packs. Most grooming of the young was done by the immigrants and natal animals, with the alpha pair grooming considerably less frequently.

Comparisons with Other Social Mongooses

Table 9 compares some parameters of ecology and social organization in the social mongooses. Clearly there is great scope for further field work; for example, field studies on species of cusimanses (*Crossarchus*) are completely lacking and only short term behavioral observations have been made on the meerkat (*Suricata suricatta*) in the field. Although the yellow mongoose (*Cynictis penicillata*) has been reported to den in colonies of up to 50 individuals (Hinton and Dunn, 1967; Roberts, 1951), and has been included in lists of the social mongooses (Gorman, 1979), Lynch (1980) points out that reports of such large colonies stem from Fitzsimmons (1919) who must have been confusing *C. penicillata* with *S. suricatta*. My brief field observations made in South Africa in 1973 and 1976 suggest that *C. penicillata* lives in pairs and family groups and Lynch (1980) found that this mongoose was usually found singly or in pairs and that larger groups (up to seven) usually consisted of an adult pair and their young.

Table 9 shows a basic similarity in ecology between the social mongooses. All are diurnal and feed primarily on invertebrates, although small vertebrates provide a secondary food source when available. With the exception of *Crossarchus,* which are reported from forest, the social mongooses live primarily in open habitats. In these characteristics, they differ from the majority of viverrids which are solitary, nocturnal, small vertebrate feeders, living primarily in forest (Gorman, 1979; Walker et al., 1975). Group cohesion is facilitated by diurnality, with visual contact in open habitats as well as contact calls keeping the group together. Invertebrate feeding is probably an important correlate of group living in small carnivores. The savanna habitat supports a high biomass of invertebrates with a high rate of renewal compared to small vertebrate prey. Thus food competition may be reduced in small carnivores which are invertebrate feeders. A high density of individuals may facilitate the evolution of a cooperating group.

Group size is largest in the banded mongoose, although the sample size is small for meerkats and virtually no data are available for cusimanses. No obvious sexual dimorphism occurs in the social mongooses which have been studied. In the dwarf mongoose this correlates with the effectively monogamous breeding system (see Kleiman, 1977), in which a dominant pair are the parents of the offspring. In the banded mongoose, several males and females mate during synchronous mating periods and up to four females regularly become pregnant and rear offspring. Field data on *S. suricatta* and *Crossarchus* spp. are lacking, but Wemmer and Fleming (1975) noted that in captive meerkats only one female per group produced young.

In the dwarf mongoose, both males and females commonly transfer between groups. Observations in Uganda on banded mongooses indicate that the rate of transfer is considerably lower in this species. No immigration was recorded in five groups monitored for over 2 years (Rood, 1975). However, evidence that groups of males occasionally operated as independent units was obtained. In one instance, six males left their pack in the early morning, crossed an airstrip into the range of a neighboring pack and sniffed and defecated at a den used by that pack the previous night. They returned to their own range in the late morning but did not rejoin their pack. After resting until the late afternoon, they again crossed into the neighboring pack's range and sniffed, defecated, and marked at the den. Twenty minutes later, the six males were joined by the four remaining males of their pack, following the same route and making loud "churr" vocalizations. After vigorous greetings, all ten ran across the airstrip and rejoined their pack females at a den before dark.

At Kirawira, two male banded mongooses separated from their natal pack and appeared to attempt to join an adjacent pack by trailing and denning near it at night, but they were chased by pack members. However, on one occasion, a pack member temporarily joined the two transient males after first chasing them and then allomarking and sniffing under their tails. The two males left the area after 8 days. Further evidence of social mobility in the banded mongoose was obtained on the Serengeti shortgrass plains although,

TABLE 9

COMPARISONS OF SOME ECOLOGICAL AND SOCIAL ORGANIZATION PARAMETERS IN SOCIAL MONGOOSES

Ecological and social parameters	Dwarf mongoose (*Helogale*)	Banded mongoose (*Mungos*)	Meerkat (*Suricata*)	Cusimanse (*Crossarchus*)
Habitat	Savanna, woodlands	Savanna, woodlands, open grassland	Open grassland, semi-arid bush	Forest
Diet	Invertebrates, small vertebrates	Invertebrates, small vertebrates	Invertebrates, small vertebrates	Invertebrates, small vertebrates
Activity cycle	Diurnal	Diurnal	Diurnal	Diurnal
*Group size				
Mean	8.7	14.2	11.6	10–24
Maximum	27	36	30	
N	83	142	34	
Sexual dimorphism	No	No	?No	
Mating system	Monogamous	?Polygamous		
Intergroup transfer				
Males	Yes	?Yes	?Yes	
Females	Yes			
Antipredator behavior				
Alarm call warns group members	Yes	Yes	Yes	

TABLE 9
CONTINUED

Ecological and social parameters	Dwarf mongoose (*Helogale*)	Banded mongoose (*Mungos*)	Meerkat (*Suricata*)	Cusimanse (*Crossarchus*)
Bunching	No	Yes	Yes	
Alloparental behavior				
Babysitting	Yes	Yes	Yes	
Feeding	Yes	Yes	Yes	
Transporting	Yes	Yes	Yes	
Grooming	Yes	Yes	Yes	
References	Dorst and Dandelot (1970); Rood (1978); Rood (1980); this study	Dorst and Dandelot (1970); Rood (1974); Rood (1975); pers. observ.	Dorst and Dandelot (1970); Ewer (1963); Lynch (1980); Smithers (1971); Wemmer, pers. comm.; Zumpt (1968); pers. observ.	Dorst and Dandelot (1970); Ewer (1973); Walker et al. (1975)

* Data used to compute mean group size in dwarf and banded mongooses were counts of regularly monitored packs in October and additional counts of packs seen only once. Dwarf mongoose counts were made in the Serengeti National Park, Tanzania. Most banded mongoose counts were also made in the Serengeti but counts made in the Mara Game Reserve, Kenya, and in Rwenzori Park, Uganda, were also included. Meerkat pack counts combine data collected by Lynch (1980) and the author (pers. observ.) in South Africa and by Smithers (1971) in Botswana.

because the packs were not marked, events could only be generally followed. One pack of 12 split into two groups of seven and five which used adjacent ranges. Two mongooses transferred from the group of seven to the five and then back again within a period of 3 months. Another pack of 24 split into groups of 13 and 11 which both attempted to stay within the original range. The group of 13 broke down into groups of eight and five and fights between these groups frequently occurred when both attempted to use the same den. On one occasion, a mongoose transferred from an adjacent pack of seven to the five and on the following day transferred back. A period of instability involving frequent aggressive interactions between subgroups of the original pack and transfers between them lasted for about 6 weeks and was resolved by most of the animals leaving the area. The remaining 10 mongooses formed a stable unit and when they were later trapped proved to consist of old animals, suggesting that the emigration was by younger pack members. As a result of the fighting which occurred during the period of instability, many mongooses had most of the hair pulled from their tails.

Evidence for intergroup transfer in the meerkat is suggested by one male which was observed trailing a pack of 12 for 3 days and denning within 50 m of them. This animal was frequently chased by the pack.

All species which have been studied have distinct alarm calls. Bunching has been described in the banded mongoose to repel potential predators (Rood, 1975), and in the Serengeti I have seen large packs bunch together and chase off ground predators up to the size of a jackal. I have also observed a pack of meerkats in the Kalahari Gemsbok National Park, South Africa, bunch and chase Cape foxes (*Vulpes chama*). Dwarf mongooses do not use bunching as an antipredator device, perhaps because their small size would render it ineffective.

The social mongooses which have been studied are particularly noteworthy for the amount of care of the young given by group members other than the parents (Table 9).

Discussion

Among group-living mammals, it is common for the males to leave their natal groups while the females remain and breed in them (Eisenberg, 1977; Wilson, 1975). Dwarf mongooses are unusual in that both males and females commonly transfer between groups and this appears to be correlated with their effectively monogamous breeding system in which a dominant breeding pair produces the offspring. In monogamous species, the reproductive advantages of emigration and intergroup transfer would appear to be similar for both sexes. Transferring dwarf mongooses increase their reproductive potential by moving from packs with greater numbers of older same-sex animals to packs with fewer. Intergroup transfer allows the maintenance of viable packs within defined home ranges despite the loss of breeding members and prevents excessive inbreeding. Several studies have indicated that inbreeding has deleterious consequences in a variety of animal species (see Greenwood et al., 1978; Wilson, 1975); this usually does not occur in dwarf mongooses because of the voluntary dispersal of relatives who are potential mates.

Long-term studies are needed to reveal how choices are made from among the diverse strategies available to individuals to maximize reproductive potential. Breeding strategies observed in this study were to: 1) emigrate and join an existing pack with fewer same-sex individuals than in the natal pack; 2) emigrate and form a new pack by joining with opposite sexed individuals; 3) remain in the natal pack and achieve breeding status through the deaths of older same-sex individuals; 4) emigrate with same-sex pack members and take over an existing pack by driving out residents of the same sex; 5) temporarily join and mate with a transient individual; or 6) transfer back and forth between two packs and move into the first available breeding opening.

With the exception of strategy 6, which was used by only one female, these strategies were employed by both sexes. Females were more likely to remain in their natal groups than males were, resulting in an adult sex ratio in the packs which was biased toward females. Males were more common in transient groups where they probably face greater predation pressure than when in packs. Breeding dispersal (see Greenwood, 1980) was recorded in alpha males following the death of the alpha females. However, no evidence was found for dispersal by alpha females. Females that attain this status probably remain within the same home range throughout their lives.

Predation has probably been the most important selective pressure favoring sociality in the social mongooses, and increased protection through the vigilance and antipredator responses of pack members is an important function of group life. Individuals which adopt strategy 3 maintain this protection and can also increase their inclusive fitness (Hamilton, 1964) by helping relatives. However, most successful breeders of both sexes in this study had emigrated. Although more vulnerable to predation while emigrating, mongooses which are able to join existing packs or form new ones achieve the benefits of earlier breeding (Rood, 1980) and avoidance of inbreeding.

Carnivores display a high degree of male parental investment compared to other orders (Kleiman and Malcolm, 1981). In dwarf mongooses male parental care is particularly important in newly formed pairs and small groups in which there are few helpers. In one pack, the alpha male was the main babysitter when the pack consisted of only three mongooses, but subsequently when five animals were present, an immigrant female stayed with the young most frequently and the alpha male was not recorded to stay alone with them (Rood, 1978). Care of young by individuals in addition to the mother is highly developed in the social mongooses. Helpers have been reported in several carnivores (Macdonald, 1979; Moehlman, 1979, this volume), but the dwarf mongoose is one of few species in which helpers unrelated to the recipients of the aid have been reported. In the B pack, the immigrant male and female were active helpers. In another pack (T pack), an immigrant female spent more time in babysitting than any of the other four related pack members (Rood, 1978).

Because kin selection cannot be operating in cases where unrelated immigrants are giving aid to the young, why do these immigrants help? Trivers (1971) developed the theory of reciprocal altruism to explain how apparently altruistic behavior might be selected for in cases where the recipient is so distantly related to the altruist that kin selection can be ruled out (for a discussion of kin selection and altruism, see Bertram, this volume). His argument may apply to the situation where the survival of the recipient of the aid is beneficial to the immigrant mongoose. The immigrant may, for example, receive aid from the young it helped to rear in the form of antipredator responses. Reciprocity is also likely to operate in care of the young. Immigrants may eventually breed in packs which they join, and young they have helped to raise may then act as helpers to their own young. This has already occurred in the T pack in which the former alpha female died in March

1979. The immigrant female helper is now breeding and receiving aid from young which she previously helped rear.

Acknowledgments

I thank the Tanzania National Scientific Research Council, the Director of Tanzania National Parks, the Director of the Serengeti Research Institute, and the Chief Park Warden of the Serengeti National Park for permission to conduct this research. Financial support was provided by the National Geographic Society, the Max-Planck-Institut fur Verhaltensphysiologie, the Harry Frank Guggenheim Foundation, and the Friends of the National Zoo. I am grateful to the many Serengeti Research Institute scientists who told me of mongooses they had seen and particularly to G. and L. Frame for recording counts of banded mongooses. I also thank D. Kleiman for her constructive criticism of the manuscript. My wife, Hazel Rood, gave invaluable assistance throughout all stages of the study.

Literature Cited

DORST, J., AND P. DANDELOT. 1970. A field guide to the larger mammals of Africa. Houghton Mifflin Co., Boston, 287 pp.

EISENBERG, J. F. 1977. The evolution of the reproductive unit in the class Mammalia. Pp. 39–71, *in* Reproductive behavior and evolution (J. S. Rosenblatt, and B. R. Komisaruk, eds.). Plenum Press, New York, 170 pp.

EWER, R. F. 1973. The carnivores. Weidenfeld and Nicolson, London, 494 pp.

FITZSIMMONS, F. W. 1919. The natural history of South Africa. Longmans Green, London.

GORMAN, M. L. 1979. Dispersion and foraging of the small Indian mongoose, *Herpestes auropunctatus* (Carnivora:Viverridae), relative to the evolution of social viverrids. J. Zool., 187:65–73.

GREENWOOD, P. J. 1980. Mating systems, philopatry and dispersal in birds and mammals. Anim. Behav., 28:1140–1162.

GREENWOOD, P. J., P. H. HARVEY, AND C. M. PERRINS. 1978. Inbreeding and dispersal in the great tit. Nature, 271:52–54.

HAMILTON, W. D. 1964. The genetical evolution of social behavior. I, II. J. Theoret. Biol., 7:1–52.

HERLOCKER, D. 1976. Woody vegetation of the Serengeti National Park. Texas A & M Univ. College Station, Texas.

HINTON, H. E., AND A. M. S. DUNN. 1967. Mongooses, their natural history and behavior. Oliver and Boyd, London, 144 pp.

KLEIMAN, D. G. 1977. Monogamy in mammals. Quart. Rev. Biol., 52:39–69.

KLEIMAN, D. G., AND J. R. MALCOLM. 1981. The evolution of male parental investment in mammals. Pp. 347–387, *in* Parental care in mammals (D. J. Gubernick and P. H. Klopfer, eds.). Plenum Press, New York, 459 pp.

KRUUK, H. 1972. The spotted hyena. Univ. Chicago Press, Chicago, 335 pp.

LYNCH, C. D. 1980. Ecology of the suricate, *Suricata suricatta,* and yellow mongoose, *Cynictis penicillata,* with special reference to their reproduction. Mem. Van Die Nasionale Mus., Bloemfontein, 14:1–145.

MACDONALD, D. 1979. "Helpers" in fox society. Nature, 282:69–71.

MOEHLMAN, P. D. 1979. Jackal helpers and pup survival. Nature, 277:382–383.

NORTON-GRIFFITHS, M., AND A. R. E. SINCLAIR (eds.). 1979. The Serengeti ecosystem. Univ. Chicago Press, Chicago, 387 pp.

RASA, O. A. E. 1977. The ethology and sociology of the dwarf mongoose (*Helogale undulata rufula*). Z. Tierpsychol., 43:337–406.

ROBERTS, A. 1951. The mammals of South Africa. Central News Agency, Cape Town, 700 pp.

ROOD, J. P. 1974. Banded mongoose males guard young. Nature, 248:176.

———. 1975. Population dynamics and food habits of the banded mongoose. E. African Wildl. J., 13:89–111.

———. 1978. Dwarf mongoose helpers at the den. Z. Tierpsychol., 48:277–287.

———. 1980. Mating relationships and breeding suppression in the dwarf mongoose. Anim. Behav., 28:143–150.

ROOD, J. P., AND D. NELLIS. 1980. Freeze marking mongooses. J. Wildl. Mgmt., 44:500–502.

SCHALLER, G. B. 1972. The Serengeti lion. Univ. Chicago Press, Chicago, 480 pp.

SMITHERS, R. H. N. 1971. The mammals of Botswana. Mus. Mem. The Trustees of the National Museums of Rhodesia, Salisbury, 4:1–340.

TRIVERS, R. L. 1971. The evolution of reciprocal altruism. Quart. Rev. Biol., 46: 35–57.

WALKER, E. P., ET AL. 1975. Mammals of the World. Johns Hopkins University Press, Baltimore, 1500 pp.

WEMMER, C., AND M. J. FLEMING. 1975. Management of meerkats. Internatl. Zoo Yearb., 15:74–77.

WILSON, E. O. 1975. Sociobiology: the new synthesis. Harvard Univ. Press, Cambridge, Massachusetts, 697 pp.

ZUMPT, I. F. 1968. The feeding habits of the yellow mongoose, *Cynictis penicillata,* the suricate, *Suricata suricatta,* and the Cape ground squirrel, *Xerux inauris.* J. South African Vet. Med. Assn., 39:89–91.

THE EVOLUTION OF MATING SYSTEMS IN PINNIPEDS

IAN STIRLING

Abstract

ECOLOGICAL factors are primarily responsible for the evolution of mating and social systems in pinnipeds, but phylogeny has also influenced some aspects. Possibly the single most important influence on the development of social and mating systems in pinnipeds has been the availability of and access to suitable habitat for parturition for the following reasons: 1) birth of the young must take place out of the water; 2) females develop fidelity to suitable pupping sites; 3) females mate shortly after parturition; 4) males are attracted by the presence of estrous females. Polygyny occurs in all otariids, the walrus, and four species of phocids and is, in general, more characteristic of temperate than polar regions. This paper discusses the major factors influencing the evolution of mating systems in pinnipeds, including availability of habitat for parturition, access to it and its stability; gregariousness of females; sexual dimorphism; the role of terrestrial and marine predators; terrestrial and aquatic mating; and, thermal considerations. Examples are given of species and ecological situations in which detailed long-term behavior studies would permit the testing of some specific hypotheses.

Introduction

The Pinnipedia is a familiar, nearly worldwide suborder, comprising three families and about 33 living species. The three families are the Odobenidae (walruses), the Otariidae (fur seals and sea lions), and the Phocidae (the true or earless seals). The question of whether or not the pinnipeds are monophyletic or biphyletic in origin has not been completely resolved but the most current views tend to favor the biphyletic view (Ray, 1976). According to the most recent studies (Ray, 1976; Repenning, 1976, 1980; Repenning and Tedford, 1977; Repenning et al., 1979) the Otarioidea first appeared in the late Oligocene and, in the early Miocene, divided into the lineages that gave rise to the present day otariids and odobenids. Other related groups are now extinct. The earliest phocids appeared about five million years later in the middle of the Miocene. It appears that the otariids originated in and dispersed from the North Pacific Basin and that phocids did so from the North Atlantic. The otariids never invaded the North Atlantic, although the present day walrus has. The phocids dispersed into the North Pacific, probably via both the Arctic Ocean and the then-submerged region of Panama.

Although pinnipeds feed at sea and some species maintain a pelagic existence for months at a time, their distribution is essentially coastal or associated with pack ice at high latitudes. This distributional pattern probably derives from the fact that pinnipeds must leave the water for parturition and that pinnipeds occasionally rest out of the water on land or ice independent of the breeding season. The two most fundamental features of pinniped life history, in terms of the evolution of their social behavior, are their marine feeding and terrestrial parturition (Bartholomew, 1970).

In most species, parturition is seasonal and highly synchronized. The pups are weaned at the birth site on either land or ice. Phocids wean the pups by 6 or 7 weeks at the most. The females of several species do not return to the water until nursing has ceased. The otariids take several months to wean pups so that, after an initial intense nursing period of a week or more, the females begin to make feeding trips to the sea to maintain their reserves for continued nursing. Some otariids may nurse their pups for over a year (for example, the Cape fur seal, *Arctocephalus pusillus*; Stirling and Warneke, 1971). In no species does the male help feed the young. Although it is not known, it does not seem likely that males of any species aid or teach young when mixed sex groups are foraging.

With the exception of the walrus, ovulation and mating follow within a few days or weeks after parturition. Thus, all the aspects of the reproductive cycle, including intrasexual competition between males for females, occur in a relatively short time period either at the birth site or in the water immediately adjacent to it. Consequently, despite the essentially marine existence of pinnipeds, much of the reproductive behavior of several species has been fairly well documented.

In the strict sense of the word, all pinnipeds are polygynous because the males of all species will mate with more than one female if possible, and there is usually considerable variance in male reproductive success. In no species are the males and females known to associate in any socially organized way other than during the breeding season. For the purposes of this paper, there are four fairly clear, albeit arbitrary, categories into which the mating systems of most species, or in some cases subpopulations of species, can be classified. These are serially monogamous, promiscuous, polygynous, and extremely polygynous.

In a serially monogamous species or population, an adult male defends a parturient female until she weans her pup and ovulates. He then mates with her, and departs to seek another parturient or lone estrous female. An interesting aspect of this behavioral pattern is that the male protects the pup of an unrelated male from conspecifics and some predators until weaning. Mating may occur either in or out of the water. Examples are the crabeater seal (*Lobodon carcinophagus*; Siniff et al., 1979) (Fig. 1), and the hooded seal (*Cystophora cristata*; Øritsland, 1964).

In promiscuous species, variable sizes of mixed sex groups occur together or at least in reasonably close proximity to one another. Although some males are undoubtedly dominant over others they are able to exert only limited, if any, control over the reproductive activities of their fellows. Although not adequately documented, females are thought to mate with more than one male with some regularity. Most, if not all, mating probably occurs in the water. Examples are the harbor seal (*Phoca vitulina*; Venables and Venables, 1957) and the Hawaiian monk seal (*Monachus schauinslandi*; Kenyon and Rice, 1959).

In polygynous species, parturient females aggregate on terrestrial colonies for breeding. Adult males are able to defend terrestrial territories (including several estrous females) from competitors, and thereby ensure reproductive access to more than one female. Mating occurs terrestrially. Examples include the California sea lion (*Zalophus californianus*; Peterson and Bartholomew, 1967) and the New Zealand fur seal (*Arctocephalus forsteri*; Stirling, 1971*a*).

In extremely polygynous species, males are able to defend terrestrial territories with large numbers of estrous females or be dominant enough in a hierarchical situation to do most of the breeding. Examples include the northern fur seal (*Callorhinus ursinus*) in which Bartholomew and Hoel (1953) recorded a mean harem size

FIG. 1. Family group of crabeater seals (*Lobodon carcinophagus*) in Antarctic pack ice; note vulnerability of habitat to disintegration and extensive scars on adult female from attacks by leopard seals.

of 39 and a maximum of 108, and the northern elephant seal (*Mirounga angustirostris*) (Fig. 2) where, in one colony, LeBoeuf and Peterson (1969) reported that only 4% of the males inseminated 85% of the females.

Because some of the more accessible (to humans) otariids and phocids of temperate latitudes are highly gregarious and mate on land at a limited number of sites, much published research has been concerned with these species and their impressive development of sexual dimorphism and polygyny (for example, Bartholomew, 1952, 1970; Bertram, 1940; McLaren, 1967; Nutting, 1891; Peterson and Bartholomew, 1967). However, the concentration of pinniped behavior studies on temperate species has tended to obscure two important points. First, the development of polygyny may be linked to phylogeny (Stirling, 1975). All otariids and apparently the walruses are polygynous, whereas only four of the 19 species of phocids appear to be. Second, there appears to be a strong geographic effect on pinniped mating systems. Although many of the species breeding in ice-free waters have evolved a significant degree of

FIG. 2. Breeding colony of northern elephant seals (*Mirounga angustirostris*); note high density of parturient females and presence of competing adult males.

polygyny, apparently only two polar species have done so (the walrus, *Odobenus rosmarus,* and the Weddell seal, *Leptonychotes weddelli*). One consequence of these two overlooked aspects has been that marked sexual dimorphism and highly developed polygyny is generally considered to be more characteristic of pinnipeds than is the case. In fact, almost half of the living species of pinnipeds do not exhibit these characteristics.

Bartholomew's (1970) model explains the evolution of terrestrial polygyny but it assumes that the gregariousness of females and terrestrial mating are already-evolved characters. The purpose of this paper is to evaluate and discuss on a broader scale the most important factors influencing the evolution of the spectrum of mating systems in pinnipeds today. I will sacrifice a certain amount of detail because the information available is voluminous and the life cycles of each species (and even of subpopulations of the same species) have their own distinct characteristics. Also, much of this paper will be speculative because of gaps in information on critical points. Consequently, wherever possible, I will emphasize unanswered questions and discussions of species in ecological contexts that favor the testing of hypotheses. Because of the wide variety of behavioral and ecological characteristics and individual species discussed, I have briefly summarized the most relevant information in Table 1.

The central thesis of this paper is that ecological factors have determined how social and mating systems have developed in pinnipeds, but that phylogeny has also been influencial. The most important factors influencing the development of social systems in seals have probably been the availability of habitat for parturition, including access to it, and its effect on mating behavior. This is because birth of the young must take place out of the water, females develop fidelity to suitable pupping sites and mate shortly after parturition, and males are attracted by the presence of estrous females. In general, the extent to which polygyny will develop is inversely proportional to the amount of breeding habitat and the factors which influence the seals' access to it.

Major Factors Influencing Behavior

Habitat for Parturition: Availability, Access, and Stability

The availability of suitable stable habitat and access to it for parturition and safe weaning of the pup varies both geographically and seasonally. For example, in areas such as the southern ocean between about 40° and 60°S, there is little ice and the land masses of the subantarctic islands are both small and widely separated. Furthermore, the shorelines of these islands tend to be rugged and exposed to heavy seas so that sheltered beaches and sites with access to other protected areas are further reduced. Consequently, it may be necessary for females to give birth on islands at high densities (for example, Fig. 2). Under such conditions males can compete for tenure in areas where females concentrate and exclude competitors. Even on small islands where there appears to be more suitable land available than is being used, it may be misleading to conclude that available space is not of vital importance. Since most species of seals that breed on islands were at one time severely depleted by man, what we see today may not be representative of the past. Similarly, the presence of marine predators, local currents, or submarine geog-

TABLE 1

SUMMARY OF BEHAVIORAL AND ECOLOGICAL CHARACTERISTICS OF SPECIES DISCUSSED IN TEXT (REFERENCES GIVEN IN THE TEXT WITH SUPPLEMENTAL INFORMATION FROM SCHEFFER, 1958; KING, 1964; HALEY, 1978; AND FAO, 1979)

Pinniped families and species	Breeding habitat	Mating system	Degree of sexual dimorphism	Synchrony of pupping and breeding	Gregariousness of females and intrasexual thigmotaxis	Terrestrial or aquatic mating
Family Otariidae						
Eumetopias jubatus (Steller sea lion)	Islands	Polygyny	Males 3–4 × weight females	3–4 weeks	Gregarious; positively thigmotactic	Terrestrial
Zalophus californianus (California sea lion)	Islands	Polygyny	Males 2–3 × weight females	3–4 weeks	Gregarious; positively thigmotactic	Mainly terrestrial, some aquatic
Neophoca cinerea (Australian sea lion)	Islands	Polygyny		Some pupping on 18-month cycle; most pupping and mating in 2 months	Gregarious; positively thigmotactic	Terrestrial
Callorhinus ursinus (northern fur seal)	Islands	Extreme polygyny	Males 3–5 × weight females	up to 2 months	Gregarious; positively thigmotactic	Terrestrial
Arctocephalus townsendi (Guadalupe fur seal)	Islands; more recently inaccessible caves	Polygyny	Males 4–5 × weight females	4–6 weeks	Gregarious; negatively thigmotactic	Terrestrial

TABLE 1
CONTINUED

Pinniped families and species	Breeding habitat	Mating system	Degree of sexual dimorphism	Synchrony of pupping and breeding	Gregariousness of females and intrasexual thigmotaxis	Terrestrial or aquatic mating
Arctocephalus pusillus (Cape fur seal)	Islands	Polygyny	Males 2–3 × weight females	3–4 weeks	Gregarious; positively thigmotactic	Terrestrial
Arctocephalus forsteri (New Zealand fur seal)	Islands	Polygyny	Males about 2 × weight females	Up to 2 months	Gregarious; negatively thigmotactic	Terrestrial
Family Odobenidae						
Odobenus rosmarus (walrus)	Pack ice	Polygyny	Males 1.4–1.5 × weight females	2 months mating period precedes 8 week pupping	Gregarious; positively thigmotactic	Aquatic
Family Phocidae						
Phoca vitulina (harbor seal)	Islands or inaccessible mainland beaches, pack ice in Bering Sea	Promiscuous on land; serial monogamy on ice	Male slightly larger	2–4 weeks	Gregarious; positively thigmotropic on land only	Aquatic unknown in Bering Sea

TABLE 1
CONTINUED

Pinniped families and species	Breeding habitat	Mating system	Degree of sexual dimorphism	Synchrony of pupping and breeding	Gregariousness of females and intrasexual thigmotaxis	Terrestrial or aquatic mating
Phoca hispida (ringed seal)	Landfast ice	Unknown; suspect polygyny	Male slightly larger	About 1 month		Aquatic
Pagophilus groenlandicus (harp seal)	Pack ice	Promiscuity or polygyny	Male slightly larger	About 3 weeks	Gregarious; negatively thigmotactic	Aquatic; on ice if forced
Erignathus barbatus (bearded seal)	Mainly pack ice; some landfast ice	Unknown, suspect promiscuity	Female slightly larger	Unknown	Not gregarious	Unknown; suspected aquatic
Cystophora cristata (hooded seal)	Pack ice	Serial monogamy	Male 20–30% heavier	About 3 weeks	Not gregarious	Aquatic; on ice if forced
Halichoerus grypus (grey seal)	Islands; some on landfast ice and pack ice	Polygyny	Male 10–20% heavier	4–6 weeks	Gregarious	Terrestrial or on ice
Monachus monachus (Mediterranean monk seal)	Islands; more recently inaccessible caves	Promiscuity	Male slightly larger	7 month span but most in 2 months	Not gregarious	Aquatic
Monachus schauinslandi (Hawaiian monk seal)	Islands	Promiscuity	Male slightly larger	Births for 6 months; mating for 3 months	Not gregarious	Aquatic

TABLE 1
CONTINUED

Pinniped families and species	Breeding habitat	Mating system	Degree of sexual dimorphism	Synchrony of pupping and breeding	Gregariousness of females and intrasexual thigmotaxis	Terrestrial or aquatic mating
Mirounga leonina (southern elephant seal)	Islands	Extreme polygyny	Males 4 × weight females	6–8 weeks	Gregarious; positvely thigmotactic	Terrestrial
Mirounga angustirostris (northern elephant seal)	Islands	Extreme polygyny	Males 2 × weight females	4–6 weeks	Gregarious; positively thigmotactic	Terrestrial
Lobodon carcinophagus (crabeater seal)	Pack ice	Serial monogamy	Female slightly larger	About 3 weeks	Not gregarious	Unconfirmed but suspected on ice
Hydrurga leptonyx (leopard seal)	Pack ice	Unknown	Female slightly larger	Unknown	Not gregarious	Unknown; suspected aquatic
Leptonychotes weddelli (Weddell seal)	Landfast ice	Polygyny	Female slightly larger	About 1 month	Gregarious; negatively thigmotactic	Aquatic

TABLE 1
CONTINUED

Pinniped families and species	Terrestrial or aquatic male defense of territories	Color of lanugo	Time to weaning	Pelagic or coastal feeders	Extent of migrations	Predators possible in breeding areas
Family Otariidae						
Eumetopias jubatus (Steller sea lion)	Terrestrial	Dark	Usually less than 1 year; some pups nursed longer	Pelagic and coastal	Long migrations; males more than females	Sharks, killer whale
Zalophus californianus (California sea lion)	Terrestrial	Brown	Usually less than 1 year; some pups nursed longer	Pelagic and coastal	Long migrations; males more than females	Sharks, killer whale
Neophoca cinerea (Australian sea lion)	Terrestrial	Light brown	Usually less than 1 year; some pups nursed longer	Coastal	Local movements	Sharks
Callorhinus ursinus (northern fur seal)	Terrestrial	Dark	4 months	Pelagic	Long migrations	Killer whale, sharks
Arctocephalus townsendi (Guadalupe fur seal)	Terrestrial	Dark	Less than 1 year	Unknown; suspect coastal	Unknown	Sharks

Table 1
CONTINUED

Pinniped families and species	Terrestrial or aquatic male defense of territories	Color of lanugo	Time to weaning	Pelagic or coastal feeders	Extent of migrations	Predators possible in breeding areas
Arctocephalus pusillus (Cape fur seal)	Terrestrial	Dark	1 year but occasionally to 3 years	Pelagic and coastal	Unknown	Sharks, killer whale
Arctocephalus forsteri (New Zealand fur seal)	Terrestrial	Dark	About 10 months	Pelagic and coastal	Unknown	Sharks, killer whale
Family Odobenidae						
Odobenus rosmarus (walrus)	Aquatic	Brown	1.5–2 years	Pelagic	Some populations long distances, some short; females more than males in Bering Sea	Polar bears
Family Phocidae						
Phoca vitulina (harbor seal)	Aquatic unknown in Bering Sea	Dark when born on land; white when born on ice	4 weeks	Coastal	Non-migratory	Sharks and killer whale near land, polar bear on ice

Table 1
CONTINUED

Pinniped families and species	Terrestrial or aquatic male defense of territories	Color of lanugo	Time to weaning	Pelagic or coastal feeders	Extent of migrations	Predators possible in breeding areas
Phoca hispida (ringed seal)	Aquatic	White	6 weeks	Coastal	Non-migratory	Polar bear
Pagophilus groenlandicus (harp seal)	Aquatic	White	10–14 days	Pelagic and coastal	Long migrations	Polar bear
Erignathus barbatus (bearded seal)	Aquatic	Grey with white blotches	12–18 days	Coastal	Non-migratory	Polar bear
Cystophora cristata (hooded seal)	On ice; suspect also aquatic	Silvery light blue	8–12 days	Pelagic	Medium length migrations	Polar bear
Halichoerus grypus (grey seal)	Terrestrial or on ice	White	16–19 days	Coastal	Non-migratory	Sharks and killer whale possible
Monachus monachus (Mediterranean monk seal)	Aquatic	Dark	6 weeks	Coastal	Non-migratory	Unknown, suspect sharks

Table 1
CONTINUED

Pinniped families and species	Terrestrial or aquatic male defense of territories	Color of lanugo	Time to weaning	Pelagic or coastal feeders	Extent of migrations	Predators possible in breeding areas
Monachus schauinslandi (Hawaiian monk seal)	Aquatic	Dark	5–6 weeks	Coastal	Non-migratory	Sharks
Mirounga leonina (southern elephant seal)	Terrestrial	Dark	23 days	Pelagic and coastal	Unknown	Killer whale
Mirounga angustirostris (northern elephant seal)	Terrestrial	Dark	28 days	Pelagic and coastal	Unknown	Sharks, killer whale
Lobodon carcinophagus (crabeater seal)	On pack ice	Dark	4 weeks	Pelagic and coastal	Unknown, suspected non-migratory	Leopard seal, killer whale
Hydrurga leptonyx (leopard seal)	Unknown	Dark	Unknown	Pelagic and coastal	Unknown; suspected non-migratory	killer whale
Leptonychotes weddelli (Weddell seal)	Aquatic	Dark	6 weeks	Coastal	Non-migratory	None in most areas; killer whale in some areas

raphy may make one part of an island or beach more suitable than another for successfully weaning a pup.

In simplistic terms, there are two types of ice on the sea in polar regions on which seals breed. Landfast ice occurs in bays and along the coastlines and is always contiguous with a continent or island compared to pack ice, which is free floating and moves at the will of the wind or sea currents. Pack ice is far more abundant than landfast ice.

The vastness of the polar pack ice offers a quite different situation from an island. In general, space on the surface of the ice, and access to it, are virtually unlimited although its actual suitability for parturition may be highly variable between areas and seasons. Factors such as the size and topography of floes, the amount of open water, and weather may strongly influence suitability (Fig. 1). Regardless of the cause, parturient females are more widely spaced on polar pack ice than terrestrial habitats. Even in species such as the harp seal (*Pagophilus groenlandicus*), which congregate in localized areas for parturition, the densities do not begin to approach those of the otariids or elephant seals (*Mirounga* sp.) (for example, see photos in LeBoeuf, 1971; Mathisen et al., 1962; Sergeant, 1965). Because the instability of the habitat favors low densities of parturient females, it is also unfavorable for the development of extreme polygyny. So far as is known, no phocids breeding in the pack ice have developed a polygynous mating system with specific site fidelity. In contrast, however, the walrus is polygynous in the same unstable habitat (Fay and Ray, 1979); some possible reasons for this will be discussed later.

Between the extremes of the reliable terrestrial habitat and the unstable pack ice is the landfast ice which is stable during the winter and usually through the breeding season before breakup in the spring. The ringed seal (*Phoca hispida*) of the Arctic and the Weddell seal of the Antarctic exploit this habitat at opposite ends of the globe (McLaren, 1958; Stirling, 1969). Although landfast ice is relatively unlimited as a site for parturition, access to it is restricted because breathing holes for hauling out must be kept open by the seals, something they do not have to do in pack ice. Because in landfast-ice areas more than one female can and does use a single breathing hole, a male can restrict the access of competitors to a group of females (Fig. 3). In the Antarctic, in the absence of terrestrial predators, polygyny has developed to a greater extent than in the Arctic, where terrestrial predators are present (Stirling, 1977).

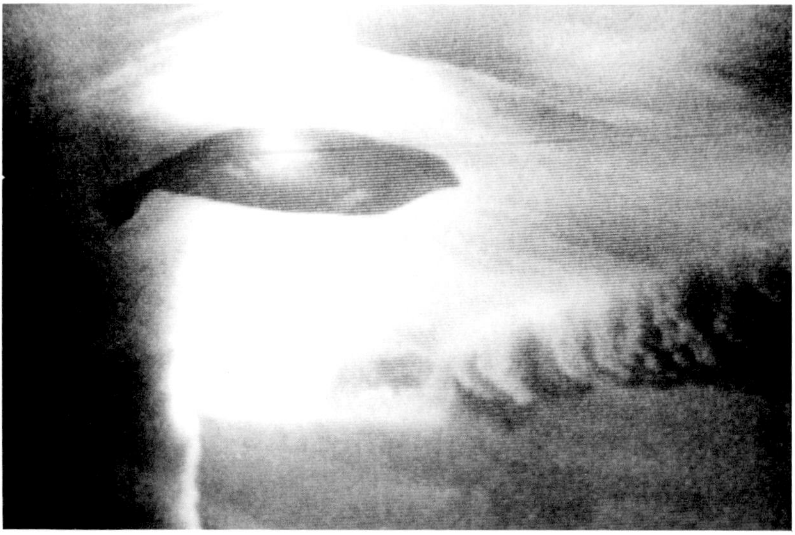

FIG. 3. Male Weddell seal (*Leptonychotes weddelli*) defending underwater territory below breathing hole.

The extent to which the location, stability, and access to suitable habitat for parturition are predictable have had a strong influence on the behavior of pinnipeds. The locations of islands are completely predictable. In ice-free waters at least, access is also predictable, although it may be influenced locally by geographic features, and thus affect the density to which parturient females will aggregate, and the distribution of terrestrial males. Consequently, pinniped females have developed a high degree of fidelity to specific pupping sites and males are faithful in the defense of specific territories (for example, Carrick et al., 1962; Johnson, 1968). In the polar landfast-ice habitat, cracks in the ice where breathing holes can be maintained occur in the same places each year although annual variations in ice conditions may affect the extent to which breathing holes can be maintained. Consequently, individual Weddell seals have a high degree of fidelity to a general area for pupping but they may move about between particular colonies between years depending on ice conditions (Stirling, 1969, 1974). In the pack ice there may be considerable variation between years in the size and type of ice floes present, the amount of open water, and the movement of the ice itself. Pinnipeds faced with such a large degree of environmental variability between years can only maintain fidelity to general areas and must be prepared to be quite flexible in choosing an actual site for parturition each year.

Synchronized Parturition and Mating

In unstable habitat such as most areas of pack ice, it is advantageous for the maximum number of females to give birth over a very short period when conditions are most favorable. Thus, most adult females are receptive at about the same time, but because they are widely dispersed, one male is unlikely to be able to defend reproductive access to more than a single female (see Fig. 1). In pack-ice phocids for which there are reasonable data (harp seal, hooded seal, crabeater seal), it appears that the period in which most parturition and mating occurs may be as short as 3 weeks (Øritsland, 1964; Sergeant, 1965; Siniff et al., 1979). However, some adult males may form a series of temporary monogamous bonds with several estrous females even within the brief breeding season.

In most stable terrestrial habitats, parturition and mating may take place safely over a relatively longer period. Consequently, in species such as the northern elephant seal, the northern fur seal, and the New Zealand fur seal, the pupping and mating period may span up to 2 months (Bartholomew, 1952; Bartholomew and Hoel, 1953; Stirling, 1971*a*). The intermediate stability of landfast ice results in pinnipeds breeding in this habitat having an intermediate duration of about a month for pupping and mating (McLaren, 1958; Stirling, 1969).

Although synchronized pupping and mating are the general rule in pinnipeds, there are at least two interesting exceptions. The Hawaiian monk seal, which appears to be promiscuous, gives birth over a 6 month period from late December through June and pairs of seals have been observed exhibiting apparent sexual behavior between early March and early July (Kenyon and Rice, 1959). Recent studies on the Australian sea lion (*Neophoca cinerea*) have shown that pups may be born throughout the year and that some females may have an 18-month breeding cycle (Ling and Walker, 1978), although more pups are born in the austral spring from October to December (Marlow, 1975; Stirling, 1972). Ecologically, it is all the more interesting that this sea lion lives in the same habitat and breeds on many of the same islands as the New Zealand fur seal which has a reasonably long, but highly synchronized pupping and breeding season (Stirling, 1971*a*). The evolutionary significance of these variations in the usual pinniped pattern are not understood.

Gregariousness of Females and Sexual Dimorphism

Although female pinnipeds may be highly gregarious, this varies among species. For example, a female New Zealand fur seal will not tolerate a seal other than her pup within about two meters (Fig. 4), whereas the Cape fur seal is positively thigmotactic (Fig. 5) (Stirling and Warneke, 1971). Elephant seals or sea lions are highly thigmotactic even when additional space is abundant (Figs. 1 and 6) (Bartholomew, 1952; Stirling, 1972). Parturient Weddell seal females hauling out on the ice from the same breathing hole or crack in the ice usually lie several meters apart (Fig. 7) and they will attack a seal that approaches more closely. Consequently, far fewer females use a particular exit hole or open crack than could do so if they were as gregarious as elephant seals. In the pack ice, parturient harp seals may haul out in densities of 700–800/km^2 (Sergeant and Fisher, 1960), whereas crabeater seals (possibly the most abundant seal in the world) in similar habitat reach densities of only about 0.7 to 1.0/km^2 during the pupping season (Øritsland, 1970; Siniff et al., 1979). Thus there are considerable differences in the extent to which female seals aggregate during the breeding season.

Large size in males and the consequent ability to fast for extended periods have been influenced by reinforcing feedback loops. In species in which ovulation occurs over several days and where females aggregate in high densities on stable habitat, aggressive males can exclude competitors from estrous females and mating is terrestrial, which should favor large size of males (Bartholomew, 1970; Stirling, 1975). However, some caution should be taken in interpreting large size of males solely in terms of reproductive advantage. As Ralls (1976) has pointed out, sexual dimorphism could be related to sex differences in niche exploitation. For example, male and female Steller sea lions (*Eumetopias jubatus*) have different migration patterns (Kenyon and Rice, 1961). In Antarctic phocids and

FIG. 4. Breeding colony of New Zealand fur seals (*Arctocephalus forsteri*); note how widely spaced the individuals are.

FIG. 5. Breeding colony of Cape fur seals (*Arctocephalus pusillus*); note close proximity of individuals.

FIG. 6. Group of Australian sea lions (*Neophoca cinerea*) sleeping; note extent of thigmotaxis despite abundance of space.

FIG. 7. Spacing of parturient Weddell seals (*Leptonychotes weddelli*) in a pupping colony.

in the bearded seal (*Erignathus barbatus*), the males are smaller than the females (Burns, 1967; Laws, 1957, 1958; Stirling, 1971*b*). This includes the Weddell seal which is polygynous with males defending underwater territories (Kaufman et al., 1975; Ray, 1967; Siniff et al., 1977).

Repenning (1976) has hypothesized that the mating systems of pinnipeds (including the development of sexual dimorphism) are a product of feeding patterns. "In the hypothesis that follows, pelagic feeders, because of their wide dispersal over the surface of the ocean, survive as a species only through evolution of a very specific homing instinct: thus designated rookeries: thus dimorphism and pinniped polygyny as reconstructed by Bartholomew. Coastal marine feeders have a ready access to land or ice; they may haul out in groups at any time of the year, but their grouping is protective rather than for birth and breeding; they have evolved relatively insigificant homing instinct and dimorphism; and they do not have designated rookeries for breeding and pupping. The exception is the walrus, a dimorphic coastal feeder whose ancestors, as we shall see, were pelagic feeders" (Repenning, 1976: 375). Essentially, Repenning's hypothesis states that pelagic feeding has imposed certain restrictions to ensure access to available breeding habitat. Alternatively, it seems to me that because pinnipeds evolved from terrestrial mammals, the distribution and availability of habitat for parturition and breeding would be more likely to influence the way in which the different species evolved to obtain maximum utility of the marine habitat, rather than the other way around. However, Repenning's hypothesis about the migrations of pelagic feeders does not seem adequate on its own to explain the degree of sexual dimorphism observed in some species. There does not appear to be a clear correlation between the degree of sexual dimorphism and the extent to which the polygynous or extremely polygynous species migrate (Table 1). The sexually dimorphic hooded seal which breeds in the pack ice appears to be a contradiction. Harp seals have one of the longest annual migrations of any pinniped species (Sergeant, 1965) and, although they have a high degree of fidelity to a breeding area, lack sexual dimorphism. Nevertheless, Repenning's hypothesis should be examined using available data on feeding habits by sex, species, and habitat.

Clutton-Brock (in press) has suggested that in ungulates, sexual dimorphism is related to whether or not males use their bodies to push their opponents during intrasexual fighting as opposed to using methods such as biting the legs of an adversary. This hypothesis should be examined in pinnipeds as well. The males of species which mate on land also fight on land. None mate in the water to a significant extent. They push each other a great deal when fighting on land and sexual dimorphism is marked. For seals in which mating, and consequently male fighting, occur in the water, attempting to push would probably be less effective than direct biting. Thus, it is interesting to note that sexual dimorphism is much reduced in those species. In the walrus and Weddell seal, both of which are polygynous and fight in the water, sexual dimorphism is reduced, and in the latter, even reversed. Thus, it may be that in polygynous pinnipeds the development of sexual dimorphism is determined by the extent to which males fight terrestrially and use their bodies to push each other.

Role of Predation

Although predation is generally acknowledged to have been a significant influence on the locations of pinniped breeding colonies (for example, most are on islands without terrestrial predators), the influence of this factor on the evolution of pinniped behavior has been inadequately addressed. Stirling (1977) compared the behavior of ringed seals, which evolved with terrestrial predators, to that of Weddell seals, which evolved without them, and found significant differences between the species in their distribution patterns, densities, behavior, morphology, birth sites, and underwater communication.

The influence of marine predators has probably been equally significant, but the interactions have been more difficult to study. Consequently, much less is known except for anecdotal accounts of seals in the stomachs of sharks and killer whales (for example, see Slijper, 1976). Carrick and Ingham (1962) discussed killer whale (*Orcinus orca*) wounds on southern elephant seals (*Mirounga leonina*). Condy et al. (1978) have documented seasonally occurring predation of killer whales on southern elephant seals at breeding colonies at Marion Island. Smith et al. (in press) documented a remarkable incident in which a pod of killer whales coordinated their behavior to hunt a crabeater seal, which indicates that such predation is probably highly evolved and of great importance to the seals. Siniff and Bengtson (1977) reported that 78% of a sample of 200 crabeater seals were scarred from leopard seal (*Hydrurga leptonyx*) attacks (for example, see Fig. 1). Condy (1977) reported similar observations from the area of the King Haakon VII Sea, Antarctica. Barlow (1972) described adult male California sea lions (*Zalophus californianus*) on the Galapagos Islands regularly going into the water to deflect sharks away from the breeding colony, apparently to protect the pups. Shark predation has been documented near breeding colonies of many species of pinnipeds (for example, see Fig. 8).

Even in the short term, it is possible to see the effects of the predator, man, on some seals. For example, both the Mediterranean monk seal (*Monachus monachus*; Bareham and Furreddu, 1975) and the Guadalupe fur seal (*Arctocephalus townsendi*; Hubbs, 1956) have been so reduced in numbers that large proportions of the remnant populations now breed in caves inaccessible to humans. Conversely, in some areas where terrestrial predators have been eliminated and human activities are controlled, populations of some species, such as northern elephant seals and Australian sea lions, are now breeding on readily accessible mainland areas (LeBoeuf

FIG. 8. Adult male Australian sea lion (*Neophoca cinerea*) with fatal shark bite that severed the spine.

and Penken, 1977; Stirling, 1972). Best and Shaughnessy (1979) documented the reoccupation of colonies by Cape fur seals on the mainland coast of South Africa after their apparent elimination by commercial sealers prior to 1829. In some instances, colonies exist today which were apparently not present in 1829. What is even more interesting, though yet unstudied, is that seal pups in at least one of these colonies are now being preyed upon by black-backed jackals (*Canis mesomelas*) while at another colony, scavenging by the brown hyena (*Hyaena brunnea*) occurs (Best and Shaughnessy, pers. comm.).

Terrestrial and Aquatic Mating

The importance of terrestrial mating has been discussed extensively by most authors who have considered the origin of polygyny in pinnipeds (Bartholomew, 1970; Bertram, 1940; Stirling, 1975). Bartholomew (1970) suggested two factors which may have contributed to the development of terrestrial copulation: 1) females stay on land until they come into estrous, and 2) the "marginal male effect" operates so that the most successful males contribute disproportionately to the gene pool. As I have discussed before (Stirling, 1975), these are both powerful arguments, but they would really apply only after terrestrial copulation had been established and the feedback loops related to gregariousness and sexual dimorphism had been initiated.

Trying to evaluate why some species mate in the water while others do so on land or ice is one of the most fundamental problems. At least three factors could be significant—phylogeny, predation, and changes in habitat.

Phylogeny may be one of the most important factors (Stirling, 1975). Although it is possible for otariids to mate in the water, few appear to actually do so. Present day ursids, which evolved from the same basic stock as the otariids, copulate terrestrially or on ice. It seems probable that their ancestors also did so at the time the fur seals and sea lions diverged from the ursids and that selection pressures have been insufficient to stimulate a change. In comparison, present day otters appear to be capable of mating either on land or in the water, but prefer the latter (Cocks, 1881; Crandall, 1964; Kenyon, 1969; Liers, 1951). If the phocids evolved from a lutrine or lutrine-like stock with a preference for aquatic mating, that might be sufficient to account for their preference for aquatic mating.

However, four species of phocids appear to mate predominantly out of the water—the northern and southern elephant seals, the grey seal (*Halichoerus grypus*), and the crabeater seal. Conversely, the walrus, which apparently diverged from the otariids at an early stage, mates in the water. If the earlier generalizations about phylogenetic preferences for mating habitat are true, the question then is, why did some species apparently change from one habitat to another for copulation while others did not? Two factors—predation and change in habitat—could have been important.

In the past, colonies of pinnipeds on land, entering the sea in restricted areas, may have attracted marine predators as they apparently still do in many areas today (for example, see Condy et al., 1978). However, even in the presence of a large number of conspecifics which could create a prey saturation effect (Taylor, 1976), pinnipeds which were preoccupied by intrasexual competition, courtship, and copulation in the water would be vulnerable to predation. Although difficult to evaluate quantitatively, marine predators may have played a role in preventing aquatic mating from developing in some species, while terrestrial predators may have aided its development in others.

In the Arctic, all species of phocids normally mate in the water, although mating on the ice has been reported in harp and hooded seals when the ice becomes so compacted that the seals cannot get into the water (Øritsland, 1964; Popov, 1966; Sergeant, 1965). I am tempted to speculate that copulation has not developed on the ice because of the presence of terrestrial predators, such as the polar bear (*Ursus maritimus*). However, in the Weddell seal at least, polygyny has developed with aquatic mating, despite males being smaller than females, because territorial males could still control the access of competing males to the habitat occupied by parturient females. The reversal in the usual pattern of sexual dimorphism in a polygynous species is also interesting to consider in relation to Repenning's (1976) hypothesis because the Weddell seal is strictly a coastal feeder but has evolved polygyny.

The selective pressure of marine predation by leopard seals and killer whales in the Antarctic pack ice may have been sufficient to stimulate the crabeater seal to mate on the surface of the ice (Siniff et al., 1979). One can only speculate about whether marine predators could have influenced the development or retention of terrestrial mating in the elephant and grey seals. Similarly, continued marine predation could have influenced the historical distribution and numbers of monk seals. These species still mate aquatically and, in Hawaii at least, shark predation is thought to be a significant mortality factor (Kenyon, 1972).

Thermal Considerations

The influence of thermal regimes on the development of social structure and mating systems is difficult to evaluate, but it may be important, particularly in warmer areas. In several species, animals may enter the water on hot days during the breeding season for thermoregulatory purposes (Bartholomew and Hoel, 1953; Bartholomew and Wilke, 1956; Gentry, 1973; Odell, 1974; Peterson and Bartholomew, 1967; Ferreira and Palerm, 1961). If breeding males and females are forced into the water often, where access to females can be less rigidly controlled, the importance of territorial or hierarchical behavior may be reduced, thus weakening the feedback loops related to sexual dimorphism and polygyny.

FIG. 9. Territorial male New Zealand fur seal (*Arctocephalus forsteri*) using a pool of water to thermoregulate, thus permitting him to remain on his territory despite high ambient temperatures.

In some species, such as the New Zealand fur seal, males that have water on their territories are not forced to abandon them to thermoregulate on hot days (Fig. 9) (Stirling, 1971*a*). The significance of this is emphasized by Gentry's (1973) data, which showed that 82% of 109 copulations observed were by males that had water on their territories.

Under cold conditions, it is difficult to separate possible thermoregulatory need from simple thigmotactic responses in seals. Fay and Ray (1968) attribute "huddling" in the walrus to conservation of heat. I have observed young fur seals, sea lions, and elephant seals lying in close contact in cold weather, but the physiological necessity or value of this behavior was not clear. The possible influence of close body contact for thermal benefit on the development of intraspecific tolerance for high density living can probably only be speculated upon.

In a recent paper, Pierotti and Pierotti (1980) suggested that the lack of "terrestrial" polygyny in ice-breeding pinnipeds, in comparison to more temperate species, has resulted because of the excessive heat loss that would be experienced by males defending territories on the ice near parturient females. Although this hypothesis appears attractive superficially, it ignores many aspects of the biology of ice-breeding pinnipeds—in particular, the instability of the sea ice habitat; the role of terrestrial predators; and the fact that the females of several Antarctic seal species (without surface predators) may spend several weeks on the ice weaning their newborn pups, which surely indicates that the males would be physiologically capable of doing so as well.

Discussion

The prototype pinnipeds were probably already adapted to aquatic feeding in lakes, rivers, estuaries, and along the coast before they began to feed offshore. As they became more dependent on marine resources and adapted to longer feeding trips, they would have become less tied to the land. Dispersal would have tended to take place along coastlines and archipelagos. Seasonal breeding in response to seasonal variations in climate and food availability was likely already present in the original fissiped stock, and delayed implantation, characteristic of the present day ursids and mustelids (Wright, 1963), may have been as well. There would have been particularly strong selective pressure on those species that began to feed far offshore to develop strong breeding-site fidelity to ensure reuniting of the sexes for breeding. The fossil evidence indicates that colonial breeding and sexual dimorphism were already developed in the odobenids by the middle of the Miocene (about 14 million years ago) and in the otariids by the late Miocene (about 8 million years ago) (Mitchell and Tedford, 1973; Repenning and Tedford, 1977). It is less clear when sexual dimorphism first appeared in the phocids, but Ray (1976) suggested that the elephant and grey seals did not evolve until sometime during the Pliocene.

As the early pinnipeds became more adapted to life at sea and less mobile on land, they probably became progressively more vulnerable to predation by terrestrial carnivores. Consequently, their shift toward breeding on islands and inaccessible points along the mainland may have begun by the late Miocene as well, although this has not been clearly established.

From the Eocene to the Oligocene there was a period of cooling, followed by a period of warming by the mid-Miocene and a gradual cooling again to the present (Repenning, 1980). For example, only about 10% of the earth's land surface is glaciated at present compared to about 30% during the glacial periods of the Pleistocene (Flint, 1947).

The formation of pack ice and landfast ice along the coast would have made the shore less accessible to seals returning there to breed. Pinnipeds could have responded to this change by moving south or by adapting their behavior to breed and feed in association with the ice. Variation in the extent of glaciation has probably had a strong influence on patterns of dispersal, particularly during the last stages of pinniped evolution (Davies, 1958*a*, 1958*b*; Ray, 1976; Repenning, 1980). There may also have been effects on speciation and behavior. Judging from the present day species, the otariids appear to be less adapted to living with ice than the phocids and odobenids.

When ice first began to occur along the coastline, the phocid seals and some of the odobenids probably gave birth on the ice, basked on it, and used it for a platform much as if it were land. Because there was unlimited space for parturition, and access to it, the females did not need to aggregate. Because estrous females probably became more spread out on ice than on land, territorial males could exert control over access to limited numbers of females at best and probably only one at a time, which would have limited the development of polygyny. In general, continuous breeding on the pack ice seems to have created selection pressures for short synchronized birth, weaning, and mating periods; a promiscuous mating system and apparently even short-term serial monogamy; reduced disparity in the sex ratio and body size of adults; and similar mortality rates of adult males and females (Stirling, 1975). Copulation continued to occur predominantly in the water.

The main exception to the generalizations made above is the walrus, which is at least distantly related to the otariids. Ancestral walruses occurred in large colonies on land, were markedly sexually dimorphic and, presumably, bred on land (Repenning, 1976). Present day walruses give birth to their young on the pack ice and may take up to two years to wean the pups (Buckley, 1958). They mate in the water in groups of several females to one male, and have a long breeding season (December to March in the Bering Sea) (Fay 1982). Aquatic mating could be an evolutionary response to the problem of a large mammal mating on variably-sized floes in the pack ice, but the gregariousness of females and the long weaning period may be phylogenetic relicts.

The first phocids to invade the pack-ice habitat probably mated in the water. A possible stimulus for mating on the surface of the ice could have been closing of the cracks in some years by wind or ocean currents when the females were in estrous (Stirling, 1975). This would require that copulation take place on the surface of the ice if it were to occur at all. As noted above, this happens occasionally in present day harp and hooded seals. Theoretically, if there were enough such years in a row, this might facilitate males controlling reproductive access to more than one female, and possibly lead to polygyny through the pathways suggested by Bartholomew (1970).

In the case of the crabeater seal, it appears that the selection pressure promoting mating on the ice has been from marine predators (Siniff et al., 1979), and of course there are no terrestrial predators in the Antarctic to counteract such an influence. If a species did develop copulatory behavior on the ice, this might easily transfer to land areas if the ice retreated. Stirling (1975) suggested that this might be the mechanism whereby the elephant and grey seals began terrestrial mating and noted that the weaning times for these two species are more like pack-ice than landfast-ice phocids. Grey seal pups, like those of most other Arctic phocids, are born in white lanugo, suggesting a previous association with both ice and terrestrial predators (Burns, 1970; Fay, 1974; McLaren, 1966, 1975; Stirling, 1977). Elephant seal pups, like pups of monk seals and Antarctic phocids, are all born with dark lanugo, suggesting no evolutionary experience with predators on ice.

It is not clear whether or not the evolution of the polar bear in the late Pleistocene (Kurten, 1964) was too recent to have influenced the presence or absence of mating on the surface of the ice or other aspects of the reproductive cycle. However, the white lanugo of the pups of most Arctic phocids is suggestive of terrestrial predators having been present long enough for the seals to have evolved protective coloration for their young. An exception is the mostly dark grey lanugo of newborn bearded seal pups which has substantial white patches dorsally (Fig. 10) suggesting that this species might presently be in the process of evolving a cryptic coat for the neonate (Stirling, 1977). As noted earlier, the weaning period of pack-ice phocids is shorter than landfast-ice phocids and this has been attributed to instability of the habitat. However, the weaning period of Arctic pack-ice phocids such as harp and hooded seals is

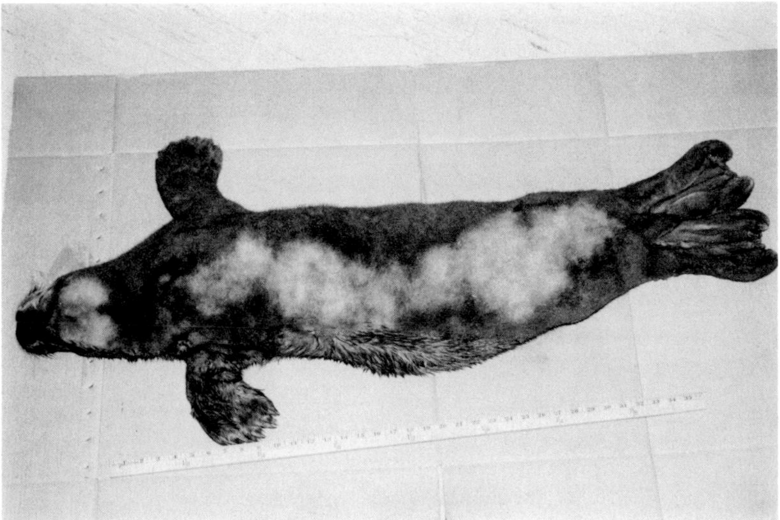

FIG. 10. Neonate bearded seal pup (*Erignathus barbatus*); note patches of white lanugo on predominantly dark pelage.

less than half the time required by the crabeater seal in the Antarctic. Could this have been influenced by surface predators?

The hooded seal has marked sexual dimorphism and secondary sexual display organs similar to the elephant seals (Berland, 1965). The hooded seal also breeds in loose aggregations in heavy pack ice where it occasionally has to maintain breathing holes (Høst, 1938; quoted by Øritsland, 1964). Mating has been observed on the ice (Rasmussen, 1952; quoted by Øritsland, 1964), although it apparently usually takes place in the water. Prior to the evolution of the polar bear, mating on the sea ice may have been more common, especially if ice conditions were heavier than they are in the present. Certainly, the evolution of the hood as a male sexual display organ (Berland, 1965) and its in-air vocalizations (Terhune and Ronald, 1973) parallel the evolution of similar features in the elephant seal. If Clutton-Brock's (in press) hypothesis is correct, it might also suggest that in earlier times hooded seals may have had a more organized social structure which involved male fighting and mating out of the water. Presumably this could not have occurred in the presence of surface predators. Also, hooded seal pups are not born with white lanugo. These features are at least suggestive of the existence of a non-aquatic mating system without surface predators at an earlier time. Although highly speculative, this may have permitted some degree of polygyny and subsequent sexual dimorphism to develop, which later broke down under pressure of predation from polar bears. The heavy pack ice preferred by hooded seals is easier for polar bears to hunt in than the lighter ice preferred by harp seals, and even now polar bears are sighted around the hooded seal whelping patch 200 to 240 km offshore in Davis Strait (Sergeant, 1974). It has also been suggested that polar bears travel into the offshore ice each year to prey upon hooded seals (Sergeant, 1977), but to date there are not data to confirm or deny this speculation. The final detrimental selection pressure on copulation on ice and polygyny in hooded seals could have been the century-long commercial sealing of primarily adult females and their young (Sergeant, 1976), because it has resulted in a surplus of males in the pupping colonies (Øritsland, 1964).

The central hypothesis, that the most important factor influencing the development of social and mating systems in seals has been the availability of habitat for parturition and access to it, and attendant hypotheses involving factors such as predation, phylogeny, and habitat change, can be tested. There are several species in ecological contexts from which detailed long-term behavioral studies could provide tests of particular hypotheses. Such studies should concentrate on natural variation in the environment, such as in species which have ecological counterparts in other areas, and species in which different subpopulations live in different ecological circumstances; or examine species in which an important influence on a population (such as harvesting) is changed or eliminated. Maximum use should be made of individually-marked seals that can be followed over a series of years. Stirling (1975) detailed several species and geographic areas in which such studies could be conducted but, to date, no long-term studies of the type required have been initiated.

The grey seal offers the greatest opportunity for testing hypotheses on the effects of different breeding habitats on mating systems. They breed in several types of terrestrial habitats as well as on ice and important differences in the mating systems of different populations have been recorded (Anderson et al., 1975; Boness and James, 1979; Curry-Lindahl, 1970; Hewer, 1960; Mansfield, 1966; and Smith, 1966). It is not clear whether or not grey seals copulate in the water when breeding occurs on ice. To date no comparative study has been undertaken.

At Signy Island, which is near the southern limit of the breeding range of the southern elephant seal, offshore ice in some years prevents the seals from reaching the beach and establishing harems on land (Laws, 1956, pers. comm.). The total population is small and apparently the development of harems on ice is similar to that on land. However, on ice, where space is unlimited, the harems are further apart. Subadult males occupy the space in between and would seem better situated to intercept females and mate with them. At South Georgia, by comparison, landfast ice does not occur. Breeding space on the beaches is therefore limited and the cows form a continuum in which the breeding bulls occupy territories. Here subadult activity is confined to swimming back and forth parallel to the beaches in the hope of intercepting a female and copulating with her in the water. Under these circumstances, the pattern of social behavior of the southern elephant seal at South Georgia appears static. This is not the case at Signy Island, however, and for this reason the effect of ice and land on social behavior should be studied in detail as opportunity permits.

In ice-free waters, the harbor seal gives birth on land and most copulations occur in the water in promiscuous social groups (Bishop, 1968; Venables and Venables, 1955, 1957, 1959). In *P. vitulina largha*, which breeds in the pack ice of the Bering Sea, "family groups" consisting of a male, a female, and a pup are formed (Burns et al., 1972) which are apparently similar to those of the hooded seal (Olds, 1950). A comparison of the details of the breeding behavior of this subspecies with more southerly forms could provide a valuable insight into the effect of pagophilic breeding on the development of social behavior. The Weddell seal defends aquatic territories and is polygynous in the stable landfast-ice habitat of continental Antarctica (Kaufman et al., 1975; Ray, 1967). At Signy Island, which is near the northernmost limit of the breeding range of the Weddell seal there may be no sea ice in some years on which females can give birth. In these seasons the cows have their pups on the beaches and apparently some even go inland (Smith and Burton, 1970). At Larsen Harbour, at the southern end of South Georgia there is a small unstudied population of Weddell seals which breeds away from regular contact with landfast ice although pack ice may occur along the coast in the austral spring. The comparative social behavior of the Weddell seal under these different circumstances should be documented in detail.

There are also several unstudied aspects alluded to in this paper which merit research, the results of which may aid us in understanding how mating systems have evolved in pinnipeds as well as in other mammals. How much could sexual dimorphism be related to habitat utilization or terrestrial fighting of males? Why are harbor seals and monk seals not more polygynous? What variables affect the degree to which parturient females will aggregate? Why does the Australian sea lion give birth throughout the year in the same habitat in which two species of fur seals are highly seasonal breeders? Why are female Antarctic phocids larger than the males, even in the polygynous Weddell seal? Why are the densities of parturient crabeater seals and harp seals so different in such a similar habitat? Why do hooded seals exhibit such a large degree of sexual dimorphism? How has man's harvesting and harassment of seals affected their social systems as we see them today?

Acknowledgments

I am particularly grateful to the following for logistic support: Canadian Wildlife Service, Polar Continental Shelf Project, National Science Foundation, New Zealand Antarctic Research Program, and the Mawson Institute for Antarctic Research. Over the years, I have benefited a great deal from cooperative field projects and discussions with R. Carrick, R. M. Laws, C. A. Repenning, D. B. Siniff and T. G. Smith. Constructive criticism of this manuscript was received from W. Calvert, C. A. Repenning, and D. B. Siniff. I thank P. D. Shaughnessy and P. B. Best for permission to use unpublished observations.

Literature Cited

Anderson, S. S., R. W. Burton, and C. F. Summers. 1975. Behaviour of grey seals (*Halichoerus grypus*) during a breeding season at North Rona. J. Zool., 177:179–195.

Bareham, J., and A. Furreddu. 1975. Observations on the use of grottoes by Mediterranean monk seals (*Monachus monachus*). J. Zool., 175:291–298.

Barlow, G. W. 1972. A paternal role for bulls of the Galapagos sea lion. Evolution, 26:307–308.

Bartholomew, G. A. 1952. Reproductive and social behavior of the northern elephant seal. Univ. California Publ. Zool., 45:369–472.

———. 1970. A model for the evolution of pinniped polygyny. Evolution, 24:546–559.

BARTHOLOMEW, G. A., AND P. G. HOEL. 1953. Reproductive behavior in the Alaska fur seal (*Callorhinus ursinus*). J. Mamm., 34:417–436.

BARTHOLOMEW, G. A., AND F. WILKE. 1956. Body temperature in the Northern fur seal, (*Callorhinus ursinus*). J. Mamm., 37:327–337.

BERLAND, B. 1965. The hood and its extrusible balloon in the hooded seal, *Cystophora cristata* (Erxl.). Norsk Polarinstitutt-Arbok, 1965:95–102.

BERTRAM, G. L. C. 1940. The biology of the Weddell and crabeater seals. Sci. Rept. British Graham Land Exp. 1934–37, 1:1–139.

BEST, P. B., AND P. D. SHAUGHNESSY. 1979. An independent account of Captain Benjamin Morrell's sealing voyage to the south-west coast of Africa in the *Antarctic,* 1828/29. Fish. Bull. South Africa, 12:1–19.

BISHOP, R. H. 1968. Reproduction, age determination, and behavior of the harbor seal, *Phoca vitulina* L., in the Gulf of Alaska. Unpubl. M.Sc. thesis, Univ. Alaska, College, Alaska, 121 pp.

BONESS, D. J., AND H. JAMES. 1979. Reproductive behaviour of the grey seal (*Halichoerus grypus*) on Sable Island, Nova Scotia. J. Zool., 188:477–500.

BUCKLEY, J. L. 1958. The Pacific walrus. U.S. Fish and Wildl. Service, Spec. Sci. Rept. Wildl., 41:1–29.

BURNS, J. J. 1967. The Pacific bearded seal. Fed. Aid in Wildlife Restoration Proj. Covering Invest. completed by Dec. 31, 1966, 3:1–66.

———. 1970. Remarks on the distribution and natural history of pagophilic pinnipeds in the Bering and Chukchi Seas. J. Mamm., 51:445–454.

BURNS, J. J., et al. 1972. Adoption of a strange pup by the ice-inhabiting harbor seal, *Phoca vitulina largha*. J. Mamm., 53:594–598.

CARRICK, R., AND S. E. INGHAM. 1962. Studies on the southern elephant seal. V. Population dynamics and utilization. C.S.I.R.O. Wildl. Res., 7:198–206.

CARRICK, R., et al. 1962. Studies on the southern elephant seal, III. The annual cycle in relation to age and sex. C.S.I.R.O. Wildl. Res., 7:119–160.

CLUTTON-BROCK, T. H. In press. Red deer: the ecology of two sexes. Princeton Univ. Press.

COCKS, A. H. 1881. Note on the breeding of the otter. Proc. Zool. Soc. London, 1881:249–250.

CONDY, P. R. 1977. Whale observations in the pack ice off the Fimbul Ice Shelf, Antarctica. South African J. Antarctic Res., 7:7–9.

CONDY, P. R., R. P. VAN AARDE AND M. N. BESTER. 1978. The seasonal occurrence of killer whales (*Orcinus orca*), at Marion Island. J. Zool., 184:449–464.

CRANDALL, L. S. 1964. The management of wild animals in captivity. Univ. Chicago Press, Chicago, 761 pp.

CURRY-LINDAHL, K. 1970. Breeding biology of the Baltic grey seal (*Halichoerus grypus*). Zool. Gart., 38:16–29.

DAVIES, J. L. 1958*a*. Pleistocene geography and distribution of northern pinnipeds. Ecology, 39:97–113.

———. 1958*b*. The Pinnipedia: an essay in zoogeography. Geogr. Rev., 48:474–493.

FAO. 1979. Mammals in the seas. Pinniped species summaries and report on Sirenians. FAO Fisheries Series No. 5, Vol. II, FAO, Rome, 151 pp.

FAY, F. H. 1974. The role of ice in the ecology of marine mammals of the Bering Sea. Pp. 383–399, *in* Oceanography of the Bering Sea (D. W. Hood and E. K. Kelley, eds.). Institute of Marine Science, Univ. Alaska, Fairbanks.

———. 1982. Ecology and biology of the Pacific walrus, *Odobenus rosmarus divergens* Illiger. USDI No. Amer. Fauna Ser., 74:1–279.

Fay, F. H., and C. Ray. 1968. Influence of climate on the distribution of Walruses, (*Odobenus rosmarus*) (Linnaeus). I. Evidence from thermoregulatory behavior. Zoologica, 53:1–14.

———. 1979. Reproductive behavior of the Pacific walrus in relation to population structure. Trans. 29th Alaska Sci. Conf., 29:409–410.

Ferreira, R. Vaz, and E. Palerm. 1961. Efectos de los cambios meteorologicos sobre agrupaciones terrestres de Pinnipedios. Revista Facultad de Humanidades y Ciencias, Univ. Rep., Montivideo, 19:281–293.

Flint, R. F. 1947. Glacial geology and the Pleistocene Epoch. Wiley and Sons, New York, 589 pp.

Gentry, R. L. 1973. Thermoregulatory behavior of eared seals. Behavior, 46: 73–93.

Haley, D. (ed.). 1978. Marine mammals of eastern north Pacific and Arctic waters. Pacific Search Press, Seattle, 256 pp.

Hewer, H. R. 1960. Behaviour of the grey seal (*Halichoerus grypus* Fab.) in the breeding season. Mammalia, 24:400–421.

Høst, P. 1938. Sel og selfangst veol New Foundland. Utdrag av fangst-møtet Oslo, den 25 oktober 1948, stesilert, 15 s, 5 fig. (not seen)

Johnson, A. M. 1968. Annual mortality of territorial male fur seals and its management significance. J. Wildl. Mgmt., 32:94–99.

Hubbs, C. L. 1956. Back from oblivion. Pacific Disc., 9:14–21.

Kaufman, G. W., D. B. Siniff, and R. A. Reichle. 1975. Colony behavior of Weddell seals, (*Leptonychotes weddelli*) at Hutton Cliffs, Antarctica. Pp. 228–246, *in* Biology of the Seal (K. Ronald and A. W. Mansfield, eds.). Rapp. P.-v. Reun. Cons. Inf. Explor. Mer., 169:1–557.

Kenyon, K. W. 1969. The sea otter in the Eastern Pacific Ocean. N. Amer. Fauna, 68:1–352.

———. 1972. Man versus the monk seal. J. Mamm., 53:687–696.

Kenyon, K. W., and D. W. Rice. 1959. Life history of the Hawaiian monk seal. Pacific Science, 13:215–252.

———. 1961. Abundance and distribution of the Stellar sea lion. J. Mamm., 42: 223–234.

King, J. E. 1964. Seals of the world. British Mus. Nat. Hist., London, 154 pp.

Kurten, B. 1964. The evolution of the polar bear (*Ursus maritimus*) Phipps. Acta Zool. Fennica, 108:1–30.

Laws, R. M. 1956. The elephant seal (*Mirounga leonina* Linn.). II. General, social and reproductive behaviour. Falkland Is. Depend. Surv., Sci. Rep., 13: 1–88.

———. 1957. On the growth rates of the leopard seal (*Hydrurga leptonyx*) (de Blainville). Saugetierk Mitt., 5:39–55.

———. 1958. Growth rates and ages of crabeater seals (*Lobodon carcinophagus*), Jacquinot and Pucherhan. Proc. Zool. Soc. London, 130:275–288.

LeBoeuf, B. J. 1971. Fighting in elephant seals. Nat. Hist., 80:83–94.

LeBoeuf, B. J., and K. J. Panken. 1977. Elephant seals breeding on the mainland in California. Proc. California Acad. Sci., 41:267–280.

LeBoeuf, B. J., and R. S. Peterson. 1969. Social status and mating activity in elephant seals. Science, 163:91–93.

Liers, E. E. 1951. Notes on the river otter (*Lutra canadensis*). J. Mamm., 32: 1–8.

Ling, J. K., and G. E. Walker. 1978. An 18-month breeding cycle in the Australian sea lion? Search, 9:464–465.

Mansfield, A. W. 1966. The grey seal in eastern Canadian waters. Canadian Audubon, 28:161–166.

Marlow, B. J. 1975. The comparative behavior of the Australian sea lions, *Neophoca cinerea* and *Phocarctos hookeri* (Pinnipedia: Otariidae). Mammalia, 39:159–230.

Mathisen, O., R. T. Baade, and R. J. Lopp. 1962. Breeding habits, growth, and stomach contents of the Steller sea lion in Alaska. J. Mamm., 43:469–477.

McLaren, I. A. 1958. The biology of the ringed seal (*Phoca hispida* Schreber) in the Eastern Canadian Arctic. Bull. J. Fish. Res. Board Canada, 118: 1–97.

———. 1966. Taxonomy of harbor seals of the western North Pacific and evolution of certain other hair seals. J. Mamm., 47:466–473.

———. 1967. Seals and group selection. Ecology, 48:104–110.

———. 1975. A speculative overview of phocid evolution. Pp. 43–48, *in* Biology of the seal (K. Ronald and A. W. Mansfield, eds.). Rapp. P.-v. Cons. Int. Explor. Mer., 169:1–557.

Mitchell, E. D., and R. H. Tedford. 1973. The Enaliarctinae: a new group of extinct aquatic carnivora and a consideration of the origin of the Otariidae. Bull. Amer. Mus. Nat. Hist., 151:205–284.

Nutting, C. C. 1891. Some of the causes and results of polygamy among the Pinnipedia. Amer. Nat., 25:103–112.

Odell, D. K. 1974. Behavioral thermoregulation in the California sea lion. Behav. Biol., 10:231–237.

Olds, J. M. 1950. Notes on the hood seal (*Cystophora cristata*). J. Mamm., 31: 450–452.

Øritsland, T. 1964. Klappmysshunnens forplantnings-biologi. Fisken og Havet (Bergen), 1:1–15.

———. 1970. Sealing and seal research in the south-west Atlantic pack ice, Sept.–Oct. 1964. Pp. 367–376, *in* Antarctic ecology (M. W. Holdgate, ed.). Academic Press, London, 1:1–1604.

Peterson, R. S., and G. A. Bartholomew. 1967. The Natural History and Behavior of the California sea lion. Spec. Publ., Amer. Soc. Mamm., 1: 1–79.

Pierotti, R., and D. Pierotti. 1980. Effects of cold climate on the evolution of pinniped breeding systems. Evolution, 34:494–507.

Popov, L. A. 1966. On an ice floe with the harp seals: ice drift of biologists in the White Sea. Priroda, 9:93–101.

Ralls, K. 1976. Mammals in which females are larger than males. Quart. Rev. Biol., 51:245–276.

Rasmussen, B. 1952. Selfangsten ved Newfoundland varen 1951. Arsberetn. vedk. Norges Fiskeries, 1951, 5:53–85.

Ray, C. E. 1976. Geography of phocid evolution. Syst. Zool., 25:391–406.

Ray, G. C. 1967. Social behavior and acoustics of the Weddell seal. Ant. J. U.S., 2:105–106.

Repenning, C. A. 1976. Adaptive evolution of sea lions and walruses. Syst. Zool., 25:375–390.

———. 1980. Warm blooded life in cold ocean currents. Oceans, 3:18–24.

Repenning, C. A., C. E. Ray, and D. Grigorescue. 1979. Pinniped biogeography. Pp. 357–369 *in* Historical biogeography, plate tectonics, and the changing environment (J. Gray and A. J. Boucot, eds.). Oregon State University Press, Newport, 500 pp.

Repenning, C. A., and R. H. Tedford. 1977. Otarioid seals of the Neogene. U.S. Geol. Surv. Prof. Paper, 992:1–93.

Scheffer, V. B. 1958. Seals, sea lions and walruses. A review of the Pinnipedia. Stanford Univ. Press, Stanford, 179 pp.

Sergeant, D. E. 1965. Migration of harp seals (*Pagophilus groenlandicus*) (Erxleben) in the Northwest Atlantic. J. Fish. Res. Board Canada, 22:433–464.

———. 1974. A rediscovered whelping population of hooded seals (*Cystophora cristata*) Erxleben and its possible relationship to other populations. Polarforschung, 44:1–7.

———. 1976. History and present status of populations of harp and hooded seals. Biol. Conserv., 10:95–118.

———. 1977. Research on hooded seals in the western North Atlantic in 1977. ICNAF Res. Doc. 77/XI/57.

SERGEANT, D. E., AND H. D. FISHER. 1960. Harp seal populations in the western North Atlantic from 1950 to 1960. Circular Fish. Res. Board Canada, 5: 1–58 (mimeo).

SINIFF, D. B., AND J. L. BENGTSON. 1977. Observations and hypotheses concerning the interactions among crabeater seals, leopard seals, and killer whales. J. Mamm., 58:414–416.

SINIFF, D. B., et al. 1977. An analysis of the dynamics of a Weddell seal population. Ecol. Monogr., 47:319–355.

———. 1979. Social and reproductive behavior of crabeater seals (*Lobodon carcinophagus*) during the austral spring. Canadian J. Zool., 57:2243–2255.

SLIJPER, E. J. 1976. Whales and dolphins. Univ. Michigan Press, Ann Arbor, 170 pp.

SMITH, E. A. 1966. A review of the world's grey seal population. J. Zool., 150: 463–489.

SMITH, E. A., AND R. W. BURTON. 1970. Weddell seals of Signey Island. Pp. 415–418, *in* Antarctic ecology (M. W. Holdgate, ed.). Academic Press, London, 1:1–1604.

SMITH, T. G., et al. In press. Coordinated behavior of killer whales (*Orcinus orca*), hunting a crabeater seal (*Lobodon carcinophagus*). Science.

STIRLING, I. 1969. Ecology of the Weddell seal in McMurdo Sound, Antarctica. Ecology, 50:573–586.

———. 1971*a*. Studies on the behavior of the South Australian fur seal (*Arctocephalus forsteri*) (Lesson): I. Annual cycle, postures and calls, and adult males during the breeding season. Australian J. Zool., 19:243–266.

———. 1971*b*. Population aspects of Weddell seal harvesting in McMurdo Sound, Antarctica. Polar Record, 15:653–667.

———. 1972. Observations on the Australian seal lion, (*Neophoca cinerea*) (Peron). Australian J. Zool., 29:271–279.

———. 1974. Movements of Weddell seals in McMurdo Sound. Australian J. Zool., 22:39–43.

———. 1975. Factors affecting the evolution of social behavior in the Pinnipedia. Pp. 205–212, *in* Biology of the seal (K. Ronald and A. W. Mansfield, eds.). Rapp. P.-v. Reun. Cons. Int. Explor. Mer., 169:1–557.

———. 1977. Adaptions of Weddell and ringed seals to exploit polar fast ice habitat in the presence or absence of land predators. Pp. 741–748 *in* Adaptations within Antarctic ecosystems (G. A. Llano, ed.). Proc. Third symposium on Antarctic biology, Washington, D.C., 1252 pp.

STIRLING, I., AND R. M. WARNEKE. 1971. Implications of a comparison of the airborne vocalizations and some aspects of the behavior of the two Australian fur seal species (*Arctocephalus sp.*) on the evolution and present taxonomy of the genus. Australian J. Zool., 19:227–242.

TAYLOR, R. J. 1976. Value of clumping to prey and the evolutionary response of ambush predators. Amer. Nat., 110:13–29.

TERHUNE, J. M., AND K. RONALD. 1973. Some hooded seal (*Cystophora cristata*) sounds in March. Canadian J. Zool., 51:319–321.

VENABLES, U. M., AND L. S. V. VENABLES. 1955. Observations on a breeding colony of the seal *Phoca vitulina* in Shetland. Proc. Zool. Soc. London, 125:521–532.

———. 1957. Mating behaviour of the seal *Phoca vitulina* in Shetland. Proc. Zool. Soc. London, 128:387–396.

———. 1959. Vernal coition of the seal *Phoca vitulina* in Shetland. Proc. Zool. Soc. London, 132:665–669.

WRIGHT, P. L. 1963. Variations in reproductive cycles in North American mustelids. Pp. 77–97, *in* Delayed implantation (A. C. Enders, ed.), Univ. Chicago Press, Chicago, 318 pp.

KIN IDENTIFICATION, MATRIARCHIES, AND THE EVOLUTION OF SOCIALITY IN GROUND-DWELLING SCIURIDS

GAIL R. MICHENER

Abstract

THE social organization of North American ground-dwelling sciurids is reviewed. The five grades of social structure identified—asocial, single-family female kin clusters, female kin clusters with male territoriality, polygynous harems with male dominance, and egalitarian polygynous harems—form a continuum of sociality, based on the mother-headed family as the fundamental social subunit. Three simultaneous trends characterize the sequence of sociality: retention of daughters near or in the mother's range (single-family clusters), relaxation of distinctions between adjacent litters (cohesive multiple-family clusters), and superimposition of male territories, maintained beyond the breeding season, over female ranges (polygynous harems).

Social structure is based upon genetic relatedness between females. Familiarity with close kin, established via socialization in amicable social contexts during the first 4 to 6 weeks of life, is proposed as the proximate mechanism determining differential behavioral responses to conspecifics that vary in their degree of genetic relatedness.

Social structure correlates with the proportion of their preadult life that immatures are simultaneously in the active, above-ground population with adults. Prolonged active seasons for adults and/or increased number of seasons required for juveniles to attain maturity permit greater coincidence in the seasonal activity of adults and immatures. The selective factors that have determined the timing of the circannual cycle are proposed to have indirectly promoted the evolution of sociality.

Introduction

Biologists are increasingly aware that by combining comparative studies of the social behavior of closely related species with the fundamental concept of natural selection acting at levels no higher than the individual, the evolution of social behavior is amenable to interpretation and to the formulation of predictive hypotheses. Discussions of the evolution of social behavior and of mating systems (for example, see Alexander, 1974; Orians, 1969; West Eberhardt, 1975; Wittenberger, 1980) have led to classificatory systems or models useful for providing functional descriptions of social systems in terms of adaptive behavioral strategies to environmental constraints (Crook et al., 1976; Emlen and Oring, 1977). When a range of social systems occurs within a taxonomic group, adaptive radiation in relation to ecology is probable (Barash, 1974; Clutton-Brock, 1974; Crook et al., 1976; Eisenberg et al., 1972). The social systems of related species can be analyzed for common, underlying social subunits and for overlying, more recently evolved social trends.

Among mammals, the ground-dwelling sciurids (Sciuridae) have attracted scientific interest because they include several highly social species that are readily observable in field situations and because most North American species are obligate seasonal hibernators. An ever-increasing body of information is available on the social organization and on the hibernation physiology of several North American species. The aims of this article are: 1) to review the current literature on social organization of North American sciurid species; 2) to classify sciurid social systems; 3) to identify common themes and evolutionary trends in social organization; 4) to discuss the role of kin identification in maintenance of the social system; and 5) to relate sociality to environmental constraints and circannual rhythmicity.

Classification

The North American Sciuridae are classified as arboreal (tree squirrels and flying squirrels) or terrestrial (chipmunks and ground squirrels). The semi-fossorial ground squirrels are further subdivided into the giant ground squirrels (subtribe Marmotina) and the ground squirrels, prairie dogs, and antelope squirrels (subtribe Spermophilina) (Moore, 1961). The antelope squirrels, which have an earlier origin than the other living species of Nearctic ground squirrels (Hight et al., 1974) and which are not obligate hibernators, are not treated in this review because of the paucity of data on their social behavior and organization. The remaining members of the Spermophilina belong to two genera, *Spermophilus* (ground squirrels) and *Cynomys* (prairie dogs); the latter are a specialized, recently evolved group whose closest affinities are to ground squirrels of the subgenus *Spermophilus* (Hight et al., 1974; Nadler et al., 1971; Seaman, 1976).

Circannual Cycles

Obligate seasonal hibernation is characteristic of the majority of species of North American ground-dwelling sciurids that occur north of 30°N. In such species each year is divided into an active season, during which some or all of the individuals enter the above-ground population daily, and an inactive (hibernation) season, in which all members of the population remain below ground continuously for several months. Mating occurs shortly after adult females emerge from hibernation, and juveniles typically enter the above-ground population 56 to 60 days after adult females first appear from hibernation (Knopf and Balph, 1977; Michener, 1977*a*, 1977*b*, 1980*a*; Morton, 1975; Murie and Harris, 1982). The emergence of juveniles from the natal burrows results in a dramatic increase in the density of the surface active population. However, the density typically begins to decline shortly thereafter with the immergence of adults. The sequence of late summer and autumn immergence into hibernation for most species is adult males, adult females that failed to wean offspring, adult females that successfully reproduced, juvenile females and, finally, juvenile males (Clark, 1977; Grizzell, 1955; Iverson and Turner, 1972; Loehr and Risser, 1977; McCarley, 1966; D. Michener, 1974; Michener, 1977*a*, 1978, 1979*a*; Morton and Parmer, 1975; Murie, 1973; Skryja and Clark, 1970). If there is a non-breeding yearling age class, these animals usually disappear after the adults but before the juveniles. Exceptions to the above sequence are Uinta ground squirrels (*Spermophilus armatus*), in which adult males immerge shortly after adult females but before yearlings and juveniles (Knopf and Balph, 1977), and Arctic ground squirrels (*Spermophilus parryii*), in which adult males remain active later than adult females, sometimes not disappearing until late autumn at the same time as juvenile males (Green, 1977; McLean and Towns, 1981). Of the ground-dwelling sciurids whose ranges extend north of 40°N, only the black-tailed prairie dog (*Cynomys ludovicianus*) is not an obligate hibernator. King (1955) and Koford (1958) noted that prairie dogs were active on all but the most inclement days during winter, with any individual rarely continuously underground for more than a couple of days.

Among the species of ground-dwelling sciurids, there are variations in the amount of time that different cohorts are simultaneously active (Table 1) and, therefore, able to interact socially. Once juvenile black-tailed prairie dogs have emerged from the natal burrow, their above-ground activity coincides with that of all other cohorts for the remainder of their lives. In contrast, for some of the hibernating species there is a period of only a few weeks when all age and sex classes are simultaneously active. Because of the early immergence of adult males in most species, the cohorts with the least amount of coincidence in their activity patterns are adult males and juveniles. Richardson's ground squirrels (*Spermophilus richardsonii*) have a short period of overlap in the activity of adult males and juveniles, with juveniles typically spending the last 10 weeks of their first summer of life out of contact with males. The atypical immergence pattern of adult male Arctic ground squirrels results in their presence throughout the entire period that juveniles are active. For those species in which reproductive maturity is delayed until years 2 or 3, coincidence in the seasonal activity of adults and immatures is high (Table 1).

Grades of Social Structure

Isolated individual ground squirrels are rare; typically aggregates of conspecifics are found wherever suitable resources (primarily food and burrows) occur. Although rarely geographically solitary, some species are characterized by social solitariness. Sociality will have evolved from ancestral, asocial aggregations if social integration and cohesiveness increased the survival and reproductive success of individuals in the group.

The social organization of North American ground-dwelling sciurids forms a continuum ranging from asocial species, such as the woodchuck (*Marmota monax*) and Franklin's ground squirrel (*Spermophilus franklinii*), to highly social species, such as the Olympic marmot (*Marmota olympus*) and black-tailed prairie dog. Between these extremes are species with intermediate degrees of

TABLE 1
SELECTED LIFE HISTORY TRAITS OF 18 SPECIES OF GROUND-DWELLING SCIURIDS RANKED BY INCREASING SOCIAL GRADE

Species	Social grade	FAFR	Size	Seasonal hetero-thermy	Subadult active season (days)	Adult-subadult seasonal coinci-dence as percent	References
→*Spermophilus franklinii*	1	Y	M	O	65–90	40–60	Iverson and Turner (1972); Murie (1973); Sowls (1948)
Spermophilus lateralis	1	Y	S	O	110–130	50–60	Cameron (1969); Ferron (1977); Kivett et al. (1976); Skryja and Clark (1970); Tevis (1955); Wirtz (1967)
→*Marmota monax*	1	Y (2)	VL	O (P)**	140–160	80–90**	Grizzell (1955); Snyder et al. (1961)
Spermophilus townsendii	1 or 2	Y	S	O	60–75	65	Alcorn (1940); Rickart (in press)
→*Spermophilus tridecemlineatus*	1 or 2	Y	S	O	85–100	40–65	McCarley (1966); Rongstad (1965)
Spermophilus elegans	2	Y	M	O	80	45	Clark (1970)
→*Spermophilus richardsonii*	2	Y	M	O	100–130	30–60	Dorrance (1974); D. Michener (1974); Michener (1977*a*, 1978, 1979*a*, 1979*c*)
→*Spermophilus beldingi*	2	Y	M	O	65–70	60	Loehr and Risser (1977); Morton and Parmer (1975); Morton et al. (1974)
Spermophilus armatus	2	Y	M	O	65	45–70	Knopf and Balph (1977)
→*Spermophilus tereticaudus*	2	Y	S	O	130	80–100	Dunford (1977*b*); Neal (1965)
→*Cynomys leucurus*	2	Y	L	O	120–140	35–55	Clark (1977); Tileston and Lechleitner (1966)
Spermophilus beecheyii	2 or 3	Y	L	P (O)#	300	50–100	Bickford (1979); Fitch (1948); Owings et al. (1977); Storer et al. (1944); Tomich (1962)

TABLE 1
CONTINUED

Species	Social grade	FAFR	Size	Seasonal hetero-thermy	Subadult active season (days)	Adult-subadult seasonal coinci-dence as percent	References
→*Spermophilus parryii*	3	Y	L	O	75–100	70–95	Carl (1971); Green (1977); McLean (1981); McLean and Towns (1981)
→*Spermophilus columbianus*	3	2	L	O	140–150	70–95	Michener (1977*a*); Murie and Harris (1982)
Cynomys gunnisoni	4	(Y) 2	L	O	140–150	80–95	Fitzgerald and Lechleitner (1974); Rayor (1980)
→*Marmota flaviventris*	(1)† 4	(2) 3[∞]	VL	O	350[∞]	95–100	Armitage et al. (1976); Kilgore and Armitage (1978)
→*Marmota olympus*	5	3	VL	O	240	90–100	Barash (1973*a*)
→*Cynomys ludovicianus*	5	(Y) 2	L	T	650	100	King (1955); Koford (1958)

* See text for descriptions of the five social grades. → indicates species discussed in detail in the text. FAFR is the typical female age at first reproduction (symbol in parentheses indicates age at which some females first reproduce): Y, yearling; 2, 2-year-old; 3, 3-year-old. Size classifies species by head plus body length: S, <200 mm; M, 200–250 mm; L, >250–350 mm; VL, >350 mm. Seasonal heterothermy: O, obligate hibernator; P, permissive hibernator; T, occasional torpor. Subadult active season is the typical total number of days immatures are active from first emergence from natal burrow to immergence in final subadult season, and hence is cumulative over several seasons for species in which reproductive maturity is not achieved until 2 or 3 years of age. Adult-subadult seasonal coincidence indicates the proportion of the number of days immatures spend above ground that adults are also present in the surface-active population. Figures represent modal values rather than extreme values.

** Anthony (1962) reported some activity by woodchucks in southern Illinois throughout winter, in which case adult-subadult coincidence is 100%.

Storer et al. (1944) reported that all California ground squirrels at 3000 m elevation are underground from December to March. At lower elevations most adults hibernate for varying time periods whereas most juveniles remain active throughout the winter months (Fitch, 1948; Tomich, 1962).

† 25–50% of adult marmots occupy satellite sites and are not members of harems (Armitage and Downhower, 1974; Svendsen, 1974).

∞ Although some 2-year-old yellow-bellied marmots do reproduce, Armitage and Downhower (1974) reported that fewer than half do so, with most first producing young as 3-year-olds; subadulthood has therefore been considered to typically extend over 3 summers. Even for those marmots which first reproduce as 2-year-olds, about 95% of subadulthood is coincident with adult activity.

sociability and group cohesiveness. The aims of this section are to describe the characteristic social and spatial organization of several well-studied species of ground-dwelling sciurids and to classify them into one of five grades of sociality. Classification necessarily breaks a continuum into discrete units, but this does not imply that the social grades are absolutely distinct from each other.

The proposed grades of sociality are:

1. *Asocial.*—Males and females do not share territories. Juveniles disperse shortly after weaning with each establishing a home range distinct from those of the mother and littermates. Social interactions, even between kin, are mainly agonistic.
2. *Single-family female kin clusters.*—After the breeding season males and females occupy individually distinct ranges. Sons disperse from the natal area, whereas daughters often remain in or near the mother's range throughout life with subsequent sharing of the ancestral range between mother and adult daughters and among littermate sisters. Juveniles from different litters do not intermingle; a mother and her offspring form a sociable, moderately cohesive group that is unsociable with and aggressive toward adjacent groups.
3. *Female kin clusters with male territoriality.*—Adult males maintain their territories beyond the breeding season, defending an area that overlaps the smaller ranges of several adult females (composing one or more kin clusters) and their offspring. Juveniles from adjacent litters may associate together after weaning, and mothers show no strong discrimination between their own offspring and those of adjacent females. Dispersal is sex-biased with more sons than daughters leaving the natal area and with males moving further than females.
4. *Polygynous harems with male dominance.*—The harem male maintains a territory throughout the active season within which several females and their offspring live. The male dominates all other harem members. Juveniles from different litters within a harem frequently intermingle after weaning. Sons leave the harem but daughters may be recruited.
5. *Egalitarian polygynous harems.*—A male and several females form a cohesive group that maintains and defends a common range. Males do not dominate females and they engage in amicable contacts with juveniles. Litter distinctions are not main-

tained following emergence from the natal burrow, and adults do not discriminate between young from different litters. Sons leave the harem whereas daughters are recruited.

The accuracy of assigning particular species to a given grade is limited by the information available in the literature. The following descriptions include only those species for which assignment can be made with reasonable confidence. Table 1 provides a more extensive list of species with their probable social grade. Although grade assignment is based on the typical or most commonly described social pattern exhibited by the species, such assignment does not imply inflexibility in the social system of that species.

Asocial

Because of the unrewarding nature of studying their social behavior and organization, little quantitative information is available on those ground-dwelling sciurids that show minimal social integration. Information is most extensive for the woodchuck, *Marmota monax* (Anthony, 1962; Bailey, 1965; Bronson, 1963, 1964; Grizzell, 1955; Hayes, 1977; Merriam, 1971; Snyder, 1976). After the breeding season, each adult woodchuck defends a den and foraging area. Juveniles disperse from the natal den shortly after first emergence above ground. Although littermates may share a burrow initially, aggression between siblings ultimately results in each animal establishing its own range. Snyder (1976) suggested that dispersal, particularly of juveniles, is sex-biased, with males moving further than females.

Limited field data indicate that Franklin's ground squirrels (*S. franklinii*) are the least social of the six species of *Spermophilus* found in Canada (Kivett et al., 1976; Sowls, 1948). Captive Franklin's ground squirrels rarely scent-mark, and they interact infrequently, maintaining interindividual spacing primarily by mutual avoidance.

Wistrand (1974) described thirteen-lined ground squirrels (*S. tridecemlineatus*) as being social only at courtship and during rearing of litters; at other times individuals tend to avoid each other. Evans (1951) noted apparent family groupings consisting of a mother and several juveniles, but McCarley (1966) reported that family units began to break up within 2 weeks of juvenile emergence. Four weeks after emergence, juveniles were living in separate burrow systems. Juvenile females usually remain closer to the natal area than juvenile males, and adult females tend to be sedentary (McCarley, 1966; Rongstad, 1965), so groupings of closely related females may occur, though this has not been documented. Because Rongstad, McCarley, and Wistrand studied thirteen-lined ground squirrels on golf courses, cemeteries, and mown park lawns, their observations on social behavior may not be typical for the species. Based on the brief study by Evans (1951) on an abandoned field, *S. tridecemlineatus* might more appropriately be placed in the next grade of social organization.

Single-Family Female Kin Clusters

Belding's ground squirrels (*S. beldingi*; Sherman, 1977, 1980, 1981*a*; Sherman and Morton, 1979), round-tailed ground squirrels (*S. tereticaudus*; Dunford, 1977*a*, 1977*b*), Richardson's ground squirrels (*S. richardsonii*; D. Michener, 1972; Michener, 1973*a*, 1973*b*, 1979*b*, 1980*b*), and white-tailed prairie dogs (*Cynomys leucurus*; Clark, 1977; Hoogland, 1979*a*, 1979*b*, 1981*a*; Tileston and Lechleitner, 1966) have similar social systems based on matrilineal kin groups. Females tend to be sedentary, remaining in or near the natal area throughout life, and ancestral homesites are ultimately occupied by daughters and granddaughters. Males typically disperse individually as juveniles or as yearlings. Consequently, males do not tend to be resident near to kin of either sex, whereas females tend to live in proximity to near and distant female relatives. Adult males typically move extensively during or after the breeding season (Michener, 1979*c*; Sherman, 1977), so fathers are rarely in proximity to offspring. Males do not defend a postbreeding territory nor do they defend resources that could be used by mate(s) and progeny. Each female rears young in isolation and defends the nest burrow from all conspecifics, including her mate and female relatives.

Round-tailed ground squirrels (Dunford, 1977*a*), Richardson's ground squirrels (Michener, 1979*b*), and white-tailed prairie dogs (Clark, 1977) show corresponding patterns of spatial and social relationships throughout the active season. All exhibit site-dependent dominance and site-attachment. The home range is not overtly defended but individuals tend to concentrate activity in one area (core area) which is unlikely to be simultaneously occupied by another animal. Clark (1977) described this as "core monopolization," and Michener (1979*b*) suggested that Brown's term "dominion" (Brown, 1975:62) would apply to such a spatial-behavioral system. Following the emergence of adult females in spring, four major phases can be identified in the spatial and social patterns among conspecifics throughout the active season. First, during the copulatory phase, interaction rates (especially between individuals of opposite sex) are high. Adult males are conspicuous and active, often sustaining numerous wounds. Males tend to dominate females, and they frequently intrude into female core areas. The second phase, pregnancy-lactation, is characterized by even dispersion of home ranges and core areas with a minimal amount of overlap. Males are relatively inconspicuous and asocial, and are dominated by females, except when within their own core area. Pregnant and lactating females are aggressive toward conspecifics, confine their activity to a relatively small area around the natal burrow, and intrude into the areas of only immediately adjacent neighbors. During the third phase, juvenile emergence, females spend more time above-ground and they range over a larger area. Overlap of home ranges is maximal but core areas continue to remain essentially distinct. Following emergence, juveniles gradually increase their ranges of activity around the natal burrow and, after about 2 weeks, start interacting with juveniles from neighboring litters. Social interactions among littermates and between mother and offspring are predominantly amicable, whereas those between non-kin are predominantly agonistic. Mothers and offspring form family units that are distinct from and aggressive toward other family units. The final phase, prehibernation, is characterized by a decrease both in the amount of time spent above ground and in the range of activity of adults. Juveniles cease sharing burrows with siblings or mother; each juvenile establishes its own core area that is essentially distinct from, although in proximity to, those of its family members. In effect, each mother and any of her offspring (primarily daughters) that do not disperse, subdivide and share the range she had previously occupied. Interaction rates are low, but kin continue to interact amicably and non-kin agonistically. Juveniles are behaviorally indistinguishable from adults both in terms of the use of space and the nature of interactions with conspecifics. By late summer, juveniles have attained 90% of adult size and are virtually indistinguishable morphologically from adults. There is no conclusive evidence that animals hibernate communally; judging from the establishment of separate core areas and the cessation of burrow sharing by family members, hibernation in isolation seems probable.

Female Kin Clusters with Male Territoriality

Arctic ground squirrels (*S. parryii*; Carl, 1971; Green, 1977; McLean, 1981, 1982, in press; Watton and Keenleyside, 1974) and Columbian ground squirrels (*S. columbianus*; Betts, 1976; Festa-Bianchet and Boag, 1982; Michener, pers. observ.; Murie and Harris, 1978; Steiner, 1970) have a social system that is based on matrilineal kin groups. However, compared with the previous category, adult males play a more prominent role in the social organization, and thier territorialism is more pronounced, with territorial defense continuing at least until juveniles emerge. The male's territory overlaps the ranges of several females and offspring.

Although Arctic and Columbian ground squirrels show similarities in social systems, there are major differences between the species in life history traits. Litter size is smaller in *S. columbianus,* with females typically weaning 2.5 young (Murie et al., in press), whereas *S. parryii* typically wean 6.1 young (McLean, 1980). Columbian ground squirrels are 10 to 12 weeks old when they first enter hibernation and they rarely breed as yearlings (Festa-Bianchet, 1981; Michener, 1977*a*; Murie et al., 1980), whereas Arctic ground squirrels are 18 to 22 weeks old at first immergence and are reproductively competent the following spring, with all yearling females and most yearling males breeding (McLean and Towns, 1981). A second summer of growth is required to attain adult size in *S. columbianus* (Boag and Murie, 1981*a*). Coincidence of the active seasons of adults and subadults is high in both species (Table 1) despite the difference in age at maturity. Although Columbian ground squirrels can reproduce as 2-year-olds, males probably do not first reproduce until 3 to 4 years old when they are capable of maintaining territories (Murie and Harris, 1978).

Male *S. parryii* and *S. columbianus* establish territories following emergence in spring. Males are aggressive and wounding results from boundary disputes. Males unsuccessful in establishing territories become "floaters" in the population and are subordinate to territorial males although tolerated by them. Others disperse and become "refugees" in suboptimal and peripheral habitat. Females establish ranges, which are smaller than and typically overlap with the territory of at least one male (Murie, 1980). A minimum of fighting and wounding is involved in interfemale disputes, probably because females are familiar kin or neighbors from the previous year. The ranges of males and females are established independently; females occupy a burrow system in spring and, incidentally, select the male within whose territory the system is located. Males do not "herd" females (Wittenberger, 1980) or attempt to prevent them from leaving the defended area. Murie and Harris (1978) hypothesized that the main function of male territoriality in *S. columbianus* is to provide access to estrous females. McLean (in press) suggested that, because the territories of male *S. parryii* are more exclusive after than during the breeding season, the main function of territorial defense by males of that species is to protect offspring born within each territory.

As with the previous grade of social structure, sciurids in this category rear litters in isolation; females defend the area around the natal burrow from the territorial male as well as other females and intruders. However, intermingling of juveniles from different litters may occur shortly after emergence from the natal burrow. In Arctic ground squirrels, up to three females who are close kin (mother-daughter, littermate sisters) may bring litters to a common burrow system following juvenile emergence (McLean, 1982). Juveniles, therefore, frequently interact with conspecifics other than their family members within a few days of initial emergence.

In Arctic ground squirrels, there is sex-biased dispersal of juveniles, commencing about 4 weeks after weaning (when young are about 8 weeks old), with more males disappearing than females. In Columbian ground squirrels, juveniles tend to remain in or near the natal area through the first summer, but then leave the mother's area as yearlings (Boag and Murie, 1981*b*). Yearlings from different litters aggregate and frequently play in early spring. Such play becomes more aggressive as the season progresses, and the yearling groups break up in mid-summer when most yearling males disperse. Yearling groups may be composed of squirrels of both sexes (Michener, pers. observ.) or primarily of males (Murie and Harris, pers. comm.). Sex-biased dispersal of juveniles (*S. parryii*) and yearlings (*S. columbianus*) and the sedentariness of females permit formation of female kin clusters. Related adult females (mother-daughters, siblings) engage in amicable social interactions more frequently than do more distantly related squirrels, and they are more likely to share space, food, and burrow systems (with the exception of the natal burrow during lactation). Combining of litters after weaning of adjacent close kin and prolonged territorial defense by adult males may represent behavioral adaptations to protect pups from predators and aggressive males (McLean, 1982, in press; Steiner, 1972). Rivett et al. (1976) suggested that the territorial adult male is a unifying force in the social system of Columbian and Arctic ground squirrels. Territorial males initiate the majority of social interactions within the social groups, they are aggressive towards neighboring conspecifics, and they scent-mark the territory more frequently and more consistently than do females.

Polygynous Harems with Male Dominance

The basic social group in yellow-bellied marmots (*Marmota flaviventris*) is a harem composed of one adult male, one to several adult females, and variable numbers of yearlings and juveniles (Armitage, 1973, 1974, 1975, 1977; Armitage and Downhower, 1974; Barash, 1973*b*; Downhower and Armitage, 1971; Johns and Armitage, 1979; Nowicki and Armitage, 1980; Schwartz and Armitage, 1980; Svendsen, 1974). These marmots have been studied at medium to high elevations, where they inhabit subalpine and alpine meadows in discontinuous habitat. Typically any one suitable location sustains only one harem.

In *M. flaviventris* the harem male is territorial throughout the active season, is behaviorally conspicuous, is socially dominant to other harem members, and is aggressive toward all other males, who tend to avoid him. The foraging ranges of harem members overlap extensively, with the adult male moving over a larger area than any individual female. The adult male and females within a harem interact infrequently except when an unfamiliar male takes over a harem. According to Armitage (1974) the advantages of territoriality to males are: creation of a socially stable environment optimal for reproductive success; enhancement of outbreeding through the presence of only one male and through the dispersal of sons born within the harem; and maximization of male fitness through access to several estrous females.

Yearlings do not reproduce and, although 2-year-olds may reproduce, most females produce their first litter as 3-year-olds. Females typically rear their young in isolation from other litters and from offspring of the previous year. Young remain in close contact with parents beyond the age of weaning, sharing summer burrows, foraging areas, and hibernacula with them. Harem females are dominant to yearlings but interactions between parents and yearlings are typically more amicable than those between adults and unrelated yearlings. Yearling males disperse, usually after the snow has melted and before juveniles emerge. As adults, males are rarely resident in the harem of birth. In contrast, half or more of the adult females within a harem were born there. Typically those females that, as yearlings, establish home ranges that extensively overlap the home range of the mother are recruited into the natal harem. Male tenure in harems is dynamic with male turnover every 2 or 3 years, whereas female tenure is more static with some females residing in the same locality for 11 years. The differential length of tenure of male and female marmots and the likelihood of daughters being recruited into the natal harem, means that harems are typically matriarchies.

Although the basic social unit is the polygynous harem, yellow-bellied marmots living alone or in pairs in suboptimal habitat are common; Svendsen (1974) classed 24% of 67 adults in his population as satellite occupants, and Armitage and Downhower (1974) first captured 52 of 98 adults and 11 of 21 yearlings at satellite sites. Such animals are resident for shorter periods than harem animals, they suffer higher mortality, and their lifetime reproductive success is lower. Individuals may vary between satellite and harem sites between years.

Egalitarian Polygynous Harems

Olympic marmots (*Marmota olympus*; Barash, 1973*a*) and black-tailed prairie dogs (*Cynomys ludovicianus*; Hoogland, 1979*a*, 1979*b*, 1981*a*, 1981*b*; King, 1955; Smith et al., 1973; Tileston and Lechleitner, 1966) live in closed social groups that are characterized by weak or non-existent dominance relationships, by amicable social interactions, by frequent "greetings" (oral and nasal contacts), by unrestricted use of space and burrows (except a female's natal burrow), and by communal use of sleeping burrows and winter burrows.

Although the two species have similar social systems, differences in life-history patterns and social demography coincide with differences in the habitat each occupies. Olympic marmots, which inhabit alpine meadows with short vegetational growing seasons, are obligate hibernators and biennial breeders. Young attain adult size and sexual maturity at 3 years of age, so the subadult cohort is composed of both yearlings and 2-year-olds. Yearlings reside with the mother; those marmots which disperse do not leave until they are 2 years old. Because of the discontinuous nature of the habitat, many harems are isolated and solitary. In contrast, black-tailed prairie dogs inhabit open, short-grass prairie where harems are tightly packed into the continuous habitat. The growing season is longer and prairie dogs are not obligate hibernators. Prairie dogs typically disperse as yearlings, first reproduce as 2-year-olds, and breed annually.

Olympic marmot harems (called colonies by Barash, 1973*a*) are typically composed of one adult male, two adult females (one of which bred the previous year, the other in the current year), the yearling offspring of one female, the juvenile offspring of the other, and several 2-year-olds. Black-tailed prairie dog harems (called coteries by King, 1955) are typically composed of one adult male, one to four adult females, and variable numbers of yearlings and juveniles. Isolated sites occupied by solitary animals have not been reported for either species. A second, subordinate adult male may be associated with a harem. Among Olympic marmots, such an additional male occupies peripheral burrows in early summer but, with a decline in aggressive chasing by the harem male, he is increasingly incorporated into group activities and he hibernates with other harem members. The subordinate male in a black-tailed prairie dog coterie is able to remain within the harem unmolested by the harem male unless he attempts to mate with females, which then results either in his eviction or in the splitting of the coterie into two groups.

Because a black-tailed prairie dog coterie is surrounded by other coteries with which it has contiguous boundaries, intracoterie versus intercoterie social relationships can be compared. Whereas intercoterie interactions are hostile, intracoterie interactions are amicable, except when they involve pregnant and lactating females (Michener and Murie, in press). Coterie members communally and cooperatively engage in burrow and mound construction, in modification of vegetative cover, in nest building, in territorial defense, and in predator defense. Coterie members also sleep communally, with the exception of pregnant and lactating females who maintain individual nest burrows that are defended against all coterie members. Following weaning of pups, females cease maintaining nesting territories and again all members have equal access to all burrows within the coterie. Although juveniles interact only with the mother and littermates during the first 6 weeks of life, mingling of litters within a coterie commences immediately after juveniles first leave the natal burrow. As a result, a juvenile may sleep with group members other than its immediate family within days of first coming above ground. Mothers show no discrimination between pups from various litters within the coterie, and all coterie members, including males, are amicable toward newly emerged pups. Consequently juveniles enter a social environment that is characterized by frequent amicable interactions with all coterie members. Juveniles are not, however, tolerated by the adults in adjacent coteries. As with older animals, intercoterie interactions involving juveniles are characterized by aggression and by defense of the mutual coterie boundary.

The prairie dog coterie is an egalitarian society. There are no dominance relationships among adult females, among juveniles, or between the harem male and females. If two adult males are resident within the same coterie one will be dominant, but such males share burrows and allogroom. All coterie members, regardless of age and sex, actively defend the coterie boundary, range over the entire coterie area, and have equal access to all but natal burrows.

Black-tailed prairie dog coteries, although described in terms of a male and associated females, are matriarchies. Because daughters usually stay within the coterie of birth, whereas sons disperse (usually as yearlings), the adult females are typically native to a coterie, whereas the adult male is an immigrant. When a coterie splits into two, the females that are nominally associated with each male continue to interact with each other amicably—as they had prior to fission. Occasionally a male incorporates an adjacent coterie with his original one, in which case the two sets of females will continue to interact agonistically—as they had prior to fusion. The constant factor in the social composition of a coterie is membership via the female line. King (1955) classed the coterie territory as a heritable resource passed "on from generation to generation and from individual to individual." Inheritance is matrilineal. Sons do not inherit the father's coterie; daughters do inherit the mother's coterie.

Although I have identified five grades of social structure among the ground-dwelling sciurids and have allocated certain species to particular grades, I must emphasize that these grades form a continuum with intergradations between each, that species placed in the same category have similar but not identical social systems, and that, because social organization can vary with habitat, different populations of the same species could be allotted to different grades. For yellow-bellied marmots, individuals within the same population can be either asocial (in satellite sites) or integrated into a harem. Furthermore, distinctions between the social systems of some species (for example, Columbian ground squirrels and yellow-bellied marmots) may be due more to habitat differences than to social differences. Because of their larger size and more discontinuous habitat, yellow-bellied marmots occur in smaller groups than do Columbian ground squirrels. Typically there is only one marmot harem per site, whereas aggregations of Columbian ground squirrels may include several hundred animals. In a more continuous habitat, the marmots may have less well-defined harems, and males may not have exclusive access to estrous females.

Trends in Sociality

Three simultaneous trends characterize the range of ground-dwelling sciurid social organization from asocial individuals to cooperative polygynous harems: 1) retention of daughters within the mother's range and subsequent sharing of this range when the daughters are themselves reproductively mature (single-family kin clusters); 2) a relaxation of distinctions between litters of adjacent females such that inter-litter interactions are amicable (cohesive multiple-family kin clusters); 3) male territoriality extending beyond the breeding season, with territories encompassing the areas used by several females (polygynous harems). Hypothetical evolutionary pathways for the utilization of space based on daughter retention and female cohesiveness, and on male territoriality are depicted in Figs. 1 and 2, respectively.

The mother-infant bond, specifically the mother-daughter bond, is the most fundamental unit of ground-dwelling sciurid social organization, and it forms the basis from which other social units have derived. Single-family female kin clusters develop from the mother-daughter bond via extended social tolerance such that a female remains in spatial proximity to both her adult female offspring and her adult littermate sisters. Ultimately such tolerance permits

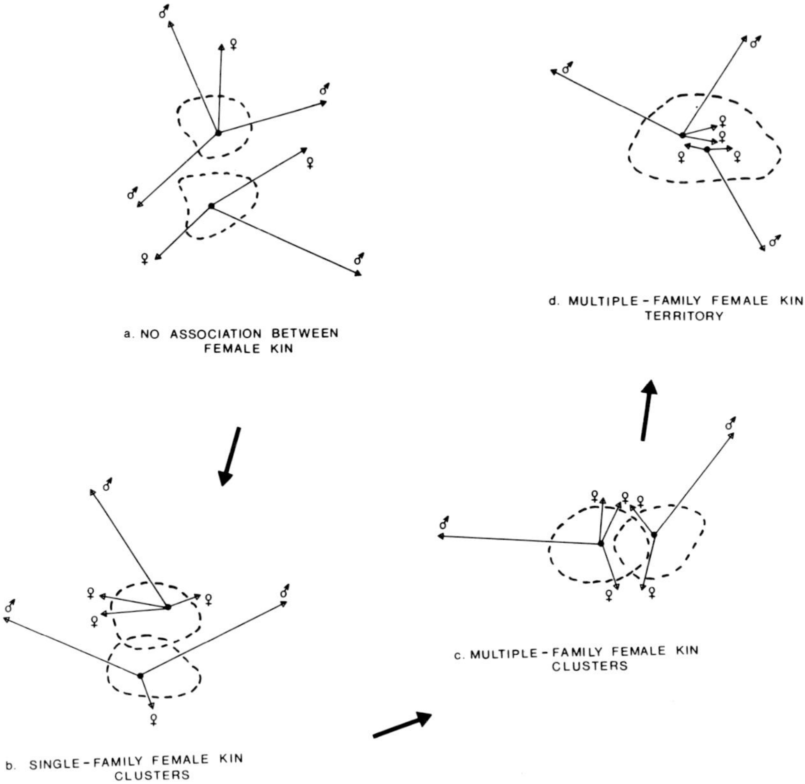

FIG. 1. Hypothetical evolutionary pathway showing increased social complexity based on retention of daughters near or in the mother's range and on relaxation of litter distinctions. Dashed lines indicate ranges of adult females. Solid circles indicate location of the natal burrow. Radiating lines indicate movement of offspring from the natal burrow. a, probable ancestral condition in which both sexes disperse from the natal area and no associations form between kin; b, sons disperse from the natal area but daughters establish ranges close to the mother's area, leading to clusters of females from the same family; c, sons disperse from the natal area, daughters establish ranges close to the mother's area, and daughters from litters of adjacent (closely related) females associate together; d, sons disperse from the natal area but daughters remain entirely within and codefend the mother's area along with daughters from other litters and their mothers.

daughters to remain entirely within the mother's range, and to cooperatively defend it with her. Multiple-family kin clusters result when litter distinctions relax, permitting daughters of adjacent females (who are themselves likely to be close kin) to associate as

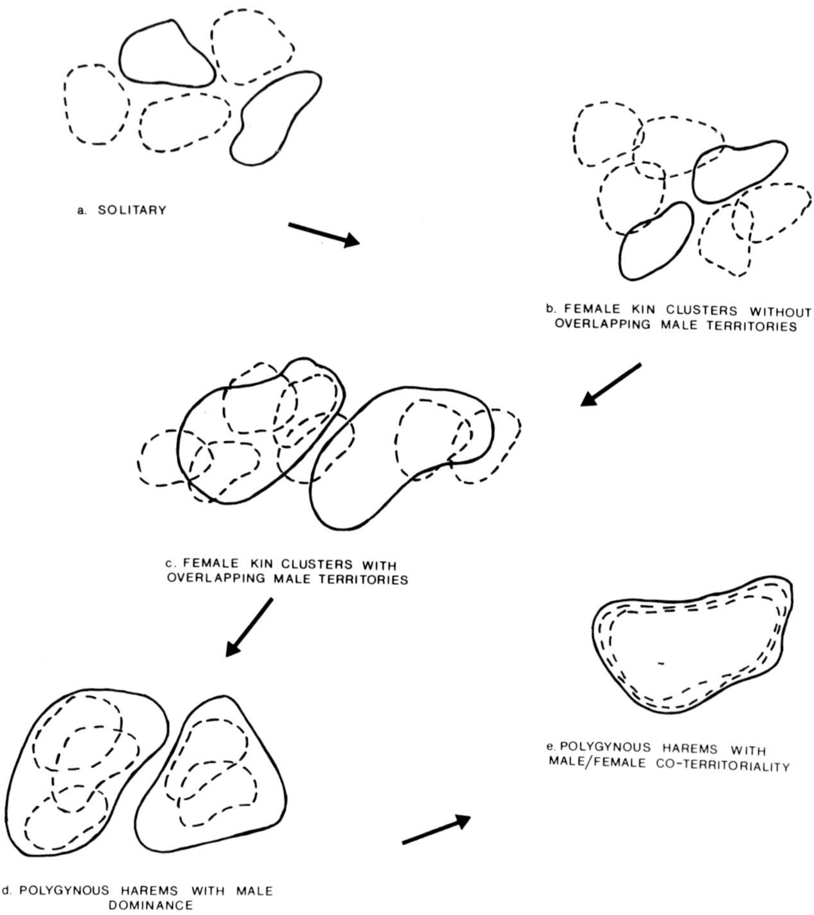

FIG. 2. Hypothetical evolutionary pathway showing increased social complexity based on superimposition of the post-breeding territories of adult males over female ranges. Solid lines indicate ranges of males. Dashed lines indicate ranges of females. a, probable ancestral condition in which both sexes maintain individually distinct ranges; b, related females occupy overlapping ranges that are distinct from the ranges of males; c, related females occupy overlapping ranges that may be overlapped by the territory(ies) of post-breeding male(s); d, related females occupy overlapping ranges that lie within the territory of the harem male; e, related females occupy and codefend a common range with the male.

adults, and, ultimately, permitting females from different litters to cooperatively defend the common area used by their mothers.

The male-female pair bond is not a fundamental unit in sciurid social organization. The association of a male with females initially derived from simultaneous choice of preferred habitat by both sexes. Males play no direct role in caring for or provisioning young, and male-offspring bonds comparable to those between females and daughters do not form.

Crook et al. (1976) suggested classifying mammalian social systems in terms of three strategies—dispersion, mating, and rearing. All three strategies vary with the trends in sociality described above. With regard to dispersion, all ground-dwelling sciurids are dependent on burrows for refuges (from predators and inclement weather), for reproduction (copulation, parturition, and rearing of altricial pups), for sleeping, and (in hibernating species) for hibernacula. Thus, in all species, the potential for mobility and for the use of space are constrained by the need to be close to burrows, which then serve as foci for above ground activity. The only variation between species is in the degree of exclusivity of the ranges surrounding burrows. This varies from little or no overlap between adjacent ranges to complete coincidence (Table 2). With increasing range overlap there is an increased probability of communal burrow use.

Typically ground-dwelling sciurids are polygynous, and no pair bond forms beyond that necessary for copulation. In some species, such as Richardson's ground squirrel, where the adult sex ratio is biased in favor of females (Michener, 1979*a*, 1980*b*; Michener and Michener, 1971), polygyny is an inevitable outcome of population demography. In others, in which the adult sex ratio approximates 1:1, polygyny results from differential male access to estrous females. Differential access could also occur in species with a biased adult sex ratio, but has not been documented. Reproductive success among male Belding's ground squirrels is asymmetric with about a quarter of the males accounting for two-thirds of copulations (Sherman, 1980; Sherman and Morton, 1979). Age is an important factor; most of the successful copulations involve males that are 3 to 4 years old. Similar asymmetric reproductive success, at least partly related to age, probably occurs in *S. columbianus* (Murie and Harris, 1978), *S. parryii* (Carl, 1971), and *M. flaviventris* (Armitage, 1974), all

TABLE 2
CLASSIFICATION OF SCIURID SOCIAL ORGANIZATION (AFTER CROOK ET AL., 1976)

Social organization	Strategies						
	Dispersion		Mating	Rearing			
Basic patterns	Refuge utilization		No pair bonding	Litters isolated in individual natal burrows, male not directly involved in rearing			
Variable patterns	Degree of range exclusivity		Type of polygyny	Duration of M-I bond† with		Male association with rearing group	
	♀–♀	♀–♂		Daughters	Sons (age at dispersal*)	Present	Duration
Asocial	High	High	Promiscuous	Short <1 summer	J	no	—
Single-family female kin cluster	Medium	High	Promiscuous	Prolonged weak assocn	J	no	—
Female kin cluster with male territoriality	Medium	Low	Limited access (resource defense)	Prolonged	J or Y	yes	To male immergence (1 or 2 summers)
Harem with male dominance	Low	None	Harem (resource defense)	Prolonged	Y	yes	To male immergence (2 summers)
Egalitarian harem	None	None	Harem (?harem defense)	Continuous	Y or 2	yes	Continuous or to male immergence (3 summers)

* J, juvenile; Y, yearling; 2, 2-year-old.
† M, mother; I, infant.

species in which the adult sex ratio approximates 1:1. Asymmetry is primarily related to the inability of some males, particularly younger animals, to establish territories in spring. In these three species, the mating strategy approaches Emlen and Oring's (1977) classification "resource defense polygyny," where the successful males monopolize a resource (burrow system) used by females. Because black-tailed prairie dogs live in continuous habitat that is partitioned among groups of female kin, the male mating strategy approximates that of "harem defense polygyny." In both types of polygynous systems the males neither control nor herd the females, but males "acquire" females by simultaneously using the areas occupied by females.

The three variables that Crook et al. (1976) considered in their classification of rearing strategies were: 1) whether or not the male was present in the rearing group; 2) the duration of the male's presence; 3) the duration of the mother-infant bond. These factors vary across the continuum of social organization exhibited by ground-dwelling sciurids (Table 2). The male is associated with female(s) and young only in those species in which the males maintain territories beyond the breeding season. In these systems, the duration of the male's association with subadults is influenced by the circannual cycle (Table 1). The duration of the mother-infant bond varies with sex of offspring and, for daughters, covaries with the extent of range exclusivity. In all species sons disperse, thereby terminating the social bond with the mother. This termination may occur as early as 3 to 6 weeks after weaning (for example, Richardson's, Belding's, and round-tailed ground squirrels), or not until the second (for example, *M. flaviventris, S. columbianus,* and *C. ludovicianus*), or third summer (for example, *M. olympus*) of life. With the possible exception of the asocial species, the mother-daughter bond is not usually abruptly terminated. Daughters typically establish home ranges that overlap with the mother's range and so the opportunities for social interactions persist into adulthood. Among *S. richardsonii,* mothers and daughters continue to interact amicably more often than agonistically but, because they spend the majority of the time in non-overlapping core areas, such interactions are not frequent. Female black-tailed prairie dogs occupy ranges that coincide with their mother's range throughout their life-time, permitting continued cohesive behavior.

The trends in the evolution of sociality among ground-dwelling sciurids pose three fundamental questions: 1) what are the selective advantages of female sedentariness and male dispersal; 2) what are the selective advantages of female clumping and range sharing; and 3) what are the selective advantages to both males and females of male territoriality.

Costs of dispersal are probably high. A male that leaves the natal area must forage, avoid predators, locate sleeping burrows and hibernacula, etc., in unfamiliar terrain, with no guarantee that he will ultimately locate a suitable environment, ecologically and socially, in which to settle. Both the proximate and ultimate causes of dispersal remain elusive for ground-dwelling sciurids (Dobson, 1979; Downhower and Armitage, 1980; Dunford, 1977*b*; Green, 1977; Michener, 1979*c*; Michener and Michener, 1973, 1977; Pfeifer, 1980; Schmutz et al., 1979; Yeaton, 1972) and for other mammals (see Bekoff, 1977, for review). One obvious result of sex-biased differential dispersal is the avoidance of inbreeding (Hoogland, 1982), which could be achieved either by male or by female dispersal. Male sedentariness, however, is the exception among mammals (see Greenwood, 1980, and Packer, 1979, for reviews), implying that advantages typically accrue to females that remain in the natal group or area to reproduce. For ground-dwelling sciurids, burrows are an essential resource for both survival and reproduction. By remaining in or near the natal area, females have an opportunity either to use burrows already proven adequate by the survival and reproductive success of the mother, or to locate nearby burrows readily before hibernation commences. Dispersing males, in unfamiliar terrain, must locate unused burrows or excavate new burrows that are of suitable quality (depth and drainage) to survive overwinter. Males could avoid the costs of dispersal and of inbreeding by being able to identify close female kin and not copulating with them. Selection has not favored this alternative in ground-dwelling sciurids, in part because the rearing of litters in isolation permits positive identification of only the mother and littermate sisters, and in part because the majority of females in and around the natal area are relatives of varying degrees of kinship, providing a sedentary male with few opportunities to locate unrelated females even if he were able to identify them.

Aside from avoidance of inbreeding, other potential benefits accruing to males that disperse include increased access to females and avoidance of aggression from members of the natal group. The eviction of subadult males (including sons) by a territorial male reduces intrasexual competition for females within the territory (Armitage, 1974; McLean, in press; Slobodchikoff, pers. comm.). Emigration by subadult males in response to conspecific aggression will have evolved only if the costs of resisting eviction outweigh the costs of leaving the natal area. To date, the factors promoting the evolution of sex-biased differential dispersal have not been adequately explained for any species of ground-dwelling sciurid. Potential advantages of differential dispersal by males include promotion of outbreeding (advantageous to both sexes), reproduction by females in a familiar area (primarily advantageous to females), location by males of more mates (primarily advantageous to dispersing males), reduction of intrasexual competition for mates (primarily advantageous to territorial males), and avoidance of intraspecific aggression (primarily advantageous to dispersing males).

Various automatic costs result from coloniality (Alexander, 1974) and group living will be maintained via selection only if individuals benefit more from membership than they suffer. The probable major advantage of female clumping and range overlap is improved predator detection; deterrence of predators (Stromberg, 1974) and statistical avoidance of becoming the predator's victim (see Bertram, 1978, for review) are additional advantages. Females spend less time alert when they are in larger groups and when they are centrally located (Hoogland, 1979*a*, 1981*a*; Svendsen, 1974). The prevalence of alarm calling and the tendency for callers to be females, especially reproductives with living family (Dunford, 1977*c*; Leger and Owings, 1978; Sherman, 1977; Schwagmeyer, 1980) further suggest that the prime advantages of daughter retention and female clumping are improved survival and reproductive success via early warning of impending danger. Armitage (1981) emphasized that retention of daughters can be viewed as a mechanism permitting continued reproductive investment by parents (particularly mothers), thereby increasing their reproductive success and that of the daughters.

Hoogland (1979*a*, 1981*a*) and King (1955) concluded that, for black-tailed prairie dogs, the most important benefit of coloniality is predator defense. The unrestricted access that coterie members have to all burrows (except natal burrows in spring) ensures proximity to a refuge; the clipping of vegetation and the general modification of ground cover through burrowing and feeding ensures greater visibility and rapid detection of approaching predators; alarm calling by conspecifics warns colony members of impending danger. However, females do not communally nurse pups, instead they actively defend a natal territory against coterie members, even though the offspring will mingle with juveniles of other litters in the coterie immediately following weaning. The costs of extending communal and cooperative behavior to rearing of pups have not been identified (Michener and Murie, in press). In general, communal nursing is rare among group-living mammals (Bertram, 1978) (but see Rood, this volume, for joint nursing in the dwarf mongoose, *Helogale parvula*).

Armitage (1981) noted that social grade correlates positively with delayed maturity, and that prolonged amicable associations between parents and offspring permit offspring to remain in the natal area until the age of dispersal (sons) or recruitment (daughters). Although it is appropriate that the male associated with a female group behaves so as to permit (his) offspring to remain with mothers, Armitage (1981) did not address the question of why harem systems evolved from female clusters originally. The selective advantages of the superimposition of male territories over female ranges are probably associated with greater assurance of access to estrous females and greater protection of offspring born within the territory (see Barash, 1975, for discussion of paternal behavior in the hoary marmot). For species in which males defend territories after the breeding season, the territorial male is probably the sire because of his preferential, if not exclusive, access to the females within his territory (Murie and Harris, 1978; McLean, in press). From the female's point of view it is unimportant which male controls the territory provided that the male tolerates her offspring and is successful at repelling conspecifics and reducing risks of predation. Consequently, there is a potential asymmetry in the value of male territorial behavior to males and females. Males are not assured of paternity because they do not control the movements of females in and out of the territory (Carl, 1971; Murie and Harris, 1978); the system will have evolved only if, on average, the male is actually protecting his own genetic investment rather than that of some other male, and if males defending territories have greater reproductive success than those without territories. Little information is available on alarm calling in the more social species; territorial males, because of association with (their) offspring, are expected to be more frequent callers than are males in the non-territorial systems. McLean (in press) noted that during the period when pups were in the burrow, male Arctic ground squirrels vocalized more than females, but once pups emerged both sexes called equally frequently. Hoogland (1981*b*) reported that adult and yearling male black-tailed prairie dogs with relatives in the home coterie alarm-called more often in the presence of a stuffed badger than did males without relatives.

One cost of group living may be increased exposure to intraspecific predation. Killing of conspecifics in free-living populations has been observed or imputed for various ground-dwelling sciurids (Townsend's ground squirrel [Alcorn, 1940]; Belding's ground squirrel [Sherman, 1981*b*]; Columbian ground squirrel [Holmes, 1977; McLean, 1978; Steiner, 1972]; Arctic ground squirrel [McLean, 1981, in press; Steiner, 1972]; yellow-bellied marmot [Armitage et al., 1979]). Where subsequent consumption of the victim occurs, one suggested benefit of killing and cannibalism is acquisition of food, particularly protein (Armitage et al., 1979; Holmes, 1977; Sherman, 1981*b*). Another benefit, reducing competition for space, has been proposed for the two species, Arctic and Belding's ground squirrels, in which infanticide (killing of unweaned young that have not yet left the natal burrow) has been documented (McLean, 1981, in press; Michener, in press; Sherman, 1981*b*). Because of the vulnerability of young animals to intraspecific predation, selection should favor behaviors promoting defense of progeny by the parent(s). One consequence of exclusion of (unrelated) conspecifics from the area surrounding a natal burrow is protection of infants. Active defense, via physical attack directed at the intruding male, of recently weaned juveniles by the mother or her female relatives occurs in Arctic ground squirrels. I have previously noted (Michener, 1973*b*) that immature animals are more vulnerable to attacks by adult males, sometimes leading to death, in species in which adult males are territorial than in species in which males have no prominent social position after the breeding season. This correlation between male aggression toward juveniles (and yearlings in species with delayed maturity) and sociality may extend to male-infant interactions; in *S. parryii* infanticide is typically perpetrated by adult males establishing new territories and includes attacks on recently weaned young (McLean, in press), whereas in *S. beldingi* infanticide is most frequently committed by adult females (that have lost their own young to interspecific predators) and ceases once infants reach weaning age (Sherman, 1981*b*). Although female Arctic ground squirrels defend their progeny, they are not as successful as the territorial male at evicting intruding males or preventing attacks from culminating in infant or juvenile mortality. As McLean (in press) suggested, male territorial behavior persisting through the period when infants are underground may represent paternal protection, necessary in a species in which adult male intruders can overcome the mother's defense and kill some or all of her offspring. Unresolved, in the evolutionary perspective, is whether male territoriality evolved in order to thwart infanticide or whether male-committed infanticide evolved in order to reduce competition for space (i.e., whether infanticide is a cause or a consequence of the social system).

Kin Identification

The social systems of ground-dwelling sciurids are derived from mother-daughter and littermate associations, and consequently tend to be matrilocal, matrifocal, and matrilineal. Kinship forms the basis of the social system, raising the question of how (and if) kin are identified.

Because litters are reared individually and in isolation from other litters in all species of ground-dwelling sciurids, the primary socialization of juveniles is always with the mother and littermates. Juveniles do not normally first encounter other conspecifics until some time after emerging from the natal burrow. Among species such as Richardson's ground squirrels, these encounters are both rare and agonistic during the first 2 to 3 weeks of above-ground activity, and do not result in amicable bonds forming with members of other families. Occasional exceptions occur in *S. richardsonii* when the mothers are littermate sisters living in close proximity; litters may amalgamate around the time of weaning to form a single supra-litter (G. Michener, 1972). Among black-tailed prairie dogs, encounters with conspecifics occur as early as the first day of above-ground activity, are amicable, and often result in juveniles sleeping with coterie members other than the mother and littermates. Because female black-tailed prairie dogs do not discriminate between juveniles from different litters within the coterie, Hoogland (1979*b*) concluded that they cannot distinguish their own young from those of other females in the coterie. However, there is a distinction between whether a mother *can* identify her offspring and whether she *does* identify them. Prairie-dog mothers may recognize their own juveniles but not behave preferentially toward them.

The quality of interactions occurring between conspecifics in black-tailed prairie dogs requires identification of conspecifics into one of two classes—coterie members versus non-coterie members. Typically this distinction coincides with (apparent) discrimination between close genetic relatives and distant genetic relatives. Although there are varying degrees of genetic relatedness between members of the same coterie (full-sibling, half-sibling, grandmother, granddaughter, first cousin), differential behavior based on differences in genetic relatedness has not been documented. Among Belding's and Richardson's ground squirrels, animals discriminate two classes of conspecifics that coincide with genetic relatedness; mother-offspring and littermate combinations (family) are differentiated from all other possible degrees of relatedness (non-family). For all species, family (or harem) membership, genetic relatedness, and familiarity via early socialization covary; observed patterns of social discrimination could be based on any of these factors.

In no sciurid has individual recognition been established. In all cases where discrimination between two classes of conspecifics has been demonstrated experimentally (Harris and Murie, 1982; Michener, 1973*a,* 1974; Michener and Sheppard, 1972; Sheppard and Yoshida, 1971), the most parsimonious explanation is discrimination of familiar versus unfamiliar animals. In natural contexts the class of familiar conspecifics is typically established through primary socialization in the natal burrow and socialization occurring during the first few days of above-ground activity. Should associations between non-family members occur during this period, recognition "errors" result; that is, an individual behaves toward a non-relative in the same manner it normally does toward its closest genetic relatives. Such "errors" have been documented in Richardson's and Belding's ground squirrels (G. Michener, 1972; Sherman, 1980). Newly emerged *S. beldingi* pups occasionally inadvertently attach themselves to a neighboring litter and subsequently treat former littermates (siblings) and mother as unfamiliar (unrelated) animals. If a *S. richardsonii* pup, up to at least 16 days old, is fostered to a lactating female in the field or laboratory, subsequent social interactions between the weaned juvenile and its foster mother are amicable, whereas those with the biological mother are agonistic. Lactating Richardson's ground squirrels do not discriminate between pups, retrieving both their own and others, but discrimination does occur by the time young are 5 weeks old (G. Michener, 1972, 1974). Pups as young as 20 to 24 days old and with unopened eyes respond differentially to the mother and to an unfamiliar lactating female (G. Michener, 1974) but, in the field, newly emerged pups sometimes fail to identify an intruding adult when that adult is in an area normally used by the mother. Within 2 to 3 weeks of emergence, juveniles correctly distinguish between the mother and all other adults regardless of the location of the interaction (Michener, 1973*a*, 1981). These various observations suggest that, in natural contexts, squirrels identify close relatives on the basis of familiarity established via amicable interactions with these relatives during the first 4 to 6 weeks of life.

Incorporation of those conspecifics that are frequently encountered in amicable contexts during the first few days of above-ground activity into the same class as mother and littermates, would account for the inclusion of coterie members in black-tailed prairie dogs and of older non-littermate sisters in Belding's ground squirrels into the class of preferred conspecifics. The social context within which such extrafamilial conspecifics are encountered influences the ultimate significance of the encounter. For instance, juvenile *S. beldingi* only behave toward older sisters as they do toward littermates if they encounter them in the presence of the common mother (Sherman, 1980). The extent to which adult prairie dogs are necessary to facilitate litter mixing and amicableness within a coterie is not known.

The coincidence between genetic relatedness and familiarity is reduced if there is multiple siring of litters in polygynous species because half-siblings could either be littermates (same mother, different fathers) or non-littermates (same father, different mothers). Female Belding's ground squirrels copulate with more than one male, and resulting litters are typically multiply-sired (Sherman, 1977; Sherman and Morton, 1979; Hanken and Sherman, 1981). Although less common, female black-tailed prairie dogs sometimes copulate with more than one male (Hoogland, 1981*b*), where the second male is from an adjacent coterie. Assuming that differential favoritism is based solely on familiarity, littermate half-siblings are expected to interact as do littermate full siblings whereas non-littermate half siblings are expected to interact as do totally unfamiliar conspecifics. If this prediction is not fulfilled, a mechanism other than, or in addition to, familiarity through primary socialization must be operating to enable unfamiliar relatives to identify each other.

Ecological Correlates

Figure 3 outlines the ranges of the species of ground-dwelling sciurids that have been discussed in detail, according to the grade of social structure. Only the more solitary species occur predominantly in woodland-field ecotones; the other species inhabit more open areas varying from short-grass prairie to alpine meadow. This difference suggests that increased visibility, both to conspecifics and to predators, may be associated with the formation of social groups. Barash (1974) and Kivett et al. (1976) suggest that group cohesion is associated with increased harshness of the environment. This trend is exhibited among species of *Marmota,* where the increasing sociality from woodchucks (low elevation) to yellow-bellied marmots (medium elevation) to Olympic marmots (high elevation) correlates with the decreasing length of the vegetative growing season and, consequently, with delayed age of dispersal and of maturation (Barash, 1973*a,* 1974). However, similar direct relationships with either altitude or latitude are not obvious among ground squirrels or prairie dogs. Slobodchikoff (pers. comm.) notes that among three species of prairie dogs (white-tailed, *Cynomys leucurus*; Gunnison's, *C. gunnisoni*; and black-tailed, *C. ludovicianus*) increasing sociality correlates with decreasing altitude, the opposite trend to that reported for marmots.

Armitage (1981) stated that the more social species of ground-dwelling sciurids are those in which the combination of large body size and a relatively short growing season results in delayed ages of dispersal and attainment of sexual maturity. As Table 1 indicates, the least social species (grades 1 and 2) are, with the exception of the woodchuck, small- to medium-sized animals that breed as yearlings and that have active seasons ranging from short to long. In the higher social grades, in which adult males play a prominent role (grades 3, 4, and 5), are species that first reproduce as yearlings, 2-year-olds, and 3-year-olds. Although these social species are all large sciurids, their active seasons range from short, to long, to continuous. What characterizes the social species is the prolonged association between adults and subadults. High adult : subadult sea-

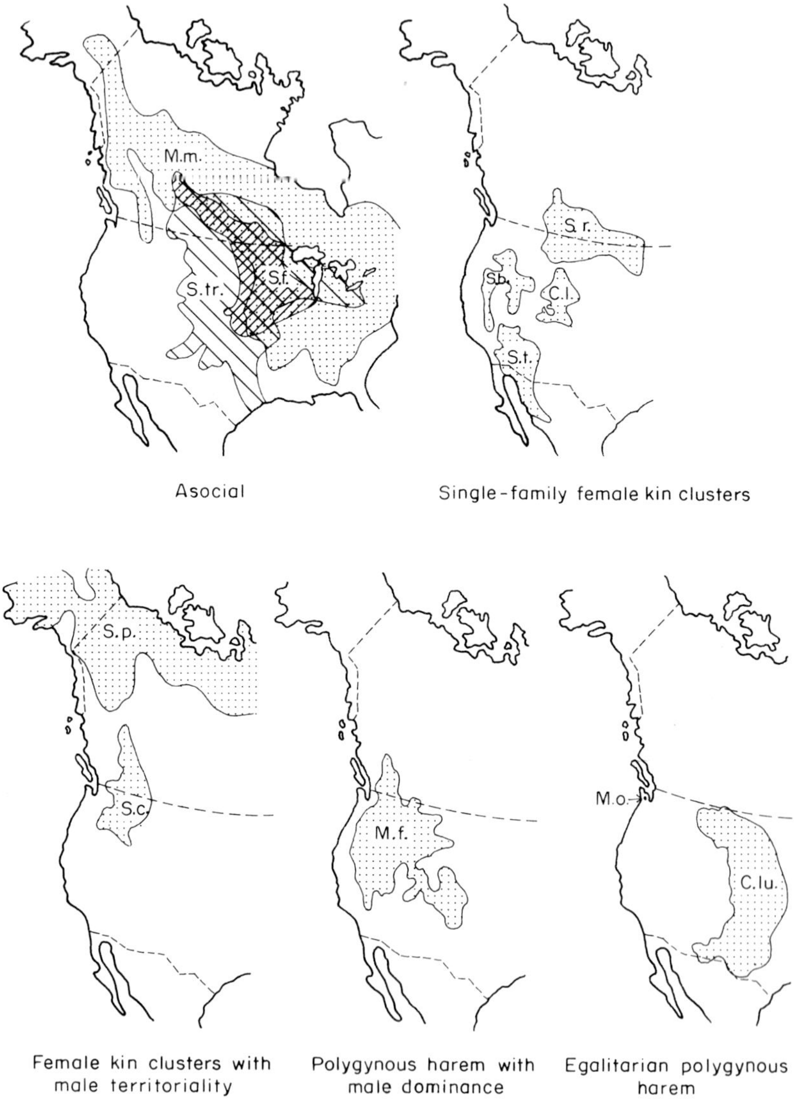

FIG. 3. Range maps of selected species of ground-dwelling sciurids grouped according to the grade of social structure exhibited by the species (maps modified from Hall and Kelson, 1959). Asocial: M.m., *Marmota monax,* S.f., *Spermophilus franklinii*; S.tr., *S. tridecemlineatus.* Single-family female kin clusters: S.r., *S. richardsonii*; S.b., *S. beldingi*; S.t., *S. tereticaudus*; C.l., *Cynomys leucurus.* Female kin clusters with male territoriality: S.p., *S. parryii*; S.c., *S. columbianus.* Polygynous harem with male dominance: M.f., *M. flaviventris.* Egalitarian polygynous harem: M.o., *M. olympus*; C.lu., *C. ludovicianus.*

sonal coincidence is achieved either by delayed adult immergence which prolongs the active season (for example, Arctic ground squirrel), by delayed age of sexual maturity which prolongs subadulthood over several years (for example, Olympic marmot), or by year-round activity (for example, black-tailed prairie dog). Among the 18 species listed in Table 1, the typical age at which females first reproduce (FAFR) is a good predictor of social grade, accounting for 70% of the variation in social grade (n = 18, r = 0.838). However, among the seven species listed with social grades higher than 2, adult : subadult seasonal coincidence (taken as the mid-point value of the range given) is a better predictor of social grade (r = .915, $P < .01$, variance explained = 84%) than is FAFR (r = .719, $P >$ .05, variance explained = 52%).

Because social species are characterized by large size and by synchrony in the active seasons of various cohorts, whereas relatively asocial species are characterized by early immergence of adults, especially males, it is relevant to know why the circannual cycle is asynchronous among cohorts in the smaller species. In those species with strongly staggered patterns of immergence of the different cohorts, juveniles continue above-ground activity and gain weight for many weeks after the disappearance of adults, implying that adult immergence is not precipitated by inadequate food supplies or inclement weather. Clark (1977) and Yeaton (1972) suggested that early immergence of adults is an adaption to reduce intraspecific competition, particularly competition for food between males and juveniles, but I reject this hypothesis for two reasons. First, early immergence by a male should be favored only if the juveniles that profit from the male's absence are his own progeny; that is, if a male resides close to his progeny. As discussed above, males often do not remain close to their mate(s) and offspring. Second, the disappearance of males is unlikely to significantly increase the food available to juveniles. Among Richardson's ground squirrels, for instance, which have an adult sex ratio of about three females per male and litter size at weaning of four or five pups (Michener, 1979*a*, 1979*c*; Michener and Michener, 1971; I recorded an average litter size at weaning of 9.1, range 7–14, for 13 litters on a pasture in southern Alberta in 1981), there are at least 12 to 15 juveniles in the population for every adult male. The disappearance of the male can have only a negligible effect in providing additional food to these juveniles. Furthermore, adult male *S. richardsonii* are socially subordinate to juveniles when they intrude into the juvenile's core area (Michener, pers. observ.), so juveniles are unlikely to be limited, either directly or indirectly, from access to food by males. Early male immergence does not benefit juveniles; the short active season of adult males is more likely attributable to survival of the male. Male *S. richardsonii* become obese during June, when daily temperatures approach the maximum for the year, and this increasing size renders them less agile and less able to squeeze down small bolt holes (pers. observ.). Fat males are presumably vulnerable to predation, hyperthermia, and water stress; early immergence represents a compromise between further activity permitting additional weight gain but increasing mortality from predation and environmental stress. Provided that the decreased mortality associated with early immergence more than compensates for the increased mortality associated with the longer period of overwinter dependence on fat stores, the short active season will be favored by natural selection. Morton (1975) suggested that, for Belding's ground squirrels, early entry into hibernation is related to predator avoidance.

Adult male Arctic ground squirrels are atypical in that they do not immerge early. In contrast to female *S. parryii* and to members of the other hibernating species, male Arctic ground squirrels hoard food and they lose relatively little body weight over winter (Green, 1977; McLean and Towns, 1981). By providing energy for thermogenesis in the form of food stores in the hibernaculum, males may be avoiding some of the costs of large and ungainly fat depots, thereby allowing them to prolong the active season.

Some of the apparent inconsistencies in correlating sociality and environment arise from the difficulty of defining "environmental harshness" (see Webb, 1980, 1981). Alpine and subalpine habitats are typically considered severe because of the shortness of the growing season and the possibility of snow through much of the year. However, such habitats are usually moist and experience a relatively narrow range of temperature extremes. Water stress, an environmental hazard unlikely to face high elevation species, is a potential threat to species inhabiting short-grass prairie. Blake (1977) and Davis (1976) suggested that seasonal hypothermia may be an adaptation to stressful conditions imposed by lack of water. Differing abilities to withstand water stress are related to the differences in hibernation patterns between white-tailed and black-tailed prairie dogs (Bakko, 1977; Pfeiffer et al., 1979). White-tailed prairie dogs, unable to tolerate water loss, are obligate hibernators whereas black-tailed prairie dogs, adapted to withstand dehydration, remain active year-round, permitting continuous social contact.

In summary, increased social tolerance correlates with increased coincidence of the active periods of various cohorts via prolonged active seasons or shortened active seasons associated with delayed dispersal. Consequently, the timing of the circannual cycle and, ultimately, the environmental parameters that have determined that timing, are factors that have promoted the evolution of sociality in ground-dwelling sciurids.

Areas for Future Investigation

Our understanding of ground-dwelling sciurid systems is currently limited by the lack of information on paternity. Not only is paternity difficult to establish but, for most species, it is not known whether females copulate with several males and, if they do, whether they produce multiply-sired litters. For the species whose social organization is based upon single-family matrilineal kin groups, the male residing nearest to a litter cannot be assumed to be the father because males move extensively during and after the breeding season. For those species in which adult males are territorial in spring beyond the breeding season, males may be associated with their mate(s) and progeny (Foltz and Hoogland, 1981). Until paternity testing is performed, the supposition that a territorial male is the father of the juveniles that emerge within his territory is only speculative.

A potential cost of the greater sedentariness of males in species with territoriality is inbreeding. In the yellow-bellied marmot, however, adult males rarely reside for more than 2 to 3 years in a given harem and because females do not commence reproduction until at least 2 years old, often not until 3 years of age, a male is unlikely to mate with a daughter. Black-tailed prairie dogs typically first breed as 2-year-olds, although yearlings have been reported to successfully reproduce in newly established colonies (Franklin, pers. comm.; Halpin, pers. comm.). Because females are sedentary and usually reproduce in the coterie of birth, inbreeding could be avoided by male emigration. Yearling males do disperse and so are unlikely to reproduce with sisters. Hoogland (1981*b*, 1982) suggested that adult males usually reside for a maximum of 2 years in a given coterie and hence are no longer present when daughters achieve reproductive maturity. Long term studies are necessary with more species to establish whether inbreeding is avoided through dispersal of adult males in the long-lived species with territorial males.

Familiarity and maternal kinship co-vary. Familiarity through primary socialization is likely to be the proximate mechanism by which kin selection can operate. Cross-fostering experiments are the most valuable technique for partitioning out the relative contributions of postnatal familiarity and kinship to observed social patterns. Such experiments are currently being conducted with Arctic and Belding's ground squirrels (Holmes and Sherman, in press, pers. comm.), and Richardson's ground squirrels (Davis, in press, pers. comm.). Preliminary indications are that, for juvenile *S. parryii* and yearling *S. beldingi,* animals reared together rarely interact agonistically, regardless of their genetic relatedness and sex, suggesting that familiarity typically plays a major role in sibling recognition in natural contexts. However, among animals reared apart siblings are less aggressive when subsequently paired than are non-siblings. This effect is greater for female-female pairs than for mixed sex or male-male pairs, and suggests that sisters do not necessarily require lengthy post-natal association to be able to identify each other. Cross-fostered juvenile *S. richardsonii* also exhibit differential responses to unfamiliar siblings versus unfamiliar non-siblings; siblings that were separated on the day of birth approach and touch each other more frequently and remain in greater proximity than do unfamiliar non-siblings. Because littermates share a common uterine and post-natal environment, prior association could account for the apparent ability of siblings to identify each other even when reared apart from within a few hours of birth. Female *S. beldingi* frequently copulate with more than one male, producing multiply-sired litters (Hanken and Sherman, 1981); interactions among yearling female littermate half-siblings are more aggressive than those between littermate full-siblings, suggesting an ability to distinguish between differing degrees of genetic relatedness even when reared together from conception (Holmes and Sherman, in press). Further cross-fostering experiments with more species will assist in assessing the relative significance of early social experience as an underlying mechanism of kin identification, kin preference, and, ultimately, of social structure.

Quantification of rates of social interactions in natural contexts between conspecifics of varying degrees of kinship need to be augmented and improved (see Michener, 1980*c*). Because kin are likely to be in greater physical proximity than non-kin, measurements of interaction rates need to be corrected for the likelihood that different classes of animals will be physically able to interact. To date, comparisons of interaction rates in the field among conspecifics of varying genetic relatedness are limited (Dunford, 1977*b*; Michener, 1973*a*, 1981; Sherman, 1980; Yeaton, 1972).

Attempts to relate social structure to species ecology require additional information on the timing of the circannual cycle under natural conditions, on the advantages of short active seasons, and on the reasons for age and sex differences in timing of immergence. Intraspecific comparisons of populations located in different habitats, and interspecific comparisons involving data from more of the species of ground-dwelling sciurids would contribute to interpretation of correlations between environment and socioecology. Valuable comparisons could be made of the socioecology and physiological ecology of Richardson's ground squirrels, white-tailed prairie dogs, and black-tailed prairie dogs. Although the latter two species are closely related phylogenetically (Nadler et al., 1971), their social systems diverge. The white-tailed prairie dog social system and circannual cycle correspond with those of the Richardson's ground squirrel despite differences in habitat and despite greater phylogenetic divergence. Richardson's ground squirrels and black-tailed prairie dogs are sympatric over portions of their ranges but they show different adaptations to xeric conditions (obligate hibernation and tolerance of water stress, respectively) and have different social systems.

Summary

The five proposed grades of social structure—asocial, single-family female kin cluster, female kin cluster with male territoriality, polygynous harem with male dominance, egalitarian polygynous harem—form a continuum of sociality based on the mother-headed family as the most fundamental social subunit. Sex-biased differential dispersal results in clustering of female kin in all species. Prolonged social association between mother and daughters and between littermate sisters, increasing overlap of the ranges used by such female kin, and relaxation of social distinctions between the litters of adjacent female kin have led to multiple-family female kin clusters that communally share a common range. The male-female pair bond is not a fundamental social unit in ground-dwelling sciurid social organization, and the association of males with clusters of female kin probably commenced as an incidental result of males and females independently choosing to live in the same preferred habitat. Increased probability of guaranteed access to estrous females and the improved survival of offspring living under the indirect protection of their father would promote male territoriality and the establishment of uni-male harems. The simultaneous development of female cohesiveness and male territoriality account for the integrated social structure of Olympic marmots and black-tailed prairie dogs.

The prime forces determining the trends in social organization are hypothesized to be increased protection from both intra- and interspecific predation and the avoidance of inbreeding. Sex-biased differential dispersal is the proximate mechanism that leads to formation of female kin clusters. It also promotes outbreeding while permitting females to hibernate and reproduce within a familiar locale already proven successful by the survival and reproductive success of the mother. The clumping of female kin permits improved survival and reproductive success via increased group vigilance, use of alarm calls to warn of danger, sharing of resources, and reduced time spent in establishing and maintaining natal territories. The association of a male (presumably the father) with female kin groups further enhances survival and reproductive success by protection of resources within the territory, by protection from intraspecific aggression, and by increased group vigilance.

Although the social structure of the ground-dwelling sciurids is based upon genetic relatedness between females, the observed patterns of differential behavior to kin and non-kin do not require that animals be able to identify kin as such. Because litters are reared in isolation from all conspecifics, the primary socialization is always with the mother and littermates. For some species these relations compose the group of conspecifics that is distinguished from all other animals. In the more social species, where juveniles engage in amicable contacts with conspecifics other than the immediate family in the days following entry into the above-ground population, the preferred group of conspecifics expands to include these individuals. Typically the additional individuals will be close kin (aunts, non-littermate sisters, cousins). In either situation it is the animals that have been met in amicable social situations during the first 4 to 6 weeks of life, and hence are familiar, that constitute the favored group. Thus kinship and familiarity covary, and familiarity through early socialization is an adequate proximate mechanism for distinguishing between kin and non-kin classes. The assumption that differential behavior to kin and non-kin is based on familiarity is amenable to testing through cross-fostering experiments and through comparison of the interactions between littermate half-siblings and non-littermate half-siblings.

Degree of sociality does not correlate meaningfully with either latitude or altitude. The two most social sciurid species, Olympic marmots and black-tailed prairie dogs, occur in divergent habitats (alpine meadow and short-grass prairie). Within a given habitat, such as subalpine meadow, social systems range from single-family kin clusters (Belding's ground squirrels), kin clusters with male territoriality (Columbian ground squirrels), to polygynous harems with male dominance (yellow-bellied marmots). Despite these apparent inconsistencies, social structure does appear to have ecological correlates if the timing and patterns of circannual rhythmicity are assumed to be evolutionary adaptations to environmental exigencies. The least socially tolerant species mature as yearlings and have the least amount of coincidence of above-ground activity between adult cohorts and immature cohorts, whereas the most tolerant species have extensive overlap in the active seasons of adults and immatures. Sociality is related to prolonged active seasons for adults and/or to increased number of seasons required for juveniles to attain maturity. Ultimate understanding of the socioecology of ground-dwelling sciurids requires knowledge of the selective factors that have determined the nature of the circannual cycle, and that have thereby determined the degree of coincidence in the seasonal activity of various cohorts.

Acknowledgments

Financial assistance was provided by the Department of Zoology, University of Alberta, by Dr. J. O. Murie via NSERC grant A-5865, and by the Friends of the National Zoo. Secretarial assistance was provided by the Department of Biological Sciences, University of Lethbridge and by EPO Biology, University of Colorado.

The following people generously took time to read and comment on an earlier version of the manuscript and to provide information on their current research with sciurids: Z. T. Halpin, J. O. Murie, I. G. McLean, W. G. Holmes, and M. A. Harris. I thank my son, Rory, and my husband, Dan, for providing me with the uninterrupted time in which to pursue my interest in sciurid social organization.

Literature Cited

ALCORN, J. R. 1940. Life history notes on the Piute ground squirrel. J. Mamm., 21:160–170.

ALEXANDER, R. D. 1974. The evolution of social behavior. Ann. Rev. Ecol. Syst., 5:325–383.

ANTHONY, M. 1962. Activity and behavior of the woodchuck in southern Illinois. Occas. Papers C. C. Adams Center Ecol. Stud., 6:1–25.

ARMITAGE, K. B. 1973. Population changes and social behavior following colonization by the yellow-bellied marmot. J. Mamm., 54:842–845.

———. 1974. Male behavior and territoriality in the yellow-bellied marmot. J. Zool., 72:233–265.

———. 1975. Social behavior and population dynamics of marmots. Oikos, 26: 341–354.

———. 1977. Social variety in the yellow-bellied marmot: a population-behavioural system. Anim. Behav., 25:585–593.

———. 1981. Sociality as a life-history tactic of ground squirrels. Oecologia, 48: 36–49.

ARMITAGE, K. B., AND J. F. DOWNHOWER. 1974. Demography of yellow-bellied marmot populations. Ecology, 55:1233–1245.

ARMITAGE, K. B., J. F. DOWNHOWER, AND G. E. SVENDSEN. 1976. Seasonal changes in weights of marmots. Amer. Midland Nat., 96:36–51.

ARMITAGE, K. B., D. JOHNS, AND D. C. ANDERSEN. 1979. Cannibalism among yellow-bellied marmots. J. Mamm., 60:205–207.

BAILEY, E. D. 1965. The influence of social interaction and season on weight change in woodchucks. J. Mamm., 46:438–445.

BAKKO, E. G. 1977. Field water balance performance in prairie dogs *Cynomys leucurus* and *C. ludovicianus*. Comp. Biochem. Physiol., 56A:443–451.

BARASH, D. P. 1973*a*. The social biology of the Olympic marmot. Anim. Behav. Monogr., 6:173–245.

———. 1973*b*. Social variety in the yellow-bellied marmot (*Marmota flaviventris*). Anim. Behav., 21:579–584.

———. 1974. The evolution of marmot societies: a general theory. Science, 185: 415–420.

———. 1975. Ecology of paternal behavior in the hoary marmot (*Marmota caligata*): an evolutionary interpretation. J. Mamm., 56:613–618.

BEKOFF, M. 1977. Mammalian dispersal and the ontogeny of individual behavioral phenotypes. Amer. Nat., 111:715–732.

BERTRAM, B. C. R. 1978. Living in groups: predators and prey. Pp. 64–96, *in* Behavioural ecology: an evolutionary approach (J. R. Krebs and N. B. Davies, eds.). Sinauer, Sunderland, Massachusetts, 494 pp.

BETTS, B. J. 1976. Behaviour in a population of Columbian ground squirrels, *Spermophilus columbianus columbianus*. Anim. Behav., 24:652–680.

BICKFORD, C. E. 1979. Aspects of the social structure of the California ground squirrel (*Spermophilus beecheyi*) in western Oregon. Unpubl. M.Sc. thesis, Oregon State Univ., Corvallis, 56 pp.

BLAKE, B. H. 1977. The effects of kidney structure and the annual cycle of water requirements in golden-mantled ground squirrels and chipmunks. Comp. Biochem. Physiol., 58A:413–419.

BOAG, D. A., AND J. O. MURIE. 1981*a*. Weight in relation to sex, age, and season in Columbian ground squirrels (Sciuridae: Rodentia). Canadian J. Zool., 59:999–1004.

———. 1981*b*. Population ecology of Columbian ground squirrels in southwestern Alberta. Canadian J. Zool., 59:2230–2240.

BRONSON, F. H. 1963. Some correlates of interaction rate in natural populations of woodchucks. Ecology, 44:637–643.

———. 1964. Agonistic behaviour in woodchucks. Anim. Behav., 12:470–478.

BROWN, J. L 1975. The evolution of behavior. Norton, New York, 761 pp.

CAMERON, D. M. 1969. The population biology and interactions of two sciurid rodents (*Eutamias amoenus* and *Citellus lateralis*) in northeastern California. Unpubl. Ph.D. dissert., Univ. California, Davis, 500 pp.

CARL, E. A. 1971. Population control in Arctic ground squirrels. Ecology, 52:395–413.

CLARK, T. W. 1970. Richardson's ground squirrel (*Spermophilus richardsonii*) in the Laramie Basin, Wyoming. Great Basin Nat., 30:55–70.

———. 1977. Ecology and ethology of the white-tailed prairie dog (*Cynomys leucurus*). Publ. Biol. Geol., Milwaukee Public Mus., 3:1–97.

CLUTTON-BROCK, T. H. 1974. Primate social organisation and ecology. Nature, 250:539–542.

CROOK, J. H., J. E. ELLIS, AND J. D. GOSS-CUSTARD. 1976. Mammalian social systems: structure and function. Anim. Behav., 24:261–274.

DAVIS, D. E. 1976. Hibernation and circannual rhythms of food consumption in marmots and ground squirrels. Quart. Rev. Biol., 51:477–514.

DAVIS, L. S. In press. Sibling recognition in Richardson's ground squirrels (*Spermophilus richardsoni*). Behav. Ecol. Sociobiol.

DOBSON, F. S. 1979. An experimental study of dispersal in the California ground squirrel. Ecology, 60:1103–1109.

DORRANCE, M. J. 1974. The annual cycle and population dynamics of Richardson's ground squirrels. Unpubl. Ph.D. dissert., Univ. Wisconsin, Madison, 150 pp.

DOWNHOWER, J. F., AND K. B. ARMITAGE. 1971. The yellow-bellied marmot and the evolution of polygamy. Amer. Nat., 105:355–370.

———. 1980. Dispersal of yearling yellow-bellied marmots. Anim. Behav. Soc. Ann. Meet. (Abstr.)

DUNFORD, C. 1977*a*. Social system of round-tailed ground squirrels. Anim. Behav., 25:885–906.

———. 1977*b*. Behavioral limitation of round-tailed ground squirrel density. Ecology, 58:1254–1268.

———. 1977*c*. Kin selection for ground squirrel alarm calls. Amer. Nat., 111: 782–785.

EISENBERG, J. F., N. A. MUCKENHIRN, AND R. RUDRAN. 1972. The relation between ecology and social structure in primates. Science, 176:863–874.

EMLEN, S. T., AND L. W. ORING. 1977. Ecology, sexual selection, and the evolution of mating systems. Science, 197:215–223.

EVANS, F. C. 1951. Notes on a population of the striped ground squirrel (*Citellus tridecemlineatus*) in an abandoned field in southeastern Michigan. J. Mamm., 32:437–449.

Ferron, J. 1977. Le comportement de marquage chez le spermophile á mante dorée (*Spermophilus lateralis*). Nat. Canadien, 104:407–418.

Festa-Bianchet, M. 1981. Reproduction in yearling female Columbian ground squirrels (*Spermophilus columbianus*). Canadian J. Zool., 59:1032–1035.

Festa-Bianchet, M., and D. A. Boag. 1982. Territoriality in adult female Columbian ground squirrels. Canadian J. Zool., 60:1060–1066.

Fitch, H. S. 1948. Ecology of the California ground squirrel on grazing lands. Amer. Midland Nat., 39:513–596.

Fitzgerald, J. P., and R. R. Lechleitner. 1974. Observations on the biology of Gunnison's prairie dog in central Colorado. Amer. Midland Nat., 92: 146–163.

Foltz, D. W., and J. L. Hoogland. 1981. Analysis of the mating system in the black-tailed prairie dog (*Cynomys ludovicianus*) by likelihood of paternity. J. Mamm., 62:706–712.

Green, J. E. 1977. Population dynamics and annual cycles of activity and dispersal in the Arctic ground squirrel. Unpubl. M.Sc. thesis, Univ. British Columbia, Vancouver, 193 pp.

Greenwood, P. J. 1980. Mating systems, philopatry and dispersal in birds and mammals. Anim. Behav., 28:1140–1162.

Grizzell, R. A. 1955. A study of the southern woodchuck, *Marmota monax monax*. Amer. Midland Nat., 53:257–293.

Hall, E. R., and K. R. Kelson. 1959. The mammals of North America. Ronald Press, New York, 1:1–546 + *79*.

Hanken, J., and P. W. Sherman. 1981. Multiple paternity in Belding's ground squirrel litters. Science, 212:351–353.

Harris, M. A., and J. O. Murie. 1982. Responses to oral gland scents from different males. Anim. Behav., 30:140–148.

Hayes, S. R. 1977. Home range of *Marmota monax* (Sciuridae) in Arkansas. Southwestern Nat., 22:547–550.

Hight, M. E., M. Goodman, and W. Prychodko. 1974. Immunological studies of the Sciuridae. Syst. Zool., 23:12–25.

Holmes, W. G. 1977. Cannibalism in the Arctic ground squirrel (*Spermophilus parryii*). J. Mamm., 58:437–438.

Holmes, W. G., and P. W. Sherman. In press. The ontogeny of kin recognition in two species of ground squirrels. Amer. Zool.

Hoogland, J. L. 1979*a*. The effect of colony size on individual alertness of prairie dogs (Sciuridae: *Cynomys* spp.). Anim. Behav., 27:394–407.

———. 1979*b*. Aggression, ecoparasitism, and other possible costs of prairie dog (Sciuridae: *Cynomys* spp.) coloniality. Behaviour, 69:1–35.

———. 1981*a*. The evolution of coloniality in white-tailed and black-tailed prairie dogs (Sciuridae: *Cynomys leucurus* and *C. ludovicianus*). Ecology, 62:252–272.

———. 1981*b*. Nepotism and cooperative breeding in the black-tailed prairie dog (Sciuridae: *Cynomys ludovicianus*). Pp. 283–310, *in* Natural selection and social behavior: recent research and new theory (R. D. Alexander and D. W. Tinkle, eds.). Chiron Press, New York, 532 pp.

———. 1982. Prairie dogs avoid extreme inbreeding. Science, 215:1639–1641.

Iverson, S. L., and B. N. Turner. 1972. Natural history of a Manitoba population of Franklin's ground squirrel. Canadian Field-Nat., 86:145–149.

Johns, D. W., and K. B. Armitage. 1979. Behavioral ecology of the alpine yellow-bellied marmot. Behav. Ecol. Sociobiol., 5:133–157.

Kilgore, D. L., and K. B. Armitage. 1978. Energetics of yellow-bellied marmot populations. Ecology, 59:78–88.

KING, J. A. 1955. Social behavior, social organization, and population dynamics in a black-tailed prairiedog town in the Black Hills of North Dakota. Contrib. Lab. Vert. Biol., Univ. Michigan, 67:1–123.

KIVETT, V. K., J. O. MURIE, AND A. L. STEINER. 1976. A comparative study of scent-gland location and related behavior in some northwestern Nearctic ground squirrel species (Sciuridae): an evolutionary approach. Canadian J. Zool., 54:1294–1306.

KNOPF, F. L., AND D. F. BALPH. 1977. Annual periodicity of Uinta ground squirrels. Southwestern Nat., 23:213–224.

KOFORD, C. B. 1958. Prairie dogs, whitefaces, and blue grama. Wildl. Monogr., 3:1–78.

LEGER, D. W., AND D. H. OWINGS. 1978. Responses to alarm calls by California ground squirrels: effects of call structure and maternal status. Behav. Ecol. Sociobiol., 3:177–186.

LOEHR, K. A., AND A. C. RISSER. 1977. Daily and seasonal activity patterns of the Belding ground squirrel in the Sierra Nevada. J. Mamm., 58:445–448.

MCCARLEY, H. 1966. Annual cycle, population dynamics and adaptive behavior of *Citellus tridecemlineatus.* J. Mamm., 47:294–316.

MCLEAN, I. G. 1978. Plugging of nest burrows by female *Spermophilus columbianus.* J. Mamm., 59:437–439.

———. 1981. Social ecology of the Arctic ground squirrel, *Spermophilus parryii.* Unpubl. Ph.D. thesis, Univ. Alberta, Edmonton, 149 pp.

———. 1982. The association of female kin in the Arctic ground squirrel *Spermophilus parryii.* Behav. Ecol. Sociobiol., 10:91–99.

———. In press. Paternal behavior and killing of young in Arctic ground squirrels. Anim. Behav.

MCLEAN, I. G., AND A. J. TOWNS. 1981. Differences in weight changes and the annual cycle of male and female Arctic ground squirrels. Arctic, 34:249–254.

MERRIAM, H. G. 1971. Woodchuck burrow distribution and related movement patterns. J. Mamm., 52:732–746.

MICHENER, D. R. 1972. Notes on home range and social behavior in adult Richardson's ground squirrels (*Spermophilus richardsonii*). Canadian Field-Nat., 86:77–79.

———. 1974. Annual cycle of activity and weight changes in Richardson's ground squirrel, *Spermophilus richardsonii.* Canadian Field-Nat., 88:409–413.

MICHENER, D. R., AND G. R. MICHENER. 1971. Sex ratio and interyear residence in a population of *Spermophilus richardsonii.* J. Mamm., 53:853.

MICHENER, G. R. 1972. Social relationships between adult and young in Richardson's ground squirrel *Spermophilus richardsonii richardsonii.* Unpubl. Ph.D. dissert., Univ. Saskatchewan, Regina, 256 pp.

———. 1973*a*. Field observations on the social relationships between adult female and juvenile Richardson's ground squirrels. Canadian J. Zool., 51:33–38.

———. 1973*b*. Intraspecific aggression and social organization in ground squirrels. J. Mamm., 54:1001–1003.

———. 1974. Development of adult-young identification in Richardson's ground squirrel. Develop. Psychobiol., 7:375–384.

———. 1977*a*. Effect of climatic conditions on the annual activity and hibernation cycles of Richardson's ground squirrels and Columbian ground squirrels. Canadian J. Zool., 55:693–703.

———. 1977*b*. Gestation period and juvenile age at emergence in Richardson's ground squirrel. Canadian Field-Nat., 91:410–413.

———. 1978. Effect of age and parity on weight gain and entry into hibernation in Richardson's ground squirrel. Canadian J. Zool., 56:2573–2577.

———. 1979*a*. The circannual cycle of Richardson's ground squirrels in southern Alberta. J. Mamm., 60:760–768.

———. 1979*b*. Spatial relationships and social organization of adult Richardson's ground squirrels. Canadian J. Zool., 57:125–139.

———. 1979*c*. Yearly variations in the population dynamics of Richardson's ground squirrels. Canadian Field-Nat., 93:363–370.

———. 1980*a*. Estrous and gestation periods in Richardson's ground squirrels. J. Mamm., 61:531–534.

———. 1980*b*. Differential reproduction among female Richardson's ground squirrels and its relation to sex ratio. Behav. Ecol. Sociobiol., 7:173–178.

———. 1980*c*. The measurement and interpretation of interaction rates: an example with Richardson's ground squirrels. Biol. Behav., 5:371–384.

———. 1981. Ontogeny of spatial distribution and social behavior in juvenile Richardson's ground squirrels. Canadian J. Zool., 59:1666–1676.

———. In press. Infanticide in ground squirrels. Anim. Behav.

MICHENER, G. R., AND D. R. MICHENER. 1973. Spatial distribution of yearlings in a Richardson's ground squirrel population. Ecology, 54:1138–1142.

———. 1977. Population structure and dispersal in Richardson's ground squirrels. Ecology, 58:359–368.

MICHENER, G. R., AND J. O. MURIE. In press. Black-tailed prairie dog coteries: are they cooperative breeding units? Amer. Nat.

MICHENER, G. R., AND D. H. SHEPPARD. 1972. Social behavior between adult female Richardson's ground squirrels (*Spermophilus richardsonii*) and their own and alien young. Canadian J. Zool., 50:1343–1349.

MOORE, J. C. 1961. The spread of existing diurnal squirrels across the Bering and Panamanian land bridges. Amer. Mus. Novitates, 2044:1–26.

MORTON, M. L. 1975. Seasonal cycles of body weights and lipids in Belding ground squirrels. Bull. So. California Acad. Sci., 74:128–142.

MORTON, M. L., AND R. J. PARMER. 1975. Body size, organ size, and sex ratios in adult and yearling Belding ground squirrels. Great Basin Nat., 35:305–309.

MORTON, M. L., C. S. MAXWELL, AND C. E. WADE. 1974. Body size, body composition, and behavior of juvenile Belding ground squirrels. Great Basin Nat., 34:121–134.

MURIE, J. O. 1973. Population characteristics and phenology of a Franklin ground squirrel (*Spermophilus franklinii*) colony in Alberta, Canada. Amer. Midland Nat., 90:334–340.

———. 1980. Social interactions and dominance relationships between female and male Columbian ground squirrels. Amer. Zool., 20:827.

MURIE, J. O., AND M. A. HARRIS. 1978. Territoriality and dominance in male Columbian ground squirrels (*Spermophilus columbianus*). Canadian J. Zool., 56:2402–2412.

———. 1982. Annual variation of spring emergence and breeding in Columbian ground squirrels (*Spermophilus columbianus*). J. Mamm., 63:431–439.

MURIE, J. O., D. A. BOAG, AND V. K. KIVETT. 1980. Litter size in Columbian ground squirrels (*Spermophilus columbianus*). J. Mamm., 61:237–244.

NADLER, C. F., R. S. HOFFMANN, AND J. J. PIZZIMENTI. 1971. Chromosomes and serum proteins of prairie dogs and a model of *Cynomys* evolution. J. Mamm., 52:545–555.

NEAL, B. J. 1965. Reproductive habits of round-tailed and Harris antelope ground squirrels. J. Mamm., 46:200–206.

NOWICKI, S., AND K. B. ARMITAGE. 1980. Behavior of juvenile yellow-bellied marmots: play and social integration. Z. Tierpsychol., 51:85–105.

ORIANS, G. H. 1969. On the evolution of mating systems in birds and mammals. Amer. Nat., 103:589–603.

OWINGS, D. H., M. BORCHERT, AND R. VIRGINIA. 1977. The behaviour of California ground squirrels. Anim. Behav., 25:221–230.

PACKER, C. 1979. Inter-troop transfer and inbreeding avoidance in *Papio anubis*. Anim. Behav., 27:1–36.

PFEIFER, S. 1980. Demographic and behavioral influences on juvenile Wyoming ground squirrel dispersal. Unpubl. Ph.D. dissert., Univ. Colorado, Boulder, 146 pp.

PFEIFFER, E. W., L. N. REINKING, AND J. D. HAMILTON. 1979. Some effects of food and water deprivation on metabolism in black-tailed prairie dogs, *Cynomys ludovicianus*. Comp. Biochem. Physiol., 63A:19–22.

RAYOR, L. S. 1980. Aspects of social structure in Gunnison's prairie dog. Guild Rocky Mountain Pop. Biol. Ann. Meet. (Abstr.)

RICKART, E. A. In press. Annual cycles of activity and body composition in *Spermophilus townsendii mollis*. Canadian J. Zool.

RONGSTAD, O. J. 1965. A life history study of thirteen-lined ground squirrels in southern Wisconsin. J. Mamm., 46:76–87.

SCHMUTZ, S. M., D. A. BOAG, AND J. K. SCHMUTZ. 1979. Causes of the unequal sex ratio in populations of adult Richardson's ground squirrels. Canadian J. Zool., 57:1845–1859.

SCHWAGMEYER, P. L. 1980. Alarm calling of the thirteen-lined ground squirrel *Spermophilus tridecemlineatus*. Behav. Ecol. Sociobiol., 7:195–200.

SCHWARTZ, O. A., AND K. B. ARMITAGE. 1980. Genetic variation in social mammals: the marmot model. Science, 207:665–667.

SEAMAN, R. N. 1976. A re-evaluation of nearctic sciurid phylogeny based upon biochemical, immunological and numerical taxonomic analyses. Unpubl. Ph.D. dissert., Colorado State Univ., Ft. Collins, 220 pp.

SHEPPARD, D. H., AND S. M. YOSHIDA. 1971. Social behavior in captive Richardson's ground squirrels. J. Mamm., 52:793–799.

SHERMAN, P. W. 1977. Nepotism and the evolution of alarm calls. Science, 197: 1245–1253.

———. 1980. The limits of ground squirrel nepotism. Pp. 505–544, *in* Sociobiology: beyond nature/nurture? (G. W. Barlow and J. Silverberg, eds.). Westview Press, Boulder, 627 pp.

———. 1981*a*. Kinship, demography, and Belding's ground squirrel nepotism. Behav. Ecol. Sociobiol., 8:251–259.

———. 1981*b*. Reproductive competition and infanticide in Belding's ground squirrels and other organisms. Pp. 311–331, *in* Natural selection and social behavior: recent research and new theory (R. D. Alexander and D. W. Tinkle, eds.). Chiron Press, New York, 532 pp.

SHERMAN, P. W., AND M. L. MORTON. 1979. Four months of the ground squirrel. Nat. Hist., 88:50–57.

SKRYJA, D. D., AND T. W. CLARK. 1970. Reproduction, seasonal changes in body weight, fat deposition, spleen and adrenal gland weight of the golden-mantled ground squirrel, *Spermophilus lateralis lateralis* (Sciuridae) in the Laramie Mountains, Wyoming. Southwestern Nat., 15:201–208.

SMITH, W. J., ET AL. 1973. Behavior of a captive population of black-tailed prairie dogs. Annual cycle of social behavior. Behaviour, 46:189–220.

SNYDER, R. L. 1976. The biology of population growth. Croom Helm, London, 227 pp.

SNYDER, R. L., D. E. DAVIS, AND J. J. CHRISTIAN. 1961. Seasonal changes in the weight of woodchucks. J. Mamm., 42:297–312.

SOWLS, L. K. 1948. The Franklin ground squirrel, *Citellus franklinni* (Sabine), and its relationship to nesting ducks. J. Mamm., 29:113–137.

STEINER, A. L. 1970. Etude descriptive de quelques activitiés et comportements de base de *Spermophilus columbianus columbianus* (Ord) II.—Vie de groupe. Rev. Comp. Animal, 4:23–42.

———. 1972. Mortality resulting from intraspecific fighting in some ground squirrel populations. J. Mamm., 53:601–603.

STORER, T. I., F. C. EVANS, AND F. G. PALMER. 1944. Some rodent populations in the Sierra Nevada of California. Ecol. Monogr., 14:165–192.

STROMBERG, M. R. 1974. Group response in black-tailed prairie dogs to an avian predator. J. Mamm., 55:850–851.

SVENDSEN, G. E. 1974. Behavioral and environmental factors in spatial distribution and population dynamics of a yellow-bellied marmot population. Ecology, 55:760–771.

TEVIS, L. 1955. Observations on chipmunks and mantled squirrels in northeastern California. Amer. Midland Nat., 53:71–78.

TILESTON, J. V., AND R. R. LECHLEITNER. 1966. Some comparisons of the black-tailed and white-tailed prairie dogs in north-central Colorado. Amer. Midland Nat., 75:292–316.

TOMICH, P. Q. 1962. The annual cycle of the California ground squirrel *Citellus beecheyi*. Univ. California Publ. Zool., 65:213–281.

WATTON, D. G., AND M. H. R. KEENLEYSIDE. 1974. Social behaviour of the arctic ground squirrel, *Spermophilus undulatus*. Behaviour, 50:77–99.

WEBB, D. R. 1980. Environmental harshness, heat stress, and *Marmota flaviventris*. Oecologia, 44:390–395.

———. 1981. Macro-habitat patch structure, environmental harshness, and *Marmota flaviventris*. Behav. Ecol. Sociobiol., 8:175–182.

WEST EBERHARD, M. J. 1975. The evolution of social behavior by kin selection. Quart. Rev. Biol., 50:1–33.

WIRTZ, J. H. 1967. Social dominance in the golden-mantled ground squirrel, *Citellus lateralis chrysodeirus* (Merriam). Z. Tierpsychol., 24:342–350.

WISTRAND, H. 1974. Individual, social, and seasonal behavior of the thirteen-lined ground squirrel (*Spermophilus tridecemlineatus*). J. Mamm., 55:329–347.

WITTENBERGER, J. F. 1980. Group size and polygamy in social mammals. Amer. Nat., 115:197–222.

YEATON, R. I. 1972. Social behavior and social organization in Richardson's ground squirrel (*Spermophilus richardsonii*) in Saskatchewan. J. Mamm., 53:139–147.

CONTRASTING SOCIOECOLOGIES OF SOUTH AMERICA'S WILD CAMELIDS: THE VICUÑA AND THE GUANACO

William L. Franklin

Abstract

A comparison is made of the socioecology of South America's wild camelids, the vicuña (*Vicugna vicugna*) and guanaco (*Lama guanicöe*). These two allopatric species are the dominant large herbivores of South American aridland environments, and are the only wild ungulates to inhabit the continent's deserts, high plateau scrublands, and grasslands. Data were collected on group size, composition, numbers of animals feeding on defined plant communities (vegetation types), hourly locations, movements, and territorial defense. Measurements were made of climate, timing and length of growing season, distribution, production, surface area and species composition of plant communities, and topsoil depth. The results summarize several field studies conducted in the Andes over the past 12 years. Populations of both vicuñas and guanacos are basically organized into family groups, male groups, and solo males. Family groups are single-male breeding units composed of one adult territorial male, females, and their offspring. Male groups are non-breeding multi-male units, whereas solo males are single adult males with or without an established territory. Uneven and localized distribution of high-producing and preferred forage types has favored the evolution of a feeding territorial system occupied by the single male family groups. For the vicuña and some guanaco populations, these are year-round permanent feeding territories. Vicuñas and guanacos differ in the degree of behavioral flexibility. Vicuñas are sedentary, grazers, obligate water drinkers, and altitudinally restricted to between 3,700 and 4,900 m elevation, whereas guanacos can be either sedentary or migratory, are both grazers and browsers, periodic water drinkers, and range altitudinally from sea level to over 4,000 m elevation. Unlike vicuñas, guanacos may form female groups (without an adult male) and mixed groups (winter aggregations of both sexes and all ages in migratory populations). Sedentary guanaco family groups and male groups appear to be socially and geographically separated. Guanacos also differ from vicuñas by having a more seasonally variable family group size and composition, less group control by the adult male, and a longer period of parental investment (two versus one growing season). Guanaco offspring are expelled as yearlings instead of as juveniles, and thus have more socialization time with the adult male. In guanacos only the adult male uses dung piles within the feeding territory instead of males and females of all ages, as in vicuñas. Forage production within guanaco feeding territories is not related to family group size; it is in vicuñas. Parallel environmental conditions offer explanations for similarities, whereas contrasts in social behavior and organization appear to be the result of differing environmental conditions. The greatest contrasts occur between sedentary vicuñas and migratory guanacos because of differences in availability of forage within feeding territories. A discussion of year-round social organizational similarities and differences for the two species is presented. Male, female, and young behavioral strategies are discussed in terms of individual survival and reproductive fitness in a dry, sparsely producing, and generally harsh environment.

Introduction

Camels in South America? Absolutely. In fact, not only are there two endemic wild camelids, the vicuña (*Vicugna vicugna*) and the guanaco (*Lama guanicöe*), but two domestic South American camels—the familiar llama (*Lama glama*), historical and current day beast of burden to the highland Indians of the central Andes, and the alpaca (*Lama pacos*), producer of fine wool. Although not clearly understood, the llama and perhaps also the alpaca are believed to be derived from the wild guanaco. Just how important are the vicuña (Fig. 1) and guanaco (Fig. 2) to the mammalian fauna of the Neotropical-South American Region? Despite the region having a high mammalian diversity (second only to the Ethiopian-African Region) and the highest proportion of endemic mammalian

FIG. 1. Adult male vicuña in a short-grassland community of the Andean altiplano.

FIG. 2. Adult male guanaco in a shrubland community of the South American Patagonian plateau.

families (43%) among the zoogeographic regions of the world, South America is impoverished when it comes to hoofed mammals. Though both the South American and Ethiopian-African Regions feature great climatic and ecological diversity, from deserts to tropical jungles, South America has only 19 species of wild ungulates, compared to 95 in Africa. Vicuñas and guanacos play a feature role in this ungulate-poor continent, not only in being the dominant large herbivores of South America's aridland environments, but because they are the only wild ungulates to inhabit the continent's deserts and high plateau scrublands and grasslands.

Thus, these two allopatric species essentially occupy most of South America's dry environments, being distributed from the northern slopes of the Andes, along the Andean chain, south across the Patagonian plateau, to the islands of Tierra del Fuego at the southern tip of the continent. The vicuña and guanaco, occupying similar environments and being closely related, pose challenging questions to students of mammalian behavioral ecology. It is under such circumstances that the dramatic influence and impact that the environment has upon shaping behavioral and social systems becomes detectable. For example, what behavioral and social organizational similarities and differences do the vicuña and guanaco have? Can contrasts in behavior be explained by contrasts in environments?

This is an exciting frontier, one which requires that we understand both the animal's behavior and its environment. Though only recently revived, such a "socioecological" approach is not especially new. Petrucci (1906) first introduced the concept of social ethology when he emphasized that the spatial disperson of animals, group composition, group size, and the relations between individuals were directly responsive to the environment; to him the environmental features acting upon an animal's social system included food resources, predation, and requirements for reproduction. Contemporary social ethologists focus upon three interdependent subjects (Crook, 1970*a*)—socioecology, sociodemography, and social systems. This paper is primarily concerned with socioecology—the study of social organization and social behavior in relationship to the environment. Social organization refers to how animals in a population are spatially and temporally organized with respect to one another (Franklin, 1978*a*; Franklin and Lieb, 1979).

The social organization (social system) of a species is not a fixed event, but is flexible and can change with season, population density, habitat, food, surrounding mammal species, and other proximate factors (Fisler, 1969). However, a species' social system has evolved just as structural and physiological systems have evolved (Eisenberg, 1966), and is adapted to a particular environmental type that confers maximum reproductive success upon that population (Barash, 1974). In other words, social behavior and social systems have evolved as a result of selection to increase the genetic fitness of individuals in specific environments. The social organization for any one species also helps control and minimize intraspecific competition, interference, and aggression for promoting maximum and efficient individual reproduction (Wynne-Edwards, 1965).

The purpose of this paper is to compare the socioecologies of the vicuña and guanaco. Specific objectives include: 1) What is the social organization of each species? 2) What similarities and differences are there in the social system? 3) Are there environmental differences that might explain contrasts in social behaviors and social systems between the two species?

The results presented are summarizations and high points of field research conducted during the past 12 years in the South American Andes. Study areas have included the Pampa Galeras National Vicuña Reserve in central Peru, where vicuña socioecology was studied (Franklin, 1974, 1978*a*); the Peruvian Calipuy National Guanaco Reserve, where the continent's northernmost viable guanaco population was surveyed (Franklin, 1975); Estancia Cameron, on the Chilean side of the island of Tierra del Fuego, where guanaco family group behavioral ecology was investigated (Jefferson, 1980); and at Torres del Paine National Park in southern Chile, where field research is underway on guanaco male group socioecology. Knowledge of guanaco socioecology is less complete than for the vicuña because guanaco studies are still in progress.

Methods

Behavioral Measurements

Vicuña studies.—A grid of 12.4 km^2 was laid out in Cupitay Valley on the northern edge of Pampa Galeras Reserve (Fig. 3). The grid covered a 4.2-km length of the valley floor, the north and

FIG. 3. Pampa Galeras National Vicuña Reserve, Peru.

south-facing slopes and the southern ridge. Rock cairns 1 m in height were spaced 200 m apart in lines at right angles to a straight road.

Four observation huts were constructed on high vantage points within the gridded study area. By means of the cairns, one could estimate and record locations within the grid to the nearest 20 m. Once every hour the area was scanned from left to right, and the following information about each vicuña group present was recorded on a map of the study area: total number of animals, composition of group, location, number of adults and crias (young less than 1 year old) feeding, and the vegetation type being consumed. Ten to 20 groups were typically under observation at the same time. Details of intergroup aggressive encounters (territorial defense) were recorded throughout the hour. As groups left the observation zone and moved onto the adjacent ridges in the afternoon, some were no longer visible. Recordings of the location, numbers, and types of domestic livestock in the observation zone were also made hourly. Disturbed movements of vicuña groups due to activities of people, livestock, and dogs were recorded. By the end of a day, a group's continuous route of movement was obtained through these hourly samples. Qualitative notes were taken on unusual events such as changes in group composition, fighting, births, matings, and predation.

The above observations were made for 27 months from February 1969 to April 1971. Two observers recorded information from sunrise until noon, at which time a third observer replaced them both and remained until dark. If it was not possible to observe the entire day, observations were made only in the morning.

A total of 3,623 h of observations were made from March 1969 to April 1971 (2,209 h in the morning and 1,414 h in the afternoon); hours observed per month averaged 134 (range 36 to 227). The seasons were divided into 3-month periods, beginning with autumn on 1 May. Year one was from 1 May 1969 to 30 April 1970 and year two, 1 May 1970 to 30 April 1971.

Complete counts were made of vicuñas within the Pampa Galeras National Vicuña Reserve in August 1968 and during a return visit in December 1971. Each census was conducted over a 3-day period from mid-morning to late afternoon with the help of two or three assistants. Monthly and seasonal population densities within the study area were based upon the mean number of animals observed per hourly scan divided by the surface area of the zone under observation.

Vicuña Units (V.U.) were used to equate animals of differing sizes when calculating habitat utilization, feeding use, and preference for vegetation types. Adults and yearlings were each considered full Vicuña Units. Crias were rated by age: 1 to 3 months old (March, April, May) = 0.2 V.U.; 4 to 6 months (June, July, August) = 0.4 V.U.; 7 to 9 months (September, October, November) = 0.6 V.U.; 10 to 12 months (December, January, February) = 0.8 V.U.

Preference indices (P.I.) were computed for vicuñas feeding on different vegetation types; the percentage of total vicuñas observed feeding on a vegetation type was divided by the percentage of the total hectares that type represented in the study area and multiplied by 100. A preference index of more than 100 indicated vicuñas spent relatively more time feeding on a particular vegetation type than would be expected from that type's availability.

During 1970 and 1971 six crias were hand-captured, ear-tagged, and released. Crias could be run down on foot only within the first 15 min after birth. For those groups without tagged animals or without individuals recognized by natural markings or defects, identification was based upon group composition and consistent use of the same site. Individual vicuña within focal study groups were identified by a combination of characteristics, including body size, coat color, scars, defects, and plastic ear markers; for example, One-Eye, Two-Dent, Large Light, Tagged Cria, and so forth.

The sex of a vicuña is difficult to determine in the field because males and females lack obvious sexually dimorphic characters. An individual could be sexed when it raised its tail during defecation, revealing the absence or presence of testes. Both sexes urinated posteriorly with the hindlegs parted and the hindquarters dropped slightly, but the male's urine was at the level of the hocks, while the female's urine originated at the level of the tail. Also, the small dark and hairless penis on males and the udder on lactating females were occasionally visible. Free-ranging adult males and females could also be distinguished by behavioral differences. Adult females were nearly always with a family group; if not, solo females were usually not aggressive when attempting to enter a family group, did not interact with other females, were treated less aggressively by territorial males, and did not defend a site as did males. Yearlings were distinguished by their smaller bodies, wool of less uniform length and more shaggy appearance, and shorter muzzles than adults.

Guanaco studies.—Research conducted on guanaco family groups on Tierra del Fuego utilized the same systematic behavioral observational techniques employed during the vicuña work. However, on Tierra del Fuego a grid of 1.5 km^2 was placed out in a bog meadow surrounded by dense forest (Fig. 4). Grid markers were 100 m apart and animals were observed from a centrally located observation hut. Family groups that occupied the meadow were usually observed from the time they emerged from the forest in the early morning until their return to the forest in late afternoon or early evening. Four to six family groups were typically under observation at the same time.

The numbers of individual animals were equated to Guanaco Units (G.U.) for analyzing group size. Adults and yearlings each equaled 1 G.U., whereas juveniles 1 to 3 months old = 0.2 G.U., 4 to 6 months = 0.4 G.U., 7 to 9 months = 0.6 G.U., and 10 to 12 months = 0.8 G.U. Collection of behavioral information at Torres

FIG. 4. Meadow and beech forest habitat of the guanaco on the island of Tierra del Fuego.

del Paine Park was essentially the same except that aerial photo maps instead of grid maps were used for studying animal movements.

Environmental Measurements

Vicuña studies.—Fifteen vegetation types were identified in the study area and were analyzed by the point-frame method to determine plant species composition (percent cover), as described by NAS-NRC (1962). A homogeneous stand of each vegetation type was chosen by its characteristics of visual uniformity. During the last month of the 1970 growing season (April), 13 vegetation types were analyzed using a total of 16,000 point-frame samples.

A vegetation type map of 431 ha with a scale of 20 m to 1 cm was made by ground reconnaissance from atop high peaks and ridges using binoculars and spotting scope. The vegetation types within each 4 ha section were drawn in proportion to their size and distribution.

Plant production of each important vegetation type was estimated during the 1970 (n = 57) and 1971 (n = 45) growing seasons by clipping plots where ungulate grazing had been excluded by wire cage exclosures. The term production refers to the amount of dry plant material produced per unit surface area per year. Each year cages were put out in January before the growing season began; sites were judged to be quantitatively and qualitatively representative of a vegetation type. After the growing season, the forage within 1-m^2 plots was clipped to ground level from inside 1.2-m^2 cages, oven dried, and weighed. For shrubs and tall coarse grasses only the current year's growth was clipped.

Total production within territories was estimated by multiplying the total surface area of each vegetation type within the territory by the mean production for that type. Plant species not used by vicuñas, namely the large unpalatable bunch grasses, were not included in these estimates of total available food resources. A total of 132 soil pits were dug in 11 different vegetation types to measure the mean depth of the top soil (horizons A_1 and A_2).

Weather data were obtained from the climatological station at the Pampa Galeras headquarters 2.5 km west of the study area. Standardized weather information was taken daily at 0700, 1300, and 1900 h. Climatological information was averaged by 2-week periods beginning with August for each annual cycle. Thermometers in standard weather houses placed on the ground measured maximum and minimum temperatures in the study area.

Guanaco studies.—Methods used for measuring environmental parameters at the Tierra del Fuego study area closely paralleled those used for the vicuña research. Vegetation types were again defined on their visual distinctness, dominant plant species, and substrate differences. Primary productivity was determined with 42 1-m^2 wire exclosures and a detailed vegetation map was made of the area (Jefferson, 1980).

Results

Vicuña Socioecology

The environment.—Vicuñas are restricted in distribution to the central Andes of southern Peru, northern Chile, western Bolivia, and northeastern Argentina (Koford, 1957). They inhabit a special biological zone regionally known as the *puna* or *altiplano,* an equatorial high-altitude grassland between 3,700 and 4,900 m. This cold, arid region is unique in being above treeline but below snowline.

The Pampa Galeras National Vicuña Reserve is in southern Peru's department of Ayacucho (14°40′S, 74°25′W), 450 km south of Lima. Established in 1966 by the Ministry of Agriculture's Directory of Forestry and Wildlife, Pampa Galeras was the world's first vicuña reserve and has greatly contributed towards saving the vicuña from near extinction.

The 6,500 ha Pampa Galeras Reserve, a treeless alpine grassland, lies on the western edge of the Andean altiplano plateau and is characterized by broad open valleys separated by flattened ridges (Fig. 3). Precipitation occurs mainly in the summer. Nightly temperatures almost always are below freezing. Plant formations are primarily Subalpine Scrub Desert and Humid Alpine Tundra, with some areas of Subalpine Humid Puna and Very Humid Alpine Tundra (Tosi, 1960).

Climate.—Two general climatological periods occur in this part of the Andes: a mildly warm, rainy growing period from December through April and a cold, dry period from May through November. The autumn (May, June, and July) is characterized by freezing nights, cool days, no rain, and drying vegetation; the winter (August, September, and October) by freezing nights, cool days, little rain and dry and dormant vegetation; the spring (November, December, and January) by a decrease in the number of freezing nights, cool days, an increase in precipitation, and some plant growth; and the summer (February, March, and April) by cold nights, cool to mild days, almost daily precipitation, and plant growth.

The mean annual precipitation in Pampa Galeras for the 5 years 1967 to 1971 was 487 mm (range 281 to 742), with 90% of the annual precipitation in the spring and summer and 81% from December through March. The mornings in the rainy season were clear, warm, and rainless until noon, when storm clouds came from the east. Precipitation was usually in the form of hail or rain but rarely snow.

Mean annual temperatures for 1968 through 1971 were 4.6°C, 5.0°C, 5.3°C, and 5.5°C. From 1967 to 1971 mean minimum and mean maximum temperatures for autumn were −5.1°C and 12.2°C; for winter, −4.6°C and 13.4°C; for spring, −1.3°C and 12.9°C; for summer, 0.5°C and 12.4°C. Nightly average minimum temperature

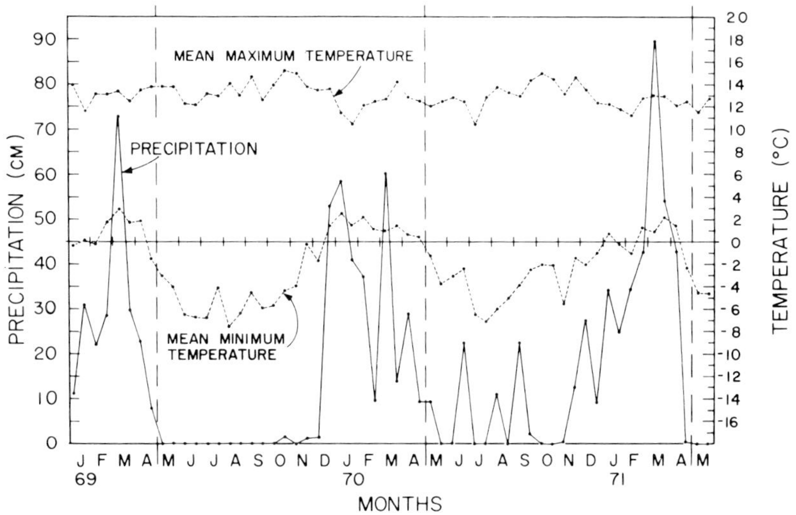

FIG. 5. Relationship between precipitation, mean minimum temperatures, and mean maximum temperatures in Pampa Galeras Reserve, January 1969 to May 1971.

was below freezing year-round except during a few weeks in late spring and summer (Fig. 5).

Cloud cover from late spring and summer storms had a profound effect upon the nighttime minimal temperatures, which in turn appeared to affect the length of the growing season and, ultimately, plant production. Nocturnal cloud cover maintained night temperatures above freezing. For example, in the summer of 1967, clouds covered 75 to 100% of the sky 75% of the time; the mean minimal temperature for this period was 1.7°C. In contrast, during the autumn of 1967, when daily maximum temperatures averaged the same as summer, the cloud cover was only 1 to 25%, most (71%) of the time and the mean minimal temperature was −4.7°C. Lapses of cloud cover in midsummer caused nighttime minima to drop below freezing.

The number of frost-free weeks when plant growth occurred varied from 8 to 17 (averaging 13 from 1967 through 1971), and appeared to have a close relationship with plant production. In 1970 the growing season was 17 weeks long. In 1971 it decreased to only 8 weeks and total production declined 40%.

The growing season at this high altitude was not necessarily "a period when sufficient moisture permitted the growth of the existing natural vegetation" (ONERN, 1971), but a period when sufficient nighttime cloud cover maintained the nightly temperature above freezing. Thus, total annual precipitation reflected, but did not fully control, the length of the growing season. Total precipitation in 1971 was 70 mm greater than 1970, yet the 1971 growing season was less than half as long as that of 1970. An unusually "wet" dry season occurred in 1971 when much of the annual moisture fell in autumn, winter, and early spring; however, cloud cover from these storms was not sufficient to bring the nightly minima above freezing (Fig. 5).

Differences in minimum and maximum temperatures within the Cupitay Valley study area were measured for the ridge (sleeping area), slope, flat, and bottomland during August 1970. Daytime mean maxima ranged from 17.1°C to 19.0°C; the ridge was the coolest of all locations. Nightly mean minima ranged from −3.3°C to −11.0°C. Temperatures on the bottomland averaged nearly 7°C colder than on the ridge sleeping area, although the ridge was only 18 to 20 m higher than the bottomland in elevation.

Topography and distribution of water.—Elevation in the Cupitay Valley study area in Pampa Galeras ranged from 4,059 m on the level bottomland at the center of the valley to 4,077 m on the ridge. The area between the bottomland and the road was referred to as the flat, whereas the area between the ridge and flat was called the slope (Fig. 6).

Fig. 6 gives the locations of surface water in the bottomland associated with the Cupitay River and of springs on the flat and slope that contained water during the first half of the dry season. By late winter, most springs had either dried up or were insufficient to satisfy the needs of vicuña.

Vegetation types.—Points of special interest on plant species composition, forage production, and utilization needed for understanding the socioecology of the vicuña are presented below for selected vegetation types. The types are given in order of abundance (Fig. 7).

Roqueño, the most common plant community, occupied 48.1% of the total 669.3 ha of area covered by vegetation (excluding rock outcroppings, sheep corrals, roads, and waterways) in the study area. Roqueno had an abundance of surface rocks of igneous origin

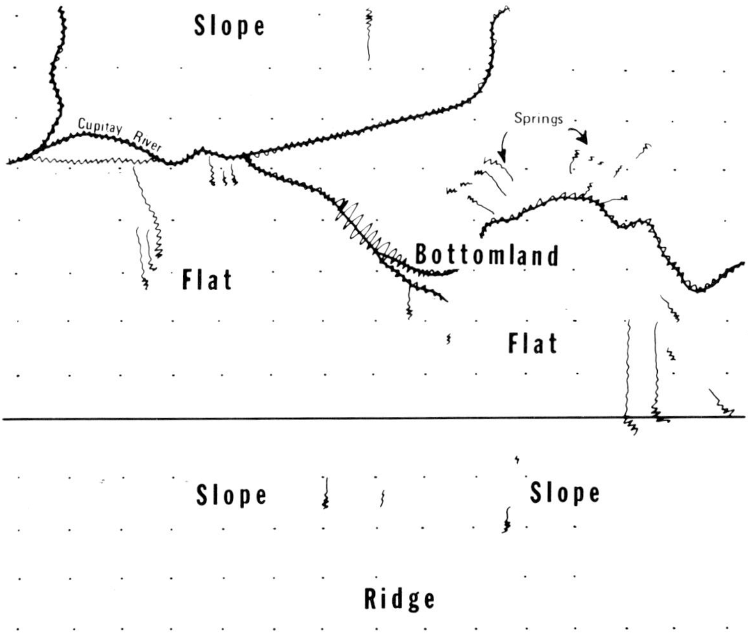

FIG. 6. Distribution of surface water during June and July 1970, Cupitay Valley study area.

and a sparsity of plants (Fig. 8). Plant production per ha was low and feeding vicuña showed no preference for this type.

Ichu, dominated by *Festuca dolichophylla,* a tall-coarse and unpalatable bunch grass, accounted for 15.5% of the area and was common on the upper dry slopes, occurring as large units throughout the Reserve. Ichu ranked second highest in production per hectare in 1970, but was the least preferred of all vegetation types (Fig. 7). Nearly 75% of the production was from *Festuca dolichophylla* in 1970.

Peccoy occurred on the flats as irregularly-shaped patches or on the slopes as linear downhill lines and covered 13.5% of the area. Sixty percent of the production in Peccoy was from a large, coarse, and unpalatable bunch grass, *Stipa ichu,* in 1970. Vicuñas showed no feeding preference for Peccoy, but fed on the small understory plants in both Peccoy and Ichu. The vicuña is mainly adapted for grazing of small forbs and small perennial grasses close to the ground and

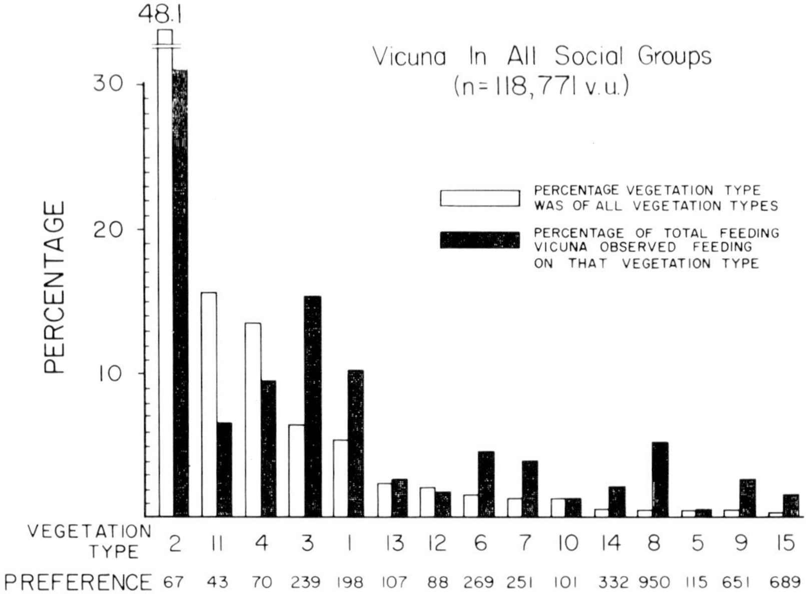

FIG. 7. Feeding use and preference for vegetation types by all vicuñas. Important vegetation types are: 1, Quisna; 2, Roqueño; 3, Chiula; 4, Peccoy; 6, Mojadal; 7, Pampa Verde; 8, Pampa Verde Claro; 9, Riparian; 10, Piedras de la Loma; 11, Ichu; 12, Quisna sin Influencia; 13, Taxona; 14, Chiula General; 15, Pampa Verde Obscuro.

is the only ungulate with open-rooted, continuously growing incisors (Miller, 1924).

Chiula, common in wet soil types on lower slopes and at the base of slopes, covered 6.4% of the vegetated area. The dominant plant was the still-culmed grass, *Festuca rigesens.* Irregularly shaped patches of Chiula were typically surrounded by large areas of Roqueno or Ichu. In Pampa Galeras, Chiula was a consequence of localized topographical runoff. Production and feeding preference was high and Chiula received the second highest use of all types by feeding vicuñas.

Quisna was the fifth most common plant community (5.3% of relative surface area); the dominant species was the perennial grass, *Calamagrostis vicunarum.* This type had the highest production of all types in 1970 (1,550 kg/ha) and was preferred by grazing vicuñas. Quisna commonly occurred as circular to oblong-shaped "put-

FIG. 8. Roqueño plant community dominated by surface rock, *Muhlenbergia peruviana*, and *Poa galeras*.

ting greens" on the flat and was surrounded by Roqueño or Peccoy (Fig. 9). Quisna was less densely vegetated on the slopes; these downhill lines had lower production and thicker topsoil than the oblong patches on the bottomland and flat. Quisna growth was influenced by vicuña excrement and urine (see below).

Mojadal, a wet and boggy plant community found on the bottomland or around perennial springs on slopes and flats, had dominant plants coarse in texture and growing close to the ground. Production was intermediate and the topsoil was the deepest of all vegetation types measured. Vicuñas preferred Mojadal, but use was low because of its small area.

Pampa Verde and *Pampa Verde Obscuro* were composed of nearly the same plant species but in different proportions. They occurred next to each other in the bottomland and took on the appearance of a densely vegetated carpet. Although use was low, vicuñas preferred both types. Alpacas (*L. pacos*) fed upon these types heavily. Although not measured, topsoil depths were equivalent if not deeper than Mojadal.

FIG. 9. An oblong patch of the preferred vegetation type Quisna dominated by *Calamagrostis vicunarum*. This type was the terminal seral stage of excrement influenced vegetation.

Piedras de la Loma was profusely littered by surface rocks and occurred on the flattened summit of ridges. Topsoil depth was the shallowest of all types, and productivity, feeding use, and preference was low.

Pampa Verde Claro occurred on the bottomland and resembled a less densely vegetated form of Quisna because of the presence of *Calamagrostis vicunarum*. This community was not influenced by vicuña excrement, but occurred in areas of lowland sediment accumulation. Total use by feeding vicuñas was low, but preference was high. Alpacas and llamas (*L. glama*) grazed heavily upon Pampa Verde Claro. Vegetation types Chiula General, Pampa Verde Claro, Chiula, and Quisna had the highest plant species diversity and were all preferred by vicuñas.

In summary, vicuña food resources were sparse. High-producing and highly-preferred vegetation types (Chiula, Quisna, and Mojadal) were scattered and patchy in their distribution, and were scarce because they covered only about 15% of the vegetated area. Thus, where good forage did occur, it was only locally abundant in a small area.

Excrement-influenced vegetation (EIV).—A striking feature of the landscape in Pampa Galeras was the alternating zones of vegetation on the slopes. Wide and sparsely vegetated strips dominated by surface rocks alternated with narrow downhill lines of densely vegetated strips dominated by Peccoy or Quisna.

The densely vegetated lines were due to the fertilizing effects of vicuña excrement and urine. Vicuñas defecated and urinated only on established and traditionally used dung piles. The downhill washing of this organic matter and nutrients by precipitation, especially after the sudden and voluminous summer rains, accelerates soil and plant succession, thus causing these narrow densely vegetated belts (Franklin, 1973*b*). Such excrement-influenced vegetation occurred also on the flats, but there the subsequent change in vegetation around the dung piles was more oblong than linear. EIV made up 18% of the total surface area in the study area.

EIV was also observed in other regions of the puna in southern Peru, where there have been vicuñas during the past decade. Inland from Pampa Galeras, its frequency and striking contrast to the surrounding vegetation decreased. Precipitation and soil fertility progressively increased eastward from Pampa Galeras, presumably moderating the fertilizing effects of dung piles.

In Pampa Galeras there were three successional stages of vegetation resulting from dung-piling behavior of vicuñas: 1) Peccoy, which was common as linear downhill lines throughout the Pampa Galeras region, irregularly-shaped patches on the flats, and occasionally as circular rings around dung piles on level ground; 2) a transitional phase between seral stages one and three; and 3) the terminal stage Quisna. Topsoil depth increased 2 cm with each succeeding stage. The soil in Pampa Galeras was poor in nitrogen and organic matter, but dung-piling behavior resulted in greater soil depth, higher plant species diversity, and increased forage production.

Effective vicuña habitat.—Forty-three percent of the 6,500 ha of the Pampa Galeras Reserve was considered to be unusable or poor vicuña habitat, including bare ground, rock outcroppings, and the vegetation types Peccoy, Ichu, and Tola, a shrub community not found in the study area. The Tola community, dominated by *Lepidophyllum tola* (=*Parastrephis lepidophylla*) and lesser amounts of *Diplostephium* spp., had invaded over the past 25 years into the lower end of Cupitay, Huisccana, and Llamaiso valleys where domestic animals, especially cattle, have heavily overgrazed the previously dominant Ichu vegetation type.

The downhill lines of Peccoy, separated by side areas of Roqueño, together covered 24% of the Reserve and were classified as fair vicuña habitat. The remainder of the Reserve (33%), which included mostly preferred vicuña vegetation types Quisna, Roqueño, Chiula, Mojadal, Quisna Sin Influencia, and Taxona, was considered good vicuña habitat. Areas classified as fair and good habitat made up 57% of the Reserve.

Quality of habitat depended not only upon good forage, but also upon the accessibility of permanent drinking water and a communal sleeping ground (Franklin, 1973*b*). Vicuñas regularly drink, and especially in the dry season, water is required daily. Streams or springs within or on the border of a territory often provided the vicuña group immediate access to water.

Social organization and behavior.—A population of vicuñas is basically divided up into family groups (FGs), male groups (MGs), and solo males (SMs) (Franklin, 1973*a*, 1974). Three types of family groups were distinguished at Pampa Galeras—Permanent Territorial Family Groups (PTFGs), Marginal Territorial Family Groups (MTFGs), and Mobile Family Groups (MFGs). PTFGs and MTFGs were the most common social units and were composed of one adult male and females with and without crias. MTFGs were primarily groups from outside the study area making brief forays into it to drink and graze. Non-territorial males formed all-Male Groups. The composition of family groups was written as an abbreviated formula—total number of animals in group (adult males-adult females-yearlings-crias), such as 19(1-10-0-8).

A *Permanent Territoral Family Group* averaged six members—one adult male, three females and two crias (Fig. 10). Fifty-one percent of PTFGs had from one to five animals, 38% six to ten, and only 11% with over 10 animals. The largest family group observed had 19 animals (PTFG 207 with one adult male, 10 adult females, and eight crias in June 1969). Vicuñas in PTFGs composed about 60% of the total population. Their territory was in two parts—a feeding territory where the group spent most of the day, and a sleeping territory located on higher terrain and used at night. A group's feeding and sleeping territory and the undefended corridors connecting the two composed its home range. The PTFG feeding territory included access to permanent drinking water. Ter-

FIG. 10. A Permanent Territorial Family Group of vicuñas in its feeding territory on good habitat.

ritories were exclusively occupied by PTFGs and defended by the adult male against all intruding outside vicuñas. Size and shape of feeding territories changed seasonally, but remained in the same basic location for the entire year. PTFGs occupied the preferred habitats that were found on the slopes, flats, and bottomlands.

Marginal Territorial Family Groups averaged one adult male, three adult females, and one cria (Fig. 11). These were found on secondary habitat surrounding the better areas occupied by the PTFGs, especially the flattened ridges and drier slopes without springs. Vicuñas in MTFGs constituted almost 20% of the total population. The size, shape, and location of their feeding territories changed more than the territories of the PTFGs, because the feeding territory commonly did not include permanent water. MTFGs had to leave the feeding territory daily during the dry season to obtain water. Family groups were categorized as PTFGs or MTFGs in order to study the differences between reproductive groups occupying good and poor habitat.

Mobile Family Groups were a temporary association of females

FIG. 11. A Marginal Territorial Family Group of vicuñas on secondary habitat.

with a non-territorial adult male. Such groups included from one to five animals, usually the male, one adult female, and two or three yearling females. MFGs were highly mobile and group size often changed. They were most common shortly after crias had been expelled from their family groups (that is, January through March). MFGs were chased out of areas occupied by PTFGs and MTFGs and were often on poor habitat.

Male Groups were bands of two to 155 non-territorial males that fluctuated in size. Three-quarters of all MGs (n = 409) had fewer than 30 animals; five to 10 was the most common group size. Large MGs were more prevalent in spring and summer when yearling males joined them. Territorial males vigorously attacked and chased MGs out of occupied habitat. As a result, MGs were often in unoccupied and non-preferred habitats and moved long distances in search of undisturbed feeding areas.

Solo Males were physically and sexually mature individuals attempting to establish territories. SMs wandered while looking for suitable unoccupied or poorly defended sites, and were aggressive toward established males that attempted to force them from zones with territories. Once a Solo Male had established his territory, it

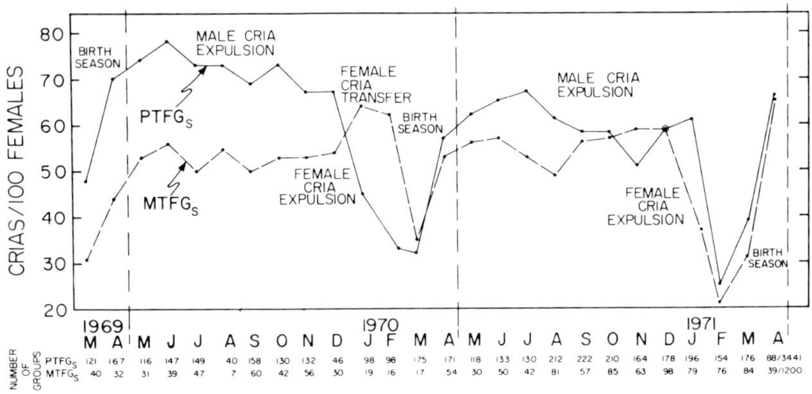

FIG. 12. The number of crias per 100 females (two years and older) for PTFGs and MTFGs in observation Zones I and II combined.

still might take him several months to obtain females. Occasionally old or injured non-territorial males were seen alone, but were not considered as solo males because of their lack of social involvement with the population.

Groups that could not be identified as a family or male group, wandering adult female-cria pairs, or solo yearlings and adult females, were referred to as Unclassified Groups.

Changes in size and composition of PTFGs and MTFGs.—The size and composition of PTFGs and MTFGs were highly influenced by births and dispersal of young. During the summer birth season, the number of crias increased (Fig. 12). Over the next 6 months, the ratio of crias to adult females declined moderately in PTFGs as adult males drove out male crias and some mortality of crias occurred. MTFGs maintained themselves at about the same size because some male crias driven from PTFGs entered MTFGs. In December and January female crias and any remaining male crias were expelled from PTFGs, with many of the female crias joining MTFGs. In February, just before the next birth season, these crias were also forced from the MTFGs; both MTFGs and PTFGs reached their lowest ratio of crias to adult females in February or March.

In most PTFGs all crias were forced out before the next birth season. However, because the data were averaged by months and because some female crias transferred to newly established groups,

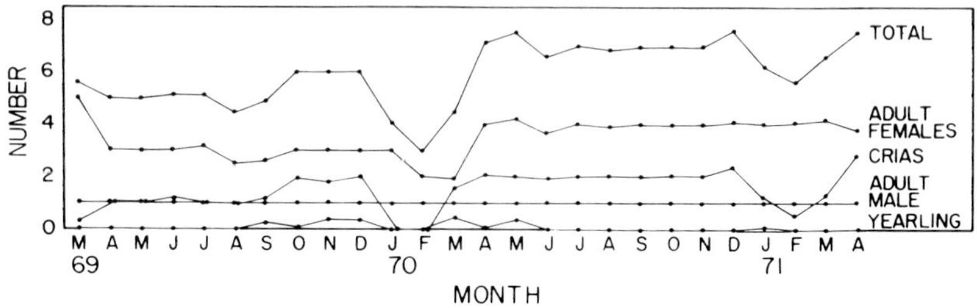

FIG. 13. Number of animals in PTFG 216.

the cria to adult female rato did not drop below 20:100 (Fig. 12). The period when crias were completely absent from most PTFGs is shown by PTFG 216 (Fig. 13).

Relative proportion of social groups.—Over two years, 59% of all animals were in PTFGs, 17% in MTFGs, and 24% in MGs. About one-third of all males older than one year held territories (Table 1).

Censuses in August 1968 and December 1971 indicated a population increase, and the number of family groups increased from 101 to 193 (Table 2). The ratio of males to females in the 1971 census was 85:100. As the population increased from 1968 to 1971, vicuñas pushed out into vacant and often less desirable range. The number of adults per family group decreased from 4.9 to 3.7, partly

TABLE 1

PERCENTAGE OF ANIMALS SEEN IN VICUÑA SOCIAL GROUPS. ABBREVIATIONS ARE: PTFG, PERMANENT TERRITORIAL FAMILY GROUP; MTFG, MARGINAL TERRITORIAL FAMILY GROUP

	Percentage		
Social group	Year one (May 1969–April 1970)	Year two (May 1970–April 1971)	Change
PTFGs	66	53	−13
MTFGs	14	20	+6
Male Groups	20	27	+7
Total	100	100	—
N	15,554	19,995	—
Percentage of all adult males territorial	32	27	−5

TABLE 2
CHANGES IN VICUÑA POPULATION SIZE AND STRUCTURE BETWEEN AUGUST 1968 AND DECEMBER 1971 IN THE 6,500 HA PAMPA GALERAS VICUÑA RESERVE

Population character	August 1968	December 1971	Percentage change
Total population	891	1,217	+37
Number of family groups	101	193	+91
Mean family group size	6.7	4.9	−27
Mean number of adults per family group (>1-year old)	4.9	3.7	−25
Number of male groups	11	27	+146
Mean male group size	19.2	17.2	−9
Percentage of groups that were family groups	90	92	+2
Percentage of population in family groups	76	81	+5
Percentage of all males territorial	32	45	+13
Percentage of population as solo males	0.2	1.1	+0.9

because newly formed family groups tended to be smaller than established ones, but perhaps also because the amount of effective habitat per group decreased.

During a 3-day field trip in 1968 to Yauriviri, southeast of Pampa Geleras, only five vicuña family groups were seen. Although a small sample, the mean group size of 9.4 at the low vicuña density in Yauriviri was the converse of the mean group size of 4.9 at high density in Pampa Galeras.

Dispersal of young.—Male and female crias were expelled from the family groups by the adult male before they reached 1 year old. Young males were forced out when 4 to 9 months old and young females when 10 to 11 months old. Most dispersed from the study area. For example, the number of 8-month-old female crias seen per h during the first week of December 1969 was 4.3, compared with 0.3 per h 25 weeks later when they were 14 months old. The 1970 female cria cohort decreased similarly from 8.0 to 2.1. The yearlings that did remain attached themselves to territorial males. Thus, in mid-April 1971, of the 20 yearling females remaining in the study area, 35% were with males that had recently established their territories, 27% were with long established PTFGs, and 35% in MTFGs.

By June 1971 only 13% of the territorial groups using the study area (n = 47) contained yearling females—two of 23 PTFGs

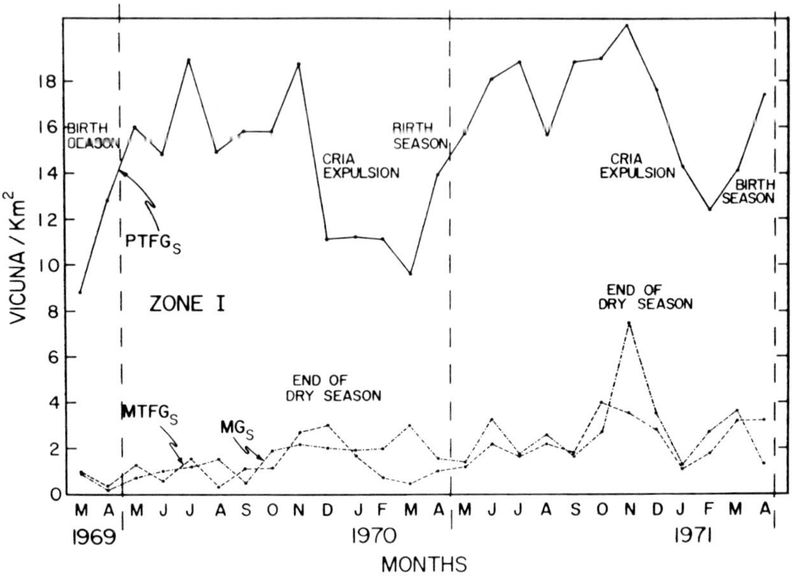

FIG. 14. Monthly density of vicuña social groups.

accepted five yearlings (two in one group and three in the other), two of 17 MTFGs accepted five yearlings (one in one group and four in the other), and two recently established territorial males out of seven took on six yearlings (three in each group). Very few territorial males were willing to take on more females. Indeed, yearling females were chased by territorial males from one family group to the next as they sought entry into a group. Thus, most yearling females dispersed from PTFGs in good habitat unless they joined groups with other yearling females. The expulsion and dispersal of young from family groups and the increase of young from births (Fig. 12) were the primary causes of seasonal changes in vicuña density (Fig. 14).

Family group cohesion, stability, and permeability.—Changes in size and composition of family groups were largely due to births, subsequent expulsion of crias, and mortality. Adult females rarely shifted from one territorial male to another. For example, the same adult females remained in family group 211 through the study except for one female joining the group after 15 months, one female dying after 20 months, and several females joining after 23 months.

In most groups, changes in the number of adult females involved only one or two individuals. In Group 207, an exception, the adult male allowed females to enter and leave at will.

During the first two weeks after birth, females with new crias were obviously nervous and would retreat to higher terrain at even the slightest sign of potential disturbance or danger. The cause of female uneasiness was often not apparent. If the group failed to follow or the male did not force them to remain, they were usually away only for a few hours. Uncommonly an adult female would be absent from her home group for several days to look for a lost or separated cria, but would usually return with or without her cria.

Territorial males did not always accept outside adult and young vicuña females attempting to enter their groups. During some 3,300 h of observation over 2 years, family group males were observed 138 times out of 240 occurrences to immediately prevent the entry of non-members into their groups. Significantly more adult females with crias were immediately rejected than other female classes (51%; $P < 0.005$; n = 70). Twenty-two percent of rejections were adult females by themselves (n = 30), 15% were single crias searching for a group (n = 12), and 12% were yearlings (n = 17) (Chi-square = 0.00; $P < 0.005$). Although the highest rejection rate of females with crias was in November, December, and January, there was no significant difference between seasons; that is, the proportion of rejection compared to acceptance was high year-round. The rejection rate of outsiders attempting to enter family groups for the four seasons beginning in the autumn was 2.2, 4.6, 6.3, and 4.0 per 100 h of observation.

In the late dry season and the months before the birth season, groups became increasingly more closed to outsiders, especially to adult females with crias. For example, on 16 September 1969, a female with cria repeatedly attempted to enter family groups, but was driven off by the territorial male of each group.

In 102 instances of females entering groups without immediate rejection, 78% entered the group only temporarily before leaving, whereas only 23% (six adult females, three females with crias, and two yearlings) permanently remained with the group. Acceptance into a group was determined by the territorial male's response. Out of 240 attempts to enter family groups, only 5% resulted in permanent acceptance.

Half of those that entered groups without initial interference were adult females. Twenty-six percent were adult females with their crias, 15% were yearlings, and 9% were crias. Sixty-three percent of the attempts by adult females to enter a group were at least temporarily successful, in contrast with 28% of attempts by adult females with crias, and 32% of attempts by single crias ($P < 0.01$). Females had the least chance of entering a group in winter ($P < 0.05$), a season when forage was most critical.

Females were prevented from leaving their family groups in 40 out of 105 attempts. These involved adult females (59%), females with crias (27%), and yearlings and crias (14%). Attempts to leave were highest in summer by flighty new mothers, females who lost their crias, and new group members. Sometimes the male forced a female back to his group after she became separated while the group was chased by another territorial male.

Vicuñas left their family groups without interference from the resident male in 56 out of 105 attempts, especially females searching for missing crias. No males were seen to allow females with crias to leave the family group and territory in summer.

Feeding and sleeping territories.—A vicuña family group feeding territory was located on a lower slope, flat, or bottomland in good habitat, whereas the sleeping territory was on some nearby flattened ridge that separated the broad valleys in an area of poor to fair habitat.

Feeding territories averaged 18.4 ha (range 1.9 to 55.8 ha; n = 98, SD = 10.8) and decreased insignificantly (6%) from year one (19.0 ha) to year two (17.8 ha) as the density of groups increased.

The size of sleeping territories was determined by the area defended around a consistent sleeping site. Sleeping territories averaged 2.6 ha in size (n = 43, SD = 1.4) and decreased insignificantly from year one (2.8 ha) to year two (2.3 ha). Territories remained in the same basic location from one season to the next although size and shape changed slightly.

Territorial borders were well defined and rigorously defended by the occupant male. Boundaries that separated territories were 1 to 3 m wide and learned by the resident family groups. Two adjacent family groups often fed tranquilly only a few meters apart, each on its side of the mutual territorial boundary. If one group wandered across the border, it was promptly chased back to its own territory by the other group's male. Some territorial boundaries conformed to natural topographical features such as streams and gullies, or man-made structures such as roads (Koford, 1957). A few groups

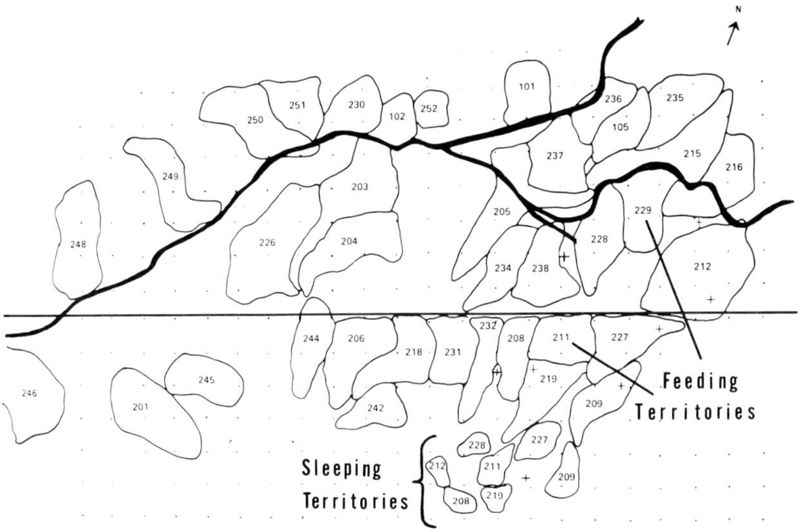

FIG. 15. Size, distribution, and shape of vicuña feeding and sleeping territories during the month of October 1970, Cupitay Valley study area. Scale is 200 m between grid dots.

seemed to use my rock cairns as visual landmarks for boundaries. Flats and upper slopes dominated by the tall perennial bunch grass communities and long gradual slopes without water supported few feeding territories, as in the vacant areas shown in Fig. 15. Vegetation types with moist deep soils at the base of shorter and steeper slopes were favored sites, especially if the site had immediate access to water.

Springs on the slopes were located between two or more feeding territories or just within the border of a single territory. Heavily used springs on the flat were often sandwiched in between territories in an undefended neutral zone.

Three important points emerged on vicuña territoral exclusiveness: 1) regular occupancy was necessary for exclusive use and continued ownership of a territory; 2) territorial exclusiveness was not honored by outsiders coming upon a vacant territory; and 3) neighbors rarely moved into an adjacent territory when the resident was absent.

Restrictions of feeding use as a result of territorial behavior.— When the study began, I arbitrarily distinguished between PTFGs

TABLE 3

COMPARISON OF VEGETATION TYPES IN TERRITORIES OF MARGINAL TERRITORIAL FAMILY GROUPS (MTFGS) AND PERMANENT TERRITORIAL FAMILY GROUPS (PTFGS), SPRING 1970–71. GOOD TO POOR VEGETATION TYPES ARE BASED UPON PREFERENCE, PLANT SPECIES COMPOSITION, AND PRODUCTION OF PALATABLE SPECIES

	Vegetation Types	Percentage of Territories in Vegetation Type			
		MTFGs	PTFGs		
		209 and 219	208 and 211	215 and 216	228 and 229
GOOD	Quisna	1	3	7	5
↓	Chiula & Chiula General	9	22	33	6
	Mojadal	0	0	1	7
	Pampa Verde Claro	0	0	5	40
	Pampa Verde	0	0	4	0
to	Quisna Sin Influencia	0	0	6	4
	Taxona	0	1	2	0
	Peccoy	11	23	5	7
↓	Roqueño	34	39	36	31
	Piedras de la Loma	11	0	0	0
POOR	Ichu	34	12	0	0
Hectares		31.6	22.7	36.3	40.0
Source of water		Semi- or permanent spring	Permanent spring	Stream	Stream

and MTFGs based upon the relative quality of habitat which the later analysis of vegetation types confirmed. Counts of surface area units showed that PTFGs had a higher proportion of the preferred vegetation types (Quisna, Chiula, Chiula General, Mojadal, Pampa Verde Claro, Pampa Verde, Quisna Sin Influencia, and Taxona) than MTFGs. In contrast, MTFGs had higher proportions of the poor vegetation types (Peccoy, Roqueño, Piedras de la Loma, and Ichu) (Chi-square = 6.82, d.f. = 1, $P < 0.01$) (Table 3).

Access to vegetation types was determined by the relative status of different social groups. The lower social status MGs and MTFGs used the common and non-preferred bunch grass communities (Peccoy and Ichu) 1.7 times more than PTFGs ($P < 0.10$ for Ichu). MGs had to rely on the non-preferred Roqueño more than other social groups ($P < 0.10$). In contrast, PTFGs fed on the highly preferred and high-producing Quisna ($P < 0.10$) and Chiula ($P <$

0.10) more than MGs and MTFGs (Franklin, 1978*a*). PTFGs also made the greatest feeding use of the highly preferred Pampa Verde Claro.

Quisna received its greatest use during the summer, especially by PTFGs, when the plant growing and birth seasons occurred. Chiula that was located on the moist to wet soil types on the lower slopes had peaks of high use by most social groups in the dry winter. MGs and MTFGs used the coarse bunch grass dominated communities (Peccoy and Ichu), particularly in the winter and spring, whereas MTFGs used the Riparian community more than other social groups in connection with their visits to the river for drinking water. Mojadal, the year-round green and marshy community, was most important to vicuña during winter when other vegetation had completely dried out and available forage was low.

During the second year of the study, when total density and the number of territorial family groups was higher, MGs had to use vegetation types less preferred by family groups. Roqueño received 17% more use in the second year than the first year by all social groups, with the greatest increase (24%) for MGs.

Relationship between group size, territory size, and forage productivity.—Koford (1957) reported that where forage was sparse and coarse, vicuña territories were larger than where forage was relatively abundant and succulent. The same was observed in this study, and larger territories also contained more animals. In both years there was a significant correlation between mean group size and mean territory size (year one, $r = 0.60$, $n = 11$, $P < 0.05$; year two, $r = 0.63$, $n = 13$, $P < 0.05$). Seasonal correlations between group and territory size for two years were $r = 0.44$ (not significant) during fall, $r = 0.48$ ($P < 0.05$) in winter, $r = 0.72$ ($P < 0.01$) in spring, and $r = 0.77$ ($P < 0.01$) in summer. The highest correlations occurred during critical seasons—spring when forage was minimal and females were in the final phase of gestation, and summer when, although forage was maximal, energy demands were high for birth and lactation.

A primary question of this study was whether the numbers of animals in feeding territories were in balance with food resources, especially during spring when forage conditions were poorest. In the spring of 1970, mean group size and total forage production were highly correlated ($r = 0.86$, $n = 12$, $P < 0.01$). The feeding territories of groups 205 and 228 were excluded from the analysis because they contained unusually high amounts of the highly preferred vegetation types (Pampa Verde, Pampa Verde Claro, or Pampa Verde Obscuro) heavily used by alpaca and sheep. The relationship between group size and territory size was nearly the same as between group size and forage production. Forage production and territory size were highly related (r = 0.98, n = 12, $P < 0.01$) in spite of the great differences in production between vegetation types.

Thus, it appears that the number of vicuña using a feeding territory is related to both the size of the territory and total forage within it. This relationship exists despite the heavy use of the area by domestic animals, suggesting that vicuñas further adjust family group size to compensate.

Territorial marking.—The vicuña is one of few ungulates that always uses a traditional dung pile for defecation and urination. Evidence suggests that defecation-urination behavior is a vital factor in the rigid territorial system and spatial distribution of vicuñas (Franklin, 1980). To summarize: 1) all age classes of both sexes eliminated only on dung piles; 2) although the density of dung piles in the sleeping territories (6.7/ha) and feeding territories (4.3/ha) did not differ, dung piles in the sleeping area (28.5 kg) were four times heavier than those in the feeding area (6.8 kg); 3) two different types of elimination behaviors were observed—Defecation-Urinations (DUs) were done by all animals primarily as routine elimination, and Defecation-Urination Displays (DUDs) were special forms of DUs done only by the territorial adult males during the defense of territories; 4) the frequency of DUDs varied with the potential threat or challenge by other social groups to the defending territorial male; 5) the male tended to perform DUDs towards the outer portion of his territory; and 6) the rate of DUs by group members dramatically increased when individuals were in potentially stressful situations, such as movement to or across the group's territorial borders.

The male's regular use of dung piles appeared to reinforce possession of a territory and functioned more to "keep insiders in" the territory through individual spatial orientation, because dung piles did not "keep outsiders out" in the absence of the resident male and his group (Franklin, 1980). Table 4 is a summary of important and relevant biological, ecological, and behavioral characteristics of the vicuña.

Guanaco Socioecology

The environment.—The guanaco is found in a wide variety of aridland environments ranging from sea level to over 4,000 m, including deserts, shrublands, grasslands, and clumped savannah (Franklin, 1975; Miller et al., 1973). Although unstudied, the occupancy of dry regions is surely related to the guanaco's ability to go for long periods without drinking water when forage moisture content is sufficiently high. They also have been observed to drink brackish and saline water, including water from ocean tidepools (Musters, 1871; Payne, pers. comm.).

The guanaco is more widely distributed than the vicuña, ranging from northern Peru on the west-facing slope and foothills of the Andes, into parts of Bolivia, the length of Chile, onto the Patagonian foothills and plains (Pritchard, 1902), and south to the islands of Tierra del Fuego and Navarino (Miller et al., 1973). Darwin (1845) observed during his visit to South America in the early 1830s that "The guanaco . . . an elegant animal . . . is the characteristic quadruped of the plains of Patagonia . . . It is very common over the whole of the temperate parts of the continent."

The guanaco's wide distribution and dominant role in South America's southern grasslands and shrublands derives from its flexible behavior and social organization. It is a versatile forager, both grazing on grasses and forbs and browsing on shrubs and trees (Franklin, 1975). Some populations are sedentary, whereas others are migratory (Raedeke, 1979).

Field studies of guanaco socioecology have been in progress in southern Chile since 1976. A sedentary population was investigated from December 1976 to January 1978 at Estancia Cameron on the 29,000-km^2 island of Tierra del Fuego. Tierra del Fuego is an extension of the Patagonian plateau, with the northern half of the island a treeless, shortgrass plain, and the southern half a picturesque mixture of beech forests, rivers, lakes, and mountains. The study area was in the interface between the northern plain and the southern forest, a region of pampas, woodlands, meadows, and bogs. The study site was a 150-ha bog meadow locally called Asseradero, 2 km west of the Russfin ranch (53°4′S, 69°55′W). It was flanked by a mature High Deciduous Beech (*Nothofagus pumilio*) forest on the north, a stream, road, and a Low Deciduous Beech (*Nothofagus antarctica*) forest on the south, and sheep fencing on the east and

TABLE 4
CONTRASTS IN THE SOCIOECOLOGY OF GUANACOS AND VICUÑAS

Characteristic	Vicuña	Guanaco
Environment	Central Andes, 3,700–4,900 m elevation; cold, dry, snow-free; growing season predictable; year-round forage available.	Patagonia and arid slopes of Andes, sea level to 4,250 m elevation; cold, dry, windy, with snow cover; predictable seasons; seasonally available forage.
Sex ratio (♂♂:♀♀)	85:100	130–160:100
Morphology	Sexes indistinguishable; 45–55 kg.	Sexes indistinguishable; 100–120 kg.
Feeding behavior	Grazer; obligate drinker.	Grazer, browser; periodic drinker.
Social system	Rigid, sedentary population.	Flexible, both sedentary and migratory populations.
Types of social groups (percent of population)	Family groups (76), male groups (24), solo males (<1).	Family groups (55), male groups (41), solo males (4); female groups and mixed groups.
Territory	Year-round (separated) feeding and sleeping territories for family groups; territory type divided into permanent, marginal, and mobile family groups. Family group zones invaded frequently by male groups. Male defense of territory. Dung piles used by all ages and sexes.	Year-round feeding territories (mainly by males) in sedentary populations; territories of family and male groups only, seasonal in migratory populations. Male groups rarely invade family group zones. Male defense of territory. Dung piles sometimes used by male.
Family group characters	Constant group size and composition, high cohesion, controlled by male. Linear hierarchy; adult male:adult females: offspring. Male leadership; low female independence.	Seasonally variable group size and composition. Low to moderate cohesion (seasonal), low control by male. Linear hierarchy less rigid; adult male:adult females:offspring. Uncommon male leadership; high female independence and aggression.

TABLE 4
CONTINUED

Characteristic	Vicuña	Guanaco
Reproduction		
Birth	Mid-growing season.	Early to mid-growing season.
Nursing	6–8 months.	11–15 months.
Parental care	Expulsion by male of juvenile males 4–9 months, juvenile females 10–11 months (1 growing season). Infrequent adult male–young interaction, only before expulsion.	Expulsion by male of yearling males and females 11–15 months (1½–2 growing seasons). Frequent adult male–young interaction, year-round.
Juvenile behavior	Play within and between groups. Exhibits submissive crouch to male (neck positioned over back) prior to expulsion.	Within-group play more common, includes yearlings. Exhibits frequent submissive crouch to male (neck near ground).
Antipredator behavior	See and flee if predator approaches; females may mob predator during birth season.	See and flee; mobbing uncommon.
Behavior-environment correlations	Male regulation of family group size correlated with forage availability.	Male regulation of family group size, but not correlated with forage availability.
FG size:territory size	Significantly related.	Not related.
FG size:forage production	Significantly related.	Not related.
Territory size:total production	Related.	Related.

west. The elevation was 490 m. An estimated 10,000 guanacos are found on the Chilean half of Tierra del Fuego (Raedeke, 1978).

A migratory population of guanacos has been under investigation since August 1979 at the Torres del Paine National Park in the Department of Ultima Esperanza, Province of Magallanes (72°55′W, 51°3′S). The 32-km² study area, located between Lakes Nordenskjold and Sarmiento, is centered around the Paso de la Muerte region and is primarily a Pre-Andean dry shrub community. Elevation of the surrounding lakes is 110 m, and the highest point in the study area is 250 m. The massive wall of the adjacent Andean mountains abruptly rises to 2,700 m, 4 to 6 km away to the north and northwest. Paso de la Muerte was the last remaining stronghold of guanacos in Torres del Paine when the national park was established and domestic stock were removed in mid-1975.

Climate, topography, distribution of water and vegetation types.—Climate at Tierra del Fuego is highly seasonal, with the summer growing season spanning from December to February. Summer days are cool, rarely above 20°C, and high winds are common. In winter snow covers the open prairie and bog meadow vegetation. The Asseradero study area was basically flat except for a gentle to steep slope leading to the forest in the northwest quarter. Water was available from the Russfin River except in winter when it was frozen. Although water was plentiful, guanacos very rarely drank, implying that their water requirements were fulfilled by forage.

Vegetation types were recognized and defined based upon plant species composition at the Asseradero study site (Jefferson, 1980). Feeding preference was shown for five out of 12 types that covered only 14% of the 152-ha meadow. Three types were particularly important. Vegetation type 4 was dominated by *Poa pratensis* grass and the forb, *Acaena ovalifolia*. It was a green belt that varied in width from 25 to 100 m along the river, and was significantly more productive per unit area than eight of the other 11 vegetation types ($P \leq 0.01$). Although it occupied less than 5% of the area and was covered by snow during the winter, vegetation type 4 was fed on more than any other type year-round (23%, n = 10,959 feeding samples), and was the most preferred forage in spring and second most preferred in summer. Vegetation type 8, occurring at the edge and extending 25 to 50 m into the *Nothofagus* forest, was the second highest producing type, accounting for 12% of total plant production. Dominated by forbs and mosses, type 8 was used mostly in the morning and late afternoon when animals emerged and returned to the forest. Vegetation type 5 was dominated by low growing shrubs (for example, *Chiliotrichium diffusum*) and only accounted for 6% of the total plant production; yet it received the highest feeding use by adult female guanacos (19%) and was highly preferred (P.I. = 557) (Jefferson, 1980).

The migratory population of guanacos in Torres del Paine inhabits a range of foothills at the base of the Andes on the western edge of the Patagonia. This desert-like brushland is dominated by the shrub, *Mulinum spinosum*, and has been classified as a "Matorral Xerofito Pre-Andino" association (Pisano, 1973). In spite of 600 mm of rainfall per year, net precipitation is low due to high evaporation caused by high-velocity dry winds coming from the Andean rain barrier to the immediate west. Numerous lagoons and ponds provide sources of drinking water in the summer. Pond littoral zones ("vegas") are frequently covered with a lush grass-forb community, an important forage type in the snow-free months. In fact, these periodic pond-bog-meadow successional sites, major habitat features of the region that provide both water and high-producing forage for guanacos, have a parallel in the range occupied by guanacos further north. Along the desolate and barren coastal plain of the rainless Peruvian desert occur scattered and temporarily vegetated "lomas" that are created by heavy winter fogs and receive seasonal use by guanacos on the steep west-facing slope of the Andes.

To summarize, guanaco habitats are characterized by: 1) highly seasonal weather, with winter snow cover (if without snow, extremely dry winters); 2) cold to freezing temperatures and moderate to high winds; 3) moderate to high precipitation coupled with high evaporation resulting in dry conditions; 4) low plant productivity; and 5) a slowly changing ecosystem. In addition, guanaco food resources are sparse; high-producing and preferred vegetation of bog-meadows are patchily abundant in a small area during the non-winter.

Excrement-influenced vegetation (EIV) is also created by dung-piling behavior of guanacos. Because not all guanacos use dung piles (see later section), dung piles are smaller and less abundant than those of vicuñas and EIV is much less common. Guanaco EIV has been observed in northern Peru (Franklin, 1975) and in the Asseradero study area on Tierra del Fuego, but not at Torres del Paine National Park. EIV on Tierra del Fuego was primarily around dung piles on shallow-soiled slopes dominated by perennial bunch grasses, but did not have an important environmental impact

FIG. 16. Guanaco family group composed of a territorial male, females, and their offspring less than 15 months old.

as it appeared to at the Calipuy guanaco population in northern Peru (Franklin, 1975). The absence of EIV at Torres del Paine, on poor soil types where they might be expected to be found, suggests that guanacos have not inhabited the area at high enough densities or for long enough periods for this process to have occurred.

Social organization and behavior.—Like the vicuña, the primary social units for the guanaco are family groups (FGs), male groups (MGs), and solo males (SMs). In addition, female groups (FMGs) can occur in sedentary guanaco populations and mixed groups (MXGs) in migratory populations.

Family Groups were composed of an adult male, females, and their young less than 15 months old (Fig. 16). Guanaco group size, largest in summer and smallest in winter, was much more variable within and between seasons than the vicuña, because adult guanaco females with and without young joined and left groups freely. Each sedentary FG had a year-round feeding territory defended and permanently occupied by the adult male. Adjacent feeding territories filled the Asseradero meadow study area, and sightings of MGs on this favorable habitat area were unusual.

Male Groups were composed of mostly immature and non-ter-

FIG. 17. Guanaco male group at Torres del Paine National Park in southern Chile.

ritorial males (Fig. 17). In the summer, yearling females, and in the winter, territorial males, were observed to temporarily join MGs in the migratory population at Torres del Paine. MGs are highly mobile and their size fluctuates from day to day. For sedentary populations, MGs seem to be socially and geographically separated from FGs with MG Zones and FG Zones; but, for migratory populations, the separation occurs only in the late spring, summer, and early autumn. Like the vicuña, *Solo Male* guanacos are mature individuals seeking a territory or with an established territory but no females.

Female Groups include adult females and yearlings and young of both sexes without a mature or territorial male. Although the territorial male and a small number of females and young used the territory almost daily, many of the females left their FGs during the winter and spent most of their time in the forest. The largest FMG seen was 14 but little is known about the size, stability, and movements of FMGs, because they infrequently appeared on the open meadow.

Mixed Groups are aggregations of both sexes of all age classes in winter migratory populations. The largest MXG seen at Torres del Paine was 176 in July 1980. The large herds of guanacos (up to several thousand animals) reported by early Patagonia naturalists were probably MXGs.

Changes in size and composition of family groups.—Guanaco sedentary family group size and composition was influenced by the mid-summer birth season, early autumn dispersal of yearlings, the loss of some adult females and their young in the fall and their return in the spring. Females and young seemed to depart voluntarily, but increased aggression from the adult male and his lack of "attentiveness" to the group may have also contributed. Sedentary family groups at Tierra del Fuego averaged 7 (n = 925, SD = 2.8) (Jefferson, 1980). Seasonal variation in family group size was typified by group 203 whose mean sizes were: winter, 2.9 (SE = 0.2, n = 21); spring, 9.2 (SE = 0.4, n = 45); summer, 12.8 (SE = 0.8, n = 39); and autumn, 9.5 (SE = 0.5, n = 50) (Jefferson, 1980). There was no within-season variation in the mean number of adult females or total family group size, but there was a significant difference in the number of females present between seasons (F = 55.6, 46.3, 31.1, $P < 0.0001$, d.f. = 3, 3; Jefferson, 1980). Family groups in the migratory population at Torres del Paine averaged 16 (n = 18) in the summer month of December 1977 and 14 (n = 20) in December 1978. Family groups and male groups combine into mixed groups on winter ranges.

Relative proportion of social groups.—Data are incomplete for the relative proportion of social groups on Tierra del Fuego because the dense forests preclude a complete census. A summer guanaco census (December 1980) in the open habitat of Torres del Paine, revealed that 54% of all animals (n = 272) were in family groups, 41% in male groups, and 4% solo males (1% miscellaneous). The ratio of males to females was 158:100. Eighty-two percent of males were in male groups, 10% in family groups (as territorial males), and 8% solo. The high proportion (44%) of solo territorial males versus territorial males with females suggested a shortage of females or lack of favorable habitat to attract females. Similarly, in northern Peru 6.4% of all guanacos were solitary males (Franklin, 1975), and on Tierra del Fuego (Raedeke, 1979) the average annual percentage of solitary individuals was about 8%. There is a marked seasonal trend in the occurrence of solitary males, with a significant tendency for males to separate from male groups during the summer breeding season (Raedeke, 1979).

The high proportion of males to females (158:100) in Torres del Paine is unexpected for a polygynous species. The mean of male: female ratios (n = 22 censuses) on Tierra del Fuego from April 1972 to March 1975 was 129:100 (Raedeke, 1979) and was 160: 100 (n = 30 censuses) at Torres del Paine from November 1979 to May 1981 (Ortega, pers. comm.). If the birth sex ratio is equal, more male guanacos survive than females for reasons not yet understood.

Dispersal of young.—Both young male and female guanacos disperse when forcefully expelled from their family group by the adult male. At the Asseradero study area on Tierra del Fuego, eight males and eight females were observed to be evicted from their family groups. There are major differences between vicuña and guanaco dispersal. Vicuña young are expelled when less than one year of age *before* the next birth season, whereas guanaco young are expelled when greater than one year of age (13 to 15 months old) *after* the next birth season. Most (75%) expulsions took place during the last 2 weeks of January and the first 2 weeks of February. Because the Tierra del Fuego birth season began in late December and ended by early February, for several weeks some reproducing guanaco females had offspring from both the former and current year at their sides. In these cases, the yearling offspring were frequently allowed to nurse. Juvenile male vicuñas are expelled from the family group before juvenile females, whereas there is no difference in the sequence of expulsion and dispersal of yearling male and female guanacos. However, yearling guanacos without chulengo siblings (that is, the mother has only the yearling at her side and no chulengo) are expelled from the family group 1 to 6 weeks *before* yearlings that have a chulengo sibling. The adult female commonly defends her yearling and fights with the attacking adult male when he is attempting to expel the yearling offspring.

The expulsion delay of yearlings with a chulengo sibling might relate to the need for additional parental investment or the necessity of a gradual decrease in the mother-yearling bond before expulsion in order to minimize the probability that the female would also leave the family group. If she leaves, the male might forfeit a subsequent mating opportunity.

Guanaco parental investment is not confined to the energetic costs of protection of the young, female lactation, and male protection of the territory, but also includes continuing parental tolerance of yearlings while they forage within a socially defined feeding territory. In effect, the parents are allowing last year's young to compete with this year's young for a limited and high priority resource—food. Although only for a short duration during the height of the plant growing season, the continued, indirect parental investment is apparently required for additional growth and/or socialization before dispersal.

Group cohesion, stability, and permeability.—Guanaco family groups are not as cohesive as vicuña family groups; they are more spread out and fragmented. When feeding undisturbed, vicuña family group members are normally in a compact unit covering an area 20 to 50 m in diameter, whereas guanaco group members are usually spaced out 50 to 200 m. Vicuña family groups move within and between territories as a tight, cohesive unit, whereas guanacos are often spaced out. At the Asseradero study area on Tierra del Fuego, it was not unusual for animals of the same group to emerge from the forest to the edge of the meadow at different times and move across the feeding territory towards the edge of the river 400 to 500 m away. Subunits of the same group might remain separated for several hours.

Guanaco groups were "semi-open" because members could come and go at will and outsiders were only occasionally allowed entrance, whereas vicuña family groups were essentially closed to outsiders and members were *not* permitted to come and go at will. Guanaco males usually chased away female non-members attempting to join their family groups.

Vicuña and guanaco territorial males differed markedly in control of their groups. Vicuña males often forced females to return if they attempted to leave the group, whereas guanaco territorial males permitted female members to leave without interference. This behavioral difference resulted in a sharp contrast in group cohesion, stability, and size between the two species. There were frequent changes in guanaco group size from day to day which would be unusual for vicuña family groups (PTFGs) whose size and composition remained unchanged for months at a time.

The relationship and interactions between family group males and offspring affected group cohesion and stability because the male eventually evicted the young. Vicuña crias uncommonly interacted with the adult male until the final few weeks before their eviction (male crias 4 to 9 and female crias 10 to 11 months old). A common cria behavior during this final period was the "submissive crouch," in which the cria walked or stood in a crouched position with the knees bent, head tucked in, ears horizontal, neck arched *posteriorly* over the back, and the tail in a full forward curl. The submissive crouch was seen when crias were threatened or attacked by the adult male and was considered an appeasement posture that might lessen male aggression and thus delay the crias' expulsion. Several types of submissive crouches were recognized—nursing, semi-submissive, full (described above), exaggerated, and forced—but the "forced-submissive crouch" was especially common in the final weeks before expulsion. In this form the cria suddenly and unexpectedly approached the male in a submissive crouch forcing him to first threaten it, then typically *avoid* the cria by turning or walking away.

In contrast to vicuña crias, guanaco chulengos begin interacting with the family group male with semi-submissive crouches (tail in forward curl, slight crouching, and slight to no change in neck posture) when only two to 6 weeks old. Full-submissive crouches were common by 2 months of age, 11 to 13 months before expulsion. Chulengo crouching occurred not only during routine feeding, but during mating and territorial defense. Guanaco chulengo submissive crouches were the same as those of the vicuña, except the long neck was arched *anteriorly* toward the ground. In the final 1 to 2 months with the group, yearling guanacos singly or in small groups of two to five were constantly going into full and forced-submissive crouches with the male who usually escaped by walking or trotting away. The adaptive value of the species difference in the timing and intensity of offspring-male interactions will be discussed below.

Feeding territories and separation of social groups.—The entire 1.5-km^2 bog-meadow at Asseradero was divided into five juxtaposed feeding territories which were vigorously defended by the resident adult males during all seasons except winter, when their continued presence probably served as a passive form of defense. Males chased trespassers out, conducted border "stand-offs" with adjacent territorial males, and exhibited DUDs on dung piles. Common boundaries between two guanaco territories were well-defined zones 5 to 10 m wide, instead of the precisely delineated "lines" 1 to 3 m wide of the vicuña.

Although territorial defense occurred within the forest where the groups slept, there were insufficient data to determine how much additional area was defended or if sleeping territories existed. The forest was not considered an important source of forage because there was no substrate vegetation in this particular mature beech forest. There is no indication thus far that separate sleeping territories exist.

Size of feeding territories was influenced by habitat quality and productivity. Feeding territores ranged from 2 to 46 ha and averaged 29.5 (n = 8, SD = 11.1; Jefferson, 1980) on the productive Asseradero meadow at Tierra del Fuego, whereas during the summer of 1977 in the desert shrubland of Torres del Paine, feeding territóries varied in size from 25 to 120 ha and averaged 65 (n = 4; Franklin, 1978*b*). Territory size was larger at the west end of Asseradero on a dry slope (dominated by unpalatable shrubs and grasses) than on the damp east end (dominated by palatable grasses and forbs) (mean = 42.2 ha, SD = 2.3, compared to mean = 21.7, SD = 3.8). The largest territory (46 ha) at the west end contained 86% and 80%, respectively, of the two least preferred forage types used by guanacos (Jefferson, 1980). This territory was occupied by a solo male for over 2 years, suggesting that the area was not desirable enough habitat to attract females.

The Asseradero meadow was a zone permanently occupied and exclusively used by territorial family groups and solo males. During 2 years of study, less than 25 observations were made of male groups in Asseradero, and then only for a few minutes before they were chased away by resident territorial males. In contrast, three to six sightings per day were common of vicuña male groups attempting to use the Cupitay Valley study area occupied by family groups.

Beyond the Asseradero study area, guanaco male groups consistently occupied the same area throughout the year. At Torres del Paine, male and family groups have been observed during three different summers to be permanently separated from one another. Thus, it appears that non-migratory guanaco male groups are geographically and socially separated from family groups into MG Zones and FG Zones, respectively. FG Zones are partitioned into family group feeding territories and unoccupied neutral areas, whereas MG Zones contain only male groups. The mechanisms for maintaining this physical and social separation are: 1) family groups do not venture into the MG Zones; 2) male groups are prevented from entering the FG Zone by resident territorial males; 3) male groups avoid the FG Zone; and 4) dominant male members of MGs may contain males within the MG Zone (Franklin, 1978*b*).

At the Paso de la Muerte study area in Torres del Paine during the summer of 1977, four family groups occupied the 250-ha FG Zone (0.14 animals/ha) and one large male group occupied the adjacent 475-ha MG Zone (0.20 animals/ha). Both zones had access to water and the highly preferred and productive "vegas" around the shores of ponds and lagoons. However, the FG Zone appeared to have a greater number of hills, ravines, and draws that might be advantageous for females with chulengos when seeking "cover" from human disturbance or the high-velocity, chilling winds typical of the Patagonian foothills. The hills also offered more observation points for territorial males for the detection of trespassing animals and predators.

The complete social and geographical separation of reproducing females from non-reproducing males may reduce competition for limited resources and prevent social harassment of females and young by immature males. Guanaco MG and FG Zones also have important conservation, management, and research implications. If guanaco male groups can be predictably found within MG Zones, then their location and observation by tourists, scientists, and wildlife managers is made easier.

Relationship between group size, territory size, and forage productivity.—Jefferson (1980) showed that as total surface area of guanaco territories in Asseradero increased, so did total plant production (r = 0.95, $P < 0.01$). Although the largest territory (107) was dominated (67%) by the two least preferred vegetation types (11 and 14) whose productivity was only low to intermediate, total plant production was high due to the large surface area of the territory. However, as the percent surface area increased of vegetation type 4, a highly preferred forage (P.I. = 477) and the highest producing type on the meadow, territory size decreased (r = −0.91, $P < 0.05$ for all seasons but winter). As production per ha of territory increased, size of territories decreased in summer and autumn (r = −0.85, $P < 0.10$) and nearly so for winter and spring (r = −0.80, $P < 0.20$).

Guanaco group size was nearly significantly correlated with territory size in the winter (r = 0.84, $P < 0.10$, d.f. = 3). However, guanaco mean group size was not correlated with total primary productivity within territories ($r = -0.21$, $P < 0.60$, d.f. = 6), nor were mean group size and productivity per ha significantly related ($r = 0.61$, $P < 0.20$, d.f. = 6). Therefore, even though the size of feeding territories defended by guanaco males proportionally declined with increasing percentage of high-producing preferred forage and with increasing production per ha of territory, the mean number of family members using the territory was not related to forage availability (production) within the territory nor to production per ha of territory. Simply stated, the number of animals using a guanaco male's feeding territory did not seem to be adjusted to the production of forage within the territory.

Territorial marking.—Dung piles were found in all the areas inhabited by guanacos from northern Peru to Tierra del Fuego. At the Asseradero study area, they occurred both on the meadow and in the adjacent forests. Compared to the vicuña, guanaco dung piles were less common (4.5/ha versus 0.5/ha) but, unlike the vicuña, single defecations (pellet groups) were also scattered throughout the area.

Only guanaco males regularly used dung piles; both sexes of vicuñas used them. At Tierra del Fuego, adult females were seen only twice in 2 years to eliminate on dung piles, compared to the thousands of DUs not on dung piles. During three summer months of study at Torres del Paine, less than 10 observations were made of females using dung piles. However, several females in family groups at Asseradero sometimes DUed in the same general vicinity, resulting in an undefined "DU-Zone" 5 to 15 m in diameter containing three to 15 pellet groups. Family group chulengos as young as 3 weeks old occasionally stopped on dung piles and intensively smelled and even consumed fecal pellets; yearlings uncommonly stopped to smell or use dung piles.

Adult males DUed on dung piles 52% ($n = 81$) of the time during the daily routine activities of feeding and moving. During aggressive interactions with other guanacos, males used dung piles 78% ($n = 90$) of the time when they stopped for a DUD (Defecation-Urination Display). Thus, dung-pile use was more prevalent during territorial defense. Although some dung piles were on or near territorial borders, they were also scattered throughout the territory.

A summary of relevant guanaco biological, ecological, and behavioral characteristics is given in Table 4.

Discussion

Behavior, Social Organization, and Environment

The evolutionary interaction between an animal's behavior, social system, and environment is obviously complex. Because we cannot delve into the past to investigate these relationships and interactions, current research (and the following discussion) employs two basic approaches: 1) the adaptive approach asks what is the adaptive significance of a particular behavior observed in nature, and 2) the correlational approach evaluates several closely related species or populations of the same species, each living in different habitats, and interprets the differences in social behavior as the outcome of natural selection. We must be cautious, however, not to use the correlational approach without proposing testable hypotheses about the specific aspects of social organization and ecology of a species (see King, 1973).

The social organization of the vicuña and guanaco are remarkably similar in that family groups and male groups are the primary social units. Yet the major and subtle differences may be explained by environmental and behavioral contrasts of the South American wild camelids. Table 4 summarizes the similarities and differences of vicuña and guanaco (sedentary and migratory) social organizations and ecology.

Throughout its historic and present range, the vicuña has been limited to the puna grasslands of the Andean altiplano. The vicuña is the only wild ungulate to prosper on this alpine grassland environment with its fluctuating and freezing temperatures, light precipitation, short but predictable growing seasons, and low plant production. Preferred and more productive vegetation types for the vicuña were very limited in area and slowly renewing, although locally abundant in small and dispersed patches. At lower elevations the grassland gives way to brushland, a habitat better suited to guanaco and huemul deer (*Hippocamelus antisensis*). Two conditions have favored sedentary habits of the vicuña—uniform rainfall patterns over extensive areas of the altiplano and the sharp transition from the altiplano grassland to the lower elevation brushland. During the dry season vicuña cannot move elsewhere to find better forage.

In contrast, the more flexible behavior of the guanaco allows it to occupy a wider array of dry environmental types and it is more widely distributed than the vicuña. Thus, its present (though in greatly reduced numbers) and historical range has included all the dry and high-plateau aridlands of the continent except for the puna grasslands of the central Andean altiplano. Whereas the vicuña is sedentary, the guanaco is both sedentary and migratory. Two conditions have favored the migratory habits of the guanaco: 1) winter snow cover or the near absence of dry winter forage, and 2) available and improved forage conditions at alternate sites. The expansive Patagonian cold desertland of shrubs and grasses has been especially favorable habitat for the guanaco and can be considered the typical or representative environment for the species.

Both the vicuña and guanaco occupy environments in which food and water are scarce; scattered patches (clumps) of preferred and productive vegetation types have favored a feeding territorial systems in both species. For such a territorial system to be feasible, the limiting resources must be defendable (Brown, 1964); they must be predictable in space and time (Wilson, 1975); and they must exist on small areas where plant community diversity and productivity fulfill the animal's living requirements (Geist, 1974). In the altiplano and Patagonia, highly productive and preferred foods are predictable in time and space even though there are marked seasonal changes in forage abundance; food is locally abundant on patchy sites; high visibility permits adult males to detect territorial interlopers; and, although visible natural landmarks for marking territories are limited (especially for the vicuña), dung piles may in part substitute for this absence.

To establish a territory is a long and costly process for male vicuñas and guanacos, involving many chases and fights. On average, 6 months elapsed before a vicuña male obtained females on his territory. The year-round need for food resources and the high cost of reestablishing a territory apparently makes it more economical to defend the territory year-round for vicuña and sedentary guanaco. Peaks of territorial defense, however, coincide with seasons when competition for forage and estrous females was highest.

The size of social groups is considered to have evolved in response to two environmental factors—the distribution and density of essential resources and the level of predator pressure (Altmann, 1974). Animals of open grasslands rely upon early detection of approaching predators, concealment within a large group, and group defense or quick escape. A group of larger size can usually more effectively detect and cope with predators (Kummer, 1971; Treisman, 1975). However, there should also be opposing selection for a smaller group size to utilize the food supply most efficiently. In the altiplano small vicuña family group size has probably been favored because slowly renewing food resources are both sparse and patchy, and a male with a small family group can defend a small but "complete" patch of the environment.

Because detection of predators is more difficult at night, one might expect nocturnal groups to be larger, as in Hamadryas baboons (*Papio hamadryas*) that feed in segregated small groups during daytime but reunite in larger groups at night (Kummer, 1968). Several vicuña family groups do sleep close together in a compact cluster of sleeping territories. Large all-male groups in both vicuña and guanaco may also benefit from improved predator detection.

The temporary merger of several vicuña family groups at water sources in neutral zones may be due to the indefensibility of this essential resource with its spatially restricted distribution. Altmann (1974) referred to such concentrated but spatially separated resources as "supermarkets." Sources of permanent water in the late dry season were under such heavy use by local and outside groups that they could not be defended by a territorial male, even if the water was within a territorial border.

The fluctuating group size of sedentary guanacos resulted from a marked seasonal difference in the distribution and availability of food. In summer there was an abundance of high-producing and preferred vegetation types *within* the feeding territories. This forage became unavailable in winter because of snow, and some animals left their family groups. Because guanaco family groups were seasonally unstable, they were also much less cohesive, females were more independent, and the adult male had less control over the group throughout the year than observed in the vicuña. Seasonal changes in guanaco food distribution and availability and changes in group size were probably responsible for the absence of a correlation between guanaco group size and total forage production of the territory.

Social Systems, Mating Strategies, and Individual Fitness

The mating systems of the vicuña and guanaco are polygynous—territorial males control access to more than one female. More precisely, it is *resource defense polygyny* because males defend food resources essential to females (Emlen and Oring, 1977). In their review of ecology and mating systems, Emlen and Oring (1977) outlined two preconditions for the evolution of polygamy. First, the environmental potential for polygamy (EPP) depends upon the degree that multiple mates, or the resources for attracting them, can be economically defended; patchy distribution of preferred and high-producing forage types in the selected Andean and Patagonian environments favors economic monopolization of several mates through resource defense. The second prerequisite is the ability to utilize this EPP; vicuña and guanaco females assume the major role of parental care, thus freeing the territorial male to defend the necessary food resources.

Territorial male vicuñas and guanacos have several strategies for enhancing their fitness. Natural selection has presumably favored individuals, both male and female, that can distinguish different habitats and especially those forage types with highest food value. The strongest males should settle on the most suitable forage types and establish the largest possible territory commensurate with the degree of competition from rivals. Such males will presumably attract more females than males forced to establish smaller territories in good habitat or larger territories in less suitable habitat.

At Pampa Galeras, male vicuñas with the largest territories usually had the most forage and the largest family groups. The correlation between group size and forage production in vicuñas suggests that territorial males were regulating family group size based on forage availability. By contrast, a correlation was absent in sedentary guanacos where group members did not depend upon the feeding territory for their year-round supply of forage, and females often left during the winter. Although both territorial vicuña and guanaco males behaviorally regulated the density of animals within their feeding territories, the pattern was weaker in guanacos.

A supplementary strategy for males is to forego the attempt to acquire a territory until older and stronger. Younger males formed all-male groups and made up 70 to 80% of all males observed during these studies. However, once having gained a territory, a male's fitness will depend upon the number of females he acquires. There should be an optimum number of females acquired by a male based on the total forage available to females and young within the male's territory, and a male should show increasing intolerance and reject females that would actually reduce his overall fitness by over-exploitation of the territory. In fact, territorial males were commonly aggressive towards unattached females attempting to join family groups.

Natural selection must also influence the amount of parental investment the male provides to his offspring (Trivers, 1972, 1974) through permitting them to remain on the territory. The male should be tolerant until the offspring can survive on their own, at which point he should evict them. In fact, juvenile vicuñas and yearling guanacos were expelled from their family groups. The later expulsion of guanaco offspring (in contrast to vicuña) that returned to the feeding territory during the second growing season may be necessary because 1) guanacos are twice the size of vicuñas and may need to reach a minimal body size before dispersal, and 2) the winter snow cover shortened the length of time preferred forage was available. Later dispersal of young occurs in the Olympic marmot (*Marmota olympus*), which occupies a harsh environment with a short growing season, than in the yellow-bellied marmot (*M. falviventris*) and woodchuck (*M. monax*), which occur in milder climates with longer growing seasons (Barash, 1974).

Strategies of adult females for maximizing fitness differ from those of males. The site and family group that a female selects may depend upon the abundance of food and water in relation to territory size, the number of competing females, security from potential predation, or the characteristics of the territorial male (size, rate of defense, and his acceptance of the female).

In species where females interact with the male primarily for breeding, such as the Uganda kob (*Adenota kob*), elephant seal (*Mirounga angustirostris*), and sage grouse (*Centrocercus urophasianus*), the female must choose mates mainly on the behavioral and physical characteristics of the males. Selection should be for size, strength, degree of activity, and experience. Because vicuña females are permanently with males in year-round feeding territories, they can base mate selection much more on the actual resources of the territory. Because the male's access to females is determined long before the breeding season, males do not intensively fight for estrous females. The minimal time and energy spent on courtship and mating may in part account for the lack of sexual dimorphism in vicuña.

Female fitness also depends upon family group size. A female joining a relatively inexperienced territorial solo male might have few to no female competitors compared to a long-established territorial male with many females. Yearling females most frequently joined solo males when there was little competition from other females, whereas adult females joined established family groups.

A resident female's fitness declines with each additional female joining her family group. Females should be intolerant of intruding females seeking group membership, as was reported by Koford (1957) for vicuñas. Though intolerance was rarely observed in my vicuña study, it was occasionally seen with guanacos. These differences might be due to differences in population density.

Female vicuñas should be familiar with and stay within the boundaries of an area that adequately provides their needs; attempts to leave will be hindered by the resident male and other territorial males will attack them. Female use of dung piles may be an important aid for orientation and assistance in staying *within* the territory (Franklin, 1980). By contrast, female guanacos do not use dung piles, which correlates with their leaving territories when unfavorable conditions exist and with their habit of showing much less attachment to an area than territorial male guanacos (which regularly use dung piles) and male and female vicuñas (which always use dung piles).

The difference between guanacos and vicuñas in the appearance and intensity of male-offspring interactions may derive from the fact that vicuña crias permanently remain with their family groups until expelled, whereas many guanaco young temporarily leave the group with their mothers during the winter and then return in the spring. It would be advantageous for guanaco young to be identifiable to the male as a former group member to ensure reacceptance into the family group-feeding territory. Frequent interaction would promote such needed familiarity and identification. Similarly, the male should be able to identify more precisely his own offspring; the greater the familiarity a male has with a female and her young, the greater the probability that the male sired that offspring.

A consequence of the one-male territorial family social system in the vicuña and guanaco is that it gives females and young guaranteed access to a source of high-producing and preferred food. Bachelor vicuña males were resident in the less favorable habitats, but could survive there presumably because they had lower energetic requirements than territorial males. Perhaps this separate foraging of all-male groups from one-male family groups allowed better allocation of limited food resources to adult females that were usually pregnant or lactating, as Kummer (1968, 1971) found for desert-dwelling Hamadryas baboons. Separate foraging also occurs under poor conditions in all-male groups of the gelada baboon (*Theropithecus gelada*) that inhabit arid mountain and alpine environments of Ethiopia (Crook, 1966, 1970*b*).

Vicuña and Guanaco Social Organization Compared to other Ungulates

Most ungulate territorial systems studied to date are based upon mating rather than feeding. Owen-Smith (1972) stated that territoriality in ungulates has evolved primarily as a system for ordering reproductive competition among males. If so, the year-round territorial feeding system of one-male groups in vicuñas and guanacos is an exception.

Territoriality is uncommon among North American ungulates, but is a major feature of the social organization of endemic pronghorn antelopes (*Antilocapra americana*). The pronghorn's high plateau habitat of short grasslands and shrublands resembles guanaco habitat in South America. The social systems of pronghorns and migratory guanacos are similar. Mature pronghorn bucks hold territories from early spring to late autumn, although the breeding season lasts for only a few weeks in early autumn (Kitchen, 1974). Recent studies have found that forage availability was a primary ecological force behind pronghorn territoriality (Bromley, 1977), and territories are located in relation to forage availability (Kitchen, 1974). In late autumn mixed herds of both sexes of all age classes combine together locally or migrate to winter grounds. Bromley (1977) believed that mixed herds formed because of a cessation of plant growth and an increase in predatory pressure.

Among African ungulates territoriality is common. In his review of African bovid social organization, Estes (1974) reported that 85% of some 70 species of bovids were territorial during at least part of the year. Vicuñas closely resemble bovids arranged by Jarman (1974) into Class D, which included primarily three genera of the subfamily Alcelaphinae—*Connochaetes* (wildebeests), *Damaliscus* (tsessebe, topi, blesbok, and bontebok), and *Alcelaphus* (hartebeests). Antelopes of this subfamily lack sexual dimorphism and have both migratory and non-migratory populations, or sedentary and nomadic phases, within one population. In many areas there is a seasonal alternation between phases; the sedentary phase (involving a mosaic of breeding territories) persisted when food was abundant, but changed to the nomadic phase (involving large mobile aggregations) when food was restricted. The migratory phase of Alcelaphinae social organization is similar to that of migratory guanacos except that African antelopes sometimes form temporary mating territories when moving herds become stationary (Jarman, 1974).

The sedentary phase of Alcelaphinae social organization has many features in common with the vicuña—a proportion of adult males defends permanent year-round territories, territories are occupied by one-male and female-offspring groups (nursery herds), and non-territorial males form bachelor herds that may be driven into marginal habitat. An aspect that differs from the vicuña is that Alcelaphinae nursery herds move within home ranges that overlap many male territories. It has been reported, however, that under some conditions female membership within territories of this group may be semi-permanent or permanent over extended periods (for example, see David, 1973; Estes, 1966, 1969; Joubert, 1972).

Whether these year-round territorial systems with one-male and permanent female groups represent feeding or breeding territories is not clear from the work done to date. However, because the arid grassland and brushland habitats of the tsessebe (*Damaliscus lunatus*) and bontebok (*Damaliscus corcas*) suggest that food and water are scarce, these species may also occupy feeding territories like vicuñas.

The strong similarities between pronghorn antelope and migratory guanaco, migratory Alcelaphinae and migratory guanacos, and sedentary Alcelaphinae and vicuñas are striking examples of evolutionary convergence in social systems by widely separated and unrelated groups. Similar ecological conditions appear to be the selective forces.

In summary, the closely related vicuña and guanaco, the dominant aridland herbivores of South America, have generally similar social systems, but important differences are observed. Parallel environmental conditions offer explanations for similarities, whereas contrasts in social behavior and organization appear to be the result of differing environmental conditions. Sedentary vicuñas and migratory guanacos differ the most, probably because of permanent availability and seasonal unavailability of forage within feeding territories, respectively. Uneven and localized distribution of high-producing and preferred forage types has favored the evolution of a feeding territorial system occupied by single male family groups. In vicuñas, non-producing males are prevented from competing with reproducing females by being relegated to marginal habitat. Males, females, and young each have evolved behavioral strategies for increasing survival and reproductive fitness in this dry, sparsely producing, and generally harsh environment. In the future many unanswered questions on South American camelid socioecology will be examined as field investigations continue.

Acknowledgments

Vicuña research was conducted through Utah State University in collaboration with the Peruvian Ministry of Agriculture. Valuable assistance from the General Directory of Forestry and Wildlife during my 3½ years in Peru made the work possible. Vicuña research was financed by The Conservation Foundation, World Wildlife Fund, Peruvian Ministry of Agriculture, and the Utah State University Ecology Center. To the many who supported and helped on the project I express my thanks. Special appreciation goes to my mentor Allen W. Stokes, to Carlos F. Ponce del Prado of the Peruvian Ministry of Agriculture, to game guard Saturnino Torres Blanco, assistant and friend Feliciano Mendoza Saiz, and to R. G. Miller, former Director of Foresta Institue.

Guanaco research was conducted through Iowa State University in collaboration with the Chilean Ministry of Agriculture's office of CONAF and the Institute of Patagonia, Punta Arenas. Guanaco research has been finanancially supported by The Organization of American States, Carnegie Museum of Natural History, National Geographic Society, Iowa State University Graduate College, World Food Institute, Institute of International Education (graduate student R. T. Jefferson, Jr.), Earthwatch, Iowa State University Research Foundation, and the Tishman Foundation. Special gratitude goes to Bob Jefferson, Isaac M. Ortega, Daniel Tishman, Mateo Martinec, Claudio Cunazza, my family, and Ona.

Literature Cited

ALTMANN, S. A. 1974. Baboons, space, time and energy. Amer. Zool., 14:221–248.

BARASH, D. P. 1974. The evolution of marmot societies: a general theory. Science, 185:415–420.

BROMLEY, P. T. 1977. Aspects of the behavioural ecology and sociobiology of the pronghorn (*Antilocapra americana*). Unpubl. Ph.D. dissert., Univ. Calgary, Calgary, 370 pp.

BROWN, J. L. 1964. The evolution of diversity in avian territorial systems. Wilson Bull., 76:160–169.

CROOK, J. H. 1966. Gelada baboon herd structure and movement. A comparative report Symp. Zool. Soc. London, 18:237–258.

———. 1970*a*. Social organization and the environment: aspects of contemporary social ethology. Anim. Behav., 18:197–209.

———. 1970*b*. The socio-ecology of primates. Pp. 103–166, *in* Social behaviour in birds and mammals (J. H. Crook, ed.). Academic Press, New York, 492 pp.

DARWIN, C. 1845. Journal of researches into the natural history and geology of the countries visited during the voyage of HMS "Beagle" round the world. Ward, Lock and Co., London, 492 pp.

DAVID, J. H. M. 1973. The behaviour of the bontebok, *Damaliscus dorcas dorcas,* (Pallas 1766), with special reference to territorial behaviour. Z. Tierpsychol., 33:38–107.

EISENBERG, J. F. 1966. Social organizations of mammals. Handbuch der Zoologie, VII (10/7), Lieferung, 39:1–92.

EMLEN, S. T., AND L. W. ORING. 1977. Ecology, sexual selection, and the evolution of mating systems. Science, 197:215–223.

ESTES, R. D. 1966. Behaviour and life history of the wildebeest (*Connochaetes taurinus* Burchell). Nature, 212:999–1000.

———. 1969. Territorial behaviour of the wildebeest (*Connochaetes taurinus* Burchell 1823). Z. Tierpsychol., 26:284–370.

———. 1974. Social organization of the African Bovidae. Pp. 166–205, *in* The behaviour of ungulates and its relation to management (V. Geist and F. Walther, eds.). I.U.C.N., Morges, Switzerland, 24:1–940.

FISLER, G. F. 1969. Mammalian organizational systems. Contrib. Sci., Los Angeles Co. Mus., 167:1–32.

FRANKLIN, W. L. 1973*a*. High, wild world of the vicuña. Natl. Geogr., 143:76–91.

———. 1973*b*. Conservation of the vicuna in Peru. Pp. 209–212, *in* World wildlife yearbook 1972–73 (P. Jackson, ed.). World Wildlife Fund, Morges, Switzerland, 311 pp.

———. 1974. The social behavior of the vicuña. Pp. 477–487, *in* The behaviour of ungulates and its relation to management (V. Geist and F. Walther, eds.), I.U.C.N., Morges, Switzerland, 24:1–940.

———. 1975. Guanacos in Peru. Oryx, 13:191–202.

———. 1978*a*. Socioecology of the vicuña. Unpubl. Ph.D. dissert., Utah State Univ., Logan, Utah, 172 pp.

———. 1978*b*. Preliminary results: behavioral ecology of guanaco male groups at Torres del Paine National Park, Chile. Earthwatch Field Rep., 7 pp.

———. 1980. Territorial marking behavior by the South American vicuña. Pp. 53–66, *in* Chemical signals: vertebrates and aquatic invertebrates (D. Müller-Schwarze and R. M. Silverstein, eds.). Plenum Press, New York, 445 pp.

FRANKLIN, W. L., AND J. M. LIEB. 1979. The social organization of a sedentary population of North American elk: a model for understanding other populations. Pp. 185–198, *in* The ecology, behavior and management of North American elk (M. Boyce and L. Hayden-Wing, eds.). Univ. Wyoming Press, Laramie, 294 pp.

GEIST, V. 1974. On the relationship of social evolution and ecology in ungulates. Amer. Zool., 14:204–220.

JARMAN, P. J. 1974. The social organization of antelope in relation to their ecology. Behaviour, 58:215–267.

JEFFERSON, R. T. JR. 1980. Size and spacing of sedentary guanaco family groups. Unpubl. M.S. thesis, Iowa State Univ., Ames, 32 pp.

JOUBERT, S. C. J. 1972. Territorial behaviour of the tsessebe (*Damaliscus lunatus lunatus* Burchell) in the Kruger National Park. Zool. Africana, 7:141–156.

KING, J. A. 1973. The ecology of aggressive behavior. Ann. Rev. Ecol. Syst., 4: 117–138.

KITCHEN, D. W. 1974. Social behavior and ecology of the pronghorn. Wildl. Monogr., 38:1–96.

KOFORD, C. B. 1957. The vicuña and the puna. Ecol. Monogr., 27:153–219.

KUMMER, H. 1968. Social organization of hamadryas baboons: a field study. Univ. Chicago Press, Chicago, 189 pp.

———. 1971. Primate societies; group techniques of ecological adaptation. Aldine-Atherton, Chicago, 169 pp.

MILLER, G. S., JR. 1924. A second instance of the development of rodent-like incisors in an artiodactyl. Proc. U.S. Natl. Mus., 66:1–4.

MILLER, S., J. ROTTMANN, AND R. D. TABER. 1973. Dwindling and endangered ungulates of Chile: vicugna, lama, hippocamelus, and pudu. N. Amer. Wildl. Nat. Res. Conf., 38:55–68.

MUSTERS, G. C. 1871. At home with Patagonians. John Murray, London, 375 pp.

NATIONAL ACADEMY OF SCIENCES—NATIONAL RESEARCH COUNCIL. 1962. Basic problems and techniques in range research. Publ. National Academy of Sciences—National Research Council, Washington, D.C., 890:1–341.

ONERN. 1971. Inventario, evaluación y uso racional de los recursos naturales de la costa: cuenca del Rio Grande (Nazca). Vol. 1, 458 pp.

OWEN-SMITH, N. 1972. Territoriality: the example of the white-rhinoceros. Zool. Africana, 7:273–280.

PETRUCCI, R. 1906. Origine polyphylétique, homotypie & non-comparabilité directe des sociétés animales. Misch & Thron, Brussels, 126 pp.

PISANO, E. 1973. La vida en los Parques Nacionales de Magallanes. Monografía Inst. de la Patagonia, Punta Areanas, Chile, 6:1–65.

PRICHARD, H. 1902. Field-notes upon some of the larger mammals of Patagonia, made between September 1900 and June 1901. Proc. Zool. Soc. London, 1:272–277.

RAEDEKE, K. J. 1978. El guanaco de Magallanes, Chile: distribucion y biologia. CONAF, 181 pp.

———. 1979. Population dynamics and socioecology of the guanaco (*Lama guanicoe*) of Magallanes, Chile. Unpubl. Ph.D. dissert., Univ. Washington, Seattle, 404 pp.

TOSI, J. A., JR. 1960. Zonas de vida natural en el Peru. Instituto Interamericano de Ciencias Agrícolas de la OEA, Zona Andina, Boletín Técnico, 5:1–271.

TREISMAN, M. 1975. Predation and the evolution of gregariousness. I. Models for concealment and evasion. Anim. Behav., 23:779–800.

TRIVERS, R. L. 1972. Parental investment and sexual selection, Pp. 136–179, *in* Sexual selection and the descent of man, 1871–1971 (B. G. Campbell, ed.). Aldine-Atherton, Chicago, 378 pp.

———. 1974. Parent-offspring conflict. Amer. Zool., 14:249–264.

WILSON, E. O. 1975. Sociobiology, the new synthesis. Belknap Press, Cambridge, Massachusetts, 697 pp.

WYNNE-EDWARDS, V. C. 1965. Social organization as a population regulator. Symp. Zool. Soc. London, 14:173–178.

PART IV.

THE ADAPTIVENESS OF BEHAVIOR: CONSTRAINTS, POPULATION MECHANISMS, AND EVOLUTION

INTRODUCTION

We have seen through the previous chapters that behavior can be studied from a wide variety of perspectives. Behavioral competence changes in predictable stages as an individual matures. Behavior may be viewed as the dynamic expression of an organism's morphology. Behavior patterns may also be analyzed from the standpoint of their survival function. In brief, behavior patterns can be viewed as adaptive mechanisms that are as much a product of natural selection as morphology.

In this section, we view behavior in the context of broader evolutionary concepts. Behavioral limits are defined by morphology. How does the simple metric of size limit behavior? Indeed, what selective pressures determine the size of an organism? Clutton-Brock and Harvey analyze the advantages and disadvantages of evolving a large body size. That such a simple measure as body size could be such a powerful tool in predicting life history strategies is not at all obvious, yet general principles and rules can be derived from just such a consideration of relative mass.

McNab offers yet another perspective. Adaptation for a specific trophic strategy has profound consequences for an organism. Adaptation for feeding on a ubiquitous but energy poor dietary substrate has an impact on optimal body size, metabolic rate, longevity, and the normative form of the social organization. Size and trophic strategy are interrelated in subtle ways and ultimately the form of the social organization is an outcome of the balance struck between energy for maintenance and energy for reproduction.

Tamarin demonstrates that social behavior must be considered not only from the perspective of the family unit involved in the rearing of young but also the interrelationship of reproductive units that interact to produce population phenomena. Why is it that some species typically show wide oscillations in numbers while others have a relatively constant number of individuals per unit of space? What behavioral mechanisms are adaptively operative in the natural regulation of animal numbers?

Finally, in this section we come to a consideration of fundamental mechanisms. How does social behavior evolve? Herein lies a problem of definition. In the lexicon on social behavior studies lies the word altruism. What is an altruistic act? Brian Bertram attempts to define altruism and then attempts to discuss the selective mechanisms to account for the evolution of the forms of altruism found within the Mammalia. Hendrichs concludes the section with a number of caveats concerning the assumptions underlying sociobiological theory. His warning serves as a reminder that ethology and sociobiology are young sciences and by no means have we achieved a complete synthesis of theory and data.

J.F.E.

THE FUNCTIONAL SIGNIFICANCE OF VARIATION IN BODY SIZE AMONG MAMMALS

T. H. CLUTTON-BROCK AND PAUL H. HARVEY

Abstract

THEORIES concerning the costs and benefits of variation in body size among mammals are reviewed. We argue that, if all other things were equal, major increases in body size would be disadvantageous because they reduce reproductive rate. However, all other things are not equal and large body size can confer a wide variety of benefits including 1) the ability to produce large neonates or litters, 2) reduction in relative heat loss, 3) the ability to catch or handle larger prey or to move longer distances in search of food, and 4) the ability to survive on qualitatively inferior food.

Introduction

As Julian Huxley once remarked, size has a fascination all of its own—but it has repercussions in virtually all other areas of biology. There are few anatomical, physiological, or behavioral parameters that are not affected by size (Alexander, 1977; Clutton-Brock and Harvey, 1979*a,* 1979*b*; McNab, this volume). As a result, understanding the functional significance of size differences is of central importance in understanding adaptation. However, because size has such far-reaching consequences, it is likely to be affected by many different selection pressures and, for this reason, identifying the functions of particular differences is difficult. Many alternative explanations can usually be proposed and close correlations between size and particular ecological variables are not to be expected.

Size and Reproductive Rate

We tend to regard the vertebrates in general and the mammals in particular as the pinnacles of evolution (for example, see Wilson, 1975). Yet compare the potential reproductive rate of a female chimpanzee with that of a bacterium. A chimpanzee can, at best, produce a single offspring every 2 years, whereas a bacterium can reproduce several times per hour. Both within particular groups of animals (see Fig. 1) and across the animal kingdom as a whole (Bonner, 1965; Heron, 1972), variables resulting in a reduced reproductive rate increase as body size increases. Thus among mammals, gestation length, weaning age, age at sexual maturity, interbirth interval, and lifespan all increase with body size (Gautier-Hion and Gautier, 1976; Leutenegger, 1976; Mace, 1979; McNab, this volume; Millar, 1977; Robbins and Robbins, 1979; Sacher, 1959, 1978; Sacher and Staffeldt, 1974; Western, 1979) (see Fig. 1). In groups with variable litter sizes, litter size also tends to decline with increased body size (Fleming, 1979; French et al., 1975; Mace, 1979; Western, 1979). As a consequence, the intrinsic rate of natural increase and birth rate decline with increased body size, while generation time increases (Bonner, 1965; McNab, this volume; Western, 1979). Explanations of correlations between size and life history variables differ and the correlations are far from perfect—in particular, there are a variety of mammals that show abnormally low rates of reproduction for their size—but there is general agreement that a reduction in potential reproductive rate is an inevitable consequence of a major increase in body size or of some other variable (such as brain size) that is closely associated with it (Bonner, 1965; Mace, 1979; Robbins and Robbins, 1979; Southwood, 1976; Western, 1979).

All other things being equal, the decline in potential reproductive rate associated with increased body size can be regarded as a serious disadvantage. Consider a hypothetical population of asexual mammals that start breeding at the end of their first year of life and thereafter produce a single offspring each year until the age of 10. Suppose that all breeding attempts are successful, but that every 10 years the population is drastically reduced by some natural calamity. Imagine introducing a morph into this population, immediately after a population crash, whose body size was twice as large but which did not start breeding until it was 2 years old and thereafter

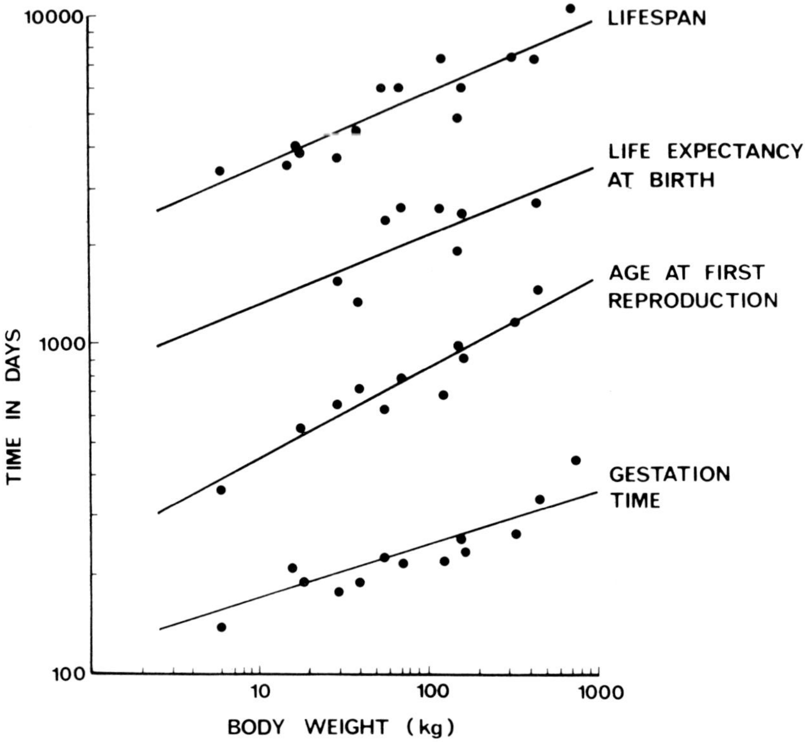

FIG. 1. Relationships between different life history variables and body weight in artiodactyls (from Western, 1979).

bred every other year. By the time of the next crash, each large morph introduced would have four daughters, six granddaughters and three great-granddaughters. In contrast, each small morph would have nine daughters, 36 granddaughters, 84 great-granddaughters, 125 great-great-granddaughters, and 121 great-great-great-granddaughters!

Our example is obviously absurd for all other things are not equal and increased body size is likely to influence survival, reproductive success, and breeding lifespan so that fitness of the small morph would be reduced and that of the large morph relatively enhanced. However, there are inevitable costs associated with major size increases and we should seek to identify the benefits of increased size in particular cases.

In some cases, the effects of body size on reproductive rate appear to constrain increases in size. For example, small-sized weasels (*Mustela nivalis*) are able to breed more than once a year in western Europe, and can consequently exploit short-term changes in rodent numbers more effectively than the larger stoats (*M. erminea*), which can only breed once a year (King and Moors, 1979). However, stoats are able to exploit a larger range of prey, and are more effective in direct (interference) competition, usually replacing weasels in such conditions. Weasels are probably able to coexist with stoats mainly due to their higher reproductive rate (King and Moors, 1979). A more straightforward example of the constraints on body size imposed by reproductive requirements is provided by northern waterfowl. Despite the possible thermodynamic advantages of large size in Arctic breeding species (Kendeigh, 1969; Mayr, 1963), geese and swans that breed in the high Arctic are *smaller* than related species or subspecies breeding further south (Lefebvre and Raveling, 1967; Owen, 1977; Syroechkovskii, 1978), apparently because high Arctic breeders must fledge young by the end of the short Arctic summer and thus cannot permit the more extended growth period which would be a consequence of increased body size.

Size and Maternal Investment

In some poikilotherms, there is evidence that large body size increases the number of eggs that a female can lay. This may contribute to selection pressures favoring increased size (Lindsey, 1966; Calow, 1978). However, in mammals increases in body size are usually negatively related to litter size (see above); consequently, selection for increased reproductive output has probably not been an important cause of gross size increases.

Nevertheless, increased size may permit a female to produce larger and stronger offspring (Ralls, 1976), although supporting evidence is contradictory. Among humans, women who are genetically short tend to be less likely to breed successfully than women who are genetically tall (Thompson, 1959), and in domestic sheep, large mothers tend to produce large lambs that grow fast (Berger, 1979). In contrast, studies of wild red deer populations have shown that maternal size (measured by jaw length) is unrelated to fecundity or to calf condition and size at 6 months (Mitchell and Brown, 1974; Mitchell et al., 1976). Interspecific comparisons show that although

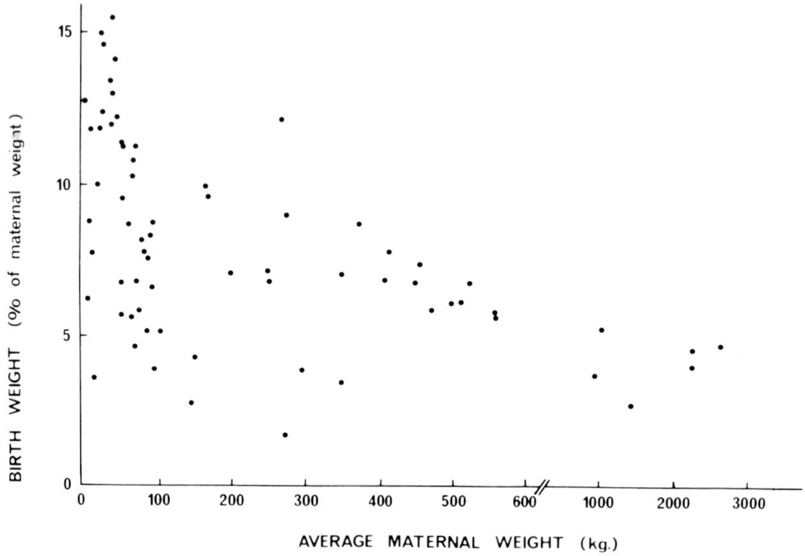

FIG. 2. Neonatal weight expressed as a fraction of maternal weight plotted against maternal weight for a variety of ungulates (from Robbins and Robbins, 1979).

absolute investment (calculated in terms of energy transfer) increases with body size, relative investment declines. For example, in ungulates over 200 kg there is a negative relationship between maternal body weight and the relative weight of neonates (Leutenegger, 1973, 1976; Robbins and Robbins, 1979)—though, for reasons that are not yet clear, the correlation disappears among the smaller species (Fig. 2).

Size and Climate

One of the first and most widely known biogeographical rules is Bergmann's (1847) Rule which states that "races from cooler climates tend to be larger in species of warm blooded vertebrates than races living in warmer climates" (Mayr, 1963). According to a more recent translation, Bergmann also intended his rule to apply to congeneric species (Barnett, 1977; James, 1970).

The validity of Bergmann's Rule has been widely discussed since Renig (1939) questioned its generality (see Irving, 1957; Mayr,

1963; McNab, 1971; Rosenzweig, 1968; Scholander, 1955). Although there are numerous exceptions, the rule does fit the distribution of size both within a variety of homeotherm species (Barnett, 1977; Brodie, 1975; Mayr, 1963; Rensch, 1936; Tomilin, 1946) and across some closely related species that have different latitudinal distributions (McNab, 1971).

Bergmann (1847) suggested that increased size led to decreased surface to volume ratios and, therefore, to decreased heat loss per unit body weight. However, although *relative* surface area is small in large animals, *absolute* surface area is increased. Indeed, using the data of Herreid and Kessel (1967), McNab showed that absolute heat loss from live mammals increases with body weight to the 0.5 power. Energy intake would have to increase with body weight at a rate equal to or greater than this for the simple thermoregulatory explanation to hold. Further criticism of Bergmann's explanation came from Scholander (1955) who argued that the surface area changes involved were physiologically unimportant compared with the thermal (insulatory) properties of body surfaces.

McNab (1971) suggested a different explanation for Bergmann's Rule. He pointed out that carnivores and granivores are the mammals most likely to conform to Bergmann's Rule, and that latitudinal changes in body size in these groups reflect changes in food size. The latter may result from larger members of a "hunting set" (sensu Rosenzweig, 1966) not occurring at higher latitudes and character displacement (or character release; Grant, 1975) causing increased body size of the other members of the hunting set, therefore allowing them to exploit wider food spectra. However, as Grant (1972) pointed out, the data are at best suggestive and "in the graphs relating body size to latitude . . . the lines drawn are biassed towards showing a change in body size corresponding with the boundary of . . . sympatry . . . and . . . allopatry." Grant suggested that further detailed study is required to substantiate the patterns. McNab's strongest case for congeneric changes in body size was the ermine or stoat (*Mustela erminea*) in the presence and absence of two other mustelids (*M. nivalis* and *M. frenata*); recent detailed analyses do not substantiate the character displacement hypothesis (Ralls and Harvey, pers. observ.). McNab's hypothesis depends on feeding ecology changing with body size and latitude, but parallel studies on birds (Behle, 1973; Johnston, 1969) and the grey squirrel (*Sciurus carolinensis*) (Barnett, 1977) reveal changes in body size with latitude that are not accompanied by changes in beak size and mandibular features. Thus, evidence for the character displacment hypothesis seems totally lacking.

A third explanation for Bergmann's Rule suggests that increased latitude results in more seasonal environments and that other correlates of seasonality are also associated with increased body size (for instance, increased altitude, continental versus coastal coimates, and savanna versus rain forest at the same latitude; see Boyce, 1979, for references). Animals with larger body size may survive better through periods of acute food shortage (see also Searcy, 1980). In a different context, Downhower (1976) formulated an appropriate model:

The rate at which large and small individuals use energy (R_L and R_S) will be proportional to body weight so that

$$R_L = aW_L{}^b$$

and

$$R_S = aW_S{}^b$$

where W_L and W_S are body weights. Because a and b are constants, R_L is greater than R_S. The length of time that large and small individuals can persist without food (T_L and T_S) will depend on the amount of food reserves ($cW_L{}^d$ and $cW_S{}^d$) divided by the rate at which energy is used:

$$T_L = cW_L{}^d/R_L = (c/a)W_L{}^{(d-b)}$$

$$T_S = cW_S{}^d/R_S = (c/a)W_S{}^{(d-b)}$$

where c and d are constants. If d is greater than b, T_L will be greater than T_S. It also follows that during periods of food surfeit, smaller individuals will be able to replenish their food reserves faster than larger individuals (R_L greater than R_S). However, size might influence the efficiency of food intake so that this last relationship may not hold.

For this model to account for Bergmann's Rule, d must be greater than b. Brodie (1975) provides evidence from cetaceans; for instance, a 48-metric-ton fin whale can last on low food intake while depleting its proportionately large food reserves for 245 days, whereas a 37-ton whale can only last 182.5 days. Studies of food reserves relative to body weight are clearly desirable and would provide valuable insight into the explanation for Bergmann's Rule offered by Boyce (1979).

Size and Predation

Major size differences are likely to have important consequences for prey species. However, data from wild populations are scarce. Extremely large size (as in the largest pachyderms) restricts the range of potential predators and appears to reduce predation rates. Both studies of the largest African herbivores and research on carnivores suggest that predation rates on elephants (*Loxodonta africana*), hippos (*Hippopotamus amphibius*), rhinoceroses (*Diceros bicornis*), giraffes (*Giraffa camelopardalis*) and, to a lesser extent, buffalos (*Synceros caffer*) are low compared to rates for antelopes (Eltringham, 1979; Kruuk, 1972; Laws et al., 1975; Schaller, 1972).

Among smaller mammals within the size range 100 g to 100 kg, size increases are associated with higher maximum running speeds (Alexander, 1977; Heglund et al., 1974; Schmidt-Nielsen, 1975), which may be an important advantage of increased size in species using open country. Moreover, the relative energetic costs of running (expressed as the energy needed to transport 1 g of body weight over 1 km) fall with increasing body size (see Fig. 3 and Taylor, 1977; Taylor et al., 1970) with the effect that larger species can sustain higher speeds for longer distances. Disadvantages of increased size in relation to predation rates probably include reduced maneuverability, less effective concealment, longer periods of juvenile dependency, and increased costs of burrowing.

Do selection pressures associated with predation help to explain the distribution of body size in mammals? Within several mammalian groups, open-country species tend to be larger than those living in closed environments, a trend which occurs across rodents (Mace, 1979), primates (Clutton-Brock and Harvey, 1977*a*, 1977*b*), carnivores (Ewer, 1973), and ungulates (Eltringham, 1979), as well as within several species in which forest forms tend to be smaller than grassland ones (examples include African buffalo, elephant, and hippopotamus). This trend could occur because forest-dwelling species rely more on concealment and less on speed to escape predators and this constrains the potential for size increase. However, these differences are also associated with variation in diet, food

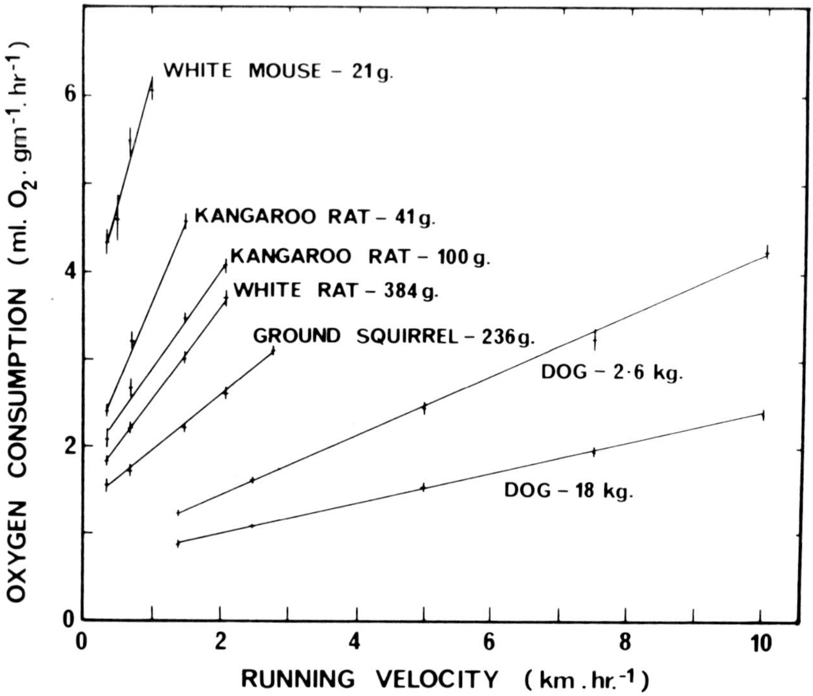

FIG. 3. The relative energetic costs of running expressed as the oxygen needed to transport 1 g of body weight over 1 km (from Taylor et al., 1970).

distribution, and zonation, and could have arisen for other reasons (see below and McNab, this volume). Moreover, there are a number of species groups in which the association between open country and increased body size is absent or even reversed, one of the most striking being the baboons of the related genera *Papio, Mandrillus,* and *Theropithecus,* in which open-country species (for example *P. cynocephalus* and *P. ursinus*) are among the smallest forms (Clutton-Brock and Harvey, 1977*a,* 1977*b*).

Also, nocturnal species of several orders tend to be smaller than diurnal or arhythmic ones (Charles-Dominique, 1974, 1975; Clutton-Brock and Harvey, 1977*b*; Mace, 1979), possibly because they, like open-country species, rely more on crypsis than flight to escape predators (Clutton-Brock and Harvey, 1977*b*).

Size and Food Handling

Body size has important consequences for the range of food items that an animal can handle effectively. For example, in granivorous birds, size differences are related to the efficiency with which they can crush seeds of various sizes (Newton, 1967); similar relationships probably exist among small granivorous mammals (Mace, 1979). Larger species of frugivores can deal with larger and tougher fruits (Struhsaker, 1978; Waser, 1977). The relationship between body size and the ability to handle large food items is probably most important among carnivores because it allows them to run down and kill larger prey; there is a close relationship between the average size of different species and the usual weight of their prey (Kruuk, 1972; McNab, this volume; Rosenzweig, 1966; Schaller, 1972; Schoener, 1969), though this breaks down in some cases (for example, in omnivorous bears).

Among herbivores, increased size has several disadvantages in relation to food handling. The larger mouths of bigger species of ungulates limit the ability to select the most nutritious parts of plants (Jarman, 1974) and, as a consequence, they cannot compensate so effectively for seasonal changes in food quality by increasing selectivity. For example, in the Serengeti, the grass grows long and rank by the end of the dry season and the proportion of leaf available is low. Impala (*Aepyceros melampus*) and, to a lesser extent, topi (*Damaliscus corrigum*), are able to pick green leaves from short grass or bushes, whereas buffalo cannot and thus eat a higher proportion of coarse grass. Thus, the quality of the food of buffalo is more strongly influenced by seasonal changes in the average quality of the standing crop (Fig. 4).

The large mouths of bigger herbivores apparently also make difficult grazing on very short-cropped swards and this, too, may restrict the ability to select high quality food during periods of shortage. Experimental evidence from sheep supports this effect. On swards with a tiller length of 10 cm, yearling sheep show a faster rate of dry matter intake than do lambs (probably because they can take larger bites), whereas on shorter swards, the intake rate of lambs is greater than that of yearlings because their smaller mouths allow them to grasp short grass more effectively (Allden and Whittaker, 1970).

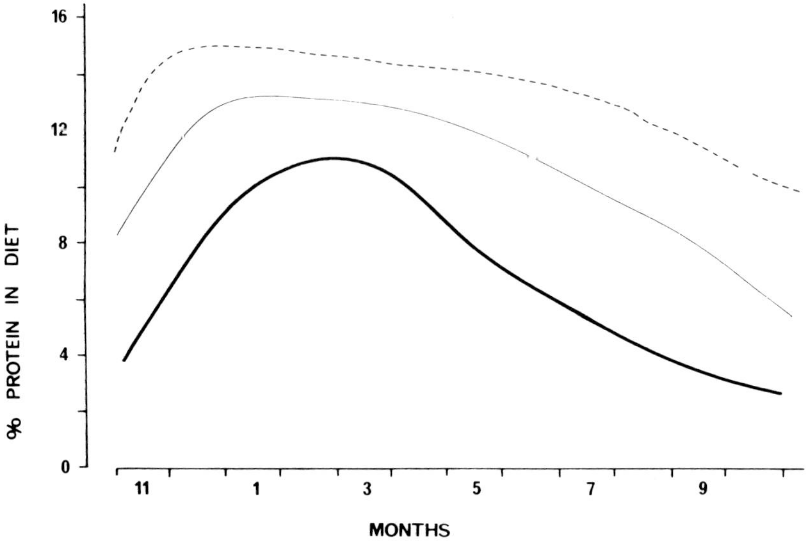

FIG. 4. Seasonal changes in the dietary protein intake of impala (broken line), topi (thin line), and buffalo (solid line) from the beginning of the rains to the late dry season (from Jarman and Sinclair, 1979).

Size and Feeding Range

Because size influences the speed of travel (see above), it has important consequences for the distance that an animal can travel each day in its search for food (Pennycuick, 1979). For example, many of the larger open-country mammals living in and on semi-arid environments travel to a water hole once a day, but collect food in other areas. As food becomes scarce during the dry season, they need to travel further from water to collect adequate supplies. The distance they can travel (their foraging radius) depends on the amount of time available for travelling and their speed of movement which, in turn, depends on size (Maynard Smith, 1968). Larger species are able to travel further from water to obtain food (Fig. 5) and frequently increase the foraging radius even further by trotting or cantering. Large size will have similar advantages for species that range from a fixed den or burrow. The relationship between size and daily travel distances also has important consequences for

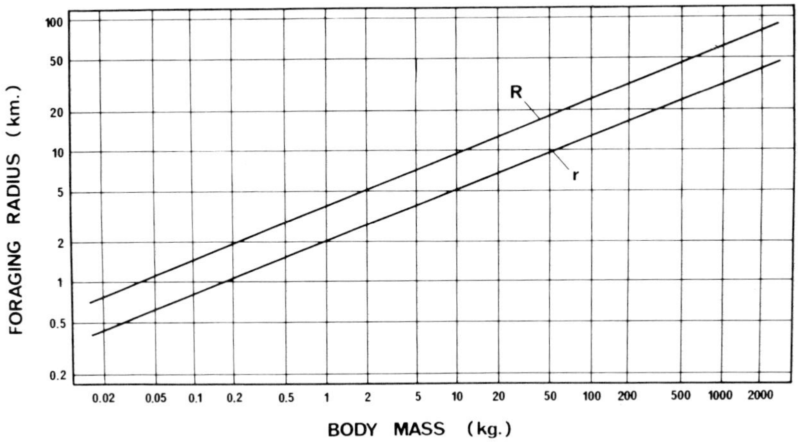

FIG. 5. Estimated foraging radius (see text) plotted against body weight (from Pennycuick, 1979). The lower line is for herbivores in dry season conditions; the upper line is the "ultimate foraging radius" obtained by neglecting expenditure on basal metabolism while travelling.

species that do not return to a fixed point each day, because it enables them to feed on food sources that are too widely dispersed to be exploited by smaller species (McNab, this volume).

Pennycuick (1979) suggested that the relationship between size and foraging radius is a potent force selecting for increased size in East African plains game. This may well be so. As we have already described, there is an association between open country and increased body size in several orders of mammals though some species of grassland ungulates (in particular, the gazelles and springbok, *Antidorcas marsupialis*) are small. Elsewhere, we have used a similar argument to explain the adaptive significance of the size differences between apes and monkeys (Clutton-Brock and Harvey, 1979*b*). Both orangutans (*Pongo pygmaeus*) and chimpanzees (*Pan troglodytes*) feed on widely and erratically distributed supplies of ripe fruit and the maximum length of their day-ranges is greater than that of forest-dwelling monkeys or of smaller apes. However, among primate species the average day-range length is not closely related to body size.

A disadvantage of increased size among arboreal species is that it reduces ease of access to the finer branches of trees and bushes. In primates, larger individuals and species spend less time feeding on the terminal twigs of trees than smaller ones (Clutton-Brock, 1977). This is probably less important among canopy-dwelling species which exploit food growing on terminal twigs by sitting on larger branches and pulling smaller ones towards them (for example, see Clutton-Brock, 1974) or by species that brachiate, but large size may play an important role in restricting the movement of larger species through smaller trees and bushes (Charles-Dominique, 1974). Both among rodents and among primates, arboreal species inhabiting the lower levels of canopy forest, scrub, or woodland tend to be smaller than those living in the canopy of rainforest (Clutton-Brock and Harvey, 1977*a*, 1977*b*; Mace, 1979). And, in primates, but not rodents, species that are partly or principally terrestrial tend to be larger than arboreal species.

Size, Energetic Requirements, and Selection for Food Quality

Among frugivores and herbivores, body size has important consequences in connection with the minimum nutritive value of exploitable foods. In a wide variety of mammalian groups, energy requirements for basal metabolism (measured as volume of oxygen metabolized per unit time) have been shown to scale approximately as body weight$^{0.75}$ (Brody, 1945; Hildwein and Gottert, 1975; Kleiber, 1932, 1961; Lasiewski and Dawson, 1967; McNab, this volume), and protein requirements scale at about the same rate (Moen, 1973). The allometric scaling of basal metabolic requirements was originally thought to be a product of the relationship between surface area and body volume but this theory has been criticized for a variety of reasons (Kleiber, 1961), of which the most important are that the exponent approximates to 0.75 rather than 0.66 (as the surface area to weight ratio would predict *if shape does not change with size*) and that a similar relationship between weight and basal metabolic requirements exists in heterotherms. McMahon (1973) points out that, for reasons connected with elastic stability and flexure, shape *does* change with size; surface area increases with weight with an exponent of 0.625. In addition, the power output of muscles is proportional to their cross-sectional area. The two findings predict a relationship between metabolic requirements and body weight with an exponent of 0.75. An earlier suggestion by Gould (1966) that an exponent greater than 0.66 is to be expected because metabolic rate will be determined by both external body surface area and circulatory system cross-sections (with the latter scaling to body weight with an exponent of 1.0) is contradicted by the heterotherm data.

Though relatively few measurements exist, active metabolic requirements show an even stronger negative allometry (for example, see French et al., 1976; Moen, 1973). Mace (1979) has reviewed measures of active daily metabolic rate (measured as the volume of oxygen metabolized *per unit weight* per unit time) for samples of rodents and other small mammals. The data clearly indicate that average daily metabolic rate (ADMR) declines more quickly with increasing body weight than basal metabolic rate (BMR) (see Fig. 6)—exponents for ADMR were −0.545 compared to −0.283 for BMR. A steeper decline in ADMR compared with BMR is to be expected on the grounds that, in terrestrial mammals, relative energy costs of locomotion and of most other activities are lower in larger species than in smaller ones (see above).

An animal's ability to digest food increases at a faster rate with increasing body size than do its metabolic requirements. The total amount of food that can be processed depends on the volume of its digestive tract and the rate of flow through the tract. In most mammalian groups, gut volume is either isometrically related to body weight or tends to show slight positive allometry. For example, in ruminants the volume of rumen plus reticulum and the weight of rumen plus reticulum contents scale with body weight with exponents of 1.26 and 1.08, respectively (Figs. 7 and 8). Theoretically, smaller species might compensate for the reduced capacity relative to the food requirements by increasing the rate of flow through the gut. Studies of throughput rate in smaller species show that they do indeed show higher rates of cellulolytic activity and faster rates of outflow from the rumen than larger species (Hungate et al., 1959; Prins and Geelen, 1971; van de Veen, 1979), but this is probably largely due to the tendency for smaller species to select food with a lower fiber fraction and to have relatively simple stomachs (Hofmann, 1973; Hoppe et al., 1977) and is partly compensated for by the reduced digestive efficiency of smaller animals. Throughput rates for animals with the same digestive system feeding on the same food probably would not vary greatly with body size (see Janis, 1976).

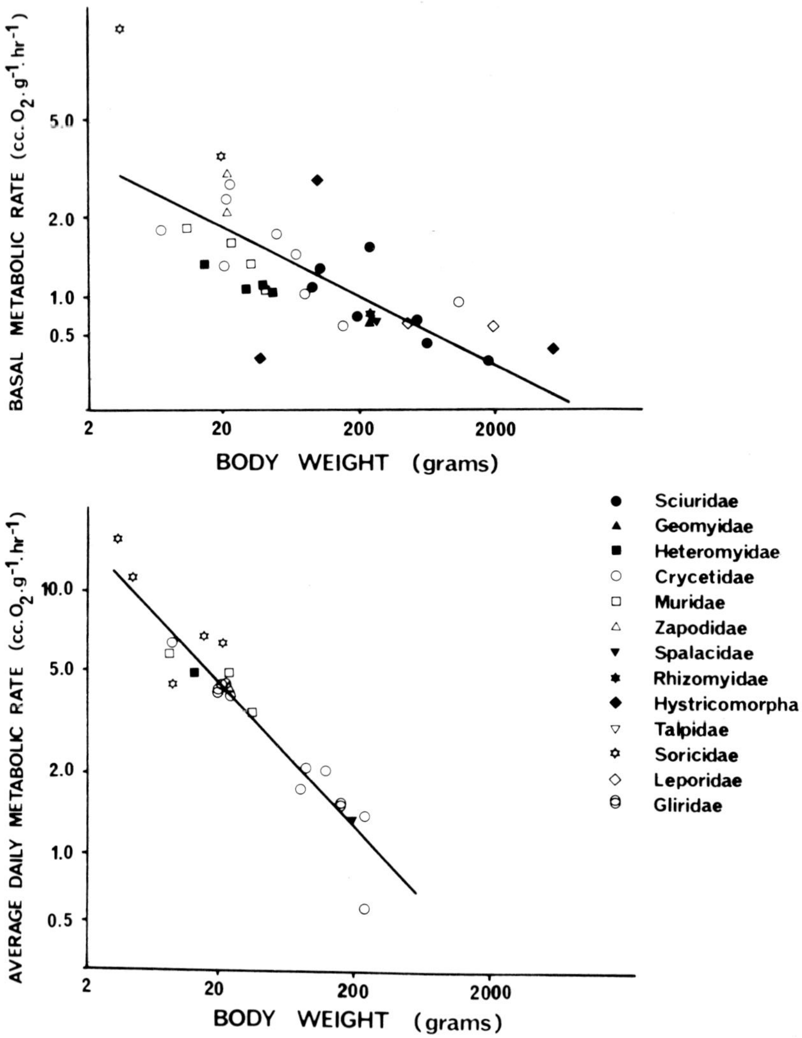

FIG. 6. Basal metabolic rate (BMR) and average daily metabolic rate (ADMR) regressed on body weight using different groups of small mammals (from Mace, 1979).

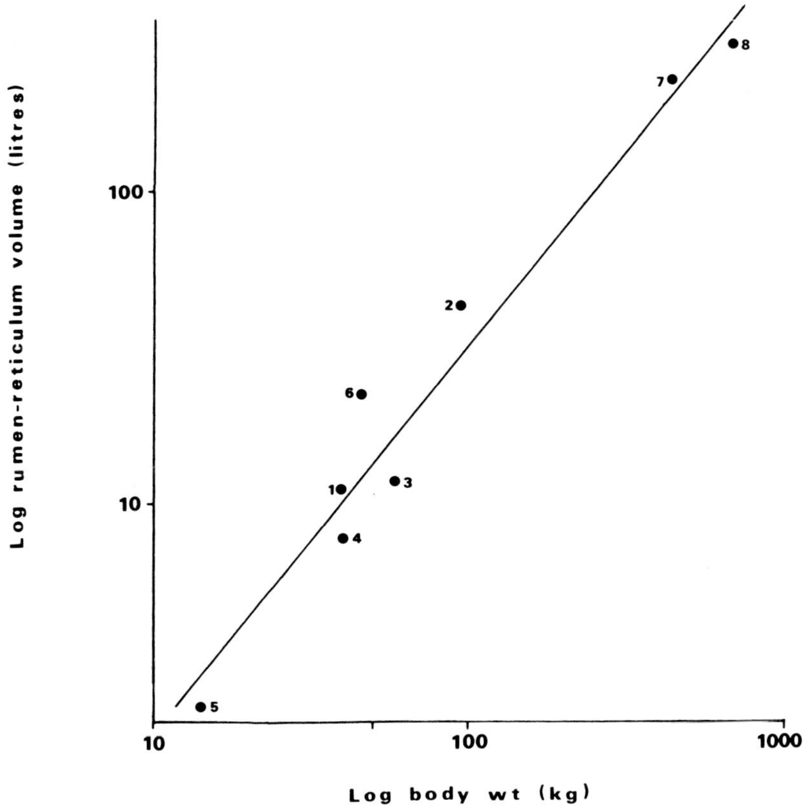

FIG. 7. Rumen-reticulum volume (liters) plotted against body weight (kg) for different species of ruminants (data from Prins and Geelan, 1971; van de Veen, 1979). 1, *Dama dama*; 2, *Cervus elaphus*; 3, *Odocoileus hemonius*; 4, *O. virginianus*; 5, *Capreolus capreolus*; 6, *Ovis*; 7, *Bos taurus*; 8, *B. americanus*.

The relevance of the difference between the scaling factors for energy requirements and digestive capability to the feeding strategies of wild herbivores was first appreciated by Bell (1969, 1970, 1971) and Jarman (1968, 1972; Jarman and Sinclair, 1979). Small species, on account of high metabolic rates, must select food with a relatively high nutritional value, whereas larger species can subsist on foods containing a lower fraction of available energy (but see McNab, this volume). Assuming that throughput rate does not change with body size, Owen-Smith (pers. comm.) has calculated

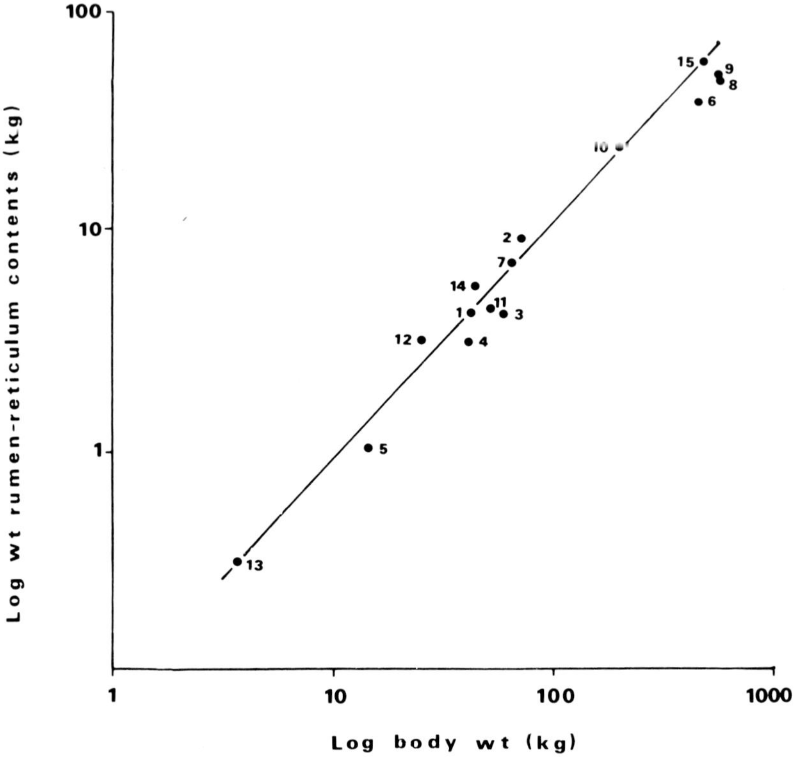

FIG. 8. Rumen-reticulum contents (kg) plotted against body weight (kg) for different species of ruminants (data from Hungate, et al., 1959; Prins and Geelen, 1971; van de Veen, 1979). 1, *Dama dama,* 2, *Cervus elaphus*; 3, *Odocoileus hemonius*; 4, *O. virginianus*; 5, *Capreolus capreolus*; 6, *Alces alces*; 7, *Rangifer taurandus*; 8, *Camelus*; 9, *Taurotragus oryx*; 10, *Connochaetes gnu*; 11, *Gazella granti*; 12, *G. thomsoni*; 13, *Nesotragus moschatus*; 14, *Ovis*; 15, *Bos taurus*.

that the minimum level of crude protein in the diet required to maintain body weight would be 5% for a 500 kg ruminant, 9% for a 50 kg animal, and 16% for a 5 kg one.

Negative relationships between measures of food quality and body size have been demonstrated for ungulates (Bell, 1970; Jarman and Sinclair, 1979; Hofmann, 1973; 1976; Hoppe et al., 1969; Hungate et al., 1969; Prins and Geelen, 1971; Owen Smith, pers. comm.). For example, the percentage of crude protein found in the stomach contents of smaller species of bovids like dik diks (*Madoqua kirkii*),

Thomson's gazelles (*Gazella thomsoni*), Grant's gazelles (*G. granti*), and impalas (*Aepyceros melampus*) is usually lower than that found in the stomach contents of medium-sized herbivores like topis (*Damaliscus corrigum*) and wildebeests (*Connochaetes* spp.); the protein fractions found in the diets of the largest species like buffalo (*Synceros caffer*) tend to be lowest of all. However, exceptions occur, especially in species whose body size allows them to gain access to food supplies that smaller species cannot reach. For example, the protein fraction of forage eaten by giraffes (*Giraffa camelopardalis*) is at least as high as that eaten by gazelles. A similar relationship occurs among cervids. The smallest species, like muntjacs (*Muntiacus* spp.) and brocket deer (*Mazama* spp.) are selective browsers and the percentage of crude protein in their diets is generally high. Medium sized species, like fallow deer (*Dama dama*), both graze and browse, whereas the largest species, including red deer (*Cervus elaphus*), graze for a substantial proportion of their feeding time. Again, there is an exception: the moose (*Alces alces*) is a specialized browser, despite being the largest living cervid. Like giraffes, its large body size enables it to reach food sources inaccessible to smaller species and its size also permits it to traverse deep snow and consequently to colonize habitats from which smaller species are excluded.

A relationship between diet quality and body size is also found among primates (Clutton-Brock and Harvey, 1977*a*). The majority of primate species are frugivorous and the protein and sugar contents of their diets are generally higher and the fiber fractions lower than in ruminants. However, the proportion of foliage eaten increases with body size, though there are many exceptions (Fig. 9). A similar trend is found within many primate families.

Similar relationships also occur within species. Red deer stags, which are about 20 kg heavier than hinds, select a qualitatively inferior diet in winter, and rumen contents contain a lower fraction of crude protein (Staines and Crisp, 1978; Staines et al., 1982; Watson and Staines, 1978). A recent study of feeding behavior in stags and hinds of different ages shows that the extent to which they differ is closely correlated with age and size (Clutton-Brock et al., in press).

The relationship between body size and food quality recurs at a higher taxonomic level. For example, differences in mean body size between rodent families are correlated with differences in diet qual-

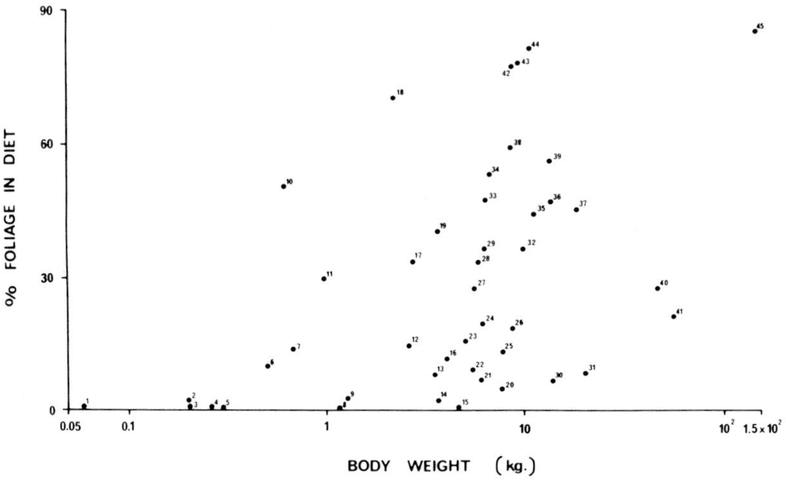

FIG. 9. Percent foliage in the diet (an inverse measure of diet quality) of different primate species (from Clutton-Brock and Harvey, 1977*a*). Species are: 1, *Galago demidovii*; 2, *Arctocebus calabarensis*; 3, *Loris tardigradus*; 4, *Galago alleni*; 5, *G. elegantulus*; 6, *Saguinus geoffroyi*; 7, *Callicebus torquatus*; 8, *Perodicticus potto*; 9, *Miopithecus talapoin*; 10, *Lepilemur mustelinus*; 11, *Aotus trivirgatus*; 12, *Cebus capucinus*; 13, *Cercopithecus cephus*; 14, *C. pogonias*; 15, *Macaca sinica*; 16, *Cercopithecus aethiops*; 17, *Lemur catta*; 18, *L. fulvius*; 19, *Propithecus verreauxi*; 20, *Cerocebus albigena*; 21, *Ateles belzebuth*; 22, *Cercopithecus neglectus*; 23, *Macaca fascicularis*; 24, *Ateles geoffroyi*; 25, *Cercocebus galeritus*; 26, *Macaca nemestrina*; 27, *Cercopithecus nictitans*; 28, *Hylobates lar*; 29, *Presbytis melalophos*; 30, *Papio hamadryas*; 31, *P. anubis*; 32, *Colobus satanas*; 33, *Presbytis obscurus*; 34, *Alouatta villosa*; 35, *Symphalangus syndactylus*; 36, *Presbytis entellus*; 37, *Theropithecus gelada*; 38, *Presbytis senex*; 39, *Indri indri*; 40, *Pan troglodytes*; 41, *Pongo pygmaeus*; 42, *Presbytis johnii*; 43, *Colobus badius*; 44, *C. guereza*; 45, *Gorilla gorilla*.

ity (Mace, 1979), and similar differences exist between subfamilies of ungulates (Jarman, 1974). However, the relationship is a relative rather than an absolute one (Mace, 1979). For example, although browsing ungulates tend to be smaller than grazing species (Jarman, 1974), they are nearly all larger than grazing rodents (Mace, 1979). Similarly, within sciurids and cricetids, folivores are generally larger than granivores and insectivores but the smallest sciurids are larger than many folivorous cricetids.

The ability to subsist on vegetation of low quality is probably one of the most important advantages of large size among mammalian herbivores. Because the abundance of different plant parts is usually inversely related to quality, large herbivores are able to use a greater fraction of the available biomass of plants than smaller species (Bell, 1970) and exploit areas that cannot support populations of smaller species. However, that large species can feed on low quality food does not explain why they do so. One might expect both large and small species to feed on the highest quality food available to them. As Jarman (1974) suggested, the larger mouths of large herbivores may limit the ability to feed selectively. Moreover, among herbivores at least, body size probably affects the rate at which they can satisfy food requirements. While the exponent of energetic requirements on body weight lies between 0.5 and 0.75, the rate of food intake (gram dry matter per unit feeding time) *for a given level of food abundance* probably increases at a considerably lower rate. The relationship between food-intake rate and body size *for a given level of food availability* is unknown, but it probably is principally determined by the relationship between body size and bite size, because bite rate varies little with size or tends to decline in larger animals (Allden and Whittaker, 1970; Hodgson and Milne, pers. comm.). The relationship between body size and bite size probably depends on the nature of the food supply. In herbivores which select discrete food items that are smaller than their mouth size (such as the tips of leaves or shoots), bite size may be unrelated to body size; in circumstances in which they feed less selectively on swards whose length is less than that of the buccal cavity, bite size may be proportional to incisor breadth, which scales as body weight$^{0.33}$ among ruminants (Fig. 10); and in those in which bite size is limited by the volume of the buccal cavity, the scaling factor may approach unity. In most grazing species, bite size is probably seldom constrained by mouth volume, and food intake may scale as, at most, body weight$^{0.33}$, whereas in circumstances in which the mechanical disadvantage of large mouth size (see above) is important, bite size may even be negatively related to body size. This is less likely to be the case among grazers that use their tongues to pull off large tufts of grass (as in the case of buffalo and domestic cows), among unselective browsers, or among frugivores in which bite size may be more closely constrained by the volume of the buccal cavity (see Belovsky, 1978*a*, 1978*b*; Chacon and Stobbs, 1976; Stobbs, 1973*a*, 1973*b*).

The suggestion that, for food of a given abundance, rate of intake

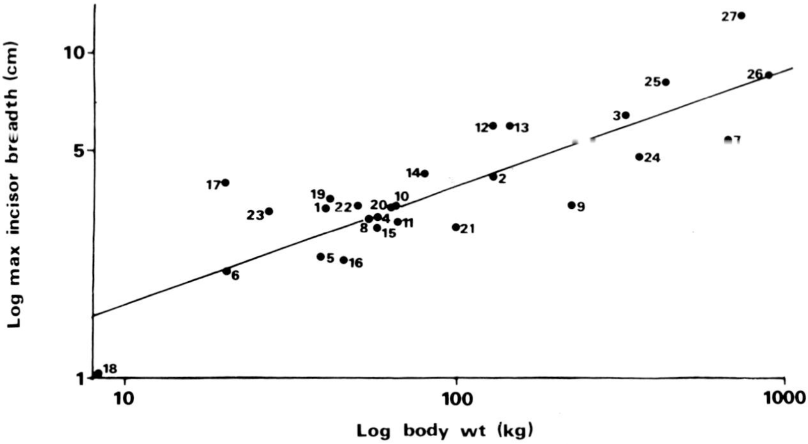

FIG. 10. Maximum breadth of the array of lower incisors plotted against body weight for 26 species of ruminants. Incisor measurements from collection of Zoology Museum, Cambridge. Body weight data from: Dorst and Dandelot (1970); Jarman (1974); Walker (1964); Whitehead (1972). 1, *Axis porcinus*; 2, *Cervus eldi*; 3, *C. canadensis*; 4, *Odocoileus hemonius*; 5, *O. virginianus*; 6, *Capreolus capreolus*; 7, *Taurotragus derbianus*; 8, *Tragelaphus scriptus*; 9, *Kobus defassa*; 10, *K. kob*; 11, *Redunca redunca*; 12, *Damaliscus korrigum*; 13, *D. lunatus*; 14, *D. dorcas*; 15, *Aepyceros melampus*; 16, *Gazella soemmerringi*; 17, *G. gutturosa*; 18, *Nesotragus moschatus*; 19, *Antilocapra americana*; 20, *Pseudois nayaur*; 21, *Capra ibex*; 22, *C. hircus*; 23, *Naemorhedus* (*Cemas*) *goral*; 24, *Ovibos moschatus*; 25, *Bos taurus*; 26, *B. americanus*; 27, *Bubalus bubalis*.

scales as around body weight$^{0.33}$ in grazers appears to conflict with Sinclair's (1977:273) demonstration that, among African ungulates, food requirements divided by feeding time scale as body weight$^{1.0}$. However, this is not the case, for larger species tend to feed on foods of lower nutritional quality but greater abundance (see above). Consequently, larger species eat a greater weight of food per unit of nutrient requirement and their rate of intake is likely to be high because they are feeding on more abundant foods. For both reasons, Sinclair's calculations do not necessarily indicate that intake rate among animals of different sizes feeding on the same swards is proportional to body weight.

If rate of food intake scales at a lower rate than food requirements, larger species should spend more time feeding. This comparison, too, is complicated by the tendency for larger species to feed more on foods with higher standing crops but lower nutritional value (see above). Despite this, in two mammal orders for which enough estimates are available for comparison to be feasible, feeding time increases with body size (Fig. 11).

Both the tendency for large grazers to be less effective at selecting small food items and the relatively slow rate at which they can satisfy their energy requirements will affect the minimum level of food abundance that they can tolerate (Bell, 1971; Sinclair, 1977). Larger herbivores may also be limited to grazing principally on communities with comparatively high standing crops in order to achieve a rate of food intake sufficient to satisfy requirements (Bell, 1969; Sinclair, 1977). This argument depends on the assumption that large grazers cannot compensate for increased food requirements by increasing the amount of time spent feeding beyond a certain point. Experiments with sheep support this (Allden and Whittaker, 1970). Although sheep will increase feeding time by as much as 100% as food availability declines, they will not graze for more than 13 h per day in circumstances in which food intake does not meet their requirements, presumably because digestive processes prevent them from further increasing grazing time. In a more natural environment with other demands (for example, predator avoidance), time available for feeding would probably be further reduced.

Size and Interspecific Competition

Increased body size confers obvious advantages in direct competition for food resources because larger animals can generally displace smaller ones from feeding sites (Clutton-Brock, 1977; Dittus, 1977, 1979; Eltringham, 1979; King and Moors, 1979; Stimson, 1973). Although direct competition is probably most important where resources are clumped and defensible, most comparisons indicate that species involved in direct competition tend to show relatively small body size (Clutton-Brock and Harvey, 1977*a*, 1978; Eltringham, 1979).

When large species require higher food density to maintain intake rate, large size may be a disadvantage in indirect competition. If two grazing species of different sizes with minimum food availability requirements of 0.5 and 2.0 kg/dm/m^2 are introduced to a fenced paddock containing a food supply with a standing crop of 3

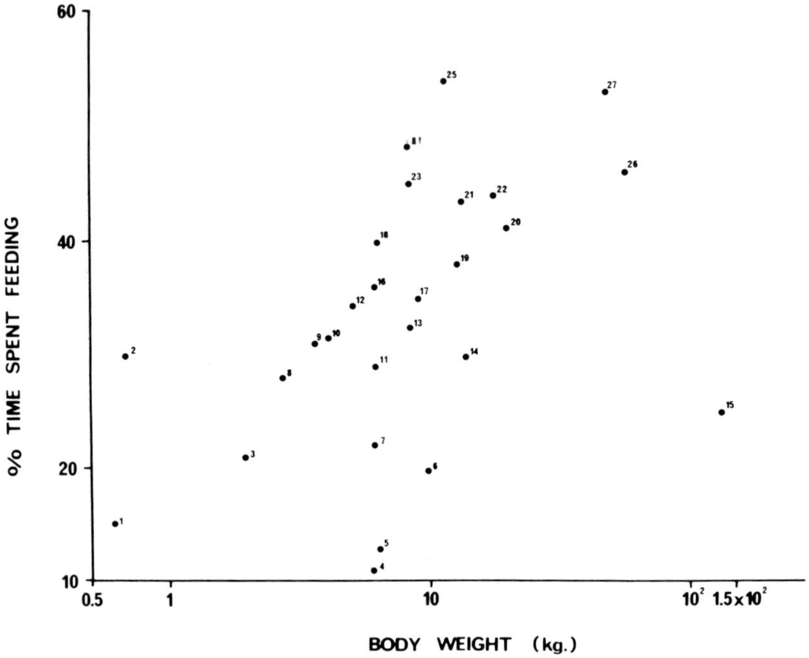

% TIME SPENT FEEDING
60
40
20
10
0.5
1
10
10^2 1.5×10^2
BODY WEIGHT (kg.)

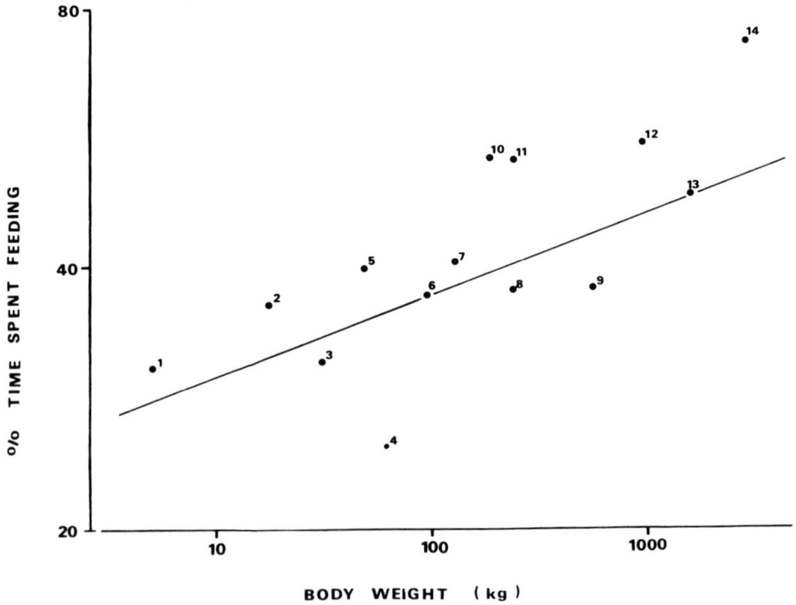

% TIME SPENT FEEDING
80
40
20
10
100
1000
BODY WEIGHT (kg)

kg/dm/m^2, both populations will graze until overall biomass is reduced below 2 kg/dm/m^2. After this, the intake rate of the larger species will not meet life-maintaining requirements unless they can exclude the smaller animals by direct interaction.

Ecological interactions as simple as this are seldom observed under natural circumstances. However, some observations suggest that where food supplies are indefensible, smaller grazers tend to exclude larger ones in times of food shortage. On the Serengeti plains, the larger herbivores are the first to leave preferred graze types and move to areas with greater food but lower food quality as standing crops decline during the dry season, whereas at the beginning of the wet season they are the last to return to feed on the first flush of annual grasses (Bell, 1970; Jarman and Sinclair, 1979). Similar sequences of use of preferred grazing areas have been recorded in other guilds of herbivores (Nicholson et al., 1970).

Evidence that smaller herbivores can exclude larger ones on a more permanent basis is also provided by other studies. When wapiti (*Cervus canadensis*) were introduced to New Zealand, they failed to spread in areas where the smaller red deer (*Cervus elaphus*) were numerous, whereas when the still smaller sika deer (*Cervus nippon*) was introduced to Ireland, it replaced natural red deer populations in parts of the country (R. Harrington, pers. comm.). In eighteenth century England, sheep were usually excluded from deer forests because, "being such close grazers, they would pick out all the finest grasses and hinder the deer from thriving" (White,

---

←

FIG. 11. Proportion of daytime (top) and the 24-h period (bottom) spent feeding by primates and ungulates. Primate data from Clutton-Brock and Harvey, 1977*a*; ungulates from Owen-Smith, in press. For top, species are: 1, *Lepilemur mustelinus*; 2, *Callicebus torquatus*; 3, *Lemur fulvus*; 4, *Ateles geoffroyi*; 5, *Alouatta villosa*; 6, *Colobus guereza*; 7, *Ateles belzebuth*; 8, *Lemur catta*; 9, *Propithecus verreauxi*; 10, *Cercopithecus aethiops*; 11, *Hylobates lar*, 12, *Macaca fascicularis*; 13, *Presbytis johnii*; 14, *Papio hamadryas*; 15, *Gorilla gorilla*; 16, *Presbytis obscurus*; 17, *Colobus badius*; 18, *Presbytis melalophos*; 19, *Indri indri*; 20, *Papio anubis*; 21, *Presbytis entellus*; 22, *Theropithecus gelada*; 23, *Cercocebus albigena*; 24, *C. galeritus*; 25, *Symphalangus syndactylus*; 26, *Pongo pygmaeus*; 27, *Pan troglodytes*. For bottom, species are: 1, *Madoqua* sp.; 2, *Gazella thomsoni*; 3, 4, *Phacochoerus aethiopicus*; 5, *Aepyceros melampus*; 6, *Damaliscus korrigum*; 7, *Alcephalus* sp.; 8, *Kobus ellipsiprymnus*; 9, *Synceros caffer*; 10, *Tragelaphus scriptus*; 11, *Equus burchelli*; 12, *Diceros bicornus*; 13, *Ceratotherium simus*; 14, *Loxodonta africana*.

1789), and in areas of Scotland where sheep and red deer compete, the latter spend little time on areas of *Agrostis-Festuca* greens for which they show a strong preference where sheep are absent (Clutton-Brock et al., pers. observ.; Osborne, 1980). Indian rhinoceroses (*Rhinoceros unicornis*) suffer from competition with domestic buffalo (Laurie, 1978), whereas African buffalo populations in the Ruwenzori National Park increased when the density of hippopotamuses (which, despite their larger size, are notoriously close grazers) was reduced (Eltringham, 1974). And increasing numbers of wildebeest (*Connochaetes taurinus*) in the Serengeti may have been partly responsible for checking the growth of buffalo populations during the last decade (Sinclair, 1979).

Comparisons of a variety of grazing species also indicate that larger individuals may be less able to tolerate long periods of acute food shortage than smaller ones. In many dimorphic north temperate ungulates, males are more likely to die than females during periods of food shortage associated with high population density (Clutton-Brock et al., pers. observ.; Flook, 1970; Grubb, 1974; Klein, 1968; Klein and Olson, 1960; Robinette et al., 1957; Tener, 1954). For example, when the reindeer population of St. Mathew Island crashed from 6,000 to 42 during a single year, only one of the survivors was male (Klein, 1968). These effects are presumably a consequence of the larger individuals being unable to gather sufficient food after fat deposits have been exhausted and need not contradict Boyce's (1979) explanation for Bergmann's Rule (see above).

Size is not the only variable that affects indirect interspecific competition. Additional factors affecting the nature of the interaction between competing species include digestive anatomy, the diversity and distribution of foods they can utilize, tolerance of water shortage, and liability to predation (Corfield, 1973; Dittus, 1977, 1979; Janis, 1976; Pianka, 1974; Sinclair, 1977; Sinclair and Norton-Griffiths, 1979). Nevertheless, if all other things are equal and food access is not determined by aggressive interaction, small species are likely to be superior competitors.

Large species may avoid competing with smaller ones by exploiting widely and erratically dispersed food sources that offer high food density for limited time. This may represent an important ecological distinction between the great apes and the lesser apes and monkeys (Clutton-Brock and Harvey, 1979).

Conclusions

What conclusions can be drawn from this brief survey of the benefits and costs of body size among mammals? Although many factors probably influence size, and these differ between ecological groupings, two common trends can be recognized. Among herbivores, frugivores, and omnivores, large-sized species usually feed on foods of lower nutritional quality than smaller species or on foods which are more widely and erratically distributed, though such differences are less apparent at times when food is abundant. In carnivores and insectivores, body size is closely correlated with prey size.

In both these cases, it is impossible to be sure that variation in body size is an adaptation to differences in feeding niche, for if differences in size had evolved for some reason unconnected with food quality or distribution, similar correlations would still be expected (Clutton-Brock and Harvey, 1979*a*). Indeed, the fact that a species' size relative to that of other species belonging to the same taxonomic group is a better predictor of its feeding niche than its absolute size, suggests that the common phylogenetic inheritance influences the feeding niches occupied by members of the same higher taxonomic unit.

The problem is partly resolved by considering the likely course of events during an adaptive radiation. Due, perhaps, to a climatic shift, a new range of habitats and foods becomes accessible. Clearly, the new resources are only available to a limited range of species. For example, the large biomass of foliage that became accessible for exploitation when the northern ice sheets drew back after the Pleistocene (see Geist, 1971) was unlikely to be exploited by bats, and was probably more easily exploited by ungulates than rodents. In this sense, an animal's size (together with other aspects of its phylogenetic inheritance) constrains the foods and habitats that it can exploit (McNab, this volume).

After the initial colonization of the habitat, a new radiation occurs—possibly quite rapidly (Gould, 1977)—and different sized forms evolve to exploit different food supplies. At this stage, the nature of the feeding niche probably determines selection pressures affecting body size. When competition occurs between members of different radiations, we would predict that the smaller species will generally replace the larger one unless the latter is able to exclude the former by direct competition, or relies on food sources which, because of the distribution, smaller animals cannot exploit.

Literature Cited

ALEXANDER, R. M. 1977. Mechanics and scaling of terrestrial locomotion. Pp. 93–110, *in* Scale effects in animal locomotion (T. J. Pedley, ed.). Academic Press, London, 545 pp.

ALLDEN, W. G., AND I. A. MACD. WHITTAKER. 1970. The determinants of herbage intake by grazing sheep: the inter-relationship of factors influencing herbage intake and availability. Australian J. Agric. Res., 32:755–766.

BARNETT, R. J. 1977. Bergmann's rule and variation in structures related to feeding in the gray squirrel. Evolution, 31:538–545.

BEHLE, W. H. 1973. Clinal variation in white throated swifts from Utah and the Rocky Mountain region. Auk, 90:299–306.

BELL, R. H. V. 1969. The use of the herb layer by grazing ungulates in the Serengeti National Park, Tanzania. Unpubl. Ph.D. thesis, Manchester Univ., 310 pp.

———. 1970. The use of the herb layer by grazing ungulates in the Serengeti. Pp. 111–123, *in* Animal populations in relation to their food resources (A. Watson, ed.). Blackwell, Oxford, 477 pp.

———. 1971. A grazing ecosystem in the Serengeti. Sci. Amer., 224:86–93.

BELOVSKY, G. E. 1978*a*. The time-energy budget of a moose. Theoret. Pop. Biol., 14:76–104.

———. 1978*b*. Diet optimization in a generalist herbivore: the moose. Theoret. Pop. Biol., 14:105–134.

BERGER, J. 1979. Weaning conflict in desert and mountain bighorn sheep, *Ovis canadensis*: an ecological interpretation. Z. Tierpsychol., 50:188–200.

BERGMANN, C. 1847. Ueber die Uerhältnisse der Wärmeökonomie der Thiere zu ihrer Grösse. Gottinger Studien, 3(pt.1):595–708.

BONNER, J. T. 1965. Size and cycle: an essay on the structure of biology. Princeton Univ. Press, Princeton, 219 pp.

BOYCE, M. S. 1979. Seasonality and patterns of natural selection for life histories. Amer. Nat., 114:569–583.

BRODIE, P. F. 1975. Cetacean energetics, an overview of intraspecific size variation. Ecology, 56:152–161.

BRODY, S. 1945. Bioenergetics and growth. Reinhold, New York, 279 pp.

CALOW, P. 1978. Life cycles: an evolutionary approach to the physiology of reproduction, development and ageing. Chapman and Hall, London, 164 pp.

CHACON, E., AND T. H. STOBBS. 1976. Influence of progressive defoliation of a grass sward on the eating behavior of cattle. Australian J. Agric. Res., 27: 709–727.

CHARLES-DOMINIQUE, P. 1974. Ecology and feeding behaviour of five sympatric lorisids in Gabon. Pp. 135–150, *in* Prosimian biology (R. D. Martin, G. A. Doyle, and A. C. Walker, eds.). Duckworth, London, 983 pp.

———. 1975. Nocturnality and diurnality, an ecological interpretation of these two modes of life by an analysis of the higher vertebrate fauna in tropical forest ecosystems. Pp. 69–88, *in* Phylogeny of primates (W. P. Luckett and F. S. Szalay, eds.). Plenum, New York, 483 pp.

CLUTTON-BROCK, T. H. 1974. Primate social organisation and ecology. Nature, 250:539–542.

———. 1977. Some aspects of intraspecific variation in feeding and ranging behaviour in primates. Pp. 539–556, *in* Primate ecology (T. H. Clutton-Brock, ed.). Academic Press, London, 631 pp.

CLUTTON-BROCK, T. H., AND P. H. HARVEY. 1977*a*. Species differences in feeding and ranging behaviour in primates. Pp. 557–579, *in* Primate ecology (T. H. Clutton-Brock, ed.). Academic Press, London, 631 pp.

———. 1977*b*. Primate ecology and social organisation. J. Zool., 183:1–39.

———. 1978. Mammals, resources and reproductive strategies. Nature, 273:191–195.

———. 1979*a*. Comparison and adaptation. Proc. Roy. Soc. London, 205:547–565.

———. 1979*b*. Home range size, population density and phylogeny in primates. Pp. 201–214, *in* Primate ecology and human origins (I. S. Bernstein and E. O. Smith, eds.). Garland Press, New York, 362 pp.

CLUTTON-BROCK, T. H., F. E. GUINNESS, AND S. D. ALBOA. In press. Red deer: behavior and ecology of the sexes. Univ. Chicago Press, Chicago.

CORFIELD, T. 1973. Elephant mortality in Tsavo National Park, Kenya. East African Wildl. J., 11:330–368.

DITTUS, W. P. J. 1977. The social regulation of population density and age-sex distribution in the toque monkey. Behaviour, 63:281–322.

———. 1979. The evolution of behaviours regulating density and age-specific sex ratios in a primate population. Behaviour, 69:265–301.

DORST, J., AND P. DANDELOT. 1970. A field guide to the larger mammals of Africa. Houghton-Mifflin, Boston, 287 pp.

DOWNHOWER, J. F. 1976. Darwin's finches and the evolution of sexual dimorphism in body size. Nature, 263:558–563.

ELTRINGHAM, S. K. 1974. Changes in the large mammal community of Mweya Peninsula, Ruwenzori National Park, Uganda, following removal of hippopotamus. J. Appl. Ecol., 11:855–865.

———. 1979. The ecology and conservation of large African mammals. Macmillan, London, 286 pp.

EWER, R. F. 1973. The carnivores. Weidenfeld and Nicolson, London, 494 pp.

FLEMING, T. H. 1979. Life-history strategies. Pp. 1–63, *in* Ecology of small mammals (D. M. Stoddart, ed.). Chapman and Hall, London, 386 pp.

FLOOK, D. R. 1970. A study of sex differential in the survival of wapiti. Canadian Wildl. Serv. Rep., 11:91.

FRENCH, N. R., D. M. STODDART, AND B. BOBEK. 1975. Patterns of demography in small mammal populations. Pp. 73–102, *in* Small mammals, their productivity and population dynamics (F. B. Golley, K. Petrusewicz, and L. Ryszkowski, eds.). I.B.P. Programme, Cambridge Univ. Press, 451 pp.

FRENCH, N. R. ET AL. 1976. Small mammal energetics in grassland ecosystems. Ecol. Monogr., 46:201–220.

GAUTIER-HION, A., AND J. P. GAUTIER. 1976. Croissance, maturité sexualle, et sociale, reproduction chez les cercopithécinés forestres africains. Folia Primatol., 20:165–184.

GEIST, V. 1971. Mountain sheep. Univ. Chicago Press, Chicago, 383 pp.

GOULD, S. J. 1966. Allometry and size in ontogeny and phylogeny. Biol. Rev., 41: 587–640.

———. 1977. Ontogeny and phylogeny. Belknap Press, Cambridge, Massachusetts, 501 pp.

GRANT, P. R. 1972. Convergent and divergent character displacement. Biol. J. Linnean Soc., 4:36–68.

———. 1975. The classical case of character displacement. Evol. Biol., 8:237–337.

GRUBB, P. 1974. Population dynamics of the Soay sheep. Pp. 242–272, *in* Island survivors: the ecology of the Soay sheep of St. Kilda (P. A. Jewell, C. Milner, and J. M. Boyd, eds.). Athlone Press, London, 386 pp.

HEGLUND, N. C. R., C. R. TAYLOR, AND T. A. MCMAHON. 1974. Scaling stride frequency gait to animal size: mice to horses. Science, 186:1112–1113.

HERNEID, C. F., II, AND B. KEISEL. 1967. Thermal conductance in birds and mammals. Comp. Biochem. Physiol., 21:405–414.

HERON, A. C. 1972. Population ecology of a colonizing species: the pelagic tunicate *Thalia democratica*. Oecologia, 10:269–293.

HILDWEIN, G., AND M. GOTTART. 1975. Standard metabolism and thermoregulation in a prosimian *Perodicticus potto*. Biochem. Physiol., 50A:201–213.

HOFMANN, R. 1973. The ruminant stomach: stomach structure and feeding habits of East African game ruminants. East African Literature Bureau, Nairobi, 32 pp.

HOFMANN, R. R. 1976. Zur adaptieven differenzierung der weidekäuer: untersuchsergebrisse auf der basis der vergleichenden funktionellen anatomie der verdauungstrakts. Der praktische Tierartz., 6:351–358.

HOPPE, P. P., S. A. QUORTRUP, AND M. H. WOODFORD. 1977. Rumen fermentation and food selection in East African zebu cattle, wildebeest, Coke's hartebeest and topi. J. Zool., 181:1–9.

HUNGATE, R. E. ET AL. 1959. Microbial fermentation in certain mammals. Science, 130:1192–1194.

IRVING, L. 1957. The usefulness of Scholander's views on adaptive insulation of animals. Evolution, 11:257–259.

JAMES, F. C. 1970. Geographic size variation in birds and its relationship to climate. Ecology, 51:365–390.

JANIS, C. 1976. The evolutionary strategy of the Equidae and the origins of rumen and cecal digestion. Evolution, 30:757–776.

JARMAN, P. J. 1968. The effects of the creation of Lake Kariba upon the terrestrial ecology of the middle Zambesi valley, with particular reference to the large mammals. Unpubl. Ph.D. thesis, Manchester Univ., 278 pp.

———. 1972. Seasonal distribution of large mammal populations in the unflooded middle Zambesi valley. J. Appl. Ecol., 9:283–299.

———. 1974. The social organization of antelope in relation to their ecology. Behaviour, 48:215–266.

JARMAN, P. J., AND A. R. E. SINCLAIR. 1979. Feeding strategy and the pattern of resource partitioning in ungulates. Pp. 130–163, *in* Serengeti, dynamics of an ecosystem (A. R. E. Sinclair and M. Norton-Griffiths, eds.). Univ. Chicago Press, Chicago, 389 pp.

JOHNSTON, R. F. 1969. Character variation and adaptation in European sparrows. Syst. Zool., 18:206–231.

KENDEIGH, S. G. 1969. Tolerance of cold and Bergmann's rule. Auk, 36:18–25.

KING, C. M., AND P. J. MOORS. 1979. On the co-existence, foraging strategy and the biogeography of weasels and stoats (*Mustela nivalis* and *M. erminea*) in Britain. Oecologia, 39:129–150.

KLEIBER, M. 1932. Body size and metabolism. Hilgardia, 6:315–353.

———. 1961. The fire of life. Wiley, New York, 358 pp.

KLEIN, D. R. 1968. The introduction, increase and crash of reindeer on St. Mathew Island. J. Wildl. Mgmt., 32:350–367.

KLEIN, D. R., AND S. J. OLSON. 1960. Natural mortality patterns of deer in Southeast Alaska. J. Wildl. Mgmt., 24:80–88.

KRUUK, H. 1972. The spotted hyena. Univ. Chicago Press, Chicago, 335 pp.

LASIEWSKI, R. C., AND W. R. DAWSON. 1967. A re-examination of the relation between standard metabolic rate and body weight in birds. Condor, 69: 13–23.

LAURIE, W. A. The ecology and behaviour of the greater one-horned rhinoceros. Unpubl. Ph.D. thesis, Univ. Cambridge, 325 pp.

LAWS, R. M., I. S. C. PARKER, AND R. C. B. JOHNSTONE. 1975. Elephants and their habitats: the ecology of elephants in North Bunyoro, Uganda. Clarendon Press, Oxford, 341 pp.

LEFEBVRE, E. A., AND D. G. RAVELING. 1967. Distribution of Canada Geese in winter as related to heat loss at varying environmental temperatures. J. Wildl. Mgmt., 31:538–546.

LEUTENEGGER, W. 1973. Maternal-fetal weight relationships in primates. Folia Primatol., 20:280–293.

———. 1976. Allometry of neonatal size in eutherian mammals. Nature, 263: 229–230.

LINDSEY, C. D. 1966. Body sizes of poikilotherm vertebrates at different latitudes. Evolution, 20:456–465.

MACE, G. M. 1979. The evolutionary ecology of small mammals. Unpubl. Ph.D. dissert., Univ. Sussex, 347 pp.

MCMAHON, T. 1973. Size and shape in biology. Science, 179:1201–1211.

MCNAB, B. N. 1971. On the ecological significance of Bergmann's rule. Ecology, 52:845–854.

MAYNARD SMITH, J. 1968. Mathematical ideas in biology. Cambridge Univ. Press, Cambridge, 152 pp.

MAYR, E., 1963. Animal species and evolution. Harvard Univ. Press, Cambridge, Massachusetts, 797 pp.

MILLAR, J. S. 1977. Adaptive features of mammalian reproduction. Evolution, 31: 370–386.

MITCHELL, B., AND D. BROWN. 1974. The effects of age and body size on fertility in female red deer (*Cervus elaphus* L.). Proc. Int. Congr. Game Biol., 11: 89–98.

MITCHELL, B., D. MCCOWAN, AND I. A. NICHOLSON. 1976. Annual cycles of body weight and condition in Scottish red deer. J. Zool., 180:107–127.

MOEN, A. 1973. Wildlife ecology. Freeman, New York, 458 pp.

NEWTON, I. 1967. The adaptive radiation and feeding ecology of some British finches. Ibis, 109:33–98.

NICHOLSON, I. A., I. S. PATTERSON, AND A. CURRIE. 1970. A study of vegetational dynamics: selection by sheep and cattle in *Nardus* pasture. Pp. 129–143, *in* Animal populations in relation to their food supplies (A. Watson, ed.). Blackwell, Oxford, 477 pp.

OSBORNE, B. 1980. Habitat use by sheep and deer on a West Highland estate. Unpubl. M.Sc. thesis, Univ. Sussex, 112 pp.

OWEN, M. 1977. Wildfowl of Europe. Macmillan, London, 358 pp.

OWEN-SMITH, N. In press. Factors influencing the transfer of plant products into large herbivore populations. *In* The ecology of tropical savannas (B. J. Huntley and B. H. Walker, eds.). Springer-Verlag, Berlin.

PENNYCUICK, C. J. 1979. Energy costs of locomotion and a concept of "foraging radius". Pp. 164–184, *in* Serengeti, dynamics of an ecosystem (A. R. E.

Sinclair and M. Norton-Griffiths, eds.). Univ. Chicago Press, Chicago, 389 pp.

Prins, R. A. and M.-J. H. Geelen. 1971. Rumen characteristics of red deer, fallow deer and roe deer. J. Wildl. Mgmt., 35:673–680.

Ralls, K. 1976. Mammals in which females are larger than males. Quart. Rev. Biol., 51:245–276.

Renig, W. F. 1939. Besteht die Bergmannsche Regel zu Recht? Arch. Naturgesch. (N.F.), 8:70–88.

Rensch, B. 1936. Studien über klimatische Parallelitat der merkmalsausprägung bei Vögeln und Sängern. Arch. Naturgesch. (N.F.), 5:317–363.

Robbins, C. T., and B. L. Robbins. 1979. Fetal and neonatal growth patterns and maternal reproductive effort in ungulates and subungulates. Amer. Nat., 114:101–116.

Robinett, W. L., et al. 1957. Differential mortality by sex and age among male deer. J. Wildl. Mgmt., 21:1–16.

Rosenzweig, M. L. 1966. Community structure in sympatric carnivora. J. Mamm., 47:602–612.

———. 1968. The strategy of body size in mammalian carnivores. Amer. Midland Nat., 80:299–315.

Sacher, G. A. 1959. Relation of lifespan to brain weight and body weight in mammals. Pp. 115–133, *in* CIBA Foundation Colloquia on Ageing. Vol. 5: The lifespan of animals. Churchill, London.

———. 1978. Longevity and ageing in vertebrate evolution. Bioscience, 28:497–501.

Sacher, G. A. and E. F. Staffeldt. 1974. Relation of gestation time and brain weight of placental mammals: implications for the theory of vertebrate growth. Amer. Nat., 105:593–615.

Schaller, G. B. 1972. The Serengeti lion. Univ. Chicago Press, Chicago, 480 pp.

Schmidt-Nielsen, K. 1975. Scaling in biology: the consequences of size. J. Exp. Zool., 194:287–307.

Schoener, T. W. 1969. Models of optimal size for solitary predators. Amer. Nat., 103:277–313.

Scholander, P. F. 1955. Evolution of climatic adaptation in homeotherms. Evolution, 9:15–26.

Searcy, W. A. 1980. Optimum body sizes at different ambient temperatures: an energetics explanation of Bergmann's rule. J. Theor. Biol., 83:579–593.

Sinclair, A. R. E. 1977. The African buffalo. Univ. Chicago Press, Chicago, 355 pp.

———. 1979. The eruption of the ruminants. Pp. 82–103, *in* Serengeti, dynamics of an ecosystem (A. R. E. Sinclair and M. Norton-Griffiths, eds.). Univ. Chicago Press, Chicago, 389 pp.

Sinclair, A. R. E. and M. Norton-Griffiths. 1979. Serengeti, dynamics of an ecosystem. (A. R. E. Sinclair and M. Norton-Griffiths, eds.). Univ. Chicago Press, Chicago, 389 pp.

Southwood, T. R. E. 1976. Bionomic strategies. Pp. 26–48, *in* Theoretical ecology (R. M. May, ed.). Blackwell, Oxford, 317 pp.

Staines, B. W. and J. M. Crisp. 1978. Observations on food quality in Scottish red deer (*Cervus elaphus*) as determined by the chemical analysis of the rumen contents. J. Zool., 185:253–259.

Staines, B. W., J. W. Crisp, and T. Parrish. 1982. Differences in the quality of food eaten by red deer (*Cervus elaphus*) stags and hinds in winter. J. Appl. Ecol., 19:65–77.

Stimson, J. 1973. The role of the territory in the ecology of the intertidal limpet *Lottia gigantea* (Gray). Ecology, 54:1020–1030.

STOBBS, T. H. 1973*a*. The effect of plant structure on the intake of tropical pastures II. Differences in sward structure, nutritive value and bite size of animals grazing *Sefaria anceps* and *Chloris gayana* at various stages of growth. Australian J. Agric. Res., 24:821–829.

———. 1973*b*. The effect of plant structure on the intake of tropical pastures I. Variation in the bite size of grazing cattle. Australian J. Agric. Res., 24: 809–819.

STRUHSAKER, T. T. 1978. Food habits of five monkey species in the Kibale Forest, Uganda. Pp. 225–248, *in* Recent advances in primatology (D. J. Chivers and J. Herbert, eds.). Academic Press, London, 980 pp.

SYROECHKOVSKII, E. 1978. Razmery lebedei, gusei i Kazarok v svyazi s adaptatsei K polyarym usloviyam. (The size of swans, geese and brant relative to adaptation to polar climes). Z. Zhurn., 57:738–749.

TAYLOR C. R. 1977. The energetics of terrestrial locomotion and body size in vertebrates. Pp. 127–141, *in* Scale effects in animal locomotion (T. J. Pedley, ed.). Academic Press, London, 545 pp.

TAYLOR, C. R., K. SCHMIDT-NIELSEN, AND J. L. RAAB. 1970. Scaling of energetic cost of running to body size in mammals. Amer. J. Physiol., 219:1104–1107.

TENER, J. S., 1954. A preliminary study of the musk-oxen of Fosheim Peninsular, Ellesmere Island, N.W.T. Canadian Dept. Natl. Parks Bd., Canada Wildl. Serv., Ottawa, 82 pp.

THOMPSON, A. M. 1959. Maternal stature and reproductive efficiency. Eugen. Rev., 51:157–162.

TOMILIN, A. G. 1946. Thermoregulation and the geographical races of cetaceans. Compt. Rend. (Doklady) Acad. Sci. U.R.S.S., 54:465–468.

VAN DE VEEN, H. 1979. Food selection and habitat use in the red deer *Cervus elaphus* (L.). Unpubl. Ph.D. dissert., Univ. Gröningen.

WALKER, E. P. 1964. Mammals of the world. 2 vols. Johns Hopkins Univ. Press, Baltimore, Maryland, 1500 pp.

WASER, P. 1977. Feeding, ranging and group size in the mangabey, *Cercocebus albigena*. Pp. 183–222, *in* Primate ecology (T. H. Clutton-Brock, ed.). Academic Press, London, 631 pp.

WATSON A., AND B. W. STAINES. 1978. Differences in the quality of wintering areas used by male and female red deer (*Cervus elaphus*) in Aberdeenshire. J. Zool., 186:544–550.

WESTERN, D. 1979. Size, life history and ecology in mammals. African J. Ecol., 17:185–204.

WHITE, G. 1789. The natural history of Selborne. Oxford Univ. Press, London, 182 pp.

WHITEHEAD, G. K. 1972. Deer of the world. Viking Press, New York, 194 pp.

WILSON, E. O. 1975. Sociobiology, the new synthesis. Belknap Press, Cambridge, Massachusetts, 697 pp.

ECOLOGICAL AND BEHAVIORAL CONSEQUENCES OF ADAPTATION TO VARIOUS FOOD RESOURCES

Brian K. McNab

Abstract

Food habits exert an important influence on the rate of metabolism of mammals independent of the influence of body mass. If they feed on foods that are seasonally unavailable, nutritionally poor, or filled with chemical poisons, placentals tend to have low basal rates, especially at large masses; otherwise they have high basal rates. The principal value to a high rate of metabolism in placentals is that it increases the population parameter r_m, the maximal intrinsic population growth constant, mainly as a result of the influence of rate of metabolism on gestation period, postnatal growth constant, and fecundity. There is at best a weak dependence of reproduction on rate of metabolism in monotremes and marsupials, which accounts for the absence of high rates of metabolism in these mammals.

Food habits and rate of metabolism also have an effect on the behavior of mammals through their effect on home-range size, extended parental care, and social behavior. Home-range size increases with body size, with the use of rare foods, and with an increase in the rate of metabolism of the consumer. The same factors influence home-range size in social species, although a troop may require a larger home range than an equally-sized solitary species due to the influence of the scaling of metabolism on body size. The pattern and timing of reproduction in mammals also depends on food habits because the various foods show a variable pattern in seasonal abundance. Extended parental care is most likely to occur in slow-growing species (that is, those that have food habits that lead to low rates of metabolism) or that are arboreal. Social behavior requires a food resource that can sustain a relatively dense population and a habitat that permits communication. The social systems of social carnivores permit the use of prey that otherwise would be unavailable; these systems may be the best models for the study of the evolution of hominid sociality.

Introduction

Mammals use most available food resources, including everything from tubers, seeds, leaves, and nectar to plankton, termites and elephants. Each food can be variously characterized in terms of nutritional and energy content; physiological cost of processing; behavioral requirements for locating, collecting, and handling; and circadial and annual availability. Clearly, foods differ quantitatively and qualitatively from each other. It remains to be seen how these differences among foods influence the mammals that use them. Some of the consequences for mammals of using one food supply or another are examined in this chapter.

Physiological Consequences of Adaptation to Different Food Resources

There are many physiological consequences of using various food resources. The morphology and enzymology of the digestive tract reflect the foods used; herbivores tend to have long intestines and carnivores short intestines. The general absence of cellulases in mammals means that most species specialized for grazing and browsing rely on pregastric or postgastric fermentation and its attendant microflora. Such physiological details may limit the ability of a phyletic line to adopt new food habits. For example, carnivores and omnivores have repeatedly given rise to herbivores in the phylogeny of therapsids and mammals, whereas true herbivores have never given rise to carnivores. However, this chapter emphasizes the relation of food habits to the energetics of mammals because of the significance of energetics to the population biology and behavior of mammals.

The principal factor influencing the daily energy expenditure of mammals is body size (McNab, 1980*a*). Unfortunately, because there are relatively few data available on the daily rate of energy expenditure in mammals, it is most useful to compare mammals in terms of basal rates of metabolism. Basal rates are standardized values in the sense that they represent measurements when mammals are postabsorptive and in thermoneutrality. Although mammals are only occasionally in this state in nature, the use of basal rates is convenient and can be justified by the correspondence that exists between basal and daily rates (see McNab, 1980*a*).

As in daily rates, the main factor dictating basal rates in mammals is body size (Kleiber, 1932, 1961), although recent evidence (McNab, in press) shows that the Kleiber scaling function must be changed at small masses or else endothermy is relinquished. Nevertheless, at all masses there is appreciable residual variation in basal rate.

Much of the residual variation in the basal rate of metabolism in mammals is correlated with food habits. This correlation is shown in monotremes and marsupials (Table 1). By placental standards almost all monotremes and marsupials have low basal rates. Most genera are in the "low" category (in which the basal rates are 60 to 89% of the Kleiber value for placentals). Some genera, however, have even lower basal rates. For example, desert carnivores (*Pseudoantechinus, Dasycercus*), desert omnivores (*Macrotis*), all ant- and termite-eaters (*Tachyglossus, Zaglossus, Myrmecobius*), and an arboreal folivore specialist (*Phascolarctos cinereus*) have basal rates that are less than 60% of the Kleiber value. The only marsupial to have an intermediate basal rate by placental standards is the aquatic carnivore, *Chironectes minimus* (McNab, 1978). Some monotremes and marsupials with certain food habits have very low basal rates, but no marsupial has a high basal rate of metabolism by placental standards, even though some marsupials have food habits that in placentals are correlated with high basal rates. This paradox shall be addressed later.

Placentals show a correlation of basal rate with food habits with even greater clarity (Table 2). Most placentals that have had their basal rates measured have intermediate (90 to 110% of the Kleiber value) or low (60 to 89%) basal rates. The difference between these two groups is that those having low basal rates tend to be desert herbivores (*Spermophilus,* Heteromyidae), fossorial herbivores (Geomyidae, Bathyergidae, *Spalax, Ctenomys,* Rhyzomyidae, *Aplodontia*), and arboreal omnivores (various prosimians, *Cebuella, Callithrix,* procyonids, palm civets). Placentals with very low basal rates (that is, basal rates less than 60% of the Kleiber value) include flying insectivores, ant- and termite-eaters (Myrmecophagidae, *Priodontes, Tolypeutes, Orycteropus, Manis*), arboreal folivore specialists (*Bradypus, Choloepus*), soil and litter omnivores (Dasypod-

TABLE 1

BASAL RATES OF METABOLISM IN MONOTREMES AND MARSUPIALS AS A FUNCTION OF FOOD HABITS

	Relative basal rates of metabolism*		
Food habits	Very low (<60%)	Low (60–89%)	Intermediate 90–110%)
Carnivores			
Terrestrial carnivores Small (insectivores)	*Pseudoantechinus*	*Antechinus, Sminthopsis, Antechinomys*	
Medium	*Dasycercus*	*Lutreolina, Dasyuroides, Sarcophilus, Satanellus, Dasyurus, Dasyurops*	
Arboreal carnivores		*Phascogale*	
Aquatic carnivores		*Ornithorhynchus*	*Chironectes*
Ant- and termite-specialists	*Tachyglossus, Zaglossus, Myrmecobius*		
Herbivores			
Terrestrial grazers		*Setonyx, Lagorchestes, Macropus, Megaleia*	
Terrestrial browsers		*Potorus*	
Arboreal folivores	*Phascolarctos*	*Trichosurus, Phalanger, Pseudocheirus*	
Omnivores			
Terrestrial omnivores		*Monodelphis*	
Arboreal omnivores		*Marmosa, Caluromys, Metachirus, Philander, Didelphis, Petaurus, Cercartetus*	
Soil and litter omnivores	*Macrotis*	*Parameles, Echymipera, Isoodon*	

* Relative to the Kleiber standard, $\dot{V}_{O_2}$ (cm^3 O_2/h) = $3.42g^{0.75}$.

TABLE 2

Basal Rates of Metabolism in Placental Mammals as a Function of Food Habits

Food habits	Relative basal rates of metabolism*				
	Very low (<60%)	Low (60–89%)	Intermediate (90–110%)	High (111–140%)	Very high (>140%)
Carnivores					
Terrestrial carnivores					
Small (insectivores)				Crocidurinae	Soricinae
Medium			*Herpestes*	*Mustela*	
Large		*Taxidea*	*Meles*	*Gulo*	*Lycaon*
Arboreal canivores				*Martes*	
Aquatic carnivores					
Medium				*Mustela vison*	*Lutra, Enhydra*
Large					Phocidae, *Phocoena, Tursiops, Rhachianectes*
Flying carnivores					
Small (insectivores)	*Noctilio labialis, Hipposideros,* Molossidae, Vespertilionidae	*Eptesicus, Molossus*			
Medium (blood)		Desmodontinae			

TABLE 2
CONTINUED

Food habits	Relative basal rates of metabolism*				
	Very low (<60%)	Low (60–89%)	Intermediate (90–110%)	High (111–140%)	Very high (>140%)
Medium (vertebrates)		*Noctilio leporinus*	*Macroderma, Tonatia, Chrotopterus*		
Fossorial insectivores			*Scalopus*		
Ant- and termite-specialists	Myrmecophagidae, *Priodontes, Cabossus, Tolypeutes, Manis, Orycteropus*	Macroscelididae, *Proteles*			
Herbivores					
Terrestrial grazers					
Small				Microtinae	Microtinae, *Sigmodon*
Medium		*Spermophilus, Procavia, Heterohyrax*	*Ammospermophilus, Sylvilagus*		*Ochotona*
Large			*Ovis, Oreamnos, Antilocapra*	*Bos, Equus*	*Taurotragus, Kobus*

TABLE 2
CONTINUED

Food habits	Relative basal rates of metabolism*				
	Very low (<60%)	Low (60–89%)	Intermediate (90–110%)	High (111–140%)	Very high (>140%)
Terrestrial browsers					
Medium		*Lepus arcticus*, *Neotoma*		*Lepus*	
Large		*Capra*	*Tragulus*		
Terrestrial spermivores	*Perognathus longimembris*	Heteromyidae, *Acomys*	*Notomys*, *Jaculus*, *Leggadina*	*Heteromys*, *Eutamias*	
Arboreal folivores	*Bradypus*, *Choloepus*, *Lemur*	*Dendrohyrax*, *Coendou*	*Alouatta*		
Arboreal spermivores			*Glaucomys*		*Tamiasciurus*
Aquatic herbivores					
Small			*Neofiber*, *Ondatra*, *Arvicola*		
Large	*Trichechus*				
Flying herbivores					
Frugivores		*Rhinophylla*	Pteropidae, Stenoderminae	*Dobsonia*, *Carollia* *Sturnira*, *Synconycteris*	
Nectarivores				Glossophaginae	

TABLE 2
CONTINUED

Food habits	Relative basal rates of metabolism* Very low (<60%)	Low (60–89%)	Intermediate (90–110%)	High (111–140%)	Very high (>140%)
Fossorial herbivores	*Heterocephalus*	*Heliophobius, Spalax, Spalacopus, Ctenomys, Tachyoryctes, Geomys, Cannomys, Aplodontia, Thomomys umbrinus*	*Thomomys talpoides*		
Omnivores					
Terrestrial omnivores					
Small			*Peromyscus, Zapus, Oryzomys*		
Medium		*Procyon, Genetta, Fennecus, Fossa, Cerdocyon*	*Canis*	*Vulpes*	
Large			*Tayassu*		

TABLE 2
CONTINUED

Food habits	Relative basal rates of metabolism*				
	Very low (<60%)	Low (60–89%)	Intermediate (90–110%)	High (111–140%)	Very high (>140%)
Arboreal omnivores	*Nycticebus, Ptilocercus, Arctictis*	*Perodicticus, Aotus, Galago, Cebuella, Callithrix, Tupaia, Potos, Eira, Hemigalus, Paradoxurus, Nandina*			
Flying omnivores		*Eonycteris, Phyllostomus*			
Soil and litter omnivores	Dasypodidae, Erinaceidae, Tenrecidae				

* Relative to the Kleiber standard, $\dot{V}_{O_2}$ ($cm^3\ O_2/h$) = $3.42g^{0.75}$.

idae, Erinaceidae), and a few desert herbivores (*Heterocephalus*). Some placentals with very low basal rates of metabolism have habits similar to monotremes and marsupials with very low basal rates.

Unlike marsupials, some placentals have high, or very high basal rates. For example, terrestrial and aquatic carnivores (Soricidae, Mustelidae, *Gulo, Lycaon,* Phocidae, *Enhydra, Lutra, Phocoena, Tursiops*), most non-desert grazers (Microtinae, *Sigmodon, Bos, Equus, Taurotragus, Kobus*), and some flying frugivores and nectarivores (*Dobsonia, Synconycteris, Carollia, Sturnira,* Glossophaginae) have basal rates greater than 110% of the values expected from the Kleiber relation. Some of the difference in allocation of placentals between the high (111 to 140%) and the very high (>140%) categories, or between the intermediate (90 to 110%) and high categories, is related to body mass; the smaller members of a set belong to the higher category, apparently reflecting the cost of continuous endothermy (McNab, in press). Thus, soricids have basal rates that are more deviant from the Kleiber relation than macroscelids; small microtines (*Microtus, Clethrionomys*) have higher relative basal rates than large microtines (*Arvicola, Neofiber, Ondatra*); and small fruit bats (*Synconycteris, Carollia, Sturnira*) have higher relative basal rates than large fruit bats (*Artibeus, Rousettus, Pteropus*). On the other hand, some small species in an ecological set have lower basal rates relative to the Kleiber standard (*Rhinophylla* among fruit bats, *Heterocephalus* among fossorial rodents, *Perognathus longimembris* among heteromyid rodents, and almost all small insectivorous bats), but in all cases these mammals are poor temperature regulators.

One of the clearest examples of the influence of food habits on the energetics of mammals is found in bats (Fig. 1). Fruit- and nectar-eaters have much higher basal rates than insect-eaters of the same mass, whereas blood-eaters and some species with mixed diets have intermediate basal rates. This correlation is independent of taxon in that it makes no difference whether the fruit bat is a phyllostomid (Microchiroptera) or a pteropodid (Megachiroptera), or whether the insectivorous bat is a vespertilionid, molossid, or noctilionid.

Certain food habits in mammals, irrespective of taxonomic affiliations, are correlated with low basal rates of metabolism—ant- and termite-eating specialists, arboreal folivore specialists, flying insectivores, and to a somewhat lesser extent desert carnivores and

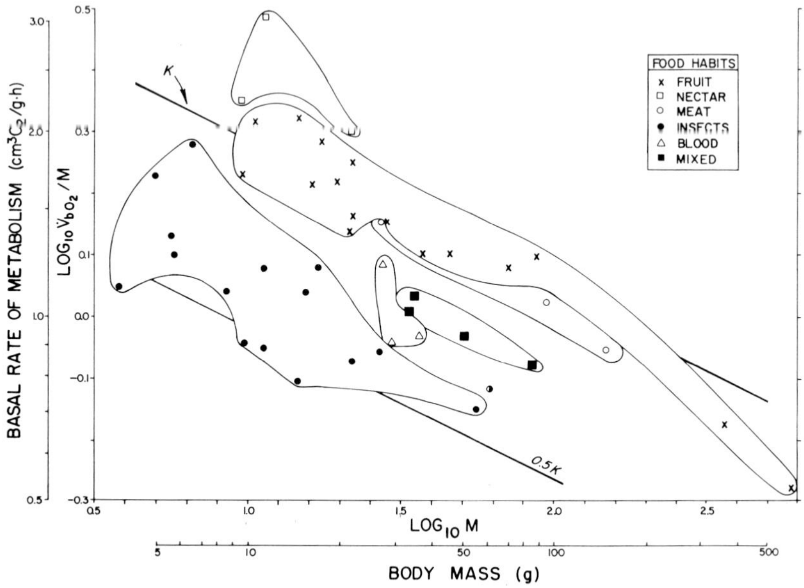

FIG. 1. Basal rate of metabolism as a function of body mass in bats as influenced by food habits (from McNab, 1980*a*). The curves labeled K and 0.5K represent Kleiber's (1932) relation for placentals and one-half of these values, respectively.

herbivores, and soil and litter omnivores. As a consequence, mammals with these food habits, regardless of whether they are monotremes, marsupials, or placentals, have low body temperatures, a reduced capacity to regulate body temperature at low environmental temperatures, and (in some instances) the capacity to enter torpor. But placentals that are terrestrial or aquatic carnivores, non-desert grazers, or flying frugivores have high rates of metabolism, high body temperatures, can usually regulate body temperature in the face of cold environmental temperatures, and (as long as they are not very small) rarely enter torpor. Monotremes and marsupials with these food habits, however, have low rates of metabolism.

The reason why there are correlations between rate of energy expenditure and food habits is, in part, perfectly obvious. Mammals with low rates of metabolism generally feed on foods that in some manner or another are sufficiently deficient (for example, are not easily digested or are widely dispersed) as to prevent their consumers from having high rates of metabolism. In other words, the low rates of metabolism of most mammals evolutionarily reflect a limitation in the food supply and not a deficiency in the mammals themselves. Thus, those foods that lead to low basal rates in mammals may be available in amounts inadequate to sustain high rates of metabolism in consuming mammals, which means that the consumers must adjust their rates down to the (maximal, as we shall see) level tolerated during the poorest period. For example, the seasonal periodicity of flying insects demands low basal rates of insectivorous bats; and the presumptive periodicity in the availability of soil invertebrates, especially in seasonally arid or cool environments, may require low basal rates of (large) soil and litter omnivores. Some foods may also lead to low basal rates in consuming mammals if they are so filled with chemical deterrents or poisons that the cost to the consumer for detoxification is great enough to reduce the value of a high rate of metabolism. For example, the anti-arthropod chemical compounds stored in many tree leaves may be a factor reducing the basal rates of arboreal folivores. Various foods may, in addition, have such poor nutritional (or energy) content that the volume that would have to be consumed to permit a high rate of metabolism in the consumer is mechanically or energetically prohibitive. For example, the leaves of trees may have such a low nitrogen content, or be so low in the amount of available energy, that they require low rate of leaf specialists; or large ant- and termite-eaters may be so much greater in size than the individual prey that these predators must resort to what is essentially filter-feeding, thereby taking in much detritus and reducing the caloric density of the ingested material.

Placental mammals seem to have as high a rate of metabolism as is possible (the reason for which shall be seen in the next section). Those with low rates of metabolism have been under selection (often by dietary adaptation) to have low rates. Marsupials conform to this picture insofar as they have low basal rates (and are of the same habits as placentals with low basal rates), but differ in that no marsupials are known to have extremely high basal rates. This is an inconsistency in the theory which is addressed later.

Populational Consequences of Adaptation to Different Food Resources

The variation in rate of metabolism in mammals with respect to the differential use of food resources has consequences for the pop-

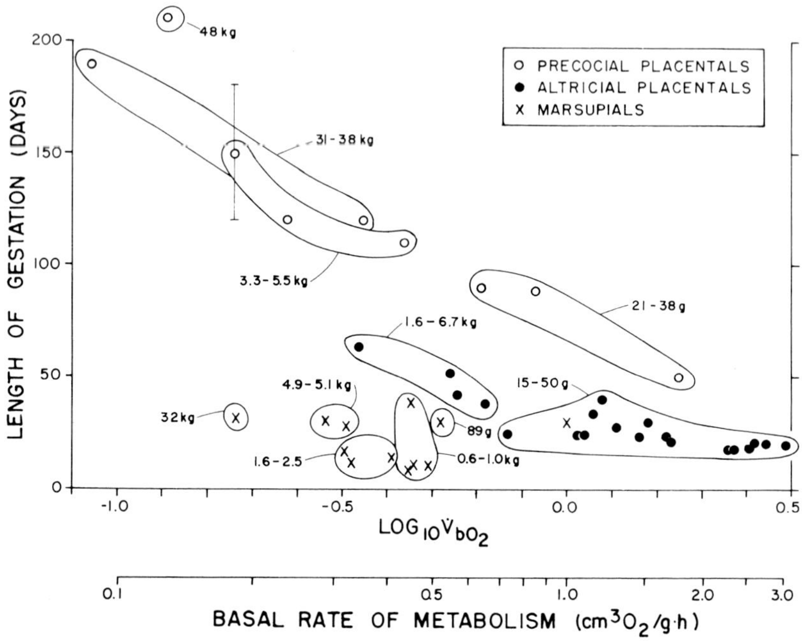

FIG. 2. Length of gestation in placentals and marsupials as a function of basal rate of metabolism for fixed mass classes (from McNab, in prep.).

ulation biology of mammals because its variation in placentals affects intrauterine and postnatal growth rates and affects fecundity (McNab, 1980*a*). Thus, in placentals 1) there is a suggestion that gestation period decreases with an increase in rate of metabolism (Fig. 2); 2) the postnatal growth constant increases with an increase in rate of metabolism (Fig. 3); and 3) the number of young produced by a female in her lifetime or in a year increases with an increase in rate of metabolism (Fig. 4). All of these functions (that is, gestation period, growth, and fecundity) also scale to body mass in placentals (McNab, 1980*a*); it is therefore necessary to demonstrate the effect of rate of metabolism independent of the influence of body size which is suggested in the scattergrams (Figs. 2–4). The production of altricial or precocial young in placentals may be unrelated to food habits; gestation period in both groups appears sensitive to rate of metabolism (Fig. 2).

Gestation period, postnatal growth constant, and fecundity inter-

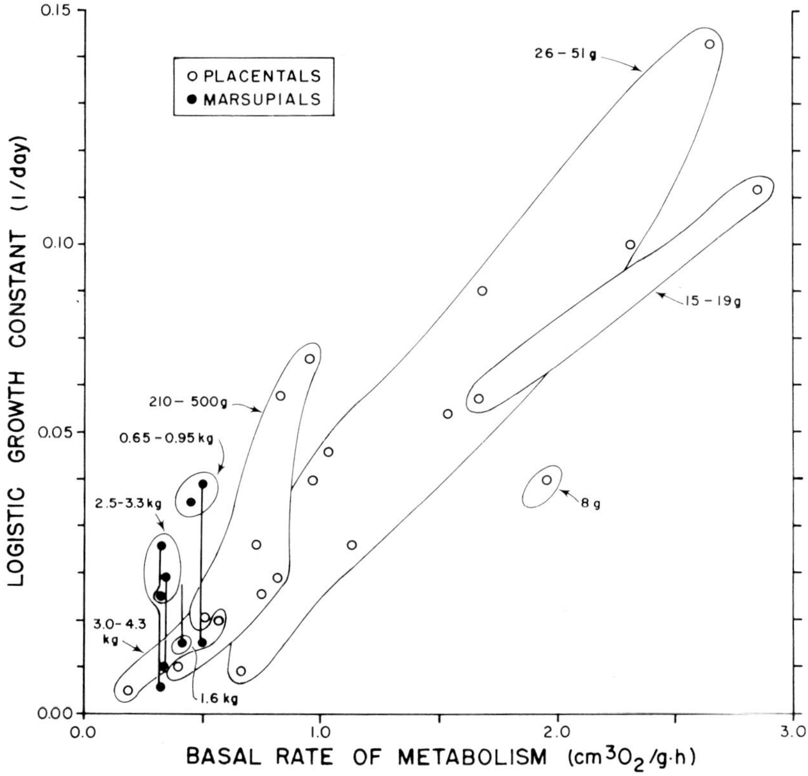

FIG. 3. Logistic postnatal growth constant in placentals and marsupials as a function of basal rate of metabolism for fixed mass classes (from McNab, in prep.).

act to influence the reproductive and populational characteristics of placentals. As a result, variation in rate of metabolism independent of mass also influences the population biology of placentals in various ways. First, placentals with high rates of metabolism have higher maximal intrinsic population growth constants (r_m) than do placentals with low rates of metabolism (McNab, 1980*a*). This correlation exists because r can be approximated by $(\ln R)/T$, where $(\ln R)$ is the natural logarithm of fecundity and T is generation time (that is, the time that it takes an individual from conception to the age of first reproduction). Thus, an increase in rate of metabolism will increase R (Fig. 4), and by reducing gestation period and by increasing the rate at which postnatal growth occurs, an

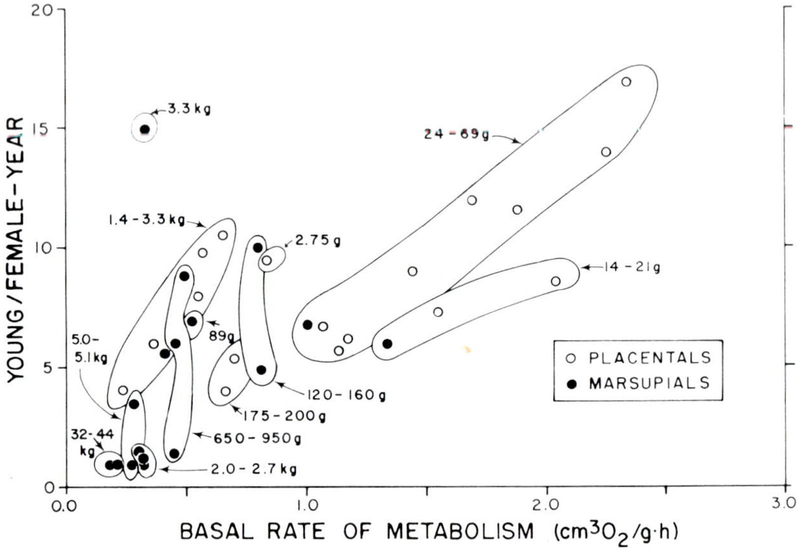

FIG. 4. Number of young produced by a female in a year in placentals and marsupials as a function of basal rate of metabolism for fixed mass classes (from McNab, in prep.).

increase in rate of metabolism will decrease *T*. The combination of these influences will greatly increase *r*, and therefore under optimal conditions r_m. Because of the correlation of rate of metabolism with food habits, mammals that feed on the leaves of trees, insects in flight, and ants and termites would be expected to have a low r_m and those that feed on grass or that are strict carnivores a high r_m. This conclusion must be conditional because so few maximal intrinsic population growth constants have been measured. The overall correlation of *r* with rate of metabolism is supported by the observations of Fenchel (1974).

Second, placentals with high rates of metabolism have larger population fluctuations under given environmental conditions than those with low rates of metabolism because of the influence of rate of metabolism on r_m (McNab, 1980*a*). Thus, the mammals most renowned for population fluctuations, namely microtine rodents (such as meadow-voles [*Microtus*] and lemmings [*Lemmus, Dicrostonyx*]), and hares of the genus *Lepus*, have high basal rates. The carnivores *Mustela, Martes,* and *Lynx* fluctuate in numbers reflect-

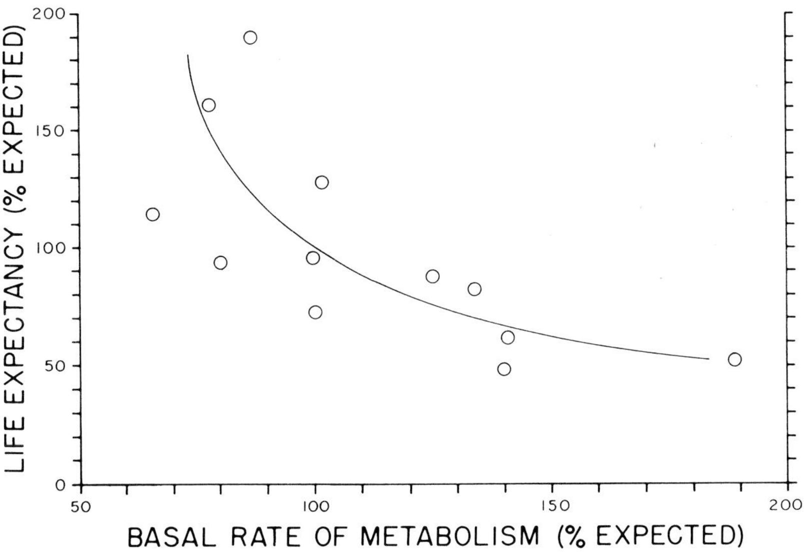

FIG. 5. Life expectancy, expressed as a percent of the mass-dependent relation of Western (1979), as a function of basal rate of metabolism, expressed as a percent of the rate expected from Kleiber's (1932) relation. Data on life span is from Western (1979) and on basal rates from other sources.

ing prey fluctuations, but *Mustela* and *Martes* also have high r_m values. At the other end of the spectrum in energetics, such mammals as ant- and termite-eaters and arboreal folivores, would be expected to show relatively constant populations (see Eisenberg, 1975). Unfortunately, there are essentially no data available on the variability of natural populations of these mammals.

Third, placentals with high rates of metabolism have shorter life spans than similarly-sized species with low rates of metabolism (Figure 5). This correlation has been used before (Bourlière, 1958) to explain why insectivorous bats have longer life spans than most other mammals of the same size, although Herreid (1964) has argued against this explanation. As a matter of fact, Western (1979) showed that life span in mammals is proportional approximately to body mass raised to the 0.23 power ($m^{0.23}$). Sacher (1959) described a similar relation. Such a function is similar to one described by McNab (1980*b*), where a time period (in this case the ability to withstand starvation) is given by the ratio of a resource (which is proportional to $m^{1.0}$) to the rate at which that resource is used ($m^{0.75}$). Thus, time here would be proportional to $m^{0.25}$. Such an analysis of life span would incorporate properly the influence of rate of metabolism (that is, an increase in rate of metabolism would decrease life span), but it does raise the question of what the "resource" would be.

The interactions between metabolism and r_m, population fluctuation, or life span, are not necessarily rigidly determined. There is always sufficient flexibility in most biological systems to permit animals to modify their behavior in such a manner as to combine apparently contradictory characteristics. An example is that some temperate insectivorous bats that hibernate in caves establish parturition and nursery colonies in warm environments exterior to caves (for example, in barns). This behavior can be interpreted to be a means of increasing r_m beyond that capable of being attained in cool caves, where body temperature and rate of metabolism would be low (McNab, 1980*a*).

It can be concluded that one reason why placentals have high basal rates of metabolism is to maximize the output of young by increasing fecundity and by reducing generation time, thereby increasing r_m. All placentals would appear to profit from such a strategy, but the maximal tolerable rates of metabolism vary radically with the food resources exploited. Even ant- and termite-eaters would appear to profit from maximizing r_m, but given the nature of their food, the maximal rate of metabolism compatible with this diet appears to be very low, resulting in a low body temperature. Mammalian ant- and termite-eaters generally are limited in distribution to the tropics; anteaters can exist in temperate regions only if they have still lower standard rates of metabolism than are found in most mammals because of the reduced abundance of these prey in temperate regions. Thus, the ant-eating habit in the temperate zone is often left to lizards.

This analysis raises the question of why marsupials are not known to have high basal rates by placental standards. Marsupials, unlike placentals, do not seem to have a close coupling of reproduction to rate of metabolism. For example, gestation period in marsupials is independent of basal rate (Fig. 2) and fecundity does not depend on basal rate independent of the influence of body mass (Fig. 4); only the postnatal growth constant may depend on basal rate in marsupials (Fig. 3), and then only slightly. Marsupials are unable to have high rates of exchange between mother and fetus due to the threat of allograft rejection; thus the gestation period must be kept short in most species. This problem is evaded in placentals by the development of a morphologically elaborate structure interposed between the fetus and mother, the trophoblast, thus permitting considerable variability in gestation length (Lillegraven, 1976). These observations on marsupials emphasize the interpretation, made earlier, that the factor mainly responsible for the high rates of metabolism in placentals is a concern with r_m, not with thermoregulation. Most marsupials are as good endotherms as are equivalently-sized placentals. The only marsupial known to have a standard placental basal rate is *Chironectes minimus,* which may reflect the aquatic habits of this marsupial and therefore may simply represent an adjustment to a thermally rigorous environment. Similar reasoning may explain the observation that *Ornithorhynchus anatinus* has a higher basal rate than *Tachyglossus. Chironectes minimus* is presumably under the same reproductive constraints as all other marsupials.

Behavioral Consequences of Adaptation to Different Food Resources

The extent to which the differential use of food resources influences the behavior of mammals is unknown. As we shall see, it has an effect on home-range size, on the pattern and timing of reproduction, and may have some indirect relationship to extended parental care and to degree of sociality, although each of these relationships is variously modified by other factors.

Size of Home Range

Food habits have a marked influence on home-range size because they dictate the comparative densities of food stuffs available within an area. For example, in a limited set of data, I suggested (1963) 1) that home-range size in mammals was approximately proportional to $m^{0.75}$ (and therefore was proportional to basal rate of metabolism); and 2) that those species that fed on seeds, fruits, insects, or vertebrates (the so-called "hunters") required at a given mass a larger home range than those that fed on grass or browse (the so-called "croppers"). Hunters generally had home ranges that were about four times those of croppers of the same size. The reason for this difference is that grass and browse (that is, the photosynthetic units of plants) are more abundant than seeds and fruits (that is, the reproductive units of plants) or insects and vertebrates (the consumers).

The relation between home-range size and body mass in mammals was re-examined recently by Harestad and Bunnell (1979). With more data available, they suggested that home-range size is approximately proportional to $m^{1.30}$ (a similar power relation seems to be found in birds; Schoener, 1968). Harestad and Bunnell agree that food habits are important in determining home-range size.

Compilation of data on home-range size by Harestad and Bunnell is of such value as to merit a detailed examination here (Fig. 6). The correlation of home-range size with mass falls between $m^{0.75}$ and $m^{2.00}$, although the trend in most individual genera or ecologically and taxonomically related sets of genera tends to fall close to $m^{1.90}$. Shrews and mustelids have the largest home ranges for a given mass, but the largest member of the genus *Mustela,* the mink (*M. vison*), has a much smaller (1/10) home range than is to be expected from mass, a condition that presumably reflects the switch from terrestrial to aquatic vertebrates as the principal prey items; on the other hand, *M. vison* forages in a three-dimensional world, but then again so does *Martes*. Otters have similarly small home ranges when surface area alone is considered. Dogs and cats have large home ranges, but they are only one-tenth of the area of equivalently-sized mustelids (for example, *Martes, Gulo*). Some of this difference may represent a specialization in prey type in *Martes* (for example, *Tamiasciurus, Lepus*) compared to *Felis* or *Lynx,* and some may reflect the omnivorous food habits of dogs. It is unclear, however, why cats do not have larger home ranges than dogs. (Nevertheless, note that one estimate of home range in *L. lynx* is much higher than the other, which emphasizes the sensitivity of home range to environmental conditions.) Herbivores have smaller home ranges, although larger *Lepus,* such as *L. californicus* and *L. alleni,* live in deserts and have proportionally larger home ranges. Equally, *Ovis canadensis* lives in arid regions and has a larger home range than equally-sized *Odocoileus.* Arid environments have lower rates of vegetative production than mesic environments (Rosenzweig, 1968), which consequently leads to larger home ranges in desert-dwelling herbivores. Omnivores have home ranges that are intermediate between carnivores and herbivores.

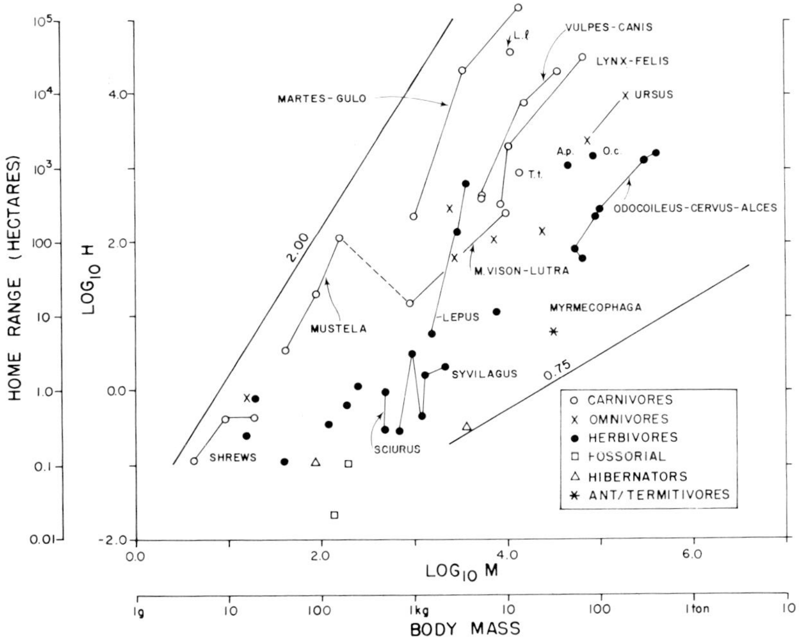

FIG. 6. Home-range size as a function of body mass in different mammals. Data taken from Harestad and Bunnell (1979) except that for *Myrmecophaga,* which came from Shaw (pers. comm.). Ecologically similar species and genera are connected with lines. The curves that represent the power functions $m^{2.00}$ and $m^{0.75}$ are indicated. The following abbreviations are used: L. l., *Lynx lynx*; T. t., *Taxidea taxus*; A. a., *Antilocapra americana*; and O. c., *Ovis canadensis*. Of the two fossorial mammals, the upper one is *Scalopus* sp. and the lower one is the gopher, *Thomomys talpoides*.

Why should home range in mammals be proportional to $m^{1.90}$ rather than to rate of metabolism (that is, $m^{0.75}$)? The answer is unknown. It may well be that the efficiency with which a consumer can find and harvest the food resources within an area diminishes with an increase in body mass and home-range area, or that the larger food particles used by larger predators occur in a sufficiently lower density to require proportionally larger home ranges than are found in smaller predators (see Clutton-Brock and Harvey, this volume).

Some mammals, including anteaters, hibernators, and fossorial mammals, have home ranges as small as, or much smaller than, grazing and browsing herbivores of the same mass (Fig. 6). All of these mammals are characterized by low basal rates of metabolism (McNab, 1979, pers. observ.). In addition, ants and termites are an exceedingly abundant food resource in many parts of the tropics. The mole, *Scalopus aquaticus,* has a higher basal rate than pocket gophers, as is reflected in a comparison of their home-range areas (Fig. 6), but the small size of a mole's home range suggests that they encounter an abundance of soil invertebrates as food. Both of these factors (low demand and high abundance of resources) would tend to reduce home-range size.

The variation in home-range size among mammals, therefore, is determined by the interaction of a series of interrelated factors. Home-range size 1) increases with the mean body mass of a species (presumably in relation to the increased energy expenditures of large individuals); 2) increases with the rate of metabolism of a species independent of body size (due either to the influence of food habits or climate); and 3) decreases with an increase in the abundance of food supplies in an area. The abundance of food may reflect either the type of food or the effect of climate on the production of various foods. An abundant food resource may compensate for a high rate of metabolism, as appears to be the case in ungulates and hares.

A similar pattern exists among home-range size, body mass, and food habits in social species. Milton and May (1976) showed that the total home range of a troop of primates divided by the number of individuals in the troop (that is, the home range corresponding to an individual) increases among species as the mean adult mass of the species increases. They further showed that the absolute size of this individual share of troop home range is smaller in folivores than in frugivores and omnivores. In fact, the size of home range within a species often varies as the diet varies in composition between leaves and fruit, the home range increasing as the proportion of fruit in the diet increases. There seems to be little evidence among primates that arboreality or terrestriality has per se an influence on home-range size, except that some terrestrial species live in arid environments and consequently have large home ranges.

One of the consequences of being a member of a troop is that an individual will traverse a much larger area than it would if it were solitary (Milton and May, 1976), which raises the question of whether such an individual pays any penalty for having a larger home range. Milton and May suggest several reasons why such a penalty may not exist: 1) locomotion may be a negligible portion of the energy budget; 2) foraging as a group may increase food-finding efficiency; or 3) a member of a social group may forage linearly, whereas a solitary individual may have to forage in a zig-zag manner. There would seem to be little difference in energetics, however, for an individual if it walked 2 km within a restricted home range or within a piece of an extended home range. The advantage of a restricted home range, namely a familiarity with sites for shelter, escape, and food gathering, may be provided in a larger area by group activity.

There is another issue concerning the relation of home range to body mass in social species: is the home-range area of a social species larger than a solitary species having the mass of the troop? Kaufmann (1974*b*) has shown this to be the case in social grazers; it can be further examined by plotting home range as a function of mass in both social and solitary species (Fig. 7). Such a plot, of course, is complicated by the complex composition of a troop, which may include one or more males, several females, and young of various ages and masses. Therefore, the mass of a troop cannot be estimated just by multiplying the number of individuals in a troop by the mean mass of a male. However, when the number of individuals belonging to a primate troop is multiplied by the average of the masses of males and females, the total home ranges of social species having large troops appear to be larger than the home ranges of social primates having smaller troops of the same total mass (Fig. 7).

Why should a troop of primates have a larger home range than a solitary species of the same mass, or why should a larger troop have a larger home range than a smaller troop? A potentially important contributor to this difference exists in the scaling of metabolism in relation to body mass. As Kleiber (1961) has shown, the total basal rate of metabolism of a mammal increases with $m^{0.75}$; therefore the summed basal rates of a group of mammals having a small individual body size is significantly greater than the basal rate of a single mammal having the same mass. The extent of the increase in the summed basal rate of the small mammals is given by

$$f = n(1/n)^{0.75}$$

where f is the dimensionless ratio of the basal rate of n individuals, each of which has a mass $1/n$, to the basal rate of a solitary individual that has a mass equal to $n(1/n) = 1$. This calculation is independent of the mass of the troop and solitary individuals, and

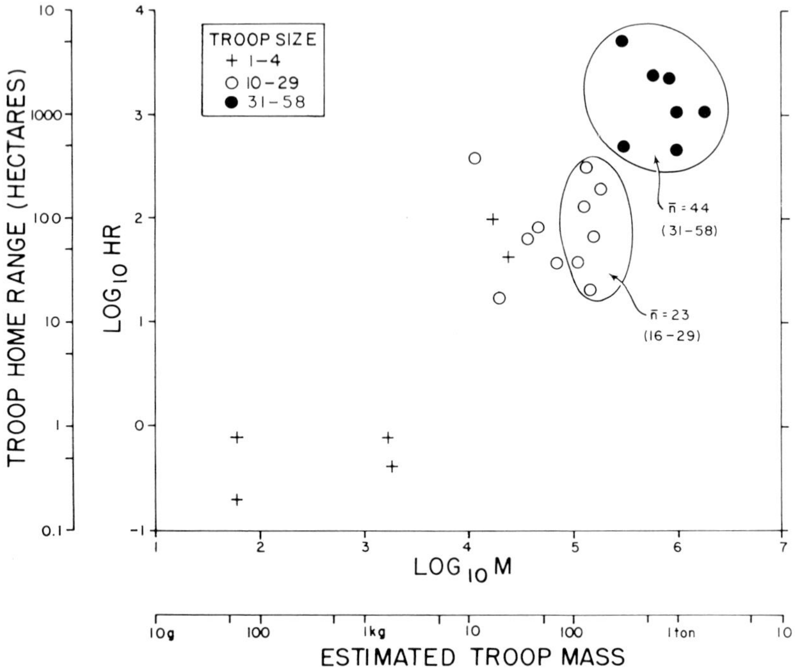

FIG. 7. Troop home range in primates as a function of estimated troop mass. Data taken from Milton and May (1976). The linear curve is calculated to indicate the presumptive influence of troop mass and numbers on troop home range (see text).

assumes for simplicity that all of the individuals that make up the troop are equally sized. The only variable in this equation is the number of individuals that make up the group. If, for example, a troop contained 23 individuals, their summed basal rates would be about 2.2 times that of a single individual having the mass of the troop; if the troop contained 44 individuals, their summed basal rates would be 2.6 times that of a single individual, or 1.18 times that of the troop having 23 individuals. In Fig. 7 the mean total mass of the troops having 44 members is about four times that of the troops having 23 members. Therefore, the factor by which home range in these troops is expected to increase as a result of the increase in troop numbers and troop mass equals $(44/23)(23/44)^{0.75} \cdot (4.19)^{1.9} = (1.18)(13.9) = 16.4$ times, which is similar to what is found (see Fig. 7).

Pattern and Timing of Reproduction

Food habits can influence the pattern and timing of reproduction through variations in the seasonal availability of various foods. These correlations are most clearly demonstrated in bats. In Central America insectivorous bats are seasonally monoestrous, frugivorous bats seasonally polyestrous, and vampire bats polyestrous with year long reproductive activity (Fleming et al., 1972). The length of the breeding season is thus affected by the seasonal availability of food: it is shortest in species that feed on flying insects, intermediate in those that use fruit or a mixture of fruit and insects, and longest in those that use the blood of endotherms.

The timing of reproduction also depends on the periodicity in the environment and on food habits. For example, in Panama the wet season starts in May, which is accompanied by a large increase in the numbers of flying insects (Fig. 8). The peak in number of species of insectivorous bats giving birth is somewhat earlier, apparently (depending on the species of bat) to time lactation, the development of young, or their early independence with the maximal amount of available food. Frugivorous bats in Panama have two peaks of reproduction, one in the dry season (when many flowers are present and just before the fruiting peak for large fruits) and the second in the wet season (during the peak of small fruits). Humphrey and Bonaccorso (1979) showed that the percentage of female fruit bats of the species *Artibeus jamaicensis* that are lactating varies in Panama with the number of trees with ripe fruits (Fig. 9), which again means that reproduction is timed in a manner such that pregnancy occurs before the peak of fruiting and lactation coincides with the peak of fruiting. Such a pattern reinforces the observation that lactation is more expensive for a female than is pregnancy.

Extended Parental Care

Some mammals show extended periods of parental care of their young. They include neotropical anteaters, pangolins, sloths, tenrecs, and most marsupials, all of which are characterized by low rates of metabolism. The occurrence of an extended period of maternal care in these species may be related to low postnatal growth rates, which in turn are associated with low rates of metabolism and particular food habits. Nevertheless, the causes of extended

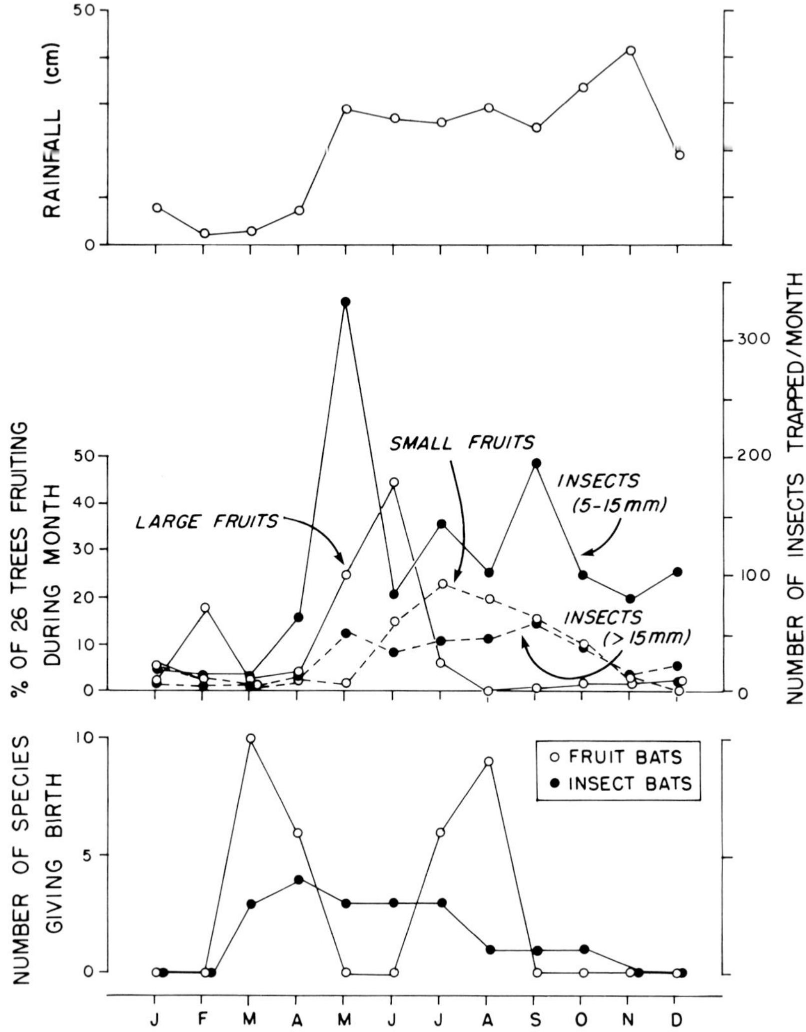

FIG. 8. Monthly variation in rainfall, the abundance of fruits and insects, and the numbers of species of insectivorous and frugivorous bats giving birth on Barro Colorado Island, Panama. Data from Fleming et al. (1972) and Wilson (1977).

periods of maternal care are undoubtedly complex. All mammals with low basal rates do not have such extended periods of care (for example, armadillos and some bats), whereas some mammals with high basal rates may have extended periods of maternal care (for

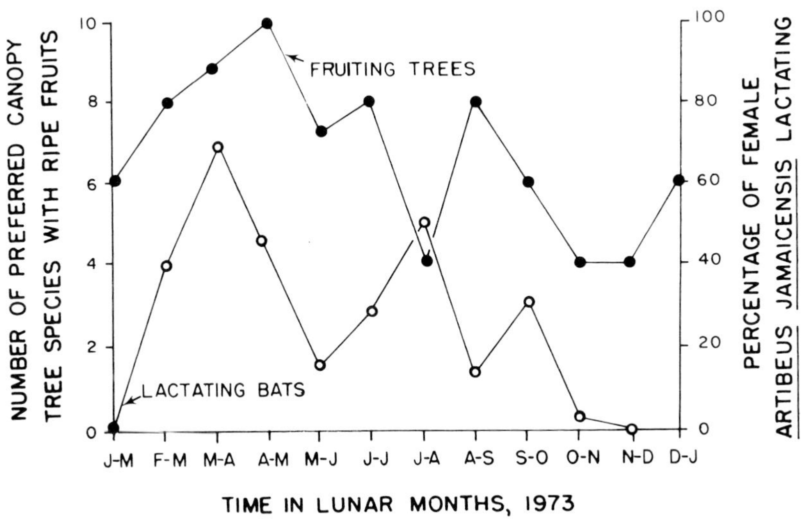

FIG. 9. Monthly variation in percent of female *Artibeus jamaicensis* that are lactating and number of preferred canopy trees with ripe fruits on Barro Colorado Island, Panama (from Humphrey and Bonaccorso 1979).

example, many primates and some ungulates). Of course, the form of reproduction in marsupials demands an extended period of postnatal care, but even beyond this influence some marsupials seem to have an extended bond established between mother and offspring. This condition is most notable in the koala (*Phascolarctos cinereus*) and various macropods. Another factor that contributes to the development of an extended period of care is life in trees (Eisenberg, 1975). Sloths, pangolins, and the koala (all of which have very low basal rates), and many primates have extended periods of care correlated in part with strictly arboreal habits.

Social Behavior

The presence of integrated social behavior is related to food habits, in part because the food must be capable of sustaining a sufficiently dense population for social behavior to occur and because the foods used are often associated with specific habitats that may place constraints on sociality. These relationships have been shown among marsupials, primates, ungulates, and carnivores. For example, most social marsupials are strict grazers. As Kaufmann (1974*a*) noted: "Being grazers, [the] abundant food supply [of whiptail wallabies, *Macropus parryi*] permits them to attain relatively dense populations overall, and to sustain groups locally. Their grazing habits and moderate size make a relatively open habitat almost inevitable, and their mobility, partly diurnal habits, and open habitat make group coordination easier." The contrast that may exist among marsupials is shown by a comparison of *Macropus giganteus* and *M. rufogriseus*. The former is social, lives in open areas, mainly eats grass, and is partly diurnal, whereas the latter is solitary, lives in closed habitats, mainly browses, and is less diurnal.

Primates have variable social systems, much of the variation being correlated with food habits (Eisenberg et al., 1972). If primates are small (under 2 kg), they tend to be nocturnal, feed predominantly on insects and fruit, live in trees, and are either solitary or have at most a parental family. Larger species, whether arboreal or terrestrial, tend to be folivorous or frugivorous, or to have a mixed diet. As has been seen, folivorous primates tend to have smaller home ranges. Folivorous, frugivorous, and omnivorous primates tend to be diurnal and to have social organizations that are based on the tolerance of various adult males in a troop.

Eisenberg and McKay (1974) and Estes (1974) showed that ungulates have a relationship between food habits and social structure similar to that demonstrated in marsupials and primates. Forest species tend to be browsers or selective feeders on leaves and fruit, to be small to medium-sized, and to be asocial, monogamous, sedentary, and territorial, whereas plains species tend to be grazers, of a large size, and to be gregarious, polygynous, and migratory or nomadic. Estes (1974) showed that of 44 browsing bovids studied, 35 (80%) were solitary, but of 26 grazing bovids studied only six (23%) were solitary. Bovids, like horses, apparently were originally forest browsers. In both groups a change in behavior accompanied a change in habitat and food preferences. With movement into open country the body size of bovids and horses increased, they (of necessity) became grazers, and sociality increased. The only forest ungulates that generally maintain large social groupings are swine, and they are omnivorous and have large home ranges.

Carnivores may also attain a high degree of sociality in relation to food habits, but in this case it is usually associated with cooperative hunting and food defense. Small species generally are solitary and feed on invertebrates and small vertebrates; there are after all few spoils to be shared. The few social viverrids, including meerkats (*Suricata suricatta*), dwarf mongooses (*Helogale undulata*), and banded mongooses (*Mungos mungos*), feed on invertebrates, are diurnal, and live on open plains or savannahs. Their sociality appears to have evolved as an antipredator strategy (see Rood, this volume). Some medium-sized to large predators, such as foxes (*Vulpes vulpes*), leopards (*Panthera pardus*), and tigers (*P. tigris*) are solitary hunters, but they either are limited to prey of a small to medium size (for example, jackals [*Canis aureus*; Schaller and Lowther, 1969], and leopard [Kruuk and Turner, 1967]), or hunt in habitats of poor visibility (for example, tigers).

Social hunters of a medium to large size increase the proportion of hunts that are successful and take significantly larger prey by hunting in packs (Fig. 10). For example, single jackals are successful in killing a fawn of Thomson's gazelle (*Gazella thomsoni*) only 16% of the time, but two jackals are 67% successful; adult Thomson's gazelles are killed by solitary lions (*Panthera leo*) about 29% of the time, but group attacks by lions on adult gazelles are 52% successful (Schaller and Lowther, 1969). Furthermore, spotted hyaenas (*Crocuta crocuta*) preferentially feed on Thomson's gazelles and juvenile wildebeest (*Connochaetes taurinus*) during the day, when hyaenas are solitary hunters, but selectively feed on adult wildebeest and zebra (*Equus burchelli*) at night when they are pack hunters (Kruuk, 1966). Lions hunting in groups are able to kill large herbivores such as buffalo (*Synceros caffer*), giraffe (*Giraffa camelopardalis*), and eland (*Taurotragus oryx*), whereas solitary lions feed on smaller prey such as gazelles (Kruuk and Turner, 1967). Wolves (*Canis lupus*) hunting in small packs generally kill prey no larger than deer (*Odocoileus*), whereas a large pack is usually required to kill large ungulates like elk (*Cervus elaphus*) and moose (*Alces americana*) (Nudds, 1978).

The differences in social behavior that may exist among carnivores are easily seen in a comparison of cats and dogs (Kleiman and Eisenberg, 1973). Felids tend to be purely carnivorous and to obtain their prey by concealment, stalking, and sudden attack. They also are usually forest-dwelling and semi-arboreal. Canids are much more omnivorous, obtain their prey by pursuit hunting, are cursorial, and generally live in open areas. Cats are generally solitary, even lacking any noticeable pair bonding; solitary hunting foxes, however, establish a pair bond. The tendency towards sociality in the lion, the only truly social cat, is much less than it is in dogs, as

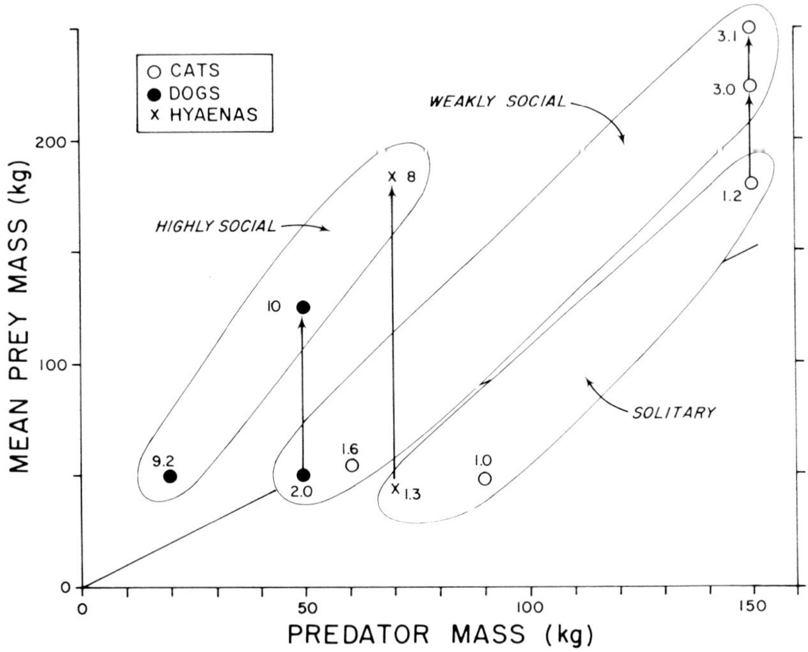

FIG. 10. Mean prey mass as a function of individual predator mass in relation to pack or pride size. Data on a particular species in relation to a variable group size are connected by arrows. Data from Kruuk (1966), Kruuk and Turner (1967), and Nudds (1978).

is shown by a reduced division of labor during hunting and by the greater level of aggression during feeding after a kill compared to the behavior of pack-hunting dogs (for example, *Canis lupus, Lycaon pictus*).

Schaller and Lowther (1969) suggested that the cooperative hunting of larger carnivores is more likely to reflect the factors responsible for the evolution of hominid sociality than is the sociality of non-hominid primates. This suggestion is based on the observations that 1) the social systems of mammals are sensitive to the ecological conditions that they face and only rarely (if ever) reflect phylogeny, and 2) the social systems of living populations of human hunter-gatherers are often similar in detail to those of carnivores in terms of group behavior and its consequences for predation, dominance hierarchies, land use systems, and cooperative hunting techniques.

Even though social hunting permits a carnivore to kill large prey

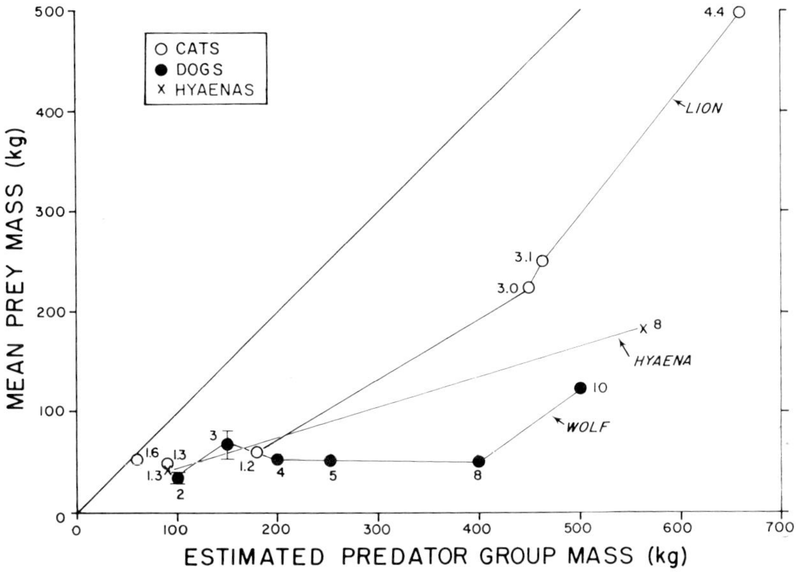

FIG. 11. Mean prey mass as a function of the estimated predator group mass. Data taken from Kruuk (1966), Kruuk, and Turner (1967), and Nudds (1978).

(Fig. 10), solitary and social hunters alike kill prey whose mass is equal to, or less than, the mass of the solitary predator or the summed mass of the social predator (Fig. 11). This observation suggests that one difficulty with a large group size in social predators is that the amount of food obtained by collective action may be inadequate to satisfy all of the members of the group. There appears to be an optimal group size judged from the amount of food provided by group hunting to the members of the group. Both Caraco and Wolf (1975) and Nudds (1978) showed that the amount of food provided per individual participating in a hunt attains a maximum at some "optimal" group size; it decreases in smaller groups, presumably because of their inability to kill large prey, and it decreases in larger groups, because of the limited amount of food obtained from a particular prey (Fig. 12). Thus, not only is a larger pack size required of wolves to kill elk and moose than is required to kill deer, the larger size of elk and moose can fulfill the energy requirements of many more wolves (up to 14) compared to deer

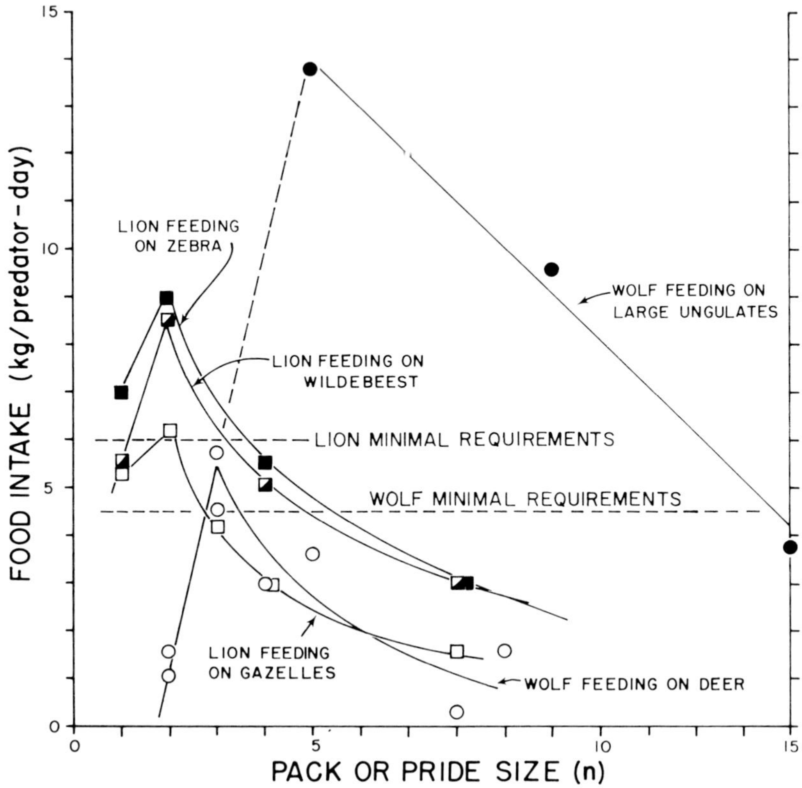

FIG. 12. Mean daily food intake of predators in relation to the size of the hunting pack or pride relative to the size of the prey. Data taken from Caraco and Wolf (1975) and Nudds (1978).

(up to four). The optimal pack sizes of wolves for these prey are six and three, respectively.

A comparison of the lion and wolf curves in Fig. 12 shows that lions maintain a positive energy balance at smaller group sizes than is the case with wolves. This difference is undoubtedly related to the larger individual size of lions (150 kg) compared to wolves (50 kg).

Finally, given the relation of social behavior to food habits and the relation of energetics to food habits, it may be asked what is the direct relation of social behavior to energetics. There is little reason to believe that either high or low basal rates are associated with sociality. Particular cases, such as the naked mole-rat (*Heterocephalus glaber*), do make one wonder, however. Unlike all other fossorial herbivores, which are normally intraspecifically hostile, *Heterocephalus* has an elaborate, cooperative social system (Jarvis, 1978; Jarvis and Sale, 1971). This species has such a low basal rate as to produce thermal incompetence (McNab, 1966, 1979). The correlation between a very low rate of energy expenditure and social structure in this animal may be coincidental. Of course, social behavior makes sense only when it is energetically feasible. As Wilson (1973) suggested: "In mammals the principal antisocial force appears to be chronic food shortage."

Conclusions

The most pervasive influences on the life of a mammal are its body size, its food habits, and the complex interactions that exist between these factors. These interactions determine in great measure the rate at which energy is expended and, consequently in placentals, the length of the gestation period, the postnatal growth rate, fecundity, and ultimately the population parameter r_m. In marsupials, reproduction appears to be more independent of energetics. Body size and food habits also influence home-range size and the extent to which a species may be social.

Acknowledgments

This chapter was written as a result of an invitation by J. F. Eisenberg, National Zoological Park, who deserves many thanks for his aid, if not for his invitation! J. H. Kaufmann and C. A. Woods, both of the University of Florida, kindly read an early draft of this manuscript and made many useful suggestions. James Shaw, Oklahoma State University, supplied the home range of *Myrmecophaga* from his studies in central Brasil.

Literature Cited

Bourlière, F. 1958. Comparative biology of aging. J. Geront., 13 (Suppl. 1):16–24.

Caraco, T., and L. L. Wolf. 1975. Ecological determinants of group sizes of foraging lions. Amer. Nat., 109:343–352.

EISENBERG, J. F. 1975. Phylogeny, behavior, and ecology in the Mammalia. Pp. 47–68, *in* Phylogeny of the primates (W. P. Luckett and F. S. Szalay, eds.). Plenum Publishing Corp., New York, 483 pp.

EISENBERG J. F., AND G. M. MCKAY. 1974. Comparison of ungulate adaptations in the New World and Old World tropical forests with special reference to Ceylon and the rainforests of Central America. Pp. 585–602, *in* The behaviour of ungulates and its relation to management (V. Geist and F. Walther, eds.). Publ. Internatl. Union Conserv. Nature, 24:1–940.

EISENBERG, J. F., N. A. MUCKENHIRN, AND R. RUDRAN. 1972. The relation between ecology and social structure in primates. Science, 176:863–874.

ESTES, R. D. 1974. Social organization of the African Bovidae. Pp. 166–205, *in* The behaviour of ungulates and its relation to management (V. Geist and F. Walther, eds.). Publ. Internatl. Union Conserv. Nature, 24:1–940.

FENCHEL, T. 1974. Intrinsic rate of natural increase: the relationship with body size. Oecologia, 14:317–326.

FLEMING, T. H., E. T. HOOPER, AND D. E. WILSON. 1972. Three Central American bat communities: structure, reproductive cycles, and movement pattern. Ecology, 53:555–569.

HARESTAD, A. S., AND F. L. BUNNELL. 1979. Home range and body weight—a reevaluation. Ecology, 60:389–402.

HERREID, C. F., II. 1964. Bat longevity and metabolic rate. Exp. Geront., 1:1–9.

HUMPHREY, S. R., AND F. J. BONACCORSO. 1979. Population and community ecology. Pp. 409–441, *in* Biology of bats in the New World family Phyllostomatidae, Part III (R. J. Baker, J. K. Jones, Jr., and D. C. Carter, eds.). Spec. Publ. Mus., Texas Tech Univ., 16:1–441.

JARVIS, J. U. M. 1978. Energetics of survival in *Heterocephalus glaber* (Rüppell), the naked mole-rat (Rodentia: Bathyergidae). Bull. Carnegie Mus. Nat. Hist., 6:81–87.

JARVIS, J. U. M., AND J. B. SALE. 1971. Burrowing and burrow patterns of East African mole-rats *Tachyoryctes, Heliophobius,* and *Heterocephalus*. J. Zool., 163:451–479.

KAUFMANN, J. H. 1974*a*. The ecology and evolution of social organization in the kangaroo family (Macropodidae). Amer. Zool., 14:51–62.

———. 1974*b*. Social ethology of the whiptail wallaby, *Macropus parryi,* in northeastern New South Wales. Anim. Behav., 22:281–369.

KLEIBER, M. 1932. Body size and metabolism. Hilgardia, 6:315–353.

———. 1961. The fire of life. John Wiley, New York, 454 pp.

KLEIMAN D. G., AND J. F. EISENBERG. 1973. Comparisons of canid and felid social systems from an evolutionary perspective. Anim. Behav., 21:637–659.

KRUUK, H. 1966. Clan-system and feeding habits of spotted hyaenas (*Crocuta crocuta* Erxleben). Nature, 209:1257–1258.

KRUUK, H., AND M. TURNER. 1967. Comparative notes on predation by lions, leopard, cheetah, and wild dog in the Serengeti area, East Africa. Mammalia, 31:1–27.

LILLEGRAVEN, J. A. 1976. Biological considerations of the marsupial-placental dichotomy. Evolution, 29:707–722.

MCNAB, B. K. 1963. Bioenergetics and the determination of home range size. Amer. Nat., 97:133–140.

———. 1966. The metabolism of fossorial rodents: a study of convergence. Ecology, 47:712–733.

———. 1978. The comparative energetics of neotropical marsupials. J. Comp. Physiol., 125:115–128.

———. 1979. The influence of body size on the energetics and distribution of fossorial and burrowing mammals. Ecology, 60:1010–1021.

———. 1980*a*. Food habits, energetics, and the population biology of mammals. Amer. Nat., 114:106–124.

———. 1980*b*. Energetics and the limits to a temperate distribution in armadillos. J. Mamm., 61:606–627.

———. In press. The scaling rate of metabolism and temperature regulation in endotherms. J. Zool.

Milton, K., and M. L. May. 1976. Body weight, diet and home range area in primates. Nature, 259:459–462.

Nudds, T. D. 1978. Convergence of group size strategies by mammalian social carnivores. Amer. Nat., 112:957–960.

Rosenzweig, M. L. 1968. Net primary productivity of terrestrial communities: prediction from climatological date. Amer. Nat., 102:67–74.

Sacher, G. A. 1959. Relation of lifespan to brain weight and body weight in mammals. Pp. 115–133, *in* The life span of animals. CIBA Foundation Colloquia on Aging. Churchill, London, 5:1–324.

Schaller, G. B., and G. R. Lowther. 1969. The relevance of carnivore behavior to the study of early hominids. Southwestern J. Anthropol., 25:307–341.

Schoener, T. W. 1968. Sizes of feeding territories among birds. Ecology, 49:123–141.

Western, D. 1979. Size, life history and ecology in mammals. African J. Ecol., 17:185–204.

Wilson, D. M. 1977. Environmental monitoring and baseline data from the Isthmus of Panama—1976, volume IV. Smithsonian Inst., Washington, D.C., 267 pp.

Wilson, E. O. 1973. On the queerness of social evolution. Bull. Ent. Soc. Amer., 19:20–22.

ANIMAL POPULATION REGULATION THROUGH BEHAVIORAL INTERACTIONS

ROBERT H. TAMARIN

Abstract

THERE is currently strong evidence to support the hypothesis that the density of animal populations is regulated by social interactions. The environment sets the ultimate limit through resource limitation, but the proximal limitation is through social organization which has evolved in a given environmental setting to maximize individual fitnesses. Watson and Moss (1970) have provided operational criteria to determine whether or not a population is regulated by social interactions.

I examine population regulation in voles, which undergo multi-annual population fluctuations. Current evidence indicates that the Watson and Moss criteria for population regulation by social interactions are supported in voles, but stronger, confirmatory data are needed. Vole mating systems range from monogamy to polygamy with female territoriality routinely found. The question of the cause of cyclic declines is still unanswered, but the heritability of aggressive behavior may play a key role. The relevant data needed to verify the Watson and Moss criteria, as well as the polymorphic-behavior hypothesis, are discussed.

Introduction

Until recently, the field of population regulation has been without a unified theory. This situation has changed over the last few years with strong evidence that social structure is important in determining densities of populations of primates, rodents, birds, and other groups of organisms (Bernstein and Smith, 1979; Cohen et al., 1980; Ebling and Stoddart, 1978).

Students of population regulation are experiencing a Kuhnian revolution (Kuhn, 1962) with regard to the paradigm: animal populations are regulated by social interactions. While not necessarily correct for all populations at all times, this model provides a method of attack within an evolutionary framework. In this paper I elucidate this model with special reference to rodents.

Historical Controversies

Past disagreements about the factor or factors that regulate densities of populations fall into several categories. Because ample historical information exists (Lidicker, 1978; Tamarin, 1978*a*), I only briefly mention these viewpoints. Basically, disagreement has centered around single versus multiple-factor hypotheses, density-dependent versus density-independent factors, and self-regulation versus imposed regulation.

At various times it has been proposed that these hypotheses were in conflict because of differences in 1) terminology among workers, 2) the choice of organisms, 3) the method of study, and 4) general point of view. All of these reasons have had validity (Tamarin, 1978*a*). Those studying insects tended to focus on density-independent population regulation (for example, climatic factors), whereas those studying birds and mammals tended to focus on density-dependent mechanisms (for example, intraspecific competition). Persons trained in theoretical techniques also tended to focus on density-dependent population regulation, emphasizing the Lotka-Volterra population growth equations. Orians (1962) pointed out that perhaps terminological differences and other apparently trivial distinctions were actually caused by a much deeper philosophical division—functional versus evolutionary approaches. Functional ecologists tended to be interested in why and how organisms died, whereas evolutionary ecologists tended to be concerned with the adaptations of organisms to their environments. I hope to demonstrate here that it is the evolutionary view that has yielded a unifying theory, one that is focused on the social structure and social interactions within an animal population.

The Model

One reason the controversy continues is the lack of a theoretical framework from which to pursue questions of population regulation. A paradigm has arisen from empirical studies and sociobiological theory; other models also fail to adequately account for population regulation. Our paradigm is: *all animal populations are regulated ultimately by a limiting resource(s), such as food, and proximally by their social behavior and organization.*

First, there is strong empirical evidence for the role of social behavior in inhibiting reproduction and limiting population density of breeding individuals from studies of confined rodent populations (Barnett, 1958; Calhoun, 1949; Southwick, 1955; Vandenbergh, this volume) and of mammals easy to observe in nature, including sciurids (Michener, this volume), ungulates (Franklin, this volume), social carnivores (Mech, 1970; Moehlman, this volume; Schaller, 1972), and primates (Altmann and Altmann, 1979; Dittus, 1977; Rudran, 1979). Similarly, the relation of territoriality to population density has been observed in many bird species (Boag et al., 1979; Krebs and Perrins, 1978, Watson, 1977; Watson and Moss, 1979).

Second, a strong theoretical framework to support the model of population regulation through behavioral interactions now exists. Wynne-Edwards (1962) suggested that social interactions among individuals in a population maintain population density below the level of mass starvation because it would be beneficial to a species to regulate its density below that level. While he erred in ascribing to "group selection" the evolution of such homeostatic mechanisms, he provided the impetus for the extensive study of the role of social behavior in population regulation. Refutation of "group selection" has resulted in a recent emphasis on demonstrating benefits to individual fitness when explaining actions that affect survival and reproduction of individuals in a population (Bertram, this volume; Davies and Krebs, 1978; Williams, 1966; Wilson, 1975).

Finally, the failure of other models of population regulation in the past had led not only to much pessimism as to the possibility of there ever being a unifying model (Peters, 1976; Tamarin, 1978*a*), but also to new attempts to uncover models of population regulation (Lidicker, 1978). However, our paradigm of population regulation may not have arisen so much as a concerted attempt at new discovery, but more as a serendipitous outcome of ethological research (Wynne-Edwards, 1962).

Criteria for Population Regulation by Social Factors

Watson and Moss (1970) suggested criteria for determining whether or not a population was regulated by social factors. 1) A substantial portion of the population does not breed either because animals are inhibited from breeding or because animals or their progeny die, presumably by socially-induced mortality. 2) The non-breeders are physiologically competent to breed, and will breed if the more dominant animals are removed. 3) The breeding animals are not themselves entirely using up some resource, like food, space, or nest sites. If a resource is depleted, it itself is limiting. 4) Behaviorally-induced mortality is compensatory with other mortality factors. Thus if other mortality factors in the environment increase their effects on a population, behaviorally induced mortality will decrease. Watson and Moss (1970) include a fifth criterion in an effort to look at the interaction of nutrition and behavior. If the nutritional level of animals influences their behavior (a well-fed animal might reduce the size of its territory), then 5) Changes in food supply cause changes in density. This demonstrates that both food and behavior are limiting.

Genetic Changes in Evolutionary and Ecological Time

We are considering a model of population regulation established over evolutionary time through social interactions and organization. Within that framework, it is possible that genetic (microevolutionary) changes occurring over ecological time also bear on population regulation. For example, Chitty (1967) has suggested that microevolutionary changes in the control of aggressive behavior can affect population size in rodents and other organisms. Other microevolutionary models have been suggested, such as Pimentel's genetic feedback mechanism (Pimentel, 1961). Until recently, it seemed that effects of natural selection would not be important over ecological time. However, we now know that microevolutionary phenomena are common (Brussard, 1978; Ford, 1975), and that natural selection acts, even within ecological time, on birth and death rates, reproductive rates, growth rates, and dispersal. The importance to population regulation of genetic changes in these parameters has not been fully documented.

Population Regulation in Voles

To discover the processes within a population that lead to density homeostasis, it is necessary to fully understand the natural history of a species, including the relationships of individuals, the kinds of interactions among individuals and their consequences for individual fitness, and the causes of mortality in a species. The goal of such research is to produce a predictive model of numerical changes in a population based on an understanding of the interactions of those factors influencing population density. For this reason, the best organisms for study should be easy to observe, which explains why much of our understanding of social interactions, mating systems, and behavior in general, has come from work on primates, social carnivores, and birds, although they are only a small part of the animal kingdom. In addition, other organisms, such as the rodents, have been the focus of intense studies of population regulation and hence an understanding of population regulation in these groups is of primary importance.

As a specific example of the role of social behavior and social organization in population regulation I discuss microtine rodents, because they have been the focus of perhaps the most intense controversy in population regulation work, and they undergo multi-annual cycles.

Vole studies have been dominated by interest in population regulation primarily because of multi-annual cycles of density, and numerous mechanisms of population regulation have been suggested, including food depletion, social interaction, and sunspots (Krebs and Myers, 1974). However, direct behavioral interactions are almost impossible to view under natural conditions because voles use underground burrows or above-ground runways hidden in grass. Thus, many conclusions regarding social structure come from laboratory manipulations, or only indirectly from field studies. However, voles are also difficult to breed in indoor colonies and they rarely use nest boxes, thus making captive observations difficult.

Most current vole population research falls into two categories. The first focuses on a multi-factorial view of cycles which emphasizes the interaction of many parameters in controlling population density (Lidicker, 1978). The second focuses on social behavior in population cycles (Getz, 1978; Krebs, 1978*a*). A compromise considers that the multi-factorial view emphasizes environmental parameters that can affect vole social behavior while the behavioral view focuses directly on social behavior as the central parameter (Tamarin, 1978*b*).

Voles are valuable study organisms because ample demographic data can be gathered because of periodic high densities, rapid turnover, and relatively short generation times. The decline and increase phases serve as readily available comparisons of regulated and unregulated populations. And, non-cyclic populations have been recently described which can be compared with cyclic populations.

There are two aspects of vole demography that require an explanation from a population regulation viewpoint: 1) why populations cease growing at a certain density levels (why a carrying capacity is reached), and 2) what causes cyclic declines. Although two independent problems, the latter has certainly dominated the literature. However, voles serve as a better model of study if our emphasis can be focused on the former problem.

The Vole Cycle

Some voles and lemmings exhibit unusual demographic changes throughout their ranges. From phases of low density when animals are rare, populations increase to levels usually in the range of one to several hundred animals per ha. Then a decline ensues and numbers return to very low levels (from extinction to several individuals per ha). The fluctuation takes about 3 to 4 years, although there is enormous diversity among different cycles (Fig. 1; Keith, 1974; Krebs and Myers, 1974). Declines may be very severe or less so, and with or without recoveries. Some populations show extensive low phases whereas others do not. Peaks vary in amplitude. All populations studied show variation from year to year in mortality and reproductive rates, body-size distribution, and dispersal patterns (Krebs and Myers, 1974). Natural selection occurs in these populations over short time periods (Krebs et al., 1973) and individual behavior varies with the phase of the cycle (Myers and Krebs, 1971).

Behavioral Hypotheses

Both major behavioral hypotheses of population regulation require that as density increases there are more interactions among individ-

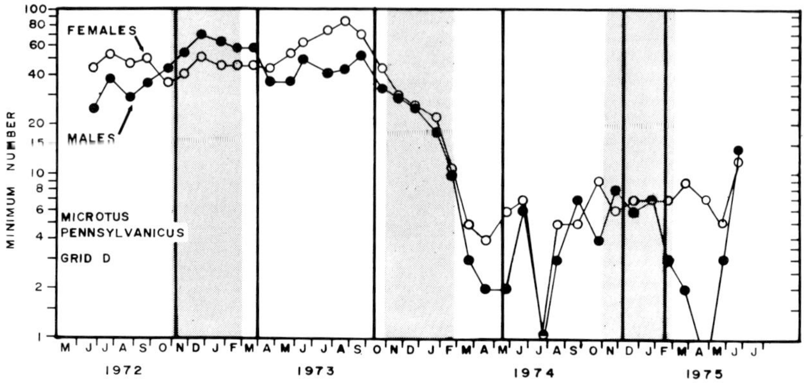

FIG. 1. Demography of a cycling vole population, *Microtus pennsylvanicus,* in southeastern Massachusetts. Shading identifies winter months. Summer breeding periods are denoted by vertical lines. (From Tamarin, 1977*b*; copyright 1977, The Ecological Society of America.)

ual voles, either through direct contacts or by indirect cues to density such as pheromonal signals (Bronson, this volume; Muller-Schwarze, this volume; Richmond and Stehn, 1976; Vandenbergh, this volume). However, a density increase does not necessarily mean that mice encounter each other more often. Pearson (1960) showed that as density increased, there was not a simple increase in runway usage; rather, mice built more runways. In fact, the relative number of active runways has been used as an indicator of density (Lidicker and Anderson, 1962). However, we will assume that increased density does lead to more interactions, or at least that mice are aware of increased density as it occurs.

According to Christian (1980), a higher rate of interaction results in the general adaptation syndrome (Selye, 1946), which causes an eventual exhaustion of the adrenal-pituitary system and leads to increased mortality, decreased reproduction, and a population decline. According to Chitty (1967; 1977), increased interaction at higher densities leads to selection for aggressive voles capable of functioning under higher density. Thus, as density increases there are more and more aggressive individuals. Eventually, however, a decline occurs because reproduction is depressed among highly aggressive individuals, and they suffer increased mortality. Both hypotheses suggest that evolutionary mechanisms affect population regulation through social behavior. Which one, if either, is correct will be discussed later.

Several current problems, including the nature of the data gathered and experimental manipulations performed intercede to prevent clear tests of hypotheses of population regulation. For example, in some studies rodents are trapped in a grid pattern on a regular basis and considerable data on numbers, positions, reproduction, weight, and general condition are amassed and treated statistically to give summary values and rates of change. The direct study of such secretive individuals is difficult. Laboratory-generated or small-enclosure data may have artifacts or be in error due to ignorance of the relevance of experimental manipulations or other parameters. Field experiments using fences are difficult to interpret because normal population demography is disrupted when a fence is erected around a population (Krebs et al., 1969).

The Watson and Moss Criteria

Given these constraints and ignoring cyclic declines for the moment, we must ask what supportive evidence exists to suggest that voles are regulated by social factors based on the criteria of Watson and Moss (1970). Only indirect evidence supports all of the criteria. First, a substantial part of cycling populations do not seem to breed, as inferred from removal experiments. When areas are cleared of animals, they are quickly repopulated, suggesting that there is an excess of individuals that respond to the vacancy (Gaines et al., 1979; Krebs et al., 1976; Myers and Krebs, 1971; Tamarin 1977*a*). However, it is uncertain whether these new individuals are non-reproductives or breeding residents drawn into the vacant habitat from other areas. Krebs (1978*a*) differentiated these two alternatives as the *behavior hypothesis,* which suggests that the colonizers are socially subordinate, and the *density-release hypothesis,* which suggests that a vacant habitat is filled more or less randomly by residents perceiving the low density. Baird and Birney (1982) and Tamarin (1977*a*) showed that few tagged residents moved into a vacant habitat from adjacent areas in the meadow vole (*Microtus pennsylvanicus*); this observation supports the behavior hypothesis and fulfills the first Watson and Moss criterion. The existence, under certain conditions, of a large number of dispersers in cycling vole populations is prima facie evidence of the existence of excess, non-breeding animals.

Support for Watson and Moss' second criterion, that surplus individuals are capable of breeding when the more dominant animals are removed, comes from pulsed removal experiments in which an area is periodically cleared of mice and new arrivals permitted to set up residency. Reproductive activity of colonizers supports the criterion if it is assumed that new animals are surplus individuals, as inferred above. Using the pulsed removal technique, Gaines et al. (1979) and Krebs (1978*a*) reported reproduction among dispersers in Townsend's vole (*M. townsendii*) and prairie voles (*M. ochrogaster*), respectively.

Krebs and Myers (1974) pointed out that the habits of voles and lemmings preclude showing the existence of surplus animals as elegantly as in birds. However, although the existence of dispersers, which may or may not be subordinate and which may be capable of breeding, supports Watson and Moss' second criterion (1970), it is not confirmatory. All animals and plants have a dispersal stage, and these dispersers may be subordinate because of age or social conditioning. It must be shown that these animals are true "floaters" (Smith, 1978); that is, they would not have bred on the area from which they came if they had not discovered the vacancy. Gaines (pers. comm.) showed that *M. ochrogaster* dispersers still seek to disperse when replaced into the population from which they came and that their reproductive success *in situ* is very low. This observation needs to be replicated.

Watson and Moss' third criterion (1970) is that "breeding animals are not completely using up some resource such as food, space or nest sites. If they are, the resource itself is limiting." Although this criterion is difficult to verify without knowledge of the total resource requirements of a species, there is strong evidence that neither quantity nor quality of food limits voles. Krebs and DeLong (1965) could not demonstrate a numerical response in voles by addition of food or by fertilization of a field. Other studies have indicated that there is no obvious food shortage during vole peaks (Chitty et al., 1968; Golley, 1960). Fencing studies have demonstrated that many more voles can "fit" into a habitat, even for short periods of time, than is found in natural, unfenced areas (Krebs et al., 1969). Fence experiments thus provide indirect evidence that food itself is not in short supply and direct evidence that physical space is not limiting. Suitable nest sites are probably not limiting because voles tunnel in the ground or build above-ground nests.

Thus, present evidence suggests that there is no limiting resource to voles in the environment. Even at peak population densities under natural conditions, voles do not seem to exhaust any resource.

Watson and Moss' fourth criterion, that behaviorally-induced mortality is compensatory, is extremely difficult to verify because little is known about vole mortality; vole studies usually involve trapping, and the disappearance of any animal may be due to mortality, dispersal (emigration), or refusal to enter traps. Mice that are no longer trapped may or may not have died.

Reproduction in voles, however, is often compensatory (Schaffer and Tamarin, 1973). Length of the breeding season and intensity of breeding (pregnancy rate) routinely change with density changes in vole populations. In fact, this response is considered part of the syndrome accompanying vole cycles (Krebs, 1970*a*). Compensatory reproduction is also seen in noncyclic populations (Tamarin, 1977*a*, 1977*b*). Because Watson and Moss (1970) define their fourth criterion in respect to both mortality and recruitment, compensatory reproduction fulfills at least a part of this criterion. Here, as density increases, reproduction (and presumably recruitment) decreases. At lower densities, reproduction and recruitment are higher.

The fifth criterion is that changes in food supply cause changes in density (which shows the interaction of both food and behavior if all other criteria have been met). In an extreme case, Krebs et al. (1969) showed that fenced populations will deplete the food resources and then decline, a finding indicating only that starved animals will die. In other experiments food has been supplemented. For example, Cole and Batzli (1978) added high quality food to a *M. ochrogaster* population and caused densities to increase 50% compared to the control (they could not prevent a population decline, however). Other studies have also noted that peak vole densities are higher in habitats with better food resources (Getz et al., 1979; Tamarin, pers. observ.). Thus food can affect density levels although it is not apparently the proximate limiting resource.

In conclusion, vole density may be regulated by social behavior although the ultimate limiting factor is probably food.

Mating Systems in Cycling Voles

Using Emlen and Oring's (1977) classification scheme of mating systems, voles should be resource-defense polygynists because voles live within a bountiful, dependable resource (usually grass), males are for the most part freed from parental care, and female sexual receptivity is asynchronous. However, recent studies of vole mating systems have suggested many differences among species.

Microtus ochrogaster and *M. pennsylvanicus* both undergo cycles, yet seem to have very different mating systems. Thomas and Birney (1979) provided evidence that *M. ochrogaster* is primarily monogamous. In the laboratory, they observed adult mice forming monogamous pairs and considerable male parental care.

Getz (1978) supported the concept of monogamy in *M. ochrogaster* but suggested that this mating system does not occur at high density. He suggested that *M. ochrogaster* and *M. pennsylvanicus* are at opposite ends of a habitat spectrum—*M. ochrogaster* lives in large, contiguous, stable habitats, and *M. pennsylvanicus* lives in small, isolated, and ephemeral or unstable habitats. *M. californicus,* an occupant of California grasslands, a habitat similar to that occupied by *M. ochrogaster,* also tends towards monogamy (Lidicker, 1979).

Field observations have not uncovered any evidence of pair bonding in *M. pennsylvanicus* (Getz, 1978). Using radio-telemetry, Madison (1980) found that *M. pennsylvanicus* females appear territorial, whereas males have a more diffuse geometry of home ranges that overlap female territories. Apparently estrous females attract several nearby males. Jannett (1978) suggested a similar spatial system in the montane vole, *M. montanus,* which lives in montane meadow habitats similar to *M. pennsylvanicus.*

We thus have some evidence that in voles the nature of the habitat affects the mating system (Getz, 1978). In large uniform habitats there is a tendency towards monogamy, whereas in patchy or less stable habitats voles tend more towards polygyny. Kleiman (1977) suggested that if a territory contains a scarce or valuable resource, or if the environment is sufficiently stringent, a male would benefit by monogamy. Thomas and Birney (1979) cited this argument, suggesting that the habitat of the prairie vole is stringent because it is kept at low resource levels by large competing herbivores. By contrast, Getz (1978) considered the prairie habitat to be stable. These contradictions indicate that additional research is needed to determine the actual nature of the habitats before we can determine why prairie voles are monogamous.

With or without a male, and hence despite different mating systems, female territoriality is a common pattern. Evidence exists for a similar system in *Microtus arvalis* (Frank, 1957) and *M. agrestis* (Viitala, 1977). Although there are several controversies as to the value of territoriality (Verner, 1977; MacLean and Seastedt, 1979), female territoriality is probably favored by natural selection to insure adequate resources for reproduction and survival. Alternatively, Krebs (1978*a*) suggested that the limiting resource is mates, although there are no supportive data. Territorial behavior may also be favored in order to secure an area for the maturation of young (Jannett, 1978) and to avoid intraspecific predation on young. These ideas are not all mutually exclusive.

Non-cyclic Species

Several vole species do not have multi-annual cycles. Townsend's vole (*M. townsendii*) in coastal British Columbia apparently undergoes annual cycles and periodic irruptions, but does not undergo long-term cycles in density (Krebs, 1979). Apparently *M. townsendii* are similar demographically to cyclic species (Krebs, 1979). The mating system is unknown, but females rather than males appear to control recruitment rates (Boonstra, 1978), which suggests that female territoriality might exist with a more diffuse spacing behavior of the males.

The beach vole (*M. breweri*) in coastal Massachusetts also does not cycle (Fig. 2; Tamarin, 1977*b*). Although it is more K-selected than its mainland counterpart (*M. pennsylvanicus*), this island species is similar demographically to the mainland species except for dispersal. The beach vole is the only vole species studied that supports Krebs' (1978*a*) density-release hypotheses of dispersal (Tamarin, 1977*a*). The mating system is unknown. Because *M. breweri* may be a conspecific of *M. pennsylvanicus* (Tamarin and Kunz, 1974), there may be parallels in social structure between these two species.

Several other non-cyclic vole populations are known. The bank vole (*Clethrionomys glareolus*) does not cycle on Crab Apple Island in Poland (Bujalska, 1970, 1975), and there is a relatively constant density of reproductive females from year to year. Other non-cyclic populations are known, but little information is available about their mating systems (Fullager et al., 1963; Lidicker, 1973; Rasmuson et al., 1977).

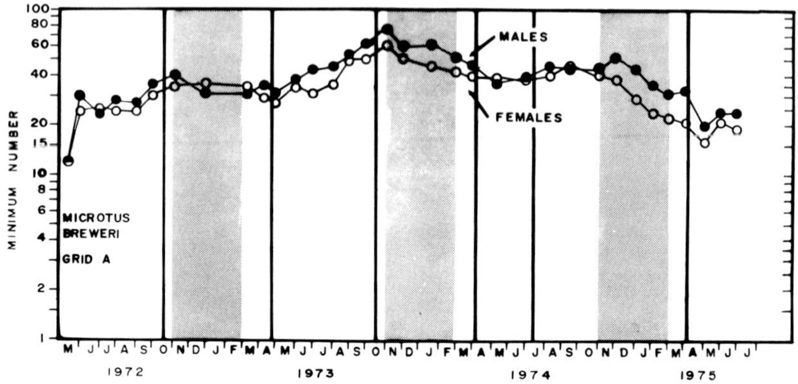

FIG. 2. Demography of a non-cycling vole population (*Microtus breweri*) on Muskeget Island, Massachusetts. Shading identifies winter months. Summer breeding periods are denoted by vertical lines. (From Tamarin, 1977*b*; copyright 1977, The Ecological Society of America.)

From this brief examination of non-cyclic species, three points emerge. First, where looked for, female territoriality seems to occur. Second, many of these populations exist on islands. And third, dispersal patterns, especially in island populations, may differ when compared with similar mainland species.

The Behavioral Hypotheses Revisited

Behavioral hypotheses, both phenotypic (Christian, 1980) and genotypic (Chitty, 1967), were mentioned above. Christian's proposed mechanism seems to work at high densities in confined populations (Christian et al., 1965), although these densities differ from those normally found in field populations. Chitty's (1967) hypothesis, however, is a viable explanation and has gained more importance in recent years. There has been a revolution in thought about the interaction of behavior and genetics (Krebs and Davies, 1978). Chitty's hypothesis spans these two areas, and precedes their merger.

Recently several alternative genetic mechanisms have been suggested for Chitty's hypothesis. Chitty (1967) suggested that natural selection acted specifically on loci that determine aggressive behavior. However, Smith et al. (1978) suggested that the level of aggression in a population derives from the level of heterozygosity which in turn is influenced by immigrants. As density increases, immigration increases, leading to increased heterozygosity and increased aggression among individuals, with the same consequences suggested by Chitty. Most recently, Charnov and Finerty (1979) suggested that aggressive levels in a population can change because levels of relatedness among neighbors will decrease as density and dispersal increase. With lowered relatedness among neighbors there would no longer be an advantage to kin to restrain aggressiveness and the aggression level in the population would increase, again with the results predicted by Chitty (1967).

In addition to the different genetic mechanisms hypothesized for a behavioral mechanism of population regulation, there are also other non-genetic behavioral hypotheses (for example, Getz, 1978; Hendrichs, 1978; Łomnicki, 1978). This profusion of variants on the general behavior theme suggests vitality in this area of study.

The role of dispersal in population regulation and in demographic processes has recently been reviewed (Gaines and McClenaghan, 1980; Lidicker, 1975). Dispersal can serve three very important roles in population dynamics. Dispersers can lower the population density, thus having an immediate "safety-valve" effect; they can prevent inbreeding in a population; and they can modify the genetic composition of the population they leave, and the one they might join, if they are not genetically a random sample of the general population. If aggressive behavior is heritable, the relative proportion of aggressive animals in a population can be greatly influenced by dispersal, which could then explain declines in vole populations (Chitty, 1967; Krebs, 1978*b*). What evidence exists linking behavior, genotype and dispersal?

Krebs (1970*b*) first showed that behavioral changes within the resident population accompany cycles in prairie and meadow voles. Myers and Krebs (1971) found behavioral differences between residents and dispersers in the same two species. Through electrophoretic analyses genetic changes have been demonstrated during vole cycles in both resident and disperser fractions of a population (Krebs et al., 1973; Myers and Krebs, 1971). However, the specific genetic changes demonstrated by electrophoresis to accompany vole cycles are probably not ecologically relevant in that they are unlikely to control agonistic, dispersal, or reproductive behavior. Rasmuson et al. (1977) demonstrated the heritability of locomotor activity in field voles (*Microtus agrestis*) in Sweden. Anderson (1976) demonstrated a lack of heritability in aggressive behavior in Townsend's vole in

Vancouver, but this species does not undergo cycles, which is consistent with Chitty's (1967) hypothesis. Hence, the available evidence suggests that dispersers are a non-random sample of the resident population, and that natural selection can act during vole cycles.

Krebs (1978*a*) suggested that cycling populations have high heritability for aggressive behavior while non-cycling populations have low heritability for aggressive behavior; this has been supported by Anderson's (1976) data for non-cycling *M. townsendii*. The lack of a cycle in some vole populations may indicate that only certain species have a high heritability for aggressive behavior. Neither the hypotheses of Smith et al. (1978) nor Charnov and Finerty (1979) require that aggression is heritable.

Hypothesis Testing in Voles

In order to demonstrate that vole populations are regulated by social interactions, four areas of research are needed, including 1) gathering more natural history data, 2) demonstration of the Watson and Moss criteria, 3) examination of the genetic components of behavior, and 4) manipulation of vole populations for confirmatory testing.

Natural History

Great gaps exist in our understanding of vole natural history. We are just now learning about mating systems and social organization of these species. One consistent observation is that females are territorial in those species studied. Wolff (1980) recently reported that the taiga vole (*Microtus xanthognathus*) may have a harem-like mating system. Virtually nothing is known about its demography; more information should be gathered on the mating system and social structure in this species as well as in other vole species and different populations of the same species.

The Watson and Moss Criteria

None of the five criteria defined by Watson and Moss (1970) has been unequivocally demonstrated. Yet some are more strongly established than others. To demonstrate the existence of excess voles or true "floaters" (that is, animals who will not breed or who will have a very low fitness if they do not disperse), animals should be marked as littermates and followed through their early life, either by trapping or with radio transmitters or radioactive isotopes. Watson and Moss (1970) would predict that these littermates will become established near the parents when density is low but not when density is high. By using radio-tagging, it should be possible to demonstrate that above a certain density subadults or young adults cannot settle into a home range, but instead continue to search for unfilled habitat. This type of study would also result in the determination of kinships, data useful for looking at kin selection.

If excess voles exist, it should also be possible to quantify the opposing selective pressures on non-dispersers and dispersers (Gaines and McClenaghan, 1980; Lidicker, 1975). Animals seem to disperse from the same area at different density levels (saturation and presaturation dispersers or Lidicker, 1975), suggesting that some animals are more prone to disperse than others. By determining the reproductive success of recruits and dispersers we can better determine the selective advantages of dispersal at different densities. Alternatively, dispersers can be returned to the home area and their reproductive success determined.

If the first criterion of Watson and Moss is firmly established, then their second criterion may already have been demonstrated. That is, if dispersers truly represent excess individuals not breeding because of social pressures, then the pulsed removal experiments described earlier have already demonstrated that these voles can and will breed successfully. Hence, excess animals have the physiological capacity to reproduce if social constraints are lifted.

Both fencing experiments and supplemental food experiments have demonstrated that voles normally neither saturate the environment nor deplete food resources. However, plant secondary compounds may be exerting an effect on the reproduction and survival of residents (Freeland, 1974; Negus and Berger, 1977; Swain, 1977). Although feeding and fencing experiments argue against this, secondary compounds could still have an effect, especially on reproduction, even within fences or with supplemental feeding. Research on the occurrence, ingestion by voles, and effects on voles of these compounds under natural conditions may be rewarding.

Compensatory reproduction occurs routinely in vole populations. Radio-telemetry may be helpful for documenting the causes of mortality of voles during a cycle because demonstrating compensatory mortality is so difficult (Hilborn and Krebs, 1976).

Finally, is there an interaction of food and social behavior in the regulation of vole population density, as seems to be the case where increases in peak density populations occurred after supplementary feeding (Cole and Batzli, 1978)? Krebs and DeLong (1965) could not prevent a decline in the California vole (*Microtus californicus*) with the addition of supplemental food and fertilizer, although it is not clear whether this population had a peak-density response to the food supplement.

The Genetic Components of Behavior

The Chitty hypothesis of population regulation in voles stands apart from other theories because of the requirement that aggressive behavior and subsequent changes within populations are influenced by natural selection operating on genetic variation. In order to demonstrate this hypothesis, the following must occur: 1) changes in aggressive behavior among voles must accompany a cycle during which there must be increased aggressiveness among residents and decreased aggressiveness among dispersers; 2) behaviors must be heritable; 3) voles from non-cyclic populations should not show a high heritability of aggressive behavior; 4) populations initiated with different genetically controlled behavioral types should undergo different population dynamics.

Inroads have been made in demonstrating that behavioral changes do occur during cycles, that dispersers and residents differ, and that a non-cyclic population does not show heritability of aggressive behavior (Anderson, 1976). It is now necessary to determine the heritability of aggression within both cycling and non-cycling populations. Despite numerous difficulties, such as the determination of relevant behaviors and their measurement, and captive breeding of voles, this area of research should have a very high priority.

Experimental Manipulation of Populations

Boonstra (1978) demonstrated that females played an important role in controlling recruitment in *M. townsendii* by artificially altering the number of adult males and females on several grids. I have recently been concerned with designing a fence that will allow manipulation of contained vole populations. In order to avoid the fence effect, in which enclosed populations reach abnormally high densities with concomitant habitat destruction, and the problems inherent in conducting unfenced experiments (swamping by immigrants, or loss of experimental animals by emigration; see Krebs, 1966), we are currently testing a design that allows dispersers to leave by having the fenced area encompass marginal habitat. Within our fences are woodlands from which voles, presumably dispersers, are continually trapped. Other fence designs have shown potential success (Beacham, 1979; Gaines et al., 1979; Getz, pers. comm.). After two years of trapping, our unfenced control and our fenced experimentals are showing the same pattern of density change. If the fenced populations exhibit a cycle similar to the unfenced control, these fences can be used to test various aspects of the behavioral hypothesis of population regulation.

Non-mammalian Species

How general is the theory that animal populations are regulated by social factors? In recent years, the importance of social structure in population regulation in higher vertebrates such as mammals and birds has been emphasized. Population regulation through social interactions may also be of paramount importance in fish and other animals with highly-developed social behaviors (Alexander, 1974; Tinbergen, 1951). There are possible examples of social structure influencing population size in insects (Milkman, 1975). Microevolutionary phenomena may also be important in population regulation in lower animals (Krebs, 1978*b*). Emphasis on density-independent population regulation in insects (Andrewartha and Birch, 1954) may be due to the relatively more r-selected nature of lower animals (Pianka, 1970). Animals that are r-selected may exist predominantly in an unregulated mode until environmental vagaries, such as the weather, act to reduce their numbers. Under these circumstances, resource limitation may prevail at high densities if there are no self-regulatory mechanisms.

Acknowledgments

The Ecology Group at Boston University provided many helpful comments on the ideas presented. I am especially grateful to Fred

Wasserman and Peter August for criticizing an earlier draft of this manuscript. My research on voles has been supported by NIH and NSF.

Literature Cited

ALEXANDER, R. 1974. The evolution of social behavior. Ann. Rev. Ecol. Syst., 5: 325–383.

ALTMANN, S., AND J. ALTMANN. 1979. Demographic constraints on behavior and social organization. Pp. 47–63, *in* Primate ecology and human origins (I. Bernstein, and E. Smith, eds.). Garland STPM Press, New York, 362 pp.

ANDERSON, J. 1976. Phenotypic correlates among relatives, and variability in reproductive performance in populations of the vole *Microtus townsendii.* Unpubl. Ph.D. dissert., Univ. British Columbia, Vancouver, 207 pp.

ANDREWARTHA, H., AND L. BIRCH. 1954. The distribution and abundance of animals. Univ. Chicago Press, 782 pp.

BAIRD, D., AND E. BIRNEY. 1982. Pattern of colonization in *Microtus pennsylvanicus.* J. Mamm., 63:290–293.

BARNETT, S. 1958. An analysis of social behavior in wild rats. Proc. Zool. Soc. London, 130:107–152.

BEACHAM, T. 1979. Dispersal tendency and duration of life of littermates during population fluctuations of the vole *Microtus townsendii.* Oecologia, 42:11–21.

BERNSTEIN, I., AND E. SMITH (eds.). 1979. Primate ecology and human origins. Garland STPM Press, New York, 362 pp.

BOAG, D., ET AL. 1979. Population regulation in spruce grouse: a working hypothesis. Canadian J. Zool. 57:2275–2284.

BOONSTRA, R. 1978. Effect of adult Townsend voles (*Microtus townsendii*) on survival of young. Ecology, 59:242–248.

BRUSSARD, P. (ed.). 1978. Ecological genetics: the interface. Springer-Verlag, New York, 247 pp.

BUJALSKA, G. 1970. Reproductive stabilizing elements in an island population of *Clethrionomys glareolus* (Schreber, 1780). Acta Theriol., 15:381–412.

———. 1975. Reproduction and mortality of bank voles and the changes in the size of an island population. Acta Theriol., 20:41–56.

CALHOUN, J. 1949. A method for self-control of population growth among mammals living in the wild. Science, 109:333–335.

CHARNOV, R., AND J. FINERTY. 1980. Vole population cycles: a case for kin-selection? Oecologia, 45:1–2.

CHITTY, D. 1967. The natural selection of self-regulatory behaviour in animal populations. Proc. Ecol. Soc. Australia, 2:51–78.

———. 1977. Natural selection and the regulation of density in cyclic and noncyclic populations. Pp. 27–32, *in* Evolutionary ecology (B. Stonehouse and D. Perrins, eds.). University Park Press, Baltimore, 310 pp.

CHITTY, D., D PIMENTEL, AND C. KREBS. 1968. Food supply of overwintered voles. J. Anim. Ecol., 37:113–120.

CHRISTIAN, J. 1980. Endocrine factors in population regulation. Pp. 55–115, *in* Biosocial mechanisms of population regulation (M. Cohen, R. Malpass, and H. Klein, eds.). Yale Univ. Press, New Haven, Connecticut, 406 pp.

CHRISTIAN, J., J LLOYD AND D. DAVIS. 1965. The role of endocrines in the self-regulation of mammalian populations. Recent Prog. Horm. Res., 21:501–578.

COHEN, M., R. MALPASS, AND H. KLEIN, eds. 1980. Biosocial mechanisms of population regulation. Yale Univ. Press, New Haven, Connecticut, 406 pp.

COLE, F., AND G. BATZLI. 1978. Influence of supplemental feeding on a vole population. J. Mamm., 59:809–819.

DAVIES, N., AND J. KREBS. 1978. Introduction: ecology, natural selection and social behaviour. Pp. 1–18, *in* Behavioural ecology: an evolutionary approach (J. Krebs and N. Davies, eds.). Sinauer Assoc., Sunderland, Massachusetts, 494 pp.

DITTUS, W. 1977. The social regulation of population density and age-sex distribution in the toque monkey. Behaviour, 63:281–322.

EBLING, F., AND D. STODDART (eds.). 1978. Population control by social behaviour. Inst. Biol., London, 304 pp.

EMLEN, S., AND L. ORING. 1977. Ecology, sexual selection and the evolution of mating systems. Science, 197:215–223.

FORD, E. 1975. Ecological genetics. Fourth ed. Chapman and Hall, New York, 442 pp.

FRANK, F. 1957. The causality of microtine cycles in Germany. J. Wildl. Mgmt., 21:113–121.

FREELAND, W. 1974. Vole cycles—another hypothesis. Amer. Nat., 108:238–245.

FULLAGAR, P., ET AL. 1963. The Skomer vole (*Clethrionomys glareolus skomerensis*) and long-tailed field mouse (*Apodemus sylvaticus*) on Skomer Island, Pembrokeshire, in 1960. Proc. Zool. Soc. London, 140:295–314.

GAINES, M., AND L. MCCLENAGHAN, JR. 1980. Dispersal in small mammals. Ann. Rev. Ecol. Syst., 11:163–196.

GAINES, M., A. VIVAS, AND C. BAKER. 1979. An experimental analysis of dispersal in fluctuating vole populations: demographic parameters. Ecology, 60:814–828.

GETZ, L. 1978. Speculation on social structure and population cycles of microtine rodents. Biologist, 60:134–147.

GETZ, L., ET AL. 1979. Comparisons of population demography of *Microtus ochrogaster* and *M. pennsylvanicus*. Acta Theriol., 24:319–349.

GOLLEY, F. 1960. Energy dynamics of a food chain of an old-field community. Ecol. Monogr., 30:187–206.

HENDRICHS, H. 1978. Die soziale Organisation von Säugetierpopulationen. Säugetierk. Mitt., 26:81–116.

HILBORN, R., AND C. KREBS. 1976. Fates of disappearing individuals in fluctuating populations of *Microtus townsendii*. Canadian J. Zool., 54:1507–1518.

JANNETT, F., JR. 1978. The density-dependent formation of extended maternal families of the montane vole, *Microtus montanus nanus*. Behav. Ecol. Sociobiol., 3:245–263.

KEITH, L. 1974. Some features of population dynamics in mammals. Proc. Internatl. Congr. Game Biol., 11:17–58.

KLEIMAN, D. 1977. Monogamy in mammals. Quart. Rev. Biol., 52:39–69.

KREBS, C. 1966. Demographic changes in fluctuating populations of *Microtus californicus*. Ecol. Monogr., 36:239–273.

———. 1970*a*. Genetic and behavioral studies on fluctuating vole populations. Pp. 243–256, *in* Dynamics of populations (P. den Boer and G. Gradwell, eds.). Ctr. Agric. Publ. Document., Wageningen, Netherlands, 611 pp.

———. 1970*b*. *Microtus* population biology: behavioural changes associated with the population cycle in *M. ochrogaster* and *M. pennsylvanicus*. Ecology, 51:34–52.

———. 1978*a*. Aggression, dispersal, and cyclic changes in populations of small rodents. Pp. 49–60, *in* Aggression, dominance and individual spacing (L. Krames, P. Pliner, and T. Alloway, eds.). Plenum Press, New York, 173 pp.

———. 1978*b*. A review of the Chitty hypothesis of population regulation. Canadian J. Zool., 56:2463–2480.

———. 1979. Dispersal, spacing behaviour, and genetics in relation to population fluctuations in the vole, *Microtus townsendii*. Fortschr. Zool., 25:61–77.

KREBS, C., AND K. DELONG. 1965. A *Microtus* population with supplemental food. J. Mamm., 46:566–573.

KREBS, C., AND J. MYERS. 1974. Population cycles in small mammals. Adv. Ecol. Res., 8:267–399.

KREBS, C., B. KELLER, AND R. TAMARIN. 1969. *Microtus* population biology: demographic changes in fluctuating populations of *M. ochrogaster* and *M. pennsylvanicus* in southern Indiana. Ecology, 50:587–607.

KREBS, C., ET AL. 1973. Population cycles in small rodents. Science, 179:35–41.

———. 1976. *Microtus* population biology: dispersal in fluctuating populations of *M. townsendii*. Canadian J. Zool., 54:79–95.

KREBS, J., AND N. DAVIES (eds.). 1978. Behavioral ecology: an evolutionary approach. Sinauer Assoc., Sunderland, Massachusetts, 494 pp.

KREBS, J., AND C. PERRINS. 1978. Behaviour and population regulation in the great tit (*Parus major*). Pp. 23–47, *in* Population control by social behaviour (F. Ebling and D. Stoddart, eds.). Inst. Biol., London, 304 pp.

KUHN, T. 1962. The structure of scientific revolutions. Univ. Chicago Press, Chicago, 172 pp.

LIDICKER, W., JR. 1973. Regulation of numbers in an island population of the California vole, a problem in community dynamics. Ecol. Monogr., 43: 271–302.

———. 1975. The role of dispersal in the demography of small mammals. Pp. 103–134, *in* Small mammals, their productivity and population dynamics (F. Golley, K. Petrusewicz, and L. Ryszkowski, eds.). Cambridge Univ. Press, Cambridge, 451 pp.

———. 1978. Regulation of numbers in small mammal populations—historical reflections and a synthesis. Pp. 122–141, *in* Populations of small mammals under natural conditions (D. Snyder, ed.). Univ. Pittsburgh, Pittsburgh, 237 pp.

———. 1979. Analysis of two freely-growing enclosed populations of the California vole. J. Mamm., 60:447–466.

LIDICKER, W., JR., AND P. ANDERSON. 1962. Colonization of an island by *Microtus californicus*, analyzed on the basis of runway transects. J. Anim. Ecol., 31: 503–517.

ŁOMNICKI, A. 1978. Individual differences between animals and natural regulation of their numbers. J. Anim. Ecol., 47:461–475.

MACLEAN, S., AND T. SEASTEDT. 1979. Avian territoriality: sufficient resources or interference competition. Amer. Nat., 114:308–312.

MADISON, D. 1980. Space use and social structure in meadow voles, *Microtus pennsylvanicus*. Behav. Ecol. Sociobiol., 7:65–71.

MECH, L. 1970. The wolf: ecology and behavior of an endangered species. Natural History Press, Garden City, New York, 384 pp.

MILKMAN, R. 1975. Specific death sites in a *Drosophila* population cage. Biol. Bull., Mar. Biol. Lab., Woods Hole, Massachusetts, 148:274–285.

MYERS, J., AND C. KREBS. 1971. Genetic, behavioural, and reproductive attributes of dispersing field voles *Microtus pennsylvanicus* and *Microtus ochrogaster*. Ecol. Monogr., 41:53–78.

NEGUS, N., AND P. BERGER. 1977. Experimental triggering of reproduction in a natural population of *Microtus montanus*. Science, 196:1230–1231.

ORIANS, G. 1962. Natural selection and ecological theory. Amer. Nat., 96:257–263.

PEARSON, O. 1960. Habits of *Microtus californicus* revealed by automatic photographic recorders. Ecol. Monogr., 30:231–249.

PETERS, R. 1976. Tautology in evolution and ecology. Amer. Nat., 110:1–12.

PIANKA, E. 1970. On r- and K-selection. Amer. Nat., 104:592–597.

PIMENTEL, D. 1961. Animal population regulation by the genetic feed-back mechanism. Amer. Nat., 95:65–79.

RASMUSON, B., M. RASMUSON, AND J. NYGREN. 1977. Genetically controlled differences in behaviour between cycling and noncycling populations of field vole (*Microtus agrestis*). Hereditas, 87:33–42.

RICHMOND, M., AND R. STEHN. 1976. Olfaction and reproductive behavior in microtine rodents. Pp. 197–217, *in* Mammalian olfaction, reproductive processes, and behavior (R. Doty, ed.). Academic Press, New York, 344 pp.

RUDRAN, R. 1979. The demography and social mobility of a red howler (*Alouatta seniculus*) population in Venezuela. Pp. 107–136, *in* Vertebrate ecology in the northern neotropics (J. Eisenberg, ed.). Smithsonian Inst. Press, Washington, D.C., 271 pp.

SCHAFFER, W., AND R. TAMARIN. 1973. Changing reproductive rates and population cycles in lemming and voles. Evolution, 27:111–124.

SCHALLER, G. 1972. The Serengeti lion: a study of predator-prey relations. Univ. Chicago Press, 480 pp.

SELYE, H. 1946. The general adaptation syndrome and the disease of adaptation. J. Clin. Endocrinol., 6:117–230.

SMITH, M., M. MANLOVE, AND J. JOULE. 1978. Spatial and temporal dynamics of the genetic organization of small mammal populations. Pp. 99–113, *in* Populations of small mammals under natural conditions (D. Snyder, ed.). Univ. Pittsburgh, Pittsburgh, 237 pp.

SMITH, S. 1978. The "underworld" in a territorial sparrow: adaptive strategy for floaters. Amer. Nat., 112:571–582.

SOUTHWICK, C. 1955. Regulatory mechanisms of house mouse populations: social behavior affecting litter survival. Ecology, 36:627–634.

SWAIN, T. 1977. Secondary compounds as protective agents. Ann. Rev. Plant Physiol., 28:479–501.

TAMARIN. R. 1977*a*. Dispersal in island and mainland voles. Ecology, 58:1044–1054.

———. 1977*b*. Demography of the beach vole (*Microtus breweri*) and the meadow vole (*Microtus pennsylvanicus*) in southeastern Massachusetts. Ecology, 58:1310–1321.

———. 1978*a*. Population regulation. Dowden, Hutchinson and Ross, Stroudsburg, Pennsylvania, 391 pp.

———. 1978*b*. A defense of single-factor models of population regulation. Pp. 159–162, *in* Populations of small mammals under natural conditions (D. Snyder, ed.). Univ. Pittsburgh, Pittsburgh, 237 pp.

TAMARIN, R., AND T. KUNZ. 1974. *Microtus breweri*. Mammalian Species, 45:1–3.

THOMAS, J., AND E. BIRNEY. 1979. Parental care and mating system of the prairie vole, *Microtus ochrogaster*. Behav. Ecol. Sociobiol., 5:171–186.

TINBERGEN, N. 1951. The study of instinct. Oxford Univ. Press, London, 228 pp.

VERNER, J. 1977. On the adaptive significance of territoriality. Amer. Nat., 111: 769–775.

VIITALA, J. 1977. Social organization in cyclic subarctic populations of the voles *Clethrionomys rufocanus* (Sund.) and *Microtus agrestis* (L.) Ann. Zool. Fennica, 14:53–93.

WATSON, A. 1977. Population limitation and the adaptive value of territorial behaviour in Scottish red grouse *Lagopus l. scoticus*. Pp. 19–26, *in* Evolutionary ecology (B. Stonehouse and C. Perrins, eds.). University Park Press, Baltimore, 310 pp.

WATSON, A., AND R. MOSS. 1970. Dominance, spacing behaviour and aggression in relation to population limitation in vertebrates. Pp. 167–218, *in* Animal populations in relation to their food resources (A. Watson, ed.). Blackwell Sci. Publ., Oxford, 477 pp.

———. 1979. Population cycles in the Tetraonidae. Ornis Fennica, 56:87–109.

WILLIAMS, G. 1966. Adaptation and natural selection: a critique of some current evolutionary thought. Princeton Univ. Press, 307 pp.

WILSON, E. O. 1975. Sociobiology, the new synthesis. Harvard Univ. Press, Cambridge, Massachusetts, 697 pp.

WOLFF, J. 1980. Social organization of the taiga vole (*Microtus xanthognathus*). Biologist, 62:34–45.

WYNNE-EDWARDS, V. 1962. Animal dispersion in relation to social behaviour. Oliver and Boyd, Edinburgh, 653 pp.

KIN SELECTION AND ALTRUISM

BRIAN C. BERTRAM

Abstract

THE distinct differences between kin selection and altruism are emphasized. Attention is drawn to the diversity of definitions of the term "altruism," and a new definition is given and justified. The possible ways by which altruistic behavior may have been selected for are outlined. Several different types of altruistic behavior are exhibited by mammals, and their occurrence in different species is discussed. It is emphasized by means of examples that a range of selective pressures was probably responsible for the evolution of any particular case of altruism.

Introduction

To be asked to write a paper on kin selection and altruism is rather like being asked to write one on electricity and motor cars—the two subjects are in many ways so dissimilar. One (kin selection) is a process, with wide ramifications, only some of which are applicable to the evolution of the second, the phenomenon of altruism. Altruism is of particular interest because one would initially think that it could not evolve. Natural selection theory has long (and easily) accounted for the spread of genes which benefit their carrier, but not for the spread of genes which are detrimental to the survival or reproduction of the animal carrying them. The process of kin selection is one of the various mechanisms responsible for altruism. Each topic is largely independent of the other. Thus kin selection operates in both the presence and the absence of altruism, and altruism can occur without the influence of kin selection. If at the end of this paper your reaction is to think "How awkward to try to link the two subjects," then I shall have achieved my aims, which are to clarify the differences between the two topics and to illustrate them with examples from mammalian behavior.

Meaning of "Kin Selection"

"The frequency of a gene in a population will be influenced not only by the effects that gene has on the survival and fertility of individuals carrying it, but also by its effects on the survival and fertility of relatives of that individual. The second kind of effect has been called 'kin selection' " (Maynard Smith, 1976). The importance of the concept, and its application to social behavior, were emphasized by Hamilton (1963, 1964, 1971, 1972), and more recently by many other authors, among them Bertram (1976), Dawkins (1976), Maynard Smith and Ridpath (1972), Trivers (1974), Trivers and Hare (1976), and Wilson (1975). The details of the workings of kin selection models have also been discussed by many, including Barash et al. (1978), Charlesworth (1978), Charnov (1977), Dawkins (1979), Hamilton (1964), and Maynard Smith (1964).

Kin selection is not in any sense an *alternative* to natural selection—it is a part of it. Likewise, kin selection is not in any sense an alternative to what we may call individual selection—rather it is complementary to, and usually less important than, individual selection. As Dawkins (1976), Wilson (1975), and others have made clear, natural selection occurs by the differential replication of genes; an animal can produce a difference in the replication of its genes not only by influencing the number of offspring it produces but also (although usually to a lesser extent) by influencing the number of offspring produced by its relatives who also carry a proportion of those genes. The whole process of natural selection is one of gene selection, and genes are selected both via offspring and via other relatives. Offspring are only one type of kin; but to widen the term "kin selection" to cover the whole of gene selection (Dawkins, 1976), while conceptually useful, deprives us of a label for a particular field of study. Within a decade or two the term will probably have outlived its usefulness, although we will still refer to kin effects in the area of gene selection or natural selection. Meanwhile, in this paper, I use the term kin selection in the same sense as Maynard Smith (1976).

Meaning of "Altruism"

Altruism has been defined differently by different authors, and this has been responsible for some of the confusion surrounding the subject. Some of these definitions are as follows: 1) "Regard for others, as a principle of action; opposed to egoism or selfishness" (Oxford English Dictionary); 2) "The principle of living and acting for the interest of others" (Chambers English Dictionary); 3) "Altruism means doing something for someone else" (Ruse, 1979: 42); 4) "Altruistic behavior can be defined as behavior that benefits another organism, not closely related, while being apparently detrimental to the organism performing the behavior, benefit and detriment being defined in terms of contribution to inclusive fitness" (Trivers, 1971:35); 5) "Altruism is self-destructive behavior performed for the benefit of others, where the self-destructive behavior can range in intensity all the way from total bodily sacrifice to a slight diminishment of reproductive powers. It is commonplace in the responses of parents towards their young" (Wilson, 1971:321); 6) "When a person (or animal) increases the fitness of another at the expense of his own fitness, he can be said to have performed an act of altruism" (Wilson, 1975:117); 7) "Altruism is here defined as any behaviour performed by an individual that benefits a recipient while incurring a cost to the donor" (Emlen, 1978:245); 8) "Altruism . . . may be described as phenotypically (or self-) sacrificing but genotypically selfish" (Alexander, 1977:294); 9) Altruism is "activity that promotes the perpetuation of an individual's genes" not "directly" but "via another individual" (Lin and Michener, 1972:133); 10) Altruism is "behavior involving detriment to individual fitness," that is, "the number of adult offspring left by an individual in the next generation in the absence of chance effects" (West Eberhard, 1975:2).

Some of the ways in which these definitions either disagree or are not explicit are outlined below, as I discuss the following definition, which I rather reluctantly propose as preferable to those above: "Altruism in biology is defined as behaviour which is likely to increase the reproductive output of another member of the same species who is not a descendant of the actor, and which at least in the short term is likely also to reduce the number of the actor's own descendants" (Bertram, in press).

The following considerations were taken into account in proposing this latter definition.

1) Because sociobiological concepts will inevitably be applied to aspects of human behavior (Trivers, 1971; Wilson, 1978), it seems advisable that the meaning of the concept of altruism in biology should on the whole be as similar as possible to its meaning in

human behavior. Of course, in each case the meaning must be clearly defined, and of course there will remain differences; in particular the question of motivation usually plays a part in "altruism" in humans (see Krebs, 1970), whereas it does not in the definition of altruism in biology. Thus, if we discover that a human altruist performed an altruistic act only in the expectation of receiving a reward later, then we no longer consider him or her to have been a true altruist. However, if he performs such an act expecting no reward but is nonetheless subsequently rewarded, then we do consider him to have behaved altruistically. If he were merely incapable of expecting or not expecting a reward, we would probably still label his act as altruistic. Most acts of altruism by animals other than human beings are probably in this latter category of "devoid of expectation."

2) The animal benefiting from the altruism must be a member of the same species as the altruist. Without this restriction, the concept of altruism becomes unmanageably broad, to no advantage. The study of interspecies interactions is fascinating, but is not helped by combining it with the study of detailed within-species interactions; the genes are not in competition in a comparable way. In human behavior too, a person who is kind to cats is not usually described as an altruist. Nor do we want to say that a deer that sneezes (and thereby attracts a predator which kills it) is behaving altruistically. Trivers' (1971) definition of altruism (4 above) could include between-species interaction, and his example of cleaner fish and their clients certainly do so. The other definitions do not specifically exclude the possibility of altruism between species.

3) Parental care is not included within the definition of altruism. There is no problem in understanding how genes responsible for parental care will spread, for they will be passed on to the offspring reared as a result of that parental care. The evolution of caring for conspecifics who are not offspring is more complex, as will be discussed. If we did include parental care within the field of altruism, we would run into further problems of definition, because such care would not always be altruistic. In almost all definitions, altruism involves a probable reduction in an animal's reproductive output. But an animal which bestows parental care on what will definitely be the last offspring it produces will not be *reducing* its reproductive output, because that was already zero. In my definition I have excluded "descendants," not just "offspring," as recipients of altru-

ism, because the evolution of caring by grandparents is a straightforward and simple extension of ordinary parental care. In a human context, a self-sacrificing mother is said to be maternal rather than altruistic. Wilson's (1971) definition (5 above) includes parental care, whereas his 1975 definition (6 above) probably does not.

4) The animal benefiting from the altruism may be a close relative (other than a descendant) of the altruist. It is mainly at this point that kin selection impinges upon altruism, as will be discussed below. As Dawkins (1976) emphasized, offspring are kin too. Nevertheless, we can make a clear nonarbitrary distinction between the classes of "offspring" and "non-offspring," whereas we cannot draw a non-arbitrary line between close relatives and less close relatives. Trivers' (1971) definition (4 above) differs from the others in that only it excluded close relatives as recipients of altruistic acts.

5) The definition of the term "altruism" should be distinct from an explanation of how altruism may have evolved. A number of different possible mechanisms by which it could have evolved have been proposed. If we were to rely on Alexander's (1977) (8 above) or Lin and Michener's (1972) (9 above) explanatory definitions, we would be storing up problems for ourselves when the explanations prove to be inadequate.

6) The cost to the altruist may be only a relatively short term one, and indeed he or his offspring may or may not benefit in the long term. We will consider such cases later.

7) The cost and benefit to altruist and recipient are measured in terms of the individual fitness (or probable reproductive output) of those individuals. Again Trivers' (1971) definition differs in this respect from those used by West Eberhard (1975) and Wilson (1975). Most of the other definitions do not specify, although several imply, that the cost and benefit are to be measured in terms of offspring probably produced or probably not produced.

In the remainder of this paper, I use the term "altruism" in the slightly cumbersome sense in which I have defined it above.

How Has Altruism Evolved?

This section outlines the main ways by which altruistic behavior may be selected for in the course of evolution. Later I examine the interaction of these different ways in some mammalian examples.

Kin Selection

Altruism has been used as a testing ground for models of kin selection, because altruism involves the relatively unusual case of kin selection operating so as to produce an effect different from that produced by individual selection alone; in most cases the two processes operate in the same direction.

Genes promoting altruism towards close relatives can be selected for in the course of evolution because, although an altruist passes on fewer of those genes to his own offspring (because he has fewer offspring), he may, by helping his relatives, ensure that they transmit a greater number of those genes; those relatives will also carry a proportion of the altruist's genes, obtained from a common ancestor. The more distantly related the relatives, the smaller is the probability that any gene in one will be identical by descent with a gene in the other; that is, the smaller the coefficient of relatedness between them. Other things being equal, genes for altruism towards a relative will be selected for if the cost to the altruist is less than the recipient's benefit devalued by the coefficient of relatedness between altruist and recipient (Hamilton, 1964; Wilson, 1975). The cost and benefit here are in terms of the individual fitness of the altruist and recipient.

Delayed Benefits

Altruistic acts, while detrimental in the short term to the donor, may in some cases benefit the donor in the long term. There are a number of different ways in which a long-term benefit may be gained. For example, in red foxes (*Vulpes vulpes*), some females do not breed but instead help the breeding vixen in the group to rear her offspring (Macdonald, 1979). They may in the long run benefit from this altruistic behavior because of the experience gained, such that they may later be more successful in rearing their own offspring. They may survive better as part of a larger group; or they may inherit the reproductive position when the breeding vixen dies.

Another mechanism by which an individual's altruistic acts may result in delayed benefits to him is reciprocity. We may distinguish reciprocity from the other types of delayed benefit by the fact that the same altruistic act is repaid. Trivers (1971) showed how some types of altruistic behavior could evolve. In a system of "reciprocal altruism," an individual trades an altruistic act for the expectation that the act will later be repaid. Such a system, based on reciprocity, depends on the benefit to the recipient being greater than the cost to the altruist, and on the recipient's reciprocating later. Trivers discussed the problem of the recipient "cheating" by failing to reciprocate, and the evolutionary consequences of this cheating.

There has been confusion about terminology. "Reciprocal altruism (is) the trading of altruistic acts by individuals at different times" (Wilson, 1975:593). It is the *system* of trading, *not* the mechanism (reciprocity) by which such trading has evolved. There is no reason why it should not take place between relatives, and indeed it often does. The essence of reciprocity is that the repayment is in the same form as the original altruistic act, such as food in return for food, rescue for rescue (Trivers, 1971), or help in fighting for help in fighting (Packer, (1977).

By contrast, many other types of delayed benefit are received in a different form (such as experience, survival, or social position) which depend less on a relationship with one particular individual (which has therefore perhaps a greater degree of uncertainty, or at least a different type of uncertainty) and more on the individual's building up his own general social and environmental position. We may usefully look on these delayed benefit systems as a type of investment by the altruist, comparable to storing food until a time when it will be more valuable to him; any investment has uncertainty attached to it, but should on average yield more benefits later. I include these "social investments" as examples of altruism because they come within the definitions given above. At least in the short term, there is a cost through spending energy, losing food, or taking risks. The donor may well benefit in the long run, but in the short term he does not.

Compulsion

It is possible for altruism to be imposed by the society (which includes the family) in which an animal lives. There are different types of compulsion, operating in different ways; how they work obviously depends on the choices available to the recipient. For example, a subordinate male impala (*Aepyceros melampus*) does more than his share of being vigilant and thereby protecting the rest of the group; but, because he is driven out from the middle to the more dangerous edge of the group, he has little alternative but to be extra vigilant for his own safety (Jarman, pers. comm.). Lionesses (*Panthera leo*) catch food that males eat. In the short term at least, the females do not benefit whereas the males do, so it would fall within the definition of altruism. The males are not related to the females (Bertram, 1976), and they do not reciprocate. It is possible that the females benefit in the long term by having healthy well-fed males with them. But it is clear that the females relinquish their food because the males are larger and stronger and compel them to do so.

It is obvious that the distinction between compulsion and delayed benefit may be blurred, because it depends both on the level at which one examines the behavior and on the choices realistically available to the actors. For example, if wild dogs (*Lycaon pictus*) were to "demand" altruistic behavior (such as feeding young) from a subordinate as the price of remaining in the pack, and if it is in the long-term interest of that subordinate to stay in the pack in spite of this price, then we would expect to see such altruistic behavior evolve; but we could not say that the mechanism for its evolution was delayed benefit rather than compulsion. The "genes for altruism" may lie in the other members of the society, not just in the altruist. Parental manipulation is probably one of the commonest forms of social compulsion resulting in altruistic behavior. The argument by Alexander (1974) that parents must always be the winners in any evolutionary conflicts with their offspring has been shown to be invalid (Dawkins, 1976). On the other hand, a parent may often be able in practice to manipulate offspring so as to compel them to be more altruistic than they otherwise would be. For example, a parent enforcing food-sharing among offspring is in effect compelling the stronger ones to be altruistic. A parent who castrates some offspring early in life (as the queens of many social insects do) is similarly compelling them to be altruistic towards their kin as their only way of reproducing (albeit by proxy).

These cases of altruism I have mentioned illustrate the problem of how specific one should be about what behavior is being considered. It makes more sense to consider the act of giving food to another individual to be an example of altruism than to include the more general remaining within a group wherein others are fed. Similarly, we should probably consider only positive acts, such as giving food, and not negative ones, such as refraining from attacking another individual. But whether an act is considered to be positive or negative may depend on the level at which it is examined. For example, in the lion case (above), supplying food for males would appear superficially to be a positive act by the females. It is only when observed closely enough that it can be seen to be the result of a negative act of withdrawal by the females when threatened by males. Such withdrawal is obviously in the females' short-term as well as long-term interests, and therefore "withdrawal" would not fall within the definition of altruism given earlier. The less specific "supplying food to males" would.

As Trivers (1971:35) pointed out, "models that attempt to explain altruistic behavior in terms of natural selection are models designed to take the altruism out of altruism." They are all designed to demonstrate ways in which an altruistic organism (or its relatives) benefits. However, the likelihood of a subsequent benefit does not in fact rule out such instances of altruistic behavior from being reasonably considered to be altruistic, as human beings apply the term to human behavior. First, there is a time lag between the benefit and the altruistic act—the donor loses before gaining later. Second, at the time of the altruistic act, the donor was not "expecting" a benefit (in the case of most non-human animals); and as argued earlier, it is the expectation of a reward, not the actual receiving of the reward, which disqualifies human acts from being classified as altruistic.

Types of Altruism in Mammals

There are a variety of different ways in which a mammal may behave altruistically towards another, and we have referred to several of these. They may be classed into several categories according to their role in defense, feeding, reproduction, and group welfare. I consider the following forms of altruistic behavior: 1) signalling to other conspecifics the presence of a predator, either visually or auditorily; 2) aiding another member of the species in encounters with predators, in cases where the donor would not otherwise be endangered; 3) helping a conspecific in the latter's hostile encounters with other conspecifics; 4) helping other individuals to find their food, and consequently getting less oneself; 5) giving food to other individuals; 6) allowing other individuals' offspring to suckle from oneself; 7) assisting other individuals to rear their offspring; 8) helping a sick or injured individual; and 9) allogrooming, that is, grooming another individual.

Obviously these forms of altruistic behavior may overlap. It is difficult, for example, in some cases to distinguish between giving food calls (4) and food sharing (5), or between helping a young animal (7) and helping a sick one (8). Nor are the different forms of behavior equal in scope or importance.

Different mammal species exhibit not only a different number of these forms of altruistic behavior but also different combinations of them. Table 1 shows roughly the differing ranges of types of altruism regularly displayed by a selection of different mammals. The table is not intended to be representative of all mammals; I have on the whole selected the more altruistic species in the various mammal orders, and I have concentrated on the carnivores and primates. I have also included for comparison a bird and a social insect.

The table demonstrates two main points. First, some species are more altruistic than others, in the sense that they regularly show more different categories of altruistic behavior. The social carnivores and the higher primates on the whole show the widest range of types of altruistic behavior, and this approaches the level reached by social insects. It is obvious that species whose members are solitary have relatively little scope for altruism. Second, different species exhibit different collections of altruistic behaviors. Thus, for example, wallabies allogroom, whereas dolphins do not. The reasons for such differences may be of several types. For example, they may be structural. Dolphins have neither fur to be groomed nor forelimbs suitable for allogrooming. They may also be ecological. For example, the food of buffaloes is not of a type or a distribution such that other buffaloes can be helped to eat it. Or they may be social. Jackals, hunting dogs, and dwarf mongooses do not regularly exhibit communal suckling (as lions do) because in those species only the one female who reproduces is likely to be capable of producing milk. Thus the number and the types of altruistic behaviors shown depends on the structure, the niche, and the social organization of the species.

Evolutionary Causes of Mammalian Altruism

In this section, I take four examples of altruism in mammals and examine the evolutionary causes of the behavior in each case. My aim is to emphasize the *variety* of evolutionary causes operating in each example.

Lion communal suckling.—A lioness usually allows the cubs of other lionesses in her pride to obtain milk from her (Bertram, 1975, 1976; Schaller, 1972). Those other lionesses are usually close relatives, as close as full cousins on average (Bertram, 1976). Thus, in allowing their offspring to suckle, a lioness is indirectly helping to propagate those of her own genes which are identical by descent in her relatives' offspring. Whether this kin-selection pressure is strong enough alone to cause this form of altruism to evolve depends on the costs to her own offspring and the benefits to those of her companions; and unfortunately we cannot yet measure these costs and benefits. However, there are also other selective pressures operating. For example, it is likely that her own cubs benefit by having the other cubs and those other cubs' mothers near them (Bertram, 1975, 1976); and her male offspring, when full grown, have much greater reproductive success if they have companions (Bygott et al., 1979). Thus, in addition to the kin-selection pressure, there are also delayed benefits to the offspring of the donor of the altruistic act.

Hunting dog not breeding.—A pack of hunting dogs generally contains one or more females who are capable of reproducing yet do not do so (Frame and Frame, 1976). These non-reproducing females, like all adult hunting dogs, help in rearing the offspring of the breeding female. They are often, but not always, close relatives of the breeding female, and thus, she may pass on some of their shared genes. Unless the breeder is likely to produce at least twice as many extra surviving offspring as the non-breeder, kin selection alone is unlikely to account for the latter's restraint; but it may contribute towards it. There are probably some delayed benefits from refraining from reproduction, such as conserving energy for possible breeding attempts later. But overriding these considerations is the fact that the breeder is dominant and prevents the subordinates from reproducing. The ways in which she achieves this are not clear in hunting dogs but are probably similar to those in wolves, in which there is reported to be both suppression of estrous and inhibition of copulation (Zimen, 1976). In addition, if a subordinate hunting dog female does nonetheless produce offspring, these may be killed by the dominant animal (Van Lawick, 1974). Hunting dogs are dependent on membership in a pack for

TABLE 1
TYPES OF ALTRUISTIC BEHAVIOR

Species	Alarm signals	Predator defence	Help in other's fighting	Help in finding food	Giving food	Suckling others	Help in rearing	Helping sick	Allogrooming	References
Whiptail wallaby, *Macropus parryi*									X	Kaufmann (1974)
Belding's ground squirrel, *Spermophilus beldingi*	X								X	Sherman (1977)
African buffalo, *Syncerus caffer*	X	X								Sinclair (1977)
African elephant, *Loxodonta africana*	X	X		X		X	X	X		Douglas-Hamilton (1972); Hendrichs and Hendrichs (1971)
Atlantic dolphin, *Delphinus delphis*				X			X	X		Pilleri and Knuckey (1969)
Blackbacked jackal, *Canis mesomelas*	X	X	X		X		X		X	Moehlmann (this volume)
Cape hunting dog, *Lycaon pictus*		X	X	X	X		X	X	X	Frame and Frame (1976); Frame et al. (1979)
Dwarf mongoose, *Helogale undulata*	X	X		X	X		X	X	X	Rasa (1977); Rood (this volume)
Lion, *Panthera leo*		X	X	X		X	X		X	Bertram (1979); Schaller (1972)

TABLE 1
CONTINUED

Species	Alarm signals	Predator defence	Help in other's fighting	Help in finding food	Giving food	Suckling others	Help in rearing	Helping sick	Allogrooming	References
Olive baboon, *Papio anubis*	X	X	X		X		X	X	X	Altmann and Altmann (1970); Packer (1977)
Chimpanzee, *Pan troglodytes*	X	X	X	X	X		X		X	Teleki (1973); Van Lawick-Goodall (1968); Wrangham (1977)
Man, *Homo sapiens*	X	X	X	X	X	X	X	X	X	Personal observation
Scrub Jay, *Aphelocoma coerulescens*	X	X		X			X		X	Woolfenden (1975)
Honeybee, *Apis mellifera*	X	X	X	X	X		X		X	Michener (1974); Von Frisch (1967)

successful hunting and reproduction, so subordinates do not have emigrating and surviving as realistic options. Thus, compulsion appears particularly important in this instance of altruism by reproductive restraint, but other selective pressures influence the response to the compulsion.

Male baboon fighting coalitions.—Packer (1977) described the process by which a male olive baboon could recruit a second male to help rob a third of a female in estrous; while the recruit was occupied in fighting the previous owner, the first male was able to gain access to the female. Packer showed that the males in these instances were generally unrelated and that they tended to reciprocate, with the second being able on later occasions to enlist the help of the first. Reciprocity appears to be the most important factor in the evolution of this particular form of altruism. However, there may be other delayed benefits from cooperation of this sort, such as the gaining of social status. And sometimes similar types of altruistic behaviors take place between relatives, in which case there will also be a kin selection pressure operating.

Chimpanzee food calling.—As Wrangham (1977) described, a male chimpanzee that finds a large food source will often advertise its presence by calling loudly; other chimpanzees gather and benefit from the caller's altruistic behavior. Some of those who come may be close relatives, so there may be a kin selection pressure operating. Wrangham considered that in the long term the male chimpanzee benefited from the presence of the others by impressing females or threatening his neighbors. The food source is a lure to induce them to gather quickly before the food is all consumed.

Conclusion

Some mammals exhibit a variety of types of behavior which with caution we may label as altruistic. Which mammals exhibit which types of altruism becomes comprehensible with knowledge of the species, its social organization, and its ecology. There are a number of ways in which altruistic behavior can be selected for in the course of evolution. It is likely that more than one of these have operated in the evolution of any particular types of altruistic behavior. Some examples of altruism may have been helped in their evolution and maintenance by the process of kin selection, and some have not. Kin selection and individual selection are the two complementary parts of the process of natural selection by which genes change in frequency in a population and only some of those genes are responsible for the performance of altruistic behavior.

Literature Cited

ALEXANDER, R. D. 1974. The evolution of social behavior. Ann. Rev. Ecol. Syst., 5:325–383.

———. 1977. Natural selection and the analysis of human sociality. Pp. 283–337, *in* Changing scenes in the natural sciences, 1776–1976 ed. (C. E. Goulden, ed.). Acad. Nat. Sci., Philadelphia, 362 pp.

ALTMANN, S. A., AND ALTMANN. J. 1970. Baboon ecology: African field research. Univ. Chicago Press, Chicago, 220 pp.

BARASH, D. P., W. G. HOLMES, AND P. J. GREENE. 1978. Exact versus probabilistic coefficients of relationship: some implications for sociobiology. Amer. Nat., 112:355–363.

BERTRAM, B. C. R. 1975. Social factors influencing reproduction in wild lions. J. Zool., 177:463–482.

———. 1976. Kin selection in lions and in evolution. Pp. 281–301, *in* Growing points in ethology (P. P. G. Bateson and R. A. Hinde, eds.). Cambridge Univ. Press, Cambridge, 548 pp.

———. 1979. Serengeti predators and their social systems. Pp. 221–248, *in* Serengeti: dynamics of an ecosystem (A. R. E. Sinclair and M. Norton-Griffiths, eds.). Univ. Chicago Press, Chicago, 389 pp.

———. In press. Problems with altruism. *In* Current problems in sociobiology (King's College Sociobiology Group, ed.). Cambridge Univ. Press, Cambridge.

BYGOTT, J. D., B. C. R. BERTRAM, AND J. P. HANBY. 1970. Male lions in large coalitions gain reproductive advantages. Nature, 282:839–841.

CHARLESWORTH, B. 1978. Some models of the evolution of altruistic behaviour between siblings. J. Theor. Biol., 72:297–319.

CHARNOV, E. L. 1977. An elementary treatment of the genetical theory of kin-selection. J. Theoret. Biol., 66:541–550.

DAWKINS, R. 1976. The selfish gene. Oxford Univ. Press, Oxford, 224 pp.

———. 1979. Twelve misunderstandings of kin selection. Z. Tierpsychol., 51: 184–200.

DOUGLAS-HAMILTON, I. 1972. On the ecology and behaviour of the African elephant: the elephants of Lake Manyara. Unpubl. Ph.D. dissert., Oxford Univ., 268 pp.

EMLEN, S. T. 1978. The evolution of cooperative breeding in birds. Pp. 245–281, *in* Behavioural ecology: an evolutionary approach (J. R. Krebs and N. B. Davies, eds.). Blackwell, Oxford, 494 pp.

FRAME, L. H., AND G. W. FRAME. 1976. Female African wild dogs emigrate. Nature, 263-227–229.

FRAME, L. H., ET AL. 1979. Social organization of African wild dogs (*Lycaon pictus*) on the Serengeti plains, Tanzania 1968–1978. Z. Tierpsychol., 50: 225–249.

HAMILTON, W. D. 1963. The evolution of altruistic behavior. Amer. Nat., 97: 354–356.

———. 1964. The genetical evolution of social behaviour, I & II. J. Theor. Biol., 7:1–52.

———. 1971. Selection of selfish and altruistic behavior in some extreme models. Pp. 57–91, *in* Man and beast: comparative social behavior (J. F. Eisenberg and W. S. Dillon, eds.). Smithsonian, Washington, D.C., 401 pp.

———. 1972. Altruism and related phenomena, mainly in social insects. Ann. Rev. Ecol. Syst., 3:193–232.

HENDRICHS, H., AND U. HENDRICHS. 1971. Dikdik und Elefanten: Okologie und Soziologie zweier afrikanischer Huftiere. Piper, Munich, 173 pp.

KAUFMANN, J. H. 1974. Social ethology of the whiptail wallaby, *Macropus parryi,* in northeastern New South Wales. Anim. Behav., 22:281–369.

KREBS, D. 1970. Altruism—an examination of the concept and a review of the literature. Psych. Bull., 73:258–302.

LIN, N., AND C. D. MICHENER. 1972. Evolution of sociality in insects. Quart. Rev. Biol., 47:131–159.

MACDONALD, D. W. 1979. 'Helpers' in fox society. Nature, 282:69–71.

———. 1976. Group selection. Quart. Rev. Biol., 51:277–283.

MAYNARD SMITH, J., AND M. G. RIDPATH. 1972. Wife sharing in the Tasmanian native hen, *Tribonyx mortierii*: a case of kin selection? Amer. Nat., 106: 447–452.

MICHENER, C. D. 1974. The social behavior of the bees: a comparative study. Belknap Press, Cambridge, Massachusetts, 404 pp.

PACKER, C. 1977. Reciprocal altruism in *Papio anubis.* Nature, 265:441–443.

PILLERI, G., AND J. KNUCKEY. 1969. Behavior patterns of some Delphinidae observed in the western Mediterranean. Z. Tierpsychol., 26:48–72.

RASA, O. A. E. 1977. The ethology and sociology of the dwarf mongoose (*Helogale undulata rufula*). Z. Tierpsychol., 43:337–406.

RUSE, M. 1979. Sociobiology: sense or nonsense? Reidel, Dordrecht, 231 pp.

SCHALLER, G. B. 1972. The Serengeti lion. Univ. Chicago Press, Chicago, 480 pp.

SHERMAN, P. W. 1977. Nepotism and the evolution of alarm calls. Science, 197: 1246–1253.

SINCLAIR, A. R. E. 1977. The African buffalo. Univ. Chicago Press, Chicago, 355 pp.

TELEKI, G. 1973. The predatory behavior of wild chimpanzees. Bucknell Univ. Press, Lewisburg, Pennsylvania, 232 pp.

TRIVERS, R. L. 1971. The evolution of reciprocal altruism. Quart. Rev. Biol., 46: 35–57.

———. 1974. Parent-offspring conflict. Amer. Zool., 14:249–264.

TRIVERS, R. L., AND H. HARE. 1976. Haplodiploidy and the evolution of the social insects. Science, 191:249–263.

VAN LAWICK, H. 1974. Solo: the story of an African wild dog. Houghton Mifflin, Boston, 159 pp.

VAN LAWICK-GOODALL, J. 1968. The behaviour of free-living chimpanzees in the Gombe Stream Reserve. Anim. Behav. Monogr., 1:161–311.

VON FRISCH, K. 1967. The dance language and orientation of bees. Belknap Press, Cambridge, Massachusetts, 566 pp.

WEST EBERHARD, M. J. 1975. The evolution of social behavior by kin selection. Quart. Rev. Biol., 50:1–33.

WILSON, E. O. 1971. The insect societies. Belknap Press, Cambridge, Massachusetts, 548 pp.

———. 1975. Sociobiology: the new synthesis. Belknap Press, Cambridge, Massachusetts, 697 pp.

———. 1978. On human nature. Harvard Univ. Press, Cambridge, Massachusetts, 260 pp.

WOOLFENDEN, G. E. 1975. Florida scrub jay helpers at the nest. Auk, 92:1–15.

WRANGHAM, R. W. 1977. Feeding behaviour of chimpanzees in Gombe National Park, Tanzania. Pp. 504–538, *in* Primate ecology (T. H. Clutton-Brock, ed.). Academic Press, London, 631 pp.

ZIMEN, E. 1976. On the regulation of pack size in wolves. Z. Tierpsychol., 40: 300–341.

ON THE EVOLUTION OF SOCIAL STRUCTURE IN MAMMALS

HUBERT HENDRICHS

Abstract

THE general attributes of mammalian social structure are described and then confronted with those theoretical concepts currently used by sociobiologists to explain the evolution of social behavior. Attention is drawn to those attributes of social structure that, though relevant to its evolution, are neglected by current sociobiological, theoretical concepts. Such current concepts include "inclusive fitness" and "evolutionary stable strategy." A specific quality of mammalian social structure is that structural attributes may in part be transmitted and altered through social learning. A general quality of living systems is their hierarchical organization and their tendency to realize new levels of integration. The center and extension of a developing social unit shifts when a higher level of cooperation is reached. These qualities of social structure contradict the assumptions of simple mechanistic models of optimizing theory as currently employed in sociobiology. The evolution of social structure cannot be fully explained by the competition of individuals or genes aimed at maximizing their fitness. To understand the function and evolution of mammalian social structures, concepts must be developed in addition to the ones currently used, thereby allowing description and analysis in a hierarchical sequence of multidimensionally related, "open" systems and their dynamics. With growing differentiation, such systems increase economy and overall requirements at the same time, and thereby become increasingly able not only to regulate their own inner state, but also to turn competition into cooperation and to realize new levels of integration together with neighboring systems of the same kind.

Introduction

The social structure of a population, or of any other more or less permanent group of conspecific animals, is the order in which the social interactions between the individuals of this population or group can be carried out and can have social consequences. We must consider the order in which social patterns, such as rank hierarchies, attachments, friendly or hostile relations, role divisions and profiles, can be established and maintained. What is this order? What does it consist of, how does it work, how is it related to its environment, and how does it develop in evolution? Following the suggestion of the organizers of this symposium on mammalian behavior to avoid pious platitudes and to be constructively critical of current theory, I try to give some answers to these questions by confronting and combining what I have learned from long-term contacts with domestic, captive, and free-living mammals with those more abstract, theoretical considerations of function and development of biological "open" systems in a hierarchical organization that have been recently developed.

To use the term evolution in the title of a paper is rather pretentious as long as we do not know more about how evolution works. Obviously only a few aspects can be treated in a short article. I do not describe different forms of social organization in mammals (which I have done elsewhere; Hendrichs, 1978), nor do I try to analyze their relation to ecological factors. This is a question to which there is still no general answer in sight, although there have been careful assessments for some mammalian groups (for example, Eisenberg et al., 1972, for primates). Instead I point out a different aspect that should be considered in addition to ecological adaptation when one looks at the evolution of social structure. Most of my comments are very general and the article should be taken as a provocative essay rather than as a scientific paper presenting data and their careful analysis.

The Reality of Social Structure

At first glance, social order seems to be the result only of the interactions of the individuals present in the population or group, whose behavior became programmed partly through phylogeny and partly through ontogeny. According to this view, the genetically transmitted possibilities for interactions, resulting in relations and patterns, provide the framework in which the single individuals can realize their specific behavior and thereby jointly create the group-specific social structure. But a closer look at a population or group and its social structure for a longer period of time reveals that, although dependent on interacting individuals, such an order is at the same time independent of specific individuals and of specific behavior shown by individuals. Within limits, individuals and behavior are exchangeable without altering the social order. The social order, resulting from the behavior of former and present individuals of the group, has a reality of its own which acts strongly on individuals and their behavior. In a sense, then, individuals are usually younger than the social structure and are formed by it to a greater extent compared with the degree to which the structure is changed by their specific individuality. This social structure, which results from the behavior of phylogenetically and ontogenetically programmed individuals, may even influence individuals up to a point where they give up maximizing their own immediate benefits for the long-term benefits offered to them and their kin by living in a group.

The Complexity of Biological "Open" Systems and Simplicity of Current Sociobiological Models

It is the aim of biology to design and test more or less general models which simulate the dynamics of *developing* structures. Biological structures are considered to be multidimensional "open" systems. The biological "open system" was contrasted with the physical "closed system" by Lotze (1882) long before Bertalanffy (1972) developed his system theory in the early thirties. The term "open" refers not only to the energy flow (which also occurs between the "closed" system and its environment) nor only to the metabolism essential for living matter, but also refers to the multidimensional relatedness of the living system and to those qualities of its environment which it can perceive and react to, thereby regulating and stabilizing its "inner milieu" (Bernard, 1966; Cannon, 1967). In my view this is the most important point.

As all objects of science, biological systems—molecules, cells, organisms, populations, ecosystems—are organized hierarchically; each system is composed of subsystems and is at the same time a component of larger systems. The disposition and behavior of a defined system is influenced by the behavior of both the composing and by that of the inclusive comprising systems. There is no biological phenomenon which has causes and consequences in only one context. When performing their defecation ceremony at a specific place on the boundary of its territory, dik diks do at least five things at once: they excrete urine and dung; they mark their territorial boundary with optical and olfactory signals; they claim their territorial ownership up to that point; they strengthen their attachment to that place; they strengthen the integration of their group.

Grooming among mammals is another example of behavior that involves costs and benefits in several dimensions and in several currencies. It has physiological effects on skin and hair. It is important in the context of social relations, where it can be a demonstration of dominance or submission—the dominant animal may be grooming or being groomed. It also has psychological and social effects on the levels of arousal, tension, and aggressiveness of the animals directly involved, as well as on the arousal levels of the other animals in the group, and thereby on the state of the group as a whole.

Every biological structure balances numerous influences from various levels of organization. A scientific model can always consider only one aspect of the real structure, thus selecting one or a few parameters for the analysis. The reduction of the multidimensional relatedness of a real structure to one or a few aspects and to a small number of observable parameters is largely dependent on the competence of the scientist and is his or her responsibility. Besides the practical limitations of scientific procedure which depend on observable events, the theoretical concepts cherished at the moment by the scientific community exert a strong influence on the selection of the parameters for study and measurement. These theoretical concepts change continuously and discontinuously, and they should develop to allow a better understanding of the phenomena under investigation.

The theoretical concepts used to describe and understand the biological evolution of social structure have, in the last decades, been strongly influenced by mathematical models, such as that of "inclusive fitness" (Hamilton, 1964) or that of "evolutionary stable strategies" (Maynard Smith, 1976; Maynard Smith and Price, 1973), which have been developed and used by sociobiologists to explain the evolution of social behavior. Through the strong impact of these models, which are brilliant and simple at the same time, the theoretical concepts used have become rather uniform (Wilson, 1975). Does this mean that social behavior has been explained and sociobiology has reached a state of perfection which is fully satisfactory, or does it mean that these theories are widely used in an uncritical manner which overlooks or suppresses serious problems?

If one looks closely at the empirical findings pertinent to the evolution of social structure in mammals, it becomes obvious that there are serious problems in applying current "sociobiological" models. Not only does the applicability of game-theoretical models seem doubtful, but so does the application of present-day mathematics in general. The tendency of a biological structure to optimize function and adaptation is no doubt an important aspect of its evolution, but it is not the only one nor one that can be easily conceptualized. A simple linear rationality, according to which isolated, independent competing units weigh cost and benefit when trying to maximize their fitness, soon comes up against serious problems.

First, biological structures always perform in several functional contexts; performance therefore represents cost and gain at the same time, which have to be measured in several contexts and different currencies that may not be convertible. In this way biological systems can increase economy and expenditure at the same time; developments allowing greater economy are often very costly and show high requirements in terms of the quantity and quality of energy. The development of pack organization in canids, deriving from the usual canid pair organization (Kleiman and Eisenberg, 1973), is one example of this. A highly-developed system of social communication allows a relatively smooth cooperation of several mature male and female animals, but not to the extent, as in many primate species, that more than one pair can breed regularly. Cooperative hunting facilitates killing of larger prey; thus, an additional food source becomes accessible and life gets easier. But at the same time, the size of the animals increases. An easier life makes it possible to sustain a larger body, which is helpful and more economical in bringing down large prey. A larger animal needs relatively less energy than a smaller one, but—and this is the other side of the coin—a larger animal always needs more in absolute terms (Clutton-Brock and Harvey, this volume). Although the system has become more economical, its total requirements have grown. As long as large prey is abundant, wolves (*Canis lupus*) have no problem, but as soon as large prey gets scarce, they run into trouble, whereas foxes (*Vulpes* spp.) survive more easily.

Second, the units that invest and the ones that gain by the investment are difficult to define and they may be different ones. The center and contour of a social unit may shift during evolution. One example is the evolutionary development of a social structure, causing integration of formerly competing individuals into a cooperative unit with a division of roles and with commonly shared costs and benefits. No longer does the gain of one individual only mean loss to another. Individual investments can result in profits to the investing animal and, at the same time, to other individuals, and finally to all animals in the group. Each element of a cooperative system acts as the organs of an organism—an element acts for itself, acts for the benefit of the other elements, and acts for the benefit of the system as a whole.

Third, changes both inside a biological system and in its environment have the effect that the expression of its adaptation is rarely "optimal" for a single context. There remains scope for an experimental approach to the solution of problems. The new solution would not have to be optimal, because the structure of a system can function well enough to provide a limited independence from the immediate ecological situation.

Lastly, the hierarchical organization and multidimensional relatedness of open living systems make it possible for developments which started independently of one another in different areas to grow together and by reciprocal stimulation generate a new temporal and structural quality of development (that is, a "breakthrough," opening-up of new possibilities). Examples of such developments are the transition to life on land, the development of temperature regulation and population regulation, sociocultural evolution, and the origin of man.

The question is, can social phenomena be simplified without distortion so that at least some aspects can be handled by present-day applied mathematics, which requires 1) clearcut action patterns with clearcut properties that 2) do not change during development and 3) whose cost and benefit can be clearly defined. The development of social phenomena should be smooth and continuous and not be disrupted by jumping to new levels of orientation and integration. Most of the current sociobiological mathematical models assume: 1) non-sexual reproduction; 2) non-overlapping generations; 3) clearcut connections between phenotype and genotype and between the unit of selection and the unit of evolution; 4) interactions between individuals that meet at random and do not know one another, or remember previous encounters; and 5) a rationality in which cost and gain can be clearly defined, evaluated, and measured in currencies convertible into fitness values.

Game-theoretical models in particular are only concerned with the stabilities of equilibria. They are not concerned with the phylogenetic development of such equilibria and are unable to describe such developments. Game theoretical payoff could be considerably increased as an evaluation of the Prisoner's Dilemma shows, if cooperative agreements or conventions were possible and could be further safeguarded against cheating.

Nature is able, as the evaluation of phylogenetical development demonstrates, to develop non-cheating programs that allow cooperation by not maximizing their own immediate benefits. Such a program does not play an "evolutionary stable strategy," but its game-theoretical payoff is considerably higher. This is the kind of program which facilitates the evolution of new qualities, and carries forward developments which reach new levels of integration.

So far sociobiological mathematical models in general have only used very simple concepts derived from the linear mathematics of population genetics. In another field of biology bordering on chemistry, very promising non-linear mathematical approaches have been developed in the last decade which can simulate "self-organization" as non-equilibrium phase transitions (Eigen and Schuster, 1979; Glansdorff and Prigogine, 1971; Thom, 1972). Mathematically oriented sociobiologists should study these approaches; they might be more suitable for the description of social development.

The Function of Social Structure

To describe the effects of the social structure, a more or less dense network of relations among the individuals, it is necessary to introduce or redefine some terms concerning the disposition of the animals. One must consider non-specific components of their inner state that together with specific endogenous and exogenous stimulation regulate the motivational balance of the animal. Each of these components can be correlated with visible behavior, and by using these behavioral indicators they can be quantified and analyzed.

The "security" of an animal is a measurement of how solidly it is embedded in the social network—that is, when none or not too many of the necessary relations pertinent to its role are seriously disturbed. Security of various degrees is a precondition for different social initiatives—parental, sexual, or agonistic. The different degrees can be correlated to specific forms of posture or movement.

The "excitement" or "arousal" of an animal is a non-specific response to stimulation. Its correlate in the physiological system has been described as "stress" (Selye, 1956). In the state of excitement an animal mobilizes its possibilities to react. Excitement can be correlated with hair-raising, frequencies of marking and excreting, and with positions of tail or ears (see also Gould, this volume).

The "tension" of an animal is a measurement of its concentration or intensity in acting. It describes the difference between playful acting and serious actions, that is, "meaning business." It can be correlated with muscle tension and with specific forms of movement.

The social "enterprise" or "aggressiveness" of an animal—the term "aggressiveness" is used in a very broad sense as introduced by Lorenz (1963)—is that component of the inner state of an animal which drives it into any social interaction (friendly or hostile) and determines its readiness to stand and even search for social situations which may induce stress. This component can be correlated with the frequency of specific behavior patterns.

Using these terms as defined above, the function of social structure can be descri ,ed as follows. 1) Social aggressiveness pushes individuals towards each other with enough impact and such regularity ιhat specific patterns of interacting and a specific network of social relations develop which influence the social program and behavior of the individuals. 2) The social network of relations moulds aggressive behavior away from the harsh and rough encounters of strangers towards calm and gentle forms of engagements between neighbors and group members, thereby lowering tension and arousal while raising the security of the individuals. 3) The social structure of a group provides a relatively stable net of relations. The acquaintance with and the attachment to this net provides security for the individuals, enabling them to regulate their states of arousal and tension and in so doing to act adequately in a given social context. 4) The social structure stabilizes and isolates groups of animals, while at the same time separating those individuals that belong to a specific net from those that belong to a different one and from migrating strangers. 5) In a developed social structure, a prolonged and highly-differentiated socialization of a slowly-maturing young animal in close contact with very few but very intimately known individuals is possible and can generate a further differentiation of the social structure. 6) Social structure reduces the turnover of individuals and energy flow by keeping those animals attached to the social net that are past their prime and whose agility, fighting, and reproductive abilities are reduced. Energy flow is thus throttled and economized while there are still more young being produced than needed in a normal situation to keep up the number. 7) The regulation of numbers as an economical achievement made possible by social structure allows an experimental orientation, an experimental strategy of the population removed from the limitations of their niche, by which more costly structures and strategies with higher investment demands can be developed, thus facilitating coevolution and the realization of new levels of integration.

The Evolution of Social Structure

In agreement with their large size and with their spatial and social programs, mammals mostly live in relatively small local populations with a moderate rate of inbreeding. This can facilitate a high rate of genetic differentiation. However, as if not satisfied with the speed of genetic differentiation, mammals have in addition developed a special kind of behavioral evolution that in speed exceeds all genetical developments. This latter process is the social or cultural evolution in which qualities of the social or cultural structure change independently of genetic changes. Young mammals mature slowly. Their long contact with a few intimately known individuals, together with the qualities of the brain and its communication channels, allow a detailed adaptation to a highly differentiated social structure. During this process of socialization, behavior patterns of individuals and the elements of the social structure can change independently of genetic changes.

Mammals are able to pass on quite a number of "cultural," social, and psychological attributes by social tradition. Besides social patterns, they can transmit knowledge of their abiotic environment, such as suitable terrain, routes, and crossings in different weather conditions. They can transmit knowledge of specific qualities of prey and predator species (for instance, where and when to be alert to a predator or a prey animal); how to hunt specific prey; and how to stalk on specific ground. Lions, rhinos, buffaloes, and elephants can learn population-specific ways to get along with each other. But mammals do not only learn about the ecological or social conditions of their specific situation and thereby become conditioned ecologically or socially to a single strategy. Instead, they are also conditioned "psychologically" in the process and they become shy or bold, peaceable or quarrelsome, quiet or fierce in relation to several specific ecological or social conditions.

A bull elephant, moving along by himself for days, makes his own decisions during that time about the temporal and spatial structure of his activity, balancing drives, and stimulations. When he then joins a group of other bulls, all well known to him, he adapts his activity to that of the group, at first influencing and then "accepting" each joint decision (for example, when to move and in what direction). For such a joint decision, competing and antagonistic tendencies of different individuals, some of which are trying to move in one, others in a different direction, have to be weighed as to their urgency and balanced with the tendencies to stay with the group. The decision can often only be reached after some "dispute," in which rank and aggressiveness of the animals play a part. In such situations mammals also learn specific ways to settle disputes or to solve general problems. This learning process not only concerns the type of possible solution but also the degree of arousal, tension, and aggressiveness involved.

The phylogenetic development of a social structure can be described as an analogy to the development of an ecosystem in successive stages leading towards its climax. The social role of the individual, its niche in the social system of the population or group, as well as the ecological role or niche of the population in a developing ecosystem, become increasingly interwoven within a network of relations which becomes denser and finer. A dense network of relations changes the competition of isolated opponents into the cooperation of members of a system, which is then able to regulate its inner conditions independent of short-term fluctuations in the environment.

Another biological development that can be seen in analogy to the evolution of social structure is the ontogeny of social development. In ontogeny the young mammal is at first only in contact with and attached to its mother. As it comes in contact with other family or group members, it builds up other relations and experiences the dynamics of the social network. By relating itself to this structure, the young gains a social identity and its "role." It becomes integrated in the group and at the same time a relatively self-reliant unit of the group. In a comparable way, the "young" social structure becomes more independent of its environment and at the same time more densely related to the environmental structure during its development towards a more differentiated network of relations.

In developing such multidimensionally-related biological systems by evaluating their overall cost and benefit, nature is not restricted to a simple, linear, "short-sighted" logic, as is assumed by current sociobiological theory. Instead, organisms are capable of using a circumspect and far-seeing rationality to design developments. Biological developments are influenced by non-linear possibilities of the developing system, which include the realization of a new level of integration. In this context, Campbell (1974) has coined the term "downward causation:" in hierarchical systems, qualities of higher levels of organization influence the evaluation of processes on a lower level; "all processes at the lower levels of a hierarchy are restrained by and act in conformity to the laws of the higher levels" (Campbell, 1974:180). I would like to go one step further in the argument and extend this concept. The tendency of biological systems to reach new levels of integration is so general that one could be tempted to translate "downward causation" on a time scale as "backward causation," where the higher level of a possible integration, even before being realized by the units competing without cooperation at the lower level, already has an influence on the development of these units, thus favoring those able to cooperate. To avoid "backward causality" only one rather reasonable assumption is necessary, one which in any case is much less restrictive than most of the ones the mathematical models have to make. This assumption is that the interrelated higher and lower levels of integration are realized in a common structure, which in general can favor not only accomplished cooperation but also (other qualities being equal) the tendency of developing units to reach new levels of integration, even when this level has not yet been reached and when the possible new cooperation has not yet produced a payoff. The potential ability to perform at a higher level of integration can then influence the evaluation of the performance at the lower level and thereby the direction of subsequent development.

Concluding Remarks

A scientific analysis of the evolution of social structure in mammals has to consider the highly complex network of multidimensional qualities and functions in a hierarchical organization. I wanted to draw attention to the complexity of the multi-dimensional interrelation which defies any monodimensional theory of selection and evolution. For an understanding of the conditions of social structure, much research is still necessary. This research will involve and combine different biological disciplines. One must consider ecology and behavior; behavior and its neural and hormonal correlates; and social behavior and the dynamics of systems in general. Although a number of approaches have been made and some promising roads have been opened up, there is no set of theoretical concepts worked out yet which would make it possible to understand the evolution of social structure. For the time being, any advice to a student would be to observe carefully and with patience and not to be influenced too much by short-lived theoretical concepts. The study of animal behavior has been handicapped for centuries by not having a theoretical basis; today there seems to be a danger from the opposite side. Too much over-simplified theory keeps the student (who is in danger of losing contact with a natural environment anyway) from observing animals and nature long enough to begin to recognize the structures and their relations.

Literature Cited

BERNARD, C. 1966. Leçons sut les phénomènes de la vie communs aux animaux et aux vegetaux. Vrin, Paris, 404 pp.

BERTALANFFY, L. V. 1972. General system theory. Foundations, development, applications. Braziller, New York, 289 pp.

CAMPBELL, D. T. 1974. "Downward causation" in hierarchically organised biological systems. Pp. 179–186, *in* Studies in the philosophy of biology (F. J. Ayala and T. Dobzhansky, eds.). Univ. California Press, Berkeley and Los Angeles, 390 pp.

CANNON, W. B. 1967. The wisdom of the body. Norton, New York, 333 pp.

EIGEN, M., AND P. SCHUSTER. 1979. The hypercycle. A principle of natural self-organization. Springer, Berlin and Heidelberg, 92 pp.

EISENBERG, J. F., N. A. MUCKENHIRN, AND R. RUDRAN. 1972. The relation between ecology and social structure in primates. Science, 176:863–874.

GLANSDORFF, P., AND I. PRIGOGINE. 1971. Thermodynamic theory of structure, stability, and fluctuations. John Wiley and Sons, New York, 306 pp.

HAMILTON, W. D. 1964. The genetical theory of social behavior. J. Theor. Biol., 7:1–52.

HENDRICHS, H. 1978. Die soziale Organisation von Säugetierpopulationen. Säugetierk. Mitt., 26:81–116.

KLEIMANN, D. G., AND J. F. EISENBERG. 1973. Comparisons of canid and felid social systems from an evolutionary perspective. Anim. Behav., 21:637–659.

LORENZ, K. 1963. Das sogenannte Böse. Zur Naturgeschichte der Aggression. Borotha-Schoeler, Wien, 415 pp.

LOTZE, H. 1882. Gründzüge der Naturphilosophie. Dictate aus den Vorlesungen. Hirzel, Leipzig, 110 pp.

MAYNARD SMITH, J. 1976. Evolution and the theory of games. Amer. Sci., 64:41–45.

MAYNARD SMITH, J., AND G. R. PRICE. 1973. The logic of animal conflict. Nature, 246:15–18.

MILLER, J. G. 1978. Living systems. McGraw-Hill, New York, 1102 pp.

SELYE, H. 1956. The stress of life. McGraw-Hill, New York, 324 pp.

SKOLIMOWSKI, H. 1974. Problems of rationality in biology. Pp. 205–224, *in* Studies in the philosophy of biology (F. J. Ayala and T. Dobzhansky, eds.). Univ. California Press, Berkeley, 390 pp.

THOM, R. 1972. Stabilité structurelle et morphogénèse. Essai d'une théorie générale des modèles. Benjamin, Reading, Massachusetts, 362 pp.

WILSON, E. O. 1975. Sociobiology. The new synthesis. The Belknap Press, Harvard Univ. Press, Cambridge, Massachusetts, 697 pp.

*INDEX**

* This index is primarily a guide to subject areas covered in the text. References to taxonomic categories include those species, genera, families, and orders that have received detailed treatment. Species briefly mentioned in the text that are part of lists or compilations used to develop general principles are not indexed.